THE LITTLE OWL

THE LITTLE OWL

Conservation, Ecology and Behavior of *Athene noctua*

DRIES VAN NIEUWENHUYSE
EHSAL Management School, Brussels, Belgium

JEAN-CLAUDE GÉNOT
Vosges du Nord Biosphere Reserve, France

DAVID H. JOHNSON
Global Owl Project, Center for Biological Diversity Virginia, USA

CAMBRIDGE UNIVERSITY PRESS
Cambridge, New York, Melbourne, Madrid, Cape Town, Singapore,
São Paulo, Delhi, Dubai, Tokyo, Mexico City

Cambridge University Press
The Edinburgh Building, Cambridge CB2 8RU, UK

Published in the United States of America by Cambridge University Press, New York

www.cambridge.org
Information on this title: www.cambridge.org/9780521714204

First published 2008
First paperback edition 2010

A catalogue record for this publication is available from the British Library

Library of Congress Cataloguing in Publication data
Nieuwenhuyse, Dries Van.
The little owl: conservation, ecology and behavior of Athene noctua /
Dries Van Nieuwenhuyse, Jean-Claude Génot, David H. Johnson.
p. cm.
Includes bibliographical references.
ISBN 978-0-521-88678-9 (hardback)
1. Little owl – Conservation. 2. Little owl – Ecology. 3. Little owl – Behavior.
I. Génot, Jean-Claude. II. Johnson, David H. III. Title.
QL696.S83N54 2008
598.9′7 – dc22 2008012037

ISBN 978-0-521-88678-9 Hardback
ISBN 978-0-521-71420-4 Paperback

Dedication

The Little Owl has always meant a great deal to me since my childhood. It was one of the first bird species that I learned to know and appreciate. At that time the Little Owl was still rather common. My favorite bird lived in an old pollarded willow nearby.

I wish to dedicate this book to Trui, my wife, and Juul and Siel, my sons, who helped me through tough times when the combination of a family life, a busy stressful job, rebuilding our house, and writing a book were extremely energy consuming. Thanks for the support, the belief that this book would succeed, and the comprehension that writing this book simply had to be done.

I also dedicate this book to De Torenvalk v.z.w., the local nature conservancy organization in which I grew up. Especially my big brother Jef "*Debaris open*" who endured his little brother during so many field trips, Marc Lievrouw "*Luizevel*", Filiep Lammertyn "*Lepus lammertinus*," and Marc De Schuyter "*Skeuterke*". Special thanks go to Friedel Nollet who started studying Little Owls first and to Maarten Bekaert. My two genial pupils substantially helped me in realizing many of my Little Owl dreams and ambitions.

Importantly, I wish to dedicate this book to the multitude of volunteers and Little Owl enthusiasts throughout its distribution range, not least to the Flemish people of Natuurpunt Studie and the many volunteers within the framework of the International Little Owl Working Group. Particular thanks go to Roy S. Leigh who laid the foundations of this wonderful group.

I wish to thank the Community of Herzele (Flanders, northern Belgium) and the Province of East-Flanders for their continuous financial support in favor of local Little Owl research.

Finally, I wish to thank Jean-Claude and David for their remarkable effort in making this book a success.

Dries Van Nieuwenhuyse

This book is a dream come true – a dream I had 20 years ago when I first became passionately interested in the Little Owl. At that time I sorely felt the absence of a synthetic monograph providing me with comprehensive and critical scientific insight into the species. I am delighted to contribute to this venture because it is the fruit of teamwork with Dries and David, and as such it is considerably richer than it might have been otherwise.

I would like to dedicate this book to Claude Kurtz, who introduced me to Little Owl study in 1983 and Michel Juillard who helped me at the beginning of my "long journey" to get insight into Little Owls. Thanks also to the Northern Vosges regional natural park, which is the framework in which I have carried out my Little Owl work since 1982. I am indebted for its enduring trust in me and the subject.

Finally, the book is unique as it draws on unpublished material furnished by members of an international group studying the Little Owl, together with the findings of varied research conducted by us in Belgium, France, and England. Following on from a thesis on the species and many other publications, the book represents the fulfilment of a personal ambition. I hope it is retained as a "reference" for many years by all those naturalists who, like myself, are fascinated by the Little Owl.

Jean-Claude Génot

I would like to dedicate this book to my wife and confidante, Shari. Your enthusiasm for learning and life fills my heart with joy and amazement every day. I would also like to dedicate this book to the members of the Global Owl Project. My life has become richer through our shared communications, explorations, addiction to owls, and the friendship I have had with you over the years. What a wonderful opportunity and privilege it has been to work with Dries and Jean-Claude: to delve into the ecology of the Little Owl and into the passion we so deeply share about owls. Finally, I truly appreciate the wonderful efforts of all of the people that have worked with the Little Owl, offering publications and perspectives, assessments and analysis, and rich cultural insights. It is upon your solid foundation that our book was possible.

David H. Johnson

Contents

Color plate section is between pages 300 and 301 (and is also available for download from www.cambridge.org/9780521714204)

Foreword

For centuries, all across the world, humans and owls have had a continuous love–hate relationship. Owls are prominent in myth, superstition, and folklore. On the one hand, owls were an omen of bad news, of doom and gloom – in Central Europe locally Little Owls are still associated with death – on the other hand, the Greeks, for example, considered them wise, especially the Little Owl. Though the relationship between man and owls can be traced back to ancient Greece and beyond, our understanding of the basic biology of owls often is still rather poor compared to other bird species. Few scientists interested in basic ecological principles would choose to study a nocturnal species breeding in low density, often in remote and inaccessible places. The situation changed in the mid twentieth century, when ornithologists observed dramatic declines in several owl species over much of Europe. Throughout Europe, the decline of Little Owls is mainly caused by habitat destruction, especially from the intensification and mechanization of agriculture. Changes in agricultural practices led to a decline of suitable nest sites and hunting grounds, and the associated population decrease resulted in the increased isolation and fragmentation of the European breeding population. These pronounced declines have attracted the attention of many conservationists and researchers, especially in Western European countries.

A first milestone, not only for conservation of Little Owls but also for detailed studies of their breeding biology, was the development of an artificial nestbox by the late Ludwig Schwarzenberg (1970) at the end of the 1960s. The original nestbox designed by Mr. Schwarzenberg was a simple wooden tube built from meter-long laths, with a diameter of about 18 cm, waterproofed with roofing felt. The nestbox was quick and easy to construct, and was quite a success. These tubes were readily accepted by Little Owls, sometimes within two to three weeks after being set up. Thousands of these nestboxes have been established in Western Europe during the last decades, mainly in Germany. In several areas population numbers increased (rapidly) after the provision of nestboxes, especially during the first few years. Later, growth rates of populations supported by nestboxes often slowed from year to year as the population density stabilized. As well as detailed habitat analyses, the high occupation rate of nestboxes provided further evidence that the lack of suitable nesting cavities was an important factor limiting Little Owl populations. The number and distribution of nest sites seems to be an ultimate limiting factor determining population densities across much of the breeding range of Little Owls in Western Europe. However, the

Figure 0.1 Little Owl frontal view (François Génot).

reduced rate of population growth associated with increasing population density indicates there are other limiting factors. Nowadays, nestboxes are widely used as a conservation tool. But, I have to point out explicitly that the provision of artificial nesting sites is not an appropriate long-term conservation strategy. In the longer term, natural breeding sites have to be re-established, e.g., through the planting of orchards and willows.

The development of an appropriate nestbox also made it much easier to study the biology of this fascinating species in detail, which otherwise breeds mainly in inaccessible cavities. Nestboxes offer the possibility to study the breeding biology from egg-laying to fledging in detail, and moreover, to catch and ring adult as well as juvenile birds with ease. An advantage of the species is that the adults are mainly sedentary, most individuals use the same day-roost, or at most a handful of alternative roosts, throughout the year, often for their entire life. Once familiar with the species and the territory, individuals are often easy to discover. Thus, besides population numbers, a lot of important demographic parameters determining population density and population regulation can be determined. During the last 20 years many studies on population biology have been carried out. In addition, recent field techniques such as radio-tracking have been applied to analyze home-range sizes and habitat use throughout the year. DNA fingerprinting has been used to analyze genetic variation within, as well as between, populations.

Along with this, in the late 1990s Little Owl researchers realized the importance of standardized monitoring methodologies, and an international co-operation, named the International Little Owl Working Group (ILOWG), was founded. At the turn of the twentieth century, after three very fruitful international conferences organized by the ILOWG, time was ripe to summarize and synthesize our present knowledge on Little Owls for an international readership. The meetings in Champ-sur-Marne, France (2000), Geraardsbergen, Belgium (2001), and Cheshire, England (2002) helped to engender a common spirit of co-operation. The ILOWG as well as a flock of Little Owl enthusiasts made countless unpublished data, notes, and observations generously available, data that previously were not available or were written in languages with which most of us are not familiar.

How welcome, therefore, is the monograph by three leading owl enthusiasts – *Owlologists* – distilling all that is known about Little Owls and providing up-to-the-minute information on current research. The authors solved the difficult task of bringing together and synthesizing often contrasting material from different sources. They address more or less all aspects of the biology of Little Owls. However, besides presenting an impressive overview of the worldwide distribution and population numbers, for example, they concentrate on two fascinating and fast-developing fields: habitat selection and factors affecting habitat selection, and population regulation. Insights from habitat analyses and population studies have proven crucial in conservation. To understand habitat suitability and habitat preferences the authors used – in contrast to most former studies – a new multi-scale approach, they analyzed a multiplicity of biotic and abiotic parameters within grids of a few hectares to several square kilometers in size. Analyzing habitat parameters at different spatial scales provides a much better insight into habitat preferences and thus environmental factors limiting population density, in particular for species that inhabit a great diversity of natural and anthropogenic landscapes. A geographic information system-based multi-scale approach enables us not only to identify limiting and important environmental habitat parameters, moreover, it allows us to select representative sampling points for large-scale surveys, a significant advance in cost-efficient monitoring.

The Little Owl monograph you hold at hand provides examples on how to use data about bird distributions, relative abundance, and habitat associations to set conservation priorities. Based on the presented data, a working structure for an owl species conservation plan has been developed to standardize research and monitoring methods across the range of the Little Owl. A harmonized monitoring program is essential for effective conservation and management, e.g., to evaluate the effectiveness of policy and mitigation measures. Though the aim of a monitoring program is also to elucidate reasons for population changes, causal factors will not necessarily be obvious from the monitoring data alone. In order to overcome these shortcomings, an indispensable tool, concomitant research, should supplement the monitoring program. More cost-efficient monitoring will save time and money. At least some of the limited resources can be redirected to other work, such as to identify the causes of declines.

Along with this, the monograph on Little Owls exemplifies in an excellent way (a) the close link between amateur and professional ornithologists, and (b) the overall importance of detailed large-scale and long-term studies by volunteers and their contributions to applied as well as to basic scientific research. The increase in knowledge during the last decades is largely through the efforts of non-professional ornithologists and non-governmental organizations (NGOs). Most of the long-term studies on Little Owls as well as large-scale surveys of their distribution were carried out by volunteers. Modern software and modeling techniques provide the theoretical framework. In addition to their role as tools for describing distribution and patterns affecting them, for example, models can also provide the means for examining and generating hypotheses about causes and consequences of habitat changes. Models can show us which parameters are important and therefore should be recorded in the future, and hence make monitoring and conservation programs more effective. To accomplish this, the integration of comprehensive empirical studies, technological advances to study the behavior of nocturnal animals, and mathematical modeling promises many new insights. The understanding of population regulation and the spatial distribution, including the understanding of dispersal patterns, will not only prove to be an interesting challenge for the future, it is also essential in constructing an effective conservation plan. Generalization of all available local insights and models across the entire geographical range of the species is another main point of attention for the future. The ILOWG provides a fertile ground and delivers the necessary platform for co-ordinated studies across Europe.

Through international co-operation and research, our knowledge has increased during the last decades. The authors have organized the existing knowledge into a coherent whole. Nevertheless gaps in our understanding of owl biology and behavior remain. In my opinion, one of the major issues is that all quantitative data available are from anthropogenic, agricultural habitats in Europe. Only qualitative descriptions are available from the primary steppe and semi-desert habitats in the core area of the species' range. To understand the decision rules that guide Little Owls in responding to their biotic and abiotic environment, and thus for a thorough understanding of the ecological and evolutionary processes, go there! I hope this monograph will stimulate ornithologists all over the world to study this fascinating species. Dries, Jean-Claude and David provide an excellent basis for sound research and conservation strategies.

Klaus-Michael Exo
Institut für Vogelforschung
"Vogelwarte Helgoland"
An der Vogelwarte 21
D-26386 Wilhelmshaven
Germany

Executive summary

In this book we provide a summary of the substantial literature and knowledge on the Little Owl (*Athene noctua*) to offer a synthesis of the current understanding of the species' rangewide ecology and conservation status. In addition to drawing from our own owl studies and experiences, and those of many colleagues, we have examined over 1900 publications and reports dealing with the Little Owl. Our key findings are briefly offered here.

Cultural aspects. Owls are prominent in myth, superstition, and folklore across the world. With its large distributional range, and an ability to co-exist commensally with many human habitations, the Little Owl has figured prominently in many cultural beliefs. Whereas individual species of owls are rarely referenced in historical mythology, there is specific evidence of Little Owls in images on coins, medallions, carvings, and sculpture, and in historical literature. In Western societies, the Little Owl is seen as a purveyor of wisdom, a perspective likely derived from Greek mythology in which the Little Owl was the favorite bird of Athena, the Goddess of Wisdom. Representations of Little Owls have been found in association with the Xian culture in Inner Mongolia, dating from 8000-7500 BCE. Images of the Little Owl were placed on silver coins minted in Greece starting about 550 BC; the same image is on the Greece *Euro* coin minted as of 2002. In other cultural uses, Little Owls were used as bait birds in catching small birds in Italy, France, and Germany up until the 19th century. In recent decades, Little Owls have appeared on postage stamps, beer labels, and corks for fine wines.

Taxonomy and genetics. The genus *Athene* includes four species; the Little Owl was formally described to science in 1769 by Giovanni Antonio Scopoli. Twelve subspecies of *A. noctua* are recognized; recent DNA data is improving our understanding of this species' complex. Michael Wink has provided a section for this book in which he examines nucleotide sequences of the cytochrome b gene of owls to infer phylogenetic and phylogeographic relationships. Within *A. noctua*, several well-defined clades are apparent that agree with recognized subspecies. However, because some of the genetic distances between the subspecies are in the range typical of that found in established species, it is likely that *A. noctua* comprises a monophyletic species complex that could be subdivided into several 'good' species. The analysis thus far covers the subspecies *vidalii*, *glaux*, *noctua*, *indigena*, *lilith*, and *plumipes*. Some of the African and Asian taxa are not yet represented. The present data reveals three phylogenetic lineages.

Morphology and diet. The Little Owl is a small (19–25 cm, 160–250 g), relatively long-legged, nocturnal, territorial, 'chunky' owl with a short tail and round head. They exhibit reverse sexual dimorphism; based on weight, wing length, tail length, bill size, and tarsus size, females are heavier and larger than males. Mean weights for males and females are heaviest in March/April (just prior to breeding season) and lightest at the end of the summer and autumn. Adult male and female owls each have a vocal repertoire of 15 call-notes, collectively comprising a total of 20 recognizable call-notes. Juveniles have 12 defined call–notes. Neighbor–stranger discrimination is used by territorial males to minimize energy expended on aggressive acts, preventing escalated contests between neighbors, and decreasing exposure time to predators. The Little Owl has a generalist diet and takes a high diversity of small-bodied prey (e.g., voles and smaller). We tallied 544 prey species of the Little Owl, reflecting 377 invertebrate species, 54 small mammals, 15 reptiles, 14 amphibians, 82 small birds, and 2 fishes. In particular, the nutrients and biomass from small mammal prey enhances reproductive success.

Distribution, population status, and trends. The Little Owl is broadly distributed, as it has been recorded from lowland areas in 84 countries in the middle and lower latitudes of the Northern Hemisphere (mainly between 22° and 51° north). Densities of Little Owls decrease with increasing latitude. Snow depths over 10–15 cm preclude hunting, and limit their distribution in both latitude and elevation. The Little Owl is considered fairly common to common in 45 countries and uncommon, rare, or a vagrant in 39 other countries. It has been introduced (and is considered common) in New Zealand and England. The distribution of the owl has increased within 8 countries, decreased in 15, remained unchanged in 30, and was insufficient for determination in 31. The population status is reported as increasing in 11 countries, declining in 14, stable in 28, and status unknown in 31 countries. There is evidence that groups of Little Owls breed in 'clusters' as compared to individual pairs separated by long distances; defended territories are non-overlapping, and territory sizes are similar for pairs inside and outside of clusters. Ringing studies indicate that the average lifespan for Little Owl females is 3.8 years (oldest female in wild was 15 years). Most (>80%) female owls nest in their first year. Clutches average 2.60–4.42 eggs, and fledgings average 1.78–2.84 young. Average first-year survival rates were 15–27%; adult survival rates average 36%. Adult owls tend to be resident; young disperse relatively short distances (0.6–4.0 km), with females dispersing about twice as far as males. Ringing and genetic data suggest that populations of Little Owls are structured as metapopulations with source-sink dynamics.

Habitat. Habitat of Little Owls has been described as open country with groups of trees and bushes, rocky country, grasslands, deserts and semi-deserts with rocks, ruins, oases, pastureland with scattered trees, old orchards, along rivers and creeks with pollarded willow and other trees, parkland, and edges of semi-open woodland, and farmsteads and urban areas with surrounding cultivated lands. Nests are in tree cavities, crevices in ruins and water wells, adobe buildings, under roofs, holes in quarries, stick nests, nestboxes, piles of stones, and animal burrows in banks. Home range sizes of breeding pairs tend to be small, averaging 14–120 ha.

Conservation. Long-term conservation of the Little Owl is complicated because habitat conditions can change rapidly and significantly due to changes in policies and management. Current threats to the Little Owl are mainly loss of habitat from human land-use practices. Threats becoming prominent across the European range include the reduction in the amount of tree-lines, deterioration of high-stem orchards, and the increase in the area of subsidized maize. Other threats include collisions with vehicles and pesticides that reduce the availability of invertebrate prey. We propose a conservation program featuring five components: increase knowledge of the species, reduce limiting factors, understand effects of landscape conditions, introduce conservation legislation and policies, and support the role of local people in conservation. We follow this with a review of the four drivers to implement the conservation program: monitoring, standardized methods, data management, and measures of success. The Little Owl is a useful ambassador of the small-scale, half-open, largely agriculturally dominated and stony-steppe landscapes. Among the features that make it a flagship of the rural environment is that the species is well known among the public, it is still present in reasonable numbers in most countries, is readily observable, easy to research, and offers relatively quick responses to habitat restoration actions. The substantial work of conservation volunteers both gathers important data for standardized and reliable databases, and broadens the social network of nature conservation. In this way, owl conservation is brought closer to the general public. Nestboxes for Little Owls are widely used in areas where nest-sites are limited, and have been effective for both research studies and conservation of this species. Nest-cameras with internet feeds have recently been placed in nestboxes, offering unique insights into the behaviors of both the owls and their excited human viewers.

Research priorities. Until now, Little Owl research has been mostly descriptive and has reflected short-term observational studies on life history, food, behavior, habitat, and general distribution. New/expanded research should be applied and focus on large-scale replicated studies, controlled experiments, and long-term studies of demography. Results that lead to effective habitat management recommendations are needed. For habitats, information is needed on specific structural features, landscape configurations, and amounts of habitat required for stable or increasing populations. We need to link habitat and resource selection to demographic performance (i.e., survival and reproduction) to ensure managers provide for quality habitat and not simply owl presence. Studies emphasizing demographic parameters (nest success, productivity, survival, dispersal) are needed to identify factors limiting populations and to contribute to understanding metapopulations dynamics (e.g., gene flow, source vs. sink populations). There is a critical need to determine the effects of various types of land use on Little Owl populations to devise effective measures to minimize or mitigate for such actions. Land uses affecting Little Owls include: livestock grazing, silviculture, recreation, fire management, oil and gas development, mining, water control and development, agriculture, roads, suburbanization, communication towers, and wind-power development. We strongly encourage further geographical information system (GIS) analyses and use of landscape scale population modeling to determine priority areas for potential habitat protection, restoration, and management. We need to test whether currently available digital maps contain enough detail to distinguish different habitat elements, to reduce the need for additional field data collection. The Little Owl could also be studied experimentally because it is easy to manipulate local population structures, it is relatively common in human-dominated landscapes, and individual owls are available from bird-care centers. We recommend conducting baseline ecological, ethno-ornithological, and distributional surveys in Middle Eastern, African, and Asian countries to estimate overall population status and aid in conservation programs in those regions. Additional DNA studies to clarify species and sub-species taxonomy are necessary and can be conducted with relatively low cost. Finally, we recognize, and strongly encourage, the significant conservation value to be achieved in co-ordinating Little Owl research with those of other agro-pastoral and steppe species like shrikes (Laniidae).

Monitoring plan. We propose a network of 30 Vital Sign demographic monitoring areas where mark–recapture studies would be conducted to locate, mark, and re-observe or recapture pairs of Little Owl adults and their offspring. Each monitoring area currently contains 50–100 pairs of Little Owls, or equivalent habitat in areas where population restoration is planned. These *Little 30* study areas are located within 20 countries across the range of the species, and overlap with monitoring efforts in places with other natural resource values (e.g., Important Bird Areas, UNESCO World Heritage Sites, National Parks, Natura 2000 sites). Our proposed program is designed to monitor the long-term status and trends of Little Owls to evaluate the success of various plans to arrest downward population trends, and provide focal areas for maintaining and restoring habitat conditions to support viable owl populations. Data from the demographic studies would be integrated in individual population and meta-population analyses, and a process for reporting to decision-makers during their periodic land use plans and policy reviews is offered. For countries without Vital Sign demographic study areas, we recommend baseline distributional surveys and ecological studies. This monitoring program is designed to produce results directly applicable to the Convention on Biological Diversity.

Bibliography. We are including a special section in this book reflecting a full bibliography of the Little Owl *Athene noctua*, and sets of queries based on key words. The Bibliography includes publications from 1769 through December 2007. This Bibliography represents all Little Owl literature (n = 1904 publications) available to us; we welcome additions and apologize for any missed citations.

Acknowledgments

The authors wish to thank François Génot for his illustrations. Jevgeni Shergalin shared with us the rich literature of the former Soviet Union for the first time through his translations. We wish to thank Michael Wink for his contribution on the genetic work that he has done on Little Owls. We are highly indebted to Roy S. Leigh and Michael Exo who were so kind to review this book for us with the utmost dedication and for writing Chapter 14 and the foreword, respectively. Reuven Yosef and Norbert Lefranc are thanked for providing some very interesting remarks on the manuscript. Thanks also go to Giorgio Dimitriadis and Marco Mastrorilli for sharing with us their insights on the importance of the species through history. Koenraad Bracke (Belgium) helped us tremendously with the collection of illustrations. Thanks go to Armand Wernet and Koenraad Bracke for translation of German literature. Special thanks go to Loïc Hardouin, Ricardo Tomé, Joan Navarro, Ronald van Harxen, Pascal Stroeken, and Marco Mastrorilli for sharing with us submitted manuscripts and even draft versions containing particularly relevant state-of-the-art results. We are also highly indebted to Jacques Bultot and his "Groupe *Noctua*" friends, Ronald van Harxen, Pascal Stroeken, Luc Vanden Wyngaert, Philippe Smets, Ronny Huybrechts, and Stanny Cerulis for making all of their historical data available to us for our analysis.

We also wish to thank Marco Mastrorilli (Italy), Hugo Framis, Inigo Zuberogoitia (Spain), Ricardo Tomé (Portugal), Anton Kristin (Slovakia), Edwin Vaassen (Turkey), Mike Toms (UK), Tibor Fuisz (Hungary), Boris Nikolov (Bulgaria), Alivizatos Haralambos (Greece), Alexandre Vintchevski (Belarus), Libor Oplustil, Libor Schröpfer, Karl Stastny (Czech Republic), Hein Bloem, Jan Van't Hoff, Niko Groen, Ronald van Harxen, Pascal Stroeken, Peter and Wies Beersma, (Netherlands), Pelle Andersen-Harild, M. B. Grell (Denmark), Karen Jenderedjian, Tigran Tadevosyan (Armenia), Tanja Sova (Serbia Montenegro), Evgeny Victorovich Vilkov (Daghestan), Sergey Volkox, Alexander Sharikov, Dr Il'yuck, Alexander Antonchikov (Russian Federation), Alexander Abuladze (Georgia), Michael Jennings (Arabian Peninsula), Michael Brown (Oman), Jevgeni Shergalin (former USSR), Armand Wernet, Sébastien Blache (France), Paul Isenmann (Algeria, Tunisia), Michel Thévenot (Morocco), Christian Meisser, Paul Schmid, and Klaus Robin (Switzerland), Hubertus Illner, Franz Robiller, Peter Haase, Hans Mohr, Ubbo Mammen, Herbert Keil, Michael Eick, Lutz Dalbeck, and Eckhard Kartheuser (Germany), Milan Vogrin, Al Vrezec (Slovenia), Rottraut Ille (Austria), Patric Lorgé and Marc Jans

(Luxembourg), Grzegorz Grzywaczewski (Poland), Alivizatos Haralambos (Greece), Lei Fu-Min (China), Otto Pfister, Chris van Orden, and Natalia Paklina (Ladakh) for their contribution on the distribution and population numbers of Little Owls in their countries.

Hein Bloem and Ronald van Harxen (Netherlands), Andreas Kämpfer-Lauenstein and Hubertus Illner (Germany), Christian Meisser (Switzerland), and Marco Mastrorilli (Italy) helped in compiling the volunteer initiatives for their countries.

Marc De Schuyter is thanked for drawing the nestbox layouts and doing all the scans of the color slides. The following photographers submitted their personal pictures for use in this book: Marc de Schuyter, Ludo Goossens, and Thierry Votquenne (Belgium), Peter Van der Leer and Rob Hendriks (Netherlands), Bruno D'Amicis and Duccio Centili (Italy), Ricardo Tomé (Portugal), Alexander Abuladze (Georgia), Amir Ezer (Israel), Girish Jathar (India), Chris van Orden and Paklina (Ladakh, India). Special thanks go Sigmund Schönn (Germany) for sharing with us his original historical black-and-white images of Little Owl habitats across its range.

We thank J. R. Duncan, E. D. Forsman, R. J. Gutiérrez, R. Kavanagh, E. Korpimäki, P. Saurola, L. L. Severinghaus, and K. Swindle for their help with literature and formatting aspects of owl dispersal, home-range, habitat, and demographic data.

We sincerely appreciate the assistance with the Little Owl bibliography (Appendix B) from Richard J. Clark and Tracy L. Fleming, who unselfishly gave us their own bibliographies on the Little Owl, in an effort to ensure that we had the fullest citation list possible.

We deeply appreciate the contributions on the Forest Owlet from Girish Jathar, Asad Rahmani, Jayant Kulkarni, Dharmaraj Patil, Prachi Mehta, and Farah Ishtiaq for their fundamental work on this critically endangered species.

We also appreciate the support of Talon Scientific and the Center for Biological Diversity for this project.

Bruce G. Marcot and Alan Sieradzki are acknowledged for helping us out with a variety of aspects.

We gratefully appreciate the contribution of Janice A. Reid (Oregon, USA), for her extra effort in providing us with materials from the Northern Spotted Owl surveys and demographic analyses. We are also indebted to the many Little Owl researchers who have contributed important information as to the locations and sizes of the recommended vital sign demographic study areas. We also appreciate the attention and support from staff of BirdLife International in the development of Chapter 13. Finally, we appreciate the formal peer reviews we have received on the monitoring chapter.

Special thanks go to Martin Griffiths, Dawn Preston and Alison C. Evans of Cambridge University Press for their patience and support in this project.

Furthermore we wish to thank Hans-Martin Berg (Austria), Matthias Korn (Germany), Jevgeni Shergalin (Estonia), Marco Mastrorilli (Italy), and Hugo Framis (Spain) for their tremendous help in getting all the references right. Thanks to Wies Beersma and Rob Hendriks, Matti Charter and Amir Ezer for sending their pictures right on time.

We sincerely wish to express our gratitude to Kathryn Pilgrem for her patience and her thorough editing of the manuscript. This allowed us to make the book unique in its style and quality.

1
Introduction

For centuries, all around the world, humans have had a continuous and strong cultural relationship with owls (Marcot & Johnson 2003), traceable back 15 000 years to caves in France. Some cultures view owls as omens of bad luck, sickness, and death, while others view them as creator beings, helping spirits, having profound wisdom, oracular powers, or the ability to avert evil. Depending on where you are within the range of the Little Owl, both of these divergent viewpoints (bad or good omens) are still held for owls. (Figure 1.1).

In the mid 20th century, drastic declines in several owl species attracted the attention of ornithologists, conservationists, and researchers. Throughout Europe the decline of Little Owls is mainly being caused by habitat destruction, especially due to the intensification and mechanization of agriculture. In order to counter this negative situation scientists and conservationists have realized the need for international co-operation through a multi-disciplinary approach.

Early literature on the Little Owl reflected general studies on the biology, distribution, diet, nesting, and habitat. In the early 1990s an excellent book was published on the species in German (Schönn *et al.* 1991). This publication gave a review of most of the literature that was available at that time. Since that time, substantial new findings on the Little Owl have been produced (in many languages!) from a growing number of countries within the range of the owl.

Recently, the International Little Owl Working Group (ILOWG) organized three international Little Owl symposia in three years: the first International Little Owl Symposium in Europe '*Little Owls and landscapes*' took place in Champ-sur-Marne, France, in November 2000, the second International Little Owl Symposium '*The Little Owl in Flanders in its international context*' in Geraardsbergen, Belgium, in March 2001, and the third in Cheshire, England, in November 2002. Participants from eight European countries attended the meetings. The ILOWG boosted international communication and the exchange of knowledge and allowed us to obtain until now unexploited information from the former Soviet Union, the Middle East, Arabia, and China. The most recent literature overview of the species was published in the Update of the Birds of the Western Palearctic (Génot & Van Nieuwenhuyse 2002).

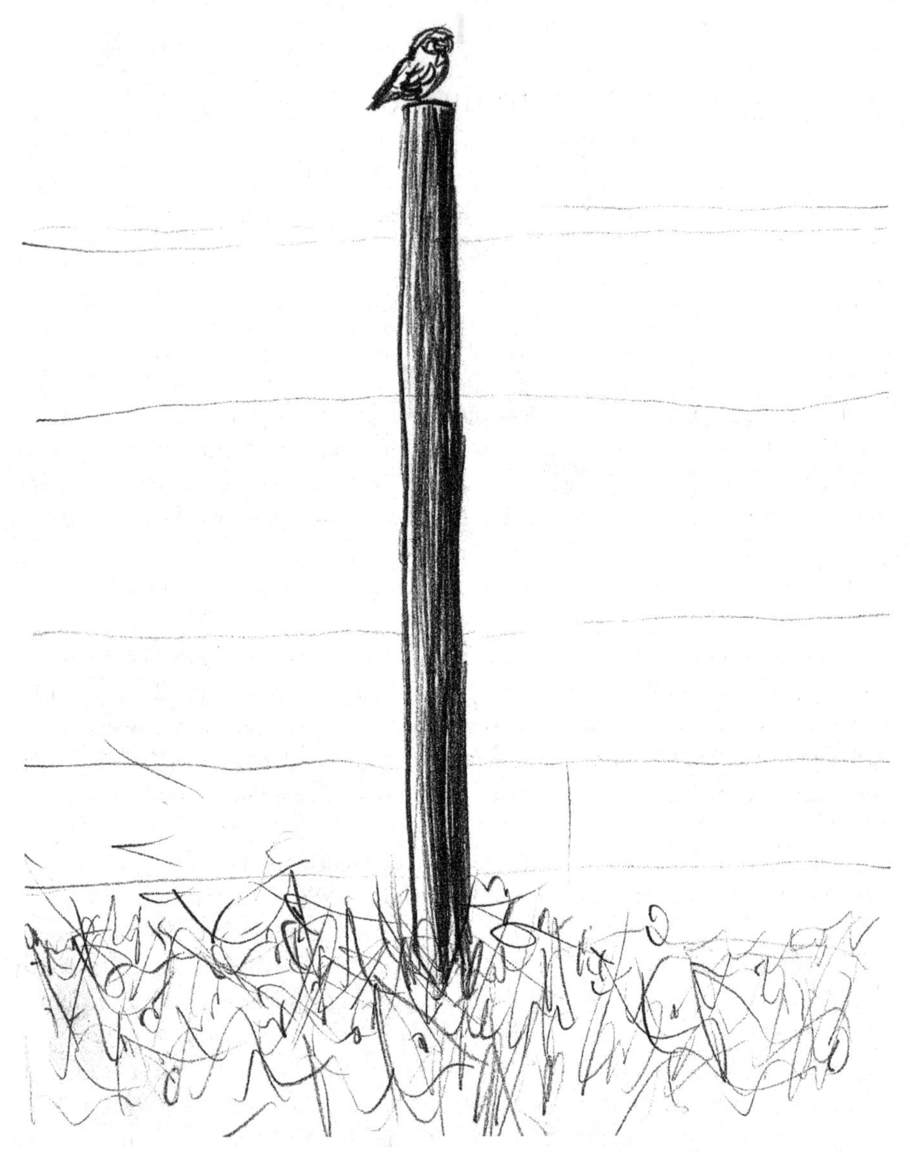

Figure 1.1 Little Owl on a fence pole (François Génot).

Researchers conducting ringing and habitat studies have made substantial findings to our understanding of the Little Owl. At the end of the twentieth century, new papers offered important data on meta-populations and the relations between the Little Owl and the landscape at different geographic scales. Growing numbers of enthusiastic volunteers want to help conserve owls. Given our field experiences with owls, the growing body of scientific literature on the Little Owl, and our desire to further international collaboration and owl conservation, we have undertaken this project.

While rooted in science, this book has the principal aim of making the information accessible to the larger international public. This follows in the spirit of the United Nations declaration of the year 2001 to volunteers and conservationists in recognition of their crucial international contributions. Recognition and appreciation of volunteer work is also of crucial importance for nature conservancy. The valuable input of volunteers not only helps to construct huge, standardized, and reliable databases, it also helps to broaden the social network of nature conservation.

The Little Owl is associated with small-scale half-open landscape (ranging from pastoral landscapes with scattered trees to stony steppe deserts). The species has a multitude of strategic features that make it an indicator of a healthy environment: the species is very well known to the public, it is still present in reasonable numbers in most European countries and is readily observed, and the Little Owl can offer us insights into the methods of implementing nature restoration. Due to its 'high cuddle factor' it is a perfect vehicle to transfer nature conservation values to the broader public.

The framework of this book reflects the complexities of the species in different social and ecological contexts (Figure 1.2). To position the Little Owl in the cultural context we look at the history and cultural traditions of the species (Chapter 2). We describe the taxonomy and genetics of the owl (Chapter 3), and then characterize the morphology and body characteristics of the species (Chapter 4). Distribution aspects, population estimates, and population trends are given in Chapter 5. Owl habitat is described and the relationships between the landscape and the species are characterized in Chapter 6. The diet of the Little Owl has been studied in many regions, and we summarize the substantial literature on diet in Chapter 7. We examine the breeding biology, nests, and foraging of the species in Chapter 8, We then describe the behavior of Little Owls (Chapter 9) mainly based upon captive breeding data but also new recent material. Chapter 10 offers insights into factors that influence the number of owls in a given geographic environment and act to regulate owl populations. Aspects of immigration, re-introduction, supplementation, emigration, and local offspring and mortality are discussed. Mechanisms that interact between local populations such as migration, meta-population functions, and the occurrence of sinks and sources place the individual parameters in a wider context. The insights obtained through population studies have proven to be crucial in conservation. After describing the main causes for declines in the species, we aummarize conservation strategies for the Little Owl (Chapter 11). In Chapter 12, we offer an overview of the most important open questions and offer recommendations for future studies summary of future research activities. In Chapter 13, we propose an ambitious monitoring program for the owl, involving a network of 30 Vital Sign demographic study areas across 20 countries. In our final section (Chapter 14) we close by focusing on the fundamental role that volunteers play in owl science and conservation.

During our work on this book, we became aware of the need to clarify the terminology used in characterizing nesting success and other aspects of owl demography. Clarification of terms is important in order to provide a scientific foundation for the accurate and consistent collection and analysis of data gathered across the range of the

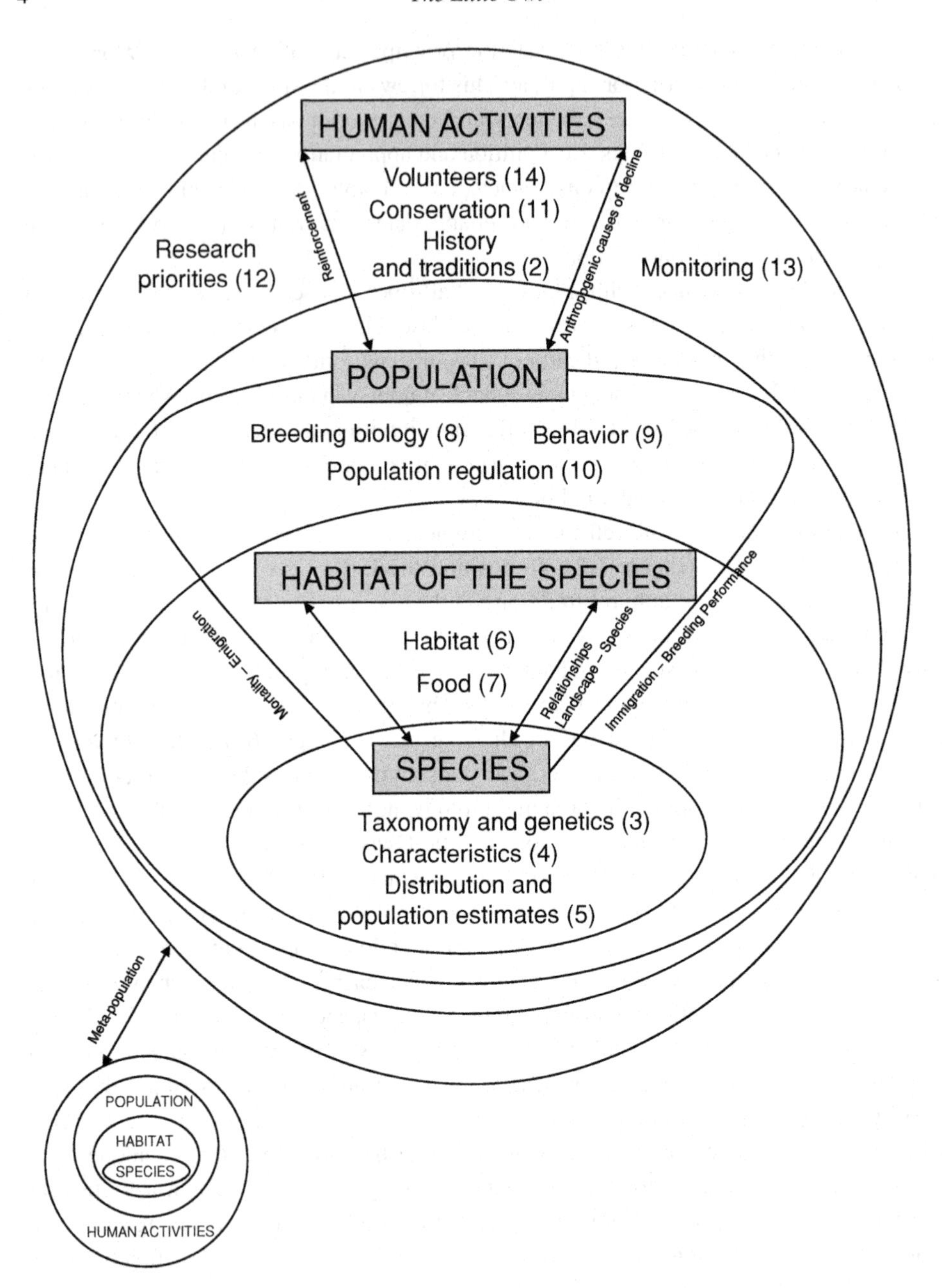

Figure 1.2 Framework of the book. (Numbers in brackets refer to the relevant chapters in the book.)

Little Owl. We offer definitions for key terms throughout the book, but particularly in the glossary.

A fudamentally important aspect of our work has been on assembling the available literature on the Little Owl. A bibliography on the species, along with queries based on keywords, is offered in a special section at the end of this book.

2

History and traditions

Throughout human history, owls have variously symbolized dread, knowledge, wisdom, creator beings, death, witchcraft, and religious beliefs in a powerful spirit worlds. See Figure 2.1. A small but important body of literature is beginning to document the important aspects of owls both across cultures and through time (e.g., Medlin 1967; Simmons 1971; Stuart 1977; Holmgren 1988; Gimbutas 1989; Weinstein 1989; Enriquez & Mikkola 1997; Marcot & Johnson 2003).

The most ancient representations of owls date from the Upper Paleolithic (13 000 BC) and are seen in two caves in France. The first site reflects three owls (generally considered Snowy Owls) in the "Gallery of the Owls", in the French cave of Les Trois Frères in Ariège. A second site, in the Cave Chauvet (Chauvet *et al.* 1996) represents a single "eared" owl that looks to be a Long-eared Owl (or Eagle Owl). Other prehistoric sites with owl art and symbolism are found in an array of locations around the world, such as the Victoria River region of northern Australia, the Columbia River region of northwestern USA, and a Mayan cultural site in what is now Guatemala (Marcot & Johnson 2003). In the east Mediterranean, archaeological digs in Syria and Jordan have recovered anthropomorphized owls in the stone or clay as early as the pre-pottery Neolithic period (8000–6500 BC)(Gimbutas 1989:190–195). Representations of Little Owls were found associated with the Xian culture (a society of agriculturists) in Inner Mongolia, dating from 8000–7500 BC (Schönn *et al.* 1991).

Owls have been viewed by human societies in many different ways. Some societies view owls as a single group of animals (e.g., species are not differentiated and any owl is a bad omen), and others view large owls as bad (harbingers of bad luck, illness, death) whereas small owls are not viewed in this way. Still others are attentive to specific owl species for medicine, religious, or hunting applications (and of course for current-day conservation purposes). Societies and societal values change, and whereas owls may have been viewed in one way or another in the past, current cultural views may differ. For example, during the Middle Ages, the owl was linked to witches and bad spirits, but nowadays in Turkmenia, the Little Owl is a sacred species and to kill one is a great sin (Shukurov in Khokhlov 1995). Another example of this change may be seen (in general) in Western culture, where newer perspectives pertain more to scientific understanding and conservation needs of owls; a specific example is how Spotted Owls have become symbols of old-growth forest protection.

Figure 2.1 Little Owl drying its feathers (François Génot).

Included in her book on the religion of the Old European Great Goddess, Gimbutas (1989) presents a pictorial "script" consisting of signs, symbols, and images of divinities. The main theme of goddess symbolism is the mystery of birth and death and the renewal of life, not only human but all life on Earth. Symbols and images cluster around the parthenogenetic (self-generating) goddess and her basic functions as giver of life, wielder of death, and as regeneratrix. Overall, owls were deeply feared, and viewed as the harbingers of death. The goddess in the guise of an owl was prominent in stone, pottery (e.g., terracotta figurines, burial urns), engravings on statues, schist plaques, bone phalanges laid in graves, drawings, amber figurines, wooden posts, and gold sculptures in this prehistoric religion, which extended from the Neolithic to the Early Bronze Age (general dates of 6500 BC to 2500 BC). Archaeological artifacts and images of owls in this religious context were generic or abstract in shape and pattern, and specific images distinguishable as Little Owls are not apparent. The features that characterize owls (round eyes and beak) can be seen on statue menhirs of southern France and the Iberian Peninsula, and in reliefs and charcoal drawings in the hypogea of the Parisian basin. A series of stelae and drawings of the Owl Goddess from Brittany and the Paris basin are depicted with breasts and one or more necklaces. Beautiful examples of owl-shaped burial urns dating from *c.* 3000 BC come from the Baden culture in Hungary, from Poliochni on the island of Lemnos, and from Troy. They have wings, the characteristic owl beak connecting arched brows, and sometimes a human vulva or a snakelike umbilical cord, symbols of regeneration. In continental Greece, gold sculptures have been found in the *tholos* tombe (Kakovatos, Pylos) and from shaft graves in Peristeria of the fifteenth century BC (Marinatos 1968: pl. 58). In spite of the gloomy aura that surrounds it, the owl has also been endowed with certain positive qualities, such as profound wisdom, oracular powers, and the ability to avert evil. However, this latter ambivalent image is a dim reflection, diffused through time, of the owl as an incarnate manifestation of the fearsome Goddess of Death. Perhaps she was respected for her grim but necessary part in the cycle of existence, as the agony of death, which we take so much for granted, was nowhere perceptible in this symbolism (Gimbutas 1989).

With its large distributional range across Europe, the Middle East, and Asia, and an ability to co-exist as a commensal with many human habitations, not surprisingly, the Little Owl has figured prominently in many cultural beliefs, and in a variety of ways. The common names given to this species across the countries are linked to its activity (Nightbird), to its voice (Kliwitt), to its morphology (Little Owl), to its food (Lark Owl or Sparrow Owl), to the beliefs (Death Bird), to its habitat (House Owl, Willow Owl, Stone Owl), and to mythology (Bird of Minerve [Roman]).

The Little Owl was first formally described to science by Giovanni Antonio Scopoli in 1769 (Scopoli 1769). The origin of its scientific name, *Athene noctua*, combines the genus *Athene*, derived from the goddess of wisdom Athena in Greek mythology, and the species *noctua*, as derived from the nocturnal characteristics of this bird of prey. An early name for the Little Owl, *Strix passerina*, was noted on Thomas Bewick's 1797 drawing. This scientific name was not used, as the name *Athena noctua* more accurately placed the Little Owl within the taxonomic naming convention of owls. While the origin of the English name, Little Owl, is probably linked to the size of the owl, the German and Dutch names – Steinkauz and Steenuil, respectively – refer to the open habitat of this species that it uses in many countries, where it breeds in piles of stones. Common names for the Little Owl in other countries include:

Figure 2.2a Greek silver tetradrachm coin showing the image of the Little Owl.

Italy: Civetta
Spain: Mochuelo Comun
Portugal: Mocho
France, Wallonia, Switzerland: Chevêche d'Athéna
Germany, Austria, Switzerland: Steinkauz
Greece: Κôυκουβάγια (Koukouvaya)
Denmark: Kirkeugle
Netherlands, Flanders: Steenuil
Turkey: Kukumav
Russia: Domovogo sycha
Georgian: Tchoti
Hungary: Kuvik
Mongolia: Chotny bugeechej
Somalia: E'yu

The relationship between the Little Owl and Greece is very long indeed, and is one of the most well-known affiliations between a particular owl species and a prominent society. This affiliation is again alive and prospering, with the new Greek one-euro coin showing the same Little Owl design as the former tetradrachm coin from 2500 years ago (Figures 2.2a and 2.2b).

In Greek mythology, Athena was the daughter of Zeus and originally a Mycenaean palace goddess. Her function later expanded to include the role of guardian of cities, war goddess, patroness of arts and crafts, and promoter of wisdom. Always shown modestly clothed and

Figure 2.2b Greek one-euro coin showing the image of the Little Owl, 2002.

often armed, the Little Owl is her special bird. Figure 2.3 shows a fifth century BC bronze sculpture of the Goddess Athena holding the Little Owl in her hand.

An emblem refers to some distinctive characteristic or activity for which the issuing city is known. The Attica tetradrachm coin bearing an obverse head of Athena and reverse owl with olive branch is perhaps the best known example of a city emblem. The Greek Goddess Athena is a punning reference to the city (in Greek, *Athenai*) that honored her as its chief protective diety. The owl as Athena's favorite bird, and the olive, which was one of the city's most lucrative exports, in time came to stand for Athens throughout the Mediterranean world. With rich silver mines at Laurium at the southern tip of Attica, the Athenians were able to export bullion for foreign exchange at a time when most Greek states restricted coinage to home use. The necessity to create a standardized Greek currency that would be widely acceptable demanded a rigid uniformity in metallic purity, type, and style seldom seen elsewhere in the Greek world. What resulted from *c*. 525 BC onward was the famous series of tetradrachms that carried an obverse image of Athena's head on one side, and a Little Owl with an olive branch and a cresent moon on the reverse. The type was kept in circulation for the next two centuries. The coins were made of silver, and their nickname "owls" become synonymous with Athen's commercial power. The *Obal, Stater, Drachm, Didrachm, Tetradrachm, Octodrachm*, and *Decadrachm* are terms for common denominations based on these weight systems. See Figures 2.4, 2.5 and 2.6.

About 77 BC, the Roman scholar Pliny the Elder assembled more information about the owl in a few chapters of Book X in his *Historia Naturalis*. He specified the Little Owl, the Eagle Owl, and the Screech Owl. His observations were laced with beliefs not entirely sound by today's zoological standards, but they were studied as gospel during the Middle Ages (Medlin 1967:20).

Figure 2.3 A fifth century BC bronze sculpture of the Greek goddess of wisdom, Athena, holding a Little Owl (from Stuart 1977).

Figure 2.4 The Little Owl on a tetradrachm coin from the second century BC (Photo Stuart 1977:7).

Figure 2.5 Roman hand-moulded oil lamp with Little Owl, probably from Corfu, *c.* AD 40–80 (Photo Weinstein 1989:116).

Figure 2.6 Greek terracotta scent bottle *c*. 650–625 BC (Photo Weinstein 1989:124).

Little Owls are also of historical importance in the Italian Peninsula since they appear in some classic fables (Mastrorilli 2005). Fedro's classic fable *The Cicada and the Little Owl* describes the capture of a cicada by a Little Owl during its diurnal hunt. The fable describes the owl's hunting strategy as a sign of its intelligence. It also demonstrates that the Little Owl was a loved species in the Roman period.

> XVI. Cicada et *Noctua*
> Humanitati qui se non accommodat
> plerumque poenas oppetit superbiae.
> Cicada acerbum *noctua*e conuicium
> faciebat, solitae uictum in tenebris quaerere
> cauoque ramo capere somnum interdiu.
> Rogata est ut taceret. Multo ualidius
> clamare occepit. Rursus admota prece
> accensa magis est. *Noctua*, ut uidit sibi
> nullum esse auxilium et uerba contemni sua,
> hac est adgressa garrulam fallacia:
> "Dormire quia me non sinunt cantus tui,
> sonare citharam quos putes Apollinis,
> potare est animus nectar, quod Pallas mihi
> nuper donauit; si non fastidis, ueni;
> una bibamus." Illa, quae arebat siti,
> simul gaudebat uocem laudari suam,
> cupide aduolauit. *Noctua*, obsepto cauo,
> trepidantem consectata est et leto dedit.
> Sic, uiua quod negarat, tribuit mortua.
> *Fedro's classic fable (Latin version).*

Leonardo da Vinci (1452–1519) wrote one fable on Little Owls and Thrushes. He described the use of the Little Owl as "callback" (in Italian "zimbello") during the hunting of Thrushes. This is an important source of evidence that Little Owls were used for hunting after the Middle Ages.

Illustrations became as important as word descriptions to naturalists and others. Conrad Gessner (1516–1565), from Zurich, Switzerland, authored one of the first printed books containing birds in the sixteenth century. His third volume, published in 1555, was the largest in a series of five books on animals (i.e., *Historia animalium*), and contained a nice image of a Little Owl ("Night Owl"); ["Von der Eul/oder Nacht-Eul/und erstlich von ihrer Gestalt. Ulula"]. See Figure 2.7.

Thomas Bewick published his authoritative *History of British Birds* in England in 1797 and 1804. Bewick rejected the copper-engraving medium that was common to book illustration at that time, in favor of incising the end grain of a hardwood block with a graver, thus achieving detail in picture and durability in printing that was previously impossible in wood. Bewick's first edition of the *History* illustrated the Long-eared Owl, Short-eared Owl, a female Horned Owl, Snowy Owl, Tawny Owl, and the Little Owl. Refinements were

Figure 2.7 Image of the Little Owl from Conrad Gessner's *Vogel-Buch* (bird book) published in 1555.

Figure 2.8 Image of the Little Owl from Thomas Bewick's 1797 *History of British Birds.*

made in the drawings, and Howard Saunders published his significant *Manual of British Birds* in 1889. See Figures 2.8, 2.9 and 2.10.

Until the beginning of the twentieth century, hunters in France enjoyed shooting the Little Owl (Delamain 1938). In the nineteenth century, despite the prejudices against this species, the Little Owl was also appreciated by people and held as a captive bird (von Risenthal 1879) and sold by fowlers in Germany (Wemer 1910). In France (Crespon 1840) and in Belgium (Dupond 1943) the Little Owl was used to hunt small birds such as larks because

Figure 2.9 Little Owl image from the *Birds of Europe* Vol 1 (Raptors), 1837, *Strix nudipes* (Nilsson), *Noctua nudipes* (Mihi) – illustrated by Edward Lear.

THE LITTLE OWL.

ATHÉNE NÓCTUA (Scopoli).

Figure 2.10 Engraving of a Little Owl from Howard Saunders' *Manual of British Birds* 1889.

the mobbing behavior of passerines towards the Little Owls was a good technique to attract them. In Italy, Little Owls were domesticated and kept (with their wings clipped) in houses or gardens, where they caught rodents, slugs, and insects (Godard 1917).

A particularly interesting situation on the cultural use of Little Owls comes from Crespina, Italy. Crespina is located 25 km from Pisa in the Toscana region of Italy. The community has a specific monument dedicated to the Little Owl called the "Place of the Little Owl Fair" (Schaaf 2005). This little town was a center for the rearing of owls in captivity to be used in a Little Owl competition. One century ago (*c.* 1900), Little Owls were taken out of their nests, reared and put in special boxes in farms of this Italian region to be used as bait for the hunting of larks. They were then sold on the Little Owl market while tied up on a roost (for an illustration of this market, see Figure 2.11). The nobility (upper-class people) commonly hunted in the countryside using Little Owls as bait. People pulled the tethers to stimulate the Little Owl to fly; larks were attracted by the owl and were themselves caught in nets. There was a *civettaio* (i.e., "owl man" or "owl-keeper") charged with tending to the Little Owls. The annual owl market of Crespina took place on September 29th of each year. At these occasions, it was possible to listen to a range of bird imitators and attend the competition of the Little Owls. This hunt reflects an ancient tradition that started in 350 BC and lasted until the twentieth century in Italy, and lasted from the seventeenth to the twentieth century in Germany.

Figure 2.11 Cover of the journal *La Domenica del Corriere* in 1903 represented the Little Owl market.

A national policy in Italy (L157/1992) and a regional policy approved in Toscana (L3/1994) have recently banned the annual owl market. The laws forbid the killing, catching, or keeping of any owl species in aviaries, or possession of dead owls or parts of them. The use of owls as bait is also forbidden. Because of the traditional aspect of the hunting method, the policy was modified to allow some rearing stations for wild animals. A Little Owl rearing station in Crespina covers two hectares and was sponsored by hunters. See Figures 2.12 and 2.13.

Little Owls were introduced in England and in New Zealand at the end of the nineteenth century and beginning of the twentieth century. After its release, the species was considered a pest by many gamekeepers and farmers in England (Dawson-Smith 1913).

Figure 2.12 Drawing from 1896 showing a seller of Little Owls and interested women.

Figure 2.13 Photo of a hunter and his wife at Crespina market, Italy.

Figure 2.14 Image of a Little Owl on the cork from a modern bottle of fine Italian wine, from Azienda Agricola (agricultural business) in Valle dell Asso, Galatina, Italy, 2001.

Figure 2.15 The Little Owl as used in marketing German beer, Kauzen Bräu.

Figure 2.16 Examples of Little Owl images used on postage stamps. (a) Belgium 1999; (b) Belgium 2007; (c) Bulgaria 1980; (d) East Germany 1982; (e) Guyana; (f) Hungary 1984; (g) Iste of Man; (h) Luxembourg 1985; (i) Poland 1990; (j) Switzerland.

Likely due to its use of vineyards, the image of the Little Owl has been printed on the cork from a modern (2001) bottle of fine wine from Valle dell Asso, Galatina, Italy (Figure 2.14). Human appreciation for the Little Owl continues to grow, and the image of the owl has recently been used for the marketing of a German beer (Figure 2.15). Many countries use the image of Little Owls on a diversity of postage stamps (Figure 2.16a–j). The Little Owl is a part of the human culture and it is one of the most widespread subjects for knick-knacks and handicrafts throughout the world. Thanks to scientific studies, conservation of the Little Owl was initiated at the beginning of the twentieth century in recognition of its diet, and it is classified as a "useful" species. Now the species is protected by law in the current European Union.

The history and traditions of the Little Owl are truly long, rich, and varied, and grow with additional recoveries of artifacts from archaeological sites, as well as evolving cultural views. In closing this chapter, we urge reviewers of owl myths, traditions, and lore to closely scrutinize the information they assemble, to determine whether the ideas and symbolism described in text and artifacts still apply in contemporary societies, or whether they are part of the colorful but quaint past. Whatever the outcome, no doubt the lure and fascination of the Little Owl will continue to make a significant impression on peoples of the world.

3
Taxonomy and genetics

3.1 Chapter summary

For biodiversity conservation, aspects of taxonomy, species and subspecies divisions and distributions are important, but inherently complex. Two main approaches are relevant here, the morphological classification for subspecies and the genetic approach. The complexity is well illustrated by the fact that both approaches can yield different results with some subspecies subsequently becoming species and vice versa. Because detailed data on the Little Owl remains to be gathered, we offer both approaches separately, without drawing final conclusions. Important work still needs to be done on the mapping of subspecies distributions. See Figures 3.5–3.7.

For the Little Owl, classical taxonomy based on the morphology and plumage classifies one species consisting of 13 subspecies, with some countries where intergrading occurs.

Phylogenetic research from nucleotide sequences of the cytochrome b gene indicates that Little Owls represent a species complex. According to genetic distances, *noctua*, *vidalii*, *indigena*, *lilith*, and *plumipes* might be considered as distinct species at least under the phylogenetic species concept if not under the biological species concept. However, before drawing far-reaching taxonomic consequences, more samples should be studied from the complete range of Little Owls, covering all distinctive populations and subspecies. In addition to a more thorough genetic analysis, vocalizations and morphological differences should be compared to corroborate the genetic findings.

3.2 Morphological approach

Genus

The genus *Athene* (Boie 1822) includes four species, one in North America (*Athene cunicularia* – formerly listed as *Speotyto cunicularia*) and three in the Old World (*Athene noctua*, *Athene blewitti*, and *Athene brama*). The Burrowing Owl (*Athene cunicularia*) has 19 subspecies (del Hoyo *et al.* 1999) with recent genetic work (Desmond 1997) helping to clarify the taxonomic complexities of this species. The Forest Owlet (*Athene blewitti* also called *Heteroglaux blewitti*) differs from the Little Owl in having a streaked abdomen. As of February 2005, Forest Owlets have been reported from six sites in central India. With their

Figure 3.1 Little Owl on a perch. (Photo Peter Van Der Leer.)

Figure 3.2 Forest Owlet (*Athene blewetti*) in India. (Photo Girish Jathar.)

recent survey work of 2001–2004, Jathar and Rahmani (2004) report Forest Owlets occurring in four locations in central India. For the breeding season of October 2003 through May 2004, they recorded 67 adults and 31 juveniles. Jayant Kulkarni (personal communication) surveyed two additional sites in February 2005, and reported four adults and two juveniles at one site (this site was also surveyed by Farah Ishtiaq in 2000 [Ishtiaq *et al.* 2002]), and three adults and three juveniles at another site.

The Spotted Owlet (*Athene brama*) ranges from south Iran, east through India and Indochina, and overlaps with the Little Owl in the Middle East. The Spotted Owlet differs from the Little Owl in nape pattern and barred underparts.

Species

The world distribution of the Little Owl ranges from the Atlantic shore of Britain into Eurasia, eastward to northeastern China and the Korean Peninsula. It extends northward through Belgium, Holland, Denmark, Latvia, regions of Pskov, Moscow, to the middle part of Meshchera, north to the 56th north latitude. It is found in the region of the Ural ridge to the 54th parallel, to the Ilek River mouth, in the north-central part of Kazakhstan and in eastern Kazakhstan to the 49th parallel, to southeastern Altai, to the Tannu-Ola ridge, the southwestern Trans-Baikalia, and the Kheiluntszyan Province of northeastern Mongolia (Figure 3.3). The Little Owl is found in 84 countries (Figure 3.4).

The species is found southward to the Mediterranean Sea, to the northern shore of the Arabian Sea, northeastern Pakistan, southern Tibet, and in central and eastern China approximately to the 35th parallel. It is found on the islands of the Mediterranean Sea, on the Arabian Peninsula, northern and northeastern Africa, in western Africa south to the 22nd parallel, to southern Sudan and northern Somalia. The owl was introduced in England with Dutch specimens (Sharrock 1976) and in New Zealand (Kinsky 1973) with German specimens.

The global distribution of the Little Owl is shown in maps published by Cramp (1985), del Hoyo *et al.* (1999) and König *et al.* (1999), and reprinted here (Figures 3.5, 3.6, and 3.7, respectively).

At the time of this writing, detailed data supporting the refined distribution maps of Little Owls offered here were only available for the following countries: France, Belgium, the United Kingdom, the Netherlands, Spain, Luxembourg, Germany, Denmark, Austria, Switzerland, Hungary, Slovakia, Arabia, and Israel. We strongly urge the refinement of other in-country distributional maps based on owl survey and habitat data. Additional survey and monitoring of the Little Owl will serve to refine these distribution maps and offer insights into range contractions or expansions of the species.

Subspecies

The main contribution on the systematics of the genus *Athene* was from Vaurie (1960, 1965) who recognized 13 subspecies. Figure 3.5 shows the subspecies distribution map from Cramp (1985) and reprinted in Schönn *et al.* (1991). Readers should expect that this map will be updated as DNA analyses and refinements to the geography are undertaken in the near future. Geographical variations are based on size measurements and plumage coloration, but there are also individual and local variations that may be due to climatic or other factors, such as the daytime activity patterns in the deserts. Further genetic

Figure 3.3 Illustration of the Altai region at the border of Siberia, Mongolia, Kazakhstan, and China.

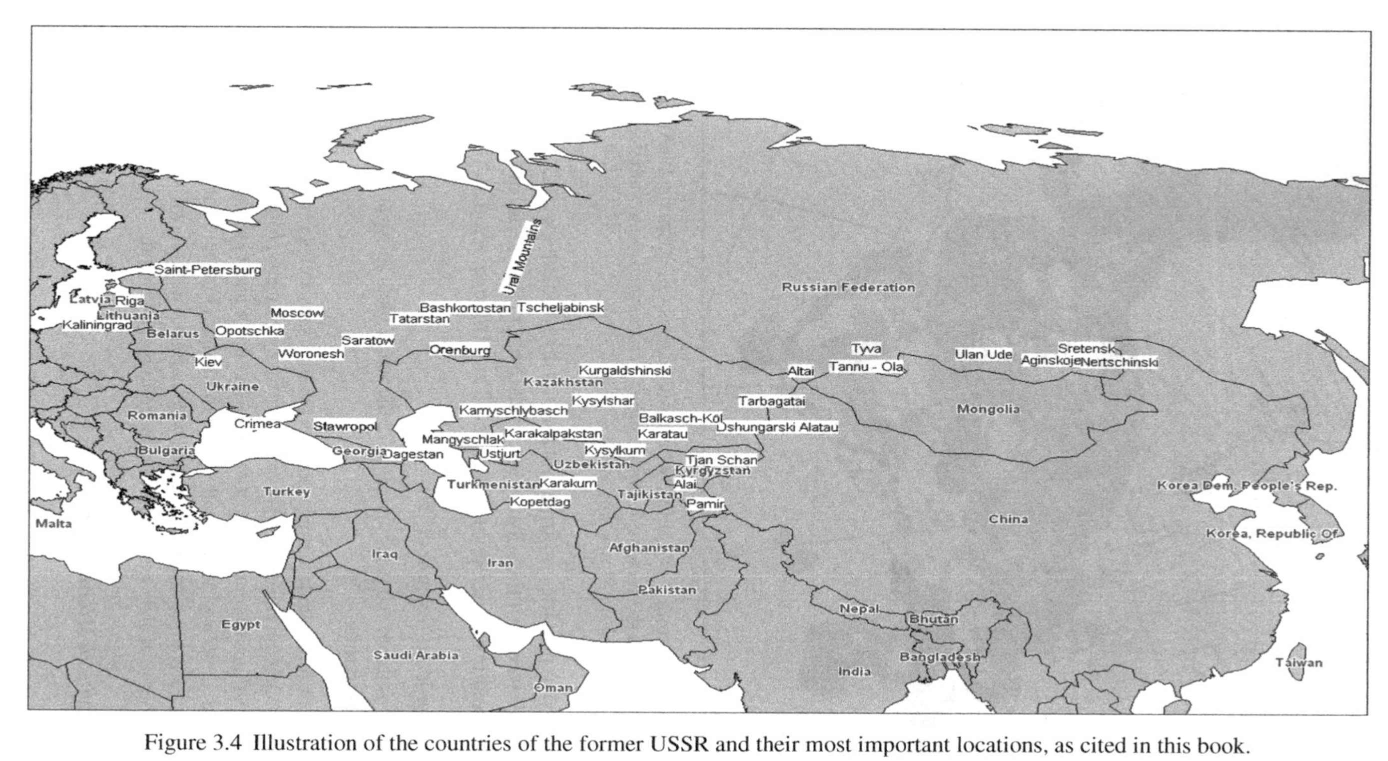

Figure 3.4 Illustration of the countries of the former USSR and their most important locations, as cited in this book.

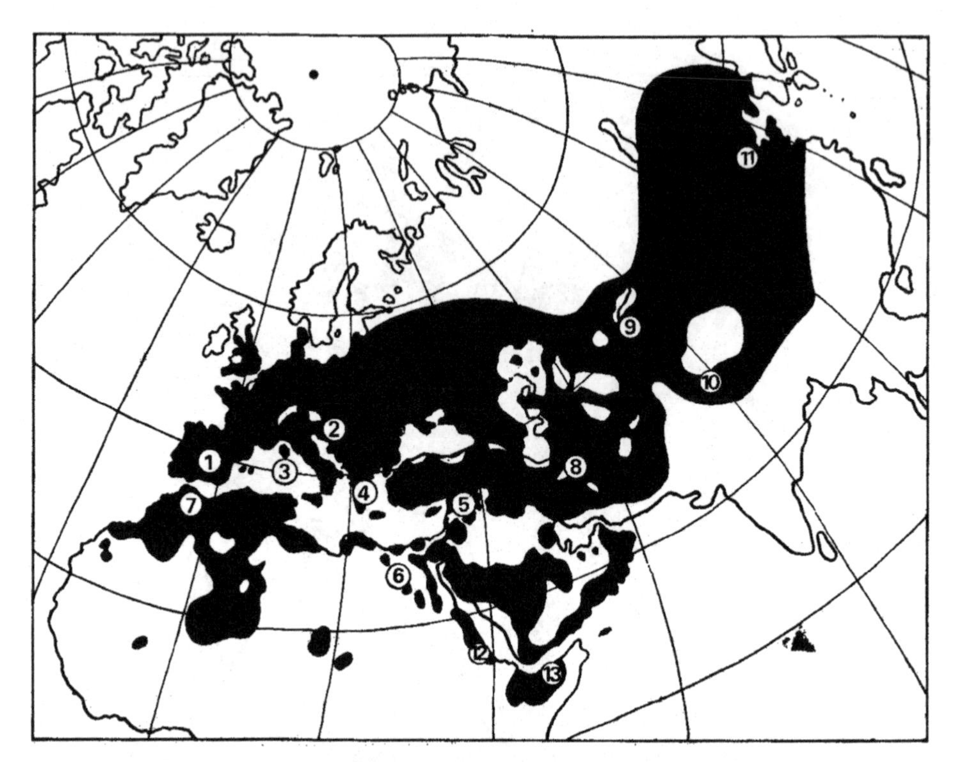

Figure 3.5 Global distribution of Little Owl and subspecies (after Cramp 1985). Subspecies 1. *A. n. vidalii* 2. *noctua* 3. *sarda* 4. *indigena* 5. *lilith* 6.*glaux* 7. *saharae* 8. *bactriana* 9. *orientalis* 10. *ludlowi* 11. *plumipes* 12. *spilogastra* 13. *somaliensis*.

testing is needed to provide confirmation for this array of subspecies. A recent genetic study confirmed the presence of the subspecies, *Athene noctua impasta*, in China (Qu *et al.* 2002). Below we present text descriptions of each of the *Athene noctua* subspecies.

Athene noctua noctua Scopoli (1769):

This race is rufous brown and paler than *vidalii*, the white spots above and the dark streaks below are somewhat less contrasting.

Range: Sardinia (Vaurie 1960) distinguished as another subspecies (*Athene noctua sarda*), Corsica, mainland Italy, southeastern Austria, Slovenia, Slovakia, Hungary, Moldova and Rumania north and west of the Carpathian Mountains, east through Denmark and Germany to Poland and the Baltic States, intergrading over a wide area with *vidalii* in southern France, Switzerland, southern Germany, Austria, the Czech Republic and Slovakia and with *indigena* in Moldova, Croatia, Bosnia; in Ukraine north of Kiev, Voronezh and Orenburg; introduced to New Zealand.

Athene noctua vidalii Brehm (1857):

The darkest race of the species. The ground color of the upperparts and of the streaks below is dark amber-brown. The tail is regularly barred with buff. The upper parts are clearly spotted with pure white.

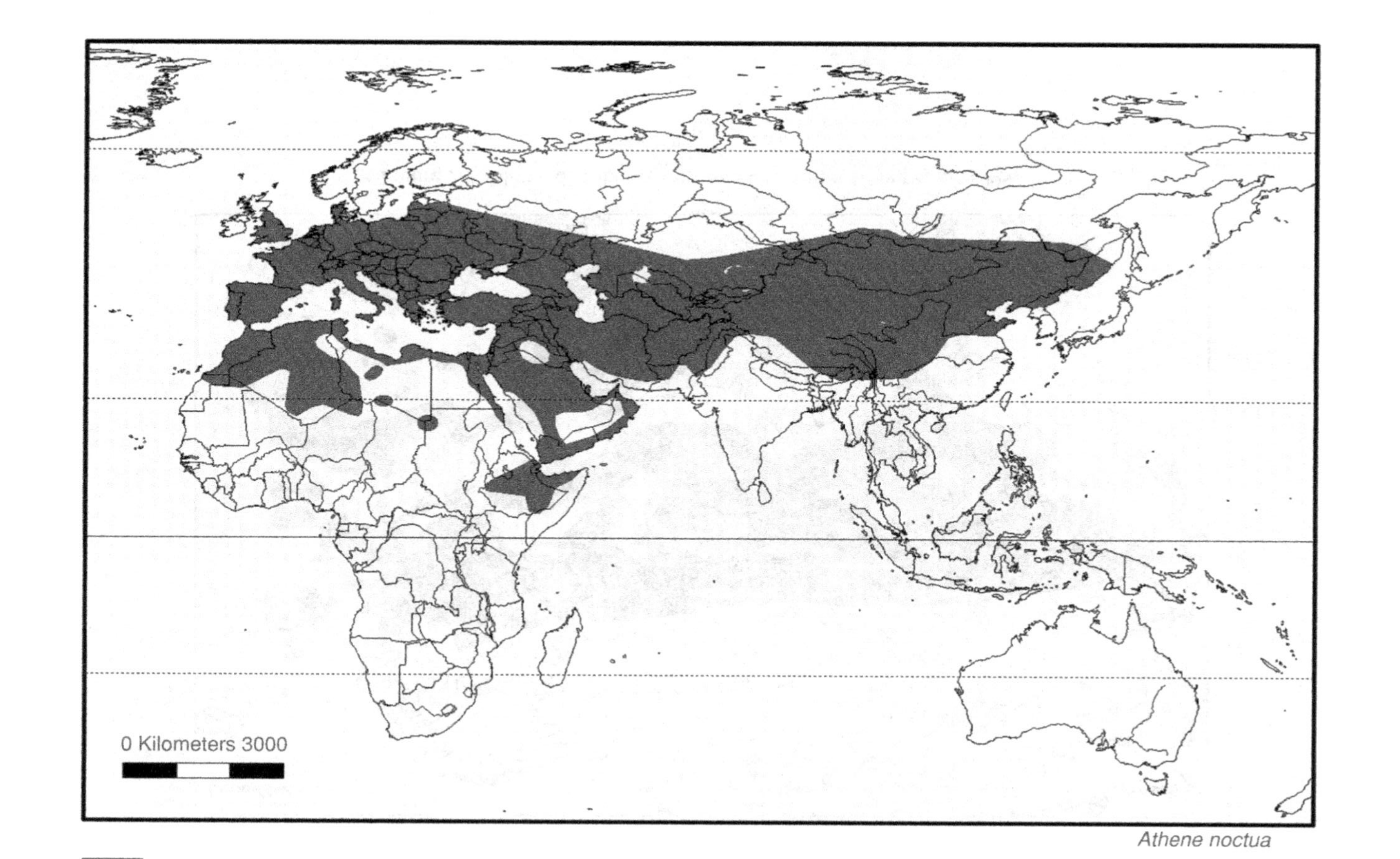

Figure 3.6 Global distribution of Little Owl (after del Hoyo *et al.* 1999).

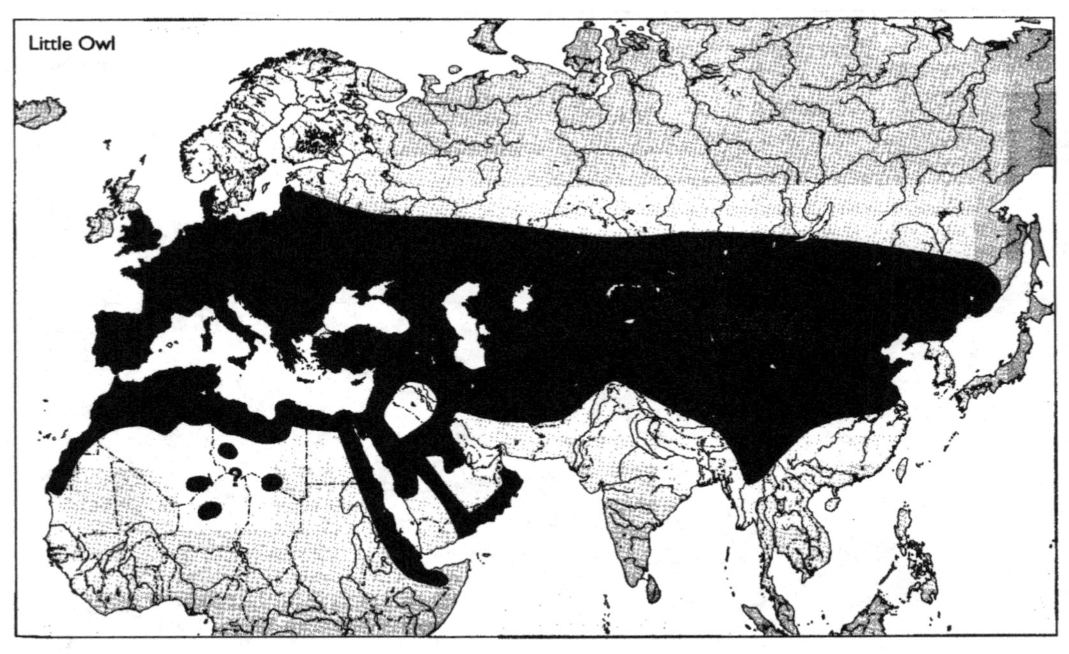

Figure 3.7 Global distribution of Little Owl (after König *et al.* 1999).

Range: Western Europe, from Netherlands and Belgium through France south to Iberia (Spain); introduced to Britain.

Athene noctua indigena Brehm (1855):

From nominate *noctua*, this race is somewhat paler, more rufous in general coloration with buff in the tail.

Range: Albania, Bosnia, southern and eastern Romania, south Moldova and Georgia, south to Crete, Rhodes, Turkey (except southeast), Levant south to about Haifa, Israel, Transcaucasia and southwest Siberia, intergrading with *bactriana* in southern Trans-Uralia.

Athene noctua glaux Savigny (1809):

A relatively dark race, the ground color of the upper parts and the streaks below are pale chocolate-brown, the tail is barred with brownish buff.

Range: north Africa and coastal Israel north to Haifa, Israel.

Athene noctua saharae Kleinschmidt (1909):

This race is similar to *glaux* but being paler and darkly streaked below, more white spotted on the crown and the entire upperparts.

Range: northern and central Saharan Desert (Africa), south of *glaux*.

Athene noctua lilith Hartert (1913):

Paler than *saharae*, very pale sand color. It is the palest race. Whiter on the crown, nape, back, and wings. Clearly less streaked below and tail more regularly barred with buff. According to genetic analysis, *Athene noctua lilith* would in fact represent a separate species of *Athene noctua*, which would be *Athene lilith* (König *et al.* 1999).

Range: Cyprus and inland Levant from Sinai to southeast Turkey, intergrading with *bactriana* in Iraq and with *saharae* in Saudi Arabia.

Athene noctua bactriana Blyth (1847):

On average larger than the preceding races with the toes more thickly feathered. Paler than *glaux* but darker than *saharae* and *lilith*. Less sandy, more greyish brown, less spotted with white than *lilith*. More greyish, less rufous brown above than *saharae*.

Range: from eastern shore of the Caspian Sea, southeast Azerbaijan and Iran east, and Ural River valley to east to eastern edge of Balkhash-Alakol hollow, foothills of Tarbagatai, Dzhungarskoe (Jungarian) Alatau, foothills of Tien Shan, foothills of Altai system. In the central and eastern part of Kazakhstan, the northern border is the 49th parallel, extending southward to the southern border of the country. Within these boundaries it inhabits lowlands, and does not occupy the upland mountains. Near the western border of its distribution (e.g., the western bank of the Ural River) it intergrades with *indigena*, and in the western and northern foothill region of Tarbagatai, Jungarian Alatau, Tien Shan, Altai system, it supposedly intergrades with *orientalis*.

Athene noctua ludlowi Baker (1926):

A relatively dark race, darker than *orientalis*, similar to *bactriana* in general coloration but better spotted with white.

Range: northern Kashmir (the foothills of the Karakoram), eastward over the Tibetan Plateau to at least the east-central region of Sikang and western Tsinghai in China.

Athene noctua orientalis Severtzov (1873):

The general coloration of the upper side of body is usually lighter than in *bactriana* (sometimes there are birds indistinct from *bactriana* in this respect), clayish-pale-grey, less brown. Dark pattern on the lower side of body is somewhat more contrasting than in *bactriana*. The extent of white spottedness on the upper side of body is more developed than in the other races.

Range: Pamiro-Altai mountain country, Tien Shan, Jungarian Alatau, Tarbagatai. In the western and northern foothill region of the marked territory it supposedly intergrades with *bactriana*, in the region of Tarbagatai supposedly with *plumipes*.

Athene noctua plumipes Swinhoe (1870):

The general coloration of the upper side of body is close to *bactriana* and insignificantly darker, more brown-grey, than in *orientalis*. The white spottedness of the upper side of body is similar to *orientalis*.

Range: Southeastern Altai (Kazakhstan – Russian Federation), Tannu-Ola (Russian Federation – Mongolia), southwestern, southern, and southeastern Trans-Baikalia (Russian Federation). Borders of the northern distribution are not yet defined. In the region of Tarbagatai (Kazakhstan) it supposedly intergrades with *orientalis*.

Athene noctua impasta Bangs & Peters (1928):

Impossible to distinguish from *Athene noctua plumipes* with morphology and plumage coloration. Only genetic analysis can separate this subspecies from *plumipes* (Qu *et al.* 2002).

Range: distributed in Qinghai and Gansu Provinces of China.

Athene noctua spilogastra Heuglin (1863) and *Athene noctua somaliensis* Reichenow (1905)

Distributed on the African coast of the Red Sea from Sudan to Somalia.

Fossil Little Owls have been described from the Pleistocene and prehistoric period from 14 sites in 11 countries across the species' range (Brodkorp 1971, Andrews 1990). See Figure 3.8.

Intergrading of subspecies

Intergrading of subspecies is important in certain countries such as Kazakhstan, Israel, China, and Russia.

Based on morphology, Kazakhstan has five subspecies (Gavrilov 1999) ranging as follows:

Figure 3.8 Little Owl on tree. (Photo Peter Van Der Leer.)

noctua: northern part of Volga-Ural inter-river area (south of the 49th parallel), and the Ural River valley eastward to the lower Ilek River valley.
indigena: southern part of Volga-Ural inter-river area northward to the 49th parallel, eastward to the lower Ural River valley.
bactriana: from the eastern shore of the Caspian Sea and Ural River valley to eastern border of Balkhash-Alakol hollow, Tarbagatai foothills, Jungar Alatau, and Tien Shan.
orientalis: Tien-Shan, Jungarian Alatau, Tarbagatai, probably foothills of Altai and Kalba.
plumipes: Altai; in the region of Tarbagatai it supposedly intergrades with *orientalis*.

Based on morphological data, Russia has five subspecies (Stepanyan 1990) with distributions as follows:

noctua: from the western border of the country eastward to the Uralian ridge and Lower Ilek valley. The northern border of the species' breeding range is in this region; the southern border is the northwestern shore of the Black Sea, Lower Dnieper River, northern shore of the Azov Sea, in the region of the Volga-Ural inter-river area to the 49th parallel. The southern border of its distribution is in the region of the Crimean Isthmus the southeastern part of Ukraine, north Caucasia; it intergrades with *indigena* at the region of the 49th parallel in the Volga-Ural inter-river area.
indigena: Crimea, Great Caucasia, Trans-Caucasia, southern part of the Volga-Ural inter-river area, eastward to the Nizhniy (Lower) Ural River area. Northward to the Crimean Isthmus, the northern shore of the Azov Sea, in north Caucasia and in the Volga-Ural inter-river area north to the 49th

parallel. It integrades with *noctua* near its northern border of distribution, in the region of the Crimean Isthmus, southeastern Ukraine, and eastwards in the region of the 49th parallel. It intergrades with *bactriana* near the eastern borders of its distribution, in the region of the western bank of the Ural River.

bactriana: from the eastern shore of the Caspian Sea and Ural River valley eastward to the eastern edge of the Balkhash-Alakol hollow, foothills of Tarbagatai, Dzhungarskoe (Jungarian) Alatau, foothills of Tien Shan, and foothills of the Altai system. Near the western borders of its distribution, in the region of the western bank of the Ural River, it intergrades with *indigena*. It supposedly intergrades with *orientalis* in the western and northern foothills of Tarbagatai, Jungarian Alatau, Tien Shan, and the Altai system.

orientalis: occupies the Pamiro-Altai Mountain country, Tien Shan, Jungarian Alatau, and Tarbagatai. It supposedly intergrades with *bactriana* in the western and northern foothills of the Tarbagatai, Jungarian Alatau, and Tien Shan region. It supposedly intergrades with *plumipes* in the Tarbagatai region.

plumipes: Southeastern Altai, Tannu-Ola, southwestern, southern, and southeastern Trans-Baikalia. Borders of the distribution on the north are insufficiently clear. In the region of Tarbagatai it supposedly intergrades with *orientalis*.

The distribution of *Athene noctua bactriana* (Hutton's Owlet) covers the area from Iran, Iraq, east northeastwards through Afghanistan and north Pakistan, Kashmir to Syr Darya. In addition, it is not uncommonly encountered in suitable habitat in western Ladakh, in the Indus Valley. The range of *Athene noctua ludlowi* (Tibetan Owlet) is much more restricted, being limited to the area from eastern Ladakh eastwards through Tibet, north Sikkim, and north Bhutan. Ladakh can therefore be considered as the zone where the two subspecies meet (Pfister 1999).

China has four subspecies (Qu *et al.* 2002) ranging as follows: *ludlowi* in Tibet, *orientalis* in Xinjiang, *plumipes* in Shanxi, and *impasta* in Qinghai and Gansu.

Serbia-Montenegro features two subspecies: *Athene noctua noctua* in the northern part and *A. n. indigena* in the southern part (Matvejev 1950).

Israel is a meeting point of four subspecies but only *lilith* occurs in its typical form. The three others (*indigena*, *glaux*, and *saharae*) represent only intergrading populations with *lilith* (Shirihai 1996). This complex situation could possibly be clarified by the genetic approach, which classifies the subspecies into the species *Athene lilith*. The three intergrading subspecies might be forms of the latter species rather than *Athene noctua* subspecies.

3.3 Genetic approach

Genetic insights

The caryotype of the species is $2n = 82$ with 18 macrochromosomes and 64 microchromosomes (Capanna *et al.* 1987). The genetic variability was studied with multi-locus enzyme electrophoresis in Italy (Randi *et al.* 1991). The polymorphism (P) of the Little Owl is

0.13 and heterozygosity (Ho) 0.026, which is similar to the Barn Owl but less than the Long-eared Owl (P = 0.23; Ho = 0.024) and Tawny Owl (P = 0.16; Ho = 0.041). Because of the low heterozygote frequency, the Little Owl could be subdivided in different and more or less isolated geographic populations, with different allele frequencies at some polymorphic loci. However, the studied samples were not large enough to allow intraspecies population analysis (Randi *et al.* 1991). In France, genetic fingerprinting was carried out to examine the basis for a supplementation experiment of a Little Owl population (Génot *et al.* 2000). Here, genetic work was conducted to establish the polymorphism within birds coming from different regions of the country and from different localities in the same region. The DNA bands were analyzed to examine the degree of genetic similarity. Technically speaking, X (X = 2Nab/Na + Nb), Na and Nb are the numbers of bands in individual a and b, Nab is the number of bands shared by a and b; if X is near one the birds are closely related genetically, if X is near 0, they are distant. The analysis of owls from different geographic origins revealed that genetic variability is not unique to each region. Two unrelated birds from the same region may be more distant genetically (X = 0.36) than two birds from different regions (X = 0.52). However, the sample size of nine individuals was too restricted to establish with any certainty the degree of genetic diversity between two unrelated birds. As far as the owls from different localities were concerned, the marked similarity coefficients indicate the existence of an important genetic proximity between the different individuals studied (X = 0.8). Such values thus reflect both the absence of great genetic diversity locally and the homogeneity of the similarity coefficients obtained.

DNA fingerprint analyses were also carried out in Germany to calculate band-sharing coefficients (0.58 ± 0.09; n = 53) between putative fathers and nestlings to check a high rate of paternal care and no extra-pair young in a Little Owl population (Müller *et al.* 2001).

Qu *et al.* (2002) showed with genetic markers that *Athene noctua impasta*, the fourth subspecies of *Athene noctua* in China, is actually a single subspecies and not a synonym of *Athene noctua plumipes*.

In France, a study (Bouchy 2004) used five polymorphic microsatellites (DNA markers), isolated *de novo*, for three aims. First, to assess the genetic structure of five French populations of Little Owls: three from the Vosges, one from the Drôme, and one from Oléron Island. Second, to confirm or rule out demographic observations obtained from mark–recapture programs, as well as to give reliable estimations of dispersion rates that could not be accurately evaluated with classic mark–recapture methods in these cases. Third, to determine if some populations, genetically close to the Vosges ones, could be used as sources of wild birds for the reinforcement program in the Vosges (instead of using the captive birds, which resulted in failure). The populations exhibited quite high internal genetic diversity, which means that the authors did not find any genetic impact of fragmentation. All analyzed populations appeared substructured. Results indicated that the dynamics of the populations were not correctly described and that further sampling efforts should be organized in light of this study. Although this genetic study on Little Owls revealed a significant genotypic differentiation between the populations studied, geographically distant populations were genetically close. Thus instead of using local birds from captivity in this reinforcement program, one can use birds from other wild and healthy populations. However, the real utility of this conservation action should be studied carefully. Indeed, some populations appeared endangered, according to their supposed dynamics. This study showed that sampling methods were not appropriate in the Drôme case and that birth and death rates were probably not correctly estimated, especially in the case of the Northern Vosges. Hence, other sampling efforts, given all the remarks made and evidence found, should be organized to improve our knowledge of these populations, prior to any conservation decision.

A mixed demographic genetic model was constructed for the subpopulation of the Northern Vosges. Deleterious mutations have an impact not only on population dynamics, but also on the rate of effective exchanges. Under a pessimistic demographic scenario, all three local Northern Vosges subpopulations are sustained by a high immigration rate from a large German population located nearby (2–3 km) in Sarre (Germany). Inbreeding depression has therefore little impact on the vital rates and mutation accumulation does not occur. Under the optimistic scenario, local demographic rates are sufficient to explain population sustainability, deleterious mutations may pose a serious threat to persistence within a 100-year period, especially if the number of German migrants is low (Bouchy 2004).

Phylogenetic and phylogeographic relationships (contributing author: Michael Wink)

Since owls are nocturnal and their morphological traits are often very similar (König *et al.* 1999) a systematic differentiation or speciation event can easily be overlooked. Together with morphological and vocal characteristics, nucleotide sequences of marker genes can help to define phylogenetic relationships, systematics, and evolution of owls.

In different studies (Heidrich & Wink 1994, 1998; Heidrich *et al.* 1995a, b; Wink & Heidrich 1999, 2000; Wink 2000; Olsen *et al.* 2002; Wink *et al.* 2004) the mitochondrial cytochrome b gene is regularly amplified and sequenced. Because this gene shows a good resolution at the genus level and since we already have quite a large database for this gene, it is used as a platform to cover all taxa of owls. About 109 taxa of the Strigidae and 21 of the Tytonidae have been studied and a phylogeny based on cytochrome b data has been published (Wink & Heidrich 1999, 2000; Wink *et al.* 2004).

In addition to *Athene* a selection of representative owls of the genera *Glaucidium, Surnia*, and *Aegolius* have been included in the analysis, since these genera form a monophyletic group together with *Athene*. A sister group to this assemblage consists of members of the genera *Bubo, Strix, Otus, Megascops, Asio*, and *Ptilopsis*. Both groups form the subfamily Strigidae that clusters as a sister group to the Tytonidae (Wink & Heidrich 1999, 2000).

The phylogeny methods maximum parsimony (MP) and maximum likelihood (ML) were employed to reconstruct phylogenetic relationships in the present data set (Figures 3.9a, b, and c).

The following clades and relationships were recognized unequivocally by all methods of tree reconstruction:

a monophyletic subfamily Tytonidae (boot-strap support 97%) with *Tyto* and *Phodilus* at the generic level
a monophyletic subfamily Strigidae (boot-strap support 68%) comprising all other owl genera
a monophyletic genus *Aegolius* (boot-strap support 100%) that clusters basal to *Athene, Surnia*, and *Glaucidium*
a monophyletic genus *Athene* (boot-strap support 91%)
a paraphyletic genus *Glaucidium* including *Surnia* (boot-strap support 69%)
a monophyletic *Bubo* complex, which comprises the former genera *Scotopelia, Ketupa*, and *Nyctea* (Wink *et al.* 2004) (boot-strap support 90%)
a monophyletic *Otus* complex of the Old World (boot-strap support 100%)
a monophyletic *Megascops* complex of the New World (formerly included into *Otus*) (boot-strap support 92%)
Ptilopsis (formerly also *Otus*) clusters as a sister to *Asio* (boot-strap support 71%)
a monophyletic genus *Strix* (boot-strap support 99%)

These relationships could be confirmed using nucleotide sequences of a nuclear marker gene (LDH b intron) (Wink *et al.* 2004).

Relationships within the genus Athene

Nucleotide sequences of the cytochrome b gene were used to infer phylogenetic and phylogeographic relationships in members of the genus *Athene*. Surprisingly, *Ninox superciliaris* turns out to be a member of the monophyletic *Athene* clade and will be regarded as *Athene superciliaris*. Within *Athene noctua* several well-defined clades are apparent that agree with recognized subspecies. A substantial differentiation might be observed in the Little Owl, *Athene noctua*, as more data becomes available. Because some of the genetic distances between the subspecies are in the range of distances that are typical for established species, it is likely that *Athene noctua* comprises a monophyletic species complex that could be subdivided into several 'good' species.

The Madagascar Hawk Owl, *Ninox superciliaris*, which occurs on Madagascar, clusters clearly in the clade of *Athene*, which would also agree with its plumage characteristics. It appears as a sister to the New World Burrowing Owl, *Athene cunicularia*, which shows some subspecific differentiation (Desmond 1997) and which has formerly been treated as a monotypic genus, *Speotyto* (Sibley & Monroe 1991).

The Spotted Owlet, *Athene brama*, clusters at the base of the Little Owl complex. Within the Little Owl complex of *Athene noctua* several geographically defined subspecies have been differentiated (del Hoyo *et al.* 1999; König *et al.* 1999), which show some overlap and some of them might not represent good taxa at all (König *et al.* 1999).

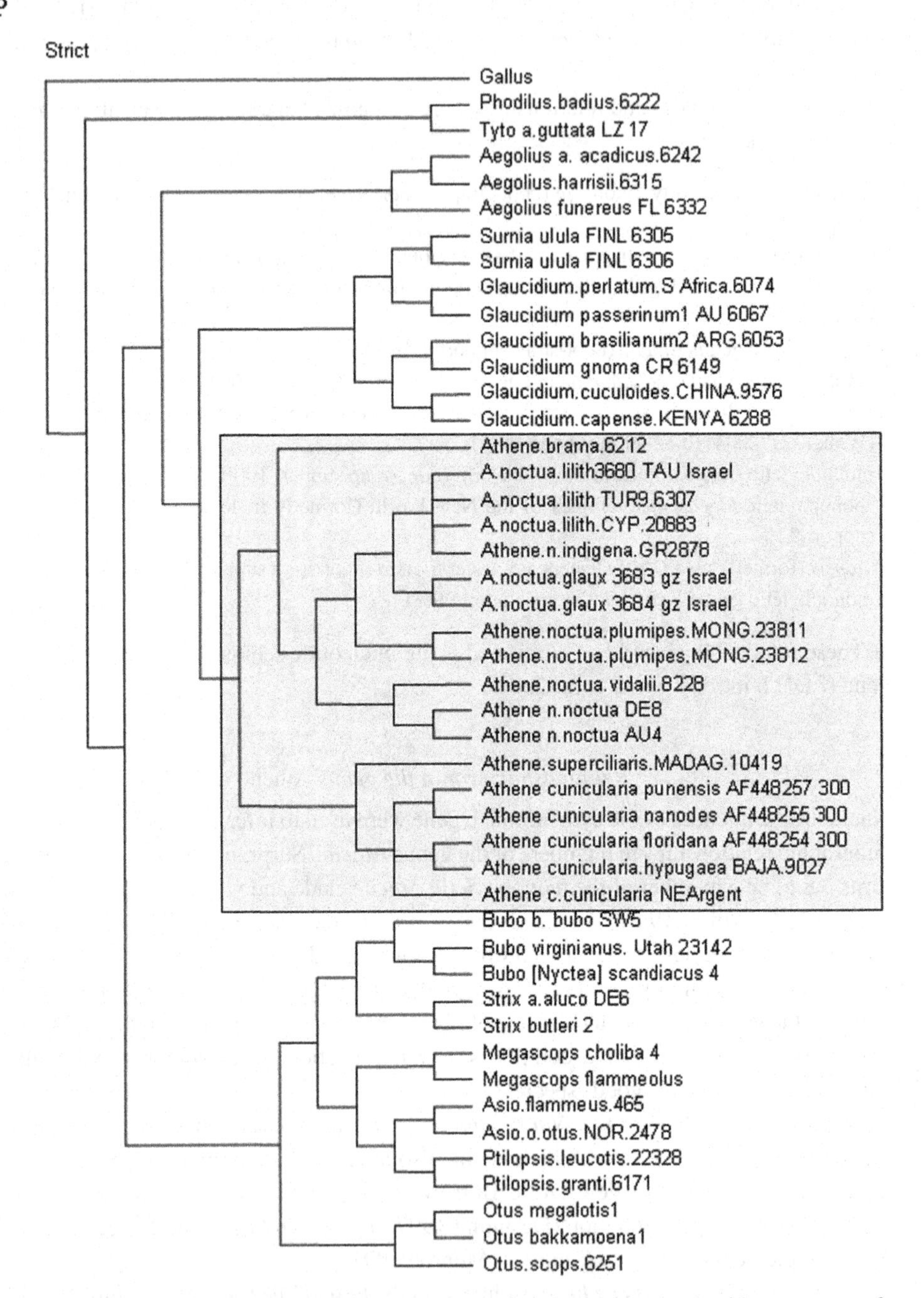

Figure 3.9a Molecular phylogeny of *Athene* and other owls inferred from nucleotide sequences of the cytochrome b gene; representation as a strict consensus cladogram of 16 equally parsimonious trees (MP).

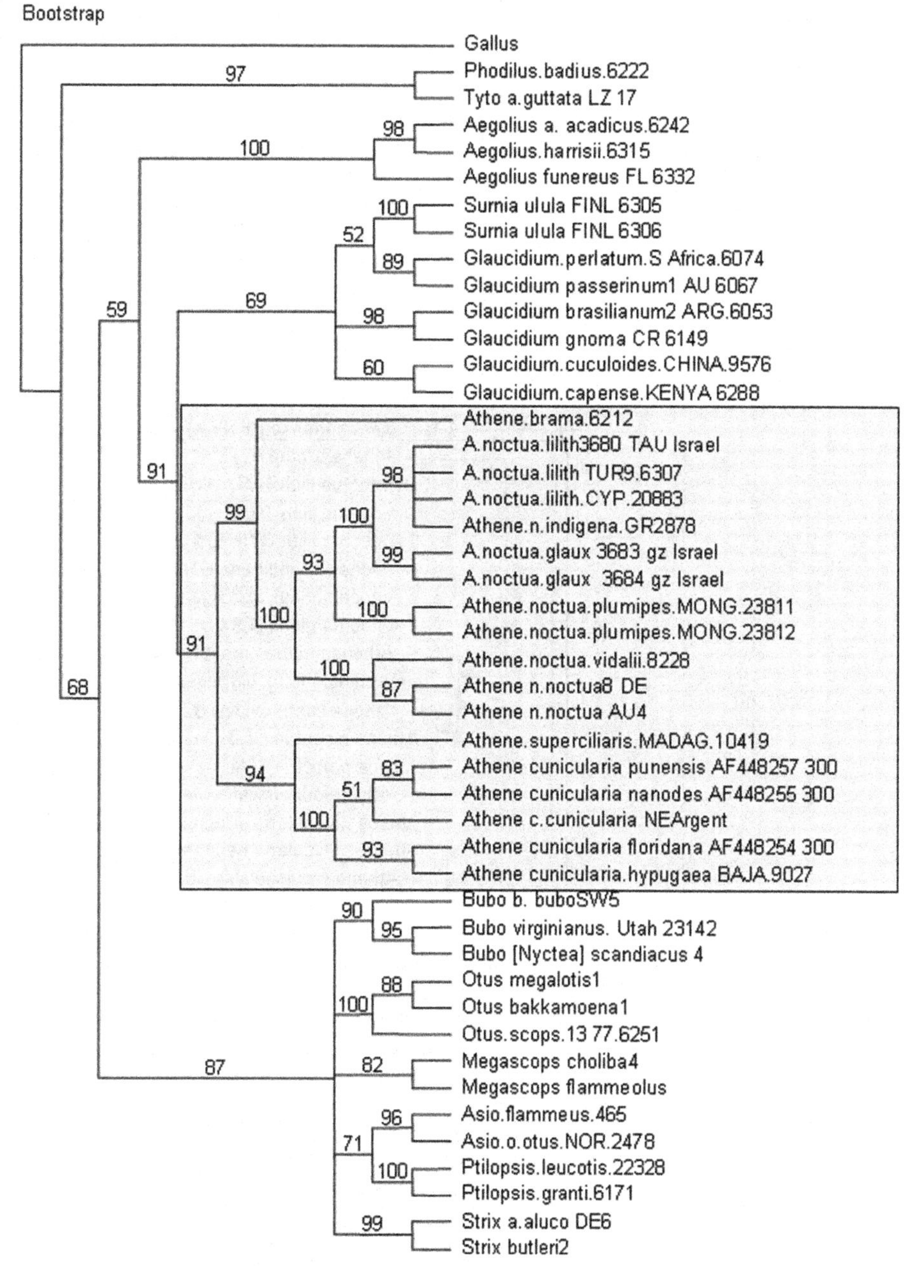

Figure 3.9b Molecular phylogeny of *Athene* and other owls inferred from nucleotide sequences of the cytochrome b gene; MP-boot-strap cladogram with boot-strap values (from 1000 replications) at branches.

MP

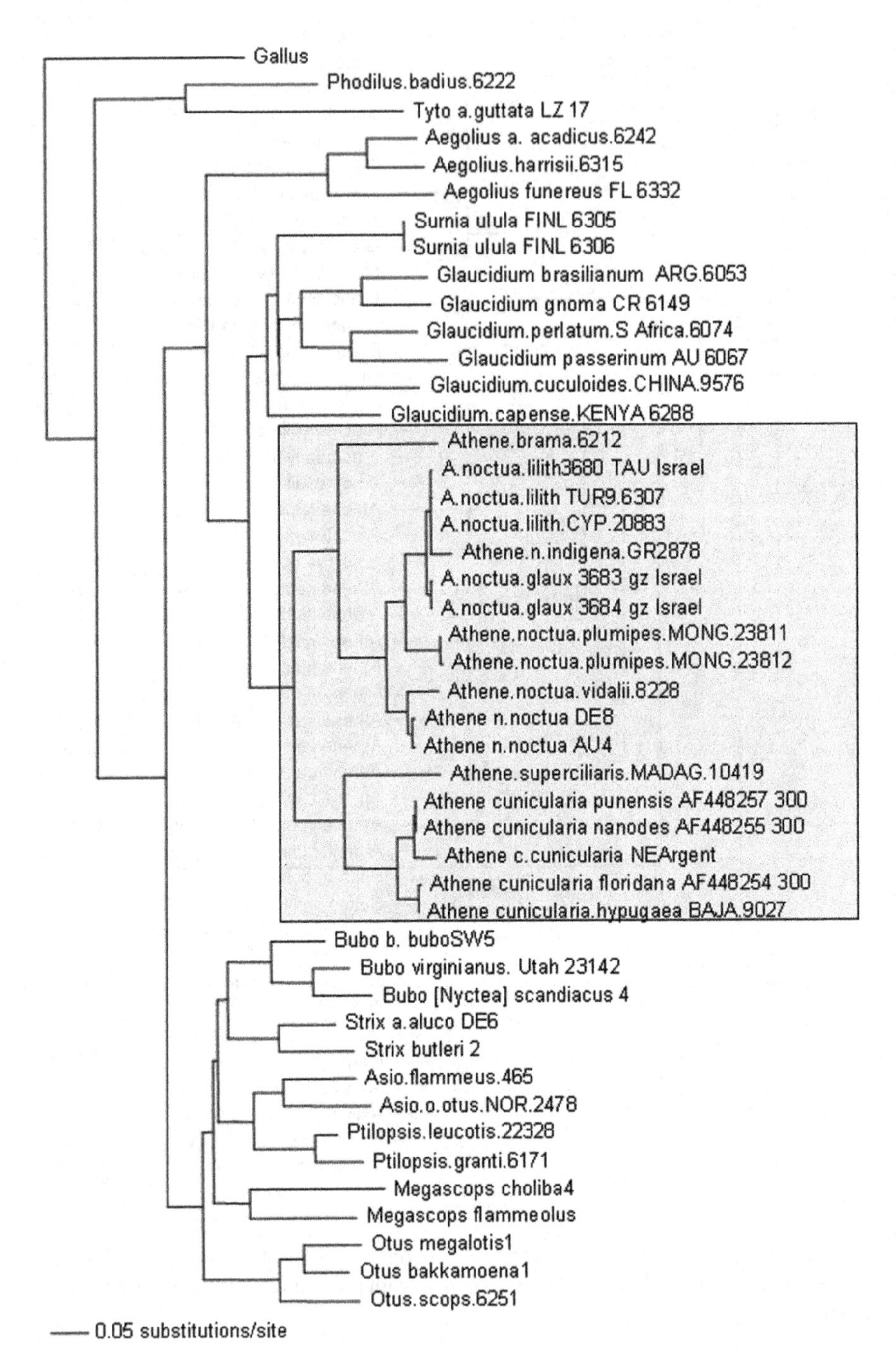

Figure 3.9c Molecular phylogeny of *Athene* and other owls inferred from nucleotide sequences of the cytochrome b gene; ML phylogram in which branch lengths reflect genetic distances.

The current subspecies listing of *Athene noctua* includes the following:

A. n. vidalii (A. E. Brehm 1857): Western Europe (introduced to England, New Zealand)
A. n. noctua (Scopoli 1769): Central Europe, south to Sardinia and Sicily
A. n. indigena (C. L. Brehm 1855): Balkans, Aegean islands south to Crete; east to Asia minor and Caspian Sea
A. n. lilith (Hartert 1913): Cyprus, inland Middle East in Turkey, Syria, and the Arabian peninsula
A. n. glaux (Savigny 1809): N Africa, coastal Israel (including *A. n. saharae* [Kleinschmidt 1909] according to König *et al.* [1999])
A. n. spilogastra (Heuglin 1869): Red Sea coast of E Africa
A. n. somaliensis (Reichenow 1905): E Ethiopia, Somalia
A. n. bactriana (Blyth, 1847): Central Asia, Iran, Iraq, Afghanistan
A. n. orientalis (Severtzov 1873): NW China, Siberia
A. n. impasta (Bangs & Peters 1928): Kokonor, W Gansu
A. n. ludlowi (Baker 1926): South central China, S and E Tibet
A. n. plumipes (Swinhoe 1870): NE China, Mongolia

The analysis of the *Athene* complex so far covers the subspecies *vidalii, noctua, indigena, lilith, glaux*, and *plumipes*. Some of the African and Asian taxa are not represented thus leaving the analysis incomplete at present. The present data set reveals three phylogenetic lineages that are supported by boot-strap values between 93% and 100%.

Athene n. noctua and *A. n. vidalii* that occur in West and Central Europe represent sister taxa that appear to be well recognized (genetic distance 4.1%) and clearly separated from *A. n. indigena* on the Balkans.

Athene n. plumipes clusters at the base of the clade leading to Little Owls of the Balkans, the Near and Middle East. Its genetic distance to the other Little Owl subspecies is between 6% and 9%.

Athene n. lilith, A. n. indigena, and *A. n. glaux* are closely related but differentiated. *A. n. lilith* from Cyprus shows an identical haplotype to Little Owls from Turkey and Israel. The genetic distances between *lilith* and *indigena* are 2.2–2.6% whereas *lilith* and *glaux* differ by 1.1% only. However, this group differs from European Little Owls by 6.8–7.6%.

Genetic distances of cytochrome b gene above 1.5–2% have been regarded as indicative for species level in owls (Wink & Heidrich 1999) and show that a gene flow has not occurred between sister taxa within the last one million years (assuming that 2% distance equals one million year divergence: Wilson *et al.* 1987; Tarr & Fleischer 1993). Within defined populations genetic distances are usually below 0.5% (also in this study). According to this argument, *noctua, vidalii, indigena, lilith*, and *plumipes* represent not only well-recognized subspecies but might even be considered as distinct species at least under the phylogenetic species concept if not under the biological species concept.

The present data could indicate that Little Owls represent a species complex, similar to the situation of the Barn Owl (*Tyto alba*). Also in the Barn Owl many geographically defined subtypes have been recognized of which many have attained species status due to morphological and genetic differences (König *et al.* 1999).

Reconstruction with maximum parsimony and maximum likelihood

Of 1143 nucleotide sequence characters analyzed, 576 are variable and 471 are parsimony informative. Base frequencies: A = 0.32, C = 0.43, G = 0.08, T = 0.15. Settings for ML correspond to the GTR + G + I model. Number of substitution types: 6; distribution of rates at variable sites: gamma distribution; shape: 0.82; rate categories: 4.

4
Morphology and body characteristics

4.1 Chapter summary

This section gives an overview of the characteristics of the Little Owl (Figure 4.1). We first look at the plumage of adult birds and juveniles as they grow. It is possible to distinguish adult and first-year birds based on their plumage. In particular, the tips of the primaries are pointed and have larger spots in first-year birds and are square with smaller spots in adult birds. The moult is described to obtain a view on the replacement of the feathers. The Little Owl has differential biometrical measures (such as length) according to the subspecies or sex (for example weight – females are heavier close to breeding time). Descriptions for taking body measurements are offered in the glossary. The species has a large vocal repertoire including 40 acoustic signals and combinations. It has retina cells similar to diurnal birds of prey. While Little Owls can differentiate several colors, the species does not see infrared light. It can hear well enough to locate small rodents with an accuracy of up to 1%. We finally examine its voice and describe specific characteristics of its flight and anatomy.

4.2 Plumage

Adult

The main color of the upper parts and the upper wings of the adult is dark chocolate-brown, usually with a slight olive-brown tinge. The forehead and crown are closely marked with cream-buff streaks, each streak often widens slightly towards the feather-tips, and the crown shows elongated spots. A distinct *white* V-mark is noticeable on the hind neck from behind the ear down to the center of the neck like a face pattern (Figure 4.2; Scherzinger 1986).

The mantle and scapulars are covered with large white or pale cream rounded subterminal spots, often joined to form a single wide subterminal bar. The back, rump, and upper tail have large rounded subterminal spots of cream-buff. The facial disc is poorly developed, and is not surrounded by a distinct ruff. The ring around the eye is white and is widest between the eye and the base of the bill.

The chin is white, bordered below by a buff-and-grey spotty band across the throat to the sides of the neck. The upper chest is also white and the lower chest, the breast, and the upper flanks are dark fuscous-brown to olive-brown. Each feather presents a paired subterminal

Figure 4.1 Little Owl on the ground (François Génot).

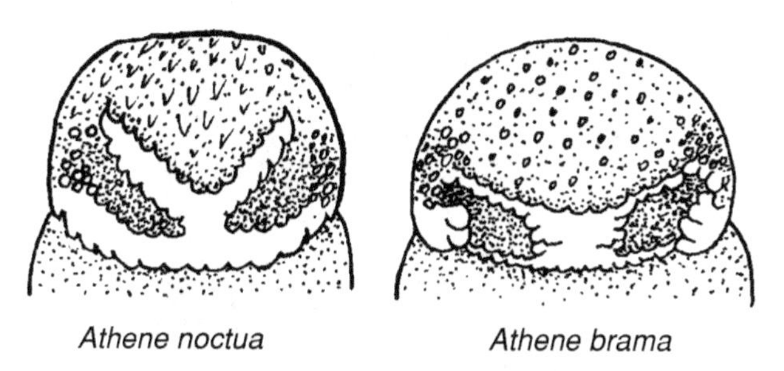

Figure 4.2 Distinct white V-mark on hind neck as seen on *Athene noctua* and *Athene brama* (after Scherzinger 1986).

pink-buff or white spot or short bar. The ground color of the tail is a similar or slightly paler olive than the upper parts of the owl; the terminal bars or spots on the tail may be lighter due to bleaching from exposure to sunlight.

The pattern of the pale bars on the tail is somewhat variable. Starting at the base of the tail feather and working onwards to the tip: one straight bar is just visible on the tail feathers at the tips of the longest upper-tail coverts. Similar second and third (and sometimes fourth) bars are found across the middle part of the tail. At the feather-tips traces of a terminal bar are often present in the form of a spot, sometimes contiguous with a white border. Thus, usually three bars, but sometimes four bars and possibly traces of a fifth bar, are visible across the tail feathers.

The flight-feathers are dark olive-brown or fuscous-brown, darkest along the shafts and outer primaries. These primaries have three to four well-spaced pale spots along the outer webs, and much larger spots or short bars are seen on the inner webs. The secondaries are similar to the primaries, but have only two to three rows of spots or bars; the tips of the secondaries have large or small, white paired spots; sometimes these spots or bars are completely absent. The upper wing-coverts, like the mantle and scapulars, have white spots smaller than those that are found on most of the primary coverts, and often restricted to one web only. The under wing-coverts and axillaries are cream-white, slightly spotted, and mottled gray. The sexes are similar, but the male tends to have a whiter face.

Juveniles and first-year adults

The first down (neoptile) of the young is short, dense, and white. It becomes mottled gray after one week when the mesoptile starts to grow. The second down (mesoptile) of the juvenile is rather feather-like. The feathers are still distinctly softer and shorter than in first-year and adult owls. The structure of the flight-feathers, tail, and greater primary coverts are as in the adult. The neoptile clings to the tips of the mesoptile for up to three to four weeks (retained longest on the crown, flanks, and thighs). The mesoptile is similar to the adult, but with a paler ground-colour and more grayish. The crown, back, rump, and upper tail-coverts are almost uniform gray-brown or with faint buff spots. The face is the same as in the adult, but grayer and less contrastingly marked. It is possible to determine the age of juveniles by measuring their growing primaries (Juillard 1979).

Figure 4.3a Difference in the shape of the tips of the primaries in first-year and adult Little Owls. Wing tip of adult bird. (Photo Jacques Bultot.)

Figure 4.3b Difference in the shape of the tips of the primaries in first-year and adult Little Owls. Wing tip of a first-year bird. (Photo Jacques Bultot.)

First-year birds are closely similar to adults (J. Bultot, personal communication). They retain their juvenile flight-feathers and tail into adulthood, with the pattern variable at all ages. The main differences are as follows.

- The shape of the tips of the primaries is rather pointed in the first year; tips are almost square in adult owls. The size of the spots on the tips of the primaries is larger in first-year birds than in adults. (Figures 4.3a and b).
- The plumage of the adult is fresh until the onset of the breeding season; the flight-feathers, tail, and in particular the tertials are slightly worn in the first autumn, distinctly worn in spring, and heavily worn in summer.
- The texture and shape of the tertials are rather soft, narrow, and taper to a tip in a first-year bird; they are broader and more square-tipped in the full adult. See Figures 4.4–4.7.

Adult and first-year owls have a lemon-yellow iris. The eyelids are a dark slate-blue. The bill is lemon-yellow or greenish-yellow. The cere is slate-gray or olive-black. The tarsus is grayish or grayish-yellow and the toes are gray-brown or olive-black. The claws are dark brown to brownish-black. At hatching, the skin under the down of the young, the cere, and

Figure 4.4 Two Little Owl primaries (François Génot).

Figure 4.5 Little Owl primary (François Génot).

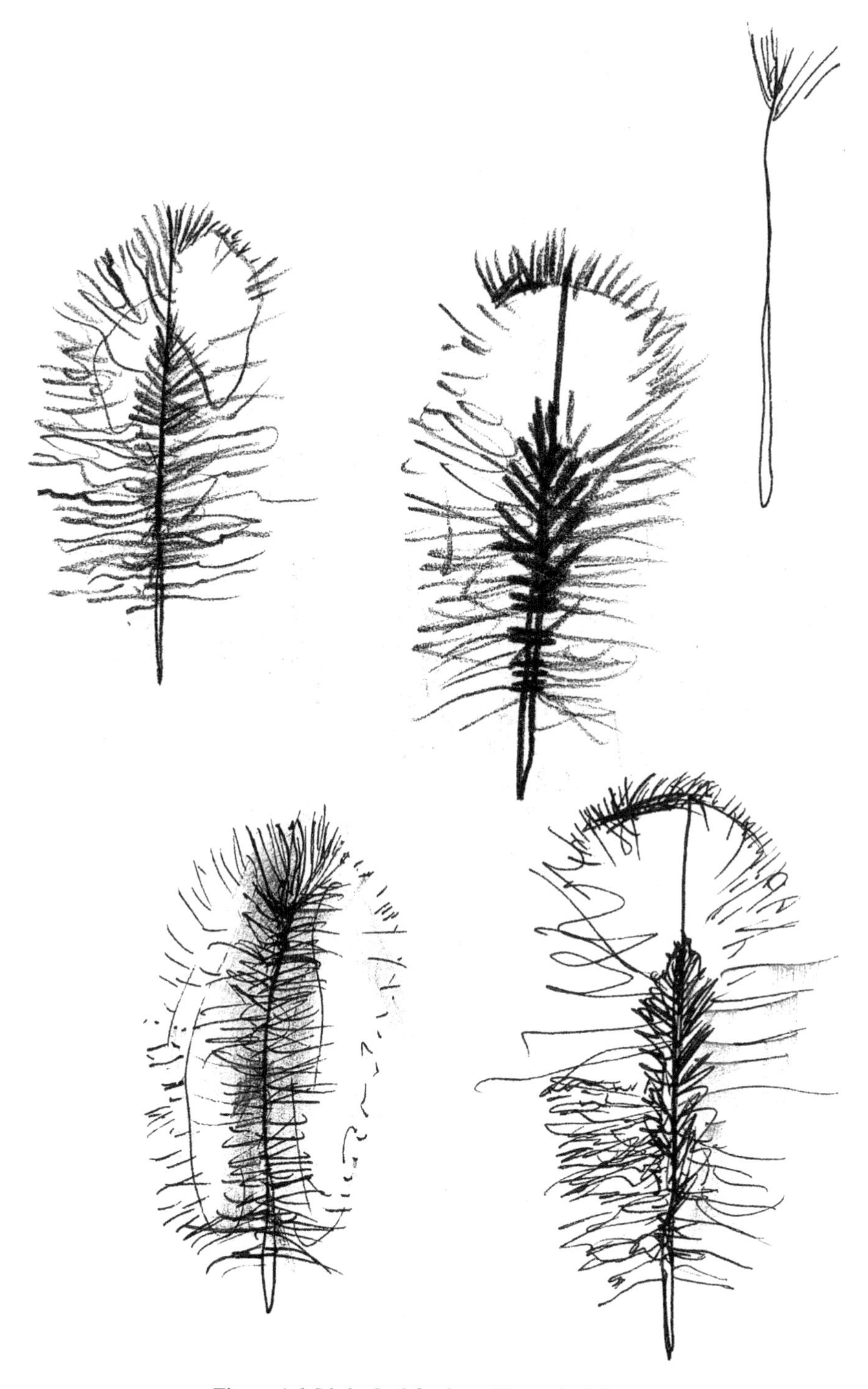

Figure 4.6 Little Owl feathers (François Génot).

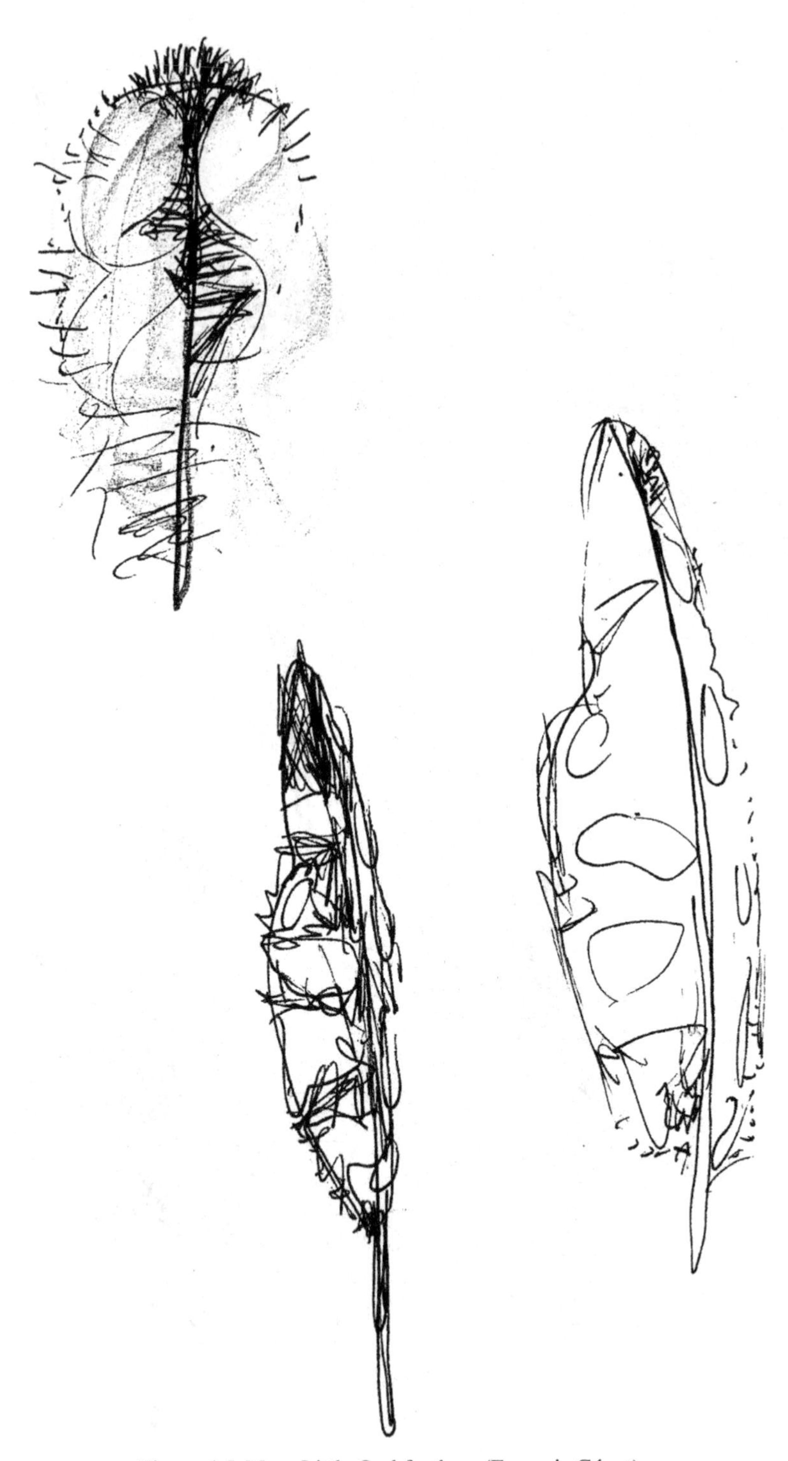

Figure 4.7 More Little Owl feathers (François Génot).

the toes are pink; the bill and claws are whitish- or grayish-pink. The eyes open on the tenth day and the iris is pale yellow at first. The skin partly darkens to gray after a few days, first on the toes and cere. At fledging, the iris and eyelids are as in the adult. The bill is olive-yellow with the tip of the culmen a pale yellow. The cere becomes slate-blue or blackish violet-green. The tarsus are yellow or yellow-gray, toes flesh-gray or dusky gray and claws black with a gray-blue base. Some anomalies can occur in the plumage as in the case of partial albinism (Bisseling 1933), leucistic (Brinzal, personal communication), or a russet form (Paris 1909).

4.3 Moult

The moult of adults starts after breeding, at the end of June or beginning of July, when the young fledge or are independent (Haverschmidt 1946; Dementyev & Gladkov 1951; Stresemann & Stresemann 1966). However, late breeders, especially those with a replacement clutch, may start moulting at about the time of hatching (Ullrich 1970), although some breeders can be found without any moult (Exo *in* Schönn *et al.* 1991). The onset date of the moult depends on the age of the owls and is earlier in older birds (Piechocki 1968; Hartmann-Müller 1973, 1974).

The moult begins with the loss of the first primary (P1) in mid-May to mid-July, and is completed with P10 in early September to early November (Stresemann & Stresemann 1966; Piechocki 1968). A different view of the moult sequence was offered by Dementyev and Gladkov (1951) who considered that replacement of the primaries goes centrifugally. There is little geographical variation in the onset of the moult, but northern birds starting from mid-May may be non-breeders, and eastern birds start earlier as in Turkmenistan (Dementyev & Gladkov 1951). It is probably due to a breeding cycle that is more extended in the year. Normally, the complete moult starts after P1 is shed (Glutz & Bauer 1980) but in other cases after a secondary or tail-feather is shed (Exo *in* Schönn *et al.* 1991).

In two captive owls, P10 was shed 98–99 days after P1 (Piechocki 1968) and three months for a male in Turkmenistan (Bel'skaya 1992).

The secondaries (S) moult from three centers: S10 starts with the shedding of P3–P4, S5 starts with P6–P7, and S1 starts with P5–P7 (Cramp 1985). S4, S7, and S8 moult last, shed at about the same time as P10 (Piechocki 1968). Most secondaries are replaced during the last stages of the primary moult (early August to mid-October).

Both sexes undergo a complete post-breeding moult, normally the male moults before the female (Martinez *et al.* 2002). The primaries commence from P1 to P10. The secondaries are shed in three groups S1–S4, S5–S7 ascending, the third group of secondaries are moulted descending from S10–S8. Tail moult occurs simultaneously usually coinciding with the period when P4 and P5 are growing. Juvenile Little Owls tend to undergo a partial body moult and will moult some wing-coverts, normally between August and November. The primaries are moulted for the first time in the owl's second summer.

The tail-feathers are shed sometimes simultaneously, but in general within a few days up to two weeks (Piechocki 1968; Exo *in* Schönn *et al.* 1991) with an irregular sequence. The

Figure 4.8 Juvenile inner secondaries. (Photo Patricia Orejas / BRINZAL.)

tail-feathers are shed 20–30 days after P1 (Piechocki 1968) or when P4–P5 are growing, and this is completed during the growth of P7 (Stresemann & Stresemann 1966).

The scheme of moult is different according to sex among Little Owls in Turkmenistan (Bel'skaya 1992). Complete replacement for females took seven months (May–November) and eight months (April–November) for males. Replacement of small feathers in females took four months (June–September) and three months (July–September) for males. Replacement of the tail-feathers in females was three months (June–August) while two months (July–August) in males.

The body moult starts with the lesser and median upper wing-coverts, soon followed by the mantle and scapulars (Cramp 1985). The skin falls out in shreds in May–June (Bel'skaya 1992).

For fledged juveniles, the moult begins in July at the age of 70–75 days, when they are separated from their parents (Bel'skaya 1992). The face feathers and lesser and median upper wing-coverts are moulted first (early July to mid-August) and largely completed six to seven weeks later (September or early October). The moult of the crown, neck, and feathering of the legs and toes occur in late November (Cramp 1985). The moult is partial for head, body, and wing-coverts.

The general moult pattern in Spain is given by Martinez *et al.* (2002) (Euring codes between brackets). (See Figures 4.8–4.16.)

1Y autumn (3) / 2Y spring (5) (bird in first calendar year to second calendar year): partial moult affecting head, body, and some wing coverts. Inner greater coverts retained. Rather worn plumage before first moult of remiges, especially noticeable in tertials.

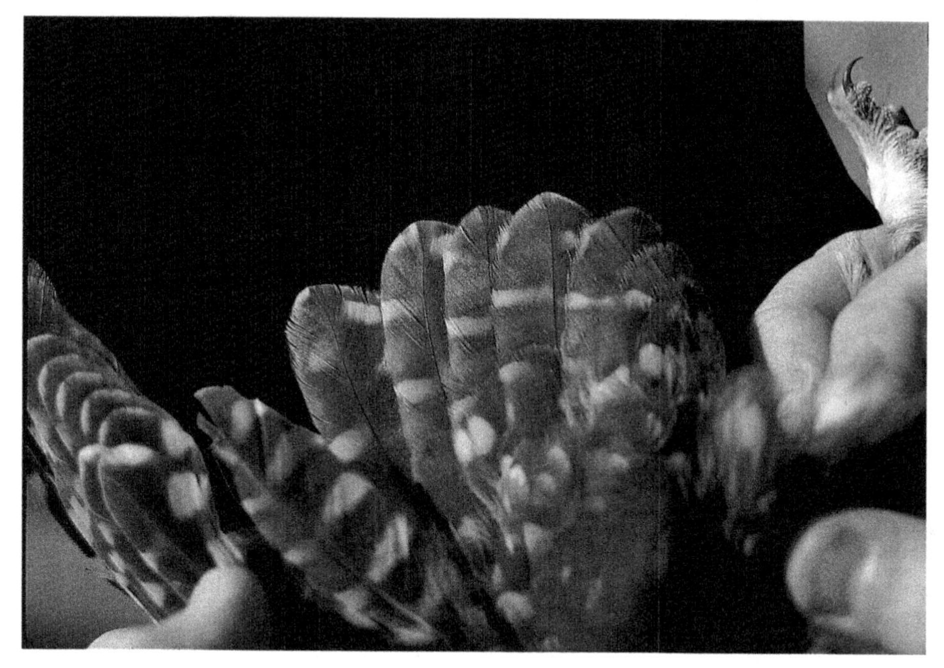

Figure 4.9 Adult inner secondaries. (Photo Patricia Orejas / BRINZAL.)

Figure 4.10 Greater coverts. Left: juvenile; right: adult. (Photo Raul Alonso / BRINZAL.)

2Y+ autumn (4) / 3Y+ spring (6) (bird in second calendar year or older to third calendar year or older): most birds undergo complete moult, but some retain a few secondaries. Adult-like tertials. If some secondary has been retained the contrast between new and old secondaries should be checked.

2Y autumn (5) / 3Y spring (7) (bird in second calendar year to third calendar year) high strong contrast between retained feathers (old juvenile) and the fresh ones.

3Y+ autumn (6) / 4Y+ spring (8) (bird in third calendar year or older to fourth calendar year or older) less contrast.

Figure 4.11 1Y spring (3). During partial moult, all greater coverts are still juvenile. (Photo Raul Alonso / BRINZAL.)

Figure 4.12 2Y spring (5). Contrast between moulted 1–9 greater coverts and not moulted inner greater coverts and primary coverts. (Photo Patricia Orejas / BRINZAL.)

Figure 4.13 3Y+ spring (6). Moulting. Two generations of adult feathers. S3–S6 not moulted. Rest of remiges moulted. (Photo Patricia Orejas / BRINZAL.)

Figure 4.14 3Y+ spring (6). All greater coverts are adult. There is no contrast between them and primary coverts or remiges. (Photo Patricia Orejas / BRINZAL.)

Figure 4.15 2Y+ autumn (4). Adult inner secondaries. Remiges of the same generation despite some variation in brightness. (Photo Inigo Zuberogoitia / E. M. Icarus.)

Figure 4.16 Moulting. Juvenile and adult remiges. (Photo Inigo Zuberogoitia / E. M. Icarus.)

Table 4.1 *General body characteristics of male and female Little Owls.*

Sex	Wing (mm)	Tail (mm)	Bill (mm)	Tarsus (mm)	Weight (g)	Author
Male	161.3 (n = 25)	82 (n = 18)		29.7 (n = 15)	143.7 (n = 32)	Simeonov *et al.* 1989
Female	164.5 (n = 21)	83.2 (n = 15)		30 (n = 11)	151.3 (n = 27)	
Male	169 (n = 4)	86 (n = 4)	13 (n = 4)	31 (n = 4)	182 (n = 4)	Khokhlov 1992
Female	170 (n = 4)	89 (n = 4)	13 (n = 4)	30 (n = 4)	220 (n = 4)	
Male	162.2 (n = 8)				153.1 (n = 10)	Blache personal communication
Female	164.8 (n = 15)				165.8 (n = 17)	
Male	162.9 (n = 20)	77.8 (n = 22)	13.5 (n = 6)	34.5 (n = 8)	160.1 (n = 14)	Mlikorvsky & Piechocki 1983
Female	166.5 (n = 26)	79 (n = 27)	13.4 (n = 13)	37.2 (n = 9)	153 (n = 13)	
Male	163.9 (n = 24)				180.1 (n = 26)	Van Harxen & Stroeken personal communication
Female	167.7 (n = 127)				200.7 (n = 233)	
Male	161.8 (n = 26)	82.1 (n = 26)		39 (n = 9)	155.3 (n = 9)	Martinez *et al.* 2002
Female	163.6 (n = 26)	82 (n = 25)		41.1 (n = 7)	185.7 (n = 7)	
Male	157.8 (n = 25)			34.59 (n = 27)	159.37 (n = 27)	Mastrorilli 2005
Female	160.5 (n = 6)			35.39 (n = 10)	176.8 (n = 9)	

4.4 Measurements

This section gives an overview of the size of the Little Owl. The measurements of the Little Owl are given in relation to the other related *Athene* species before focusing on the differences between the subspecies. In Tables 4.1–4.6 we have included detailed body measurements from a number of publications. Included in these data are references to subspecies designations of the Little Owl. Readers should be cautioned that these designations often

Table 4.2 *Variation in male and female Little Owl,* Athene noctua, *weights during the breeding season (van Harxen & Stroeken personal communication).*

	March–April	Incubation	Young 0–10 days	Young >11 days
Male	198.2 (n = 10)	170.9 (n = 15)	138 (n = 1)	
Female	220.1 (n = 8)	209.2 (n = 136)	190.6 (n = 52)	179.6 (n = 37)

pre-date findings from recent genetic studies, and that these and additional genetic data may suggest a somewhat different taxonomic structure in the Little Owl complex and the associated array of body measurement data.

Interspecific differences

According to tarsus (t) length and wing (w) length (shown as t; w respectively), the categorization between the different species of *Athene* is as follows (Cramp 1985): *Athene blewetti* (26; 148), *Athene brama* (28; 152), *Athene noctua* (32–35 and 156–162), *Athene cunicularia* (44; 164) (Schönn *et al.* 1991). There appears to be a relationship between the habitat and the size of the species. The more open the habitat, the taller the species, thus *Athene cunicularia*, the biggest *Athene*, has the longest legs because it lives on the ground in areas of steppe vegetation (Scherzinger *in* Schönn *et al.* 1991).

Body characteristics

The following measurements are most commonly used: weight (g), wing length (mm), tail length (mm), bill size (mm), and tarsus length (mm). A general overview of these measurements is given in Table 4.1. Martinez *et al.* (2002) also report mouth size (mm), back claw (mm), front claw (mm), talon length (mm), P8 length (mm), wingspan (mm) and total length (mm). Used terms are explained in detail in the glossary.

Weight

As with the majority of owl species, female Little Owls are heavier than their male counterparts. According to the mean weight over the year, the greatest difference in weights occurs during the breeding season, and the smallest during the end of the summer and the autumn. Data recorded during a ringing study in northeastern France (from March to July) gave the average weights for males at 164 g (n = 36) and 181 g for females (n = 64) (Génot unpublished data). The weight was very variable according to the annual cycle, as shown in England with an average weight of 164 g in the winter (n = 61) and 178 g in the spring

Table 4.3 *Comparison of wing lengths* (in mm) *of male and female Little Owls.*

			Males						Females						
Subspecies	Region	Origin	Average	+/–	sd	Min	Max	n	Average	+/–	sd	Min	Max	n	Author(s)
A. n. noctua	Northern Italy & Sardinia	stuffed	158.00	+/–	2.2	156	162	8	161	+/–	3.4	156	166	10	Cramp (1985)
A. n. noctua	Northern Italy	stuffed	153.00	+/–	3.85	147	158	6	152	+/–	1.41	151	154	4	Taranto pers. comm.
A. n. noctua	Central Italy	stuffed	144.83	+/–	16.43	111	160	7	145.29	+/–	19.45	105	160	7	Taranto pers. comm.
A. n. noctua	Southern Italy	stuffed	158.00	+/–				1		+/–					Taranto pers. comm.
A. n. noctua	The Netherlands	juv/stuffed	160	+/–	3	155	166	35	163	+/–	4.1	157	171	39	Cramp (1985)
A. n. noctua	The Netherlands	adults/stuffed	163	+/–	3.7	158	169	13	166	+/–	3.9	161	173	13	Cramp (1985)
A. n. noctua	Northrhein-Westfalen (Germany)	stuffed	159.9	+/–	3.9	154.5	168	13	164.9	+/–	3.6	161.5	172	10	Glutz & Bauer (1980)
A. n. noctua	Northrhein-Westfalen (Germany)	juveniles	161	+/–	1	160	162	2	166.1	+/–	4.5	157	172	15	Glutz & Bauer (1980)
A. n. noctua	Northrhein-Westfalen (Germany)	adults	164.1	+/–	2.7	160	170	15	169.6	+/–	4.3	159	177	31	Glutz & Bauer (1980)
A. n. noctua	Unterer Niederrhein (Germany)	adults	166.3	+/–	4.2	158	176	16	169.7	+/–	3.2	162	177	38	Exo (*in* Schönn *et al.* 1991)
A. n. noctua	Mittelwestfalen (Germany)	adults	163.9	+/–	3.8	154	172	64	167.8	+/–	3.4	160	176	149	Kämpfer & Lederer (*in* Schönn *et al.* 1991)
A. n. noctua	Schwäbisches Albvorland (Germany)	adults	166	+/–	3.5	158	173	44	167.8	+/–	3.3	158	174	43	Ullrich (*in* Schönn *et al.* 1991)
A. n. noctua	Bodensee (Germany)	adults	161	+/–	4.3	156	168	10	162.8	+/–	3.2	157	166	10	Knötzsch (*in* Schönn *et al.* 1991)

(*cont.*)

Table 4.3 (*cont.*)

			Males						Females						
Subspecies	Region	Origin	Average	+/–	s	Min	Max	n	Average	+/–	s	Min	Max	n	Author
A. n. noctua	Basel region (Germany)	adults	161.5	+/–	2.7	157	165	10	170.4	+/–	3.7	165	178	10	Bauer (*in* Schönn *et al.* 1991)
A. n. noctua	Eastern Germany		162.9	+/–	5.3			20	166.6	+/–	5.4			26	Mlikovsky & Piechocki (1983)
A. n. noctua	North-West Austria	stuffed	161	+/–	2.2	158	164	9	164	+/–	4.1	160	172	7	Cramp (1985)
A. n. noctua	West Slovakia	stuffed	160.3	+/–	5.2	151	167	9	162.3	+/–	5.4	155	178	9	Keve *et al.* (1960)
A. n. noctua	East Slovakia	stuffed	163.2	+/–	4.4	158	170	5	165	+/–	4.4	159	170	7	Keve *et al.* (1960)
A. n. noctua	Hungary	stuffed	161.5	+/–	4.1	155	165	6	161.7	+/–	4.7	152	168	7	Keve *et al.* (1960)
A. n. noctua	central Eastern Europe	stuffed	160.2	+/–		152	169	37	168.4	+/–		158	177	26	Dementyev & Gladkov (1951)
A. n. noctua	Italy	living birds 2/4 Euring age	157.85	+/–	2.93	154.5	160	10	160.2	+/–	2.68	158	164	5	Mastrorilli (2005)
A. n. noctua	Italy	living birds 3/5 Euring age	157.76	+/–	4.52	149	168	15	162	+/–		162	162	1	Mastrorilli (2005)
A. n. noctua	Italy	living birds all ages	157.8	+/–	3.69	149	168	25	160.5	+/–	2.56	158	164	6	Mastrorilli (2005)
A. n. noctua	Achterhoek (the Netherlands)	living birds	163.9	+/–				24	167.7	+/–				127	van Harxen & Stroeken pers. comm.
A. n. noctua	Vlaams Brabant (Belgium)	living birds	161.19	+/–	3.7	152	170	63	164.9	+/–	3.7	155	174	89	Smets, Huybrechts & Cerulis pers. comm.
A. n. noctua	Wallonia (Belgium)	living birds	157	+/–	3.24	151	165	41	161	+/–	4.5	153	174	88	Bultot pers. comm.
A. n. vidalii	Iberian Peninsula	stuffed	158	+/–	2.4	154	161	9	161	+/–	3.1	157	166	13	Cramp (1985)

A. n. vidalii	Iberian Peninsula	living birds	161.8	+/–	5.4	145	170	26	163.6	+/–	4.9	153	170.5	26	Martinez *et al.* (2002)
A. n. indigena	Greece, Albania, Romania	stuffed	164	+/–	4.1	158	171	12	167	+/–	3.9	162	174	13	Cramp (1985)
A. n. indigena	Balkans	stuffed	162.4	+/–		163	168	27	164.3	+/–		156	172	33	Dementyev & Gladkov (1951)
A. n. indigena	Crimea	stuffed	166	+/–	3.9	160	170	6	172	+/–		171	173	2	Cramp (1985)
A. n. indigena	West Asia Minor	stuffed	166	+/–	4.4	160	173	9	169	+/–	2.7	166	173	6	Cramp (1985)
A. n. bactriana	Iran, West Pakistan, Afghanistan, Kazakhstan	stuffed	166	+/–		159	174	40	169	+/–		159	177	21	Cramp (1985)
A. n. orientalis	Turkestan, Pamir, Tien Shan, Fergana	stuffed	168.1	+/–		165	178	10	175.8	+/–		170	181	5	Dementyev & Gladkov (1951)
	Sinkiang	stuffed	162	+/–		152	156	10	167.6	+/–		166	169	3	Dementyev & Gladkov (1951)
A. n. plumipes	Altai, Mongolia, North China, Korea	stuffed	163.3	+/–		158	170	23	173.7	+/–		167	178.5	9	Dementyev & Gladkov (1951)
A. n. lilith	Syria, Lebanon, Sinai, Israel	stuffed	160	+/–	3	154	164	17	162	+/–	4.1	157	172	14	Cramp (1985)
	Cyprus	stuffed	157	+/–	3.1	152	162	15	156	+/–	2.1	153	160	11	Cramp (1985)
	Egypt	stuffed	159	+/–	3.6	157	165	8	159	+/–	2.8	156	163	7	Cramp (1985)
A. n. glaux & saharae	Algeria, Tunisia, Tibesti, Libya, North-West Egypt	stuffed	154	+/–	3.6	146	161	34	157	+/–	4.1	151	165	23	Cramp (1985)
A. n. glaux	North-West Morocco	stuffed	161	+/–	2.9	157	164	4	161	+/–	2.7	157	164	6	Cramp (1985)

Table 4.4 *Comparison of tail lengths* (in mm) *of male and female Little Owls.*

			Males					Females					
Subspecies	Region	Origin	Average +/−	sd	Min	Max	n	Average +/−	sd	Min	Max	n	Author
A. n. noctua	The Netherlands	adults / stuffed	75.90	2	74	79	11	79.6	2.1	77	83	9	Cramp (1985)
A. n. noctua	The Netherlands	juveniles stuffed	73.1	2.9	69	78	29	74.1	2.7	71	80	27	Cramp (1985)
A. n. noctua	Nordrhein-Westfalen	stuffed	75.1	3.4	70	82	12	75.1	2.2	70	78	10	Glutz & Bauer (1980)
A. n. noctua	Unterer Niederrhein	adults / stuffed	75.3	3.2	69	82	18	75.8	3.1	69	84	35	Exo (*in* Schönn *et al.* 1991)
A. n. noctua	southeastern Germany	living birds	77.9	3.6			22	79	3.7			27	Mlikovsky & Piechocki (1983)
A. n. noctua	NW Austria	stuffed	74.4	2.4	70	77	9	74.4	1.6	73	77	7	Cramp (1985)
A. n. noctua	Northern Italy & Sardinia	stuffed	75.5	2.3	73	79	8	77	1.8	75	80	8	Cramp (1985)
A. n. vidalii	Iberian Peninsula	stuffed	75.5	2.3	73	79	8	77	1.8	75	80	8	Cramp (1985)
A. n. vidalii	Iberian Peninsula	living birds	82.11	7.5	75	90	26	82.08	3.3	76	90	25	Martinez *et al.* (2002)
A. n. indigena	Greece, Albania, Romania	stuffed	80	4.5	75	87	6	81.6	4	76	89	7	Cramp (1985)
A. n. lilith	Cyprus	stuffed	74.6	3	71	78	4	74.9	2.2	72	78	6	Cramp (1985)
A. n. glaux & saharae	Tunisia, Algeria, Libya, Tibesti, NW Egypt	stuffed	72.8	2.6	69	77	14	74.2	3.6	70	79	12	Cramp (1985)
A. n. glaux	Tunisia	stuffed	82.5		80	85	2	89		88	90	2	Erlanger (1898)

Table 4.5 *Comparison of bill lengths* (in mm) *of male and female Little Owls measured from the tip of the bill to the start of the cere.*

			Males						Females						
Subspecies	Region	Origin	Average	+/–	sd	Min	Max	n	Average	+/–	sd	Min	Max	n	Author
A. n. noctua	Unterer Niederrhein	adult	11.9	+/–	0.5	11.1	13	20	12	+/–	0.5	11.2	13.7	37	Exo (*in* Schönn *et al.* 1991)
A. n. noctua	Mittelwestfalen	adult	12.2	+/–	0.6	11.2	13.4	64	12.4	+/–	0.6	10.7	13.9	150	Kämpferer & Lederer (*in* Schönn *et al.* 1991)
A. n. noctua	Jülicher Börde	adult	11.9	+/–				1	12.4	+/–	0.4	12	13.1	6	Gassmann (*in* Schönn *et al.* 1991)
A. n. noctua	Northern Italy	stuffed	13.50	+/–	0.55	13	14	6	13.93	+/–	0.76	13.3	15	4	Taranto pers. comm.
A. n. noctua	Central Italy	stuffed	13.08	+/–	0.98	10	14.5	6	13.67	+/–	0.61	13	14.5	6	Taranto pers. comm.
A. n. noctua	Southern Italy	stuffed	15.00	+/–				1		+/–					Taranto pers. comm.

Table 4.6 *Comparison of tarsus lengths* (in mm) *of male and female Little Owls for different subspecies.*

Subspecies	Region	Origin	Males						Females						Author
			average	+/–	s	min	max	n	average	+/–	s	min	max	n	
A. n. noctua	The Netherlands	stuffed	34.80	+/–	1.2	33.4	36.3	21	35.6	+/–	1.1	34	36.8	20	Cramp (1985)
A. n. noctua	Nordrhein-Westphalia	stuffed	36.7	+/–	2.1	33	40	11	39.4	+/–	1.6	38	42	6	Glutz & Bauer (1980)
A. n. noctua	southeastern Germany		34.5	+/–	1.9			8	37.2	+/–	4.6			9	Mlikovsky & Piechocki (1983)
A. n. noctua	NW Austria	stuffed	34.5	+/–	0.9	33.5	35.6	8	35	+/–	1.3	33.4	36.2	7	Cramp (1985)
A. n. noctua	Northern Italy & Sardinia	stuffed	31.9	+/–	1.1	30.6	33.6	8	31.8	+/–	1.5	29.4	33.3	8	Cramp (1985)
A. n. noctua	Northern Italy	stuffed	25.92	+/–	2.5	23	30	6	27.67	+/–	1.84	25.5	30	4	Taranto pers. comm.
A. n. noctua	Central Italy	stuffed	24.33	+/–	3.69	19	31	7	24.64	+/–	3.7	20	30	7	Taranto pers. comm.
A. n. noctua	Southern Italy	stuffed	15.00	+/–				1		+/–					Taranto pers. comm.
A. n. noctua	Italy	living birds 2/4 Euring age	34.61	+/–	1.38	33	36.9	12	35.3	+/–	1.26	33.6	37.3	9	Mastrorilli (2005)
A. n. noctua	Italy	living birds 3/5 Euring age	34.57	+/–	1.24	32	36.1	15	35	+/–				1	Mastrorilli (2005)
A. n. noctua	Italy	living birds all ages	35.59	+/–	1.28	32	36.9	27	35.39	+/–	1.18	33.6	37.3	10	Mastrorilli (2005)
A. n. vidalii	Iberian Peninsula	stuffed	34.6	+/–	1	33	35.9	8	34.8	+/–	1.1	32.8	36.3	12	Cramp (1985)
A. n. vidalii	Iberian Peninsula	living birds	39	+/–	2.5	36	43	9	41.14	+/–	2.34	38	44	7	Martinez *et al.* (2002)
A. n. indigena	Greece	stuffed	33.2	+/–	0.7	32.1	34.1	6	32.7	+/–	1	31.8	34.4	7	Cramp (1985)
A. n. lilith	Cyprus	stuffed	30	+/–	0.8	29.3	32.1	4	30.2	+/–	0.8	29.2	31.6	6	Cramp (1985)
A. n. glaux & saharae	Algeria, Tunisia, Libya, Tibetsi, Egypt	stuffed	32	+/–	0.8	31.1	33.4	10	32.3	+/–	1.5	29.6	34.6	14	Cramp (1985)

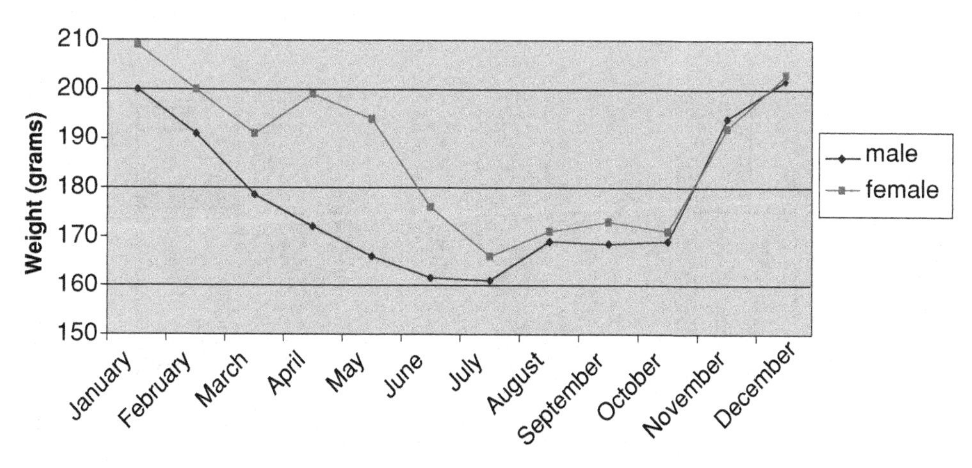

Figure 4.17 Seasonal changes in weight of male and female Little Owls in captivity during the annual cycle in 1995–1999 (26 pairs weighed; Génot & Sturm 2001).

(n = 61) (Schönn *et al.* 1991). These data were also reported from owls in captivity (see Figure 4.17) where the most important weight difference between males and females occurred in the March–June breeding season, with a peak in April and May (28 g), the laying and incubation time. The smallest weight difference (1–2 g) occurred at the end of the summer and in the autumn (Génot & Sturm 2001).

The average weights of 57 males and 94 females in northeastern Netherlands were 167 g and 185 g, respectively; the difference between the averages was significantly different ($P < 0.001$) (Figure 4.18). During the breeding season in northeastern Netherlands the weights of both male and female owls actively engaged in nesting followed a pattern of decline until the young dispersed (Table 4.2, van Harxen & Stroeken, personal communication).

Maximum average weights were 200–240 g for males and 220–250 g for females (Schönn *et al.* 1991). Minimum average weights were 160–170 g for males and 170–180 g for females (Schönn *et al.* 1991). In midwinter, the adults were 20–40% heavier than at the end of the breeding season (Schönn *et al.* 1991). Figures 4.19a–e show the seasonal changes in weight in different German study areas.

Some loss of weight can occur between the beginning and the end of winter even when prey is abundant. This was shown in Belgium where owls were weighed on two occasions. One male went from 225 g in December 2002 to 185 g in March 2003; a female went from 260 g in December 2002 to 205 g in March 2003; a male went from 240 g in November 2001 to 205 g in February 2002; a male went from 230 g in December 2002 to 165 g in February 2003 (Bultot, personal communication). There was also a difference of weight between males and females in the pre-breeding seasons (November–March: males 186.2 g, n = 45; females 204.9 g, n = 59) and at the time during which nestlings are present (May–June: males 159.3 g, n = 7; females 171.1 g, n = 56).

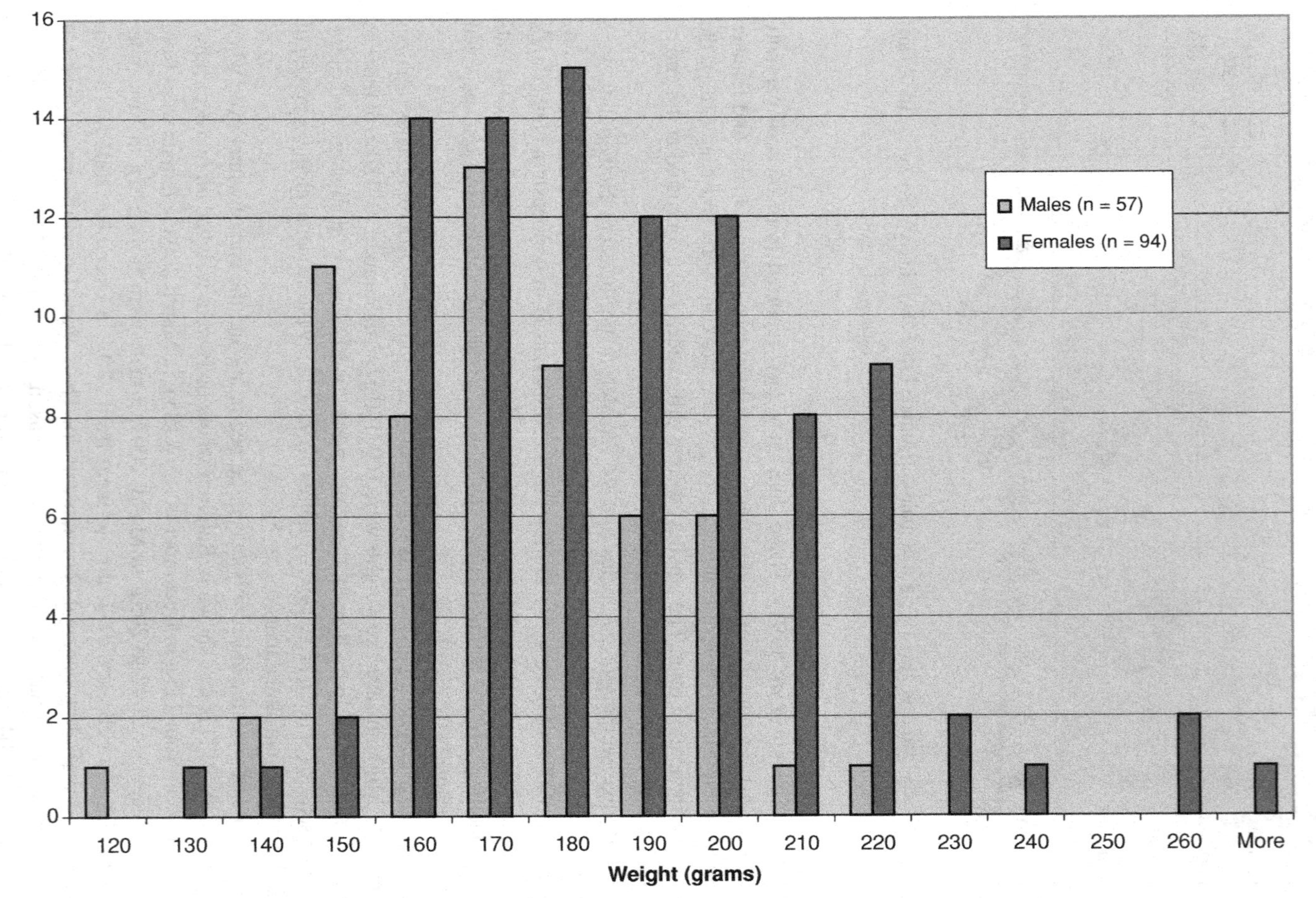

Figure 4.18 Weights for male and female Little Owls, Achterhoek, the Netherlands, 1998–2002 (R. van Harxen & P. Stroeken, unpublished data).

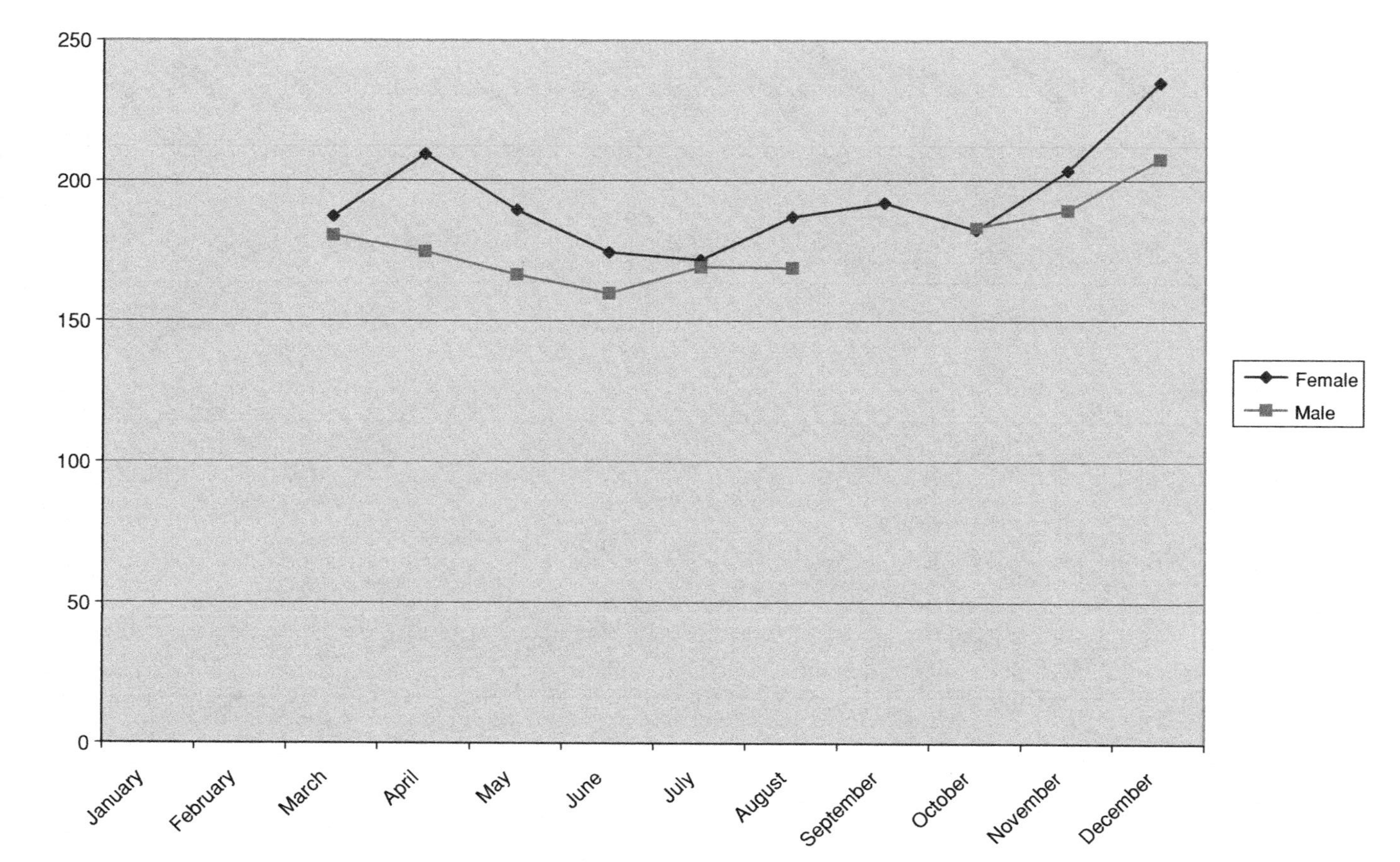

Figure 4.19a Seasonal changes in weight (in grams) of males and females in different regions of Germany (after Schönn *et al.* 1991). Mittelwestfalen.

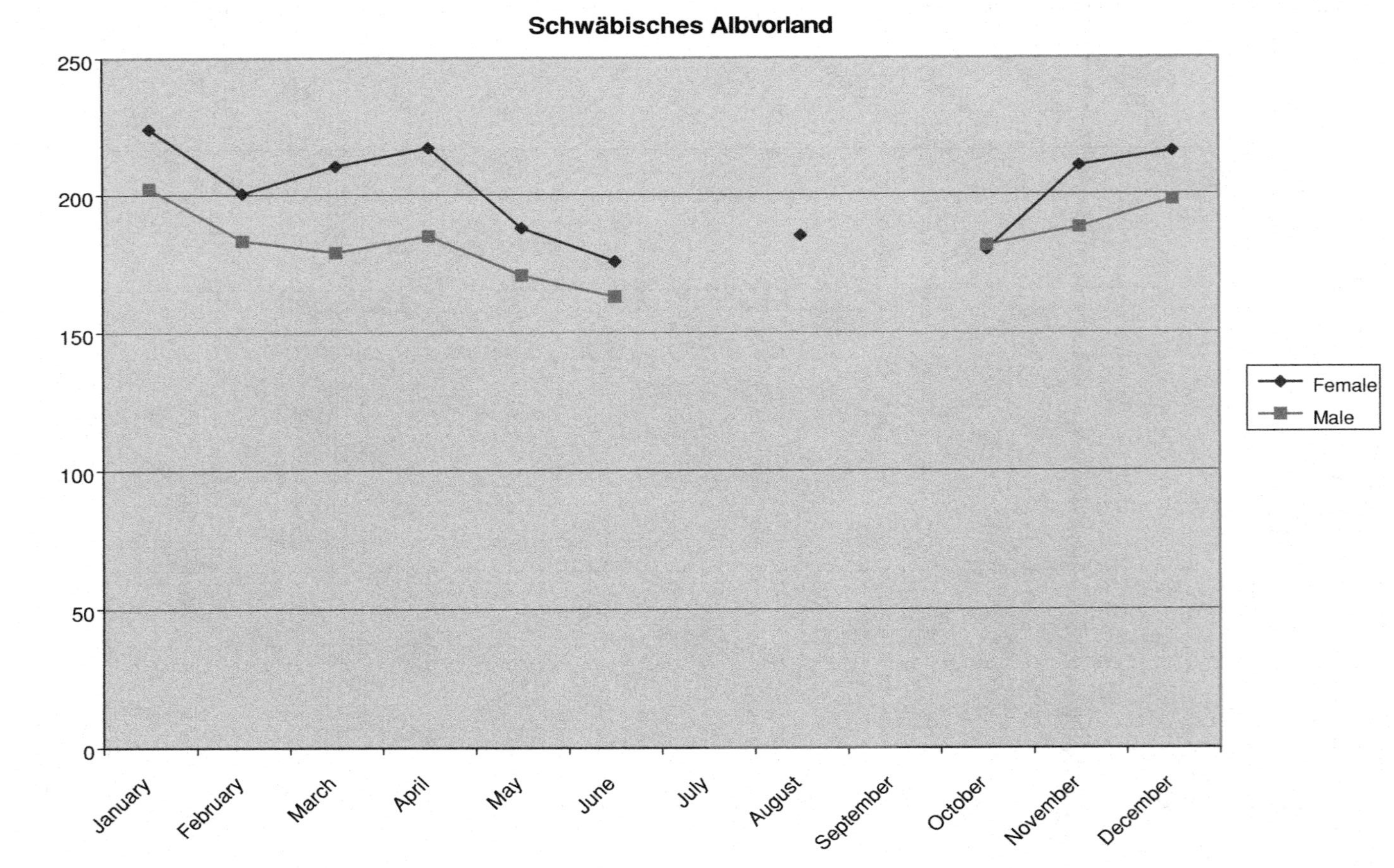

Figure 4.19b Seasonal changes in weight (in grams) of males and females in different regions of Germany (after Schönn *et al.* 1991). Schwäbisches Albvorland.

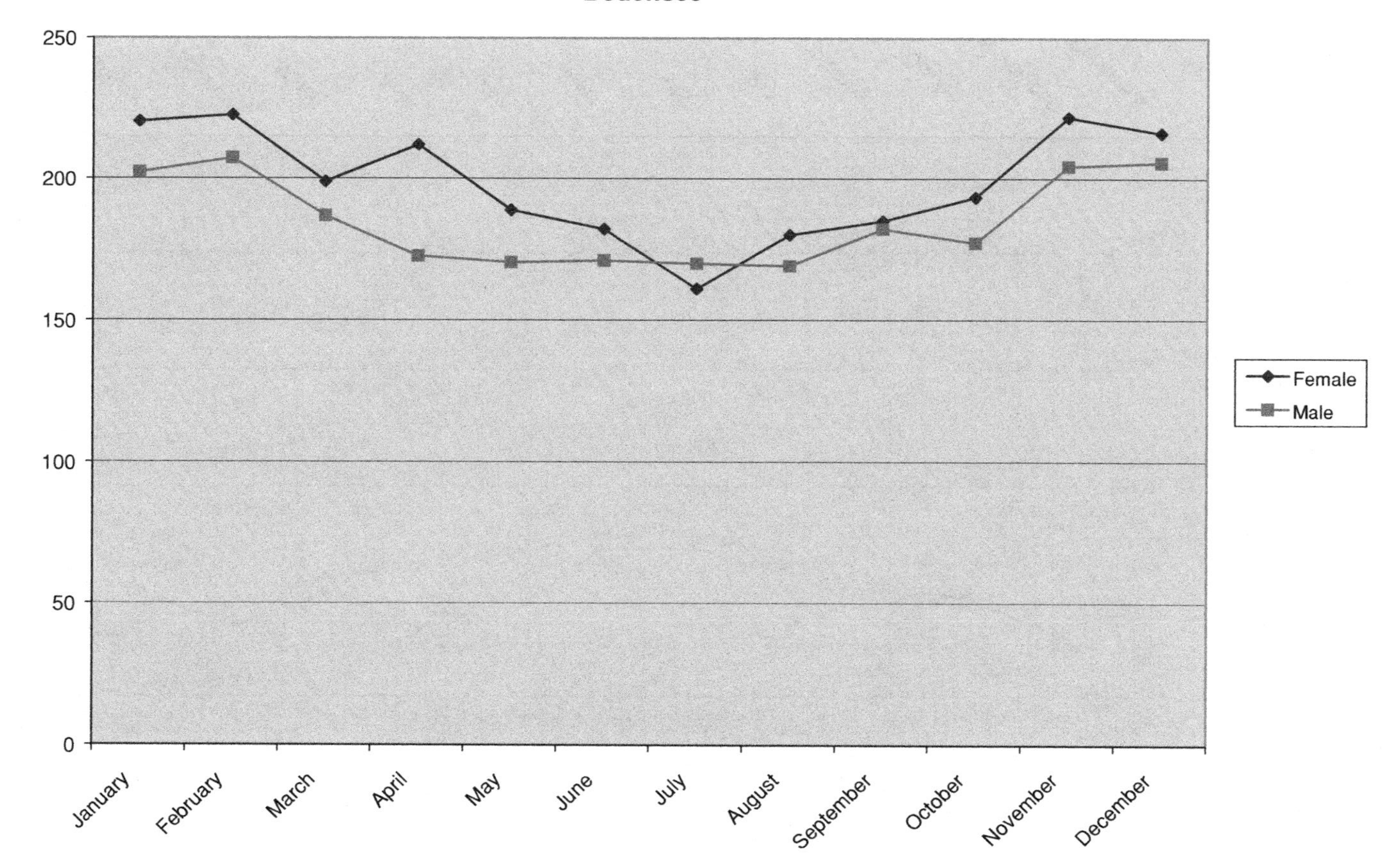

Figure 4.19c Seasonal changes in weight (in grams) of males and females in different regions of Germany (after Schönn *et al.* 1991). Bodensee.

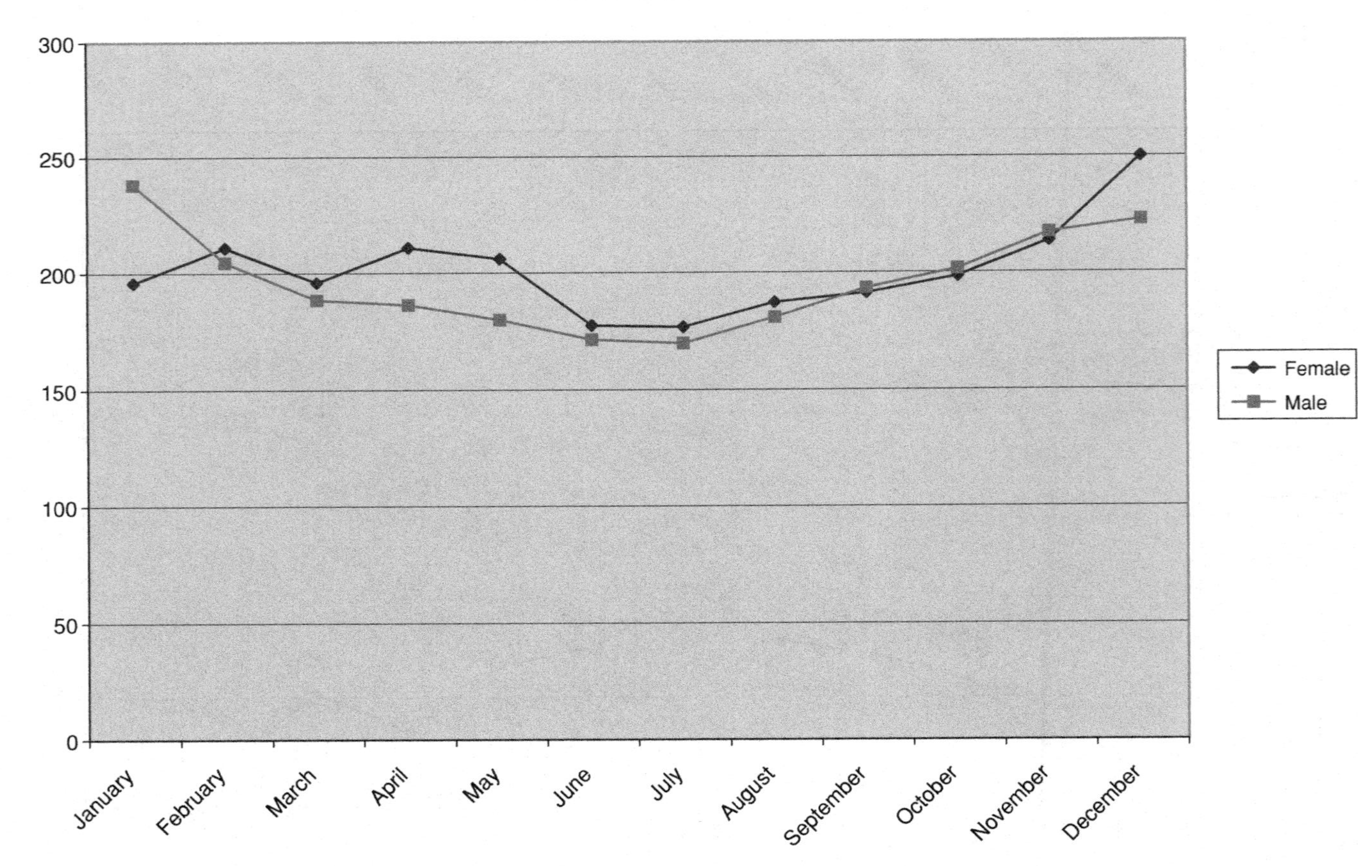

Figure 4.19d Seasonal changes in weight (in grams) of males and females in different regions of Germany (after Schönn *et al.* 1991). Unterer Niederrhein.

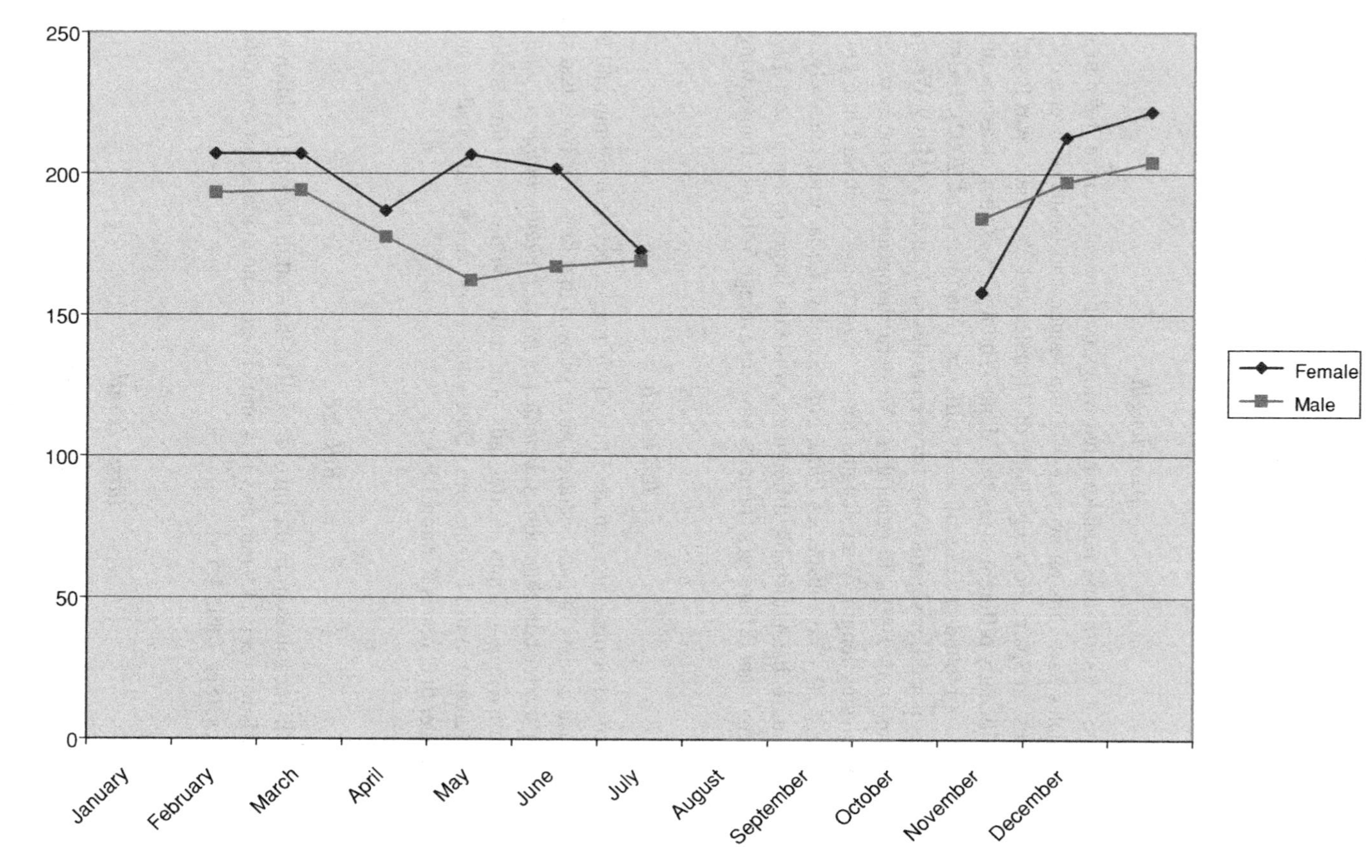

Figure 4.19e Seasonal changes in weight (in grams) of males and females in different regions of Germany (after Schönn *et al.* 1991). Jülicher Börde.

The lowest weight recorded from a wild owl in nature was 98 g (Morbach *in* Schönn *et al.* 1991). Génot (personal observation) found a recently dead 110 g adult during a snowy winter in eastern France. Glutz and Bauer (1980) consider that a weight of 120–125 g is the lower limit when an owl is hungry – death is likely imminent.

The heaviest weights recorded were 270–280 g in the wild and 300–350 g in an aviary (Schönn *et al.* 1991).

Wing length

The data in Table 4.3 show that females have wings that are on average 3–5 mm longer than males. Females have tails on average 1–2 mm longer, and tarsus on average 2.7 mm longer, than males. In Spain, the wing length of 24 males and 25 females was found to be statistically significantly different using the Mann-Whitney test. Other measures of males and females were not found to be significantly different (Martinez *et al.* 2002). The average wing length does not differ significantly between males and females (Mittel Westphalia, Germany, Schönn *et al.* 1991). In general for Western European birds the sexes are not distinguished by wing length. Wing lengths for 44 male and 43 female Little Owls are shown in Figure 4.20. The subspecies *lilith*, *glaux*, and *saharae*, with relatively shorter wings, do not feature distinguishable differences, while the larger *orientalis* and *plumipes* do show more important differences (females have on average 5–10 mm longer wings).

Tail length

The tail length of adult male and female Little Owls ranges between 69 mm and 90 mm (Table 4.4). Measurements of Dutch stuffed birds showed that first-year birds had significantly (3–5 mm) shorter tails than adults. Females of *noctua*, *vidalii*, *indigena*, *glaux*, and *saharae* have on average 1–2 mm shorter tails than males. While some time series data show no sex difference in tail length, some Dutch data show that females had on average tails 3.7 mm longer than males (Cramp 1985).

Bill size

The size of the bill measured from the tip of the bill to the start of the cere (culmen length) varies in live birds between 11.9 mm and 12.4 mm. There is no evidence of any difference between the sexes. (See Table 4.5.)

Tarsus length

The length of the tarsus shows significant differences between the subspecies (Table 4.6). The smallest average tarsus lengths were observed in birds from Cyprus (males 30.0 mm; females 30.2 mm) and are on average 20% smaller than in the birds measured in Nordrhein-Westphalia (Germany) (males 36.7 mm; females 39.4 mm) (Schönn *et al.* 1991).

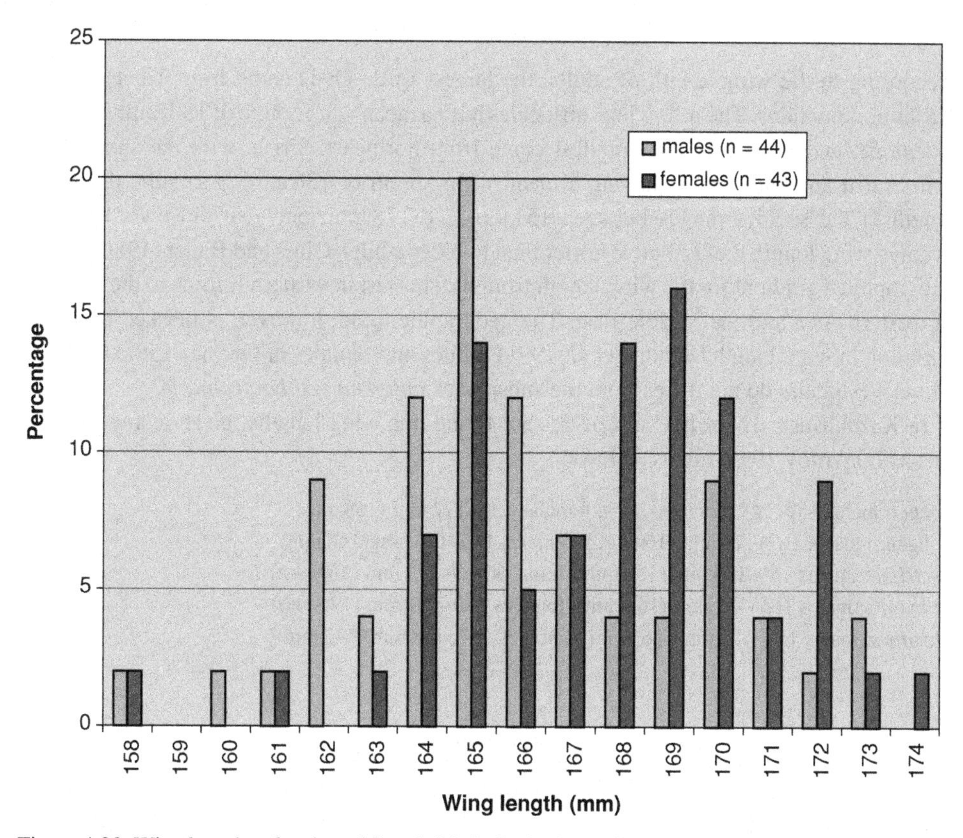

Figure 4.20 Wing lengths of male and female Little Owls (Kämpfer & Lederer in Schönn *et al.* 1991).

The Mediterranean populations, except for the Iberian birds, have relatively small tarsuses. The important spread in average measurements within the *noctua* subspecies shows important local variation. The length of the tarsus was found to differ between the sexes in two German studies (Schönn *et al.* 1991) with females having on average 2.7 mm longer tarsuses than males. All other studies showed a difference in length of less than 1 mm (see Table 4.6).

Other measurements

Besides the most common body measurements given above, Martinez *et al.* (2002) also report the mouth size (males: 18.33 ± 1.6 mm; females:19.60 ± 0.55 mm), back claw (males: 9.67 ± 1.7 mm; females: 9.31 ± 1.04 mm), front claw (males: 10.72 ± 1.10 mm; females: 11.64 ± 0.85 mm), talon length (males: 43.56 ± 1.40 mm; females: 43.57 ± 1.90 mm), P8 length (males: 117.00 ± 3.20 mm; females: 119.12 ± 3.88 mm), wingspan (males: 557.67 ± 11.90 mm; females: 565.64 ± 13.15 mm), and total length (males: 234.44 ± 11.00 mm; females: 243.57 ± 10.29 mm).

Differences between subspecies

According to the wing length of adults, the largest Little Owls come from Tibet and the Kashmir mountains. The following subspecies have a mean wing length of 180 mm: *ludlowi*, *orientalis*, and *plumipes*. The smallest come from northeast Africa, with the subspecies *spilogastra* and *somaliensis* having a mean wing length of 136 mm. Normally, the wing length of the species ranges between 151 mm and 178 mm; one-year-old owls have an average wing length that is 3 mm shorter than in older adults. Glutz and Bauer (1980) saw a geographical gradient for the wing length from the shortest in western Europe to the longest in western Asia and the Middle East. The geographic trend, however, is masked by local variation in wing length (Schönn *et al.* 1991). Males and females of German Little Owls in MittelWestphalia do not differ from the subspecies *indigena* and *bactriana*.

In Kazakhstan, where five subspecies are found, the wing lengths, given as ranges and means (Gavrilov 1999) are as follows:

noctua: males 152–169 mm (160 mm), females 158–177 mm (168 mm)
indigena: males 163–170 mm (164 mm), females 166–173 mm (168 mm)
bactriana: males 159–174 mm (166 mm), females 159–177 mm (169 mm)
orientalis: males 165–172 mm (168 mm), females 170–181 mm (176 mm)
plumipes: males 158–170 mm (163 mm), females 167–178 mm (173 mm)

4.5 Voice

Most studies are carried out through the recording of owl vocalizations in the field, with results offered as descriptions of sound spectrograms – most commonly referred to as sonagrams (Glutz & Bauer 1980; Exo 1984; 1990; Scherzinger 1988; Exo & Scherzinger 1989; Schönn *et al.* 1991). Exo and Scherzinger (1989) recorded 22 distinct call-notes (23 including non-vocal bill snapping) (Table 4.7). Two of them are specific to young owls and disappear during ontogenetic development. The juvenile repertoire contains 12 defined call-notes, which can be summarized by three or four basic call-notes. Males and females each have 15 call-notes out of a total of 20 altogether. Only three for the male and two for the female are sex-specific. The final repertoire includes 40 acoustic signals with mixtures and combinations. The most common are indicated here and are based on Exo and Scherzinger (1989) and Cramp (1985).

Calls of adults

- *Song* (Cramp 1985) or *hoot* (Hardouin, pers. comm.). This is the primary call used in pair formation and territorial defense. It is produced by both adult male and female owls. (Figure 4.21a.)

  The male's call sounds like "*goooek*"; loud, questioning "*huui*" (Haverschmidt 1946) or "*ghu(k)*", sometimes given in crescendo series; towards the end of these series it changes to an excited "*guiau*", or "*kwiau*" and ends abruptly with a shrill "*hoo-ee*", or "*miju*" (Haverschmidt 1946; Glutz & Bauer 1980; Exo & Scherzinger 1989).

Table 4.7 *Different calls of Little Owls according to age and sex (after Exo & Scherzinger 1989).*

Domain	Relative volume	Sound	Juveniles	Females	Males
Contact calls	1	Contact call	x	x	x
Feeding call	5	Begging-call juvenile	x		
Feeding call	5	Begging snoring	x		
Feeding call	5	Feeding call		x	(x)
Feeding call	5	Luring/tempting	(x)	x	x
Feeding call	5	Feeding twittering		X	
Advertizing calls	7	Begging-call adult		x	x
Advertizing calls	7	Begging snoring		x	
Advertizing calls	7	Luring/tempting		x	x
Advertizing calls	7	Chugging		?	x
Advertizing calls	7	Nest showing			X
Advertizing calls	7	Yelping		x	?
Advertizing calls	7	Copulation call			X
Enemy call and defence	6	Alarm call	x	x	x
Enemy call and defence	6	Chattering/cackling	x	x	x
Enemy call and defence	6	Peeping/cheeping juvenile	x		
Enemy call and defence	6	Defence screeching	x	x	?
Enemy call and defence	6	Chirping	x	x	x
Enemy call and defence	6	Hissing/blowing	X	?	?
Enemy call and defence	6	Bill snapping	(x	x	x)
Intraspecific aggression	6	Trembling	X		
Intraspecific aggression	6	Peeping	x	x	x
Intraspecific aggression	6	Excitement calls	x	x	x
Intraspecific aggression	6	Aggressive calls		?	x
Intraspecific aggression	6	Whispering calls	?	x	x
Intraspecific aggression	6	Copulation call			x
Song	2	Song, ordinary		x	x
Song	2	Glissando song			X
total			12	15	15

The female's call is shorter than the male's and the tone is generally higher in pitch. Most calls are louder and clearer when given by the male and repeated monotonously in a varied group of notes, while the notes occur singly when uttered by the female (Exo 1984). (Figure 4.22.)

- The *kiew-call* (Cramp 1985) or *chewing call* (Hardouin, pers. comm.) is the most commonly given call. It is a clear "*(k)weew*", "*huu*" or "*gwauu*"; sharp, complaining, "*kee-ew*"; repeated irregularly. (Figure 4.21b.) It is used in many cases of social contact such as mating, feeding young, sometimes at the end of the ordinary song during courtship behavior, during copulation and nest-showing, or during disturbance at the nest-site, occurrence of predators (e.g., feral cats) and territoriality.

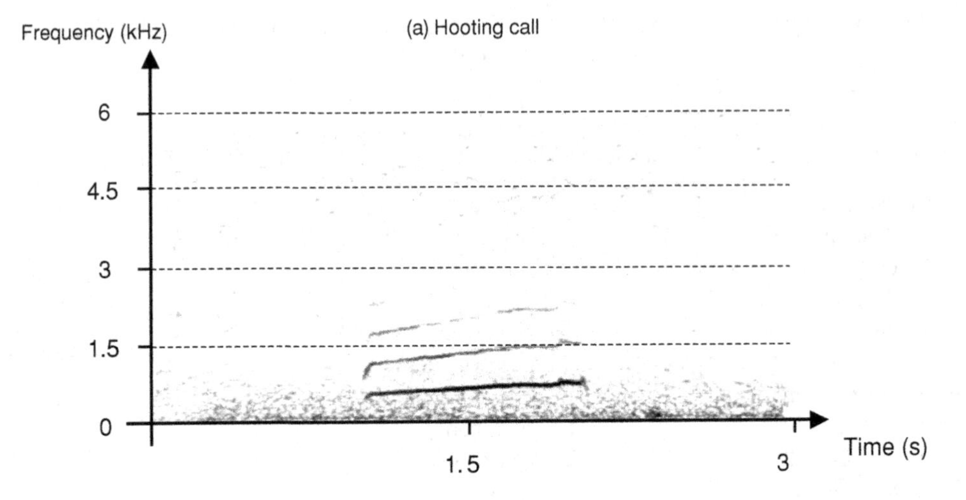

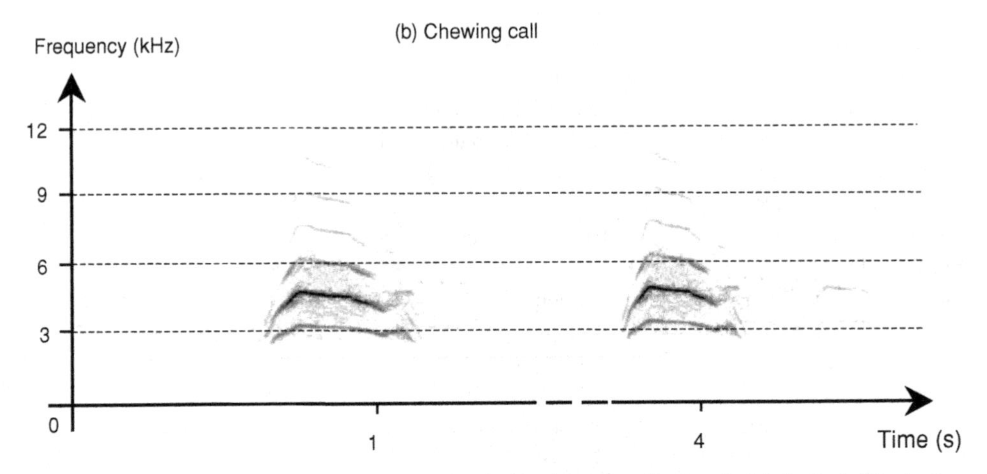

Figure 4.21 (a) Hooting call of male Little Owl; (b) chewing call of male Little Owl (Hardouin, personal communication).

- The *whispering-call*, similar to "*shrie*", is used as a contact call. It is prominently given by the female during courtship (Glutz & Bauer 1980).
- *Begging-calls* of the female are extentions of that from juveniles, like "*tsiech*", "*schräää*" and "*siej*" sounds (Glutz & Bauer 1980; Exo & Scherzinger 1989).
- The *feeding-trill* from the female is used when feeding young. It is a rapid, nasal "*gek-gok-gok-gok*", or hoarse cackling; sometimes interrupted by "*uuh*" as kiew-call (Glutz & Bauer 1980; Exo & Scherzinger 1989).
- The *excitement-call* is variable; it is described as "*jau*", "*mija*", "*kwiau*", "*miji*", "*iwidd*", and "*kuwitt*" (Exo & Scherzinger 1989) (Figure 4.23). The final shrill call "*kuwitt*"can be confused with the call of Tawny Owl *Strix aluco* (Bettmann 1951; Runte 1951). A fast call, it has been recorded at

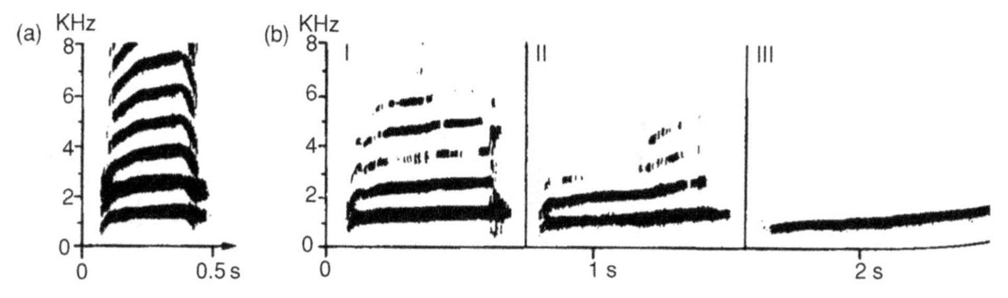

Figure 4.22 Territorial song of Little Owls: (a) female song, (b) male song I and II short calls; III glissando song of up to one second with increasing frequency (after Exo & Scherzinger 1989).

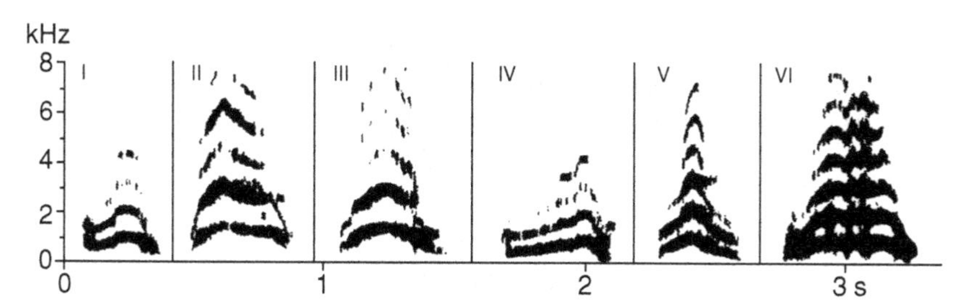

Figure 4.23 Excitement calls of Little Owls: I single "*jau*", II strong "jau", III "*mija*", IV "*mlijau*", V "*mijau*", VI "*mlji.jau*" (after Exo & Scherzinger 1989).

28–36 calls/min (Glutz & Bauer 1980). It is mainly given in association with copulation and other aspects of heterosexual behavior, but is also given in fights between rivals.

- The *nest-showing* call of the male is similar to the "*zick zick*" of the Kestrel, *Falco tinnunculus*, or "*tjuck tjuck*" of domestic chicken given during nest-showing; it is also occasionally given during mating (Glutz & Bauer 1980; Exo & Scherzinger 1989).
- The *copulation-call* is a soft "*oo oo*" given by birds sitting close together, before or after copulation. The female shrieks during copulation. The male often sings "*goooek*" before mating and sometimes during mounting (Haverschmidt 1946; Exo & Scherzinger 1989). Also the male uses the kiew-call, the excitement-call, and nest-showing call.
- The *alarm-call* is used as an alarm and given during times of anxiety and for warning. It is a loud, chattering "*kek kek*" when an owl is disturbed at its nest (Haverschmidt 1946; Exo & Scherzinger 1989). Calls based on "*queb*" and "*keck*" sounds express fear; short "*kja*" or "*kju*" are given as a warning to a rival or predator; "*quip*" express anxiety; "*quijep*" given as warning to young at fledging age. Snoring or screeching sounds are given in displeasure or fear (Glutz & Bauer 1980). The bill-snapping is a typical alarm noise of the Strigiformes family (Schönn *et al.* 1991). Little Owls can produce this sound after ten days of age (Exo *in* Schönn *et al.* 1991). Some authors (Gooch 1940; Haverschmidt 1946) have mistakenly interpreted this sound as tongue clicking. (Figure 4.24.)
- The *hissing-call* is given in threat and is often linked with the alarm-call (Schönn *et al.* 1991).
- Other calls: a squeaky "*uik*" is given during allopreening, which is a rare behavior (Scott 1980). A hissing or rasping sound is produced by males and females, probably in a begging- and

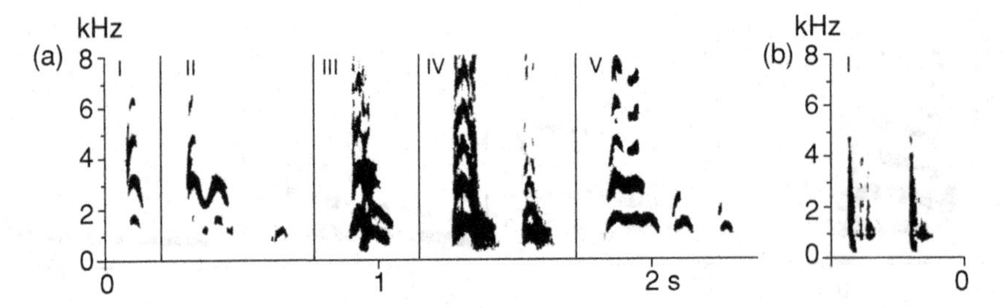

Figure 4.24 Alarm-call and bill-snapping sound of Little Owls: (a) Alarm-call – I "*kek*", II "*kju.kek*", III "*kau*", IV "*kju.kau*", V "*kju.i.pep.pep*" (I, II and V: adult birds; III and IV fledged birds); (b) bill-snapping (30-day-old juvenile birds).

contact-calling context (Glutz & Bauer 1980). A curious snoring is sometimes given by day in spring, sounding like the exhalation of a person in deep sleep (Witherby *et al.* 1938). Owls also give various shrieking, yelping, grumbling, and rasping calls (Glutz & Bauer 1980; Exo & Scherzinger 1989).

Calls of young

A food-begging call is given from unhatched (still within egg) and small young. It is a monosyllabic "*psiep*", "*szip*", or "*srie*" and a disyllabic "*uiief*". In the second week after hatching it becomes a harsh, "*chsij*", "*chriie*", or hissing "*schwo*"; in the fourth week a rasping snoring like that of the Barn Owl, *Tyto alba*, is given. Several adult calls, e.g., "*gjuu*" as the kiew-call, "*kek*", snoring (alarm-call), or hissing develop while the young are still in the nest or shortly afterwards. The typical territorial song is given during the first autumn (Ullrich 1973; Glutz & Bauer 1980).

Exo (1990) showed a geographical variation in the territorial song from eastern England where the notes last significantly longer than from the owls in northwestern Germany. The tone was higher in Little Owls from England than in Germany.

A comparative study of the vocalization repertoires within the genus *Athene* (Scherzinger 1988) showed a strong similarity of calls among the Little Owl, *Athene noctua*, Spotted Owlet, *Athene brama*, and the Burrowing Owl, *Athene cunicularia*. This study supported the placement of the Burrowing Owl in the genus *Athene*. (Figure 4.25.)

Singing behavior

The daily pattern of vocal activity includes one peak following sunset and one preceding sunrise. The length of the peaks ranged from 0.5 h up to 1.5–2 h. The shortest values were recorded from November to January, the time of the yearly minimum, with the longest during the courtship period in March to April. Rain and wind (>3 Beaufort) inhibited vocal activity (Exo 1989). In Italy, it seems that activity peaks were found to vary in different

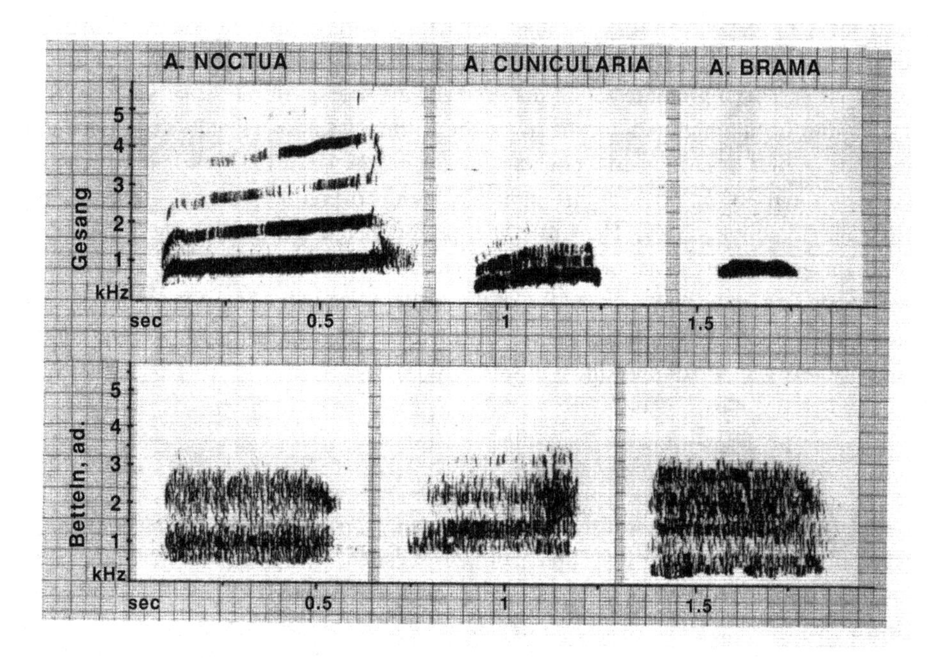

Figure 4.25 Sonagrams of three different *Athene* species (after Scherzinger 1988).

regions: early afternoon (Continental region) and early morning (Mediterranean region especially in Sardinia) (Calvi *et al.* 2005).

Hardouin (2006) studied the singing behavior of owls in relation to ambient temperature and time of night. He found that the willingness to respond to taped playback decreased with temperature. However, while the response rate of owls was lower during evenings with colder temperatures, all replied in the morning when it was even colder, suggesting that at cold temperatures the owls may need to first feed and later defend their territories. Hardouin *et al.* (2006) also investigated neighbor–stranger discrimination in a year-round territorial Little Owl. The authors used playback at the usual location for the neighbor or at an unusual location. Male Little Owls responded significantly less to their neighbor's hoots played back from the usual location. However, responses to playback of a neighbor from an unusual location were similar to responses to playback of a stranger's hoots from either location. Thus, neighbor–stranger discrimination is an economic system of territory defense that potentially saves physiological costs by minimizing the energy expended on aggressive acts, prevents escalated contests between neighbors, and decreases time lost and predation risk. This discrimination could develop with increasing population density and with experience.

To defend its territory and to attract a female, the male uses the song that is projected as far as possible. Some measures of the acoustic force were made with 21 singing males on 620 individual songs. At 10 m, the songs reach 62.1 dB (Schönn *et al.* 1991). In the field, the song of Little Owls can be heard at a distance of up to 600 m. The greatest distance to hear the song depends on the weather, especially during daylight and twilight when other species are silent. Ambient conditions of calmer and cooler air and relative humidity levels of around 60% promote sound dispersal. The dispersal of the song is better if the owl is calling from a high perch. Moreover, the Little Owl often changes the direction it faces when calling, assuring better coverage of its territory (Exo 1987). When the reach of the song is about 500 m, it represents an area of 79 ha covered by the call – several times the average home-range size. The Little Owl prefers perches located at the limits of its territory rather than singing from the center (Finck 1989). In Italy (Calabria), on 27 calling males responding to the playback, 33.8% of the replies were in the first minute of playback emission and 91.6% did not last for more than two minutes, indicating low territoriality; the density was 1.35 territories/km^2 (Giuseppe 2005).

4.6 Flight

The flight of the Little Owl is direct, swift, easy, and wave-like, very similar to a woodpecker's flight. As with other owls, the Little Owl is silent in flight. The owl's wing-beats are as silent at high frequencies as they are at low ones, and have no ultrasonic noise (Thorpe & Griffin 1962). In certain cases, in particular when hunting, Little Owls display hovering flight as in Kestrels (Martin & Rollinat 1914; Malmstigen 1970; Wahlsted 1971). The hovering takes place 2–3 m or sometimes 20 m above the ground (Gyllin 1968) with the owl finally closing its wings and dropping like a stone on the prey (Gregory 1944; Tayler 1944).

4.7 Physiology and anatomy

Most of the anatomy studies have focused on the vision of the Little Owl. Ille (1983) showed that the perception of prey is linked to sight. The big eye with large cornea has rods in the retina. The length and number of rods explain the good sight of the species during twilight (Rochon-Duvigneaud 1934). Further investigations revealed that the night vision of the owls is linked to the organization of the outer plexiform layer of the retina, specifically to the stratification of the synoptic bodies of the single and oblique cones. The adaptation to nocturnal light could be due to the lack of oblique cones (Gallego *et al.* 1975). The neuronal cells of the Little Owl's retina are divided into 20 types of simple and complex synapses (Yew 1980). Yew speculates that the contact between certain cells, in particular the amacrine–ganglion contact, are related to vision in dim light. Morphological differences in the retina cells of the Barn Owl (*Tyto alba*) and Little Owl showed that the latter has retina cells more similar to diurnal birds (Tarres *et al.* 1986). But the visual acuity of the Little Owl is the poorest in comparison with other birds of prey (Gaffrey & Hodos 2003). Binocularity is an advantage for the Little Owl (Porciatti *et al.* 1990), especially under low light conditions.

Through experiments, Little Owls can see yellow, green, blue, red (although less well), and also differentiate the gray colours (Meyknecht 1941). The presence of a greater number

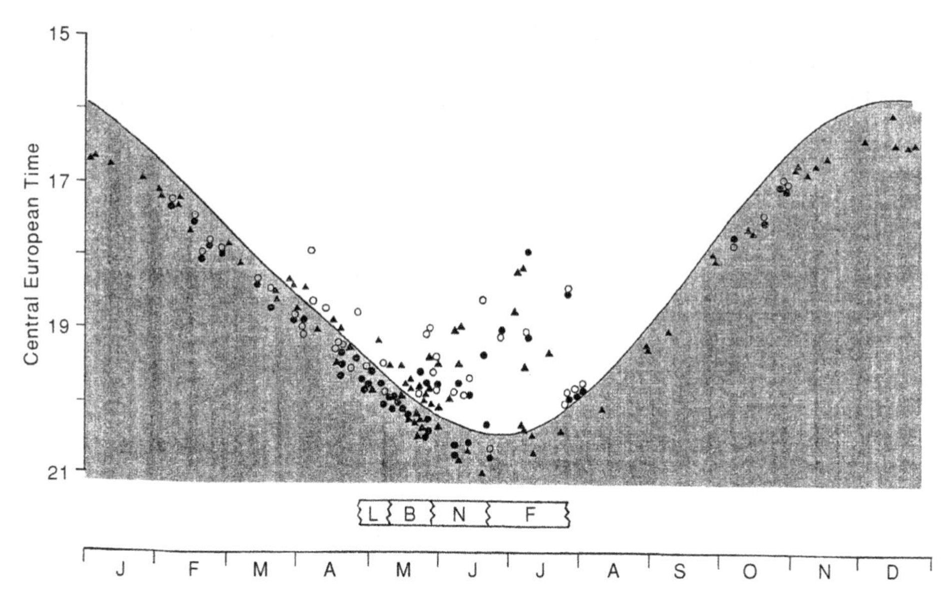

Figure 4.26 Seasonal variation in the onset of daily activity in the Little Owl. The time the bird left the roost was defined as the start of activity (after Schönn *et al.* 1991). White circles: males; black circles: females; triangles: unknown gender. L: egg-laying period; B: incubation period; N: nestling period; F: fledgling period.

of rods than cones in the retina indicates an increased ability to discrimine shades of gray, which would allow an orientation under low light levels at night (Kopystynska 1962). Kopystynska showed that the Little Owl does not see infrared rays.

The highest activity of the Little Owl in an aviary was recorded with a luminescence of about 150 Lux (measure of light intensity) (Erkert 1967). But in nature, the main activity throughout the year occurs when the light intensity is under one Lux, and it is exceptional if an owl leaves its day roost when there is a light intensity as much as 50 Lux (Exo 1989). (Figure 4.26.)

The frequency sensitivity of hearing in the Little Owl is 3–4 kc/s. It is correlated with the morphology of the hearing apparatus and the way of life of the birds. The Little Owl can locate small rodents with an accuracy of up to 1% thanks to their noises, rustling, and squeaking (Golubeva *et al.* 1970). The hearing apparatus of the different owls are compared in Schönn *et al.* (1991). Norberg (2002) details the morphology, function, and selection in outer ear asymmetry in seven owl lineages (i.e., *Tyto*, *Phodilus*, *Strix*, *Rhinoptynx*, *Asio*, *Pseudoscops*, and *Aegolius*).

Under the context of a taxonomical study, the chemical composition of the uropygial gland waxes of the Little Owl was compared to those of diurnal birds of prey and confirmed that owls seem to be an isolated order (Jacob & Hoerschelmann 1984).

Some data on the digestive organs were collected in Turkmenistan (Broun 1986). The relative weight of the glandular stomach is 0.54% of the overall bodyweight. The relative weight of the stomach muscle is 2% and it has comparatively thin walls. The bowel is comparatively short, being 198% of the body length. There are blind appendices between thin and thick parts of the bowel, representing 14.5% of the total length of the bowels. Histomorphological observations of the digestive apparatus showed that the third blind intestine is absent in the Little Owl and the developed glands are plenty. The muscular tunics and the glands of the duodenum and the organ of Meckel are well developed compared to those of granivorous birds (Richetti *et al.* 1980). A comparitive neurohistological study of the esophagus in granivorous, omnivorous, and carnivorous birds showed that carnivorous birds have one remarkable development in the innervation of the esophagus, which is more advanced than that of granivorous and omnivorous birds. The nervous plexus of the esophagus is supplied with many nervous cells and the muscularis mucosae is well developed (Costaglia *et al.* 1981).

In another field, the Little Owl has been found to be easily hypnotized in twilight or night conditions (Reisinger 1926b).

5

Distribution, population estimates, and trends

5.1 Chapter summary

Thus far, we have focused on the Little Owl itself (Figure 5.1). We have examined its morphology, the heterogeneity across subspecies, and across its geographic distribution. However, owls only survive within viable populations that are regulated in one way or another. In this chapter, we examine aspects of distribution and population in the Little Owl for its global range, which overlaps 84 countries. In very general terms, the data suggest that the distribution area of the owl has increased in 8 countries, decreased in 15, remained unchanged in 25 and was insufficient for determination in 35 countries. For population numbers, the data suggest that the number of owls increased in 11 countries, decreased in 9, remained unchanged in 30, and was insufficient for determination in 31 countries. For each country we focus on the population estimates of currently existing populations. In Table 5.1 we give details by country on the changes of the distribution and the population estimates. Except for the countries of western and central Europe, where the Little Owl has decreased during the last 40 years due to land-use changes, the species remains common or relatively widespread in the Mediterranean countries from southern Europe and Middle East to North Africa and in many Asiatic republics of the former Soviet Union where natural habitats of the Little Owl still occur. Changes in Little Owl numbers reflect the effects of regulatory factors acting on the populations (see Chapter 10).

5.2 Distribution

A map showing the rough approximation of Little Owl subspecies was discussed earlier in Chapter 3 (Taxonomy and genetics) and was shown in Figure 3.5. The aim of this section is to give an overview of the species distribution within individual countries. We have assembled within-country maps of Little Owl distributions for 26 countries (Figures 5.5–5.10 and Figures 5.12–5.22). An overview of all the available maps is given in the map shown in Figure 5.2. The individual maps are shown under each of the respective countries as they are discussed in this section. Factors that limit the local distribution of the species are given for each location whenever available.

Figure 5.1 Little Owl adopting upright position. (Photo Vildaphoto/Ludo Goessens.)

Table 5.1 *Overview of the recent changes in the distribution and population estimates of the Little Owl across its global range by country.*

	Distribution				Population trends				Remarks	Detailed map available
	up	down	no change	no data	up	down	no change	no data		
Afghanistan				✓				✓		
Albania		✓				✓				
Algeria			✓				✓			
Andorra				✓				✓		
Armenia			✓				✓			
Austria			✓		✓					✓
Azerbaijan			✓				✓			
Bahrain				✓				✓		
Belarus			✓				✓			
Belgium			✓				✓			✓
Bhutan				✓				✓		✓
Bosnia & Herzegovina				✓				✓		
Bulgaria			✓				✓			
Chad				✓				✓		
China				✓				✓		✓
Croatia		✓				✓				
Cyprus			✓				✓			
Czech Republic		✓				✓				✓
Denmark		✓				✓				
Djibouti				✓				✓		
Egypt				✓				✓		
Eritrea				✓				✓		
Estonia			✓				✓		vagrant	
Ethiopia				✓				✓		
France			✓				✓			✓
Georgia			✓				✓			
Germany		✓			✓					✓
Greece			✓				✓			
Hungary		✓				✓				✓
India			✓				✓			✓
Iran				✓				✓		
Iraq				✓				✓		
Israel			✓				✓			✓
Italy			✓				✓			✓
Jordan				✓				✓		
Kazakhstan				✓				✓		
Kuwait	✓				✓					
Kyrgyzstan				✓				✓		
Latvia		✓				✓				
Lebanon				✓				✓		
Libya				✓				✓		
Liechtenstein				✓				✓		
Lithuania			✓				✓			
Luxembourg		✓				✓				✓

(*cont.*)

Table 5.1 (*cont.*).

	Distribution				Population trends				Remarks	Detailed map available
	up	down	no change	no data	up	down	no change	no data		
Macedonia			✓				✓			
Mali				✓				✓		
Malta			✓				✓			
Mauritania				✓				✓		
Moldova	✓				✓					
Mongolia			✓					✓		
Monaco				✓			✓			
Montenegro			✓				✓			
Morocco			✓				✓			
Nepal				✓				✓		✓
Netherlands		✓				✓				✓
New Zealand			✓				✓			
Niger				✓				✓		
North Korea			✓				✓		vagrant	✓
Oman	✓				✓					✓
Pakistan				✓				✓		✓
Poland		✓				✓				
Portugal			✓				✓			
Qatar	✓				✓					✓
Romania	✓				✓					
Russian Federation		✓				✓				
Saudi Arabia	✓				✓					✓
Serbia			✓				✓			
Slovakia		✓				✓				✓
Slovenia		✓				✓				
Somalia				✓				✓		
South Korea			✓				✓		vagrant	
Spain		✓			✓					✓
Sudan				✓				✓		
Switzerland			✓		✓					✓
Syria				✓				✓		
Tajikistan				✓				✓		
Tunisia			✓				✓			
Turkey		✓				✓				
Turkmenistan				✓				✓		
Ukraine			✓				✓			
UAE	✓				✓					✓
UK			✓				✓			✓
Uzbekistan				✓				✓		
Yemen	✓				✓					✓
	up	down	no change	no data	up	down	no change	no data		
Summary	8	15	30	31	11	14	28	31		24

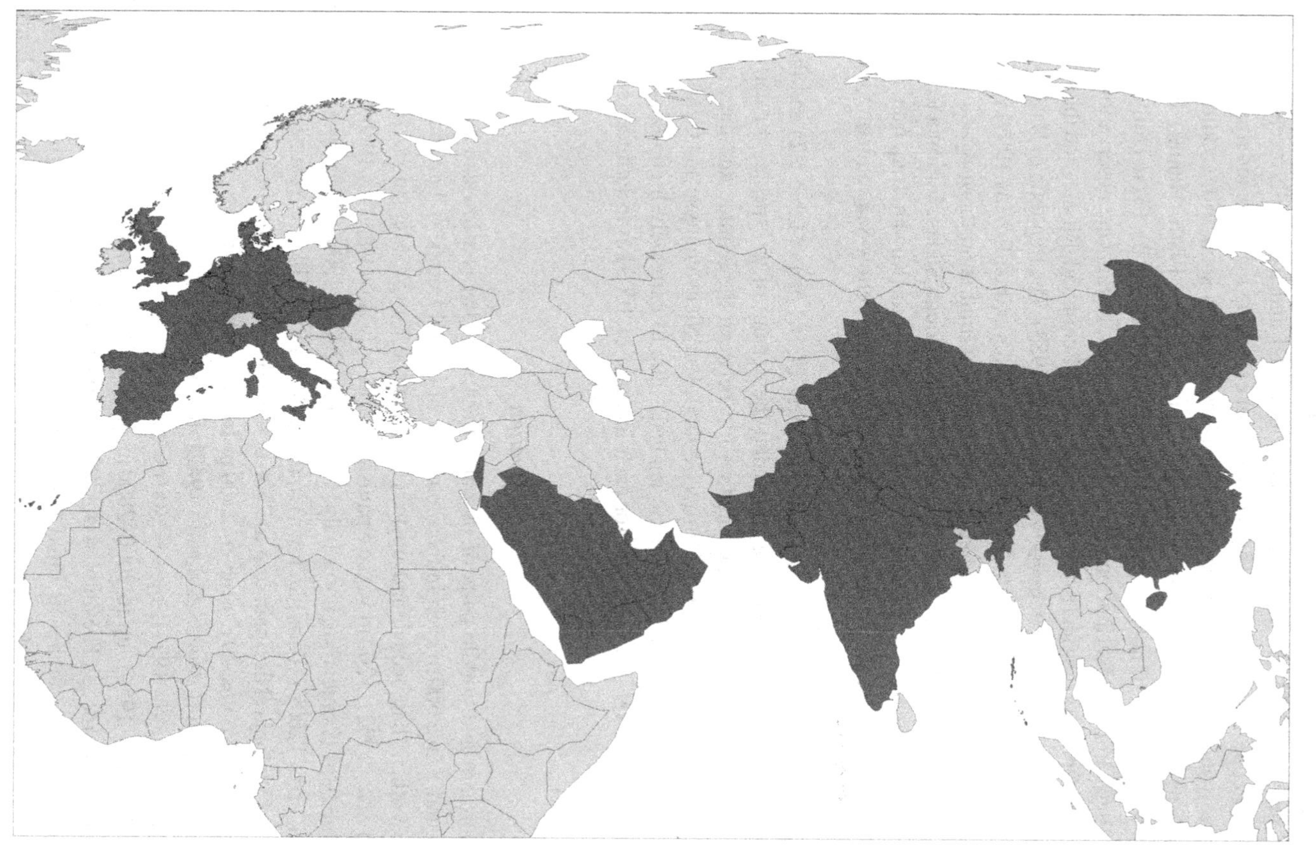

Figure 5.2 Overview of 24 countries for which a detailed distribution map is available and presented in this book.

Latitude

The species occurs in middle and lower latitudes of the Northern hemisphere (mainly between the 22nd and 51st parallels), both continental and marginally oceanic, mainly temperate, steppe, and Mediterranean biotopes, but extending to boreal and tropical. While adapting to windy and rainy climates, it favors warm, even semi-arid, conditions and is vulnerable to severe frosts and snow cover (Glutz & Bauer 1980). Hunting by the owl becomes very difficult, if not impossible, with a snow cover over 10–15 cm (Juillard 1984a; Schönn 1986), limiting distribution in both latitude and elevation. One observational record exists for the Opochka district (Pskov Region, north of Leningrad, Russia) and the species is extremely rare in Estonia (Malchevskiy & Pukinskiy 1983). The Republic of Bashkortostan forms the northern limit around the Ural mountains (Karyakin & Kozlov 1999) where only a few breeding pairs have been recorded. The southern limit is northern Africa where the species is found north of the Sahara Desert and south through Sudan, Ethiopia to Somalia. Further eastward the species is found in the Arabian Peninsula, Kuwait, Iraq, Iran, Pakistan, Kashmir, the Tibetan plateau, and north Szetschwan to the Yellow Sea south of Xinhailian, China.

Using European abundancy data on the numbers of owl pairs (Table 5.2) and the total area of each country in Europe, we calculated the density of the species for earlier and current data. We found a significant relationship between the latitude and the maximum observed densities (European Bird Census Council; Van Nieuwenhuyse and Génot unpubl. data) when removing Portugal from the analysis (due to extremely high densities). The density drops by 0.09 pairs per km^2 for each increase of 1000 km in latitude (see Figure 5.3).

Elevation

In the northern and middle parts of its range the Little Owl is a lowland species that is rarely found above 600 m even in the mountains of central Europe (Glutz & Bauer 1980). Further south, it has been found up to 1140 m in the Causse Méjean (Lozère, France) (Juillard *et al.* 1992); 1200 m in the Spanish Pyrenees (Kostrzewa *et al.* 1986); 1230 m in the Orobie Alps (Lombardia, north Italy) (Mastrorilli 2001); 1900–2000 m in Georgia and Armenia; up to 1260 m (Parashka town) in the Lviv Region (Kijko & Yakubenya 1995); at 1600 m in Catalonia (NE Spain); up to 2300 m in favorable territories in the Sierra Nevada (southern Spain) (Olea 1997); up to 2300 m in Bulgaria (Simeonov *et al.* 1989); up to 2000 m in Armenia (Lyaister & Sosnin 1942); up to 2200–2300 m in Uzbekistan (Ivanov 1969; Abdusalyamov 1971); up to 2300 m, probably even higher in Gissar-Karategin, Tajikistan (Popov 1959); up to 2000–2800 m in Altai; up to 2800 m on southern side of Todra gorges, Morocco (Thévenot *et al.* 2003); 4200 m in Pamir (Dementyev & Gladkov 1951) and sometimes higher in Tajikistan (Beik at 4260 m, Zor-Kul and Bash-Gumbez at 4200 m) (Abdusalyamov 1971); and 4400 m in Turkey (Kasparek 1992). In central Europe the population density tends to decrease with elevation in Germany (Loske 1986). See Figure 5.4.

Table 5.2 *Historical and current population estimates (after Manez 1994; BirdLife International 2004; this project).*

Population estimates		Manez (1994)					own results (2004)		BirdLife International (2004)						Densities (Manez 1994)		Densities own results		Densities BirdLife International (2004)		Trend in absolute numbers (own results 2004 – Manez 1994)		Trend in % of 1994 estimates (own results 2004 – Manez 1994)		Population estimates
Country	Surface (km²)	Pairs		Year	Trend	Breeding range trend	Min	Max	Min	Max	Trend	Data quality	% of population (excl Turkey)	% of population (excl Turkey)	Min	Max	Min	Max	Min	Max	Min	Max	Min	Max	Country
Albania	28 890	5000	10 000	1962	0	0	4000	8000	4000	8000	–	normal	1.56	1.20	0.17	0.35	0.14	0.28	0.14	0.28	–1000	–2000	–20%	–20%	Albania
Armenia							800	1500	800	1500	0	normal	0.31	0.22							800	1500			Armenia
Austria	83 790	40	60		–2	–2	60	60	70	100	0	reliable quantitative data	0.03	0.01	0.00	0.00	0.00	0.00	0.00	0.00	20	0	50%	0%	Austria
Azerbaijan							2000	10 000	2000	10 000	0	poorly known	0.78	1.50							2000	10 000			Azerbaijan
Belarus	199 900	2000	4000	1990	–1	0	400	1000	400	1000	0	normal	0.16	0.15	0.01	0.02	0.00	0.01	0.00	0.01	–1600	–3000	–80%	–75%	Belarus
Belgium	30 710	4500	6600	1981–90	–1	–1	9000	14 200	12 500	14 000	0	normal	4.89	2.10	0.15	0.21	0.29	0.46	0.41	0.46	4500	7600	100%	115%	Belgium
Bulgaria	111 400	4000	10 000		0	0	4000	10 000	5000	8000	0	reliable quantitative data	1.95	1.20	0.04	0.09	0.04	0.09	0.04	0.07	0	0	0%	0%	Bulgaria
Croatia	52 900	6000	8000		0	0	6000	8000	500	1000	–	poorly known	0.20	0.15	0.11	0.15	0.11	0.15	0.01	0.02	0	0	0%	0%	Croatia
Cyprus	8531	2000	4000		–1	0	2000	4000	5000	15 000	0	poorly known	1.95	2.25	0.23	0.47	0.23	0.47	0.59	1.76	0	0	0%	0%	Cyprus
Czech Republic	78 260	700	1100		–1	–1	500	1000	200	400	–	reliable quantitative data	0.08	0.06	0.01	0.01	0.01	0.01	0.00	0.01	–200	–100	–29%	–9%	Czech Republic
Denmark	42 800	150	150		–1	–2	175	200	100	200	–	normal	0.04	0.03	0.00	0.00	0.00	0.00	0.00	0.00	25	50	17%	33%	Denmark
France	546 900	10 000	50 000	1990	–1	0	10 330	35 000	20 000	60 000	–	normal	7.82	8.98	0.02	0.09	0.02	0.06	0.04	0.11	330	–15 000	3%	–30%	France
Georgia	85 010						3500	4000	present		0	poorly known			0.00	0.00	0.04	0.05			3500	4000			Georgia
Germany	355 300	5000	10 000	1985	–2	–1	7000	7600	5800	6100	0	normal	2.27	0.91	0.01	0.03	0.02	0.02	0.02	0.02	2000	–2400	40%	–24%	Germany
Greece	131 500	5000	10 000		–1	0	5000	1000	5000	15 000	0	poorly known	1.95	2.25	0.04	0.08	0.04	0.01	0.04	0.11	0	–9000	0%	–90%	Greece
Hungary	92 890	1500	2000		–1	–1	2000	2500	1500	2500	–	normal	0.59	0.37	0.02	0.02	0.02	0.03	0.02	0.03	500	500	33%	25%	Hungary
Italy	302 000	10 000	50 000		0	0	10 000	30 000	30 000	50 000	0	poorly known	11.73	7.49	0.03	0.17	0.03	0.10	0.10	0.17	0	–20 000	0%	–40%	Italy
Latvia	63 840	10	30		F	F	10	30	10	30	–	poorly known	0.00	0.00	0.00	0.00	0.00	0.00	0.00	0.00	0	0	0%	0%	Latvia
Lithuania	77 810	10	50	1985–8	0	0	5	10	5	10	0	poorly known	0.00	0.00	0.00	0.00	0.00	0.00	0.00	0.00	–5	–40	–50%	–80%	Lithuania
Luxembourg	2668	80	150		–1	0	45	85	40	80	–	reliable quantitative data	0.02	0.01	0.03	0.06	0.02	0.03	0.01	0.03	–35	–65	–44%	–43%	Luxembourg
Macedonia							1400	2400	1400	2400	0	normal	0.55	0.36							1400	2400			Macedonia
Moldova	49 440	5000	7000	1988	–1	–1	3200	4200	3200	4200	+	reliable quantitative data	1.25	0.63	0.10	0.14	0.06	0.08	0.06	0.08	–1800	–2800	–36%	–40%	Moldova

(*cont.*)

Table 5.2 *(cont.)*

Population estimates		Manez (1994)					own results (2004)		BirdLife International (2004)						Densities (Manez 1994)		Densities own results		Densities BirdLife International (2004)		Trend in absolute numbers (own results 2004 – Manez 1994)		Trend in % of 1994 estimates (own results 2004 – Manez 1994)		Population estimates
Country	Surface (km²)	Pairs		Year	Trend	Breeding range trend	Min	Max	Min	Max	Trend	Data quality	% of population (excl Turkey)	% of population (excl Turkey)	Min	Max	Min	Max	Min	Max	Min	Max	Min	Max	Country
Netherlands	34 760	9000	12 000	1979	−1	−1	5500	6500	5500	6500	–	reliable quantitative data	2.15	0.97	0.26	0.35	0.16	0.19	0.16	0.19	−3500	−5500	−39%	−46%	Netherlands
Poland	308 600	1000	3000		−2	0	1000	1500	1000	2000	–	normal	0.39	0.30	0.00	0.01	0.00	0.00	0.00	0.01	0	−1500	0%	−50%	Poland
Portugal	88 830	10 000	100 000	1989	0	0	50 000	150 000	50 000	150 000	0	poorly known	19.54	22.46	0.11	1.13	0.56	1.69	0.56	1.69	40 000	50 000	400%	50%	Portugal
Romania	237 300	20 000	40 000		0	0	20 000	40 000	40 000	60 000	+	normal	15.63	8.98	0.08	0.17	0.08	0.17	0.17	0.25	0	0	0%	0%	Romania
Russian Federation	16 620 000	10000	100 000	1975–90	0	0	11 500	103 300	10 000	100 000	–	poorly known	3.91	14.97	0.00	0.01	0.00	0.01	0.00	0.01	1500	3300	15%	3%	Russia
Serbia & Montenegro	102 400						14 000	22 000	10 000	15 000	0	normal	3.91	2.25	0.00	0.00	0.14	0.21	0.10	0.15	14 000	22 000			Serbia & Montenegro
Slovakia	49 030	800	1000		−1	−1	800	1000	800	1000	–	normal	0.31	0.15	0.02	0.02	0.02	0.02	0.02	0.02	0	0	0%	0%	Slovakia
Slovenia	20 560	500	800		−2	−2	200	300	150	200	–	normal	0.06	0.03	0.02	0.04	0.01	0.01	0.01	0.01	−300	−500	−60%	−63%	Slovenia
Spain	505 600	50 000	65 000		−1	−1	36 643	100 000	20 000	100 000	–	poorly known	7.82	14.97	0.10	0.13	0.07	0.20	0.04	0.20	−13 357	35 000	−27%	54%	Spain
Switzerland	40 830	30	40	1986–91	−2	−2	91	101	60	70	–	reliable quantitative data	0.02	0.01	0.00	0.00	0.00	0.00	0.00	0.00	61	61	203%	153%	Switzerland
Turkey	782 700	5000	50 000				5000	50 000	300 000	600 000	–	poorly known			0.01	0.06	0.01	0.06	0.38	0.77	0	0			Turkey
Ukraine	615 100	11 000	12 000	1988	+1	+1	15 000	22 000	15 000	22 000	0	normal	5.86	3.29	0.02	0.02	0.02	0.04	0.02	0.04	4000	10 000	36%	83%	Ukraine
United Kingdom	244 000	6000	12 000		−1	0	4000	8500	5800	11 600	0	reliable quantitative data	2.27	1.74	0.02	0.05	0.02	0.03	0.02	0.05	−2000	−3500	−33%	−29%	United Kingdom
Total incl. Turkey		184 320	568 980				235 159	649 986	555 840	1 267 900			100	100							50 839	81 006	28%	14%	Total incl. Turkey
Total excl. Turkey		179 320	518 980				230 159	599 986	255 840	667 900											50 839	81 006	28%	16%	Total excl. Turkey
Total excl. Turkey and Russian Federation		169 320	418 980				218 659	496 686	245 840	567 900											49 339	77 706	29%	19%	Total excl. Turkey and Russia

F: fluctuating populations.

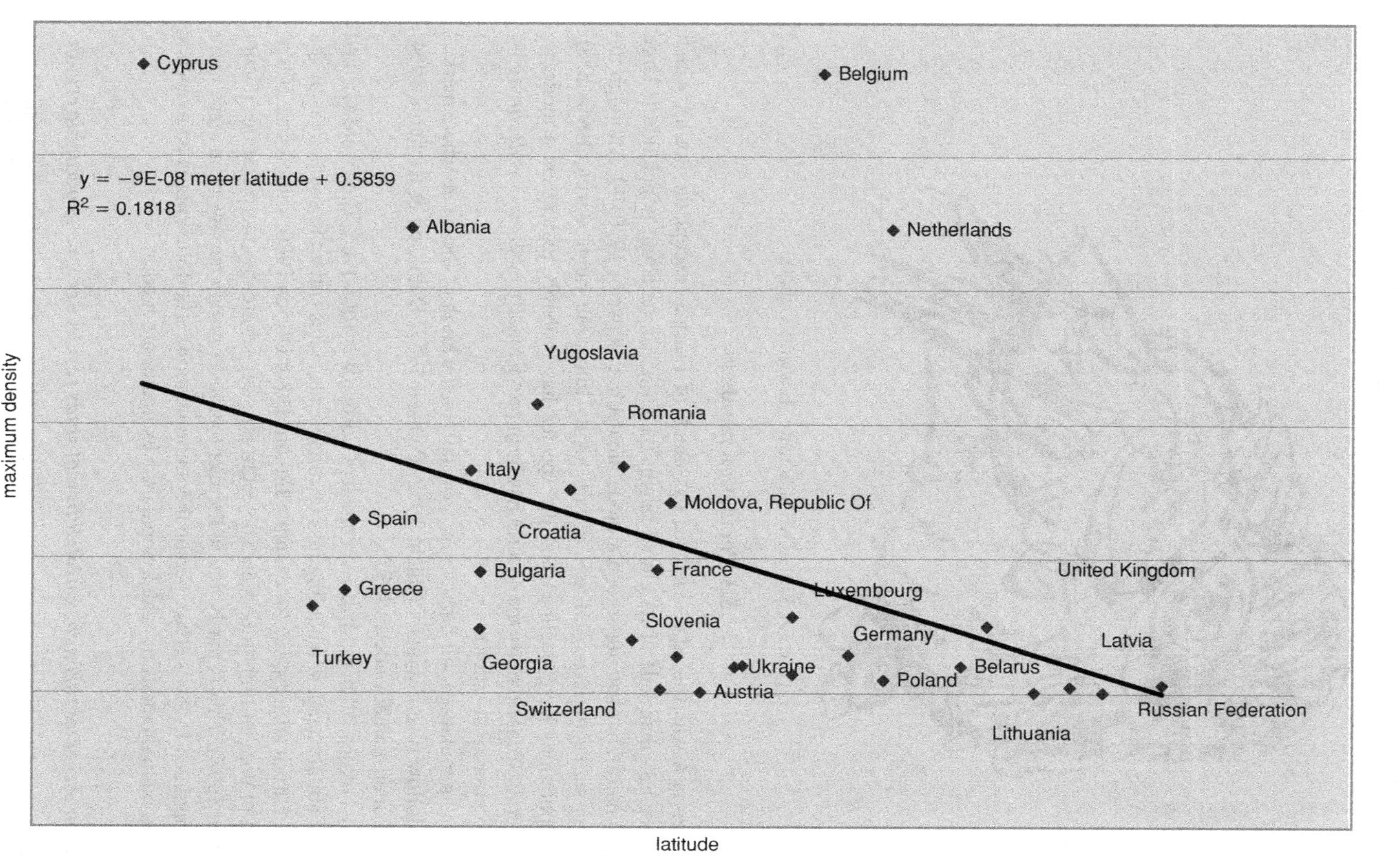

Figure 5.3 Relationship between latitude and population densities of Little Owls (densities calculated as population estimates divided by the area per country, data after Manez 1994).

Figure 5.4 Little Owl silhouette (François Génot).

5.3 Population numbers

The species is a widespread resident across much of Europe (except the north), which accounts for less than half of its global range. Its European breeding population is large (>560 000 pairs), but has undergone a moderate decline between 1970–90. Although the species was stable or increased across parts of its European range during 1990–2000, several populations have declined and the species has probably undergone a moderate decline overall. Consequently the species has been provisionally evaluated as *declining* by BirdLife International (2004).

Little Owl populations are fluctuating, especially in the north of the range where marked decreases follow severe winters (Büchi 1952; Poulsen 1957; Dobinson & Richards 1964; Kämpfer-Lauenstein & Lederer 1995).

Recent decreases, often marked, over much of Europe were observed. They were ascribed mainly to habitat changes, including loss of suitable nest sites (offset in some areas by the provision of artificial sites, see for example Juillard 1980; Ullrich 1980; Bultot *et al.* 2001). Estimation of population numbers are subject to possibly severe misinterpretations (Génot & Lecomte 1998). This is partly due to the fact that most researchers work in areas more densely populated by owls (Exo, personal communication). The following is an overview of the three main methods by which population estimates are made.

Population estimates through consolidation of study results across a geographic region

Consolidation of geographic results is the simple addition of locally based owl numbers. Using geographic population numbers has several drawbacks, i.e., data are never recorded in one single year, sometimes over a spread of more than ten years, and during this time population numbers may fluctuate substantially. Further, data for these estimates are generally obtained through the use of inconsistent methods, i.e., some observers record singing males while others record singing individuals or confirmed breeding pairs. Finally, the habitats vary from one place to another and even within regions, making extrapolations error prone. The main advantage of this type of estimate is that a minimum population is obtained without the risk of overestimation.

Population estimates through partial densities in different habitats

These estimates are derived through main habitat associations, whereby owl numbers and habitats within known study areas are recorded, and then extrapolated to areas without owl surveys but in which the main habitats are mapped. These estimates assume that strong habitat preferences exist. Previous work on the Little Owl has identified a potential correlation between owl numbers and the percentage of grassland (Exo 1983; Loske 1986; Génot & Wilhelm 1993; Zuberogoitia 2002). On a larger scale, similar observations were made (Van Nieuwenhuyse *et al.* 2001c). However, a recent study in Belgium was not able to find any association of the species with grassland (Van Nieuwenhuyse & Bekaert 2002). In France, no correlation between average population densities and the area of grassland was found (Ferrus *et al.* 2002). Thus researchers are cautioned against using grasslands as an estimator for the population of Little Owls. Using average densities of the species to calculate population numbers only makes sense when habitat-specific calculations are made. Further, this is only reasonable if an indication of the small-scale mosaic-like landscape structure is taken into account, e.g., using the length of parcel perimeters instead of areas (Van Nieuwenhuyse *et al.* 2001c). An additional limitation to this method is that it does not take the clustered distribution of the species into account, as seen when apparently unsuitable habitats are occupied and apparently suitable habitats remain empty (Génot & Lecomte 1998).

Population estimates through habitat modeling

A final method to estimate population numbers is through statistical modeling using habitat characteristics of grid cells that are, in turn, based on the area used by a breeding pair (i.e., 25 ha) (Van Nieuwenhuyse *et al.* 2001c). After a complete survey of a region covered by 25 ha grid cells, habitat models are estimated through logistic regression analysis. These models predict the probability that, based on the habitat conditions within the grid cell, a cell will harbor Little Owls. Following this, all of the cells containing a probability of 50% or more are counted and summed for the region. A range of the population numbers is obtained by estimating different models (stratified or not). Estimates through habitat modeling require both a large number of volunteers, a consistent survey method (see survey protocol), and reliable digital descriptions of the landscape, which are seldom available.

However, more and more countries are assembling national working groups that organize volunteers, and more digitized geographic data sets are becoming available.

Population estimates and trends

Despite the shortcomings of all these methods, it is still possible to give a rough overview of population numbers around the distribution range. It should be stressed that the data quality that we present is very heterogeneous from one country to another. For the European countries that were included into the EBCC-atlas (Génot, Juillard & Van Nieuwenhuyse 1997) the data were collected with similar methods across Europe. The historical and current population estimates are given in Table 5.2.

The recently observed trend in maximum estimates in Little Owl populations is negative throughout Europe for countries that had details available in 1994 (Manez 1994; BirdLife International 2004; this project).

Some central European countries have undergone a well-documented negative evolution between 1994 and 2004 (i.e., Czech Republic, 9%; Germany, 24%; Poland, 50%) probably due to a severe winter of 2001–2 (G. Grzywaczewski, personal communication). Other countries have undergone a decrease due to a reduction in reproductive offspring (i.e., the Netherlands, 46%; Willems *et al.* 2004) or intensification of agriculture (i.e., Luxembourg, 43%; United Kingdom, 29%; Slovenia 63%). Switzerland featured a recent increase of 153% of its few small well-protected and managed populations. For other countries some important adjustments to the maximum population estimates were carried out compared to Manez (1994), either upwards (Belgium, 115%; Hungary, +25%; Portugal, +50%; Spain, +54%; Denmark, +33%) or downwards (France, −30%; Greece, −90%; Italy, −40%). Recent minimum population estimates report stable populations for the United Kingdom, Germany, Portugal, Belgium, Austria, Greece, and Italy.

In Groningen (the Netherlands) increasing fragmentation of a decimated population featured a negative population trend that more significantly affected territories that were isolated compared to those clustered despite their more favorable habitat composition (Van't Hoff 2001).

5.4 Densities

The local density of a wildlife population can have a strong "feedback" effect on the demographic performance of that population. For example, the higher the local density, the fiercer the competition (for food and territory) might be and the higher the subsequent impact on the breeding success will be (Newton 1998). With this in mind, we focus in this section on examining Little Owl densities that have been observed in the breeding range of the species. Densities of owls are best determined through the use of radiotelemetry, with detailed locations over a full year (or more) on both male and female owls, yielding an accurate determination of the area (ha) being utilized by a pair of owls. More frequently, though, densities are typically measured as the number of observations (e.g., pairs, calling males, calling individuals) per km^2. Also too, densities are given by quantifying the distances between neighboring owls (i.e., the "nearest neighbor distance").

Among the publications we reviewed for this book, numbers of Little Owls have been reported in several different ways, making comparisons difficult. Some authors report owl numbers obtained by playback; however, these are often (incorrectly) interpreted and reported as confirmed breeding pairs. In some other cases calling males are reported, and in others all calling individuals are reported, no matter what the sex. Some authors denote the density of owls as the number of territories, or the number of breeding pairs, for a given study area size. While this type of density estimate often reflects the number of territorial pairs within the studied area, it may not reflect the actual areas used by the pairs within it (e.g., owls may have used habitat outside of the study area, or, conversely, there may have been substantial areas within the study area that were not used by owls). In particular, while small study areas (e.g., <2500 ha), might seem to have very high apparent densities, in actuality these estimates are biased and ecologically inappropriate for the species because of the substantial use of habitat outside the study area by the owls. Density estimates determined from large study areas are most appropriate, so long as the entire study area was thoroughly and consistently surveyed for owls. As a common measure for our review on owl densities, we use vocal observations as the basis for comparisons across all locations.

Study areas are seldom chosen at random. Most researchers favor areas with dense populations of owls; however, such areas yield biased views on owl numbers,with serious overestimations possible if extrapolating local densities to larger areas. Local extremes in densities are reflected in the following examples: 41.7 territories/km^2 on 12 ha (Exo 1988); 11.6 territories/km^2 on 86 ha (Estoppey 1992); and 15–20 territories/km^2 (Glutz 1962; Visser 1977). Owl densities are also measured by quantifying the nearest neighbor distances and by characterizing the home-range area (through radiotelemetry or detailed observations of known individuals). These measures correlate rather well since the average nearest neighbor distance decreases fairly constantly with increasing number of observed owls in a given area (Bultot *et al.* 2001). The higher the densities, the shorter the nearest neighbor distance, and the smaller the home range of the owls would be expected. We illustrate the heterogeneity of the densities first by describing the number of observed individuals per square kilometre (km^2), and conclude with descriptive statistics of nearest neighbor distances. Since densities tend to be different for different regions, we give an overview of observed densities per country where data is available.

In the country-specific accounts we do not offer vocal observations as a basis for comparisons but rather offer a wide mix of numbers (calling individuals, pairs, calling males, males), hence we are reporting the results as reported in the papers or through personal communication by the observers. (See Table 5.3.)

Most nearest neighbor distances are calculated from relatively small samples (Table 5.4), and tend to confirm the clustered nature of local populations, rather than overall densities. The closest distances between nests in Britain were 240 m and 320 m (Glue & Scott 1980); closest recorded nests in Switzerland were 50 m (Glutz & Bauer 1980). The closest calling

Table 5.3 *Densities of Little Owl,* Athene noctua *in western European anthropogenic landscapes (after Génot & Van Nieuwenhuyse 2002).*

Reference	Surface in km^2	Density	Region (country)	Number of territories	Year
Exo 1988	0.12	41.67	Niederrhein (Klein-Esserden, locally)	5	1984
Estoppey 1992	0.86	11.63	Po (Italy) locally	10	1987–90
Fajardo *et al.* 1998		8.5	Sevilla (Spain)		1996–7
Coppée *et al.* 1995	1	7	Grand-Leez (Belgium)	7	
Coppée *et al.* 1995	4	4.75	Flobecq (Belgium)	19	
Own observations	52	4	Herzele (Belgium)	209	1998–9
Fajardo *et al.* 1998		3.5	Sevilla (Spain)		1996–7
Own observations	91	3.4	Geraardsbergen (Belgium)	310	1998–2000
Own observations	35	2.17	Meulebeke (Belgium)	76	1988–9
Fuchs 1986	10.3	2.04	Betuwe (The Netherlands)	21	1972–85
Coppée *et al.* 1995	3	2	Mortier (Belgium)	6	
Coppée *et al.* 1995		0.7–2	Pays de Herve (Belgium)		
Coppée *et al.* 1995	20	1.7	Thundinie (Belgium)	34	1988–9
Exo 1988	20.7	1.7	Niederrhein (Kreis Kleve)		1974–84
Visser 1977	473	1.7	Nijmegen (The Netherlands)		1974–6
Boitier, pers. comm.	43	1.65	Livradois (France)	71	2004
Audenaert 2003	14.75	1.4	Sint-Pauwels (Belgium)	21	2003
Fajardo *et al.* 1998		1.35	Sevilla (Spain)		1996–7
Zuberogoitia & Campos 1997a	39	1.31	Biscay (Spain)	51	1992–6
Galeotti & Morimando 1991		1.25	Pavia (Italy), urban areas		
Lucas 1996	33	1.21	Pays de Herve (Belgium)	40	1995
Coppée *et al.* 1995	20	1.1	S-O Beaumont (Belgium)	22	1988–9
Centili 1996		1.02	Talfa Mountains (Italy)		1994–5
Barthelemy & Bertrand 1997	33	1	Bouches-du-Rhône (France)	32	1997
Coppée *et al.* 1995	50	1	N-E Charleroi (Belgium)	50	1988–9
Coppée *et al.* 1995	30	1	S-O Anderlues (Belgium)	30	1988–9
Juillard 1984a	275	1	Ajoie (Switzerland)		1973–80
Olea 1997		1	Cataluna (Spain)		
Galeotti & Sacchi 1996		1	Po plain, Lombardy (Italy)		
Fajardo *et al.* 1998		0.99	Sevilla (Spain)		1996–7
Blache 2004	48.5	0.87	Drôme (France)	42	2002
Launay & Calvet, pers. comm.	45	0.8	Haut-Languedoc (France)	36	2004

(cont.)

Table 5.3 *(cont.)*

Reference	Surface in km^2	Density	Region (country)	Number of territories	Year
Fajardo *et al.* 1998		0.8	Eucalyptus Sevilla (Spain)		1996–7
Petzold & Raus 1973	200	0.7	Soest (Germany)		1971–2
Mastrorilli 1997	23	0.7	Bergamo (Italy)	16	1995–6
Arson & Ranvier, pers. comm.	117	0.66	Brotonne (France)	78	2004
Lucas 1996	43	0.6	Condroz (Belgium)	26	1995
Hegger 1977	105	0.6	Viersen, Kempen (Germany)		1976
Hameau, pers. comm.	48	0.56	Luberon (France)	27	2004
Galeotti & Sacchi 1996		>0.5	Low Alps, Lombardy (Italy)		
Müller 1999		0.49	Kreis Viersen (Germany)	250	1998
Cesaris 1988		0.45	Tocino (Italy)		
Lorthois, pers. comm.	100	0.43	Scarpe-Escaut (France)	43	2004
Pirovano & Galeotti 1999	60.8	0.4	Pavia (Italy)	21	1995–7
Coppée *et al.* 1995	270	0.37	Famenne (Belgium)	100	
Kämpfer and Lederer 1988	240	0.35	Lippstadt (Germany)		1976–87
Fajardo *et al.* 1998		0.35	Marchland Sevilla (Spain)		1996–7
Dalbeck *et al.* 1999	950	0.26	Jülicher Börde (Germany)	246	1989–92
Fajardo *et al.* 1998		0.25	Urban areas Sevilla (Spain)		1996–7
Olea 1997		0.2	Granada (Spain)		
Ziesemer 1981	108	0.2	Norderstedt (Germany)		1978
Gon & Goa, pers. comm.	244	0.17	Normandie-Maine (France)	42	2004
Ille 1996	60	0.15	Marchfeld (Austria)	9	1991–4
Dombrowski *et al.* 1991	80	0.14	Mazowsze lowland (Poland)		
Bernard, pers. comm.	120	0.12	Cévennes (France)	100	1991
Danko *et al.* 1994	950	0.1	Michalovce district (Slovakia)		
Petzold & Raus 1973	300	0.1	Soest (Germany)		1971–2
Ziesemer 1981	100	0.1	Bergenhusen (Germany)		1975–8
Coppa, pers. comm.	140	0.08	Montagne de Reims (France)	12	2004
Mangin & Génot, pers. comm.	437	0.08	Vosges du Nord (France)	36	2004
Kowalski *et al.* 1991	157	0.06	Kampinos National Park (Poland)		

(cont.)

Table 5.3 *(cont.)*

Reference	Surface in km^2	Density	Region (country)	Number of territories	Year
Vogrin 1997	210	0.05	Dravsko polje (Slovenia)	10	1988
Schröpfer 1996	730	0.05	West-Bohemia		1993–4
Fronczak & Dombrowski 1991	80	0.04	South Podlasie (Poland)		
Šálek 2004	195	0.04	Ceské Budejovice and Pisek Region (Czech Republic)	8	1992–2004
Olea 1997		0.03–0.11	Galicia (Spain)		
Pykal *et al.* 1994	720	0.02	Southern Bohemia (Czech Republic)		
Ille 1996	870	0.02	Northern Burgerland (Austria)	17	1991–4
Renner, pers. comm.	500	0.01	Lorraine (France)	8	2004
Ziesemer 1981	250	0	Westensee, Preetz (Germany)		1974–8

pers. comm. = personal communication

males in Flanders, Belgium: Herzele – min.; 42 m; average; 210 m; max., 862 m; n = 210; Geraardsbergen – min., 157 m; average, 330 m; max., 942 m; n = 310; Meulebeke – min., 173 m; average, 388 m; max., 811 m; n = 78 (D. Van Nieuwenhuyse, personal observation). Breeding sites of three pairs in Ankara (Turkey), were situated at 120 m from each other around 32 apartment blocks (11 ha), with males hunting within 20 m of each other when using electricity poles as perches (Vaassen 2000). In southern Portugal the nearest neighbor distances of breeding pairs in an area of 16 km^2 of steppe-like habitat were: min., 97.5 m; average, 313.7 m; max., 555 m; n = 37; and in a woodland area of 6 km^2: min., 67.5 m; average, 447.5 m; max., 1230 m; n = 39 (R. Tome, unpublished data). Minimum distances between nests in the Mediterranean climates of Israel were 200 m at Jerusalem and at Marj Sanur, 300 m in the desert climates of Nizzana, and 500 m in northwestern Negev (Shirihai 1996). In the United Arab Emirates three nests have been recorded at a spacing of 500 m, although occupied nests were not less than 1000 m apart in Kuwait. At one farm in central Arabia, which was rich in rodents but poor in Little Owl roosting and nesting sites, 12 adults were seen on a line of rock heaps within 1 km (Génot & Van Nieuwenhuyse 2002). In the lower Zarafshan area of Uzbekistan, six Little Owls and three nests were registered during the breeding period during a road survey of 15 km. In the sandy desert of the lower parts of Kashka Darya (Uzbekistan), on a road survey of 20 km (50–60 m along each side of the road surveyed) seven pairs were found (Sagitov 1990). In Badkhyz, Turkmenia the

Table 5.4 *Nearest neighbor distances of Little Owl,* Athene noctua *in western European anthropogenic landscapes (after Génot & Van Nieuwenhuyse 2002).*

Reference	Region (country)	Number of territories	Year	Nearest neighbor distances		
				average (m)	min (m)	max (m)
Genót & Van Nieuwenhuyse 2002	Lorraine (France)			8116		
Loske 1986	Baden-Württemberg (Germany)	10	1976–84	3470		
Genót & Van Nieuwenhuyse 2002	Montagne Reims (France)			2834		
Loske 1986	Rheinland-Pfalz (Germany)	11	1976–84	2781		
Loske 1986	Niedersachsen (Germany)	5	1976–84	2070		
Loske 1986	Münsterland (Germany)	76	1976–84	1950		
Genót & Van Nieuwenhuyse 2002	Vosges du Nord (France)			1776	300	
Loske 1986	Schleswig-Holstein (Germany)	23	1976–84	1145		
Genót & Van Nieuwenhuyse 2002	Causse Méjean (France)	36		1075		
Genót & Van Nieuwenhuyse 2002	Brotonne (France)	23		1045		
Loske 1986	Saarland (Germany)	11	1976–84	682		
Lucas 1996	Pays de Herve (Belgium)	40	1995	680	300	1700
Loske 1986	Jülicher Börde (Germany)	12	1976–84	596		
Loske 1986	Essen (Germany)	13	1976–84	585		
Müller 1999	Kreis Viersen (Germany)	250	1998	556	250	925
Lucas 1996	Condroz (Belgium)	26	1995	460	225	950
Loske 1986	Niederrhein (Germany)	26	1976–84	456		
Loske 1986	Mittelwestfalen	29	1976–84	435		
Genót & Van Nieuwenhuyse 2002	Scarpe-Escaut (France)	14		403		
Genót & Van Nieuwenhuyse 2002	Meulebeke (Belgium)	78	2000	388	173	811
Genót & Van Nieuwenhuyse 2002	Geraardsbergen (Belgium)	310	1998–2000	330	157	942
Genót & Van Nieuwenhuyse 2002	Herzele (Belgium)	210	1998–9	210	42	862

highest number is recorded on clayish precipices of the Kyzyldzhar ravine, where there were 1.6 individuals per km of precipices (Simakin 2000). In Tajikistan the species can be considered as common everywhere, but nowhere in the Republic at such considerable concentrations as the one precipice, stretching almost uninterrupted along several kilometers in the "Tigrovaya Balka" Nature Reserve. Three pairs nested at a distance of 1–1.5 km from each other. As a rule, the separate breeding pairs are situated at 8–10 km from each other (Abdusalyamov 1971). In the Stavropol region, Little Owl density was two individuals per km of road surveyed (Khokhlov *et al.* 1998).

5.5 Overview by country

Belgium

Distribution

The species was found in 86% of the country except the snow-rich High-Belgium and extensively wooded areas like the Ardennes and Belgian Lorraine in the south of the country (Delmée 1988). The highest densities were observed in southern West-Flanders, East-Flanders, and Limburg (Vercauteren 1989). The Leemstreek region holds 40% and the Zand-Zandleemstreek region 46% of the total population, the Kempen and Polders regions were reported to have limited numbers in 1998–2000 (Van Nieuwenhuyse *et al.* 2001c). The Little Owl in Flanders (northern half of Belgium) can be characterized along three gradients: (1) the gradient between open and closed landscapes, (2) the gradient between dry and wet soils, and (3) the gradient between sandy and loam soils. The species avoids extremely open (polders, lakes) and extremely closed (cities, forests) landscapes, avoids soils with "excessive" and "bad" drainage, and prefers loam soil to sand. A mosaic-like landscape structure was found to be more important than the actual land-cover types. In Wallonia, the species is still found in all types of habitat, even in the Ardennes and Lorraine. Some sectors seem unoccupied for a long period of time, e.g., some empty grid cells were already present in 1973–7 (Delmée 1988) around Saint-Vith and a part of the Luxembourg Ardennes. In the Walloon Atlas (2008) project a significant proportion of the surveyed grid cells were no longer occupied, with population fragmentation and a reduction in the owl's distribution. The Little Owl may possibly not have recovered from the severe winter of 1996–7. Nevertheless, some disappearing pairs continue to be registered since 2000, at least in Lorraine.

Population estimates

The species has decreased markedly, probably due to habitat destruction and pesticides. Estimates indicate there were around 12 000 pairs in 1950, around 4000 pairs in 1972 (Lippens & Wille 1972), around 7300 pairs in 1973–77 (Delmée 1988), 4500–6600 pairs during 1981–90 but decreasing (Manez 1994).

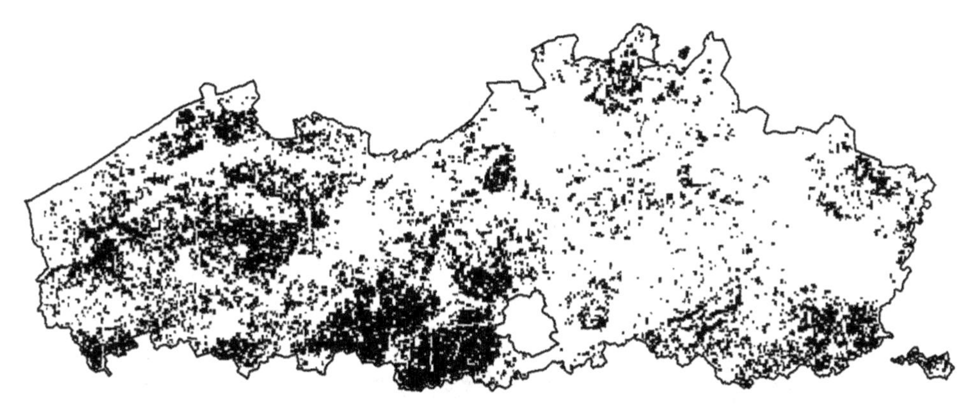

Figure 5.5 Little Owl distribution in Flanders (northern Belgium) as of 2001 (Van Nieuwenhuyse *et al.* 2001c). Dark areas indicate grid cells with modeled probabilities of Little Owl presence ≥50%.

Flanders

Van Nieuwenhuyse *et al.* (2001c) developed recent population estimates for Flanders. Here, they estimated the number of occupied grid cells (500 × 500 m) with a probability of Little Owl presence ≥50%. Their results indicated a range of 12 527 (see Figure 5.5) to 16 046 occupied grid cells for the models without regional stratification. When summing all of the minimum and maximum estimates of stratified models respectively, a range of 13 646–17 361 occupied cells was obtained. Both estimates indicate two main regions in absolute numbers, i.e., 40% of the Flemish Little Owl population is situated in the Leemstreek ecological region and 46% are in the Zand-Zandleemstreek region, while the Kempen and Polders regions have only limited owl numbers. The highest average probabilities calculated are observed principally in the Leemstreek. The total Flemish Little Owl population is estimated at 6 000–10 000 pairs (Van Nieuwenhuyse 2004).

Wallonia

Large differences are found between different sectors. The species is still rather common in Middle Belgium, Condroz, and in the Pays de Herve and Fagne-Famenne. The most observed abundance classes are 11–20 and 21–40 pairs/40 km^2; with two grid cells with 41–80 pairs. The species is rare along the Meuse valley with most grid cells having only 6–10 pairs/40 km^2. It is still present but very rare (between 1–5 and 6–10 pairs/40 km^2) around the large urban centers (Liège, Huy, Namur, Charleroi, La Louvière, Mons). The urban centers include some meadows and gardens that still offer some habitat. In the Ardenne and Lorraine the species is very rare with densities around 0 up to 1–5 pairs/40 km^2, and a few grids with 6–10 pairs/40 km^2.

The total estimate for Middle Belgium is 1800–2400 pairs; Condroz, Fagne-Famenne-Pays de Herve 950–1300 pairs; the Ardennes 200–400 pairs; and Lorraine 30–60 pairs. The total Walloon population is thus estimated at 3000–4200 pairs (J. P. Jacob, personal communication). (See Figure 5.6.)

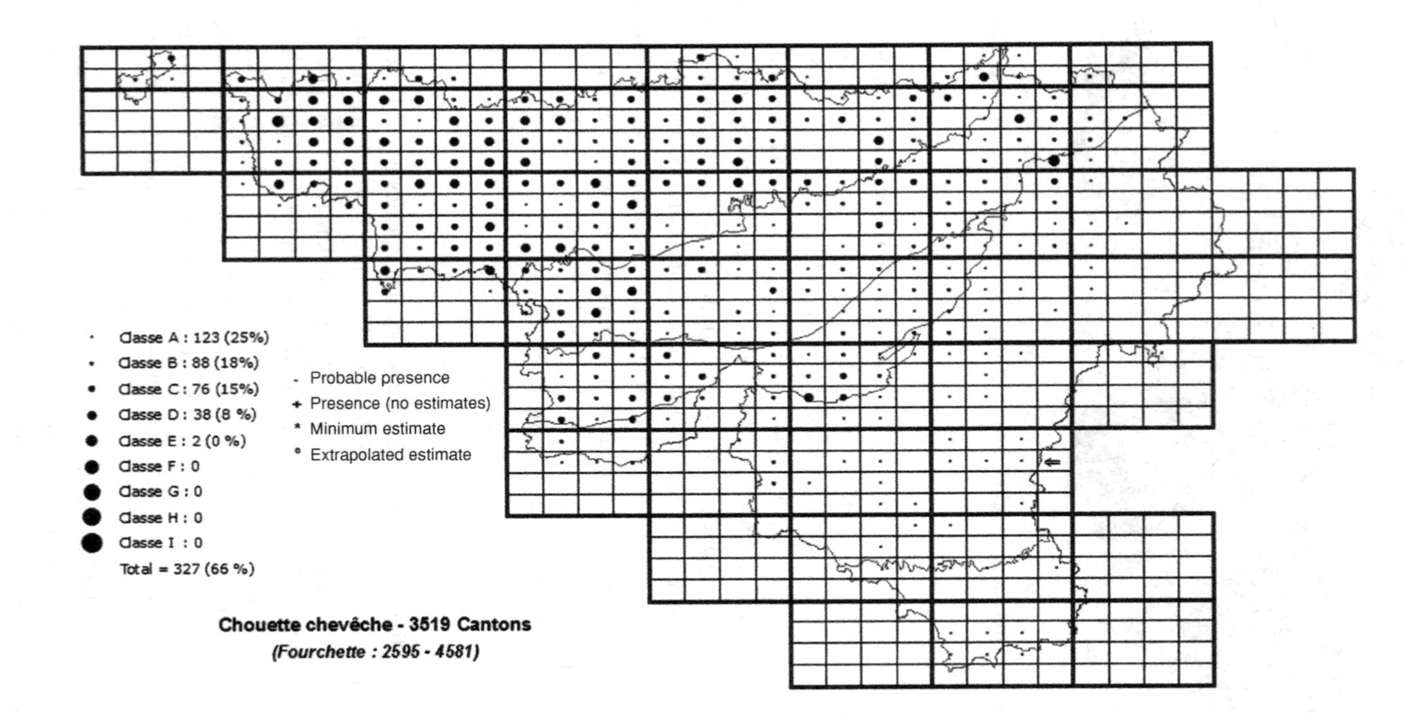

Figure 5.6 Little Owl distribution in Wallonia (southern Belgium) in 2000–3 (www.aves.be).

Thus, as of 2003, the total Belgian population is estimated at 9000–14 200 pairs (12 500–14 000 pairs BirdLife International 2004).

Densities

High densities of Little Owls have been found on larger areas in or near Belgium: e.g., 3.4 calling individuals/km^2 in Geraardsbergen with 42 out of 91 km^2 squares with at least 4 individuals; 4 calling individuals/km^2 in Herzele with 24 out of 45 km^2 squares with at least 4 individuals; 2.17 calling individuals/km^2 in Meulebeke with 8 out of 29 km^2 squares with at least 4 individuals (own observations). Locally high concentrations are also found in Condroz, with 2 out of 33 km^2 squares with 4 calling males (Lucas 1996). In the Liège region, 25 calling males in 50 km^2 with meadows and old orchards (S. Pirotte, personal communication).

The Netherlands

Distribution

The Little Owl was observed in half of the breeding bird atlas surveyed grid cells between 1998–2000 (Groen *et al.* 2002). In large parts of the three northern provinces (Groningen, Friesland, Drenthe) the species was completely absent. In the western part of the country some population clusters remain, i.e., the former fruit growing areas West-Friesland, the Beemster, the Green Heart, and Hoekse Waard. In Zeeland the species is still widely spread in Zeeuws-Vlaanderen, the part of the Netherlands bordering Flanders (De Smet 2002). The species has not been observed in Walcheren, while the species is found sporadically in Zuid-Beveland. The stronghold of the population is still found in the small-scale landscape conditions of the central and eastern river area (Meuse and Rhine) joined by the Liemers, the Achterhoek, and the delta of the Ijssel River. Locally, some clusters with higher densities still occur. Remarkable is the scattered distribution in Twente en Salland. The species is absent on the Wadden Isles, in the Ijsselmeerpolders, and the vast woods of the Veluwe.

Population estimates

The species has declined since around 1935, mainly due to habitat changes. The first estimates of 6000–8000 pairs 1973–7 (Teixeira 1979) were probably too low. In the second Dutch atlas (SOVON 1987) a new estimate of 8000–12 000 pairs was given for 1978–83 (Bekhuis *et al.* 1987); Osieck and Hustings (1994) estimated 10 000–14 000 pairs for 1979–85 and 9000–12 000 pairs for 1989–91. The lastest estimates (1998–2000) come from the new atlas: 5500–6500 pairs (SOVON 2002; BirdLife International 2004). This means a decrease between 35% and 60% in the past 20 years. Recently well-researched regions, e.g., Zeeuws-Vlaanderen, Achterhoek, and Liemers feature the highest current densities. Noord-Nederland (Groningen, Friesland) feature the lowest densities. The lower densities observed in Drenthe, Overijssel, Brabant, and Limburg are probably due to incomplete

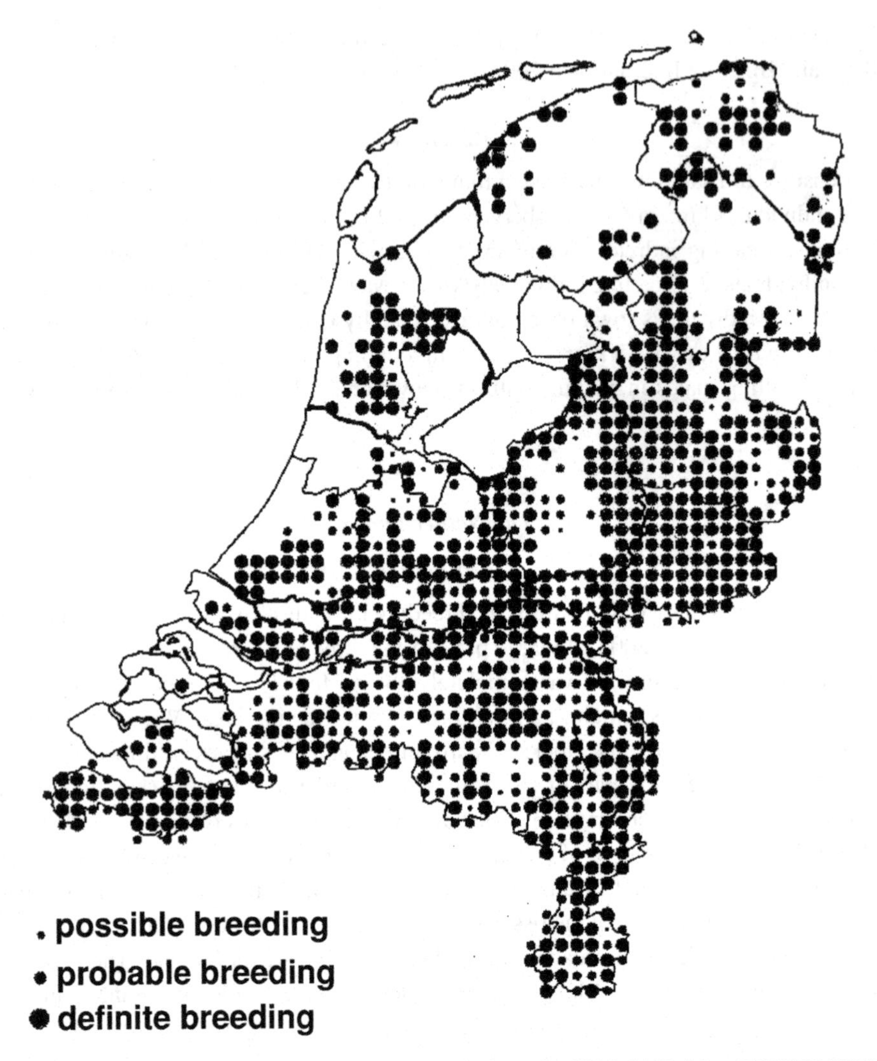

Figure 5.7 Little Owl distribution in the Netherlands from the 1988–2000 Atlas (SOVON 2002).

survey work (SOVON 2002). Population changes between the atlas periods 1978–83 (SOVON 1987) and 1998–2000 (SOVON 2002) are very clear in northern and western Holland where the species has disappeared. In 330 atlas squares the species is now absent (a decrease in atlas squares of 30%, which equates to a geographic range reduction of 8250 km^2). The situation is striking in the northern provinces, such as in Drenthe, where the species is absent in the center of the province. The number of pairs in Friesland (n = 10–20) and Groningen (n = 20) is only a fraction of what they were. In Groningen there were up to 300–400 pairs in the 1960s (Van't Hoff 2001). The presence of the species in 101 new squares in the latest atlas work is probably due to more intensive survey work rather than an expansion of the owl's range. Poor reproduction in the species is suspected to be a cause of the decline of 40% over 17 years, or a yearly decrease of 3% (Willems *et al.* 2004). A significant decline in the average clutch size has been observed since the 1980s and this has probably been occuring since the 1970s (n = 3149 nests measured from 1977–2003). A non-significant negative trend is observed in breeding success.

Densities

Within the Netherlands, owl densities differ markedly. In the northwestern part of the country, densities vary from 0.04–0.4 pairs/km^2. In the half-open cultivated areas in the southeast, densities can reach 1.0–1.5 pairs/km^2 over larger areas (Stroeken *et al.* 2001). A density of 2.04 pairs/km^2 was found in Betuwe by Fuchs (1986). See Figure 5.7.

Luxembourg

Distribution

The species is limited to a small population in the northern part of the country and a larger one in the southern part. Dramatic declines took place between 1990–2004 in the western and northern part of the country. The main population is believed to be found in the southeastern part of the country (P. Lorgé, personal communication).

Population estimates

In 1960, the population was estimated at 3400–4200 pairs (Hulten & Wassenich 1961). The *Atlas of Breeding Birds in Luxembourg* (Melchior *et al.* 1987) estimated the population at 300–500 pairs between 1976 and 1980. As of 2003 the population was estimated at 45–85 pairs, with only 15 pairs left in the northern part of the country. Dramatic declines have taken place in the last 10–15 years in the western and northern part of the country. The best remaining populations are reported in the southeastern part. In 2004 there were several conservation initiatives, with intensive monitoring and nestboxes, yielding only ten confirmed territories. The population hence is estimated in 2006 at 15–20 pairs. This means that the species is considered to be on its way to disappearing from Luxembourg (Lorgé 2006). See Figure 5.8.

United Kingdom

Distribution

Previously a rare vagrant to the British Isles from the continent, the Little Owl was introduced into England in the 1870s and 1880s and established a feral breeding population (Marchant *et al.* 1990). Initial releases were made in Kent and Northamptonshire, and by the close of the 1800s there were two separate breeding groups, in southeast England and the east Midlands. The spread was subsequently very rapid, aided by further releases, and soon Little Owls were present west to the River Severn and north to the River Trent. By 1925 they had colonized much of Wales and expanded northwards to the central parts of Lancashire

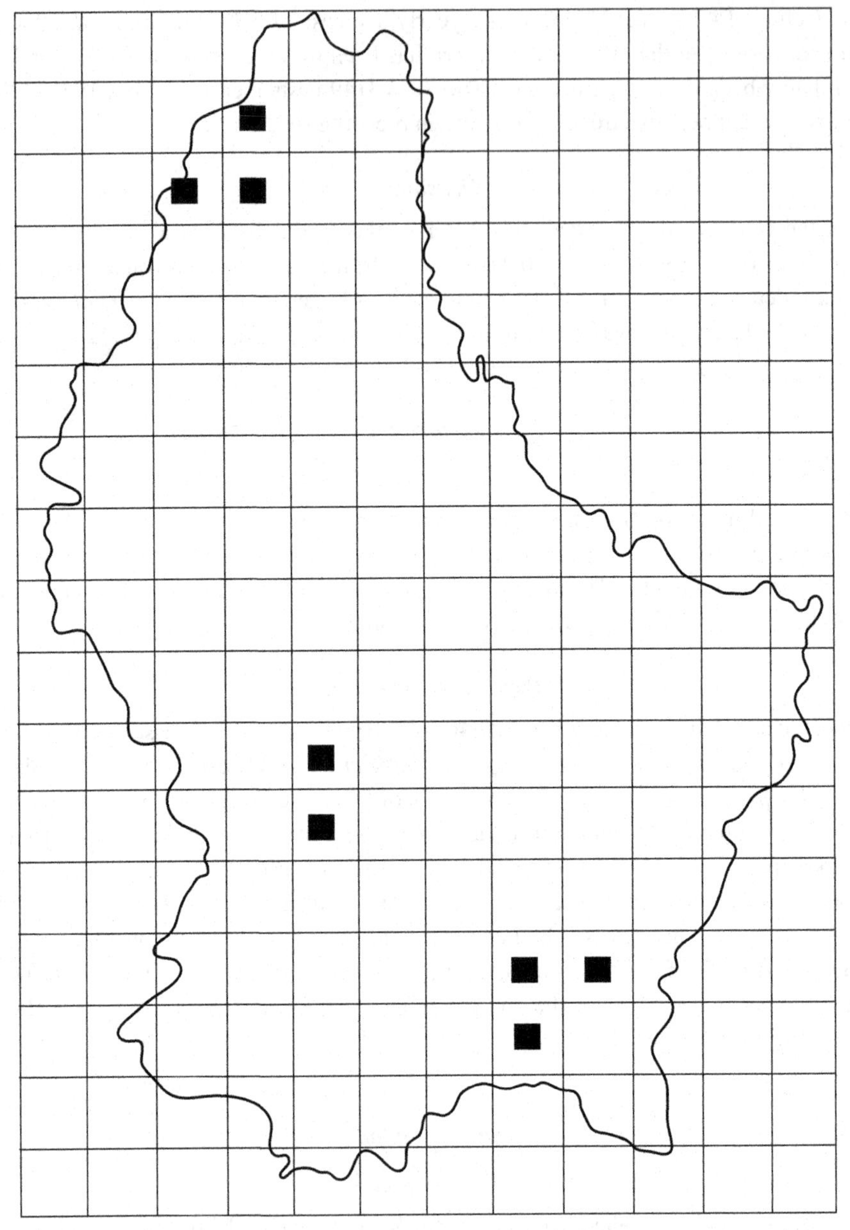

Figure 5.8 Little Owl distribution in Luxembourg (P. Lorgé, personal communication). Black squares indicate occupied grid cells as of 2005.

and Yorkshire. Durham and Northumberland were reached in the 1930s, the Lake District in the 1940s, southeast Scotland (the Borders and Lothian) from 1958, then Dumfries and Galloway from the 1970s (Marchant *et al.* 1990). However, the Little Owl remains extremely rare in Scotland and has not successfully penetrated north of the Edinburgh–Glasgow line. The first confirmed breeding of the Little Owl in Scotland was in 1958 (Thom 1986). The current Scottish population of Little Owls is unlikely to exceed a few tens of pairs (Park *et al.* 2005) and may be less than ten pairs (Greenwood *et al.* 2003). See Figure 5.9a–c.

Population estimates

Though peripheral expansion continues, the initial explosive phase after the introduction had slowed by the mid-1930s and was followed by reports of decreases in density over large parts of central and southern England. A run of severe winters in the 1940s might have adversely affected the Little Owl. Some improvement occurred in the early 1950s, but there was a further decline during 1955–63, which coincided with the organochlorine pesticide era and the cold winters of 1961–2 and 1962–3 (Parslow 1967; Sharrock 1976). The Common Bird Census (Marchant *et al.* 1990) showed fairly regular spacing of peaks and troughs between 1964 and 1988. The peak years have been at intervals of three to five years and an average of four years. This has all the appearance of being cyclical. It seems possible that reports of Little Owl declines after the mid-1930s included such cyclical oscillations reflecting natural variations in population levels and may have contributed to other low points mentioned in the literature. Superimposed onto the cyclical pattern has been some longer term variation in Little Owl densities, though it cannot yet be said that an overall trend has been confirmed. After recovery in the early 1960s, the average population level varied little until 1975, but an increase in the late 1970s and early 1980s produced the highest ever index figure in 1984. This overall increase occurred despite two very cold winters, 1978–9 and 1981–2, which seemed to have had little effect; 1982 was a natural low point in the cycle. In 2000 the British Trust for Ornithology and the Hawk and Owl Trust completed a three-year national survey of Barn Owls within the United Kingdom (Toms *et al.* 2000). As part of this project they also gathered information on breeding Little Owls, and produced a national population estimate of pairs. The Little Owl population estimates were calculated using a stratified random sampling approach based on 1100 tetrads (2 by 2 km squares) spread across the UK. A boot-strapping procedure was used to calculate 95% confidence intervals for the estimates produced from the sampling procedure. Individual national estimates were calculated for each of the three survey years to allow for any short-term climatic effects on breeding performance. The estimates of Little Owl pairs were found to be similar across each of the three years (numbers of pairs [range]: 1995: 7103 [5136–9136]; 1996: 6142 [4075–8301]; 1997: 5865 [4337–8186]). The estimates of the models using landscape data were comparable: 1995: 7468; 1996: 6653; 1997: 6253. The estimates were also recalculated for the known breeding range of the species within the UK (rather than for the whole of the UK). These data suggest a population estimate of 4000–8500 breeding pairs and represent the first reliable and repeatable estimate for the UK.

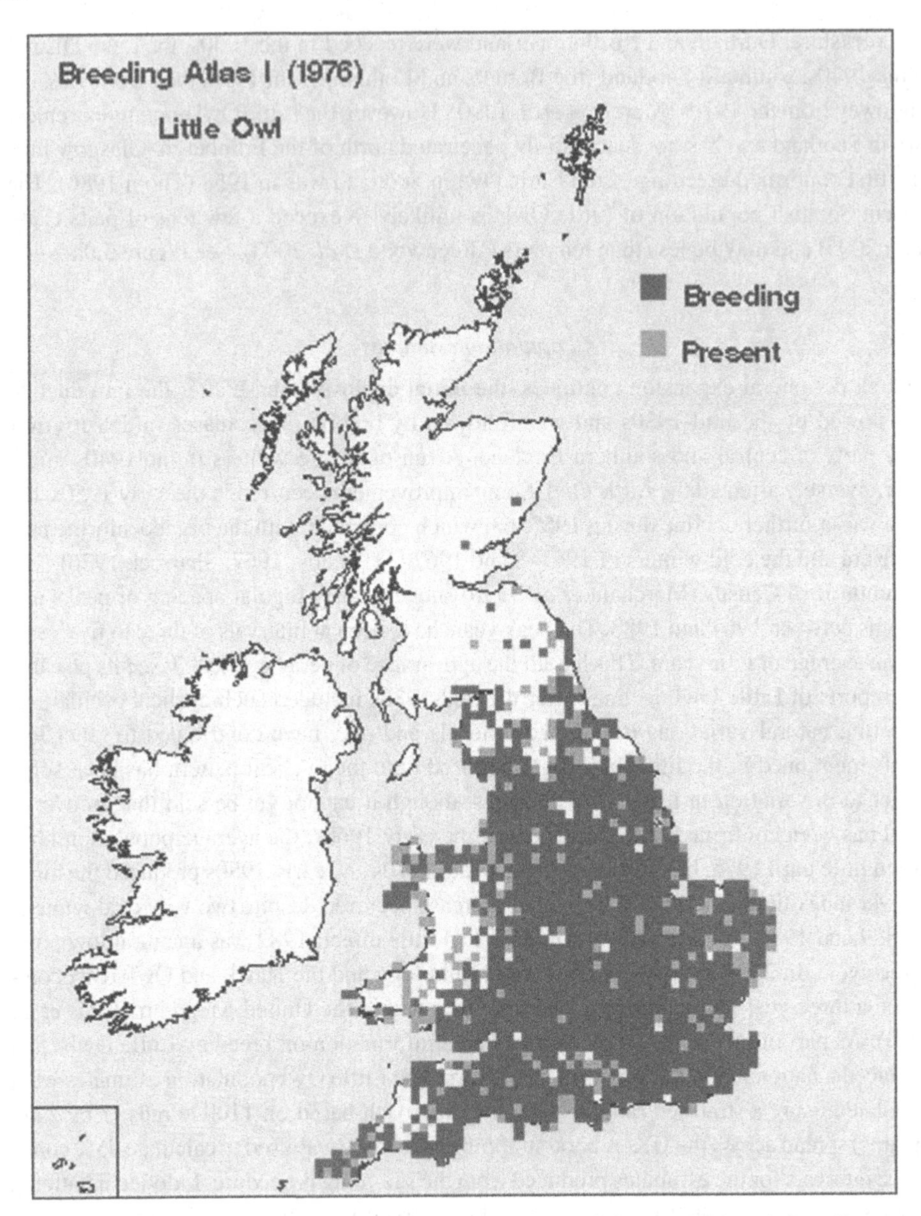

Figure 5.9a Little Owl distribution in the United Kingdom in 1976 (www.BTO.org); Breeding Atlas I (Sharrock 1976).

The previous estimate of 6000–12 000 breeding pairs (Gibbons *et al.* 1993) was derived from an atlas survey for all breeding species rather than specifically looking at Little Owls, and may be based on less accurate mean breeding densities. BirdLife International (2004) estimates the total population at 5800–11 600 pairs.

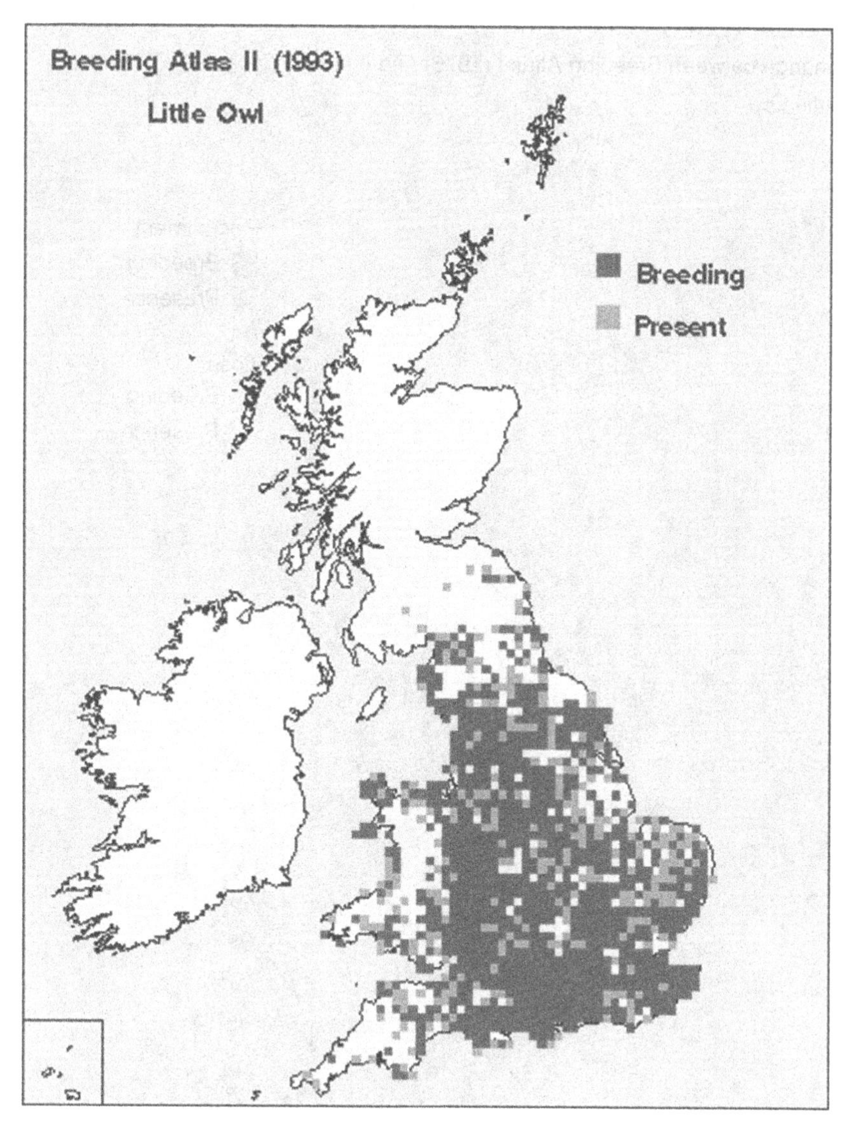

Figure 5.9b Little Owl distribution in the United Kingdom in 1993 (www.BTO.org); Breeding Atlas II (Gibbons *et al.* 1993).

Federal Republic of Germany

Distribution

The Little Owl is a rare to very rare breeding bird in Germany. The highest population densities occur in northwest Germany with Northrhine-Westphalia as the most important stronghold, followed by Hessen in Central Germany. In the southern and eastern part of Germany, the species is very rare (Bauer *et al.* 2002).

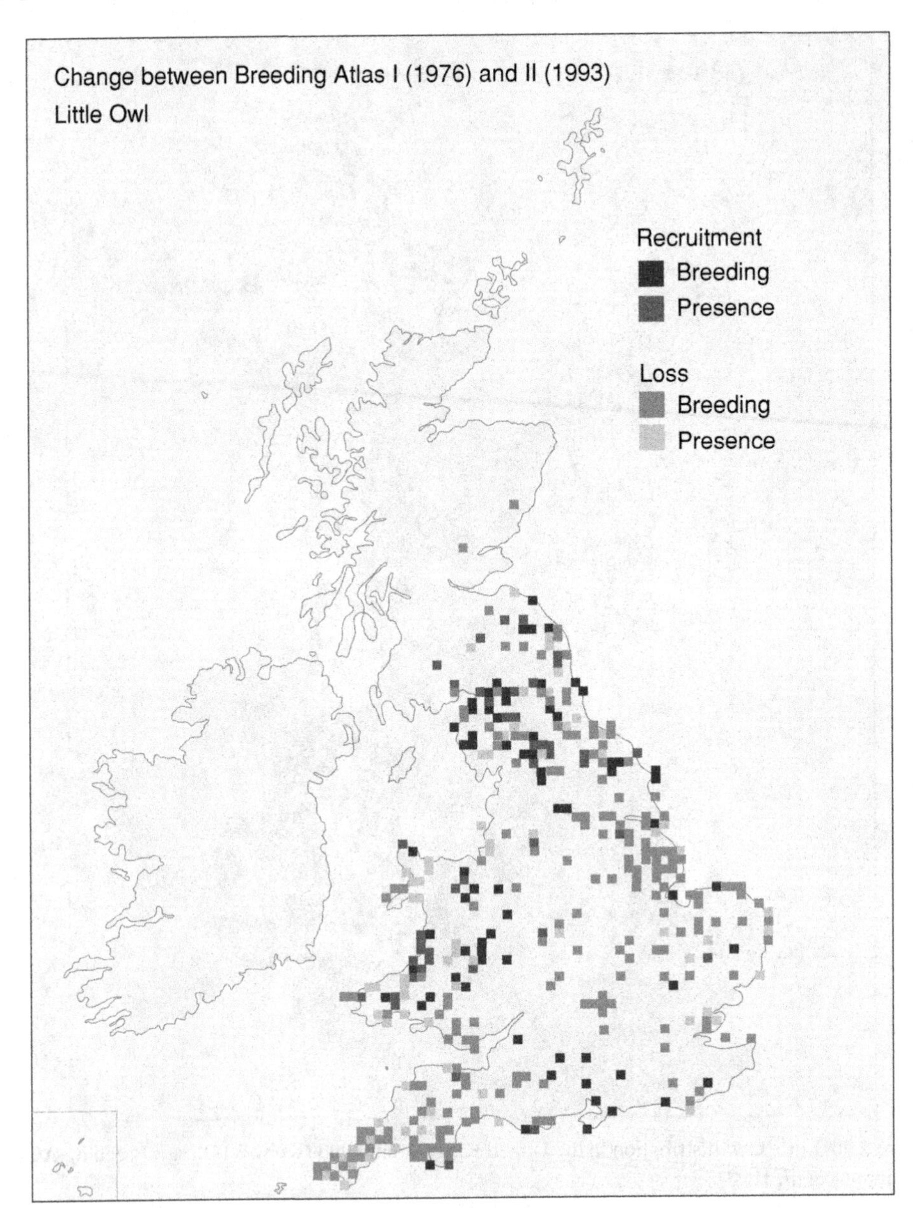

Figure 5.9c Little Owl distribution in the United Kingdom in 1993 (www.BTO.org); change between Breeding Atlas I (Sharrock 1976) and II (Gibbons *et al.* 1993).

Schleswig-Holstein has very low population levels due to its northern situation. In Niedersachsen, the species is found regularly but is very widely scattered. In Northrhine-Westphalia the species can be found everywhere except the industrial Ruhr region and the mountainous areas. The population is contiguous from the north of Nordeifel, Ahr-Mountains and

Sauerland on. Three concentration zones exist, i.e., the Niederungsgebieten in Middle Westphalia, rich in pollarded willows, the willow-rich Niederrhein, and in the pastured high-stem orchards in the western Jülicher Börde (Schönn *et al.* 1991). Rheinland-Pfalz has two strongholds, i.e., the Bachtäler Rheinhessen, and around Trier and Bitburg. In Hessen, the strongholds are situated in the hilly regions at around 350–400 m elevation. Saarland has a small concentration in the southeastern part of the region. In Baden-Württemberg, the species has only minor isolated populations mainly around the Black Forest, Rheinebene (Freiburg/Offenburg), the eastern bank of the Bodenlake, and Albvorland. In Bayern the species is very rare and mostly absent. It still occurs sporadically below 700–800 m elevation in Mainfranken and Gräulandschaft, otherwise only isolated pairs are found. In the former East Germany the species is found to be a very rare breeding bird. Mecklenburg and the Sachsen-Thüringer middle mountain range feature unfavorable conditions (northern border, vertical distribution limits) for the owl. In Brandenburg the species only breeds in isolated pairs. Most of the population of East-German Little Owls are found in Sachsen, Thüringen, and Sachsen-Anhalt. See Figure 5.10.

Population estimates

The Little Owl has shown a marked decline in Germany, probably due to habitat changes. Numbers have been severely affected by hard winters (Glutz & Bauer 1980; Bauer & Thielcke 1982). Estimates for the former East Germany were 250–300 pairs for 1985–6 (Schönn 1986), 1035 pairs in North Rheinland for 1999 (M. Wink, personal communication), 5000–10 000 pairs in 1985 in the entire country (Manez 1994). The total population was estimated at between 5800–6100 in 1996–9 with only a minor decrease between 1999–2002 to 6000 pairs (see Table 5.5; Bauer *et al.* 2002). A detailed meta-analysis using all possible data sources yielded a population estimate of 7000–7600 pairs (H. Illner, personal communication). BirdLife International (2004) estimates the total population at 5800–6100 pairs.

The population stronghold is situated in Nordrhine-Westphalia, which harbors 75% of the national population. Table 5.5 gives details of the population estimates across the country.

Trends

See Table 5.6. The highest rates of decline (−3.0 to −4.3% per year) were found in the federal states north and east from the stronghold in Northrhine-Westphalia; moderate rates of decline were found in the southern federal states (−1.2 to −1.9%). The central state Hessen, which neighbors Northrhine-Westphalia, is the only state with an apparent increase over the last 26 years. A common feature is also that all of the federal states with yearly decline rates of more than 1.7% lost significant parts (>about 50%) of the breeding range of the species, even Hessen (with a population increase assumed) shows a loss of about one-third of the breeding range that was occupied in the 1970s. Even in Northrhine-Westphalia some parts of the former breeding range were lost, e.g., the higher parts in south Westphalia and parts of the industrial area Ruhrgebiet. Southwest of the Palatinate (Rhine-Palatinate,

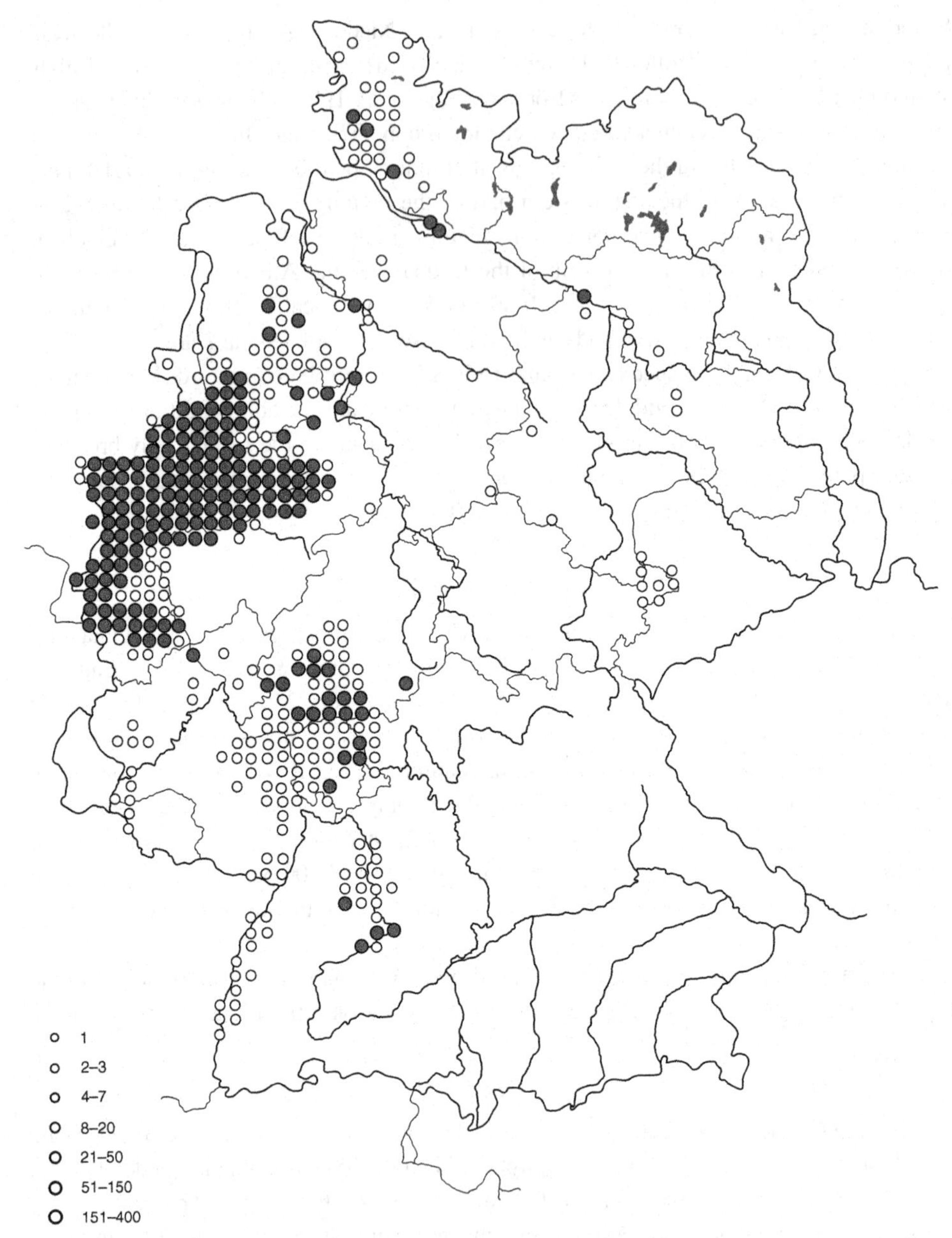

Figure 5.10 Little Owl distribution in the Federal Republic of Germany (after Jöbges 2004).

Table 5.5 *Little Owl in the Federal Republic of Germany: number of breeding pairs and territories, respectively, in the time period 1999–2003 (H. Illner personal communication).*

	1995/1996 Mädlow & Model 2000	1999 Bauer *et al.* 2002	(1990) 1999–2003 Recent publications or sources (year: pair number, sources)
Northwest Germany			
Schleswig-Holstein	77–101	122	1999: 110–30, Berndt *et al.* 2002
Hamburg	0?	1	1997–2000: 1–3, Mitschke & Baumung 2001
Niedersachsen	*c.* <150	*c.* 150	1999: at least 120[a], 2002/2003 150–200[b]
Nordrhein-Westfalen	*c.* 4500	*c.* 4500	(1993–) 2003: 5300–800 questionnaire S. Franke, March 2004[c]
Subtotal	*c.* 4727–4751	*c.* 4773	(1993–) 1999–2003: 5561–6133 (mean: 5847)
Central (west) Germany			
Hessen	≥556–582	524	1999: ascertained 625, Burbach 1997
Rheinland-Pfalz	<300	*c.* 300	1990–2000: about 300, C. Dietzen, pers. comm. February 2003[d]
Saarland	30–50	*c.* 40	2001/2002: about 40, W. Stelzl pers. comm. February 2003
Subtotal	*c.* 886–932	*c.* 864	(1990) 1999–2002: 965
Southern Germany			
Baden-Wuerttemberg	*c.* 200	*c.* 150	1999: 250, estimated by Hölzinger 2001[e]
Bayern	*c.* 150	*c.* 125	2002: 100–50, S. Hartlaub pers. comm. February 2003
Subtotal	*c.* 350	*c.* 275	1999–2002: 350–400 (mean: 375)

(*cont.*)

Table 5.5 *(Cont.)*

	1995/1996 Mädlow & Model 2000	1999 Bauer *et al.* 2002	(1990) 1999–2003 Recent publications or sources (year: pair number, sources)
East Germany			
Mecklenburg-Vorpommern	7	(7, Bauer *et al.* 2002)	
Brandenburg-Berlin	7–12	13	2000: 12, Ryslavy 2002; 13, Haase 2001
Sachsen-Anhalt	*c.* 5	15	1999: 5–20, Nicolai 2000: 10–30, J. Schroeder, pers. comm. Jan. 2004
Sachsen	10–30	17	1999/2000: 5–10, S. Schönn pers. comm. Feb. 2003
Thueringen	5–25	17	1999: about 15, J.Wiesner pers. comm. Feb. 2003
Subtotal	*c.* 27–72	69	1999–2000: 50–75 (mean 63)
Total[f]	***c.* 5990–6105**	***c.* 5981**	**6926–7573 (mean 7250)**

[a] questionnaire by the Niedersaechsisches Landesamt fuer Oekologie, pers. comm. C. Stange, March 2003.

[b] NABU Altkreis Lingen e.V. (ed.) (2003): Steinkauz. (brochure, 16pp).

[c] about 92% of the data are from the year 2003; 49.7% of these breeding pair data refer to good estimates (combination of census data and estimates for whole "Landkreise"/district censuses) and 50.3% to pure census data (74% of all "Landkreise").

[d] results of a semi-quantitative atlas study 1990–2000 (census units 32 km^2, abundance classes 1–10, 11–100); a geometric mean of the census data gives a total population of about 280 (calculations of H. Illner).

[e] 1999: ascertained 160, see Hölzinger 2001 and additional data reported by C. Stange (pers. comm. March 2003): at least 10 pairs in the eastern Rhine valley and the northern Kaiserstuhl region; in 2001/2002 the population of the eastern Rhine valley and the northern Kaiserstuhl region increased to 128 pairs.

[f] Total for 1999 estimated by Bauer *et al.* 2002: 5800–6100 (mean 5950).

pers. comm. = personal communication.

Illner pers. comm.).

	1970–80	1999–2003	Trend	Trend
	Year: number, reference		% in *n* years	%/year
Northwest Germany				
Schleswig-Holstein	1976: about 150–250, Schönn *et al.* 1991	120	−39% in 23 years	−1.7%
Hamburg	1980: about 5–10, Schönn *et al.* 1991	2	−73% in 19 years	−3.9%
Niedersachsen	1975–80(83): about 300–600, Heckenroth & Laske 1997	*c.* 175	−64% in 21 years	−3.1%
Nordrhein-Westfalen	1973/4: 7817 (a decline of 29% was calculated for 25 years[a])	5550	−29% in 25 years	−1.1%
Subtotal	**8272–677 (mean: 8475)**	**5847**	−31%	−1.2%
Central (west) Germany				
Hessen	1977: 250–750 (about 250–300, J. Weiss in Glutz & Bauer 1980)[b]	625	25% in 22 years[c]	1.2%
Rheinland-Pfalz	1978–81: 325–400, Simon 1982	*c.* 300	−17% in 20 years	−0.8%
Saarland	1979: about 100, Bauer & Thielcke 1982	*c.* 40	−60% in 20 years	−3.0%
Subtotal	**675–1250 (mean: 963)**	**965**	0%	0%
Southern Germany				
Baden-Wuerttemberg	1960–70: 500–800 Hölzinger 2001[d]	*c.* 250	−61% in 34 years	−1.8%
Bayern	*c.* 1970–83: 150–300[e]	*c.* 125	−44% in 23 years	−1.9%
Subtotal	**650–1100 (mean: 875)**	**375**		−1.9%
East Germany				
Mecklenburg-Vorpommern	*c.*1980: 20–30, Labes & Patzer 1987	7	72% in 19 years	−3.8%
Brandenburg-Berlin	*c.* 1980: 10–100, Feiler & Litzbarski 1987 (geometric mean: 32)	13	−59% in 19 years	−3.1
Sachsen-Anhalt	*c.* 1980: *c.* 170–370, Nicolai 2000	20	−93% in 19 years	−4.9%
Sachsen	*c.* 1980: 50–150, S. Schönn, pers. comm. Feb. 2003	8	−92% in 19 years[f]	
Thueringen	*c.* 1980: *c.* 50, J. Wiesner, pers. comm. Feb. 2003	15	−70% in 19 years[f]	
Sachsen + Thueringen	*c.* 1980: *c.* 100–200	21	−85% in 19 years	−4.4%

(*cont.*)

Table 5.6 *(cont.)*

	1970–80	1999–2003	Trend	Trend
	Year: number, reference		% in *n* years	%/year
Subtotal[g]	**322–632 (mean: 477)**	**63**	−87% in 19 years	−4.6%
Total [h]	**9919–11 659 (mean: 10 789)**	**7250**	**−32.8%**	**−1.3%**

[a] Long-term data sets available from west to east four out of five showing declines, one was stable; the average decline rate was about 29% in 25 years:
(1) Dueffel area (Rhine valley), 58.5 km^2, 100 territories (1978); 37 territories (1998) (trend: −63%).
(2) Emmerich-Rees (Rhine valley), 28.5 km^2, 47 territories (1976–8, Exo 1983), about 40–50 territories at about 2000 (Schwöppe pers. comm. March 2004) (trend: 0% = no change).
(3) Kreis/district Dueren (some parts of neighboring districts) (lower Rhine, bay of Cologne), 250 km^2; 123 territories (1975), 51 territories (2001/2002), source: W. Bergerhausen, pers. comm. March 2003 (trend: −58,5%).
(4) Area at Werl, Kreis/district Soest (Westphalia)(125 km^2), 65 territories (1975), 56–62 territories (1999/2001); source H. Illner, unpubl. data, (trend: −9.4%).
(5) Area at Lippstadt, Kreis Soest (Westphalia) (127 km^2), 80 territories (1974), 67 territories (2000); source A. Kaempfer-Lauenstein & W. Lederer, pers. comm. (trend: −16.3%).

[b] Higher range estimate by H. Illner as the range estimated by J. Weiss may be too low because of limited knowledge about population levels at that time (Burbach 1997).

[c] Indications that the population decreased in spite of some strong increases from 1975–95 in some parts of Hessen (Regierungsbezirke Gießen und Darmstadt, from 87 breeding pairs in 1975–8 to 159 pairs in 1995 = 84% increase) is due to the fact that the northern part of the breeding range in Hessen (Regierungsbezirk Kassel) was completely lost by 1993–1995; some old data unsystematically taken show at least 20 pairs were living there during 1975–8 (Burbach 1997). Even in the area actually occupied a data set exists from the Altkreis Dieburg (Diehl cited in Burbach 1997) showing a long-term population decline from about 150 pairs in 1950 to 14–18 pairs in 1992–5.

[d] One long-term data set available shows a population decline. The Northern Kaiserstuhl region at Wyhl held 23 territories in 1970/1 (Hölzinger 2001, C. Stange, pers. comm. March 2003), two pairs in 1994 and 14 pairs in 2002 (C. Stange, pers. comm. March 2003).

[e] Conservative estimate by H. Illner. Two sources were exploited: 1) At least about 110 breeding pairs at different places are listed by Wüst 1986 (p. 779–802) for the period 1970–80 (no systematic survey); (2) The breeding bird atlas of Nitsche & Plachter 1987 shows certified breeding in 45 squares of 100 km^2 and probable and possible breeding in a further 38 squares for the period 1979–83; assuming a low average density of five pairs per 100 km^2 unit with certified breeding would translate to 275 breeding pairs. Further evidence for declines: (a) In an area at Bad Winsheim/Uffenheim (Westmittelfranken) about 30 pairs were found in 1980, none in 2000 (H. Klein 1992 unpubl. data and S. Hartlaub, pers. comm. Feb. 2003); (b) in an area at Steigerwald in 1980 about 25 pairs were found, in 2000 3–4 pairs (Dr Lindeimer, pers. comm. Feb. 2003).

[f] In 1999 a part of Thüringen with a significant population of Little Owls formally belonged to Sachsen. So only the combined data for Thüringen and Sachsen represent the real trend.

[g] Subtotal estimated for East Germany 1985/6: *c.* 250–300 pairs, Schönn 1986.

[h] Total estimated for 1999 by Bauer *et al.* 2002: 5800–6100 pairs.

pers. comm. = personal communication.

Germany) the population increased from 6 pairs in 1987 to 34 pairs in 2004 due to the placement of 180 nestboxes in the area (Heilig & Stahlheber 2004). See Figure 5.11.

Densities

In Germany, densities recorded below one territory/km^2 were reported from Saxony with 12 pairs on 20 km^2 (Augst & Manka 1997). While higher densities were once found in Northrhine-Westphalia (Exo 1992) the situation has changed in the Aachen region where only 34 calling males were recorded on 160 km^2 (Toschki 1999). In Bönningheim, Landkreis Ludwigsburg, densities up to one pair/km^2 were recorded locally (Eick 2003) with an average of 0.17 pairs/km^2 for the entire region (120 pairs across 690 km^2 [Keil 2001]). In the orchards of southwest Germany there was a density of 3.5 breeding pairs/km^2 with a nestbox density of one box per 15 ha (Otto & Ullrich 2000). In Northrhine-Westphalia a density of 0.14 calling males/km^2 was reported in a study area of 260 km^2 (Zens 2005).

Denmark

Distribution

The major part of the Danish Little Owl population breeds in north and west Juttland. The most important areas are Himmerland (between Ålborg and Mariager), Salling (north of Skive), and the central part of Ringkøbing County. Outside these areas small populations are found in Vendsyssel (north of Ålborg, between Kattegat and Fjerritslev, north to Hjørring), in the region between Randers Fjord and Mariager Fjord, on Djursland (between Randers/Århus and Grenå) and in the western part of Ribe County. In the counties of Sønderjylland and Vejle only a few scattered breeding records exist. In Fyn only one confirmed breeding record and one possible breeding record has been found. The Little Owl was found breeding in Falster in 1937 but has probably not bred on the islands east of Store Bælt (Great Belt) since the 1940s (Grell 1998).

Population estimates

The latest estimate for Denmark of 175–200 breeding pairs dates from 1993–6 (Grell 1998). It is probable that the population is lower today since this species still seems to be decreasing in many regions of Denmark (M. B. Grell, personal communication). Fewer grazed pastures close to appropriate buildings for nesting together with the general decrease in biological diversity caused by agricultural intensification, possibly pesticides and over-fertilizing in open country, have been proposed as some of the main causes of decline. The species also seems to be quite vulnerable to the increasing levels of faster motorized traffic over the years (Glue 1971). Lack of good nesting sites due to better maintenance of buildings and predation by Stone Martens (*Martes foina*) are also possible problems. BirdLife International (2004) estimates the total population at 100–200 pairs.

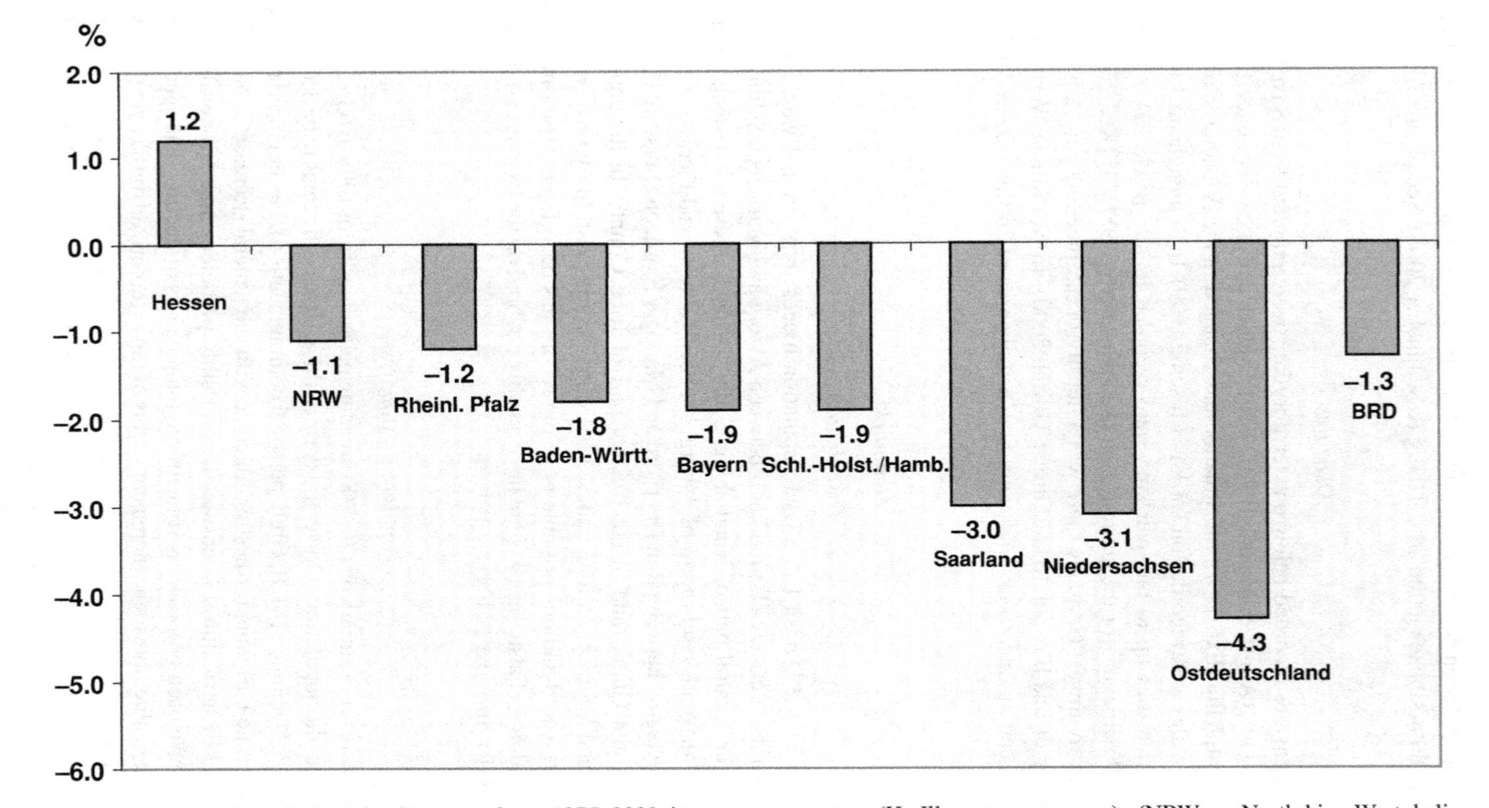

Figure 5.11 Population changes in Germany from 1975–2000 in percent per year (H. Illner, pers. comm.). (NRW = Northrhine-Westphalia; BRD = Bundesrepublik Deutschland).

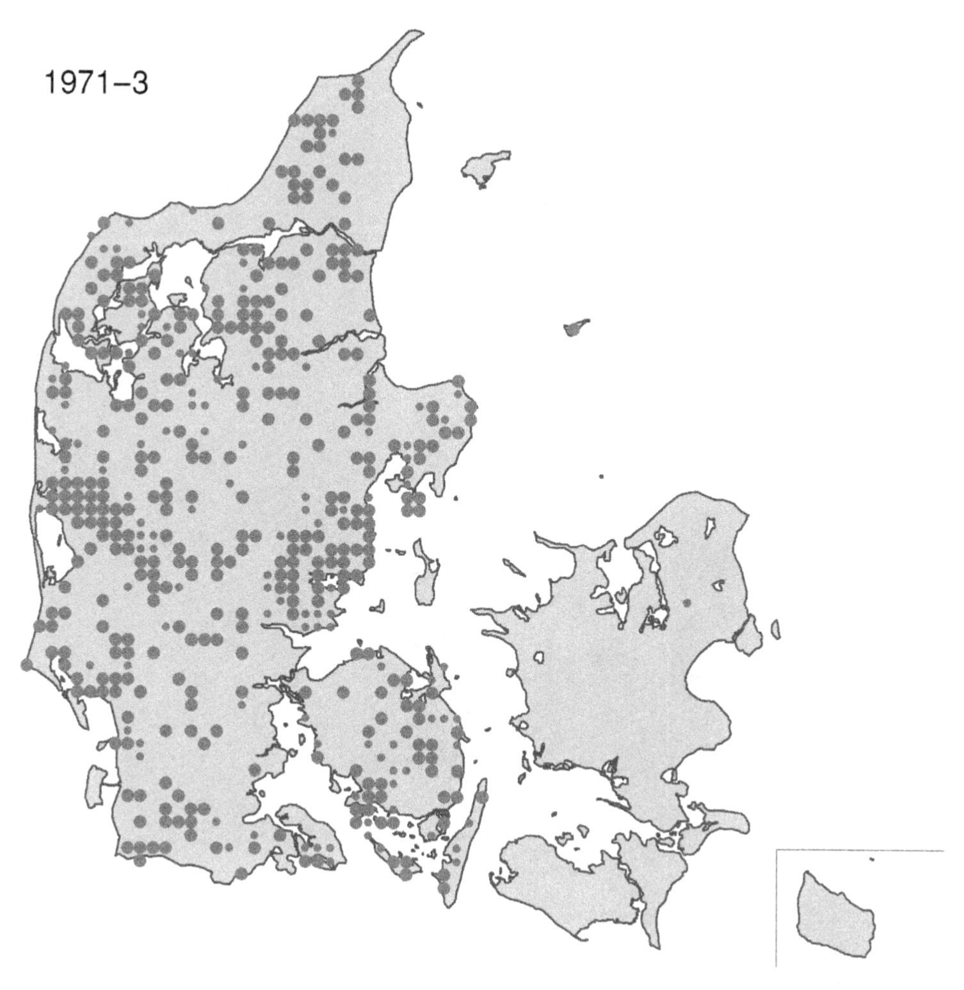

Figure 5.12a Historical distribution of the Little Owl in Denmark in 1971–3 (after DOF, BirdLife Denmark, and the Museum of Zoology, Copenhagen). Large dots represent confirmed breeding, small dots possible breeding.

Trends

An extreme restriction in the distribution area of the Little Owl has taken place during the last two decades of the twentieth century. The total number of breeding records has been reduced by 65% during 1993–6 compared with the results of the first atlas project in 1971–4. Approximately 300 atlas-squares distributed throughout the area of Juttland and Funen that reported breeding in 1971–4 did not report any breeding activity during 1993–6. In particular, a high proportion of the deserted breeding areas were observed in eastern and southern Juttland and Funen. But the species has also apparently disappeared from Thy (north of Limfjorden – west of Fjerritslev) and on Mors, in western Vendsyssel, and in middle and western Juttland (Grell 1998). See Figures 5.12a–c.

Figure 5.12b Historical distribution of the Little Owl in Denmark in 1993–6 (after DOF, BirdLife Denmark, and the Museum of Zoology, Copenhagen). Large dots represent confirmed breeding, small dots possible breeding.

France

Distribution

In France, the Little Owl is a sedentary species, present across the whole country except the mountains. Yeatman (1976) indicated that the mountains where the snow stays for a long time were not suitable for the species. The species breeds up to 1080 – 1100 m on the Causse Méjean in Lozère and up to 1155 m in Haute-Loire (Juillard *et al.* 1990). The Little Owl is absent from the islands of Brittany (Guermeur & Monnat 1980). It is very common on the island of Oléron (Bavoux & Burneleau 1983; Bretagnolle *et al.* 2001). The species is very rare in Corsica where just a few cases of breeding were recorded in the last century, in the middle-east plain, near Aleria (Thibault 1983; Thibault & Bonaccorsi 1999) and more

Figure 5.12c Historical distribution of the Little Owl in Denmark in 2003 (after DOF, BirdLife Denmark, and the Museum of Zoology, Copenhagen). Large dots represent confirmed breeding, small dots possible breeding.

recently in 1998 in the same area (Mastrorilli 2000); a dead adult was found in 2004 (Faggio 2005). The last breeding record for Corsica was in 2000 (Mastrorilli 2000). The species is believed to compete with the Scops Owl (*Otus scops*) on Corsica, and is present on the island of Frioul, not far from Marseille (Cheylan 1986). Another population stronghold is found around the Loire River with up to 11% of the total estimated French population. The distribution in France reflects the specific situation where Mediterranean populations are connected to populations in the neighboring countries of Spain, Italy, and Belgium. For most of France, the Little Owl is distributed in population clusters, sometimes irrespective of the apparent quality of the habitat (Génot & Lecomte 1998). See Figure 5.13.

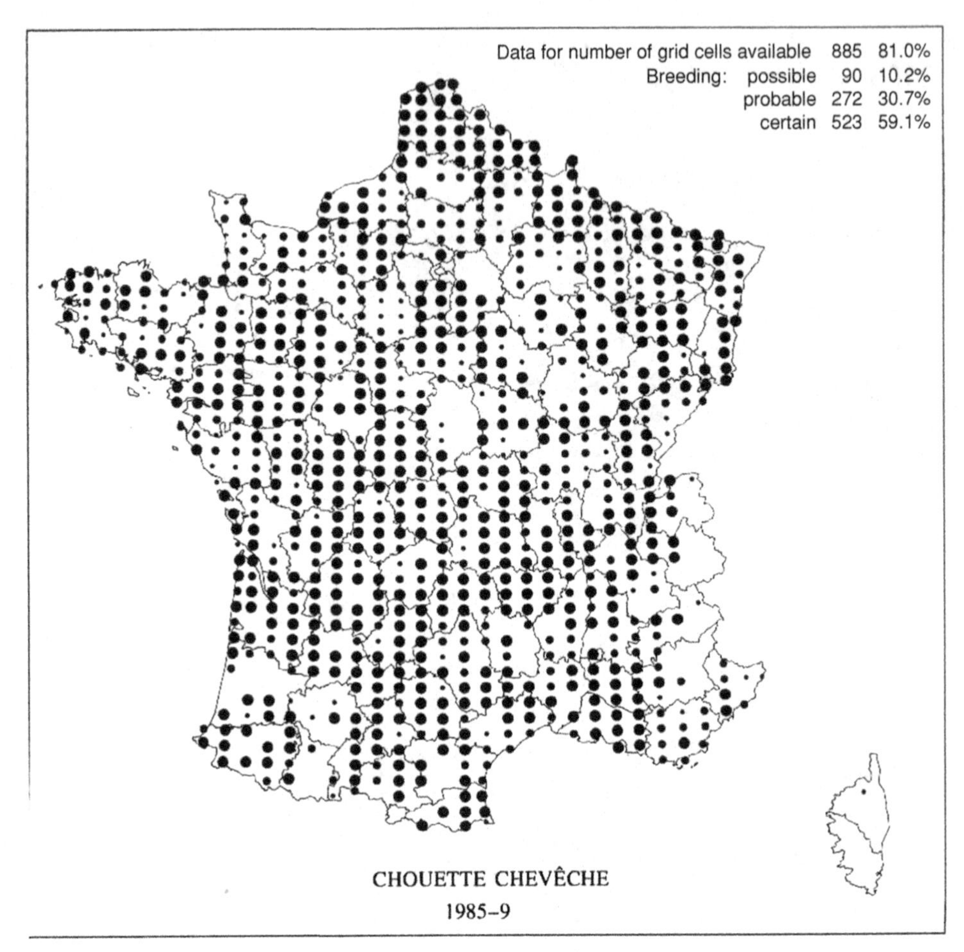

Figure 5.13 Little Owl distribution in France (after Génot 1994).

Population estimates

The French population of the Little Owl was estimated at 5000–50 000 pairs in the beginning of the 1990s (Génot 1994); and 11 000–35 000 pairs in 1998 (Génot & Lecomte 1998). A detailed analysis was done by Génot and Lecomte (1998) yielding population estimates per administrative district (Table 5.7). The total minimal population at the end of the 1990s was estimated at 10 330–35 000 pairs by summing partial estimates (Génot & Lecomte 1998). BirdLife International (2004) estimates the total population at 20 000–60 000 pairs.

Densities

In 1990, Athanaze (Blache 2003) found in farmland areas of 2.7 km^2 and 6 km^2 of Rhône, respectively, 2 and 12 pairs. In 1993, Geslin (unpublished data) recorded 24 calling males in an area of 12 km^2. In 1994, Geslin (unpublished data) recorded 2 males/km^2

Table 5.7 *Population estimates of the Little Owl* Athene noctua *for France (after Génot & Lecomte 1998).*

Region	pairs, lower estimate	pairs, upper estimate	period
Alsace	200	300	1995
Aquitaine	100	100	1987
Auvergne	1900	3100	1992
Bourgogne	500	2000	1999
Bretagne	60	60	1994
Centre	100	100	1994
Champagne-Ardennes	150	150	1989
Franche-Comté	300	300	1993
Ile-de-France	300	400	1994
Languedoc-Roussillon	1630	1630	1990–6
Limousin	200	200	1993
Lorraine	500	1000	1996
Midi-Pyrénées	350	350	1996
Nord-Pas-de-Calais	1800	2400	1994–6
Normandie	130	130	1993–6
Pays de la Loire	1100	1100	1990–4
Picardie	70	70	1991–5
Poitou-Charentes	200	200	1991–5
Provence-Côte d'Azur	530	530	1995–7
Rhône-Alpes	410	410	1999
	10 530	14 530	

in the Loire valley. In 1993, Gailliez (unpublished data) noted 136 calling males/133 km^2 (1.02 males/km^2) in the orchards and hedges of northern France. In 1997, Camuzat and Faucon (unpublished data) found 70 calling males in an area of 44 km^2 (1.6 males/km^2). In 1998, Vivier and Telle (unpublished data) recorded 115 calling males in an area of 95 km^2 (1.2 males/km^2) in meadows with pollarded willows in northern France. One territory/km^2 was recorded in Bouches-du-Rhône (by Barthelemy and Bertrand 1997). Signoret (2002) recorded 157 calling males/320 km^2 in 1999–2000 in western France, with the highest average density at the edge of the "bocage" (0.62/km^2) and the lowest average density in the sandy dunes area (0.05/km^2). Crespon and Chretienne (unpublished data) recorded 88 calling males/1175 km^2 in the natural regional park of Perche, with average densities from 0.07–0.83/km.^{2} Locally high densities of 6.7 calling males/km^2 were found on Île d'Oléron (18 km^2) (Bretagnolle *et al.* 2001). Blache (2003) recorded 6 pairs/km^2 in Plaine de Valence. Monitoring in ten regional parks revealed densities ranging from 57 calling males/44 km^2 in Livradois-Forez (1.3 males/km^2) to 7 calling males/200 km^2 in Lorraine (0.035 males/km^2) (J.-C. Génot unpublished data).

Spain

Distribution

The Little Owl nests almost all over the Iberian Peninsula, but is absent from the Balearic and Canary archipelagos, where it has only been recorded in winter in Palma de Mallorca (Olea 1997) suggesting migration of the species to that region. The species avoids excessively wet zones, dense forests, and high mountain habitats (Muntaner *et al.* 1983; Olea 1997). The Little Owl is less common in the Euro-Siberian Region than in the Mediterranean Region of the Iberian Peninsula. This could be indirectly related to the weather, e.g., the oceanic weather favors forests that are avoided by the species (Zuberogoitia & Martínez 2001). However, there are high-density populations of Little Owls in places with traditional agriculture and poultry breeding, and their associated fields and grasslands. It is not common in mountainous areas, although there are populations established in more open areas at higher elevations in the Pyrenees (1600 m) and Sierra Nevada (2300 m).

Muntaner *et al.* (1983) and Urios *et al.* (1991) observed that the winter temperature was also a limiting factor. In fact, Muntaner *et al.* (1983) established the distributional limit as the 1°C isotherm of January in Cataluña and Andorra, and Urios *et al.* (1991) identified the 3°C isotherm as the distributional limit in Valencia. This can be correlated with other variables, such as the persistence of snow cover during winter that makes it difficult for the owls to find adequate prey.

Population estimates

According to the *Atlas de las Aves de España* (Olea 1997) there were 50 000–65 000 estimated breeding pairs in Spain during the period 1975–95. The most recent estimated data, compiled in the *Atlas de las Aves Reproductoras de España* (Martí & Moral 2003), suggests the Spanish Little Owl population to be at least 36 000 pairs, although it is necessary to point out there is no abundance data for almost 25% of the squares (each square being 625 km^2) where the species was recorded (Garcia & Muñoz 2003). A reduction of more than 20% of the population has been observed over the last 30 years, following the European trend. However, the short-term population tendency for the species was considered as positive during 1998–2001. Furthermore, the Little Owl is reported as the most abundant nocturnal raptor in Spain (SEO/BirdLife 2002). Changes in agricultural practices and land use are the main factors responsible for the decrease of the species. The urban development plans are a real threat, especially in the eastern coastal semi-arid habitats and in the Basque countryside on the Atlantic coast (northern Spain) (Zuberogoitia 2002; Martínez & Zuberogoitia 2004a). The reduction of dry-farmed tree crops, in particular the carob tree (*Ceratonia siliqua*), has caused severe declines in the eastern coast. Pesticides diminish prey availability, land re-allotment schemes reduce suitable sites for nesting and hunting, and the reduction of hedges and trees along roads increases the risk of collisions with vehicles (Olea 1997; Garcia & Muñoz 2003). Likewise, the enlargement of forested areas, especially in the Pre-Pyrenees has lead to a reduction of suitable habitats (Martínez & Zuberogoitia 2003; J. Estrada, personal communication). BirdLife International (2004) estimates the total population at 20 000–100 000 pairs.

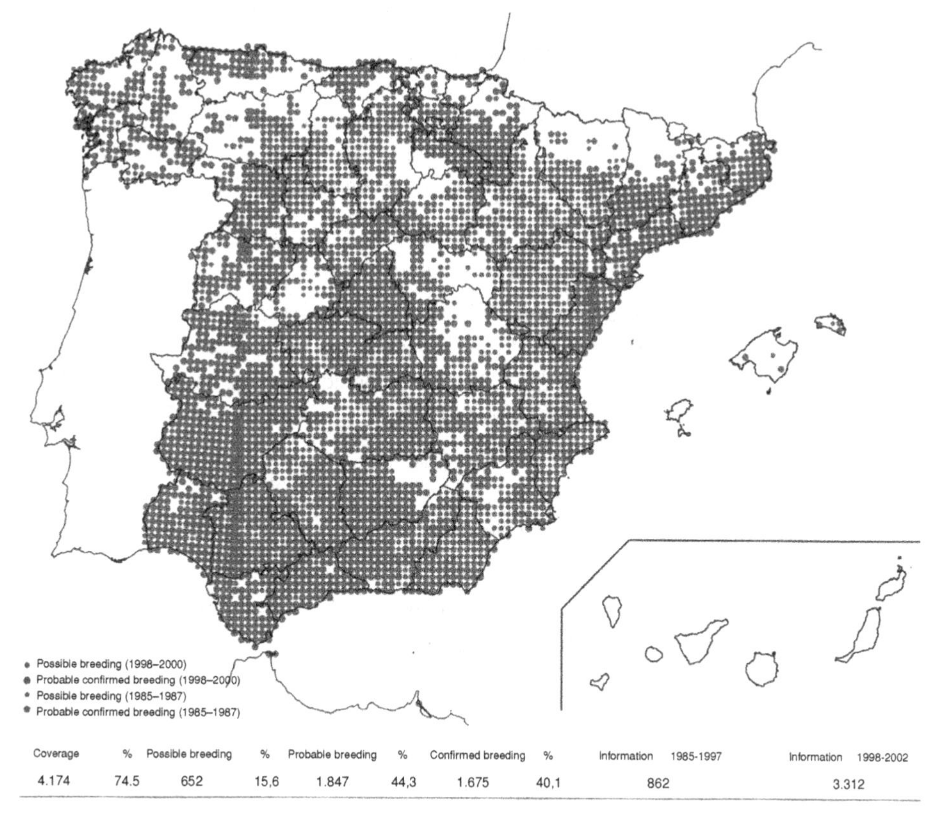

Coverage	%	Possible breeding	%	Probable breeding	%	Confirmed breeding	%	Information 1985-1997	Information 1998-2002
4.174	74.5	652	15,6	1.847	44,3	1.675	40,1	862	3.312

Figure 5.14 Little Owl distribution in Spain (after Martí & Moral 2003).

Densities

There is limited data on Little Owl abundance in the Iberian Peninsula: 0.2 individuals/km^2 in Granada (southern Spain), up to 1 pair/km^2 in Catalonia, and 0.03–0.11 pairs/km^2 in Galicia (northwest Spain) (Olea 1997). Fajardo *et al.* (1998) recorded Little Owl numbers during road surveys in the province of Seville (southern Spain). During the breeding season, they recorded 17 individual owls/km^2 in olive groves and 7 individuals/km^2 in sunflower fields, respectively. During July, they recorded up to 86.8 individuals/km^2. Zuberogoitia and Campos (1997a) found 1.3 territories/km^2 in Biscay; Olea (1997) reported 1 pair/km^2 in olive plantations and *Pinus halepensis* forests in Catalonia. See Figure 5.14.

Portugal

Distribution

The Little Owl is probably the most common and widespread owl in Portugal. Iberian breeding populations are sedentary, and young birds undertake local movements. There is evidence that some peninsular birds could be wintering in Iberia (Díaz *et al.* 1996). The species is more common south of the River Tagus, mainly in the provinces of Alto Alentejo and Baixo Alentejo (Rufino 1989) in southern Portugal. The species is scarce or absent in the extensive forest areas of central and northern Portugal. It occurs in a great variety of habitat types including regions from sea-level to over 1000 m. It is frequently found in parks and gardens in cities, small towns, and villages. It is a common species in Lisbon, frequenting the great urban parks and some gardens. Although uncommon, it can sometimes be found using the upper parts of buildings (L. Reino, personal communication).

Population estimates

As of 1989, the estimate for the Portuguese Little Owl population was between 10 000–100 000 pairs with populations considered stable (Manez 1994). The latest estimate for the Little Owl population in Portugal was 50 000–150 000 breeding pairs, and probably closer to 90 000 pairs (R. Tomé, personal communication). BirdLife International (2004) estimates the total population at 50 000–150 000 pairs.

Densities

In southern Portugal (Bixo Alentejo) densities of up to 2.36 pairs/km^2 were recorded in 16 km^2 of steppe-like habitat, and 6.95 pairs/km^2 in an area of 6 km^2 of open Holm-oak, *Quercus rotundifolia* (R. Tomé, unpublished data).

Italy

Distribution

In Italy the Little Owl is a sedentary species, being widespread across the mainland, but can also be found on the isles of Sardinia, Sicily, and others (Casini *in* Meschini & Frugis 1993). Dispersal and movement of owls occurs in autumn and winter, and some autumn movements can be considered as migration (M. Mastrorilli, personal communication).

It is a well-distributed bird of prey in lowland areas and hillsides, with rare breeding records in the lower mountains. This explains the gaps in the distribution along the Alpine Arc. Nevertheless, some breeding pairs have been recorded at higher elevations in recent years: at 1000 m in Liguria (Andreotti 1989), at 1100 m in Piedmont (Mostini *in* Mingozzi *et al.* 1988), and the highest in western Europe has been recorded in the Orobic Alps, Lombardia at 1230 m (Mastrorilli 2001).

Between 1988–2004 the Little Owl has been recognized as the most common bird of prey in Italian towns due to the high number of urban parks and wooded boulevards (with owls showing a preference for species such as Poplar [*Populus* sp.], Plane [*Platanus* sp.], and Horse-chestnut [*Aesculus hippocastanum*]), and due to the urbanization of country buildings and to the favorable characteristics of historical city centers (old buildings with plenty of hollows). Roof air spaces in recently built industrial buildings are often used as breeding sites (Mastrorilli 1999a). During several urban surveys, Little Owl presence has been observed in 52 main Italian towns (Table 5.8).

Table 5.8 *Population estimates of the Little Owl,* Athene noctua, *for main Italian cities (M. Mastrorilli, personal communication).*

Urban area	Region	Years of study	Study area (km^2)	Pairs	Density (pairs/km^2)	Reference
Bergamo	Lombardia	1997–2000	23	34	1.48	Mastrorilli 2001
La Spezia	Liguria	1994–5	51	11	0.22	Biagioni *et al.* 1996
Livorno	Toscana	1992–3	38.1	30	0.79	Dinetti 1994
Firenze	Toscana	1986–8	102.4	52	0.51	Dinetti & Ascani 1990
Pavia	Lombardia	1997–8	33	33	1.00	Bernini *et al.* 1998
Cremona	Lombardia	1990–3	10.23	17	1.66	Groppali 1994
Roma	Lazio	1989–3	360	217	0.60	Cignini & Zapparoli 1996
Biella	Piemonte	1998	30.75	8	0.26	Bordignon 1999
Napoli	Campania	1990–4	117.3	48	0.41	Fraissinet & Milone 1995
San Donà d. Piave	Veneto	2000	21.05	23	1.09	Sgorlon 2003
Marcon	Veneto	1988	25.39	15	0.59	Stiva 1990
Cossato	Piemonte	1995	34.75	10	0.29	Bordignon 1997
Crema	Lombardia	2000–1	22.75	18	0.79	Mastrorilli in press

Due to the different weather conditions along the Peninsula, there are several areas that can be colonized by the species. In the Po Plain the species' distribution is strongly affected by severe winters (i.e., snow and relatively low temperatures) causing significant population decreases. In central and southern Italy the environment and weather conditions are typically Mediterranean, thus more suitable to the owl's ecology and for this reason populations are more stable there. The most favorable conditions for Little Owls (suitable weather, food availability, suitable habitat) are encountered in southern Italy (M. Mastrorilli, personal communication).

Population estimates

A large decrease was observed during 1960–80, followed by a rapid recovery (Brichetti 1997), yielding stable populations recently estimated at 10 000–50 000 pairs (Manez 1994) and 10 000–30 000 pairs in 1996 (Brichetti 1997). BirdLife International (2004) estimates the total population at 30 000–50 000 pairs. See Figure 5.15.

Densities

The following densities have been reported in various areas of Italy: in mixed farmland (Ticino Park, Pavia, Northern Italy) 1.1 territories/km^2 (Cesaris 1988); ricefield (Pavia) 0.4 territories/km^2 (Galeotti & Pirovano unpublished data); rural areas with orchards (Modena) 0.34 territories/km^2 (Selmi *et al.* 2005); Province of Cuneo 0.45 to 1.07 territories/km^2 (Toffoli & Beraudo 2005); rural areas with ruins in Campi Flegrei (southern Italy) 1.05 territories/km^2 (Giannotti *et al.* 2005); urban areas (Pavia) 1.25 territories/km^2 (Galeotti & Morimando 1991); and based upon the results of a census in 15 cities an average of 0.59 (minimum 0.22 and maximum 1.48 territories/km^2) (Mastrorilli *et al.* 2005). Intensive farmland on the Po Plain Lombardy, northern Italy, produced 1 territory/km^2 (Galeotti & Sacchi 1996), and Low Alps (Lombardy) >0.5 territories/km^2 (Galeotti & Sacchi 1996). Table 5.8 shows density data of the owl in some Italian towns. Table 5.9 reflects densities of Little Owls in an array of Italian countryside locations.

Switzerland

Distribution

In 2003, the species was present in very low numbers in three small regions: (1) Ajoie, north of the Jura canton (four villages between Porrentruy and the northern border of the country), (2) in the Canton de Genève, and (3) near Tessin. A conservation initiative south of Basel is trying to restore a link with the neighboring German population through improvement and restoration of habitat, without a positive result so far (C. Meisser, personal communication).

Population estimates

The species has undergone a marked decrease, due especially to habitat changes. The population was estimated at 185 pairs in 1980 (Juillard 1980), 30–40 pairs in 1986–91

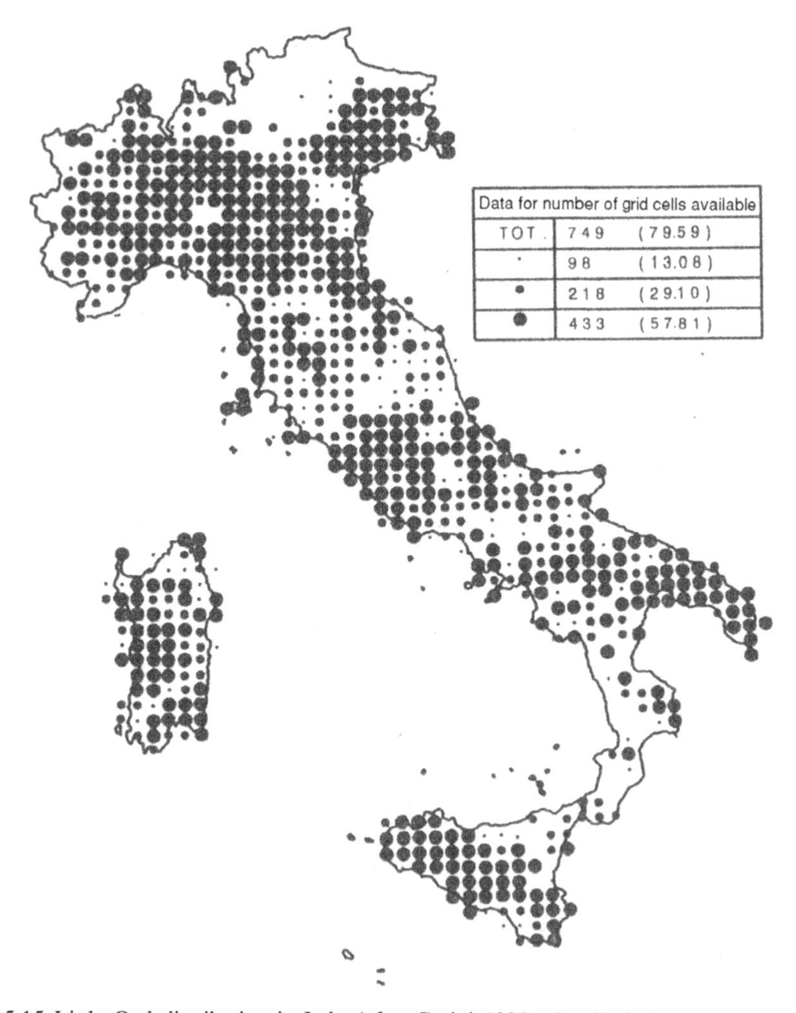

Figure 5.15 Little Owl distribution in Italy (after Casini 1993). Small circles: possible breeding; medium circles: probable breeding; large circles: definite breeding.

(Manez 1994), and 70 pairs in 1992–6 (Meisser & Juillard 1998). Around Genève, 30–40 pairs remained in 2000 (Meisser & Albrecht 2001); there were 34 territories in 2002 and 35 territories in 2003 (C. Meisser, personal communication). Tessin had less than 10 pairs and Ajoie at least 17 pairs in 2000 (Meisser & Albrecht 2001); there were 16 territories in 2002, and 12 territories in 2003 (C. Meisser, personal communication). In 2004, there were 19 territories in Ajoie (Jura), 41 near Geneva, and 5 near Tessin, totalling 65 territories with at least 28 pairs near Geneva and 13 in Ajoie. The trend for 2005 for Geneva and Ajoie is stable with a moderate increase (C. Meisser, personal communication). BirdLife International (2004) estimates the total population at 60–70 pairs.

Table 5.9 *Densities of the Little Owl,* Athene noctua, *in the Italian countryside (after M. Mastrorilli personal communication).*

Area of study	Province	Region	Density (pairs/km^2)	Years of study	Reference
Plain of Modena	Modena	Emilia Romagna	9.3–11	1987–90	Estoppey 1992
Plain of Bergamo	Bergamo	Lombardia	0.69	1995–6	Mastrorilli 1997
Regional Parks of Ticino	Pavia	Lombardia	1.1	1975–8	Cesaris 1988
Castel Porziano	Roma	Lazio	3.14–4.62	1997–8	Tomassi *et al.* 1999
Tolfa Mountains	Roma	Lazio	0.55	1994	Centili 1995
Plain of Pavia	Pavia	Lombardia	0.4	1995–7	Pirovano & Galeotti 1999
Province of Cuneo	Cuneo	Piemonte	0.8	2000–3	Toffoli & Berando 2005

Densities

Near Geneva, densities range from 0.5–1 calling males/km^2. The highest densities were 2.5 territories/km^2 (13–16 territories between 1993 and 1997 on 5.9 km^2) (Meisser & Albrecht 2001).

Austria

Distribution

The Little Owl was a very common species in Austria until the 1970s. For this reason systematic data about its density and distribution were only rarely and randomly collected until recently. Intensification of agriculture, and extension of roads and settlements, led to a decrease of appropriate habitats (e.g., cultured grasslands) followed by a continuous decline of the owl population in Austria. Monitoring studies began in 1992 in the eastern part of Austria (Ille 1992, 1995). During the 1990s the population was restricted to the east of the country, in an area that is climatically favorable with respect to rainfall and temperature. The species settles in cultural landscapes consisting of open and rather dry lowlands. Results of a research project on its habitat preferences performed between 1997–9 (Ille & Grinschgl 2001) showed that its density varied within this kind of landscape. The population fluctuated substantially during this study; pairs disappeared from traditional territories while new areas were colonized. One of the areas where owls disappeared during recent years was around "Neusiedler See". The reason for this is probably a lack of appropriate nest sites, as there are only a few trees and buildings that can be used for breeding. Another region with decreasing numbers was the eastern part of the "Weinviertel". In this area a lack of hunting areas was the main reason for the decrease. In contrast, there was an increase in the northwestern part of the "Weinviertel", along the Czech Republic border. Here, the villages are characterized

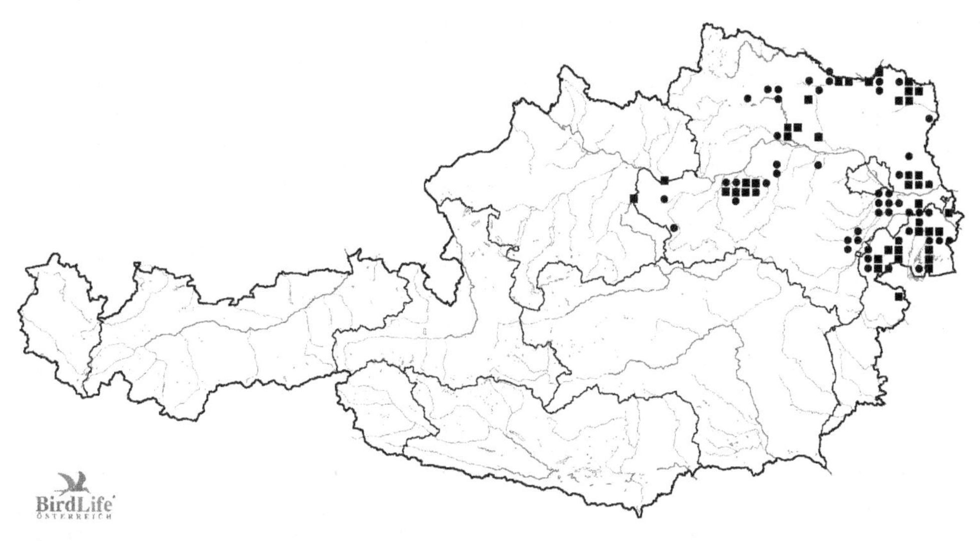

Figure 5.16 Little Owl distribution in Austria as of 2007 (BirdLife Austria, personal communication).

by vinicultural areas and alleyways with traditional wine cellars that are predominantly used by the owls for breeding. In the study, only three territories were found on the Czech side of the border, suggesting that owls are not likely dispersing into Austria from this area. A new region colonized by owls between 1997–9 was situated northeast of "Mostviertel" and characterized by walls with many natural and man-made holes and cavities.

Population estimates

The species is decreasing (Glutz & Bauer 1980) and 40–60 pairs were counted in the 1990s (Manez 1994) and 60 pairs in 1997 (Ille & Grinschgl 2001). Based on systematic searches, the population size between 1996 and 2001 was estimated at about 50 pairs and 20 single individuals (nearly exclusively males) (R. Ille, personal communication). BirdLife International (2004) estimates the total population at 70–100 pairs. See Figure 5.16.

Densities

Results of a research project on habitat preferences performed in 1997–9 showed that density varies with the kind of cultural landscape. Density was highest within a region characterized by extensively cultivated fruit meadows rich in insects and mammals ("Mostviertel"; 0.2 pairs/km^2) followed by a region dominated by vinicultural areas ("Weinviertel"; 0.09 pairs/km^2). In the other settled areas average density was only 0.03 pairs/km^2 or fewer (Ille & Grinschgl 2001).

Czech Republic

Distribution

The Little Owl was the most abundant and widespread owl in Bohemia in the 1940s (Stastny *et al.* 1996). However, in the last decades its distribution range has clearly been reduced. Between 1973–7 and 1985–9 the number of occupied squares decreased by 43% (Stastny *et al.* 1996); each survey square was 132 km^2 in size. The Little Owl's highest abundance is recorded in areas up to 500 m elevation. The main populations live south of Plzen town, in the Labe valley near Decin town, and in Ohre valley near Lovosice town. In southern Moravia, which was quite densely inhabited by owls in the past, just a few pairs remain. The Little Owl breeds mostly in farm buildings. Breeding in tree hollows is quite rare.

Population estimates

Owl numbers have been declining since the 1950s. Between survey efforts conducted in 1973–7 and 1985–9, a lack of confirmed breeding records occurred in 43% of the surveyed squares (Stastny *et al.* 1996). Between 1985–9 the population size was estimated at 700–1100 pairs (Stastny *et al.* 1996). An abundance and distribution survey was carried out in 1993–5 in the Czech Republic. On the basis of the results from 16 study areas the population size was estimated at 1000–2100 pairs (Schröpfer 1996). The same area was again surveyed in 1998–9. A noticeable decline to 500–1000 pairs was recorded (Schröpfer 2000). The main reasons for the decline since 1993–5 were most probably the severe winters of 1995/96 and 1996/97, with long-lasting snow cover. BirdLife International (2004) estimates the total population at 200–400 pairs.

Densities

In total, 16 of 27 surveyed study areas were not occupied; in the remaining 13 areas the average density was 0.012 pairs/km^2. The highest densities were recorded in western Bohemia south of Plzen town, where the density was 0.145 pairs/km^2 (Schröpfer 2000).

Slovakia

Distribution

The Little Owl is a common breeding species in lowlands and basins of south Slovakia (Danko *et al.* 2002). The species is distributed along rivers in north Slovakia too, in several areas up to 600 m elevation. In lowlands it usually occurs between 100–300 m. Its typical habitats in Slovakia include small field patches surrounding barns and other agricultural buildings and villages. These habitats are usually situated at lower elevations – in lowlands and hilly areas. The species avoids forests and big cities. The Little Owl can sometimes be found at the margins of industrial zones surrounded by fields. The highest densities in Slovakia were recorded in the lowlands of east Slovakia, the lowest in the basins and hills of north Slovakia.

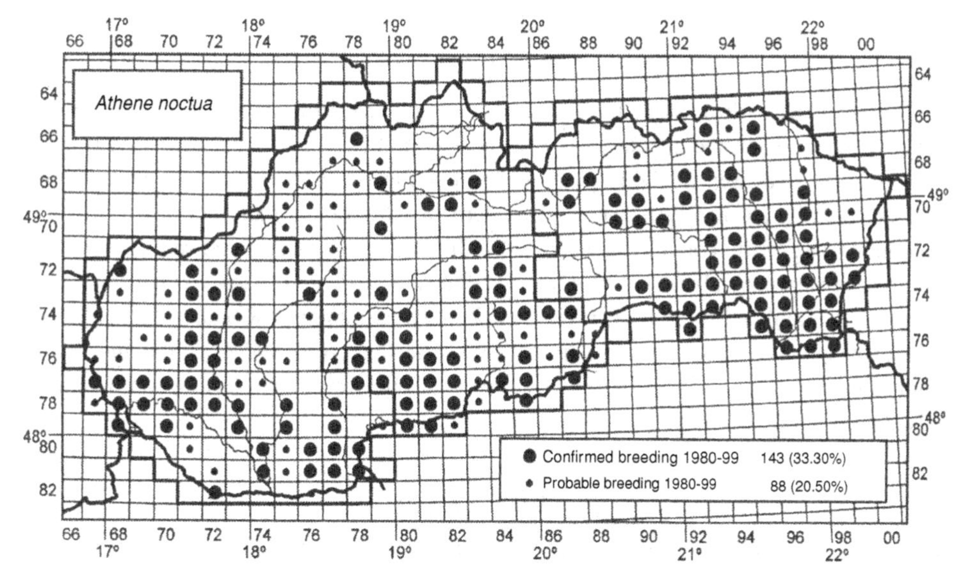

Figure 5.17 Little Owl distribution in Slovakia (after Danko *et al.* 2002).

Population estimates

Between 1980–99 the total number of breeding pairs in Slovakia was estimated at 800–1000 pairs; the number of wintering Little Owls between 1980–99 was estimated at 2000–3000 individuals. The population is slightly decreasing and the geographic distribution has decreased by 20% (Danko *et al.* 2002). BirdLife International (2004) estimates the total population at 800–1000 pairs.

Trends

A slight long-term decline of the total population is evident (Danko *et al.* 2002).

Densities

The highest densities in Slovakia were recorded in the lowlands of east Slovakia (12–15 pairs/121 km^2 or 0.09–0.12 pairs/km^2); the lowest in basins and hills of north Slovakia (1–3 pairs/121 km^2) (Danko *et al.* 2002). See Figure 5.17.

Croatia

As of 1991, the population of Croatia was estimated between 6000 and 8000 pairs (although the data quality for this estimate was not provided; Manez 1994). More recently, BirdLife International (2004) estimated the 2002 population at 500–1000 pairs.

Slovenia

Distribution

In 1769, J. A. Scopoli described the Little Owl to science from a specimen collected in Slovenia. In the past, the species was very common and widely distributed in Slovenia (Ponebšek 1917; Reiser 1925). Later a marked decline was noticed and nowadays Little Owls are distributed mainly in the southwestern (Mediterranean) and northeastern (sub-Pannonian) part of the country (Geister 1995). In the southern part of Slovenia there is a hybrid zone between subspecies *A. n. noctua* and *A. n. indigena* (Keve *et al.* 1960). The Little Owl is mainly a lowland species in Slovenia, being found below 630 m (Tome 1996a, b). However, it has also been found at higher elevations (e.g., from 690–900 m) in the alpine region (Jančar 1997; Vogrin 2001), but these reports were viewed as reflecting non-breeding individuals (Božič & Vrezec 2000).

Population estimates

In the second half of the twentieth century the Little Owl breeding population was estimated at 500–800 pairs (Geister 1995). A strong negative trend was noticed in the later part of the twentieth century, especially in northeastern Slovenia (Vogrin 1997). The species is now probably extirpated from the central part of Slovenia. According to recent surveys and population estimates made in southwestern and northeastern Slovenia (e.g., Vogrin 1997; Polak 2000; Denac *et al.* 2002; Štumberger unpublished data) the population is estimated at 200–400 breeding pairs, with a bulk of the population concentrated in the Mediterranean region of southwestern Slovenia (e.g., Karst area). In the Red Data List of Slovenia, the Little Owl is listed as an endangered species with intensification of agriculture and traffic collisions listed as the main reasons for the decline and continual threat to its population (Geister 1998; Jančar *et al.* 2001). BirdLife International (2004) estimates the total population at 150–200 pairs.

Densities

The density of Little Owls in Dravsko polje south of Maribor town was 0.048 pairs/km^2 (total area 210 km^2) (Vogrin 1997). In Central Slovenia, on Ljubljansko barje in an area of about 160 km^2, Trontelj (1994) estimated the population of Little Owls at 5–10 pairs or 0.03–0.06 pairs/km^2.

Serbia and Montenegro

Distribution

The species occurs in Serbia in the Pannonian part, the Peri-Pannonian part (south of the rivers Sava and Danube) and in the mountain regions with basins. The Little Owl is adapted to urban areas as in Jagodine (150 km south-east of Belgrade) where 24 individuals lived in 1994 (Stankovic 1997). In Montenegro (Crna Gora) it is found in Maritime Montenegro and the mountain regions with basins (Stevanovic & Vasic 1995).

Population estimates

Recent population estimates are 10 000–15 000 pairs for Serbia and Montenegro (BirdLife International 2004). Estimates for Vojvodina (northern part of Serbia – Pannonian Plain) are 1500–2000 pairs. Estimates for the central part of Serbia, excluding Vojvodina and Kosovo, are 2500–5000 pairs. In the Kosovo region (Kosovo & Metohija) the population is estimated at 4000–6000 pairs; in Serbia (including Vojvodina, Central Serbia, and Kosovo) at 8500–13 000 pairs; and in Montenegro at 1500–2000 pairs (S. Puzovic & D. V. Simic, personal communication). BirdLife International (2004) estimates the total population at 10 000–15 000 pairs.

Albania

An estimate of 5000–10 000 pairs was made for 1962 (Manez 1994). BirdLife International (2004) estimates the total population at 4000–8000 pairs.

Greece

Distribution

The species is a fairly common and widespread resident owl in Greece. Little Owls have been familiar birds to people living in Greece for at least 2500 years. They appear in ancient Greek art and in classical times they were associated with the goddess Athena. They still breed locally within Athens and other urban centers, e.g., Lamia, Chalkis, Eubä, Parnas, Zygos mountains, and Peleponesos, though they are more typically found in rural areas. They occur widely over the whole mainland and on most islands, i.e., the Ionian and Aegean Isles, the Cyclads, and Naxos (Schönn *et al.* 1991), in a wide variety of open habitats, including farmland, phrygana, and the surroundings of villages and shepherds' huts (Eakle 1994). Little Owls have been recorded as high as 1650 m on Mount Oiti (Handrinos & Akriotis 1997) and occur at relatively high elevations in other areas, provided suitable open habitat is available. It appears that Little Owls have declined since around 1970 (Cramp 1985). This is most evident in agricultural areas, where the increased use of insecticides has reduced their food supply, and in coastal areas, where urbanization has resulted in increased disturbance and loss of habitat such as derelict buildings.

Population estimates

Although they have been a protected species for many decades, they are disliked and persecuted in many rural areas due to a superstitious belief that their call is a bad omen. However, despite this level of undocumented background impact, the owls remain quite numerous and widely distributed. Their total population probably lies between 5000–15 000 pairs

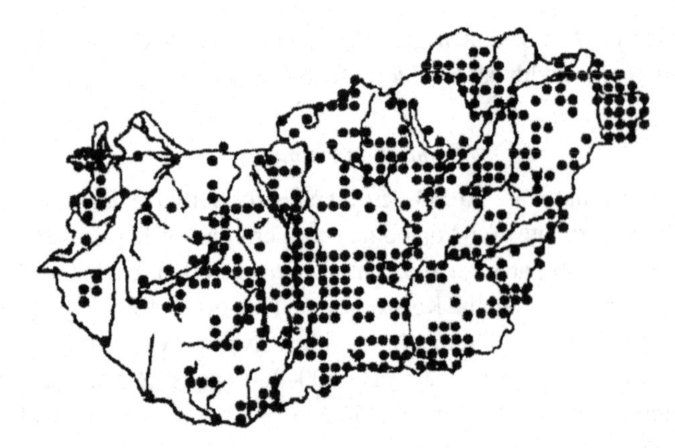

Figure 5.18 Little Owl distribution in Hungary (after Gorman 1995).

(Handrinos & Akriotis 1997). BirdLife International (2004) estimates the total population at 5000–15 000 pairs.

Cyprus

Little Owls are considered a common but decreasing species, with 2000–4000 pairs as of 1988 (Manez 1994). BirdLife International (2004) estimates the total population at 5000–15 000 pairs.

Hungary

Distribution

The Little Owl can be found throughout the entire country, especially to the east of the Donau River. In cities the species often inhabits railway stations (e.g., Budapest) (Schönn *et al.* 1991). The population has declined mainly due to agricultural intensification. The core population inhabits lowland areas though some pairs breed in hilly areas where nesting cavities can be found in trees or buildings (Gorman 1995). See Figure 5.18.

Population estimates

The population was estimated at 2000–2500 pairs as of 1994, and is probably stable again after a decline (Gorman 1995). BirdLife International (2004) estimates the total population at 1500–2500 pairs.

Bulgaria

Distribution

The Little Owl is the most widespread and numerous species among owls in Bulgaria (Nankinov 2002). During the breeding season the species occurs all over the country up to

2300 m (Simeonov *et al.* 1989). During autumn and winter its distribution is restricted to the lowlands and hilly country up to 900 m, indicating elevational migration (Simeonov 1983). It is widely distributed in all types of human settlements, especially villages – almost every village in Bulgaria has at least one or several pairs. Modern urban landscapes still provide supportive conditions for the species. In towns the species is also very well represented, in Sofia it is common even in the large apartment districts (Iankov 1983).

Population estimates

Nankinov (2002) characterized the Little Owl population in Bulgaria as exceeding 10 000 pairs. In 1982 about 140 pairs were encountered in Sofia (Iankov 1983). According to Kostadinova (1997) the breeding population was estimated at 4000–10 000 pairs. BirdLife International (2004) estimated the total population at 5000–8000 pairs.

Romania

The Little Owl is the most common owl species in the country. It can be found throughout the entire country with the exception of the forested mountainous areas. As an adaptable species, the owl can be found regularly in villages and towns. In Bucharest the species breeds in the Opera building. The species occurs up to 800 m (Schönn *et al.* 1991).

Population estimates

The population is estimated at 20 000–40 000 pairs and was considered stable as of 1988 (Manez 1994). BirdLife International (2004) estimates the total population at 40 000–60 000 pairs.

Moldova

Distribution

The species is one of the most widely distributed owl species in Moldova. It is also considered the most numerous of the nocturnal raptors (Ganya & Zubkov 1975). In parks of Askania-Nova, Little Owls willingly occupy empty nests of Magpies, *Pica pica*, in insular forests in waterless valleys with groves of the white acacia far from human dwellings. On the plot at the site "Florike", on an area of about 50 ha, about 100 Magpie nests were counted and in 42 of them Little Owls nested, as if nesting in a colony (Uspenskiy 1977).

Population estimates

The population was estimated to be 5000–7000 pairs as of 1988 (Manez 1994). BirdLife International (2004) estimated the total population at 3200–4200 pairs.

Turkey

Distribution

The Little Owl is a common and widespread species throughout the country with high densities in natural steppes and villages throughout central and eastern Turkey (Eakle 1994). It is less common to rare in forested areas, e.g., along the Black Sea coast. Due to its common status the species has not been mentioned in recent bird reports from Turkey (Ornithological Society of the Middle East and Central Asia, OSME). However, Kasparek (1992) reviews the species as a widespread, common breeding resident in nearly the whole of Turkey, only local in the Black Sea region and other coastal regions. Further, it is mentioned that the species is rare in alpine areas, though recorded up to 4400 m. In Aegean and Mediterranean coastal areas the species has been recorded scarcely and is mainly confined to villages with surrounding agriculture. However, it is again more commonly found at higher elevations where rocky habitat, macchia, and extensive agriculture are prevalent. These studies also revealed that the Little Owl is certainly much more common than estimated (E. Vaassen, personal communication). Local populations of the Little Owl as well as the other insect-eating Eurasian Scops Owl *(Otus scops)* have nearly disappeared from some habitats (villages and ruins) in the Göksu and Çukurova Deltas in southern Turkey, probably due to the massive application of insecticides (E. Vaassen, personal communication).

Population estimates

Kasparek and Bilgin (1996) estimate the total Little Owl population at 5000–50 000 pairs in their *Checklist of Turkish Vertebrates*. Recent field studies revealed that the owl is certainly much more common than estimated and therefore a population of 30 000 pairs for the whole of Turkey (782 700 km^2) is likely to be the absolute minimum (E. Vaassen, personal communication). Estimates for Turkey may prove very low given the density of 500 pairs in 80 km^2 for Emirdag. The national population may reach 200 000–250 000 pairs (E. Vaassen, personal communication). BirdLife International (2004) estimates the total population at 300 000–600 000 pairs.

Densities

Recent field studies have revealed that the species is locally very common with densities of 8–10 pairs/km^2 in rocky highland areas with marginal cultivation. Similar densities have been located in the surrounding districts of large cities like Ankara and Konya, where building sites and patches of wild-grown steppe are prevalent (Vaassen 2000).

Poland

Distribution

Until the end of the 1960s, the Little Owl was one of the most common owls in Poland (Taczanowski 1882; Sokolowski 1953). In the 1960s, one or two pairs were nesting in almost every village in the area of Legnica and Leszno in southwestern Poland (Tomialojc 1972; Bednarz *et al.* 2000). The first data concerning the decrease in population numbers of the Little Owl appeared in the 1970s and 1980s when it was considered as occurring generally in small numbers, and only locally in medium numbers (Tomialojc 1990). The disappearance of this species started in western Poland. In the Gorzów Wielkopolski Region of western Poland, this bird was relatively frequently observed in the 1970s, whereas at present it is practically extirpated (Jermaczek *et al.* 1995; A. Jermaczek, personal communication). A similar situation is observed in the Poznań Region of west-central Poland and the Silesian Region of southwestern Poland, where the Little Owl currently occurs in very small numbers (Dyrcz *et al.* 1991; Bednarz *et al.* 2000). The tendency towards the gradual disappearance of the species from western Poland was noted in the West Pomeranian Region of northwestern Poland, where at present there probably remain only two nesting sites (R. Kościów, personal communication). In the nineteenth century it occurred in Poland in about 130 larger and smaller towns, and recently, from 1990–2000, it inhabits over 60 towns, mainly in southern and southeastern Poland (Grzywaczewski & Kitowski, unpublished data). In Poland, the Little Owl mainly occurs in cities and upland areas, i.e., the Uplands of Lublin, Cracow, Cracow-Czestochowa, and the Sub-Carpathian Highlands. The cities and rural areas located in these regions of south and southeastern Poland can be considered the last strongholds for this species (G. Grzywaczewski, personal communication).

Population estimates

The species is scarce in most areas, with marked declines after severe winters (Tomialojc 1976). Based on the available information, the number of Little Owls in Poland may be estimated at *1000–1500 breeding pairs* as of 2003 (G. Grzywaczewski, personal communication). Little Owl estimates in different regions of the country are given in Tables 5.10 and 5.11. Detailed data in the 1990s indicated 10–20 pairs in the Bialystok Region in northeastern Poland (M. Polakowski, personal communication); 150–200 pairs in the Lublin Region (unpublished data); 10–100 pairs in the Łódź Region of central Poland (R. Wlodarczyk, personal communication); 50–200 pairs in the Cracow Region of southeastern and southern Poland (K. Kus, personal communication); and 120 pairs in the Radom and Kielce Regions of south-central Poland (J. Tabor, personal communication). BirdLife International (2004) estimates the total population at 1000–2000 pairs. Monitoring activities between 1982 and 2005 revealed the disappearance of a population in an area of 12 km^2 in central Poland, due to a decrease of nesting places and food availability (Zmihorski *et al.* 2006).

Densities

In the 1980s, the average density of the Little Owl population in the agricultural landscape was 0.17 territories/km^2. A relatively rapid change took place in the 1990s, when an at least 2.5-fold decrease in population density was noted, with densities dropping to 0.07

Table 5.10 *Population estimates and density of the Little Owl,* Athene noctua, *in cities in Poland (G. Grzywaczewski, personal communication).*

Number	Name of area	Study area (km^2)	Territories (pairs)	Density (territories/10 km^2)	Year	Reference
1	Gliwice (S Poland)	136	19–20	1.4	1988	Dyrcz *et al.* 1991
2	Jasło (S Poland)	36.7	12–15	3.3–4.1	1990	Stój & Dyczkowski 2002
3	Tomaszów Lubelski (SE Poland)	13.33	6	4.5	1993–8	Grzywaczewski & Kitowski 2000
4	Hrubieszów (SE Poland)	32.79	7	2.1	1993–8	Grzywaczewski & Kitowski 2000
5	Biłgoraj (SE Poland)	20.85	4	1.9	1993–8	Grzywaczewski & Kitowski 2000
6	Tarnogród (SE Poland)	10.86	1	0.9	1993–8	Grzywaczewski & Kitowski 2000
7	Szczebrzeszyn (SE Poland)	29.04	2	0.7	1993–2000	Grzywaczewski & Kitowski 2000
8	Zamość (SE Poland)	31	5–6	0.5–0.6	1994–9	Grzywaczewski & Kitowski 2000
9	Chełm (SE Poland)	35.72	14–19	0.39–0.53	1998–2000	Kitowski & Grzywaczewski 2003
10	Kraśnik (SE Poland)	25	8	3.2	2000	Frączek & Szewczyk 2000

Table 5.11 *Population estimates and density of the Little Owl,* Athene noctua, *in agricultural landscapes in Poland (G. Grzywaczewski, personal communication).*

Number	Name of area	Study area (km^2)	Territories (pairs)	Density (territories/10 km^2)	Year	Reference
1	Ogrodniki	11	2	1.8	1980	Dombrowski *et al.* 1991
2	Golice	22	1	0.5	1980	Dombrowski *et al.* 1991
			1	0.5	1986	
3	Paprotnia	18	3	1.7	1982	Dombrowski *et al.* 1991
			2	1.1	1984	
			2	1.1	1985	
			2	1.1	1986	
4	Łomianki	11.7	7	6	1982	Dombrowski *et al.* 1991
			8	6.8	1986	
			2	1.7	1988	
5	Września	20	4	2	1983	Tomiałojć 1990
6	Sobienie	10	2	2	1984	Dombrowski *et al.* 1991
7	Kampinowski National Park	62.6	9	1.4	1984	Kowalski *et al.* 1991
8	Przesmyki	80	3	0.4	1984	Fronczak & Dombrowski 1991
9	Sulechów	17.5	0	0	1989	Jermaczek *et al.* 1990
10	Sulęcin	42	0	0	1989	Jermaczek *et al.* 1990
11	Zbąszynek	26	0	0	1989	Jermaczek *et al.* 1990
12	Valley River Warta	322	1	0.03	1992	Winiecki *et al.* 1997
13	Potęgowo	50	1	0.2	1993–1994	Antczak *et al.* 1995
14	Kotlina Dzierżoniowska	44	2	0.5	1994	Wuczyński 1994
15	Żurawica	96	8–9	0.8–0.9	1994	Hordowski 1999
			5–6	0.5–0.7	1998	
1.+3.	Ogrodniki and Paprotnia	39	8	1.3	1995–6	Golawski unpublished data
16	Bystrzejowice-Skrzynice-Głusk	80	10–12	1.3–1.5	2001	Grzywaczewski & Gustaw, pers. comm.
			4.5	0.5–0.6	2002	
17	Near Siedlce town	220	9	0.4	2002	Kasprzykowski & Goławski in press
18	Małopolski Przełom Wisły	80	1	0.1	2002	Rudolf & Grzywaczewski unpublished data

territories/km^2. In 2004, in urban areas, especially in southern and southeastern Poland, an increase was observed in the owl population density to 0.26 territories/km^2. In Chełm, southeastern Poland, the density of Little Owls is among the highest in Poland, reaching on average 0.4 territories/km^2. However, after the snowy and frosty winter of 2001/2002 the density dropped to 0.22 territories/km^2 (G. Grzywaczewski, personal communication).

Russian Federation

There have been substantial changes in the political boundaries of the former USSR, complicating analysis of the older literature on the Little Owl. Below, we briefly describe the owl distribution in the former USSR, and then denote the owl situation in the respective countries. In the European territory of the former USSR the Little Owl is distributed mainly in anthropogenic landscapes, in southern Kazakhstan and Middle Asia mainly in the deserts, dry steppes, and in the mountainous regions. From the western border the species spreads eastward to the Zabaikalie (Trans-Baikalia) (Sretensk). To the north it reaches Lithuania and Latvia (in Estonia the species is a vagrant bird), Pskov, Moscow, Ryazan Region, Kazan, Buguruslan, Orenburg, from where the border stretches via central Kazakhstan to Zaisan Lake. Further eastwards the species occurs only in isolated populations in the southeastern Altai (Chuiskaya steppe), in Tuvinskaya (Tyva Autonomous Soviet Socialist Republic) and in Trans-Baikalia (Selenginskaya Dauria, Aginskaya, Borzinskaya, and Nerchinskaya steppes). In Tien-Shan and in the Pamir the Little Owl breeds at high elevations (up to 4200 m). The southern border of the Little Owl's range is situated outside the borders of the former USSR (Ivanov 1976).

In Russia the Little Owl is resident or, rarely, migratory. The total number in European Russia is estimated to be 10 000–99 999 pairs (Mishchenko 2004). BirdLife International (2004) estimates the total population at 10 000–100 000 pairs. While there is some information on owls moving northwards along human-transformed areas, fluctuations in owl numbers are difficult to determine and some researchers report that the population is currently decreasing in Russia. The population in the Republic of Bashkortostan (Karyakin 1998) is estimated at 10 breeding pairs, with winter numbers at 10–50 individuals. At least 10–15 pairs breed in the whole of the south of the Kaliningrad Region (Grishanov 1994). The Saratov region of 100 000 km^2 likely supports about 200–300 pairs (A. Antonchikov, personal communication). At present the species is comparatively stable in the Republic of Dagestan, and in some places may even increase due to an increase in human-transformed landscapes in the lowland regions of the Republic and restoration of formerly abandoned cattle-farm complexes. In the Lower Cis-Volga River area, Zav'yalov *et al.* (2000) illustrated the intra-year dynamics of abundance for two different environments of Saratov city. For 1993–8, they found Little Owl densities in old apartment buildings and the forest-park zone to be 0.2 and 0.05 individuals/km^2 in winter; 0.1 and 0.2 in pre-spring; 0.3 and 0.3 in the pre-nesting period; 0.6 and 0.4 during nesting; 0.5 and 0.4 post-nesting; and 0.3 and 0.2 for the migration period, respectively (Zav'yalov *et al.* 2000).

Kaliningrad Region (Russian Federation)

As a whole for eastern Prussia the species is considered as a rare breeding species, whose numbers decrease markedly as a result of severe winters (Grishanov 1994). Population densities increase slightly from the northwestern to the southeastern part of the region (Tischler 1941). At present the status in the Kaliningrad Region looks similar to that occuring before 1940, when the Little Owl was extremely rare to the south of the Neman River, east of the Gulf of Kursiu (Courish Gulf), on Samland Peninsula, and was slightly more common in the central part of the region. In the south, the species occurs most frequently along the border with Poland.

Republic of Dagestan (Russian Federation)

Distribution

The Little Owl is resident in the Republic of Dagestan, spread widely from the shore of the Caspian Sea and lowlands reaching to 2500 m in the mountain-alpine regions. Due to the arid climate, lowland (northern) desert regions, and insignificant forest coverage (7–8% of total area) of the region, the distribution of the owl is non-continuous. The species inhabits mountain steppes in small numbers being scattered among other types of landscapes, where it prefers slopes of southern exposures with rocks, big and small fragment screes, and outcrops of stones, which provide vertical topography. The species often settles near humans that raise cattle and poultry, because of the concentration of insects and small rodents that are supported. In the open landscapes of northern (lowland) Dagestan, including the regions of the central and southern Caspian Sea, it is especially attracted by small human settlements and cattle farms. In the lowlands it inhabits the floodland forests along rivers such as the Terek and Sulak, and the forest belts of the tape-grove type. Part of the population inhabits big cities and suburbs, nesting under the roofs of old buildings. While the major part of the population is resident, birds that inhabit the mountain-alpine plots migrate in the autumn–winter period to the flat foothill regions (E. V. Vilkov, personal communication).

Densities

Densities of the Little Owl have been reported at 16–22 individuals/km^2 in the open landscapes of the northern (lowland) Dagestan, including areas of the central and southern Caspian Sea region (Vilkov 2003).

Republic of Bashkortostan (Russian Federation)

At the northern border of its distribution, the Little Owl is a very rare breeding species in Bashkiria. Only one breeding site of this species is known in the rocky massif of Belaya

River near Kuznetsovskiy khutor. However, it is likely that the number of Little Owls is underestimated because settlements of the steppe and forest-steppe zones of the Republic have not been investigated (Karyakin 1998).

Republic of Tatarstan (Russian Federation)

In Tatarstan, the Little Owl is recorded mainly in the western and southern districts, i.e., Apostovskiy, Buinskiy, Tetyushskiy, Zelenodolskiy, and also in Leninogorskiy, Aznakievskiy, and Sarmanovskiy districts (Rakhimov 1995). Recent research points at some shift of the border of the Little Owl breeding range in Tatarstan in a northerly direction (Rakhimov & Pavlov 1999). This northerly shift seems connected with the rapid rate of forest logging in the area. Following humans, the species spreads into anthropogenic landscapes in the northern half of the Republic.

Republic of Ossetia (Russian Federation)

Komarov (1998) reported winter bird population counts of up to 4.1 individuals/km^2 in the settlements of the Ossetia plain.

Tyva (formerly Tuva) (Russian Federation)

The Little Owl is a rare, probable breeding species in Tyva. In 1983, two specimens were sighted in the upper parts of Barlyk River at Kurgak-Sai. In August of 1984, the species was observed at the Khapshi site and on a macro-slope of the Tsagan-Shibetu mountain ridge at Oruktyg and Kedyrorug. It is a rare wintering species in Tyva, observed at the Semigorka site (Popov 1991).

Latvia

Distribution

The Little Owl is very rare in Latvia (Lipsberg 1985), locally distributed in different regions of the Republic, most often in the western part. The northern border of the species' breeding range crosses the territory of the country. According to most authors, in the nineteenth century the species nested more often in Kurland than in Livland (Meyer 1815; Russow 1880). A specific nesting place was found in 1896 and several subsequent years in the Rembate settlement (not far from the Lielvarde settlement) (Sawitzky 1899). In subsequent publications authors confirm the more frequent occurrence of the species in Kurland than in the rest of Latvia. Non-breeding season records in the country are from areas surrounding the towns of Riga, Elgava, Dobele, and also the settlements Inchukalns, Rembate, Puze Misa, Olainr, Ropaji, and Burtnieki (Graubitz 1983). In the first half of the twentieth century the species was described as a very rare breeding species (Loudon 1909; Grosse & Transehe 1929; Transehe & Sinats 1936; Taurinsh & Vilks 1949) and found in Kurzem. Breeding was confirmed only at two sites in Rembate, not far from Lielvarde (around 1896 and in several subsequent years) (Sawitzky 1899) and near Elgava town (in 1949) (Lipsberg 1985). All other references deal with birds found outside the breeding season and without clear breeding evidence.

Population estimates

The Little Owl is considered rare (Baumanis & Blums 1969), with an estimate of 10–30 pairs as of 1988 (Manez 1994). BirdLife International (2004) estimates the total population at 10–30 pairs.

Lithuania

Distribution

The Little Owl is a rare resident breeder in Lithuania (Zhalakevichius *et al.* 1995) and is considered to have a decreasing breeding population. It is assumed that only isolated pairs breed.

Population estimates

Manez (1994) gave an estimate of 10–50 pairs for the period of 1985–8, and considered the population stable. BirdLife International (2004) estimates the total population at 5–10 pairs.

Estonia

Though the breeding range of the Little Owl reaches western Latvia it has been recorded only once in Estonia. Around 1880–5 one specimen was shot at Kolovere, Laanemaa district (Stall 1904; Lilleleht & Leibak 1992).

Belarus

The Little Owl is a rare resident breeding species throughout all of Belarus (Nikiforov *et al.* 1997) and is estimated at 400–1000 pairs with some decline in 1970–80. BirdLife International (2004) estimates the total population at 400–1000 pairs.

Ukraine

Distribution

The species is a resident bird in Crimea. Some short-distance winter migrations have been observed. On the southern coast of the Crimea the species probably occurs to the east of Alushta town. Numbers in steppe and foothills are moderately high and distributed relatively evenly. Along the southeastern shore (south coast) it is spread sporadically. The owls breed in walls of the fortress on Arbat(skaya) Spit (Kostin 1983).

Population estimates

Manez (1994) estimates 11 000–12 000 pairs as of 1988, and considers the population to be increasing. BirdLife International (2004) estimates the total population at 15 000–22 000 pairs.

Armenia

In Armenia, the Little Owl occurs in semi-desert, steppe, and mountain-meadow zones. Dementyev and Gladkov (1951) consider the species as one of the typical birds for the Caucasia. *Athene noctua indigena* is characteristic of different, but predominantly open, habitats, which are situated in the piedmont climatic zone of both river basins Arax and Kura (K. Jenderedjian, personal communication). Lyaister and Sosnin (1942) mention the species as a resident bird, occuring mainly in places of lower elevations and in the foothills (though up to 2000 m sporadically). It breeds in ravines and gorges, and also in stony ruins of disused settlements and houses. The species is also observed in gardens. The species has been recorded in the Oktemberyan district, in the surroundings of Novyi Shakhvarut, in typical semi-desert, at an elevation of about 900 m (Sosnikhina 1950). It was discovered in the caves of the lava mountain Karkhan (Sosnikhina 1950). The species is distributed in three state nature reserves in Armenia (Khosrov, Erebuny, and Shikahogh). It has also been recorded in rock outcrops of the Goravan sands sanctuary, and in the vicinity of Yerevan (Sokolov & Syroechkovsky 1990). BirdLife International (2004) estimates the total population at 800–1500 pairs.

Georgia

Distribution

The Little Owl is a year-round resident in the lowlands, plains, plateaus, foothills, and low- and middle-mountain belt of Georgia. It is uncommon in central parts of the country, and common in eastern and southeastern areas. It is more common in the Iori plateau, Karthli Plain, Alazani Plain, Gardabani Lowland, in lower parts of the Iori, Alazani, Khrami Rivers' valleys, Shiraki and Eldari semi-deserts, Udabno and Kvernaki ridges, and in some other points in the Kura River basin. The western boundary of its breeding range remains uncertain. It prefers dry, open and semi-open habitats with ravines, precipices, and rock outcrops. The species often utilizes older buildings and ruins to breed in. In semi-desert areas, nests were found in holes and in cracks in rock precipices. Owls sometimes nest in settlements, including the environs of Tbilisi (A. Abuladze, personal communication).

Population estimates

The present Georgian population is estimated at several thousand breeding pairs (4000–4500 pairs at the end of the 1970s; about 3500 pairs in the mid-1980s; 3500–4000 pairs in the middle to end of the 1990s). During the last 25 years no alarming trends have been detected. Numbers fluctuate over years. A slow decline was registered in some areas in the first half of the 1980s, but at present the population as well as the distribution is as stable as during the twentieth century (A. Abuladze, personal communication).

Kazakhstan

The Little Owl is spread unevenly in Kazakhstan, and occurs sporadically at certain sites. The highest breeding densities are observed in the desert zone of south Kazakhstan, in the regions of south Pribalkhashie, in the foothills of Karatau, in the Syr Daria River valley, along the Aral Sea shore, and on Mangyshlak. The northern border of Little Owl nesting in Kazakhstan is roughly the 49th parallel, and only in the Ural River basin does it extend up to the 51st parallel. It breeds in Ustyurt, where it is common in breaks in this plateau, but being seldom sighted in the central parts. In Mangyshlak it is common, but further northwards, in the northeastern part of the Near-Caspian Lowland, data regarding its presence (or absence) is missing. It is found in the northern Cis-Caspian Sea area, in the lower parts of the Ural and Volga rivers and further north, and in the Volga-Ural steppes near Urda. Along the Ural River valley it is spread north to the Ilek River mouth. It is extremely rare north of Ustyurt, in the desert steppes of the Emba River basin, and in the region of Mugodzhary. The species is well distributed and breeds in high numbers around the Aral Sea. It has been found considerably northwards, i.e., near Chelkar Station and in the Irgiz settlement (Gavrin 1962).

It is common, and at certain sites even numerous, in Kyzyl-Kumy, Syr-Daria River valley, and in the foothills of Karatau. In the plains of southern Pribalkash'ie, the Little Owl is spread more evenly and at certain sites it is a common bird. It is also common in the deserts of northern Pribalkhash'ie (Balkhash Lake area), where it normally breeds north of the 48th parallel. Northwards, in the region of the Kazakh small-knoll area, the Little Owl is very rare and is spread extremely sporadically. In the Kyzylrai mountains it was found in the stony knolls of Kotur-Kyzyltau (49′20″ N. L.), which is the northernmost nest record of the species in eastern Kazakhstan. In the Tien-shan mountain region, the Little Owl breeds on the slopes of Ugam ridge, and is also found to the south of Chimkent on the Kazgurt ridge.

Uzbekistan

The Little Owl is widely spread in Uzbekistan. It occurs in oases, in mountains, in semi-deserts and deserts (Sagitov 1990), in settlements, in cemeteries, in tugais, in sands, on rock outcrops, and 'kyr's of Ustyurt (Kostin 1983). The species is considered as a resident bird in Karakalpakia (Ametov 1981). On the northern slopes of the ridge in the lower belt of mountains the species occurs in low densities (Meklenburtsev 1936). According to Salikhbaev *et al.* (1970), the species is common in the foothill plains or *adyr's*. It is also common in the entire Zarafshan valley (Maslov 1947; Bogdanov 1956; Bakaev 1974). Population densities depend mainly on the habitat; in the Kashkadarya valley the owls are more numerous in the foothills, and in clayish and sandy deserts (Meklenburtsev 1958). In the Nuratau Ridge, it is common on *adyr's* and in the foothill plain. It is found in lower numbers in the lower belt of mountains (Salikhbaev *et al.* 1970).

Turkmenistan

The Little Owl is spread throughout the country and occurs all year round. It breeds in rock holes, in rodent burrows, on clayish precipices, in tree hollows, in walls and under roofs of buildings, and in cornices. The distribution and population densities of the species are determined by the available food that mainly consists of small mammals (Bel'skaya 1992). In the western part of Turkmenistan, the Little Owl is a resident species of the southeastern Cis-Caspian Sea area. It has been recorded during mid-January in Krasnovodsk town, below the Nebit-Dag mountain, and in the Janga settlement (Khokhlov 1995). In Badkhyz, the Little Owl is a common species, distributed throughout the region. The highest numbers are recorded on clayish precipices of the Kyzyldzhar ravine. It is rarest in the desert steppe (Simakin 2000). This species is found throughout the area north of the river-bed of Zapadnyi (western) Uzboi. The natural and northern border of the region are the precipices (cliffs) of the Ustyurt Plateau (Kaplankyr chink). The western boundary is the coast of Kara-Bogaz-Gol Gulf and from there, south along the Caspian Sea to Iran (Zarkhidze & Loskutova 1999).

Densities

In northwestern Turkmenistan *Athene noctua bactriana* occurs everywhere. In 1989–91 the population density in different habitats ranged from 0.4–5.0 individuals/km^2 (Zarkhidze & Loskutova 1999). In the Repetek Nature Reserve, fluctuating densities were found in Black Saxaul thickets (0.7–1.8 individuals/km^2), and in White Saxaul thickets (0.3–1.2 individuals/km^2) (Ataev 1977).

While the species is widely spread, it is mainly associated with desert landscapes. In the northern part of the Republic it inhabits suitable habitats of Mogol-Tau, in *adyr's* of the Kuraminskiy mountain ridge, in the Syr-Dar'ya River valley (in the vicinity of Leninabad, Kairak-Kum, and further to the border with Uzbekistan), and on the northern slope of the Turkestan ridge (surroundings of Shakhristan, the valley of Ok-Tanga at 2400 m). It also occurs on the southern slope of the Turkestan mountain ridge. The species was found in the Iskander-Kul' Lake area (Ivanov 1940), and in Gissar-Karategin (Popov 1959) where it is most commonly found in the foothills. It is found throughout the Gissar Valley occupying habitats suitable for nesting (Ivanov 1940; Popov 1959). In southern Tajikistan the Little Owl nests in the mountains of Rangin-Tau, Aruk-Tau, Ak-Tau, Buri-Tau, Kara-Tau, and Khazratishokh and from the banks of the Pyandzh River to the foothills of the Darvaz

Mountain ridge. The species has been sighted very often in the valleys of the Vakhsh and Kafirnigan Rivers, and on the lower portions of the Yakh-Su and Kizyl-Su streams (Ivanov 1940). In Badakhshan, the species has been found in the upper parts of Shakhdara (Meklenburtsev 1946), in the upper parts of Vakhan-Darya (Bazai-Gumbez) (Ivanov 1940), in the Drukhmatz River valley, and in the surroundings of the Pamir Botanical Garden (A. V. Popov, personal communication). In all probability, the Little Owl is widely distributed in Badakhshan, but due to its small numbers it remains unrecorded by ornithologists in other areas. The species occurs in the southern part of Pamir too, i.e., in the Beik River valley, Ak-Su tributary (Ivanov 1940), in the surroundings of Zor-Kul Lake and Bash-Gumbez site (Abdusalyamov 1971), and near the Soviet farm "Pamir" on Bulun-Kul (Meklenburtsev 1936).

India (Ladakh)

Distribution

Ladakh is an area through which the line of control (as per the 1972 SIMLA Agreement) between Pakistan and India passes. Although a major part of Ladakh lies in Indian-held Kashmir, a sizeable part is present on the Pakistani side of Kashmir. Further in the north, above Saltoro Range, lies the Siachen Glacier where the two armies are fighting each other at the highest battlefield in the world. Readers can see a map of the situation at www.un.org/Depts/Cartographic/map/dpko/unmogip.pdf. The region holds all the Little Owls in India. It lies above 4000 m and includes some extremely desolate areas (N. Paklina & C. van Orden, personal communication).

Population estimates

The region of Ladakh holds some 120 pairs as of 2003 (N. Paklina & C. van Orden, personal communication).

The global distribution of the Little Owl indicates a presence in the following countries, but no data on the population was available for this project. However a map of the Little Owl's distribution in India, Pakistan, Nepal, and Bhutan, from Grimmett *et al.* (1999) is shown in Figure 5.19.

Pakistan
Nepal
Bhutan

Mongolia

The species occurs in natural habitats and anthropogenic environments, and is considered rare everywhere in Mongolia. Breeding has been observed in the Chentej Mountains near Boroo, near Chudshirt, and in the Gobi desert. Other reported observations are from Somon,

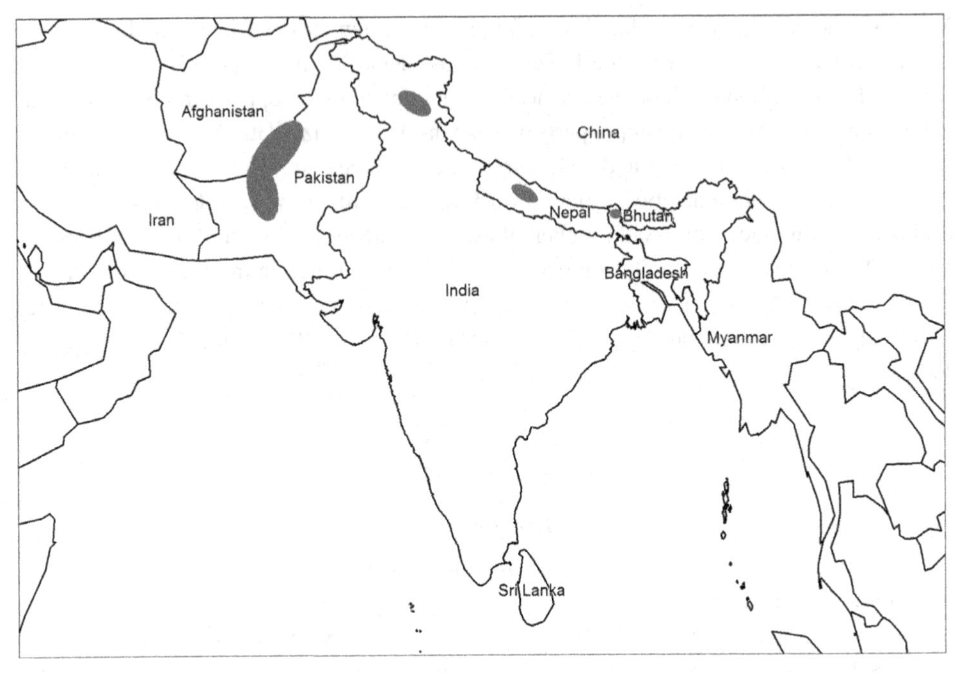

Figure 5.19 Map of Little Owl distribution in India, Pakistan, Nepal, and Bhutan (after Grimmett *et al.* 1999).

Buucagaan, Jarantaj (border with China), Conocharajch-gol, Ulaan-uul, Chvod, and Ajmak (southern Gobi) (Schönn *et al.* 1991). In Mongolia, the Little Owl is considered a rare south-Eurasian species, occuring as a breeding species in the desert mountains and stony deserts of northwestern Mongolia and southeastern Altai (Rogacheva 1988). The Little Owl is probably a vagrant species. The only specimen observed within the borders of the Krasnoyarsk Territory was in 1908 in the Us settlement. Overall, the owl breeds in stony deserts, semi-deserts, and dry steppe with separate rocks or stony ridges; in the mountains, it occurs up to 2800 m (Sushkin 1938).

China

The northern border of the Little Owl distribution range crosses northeastern China, the southern border goes through Tschajul, Yaan, Daxian, Nanyang, to the south of Xinhailian, and the Yellow Sea. Qingdqo was registered as the most eastern point of its occurrence (Schönn *et al.* 1991).

Little Owls are mainly distributed in foothills near the forest, prairie, and hillside fields. The most suitable habitats in Qishan (34.4 N, 107.6 E) are found to be open country with scattered trees, earth banks, or crevices. The population density differs according to the season; the average density is 0.04 pairs/km^2 in winter (January and February), and 0.21

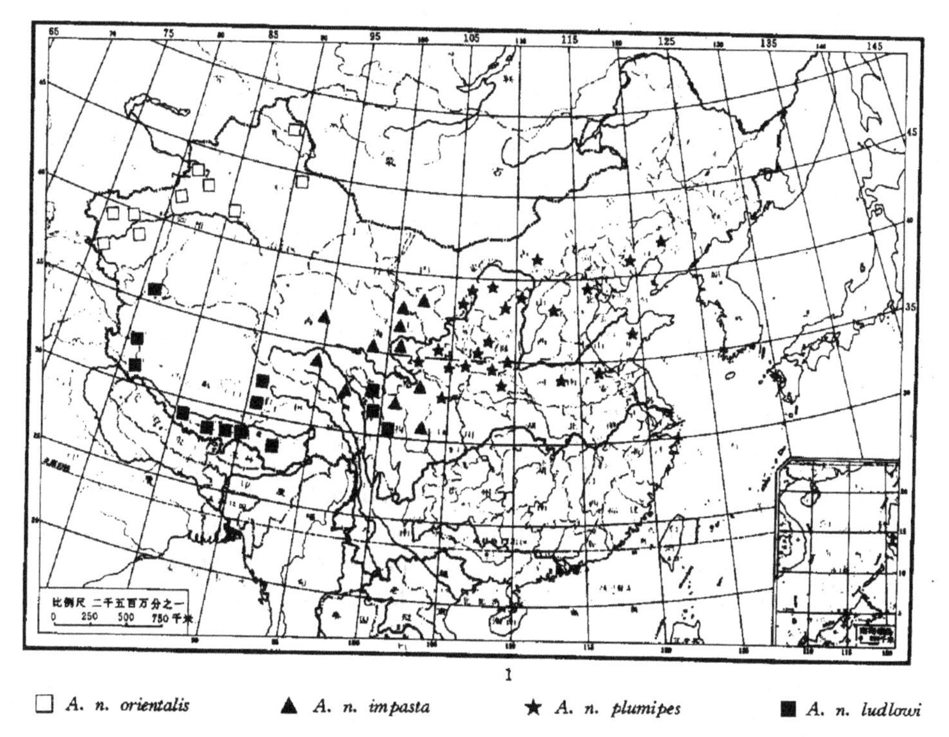

Figure 5.20 Distribution of Little Owl subspecies in China (after Lei *et al.* 1997).

pairs/km^2 in summer (June). The average distance between nests is 189 m, and the shortest distance between two nests is less than 100 m. However, they are now rarely found in the Weinan, Meixian, and Huxian counties of the Shaanxi Province, and in the Laishui and Yixian counties of the Hebei Province. See Figure 5.20.

Tibet (China)

Distribution

The Little Owl is common in a large area of northern China (Fu-Min Lei, personal communication), and the Tibetan Plateau features some isolated breeding populations above 4000 m in extremely desolate areas (N. Paklina & C. van Orden, personal communication).

Population estimates

As of 2003, the population of the Tibetan Plateau was estimated at 50 000 pairs after extrapolation of the density of Ladakh (N. Paklina & C. van Orden, personal communication).

North and South Korea

The most recent record of the Little Owl dates from 1960 when a dead Little Owl was found by one of the employees of the Army Post Exchange of Seoul, South Korea. It was said to have been found in the folds of a tarpaulin on a supply truck that had arrived that morning from the Port of Inchon on the Yellow Sea, approximately 40 km west of Seoul. According to Austin (1948) only three records of the species exist for Korea, all based on specimens taken by Won in North Korea (Pyongan Namdo) in November 1931 and February 1932.

Syria

The Little Owl can be observed in most parts of Syria. In and around Damascus the species is relatively rare; it could be absent from the intensively farmed land of Ghouta. In the fertile regions around Homs, Hama, Idlop, and Aleppo, the species is common and occupies houses, barns, and stables. Historical ruins (e.g., Palmira) are favorite breeding places. In the Euphrates River valley and in the desert, the Little Owl breeds in stone piles (Schönn *et al.* 1991).

Jordan

Distribution

A fairly common resident in the western parts of Jordan, the Little Owl is scarce to rare in the east. It is found in the highlands, rift margins, and parts of the Jordan Valley, and in true desert areas. In the highlands, it is fairly common in open habitats with rocks and low vegetation, never in woods or deep gorges.

In the rift margins the species is a common resident from the northern areas, southward to the slopes that face the Dead Sea, where it becomes scarce. It is also rather uncommon in the rift margins/slopes facing Wadi Araba, the Aqaba mountains, and Wadi Rim. In the eastern desert, the Little Owl is common in the areas just east of the highlands, which slope gently towards the eastern desert plateau. This area is characterized by open rolling country with rock boulders and low steppe vegetation. In the eastern desert it is only locally present, usually in the few limestone hills in the flat hammada/flintstone desert and around farms. It is more common and widespread in the northeastern desert, which is covered by basalt stones and rocks (Andrews 1995; Khoury, personal communication).

Population estimates

The population size is estimated at several thousand pairs (Andrews 1995; Khoury, personal communication).

Israel

Distribution

The Little Owl is resident in most areas, and territorial pairs are found during the breeding season. Outside the breeding season it normally occurs alone, but some pairs remain together throughout the year. Immature owls wander and disperse short distances. The most continuous and dense population occurs along the central mountain range, including Golan, Galilee, Carmel, Shomron, and Judea. In Hermon, it is a rare breeder up to 500 m. There is quite a dense population in the Jordan River Valley, especially on the cliffs along the Jordan River itself. It is more widely dispersed in the northern valleys and along the Coastal and Judean Plains (in Dan region mainly in the eastern hills). It is common throughout much of the Judean Desert, mainly northwestern parts, near the Judean Hills. It is almost totally absent in the low-lying arid areas of the Judean Desert, in the Dead Sea Depression, and along Arava. In Negev it occurs continuously, mainly in mountainous regions north of Nahal Paran, in the Makhteshim, Nahal Zin area, and in northeastern Negev between Arad and Yeroham. The densest concentration in the desert and semi-desert regions occurs in northwestern Negev, Holot Halutza, Nizzana, Shivta, and eastwards to Mashabbe Sade. In the Eilat mountains, and at Eilat town, the species appears occasionally as a wanderer or migrant, though a few breeding pairs are scattered in *wadis* (an annual dry riverbed or valley) north of the Eilat mountains and northwards in the area of Biq'at Sayyarim and Shizzafon. In all parts it breeds mainly outside human settlements or on fringes of villages and towns. A few breed in local enclaves within cities such as Jerusalem, where 30–40 pairs were found within 68 km^2 (0.44–0.59 pairs/km^2) in the 1980s, and a similar number in Haifa (Shirihai 1996). See Figure 5.21.

Population estimates

The total population in Israel was estimated in the 1980s at several thousand pairs (Shirihai 1996).

Arabian Peninsula

Distribution

The Little Owl is a widespread, locally common resident in the deserts of Arabia. There is no information to suggest seasonal or other movements and it is generally thought of as sedentary, occuring in all states except Bahrain. It has occurred on Das Island in the middle of the Arabian Gulf, evidence that some dispersal or movement is taking place. There are a number of records from the periphery of the Empty Quarter, but it appears to be absent from the core area. This is likely due to the lack of suitable roosting/nesting sites rather than the species itself not being tolerant of hyper-arid environments. The owl is scarce in the highland areas of the southwest, except for the drier eastern fringes. It is a resident on Farasan Island, but generally scarce at other coastal sites and not known on Masirah or

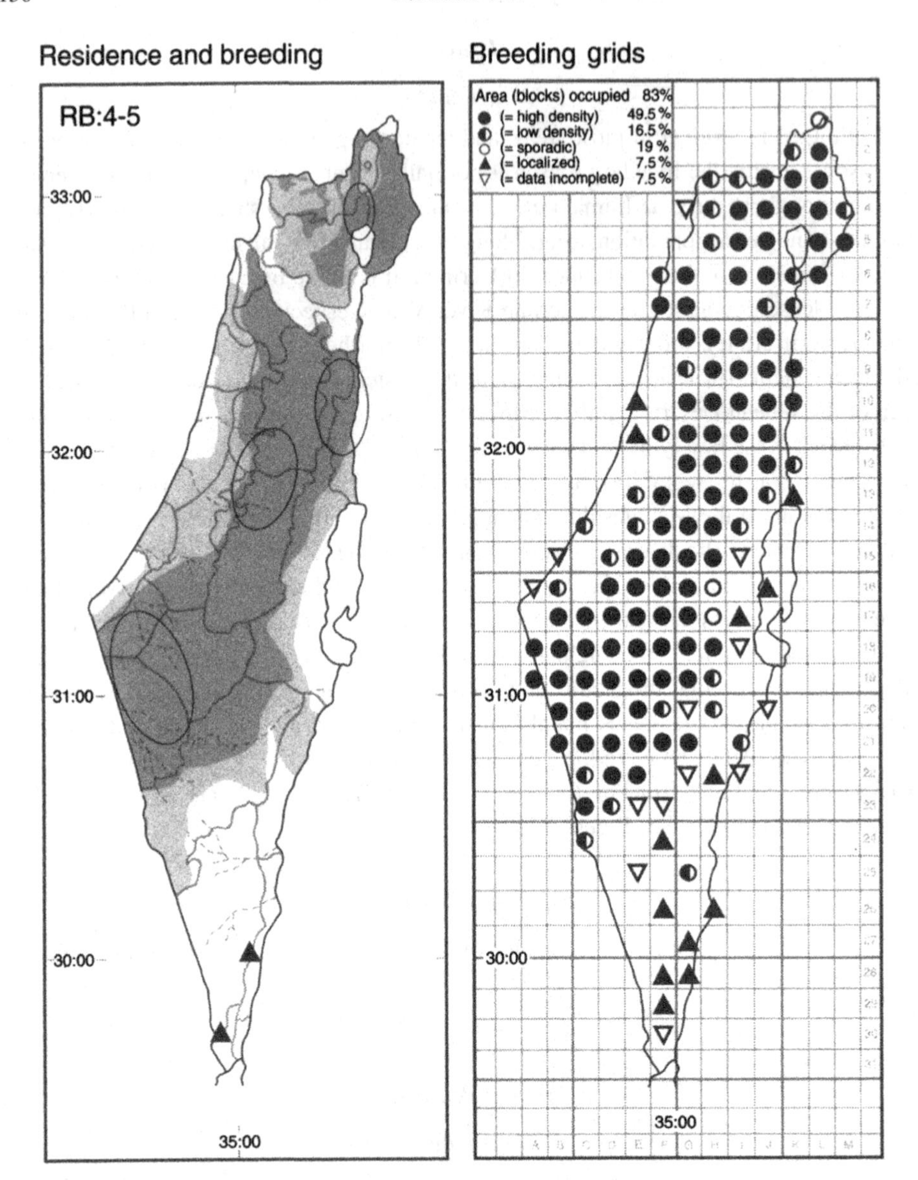

Figure 5.21 Little Owl distribution in Israel (after Shirihai 1996).

Socotra Islands. It is a common bird in the United Arab Emirates where it features a much higher breeding density than is likely in other states except possibly northern Oman (M. Jennings, personal communication). See Figure 5.22.

Population estimates

The total Arabian population may range in the order of 5000–6000 pairs. There is no evidence to suggest that populations are changing in any way, but the huge expansion in

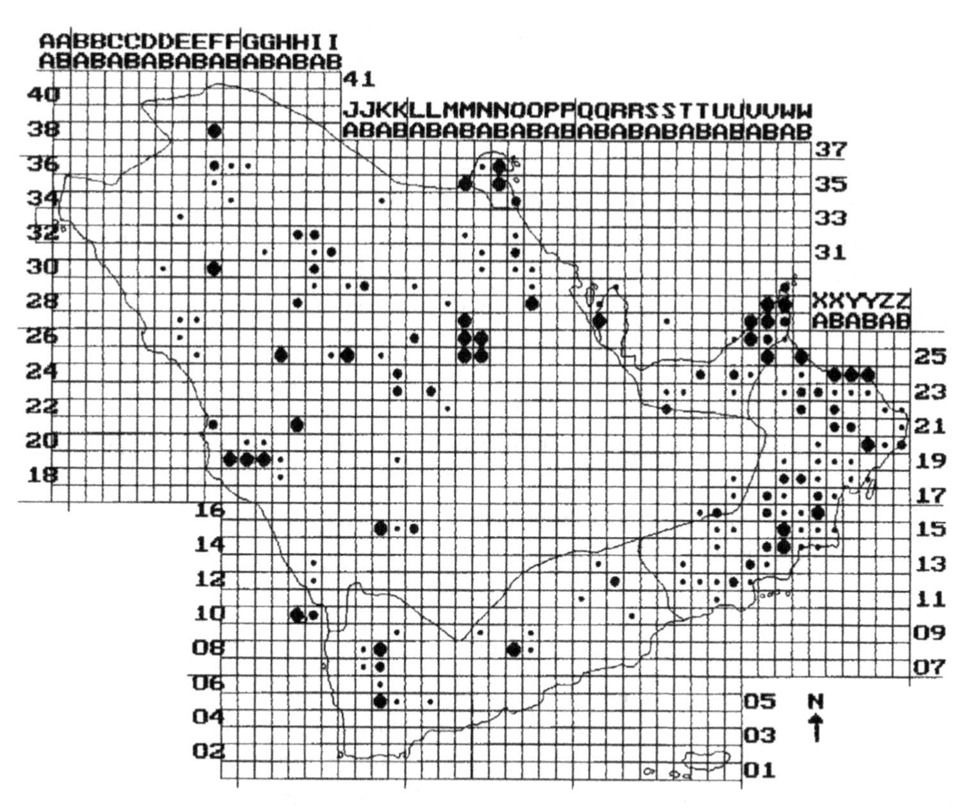

Figure 5.22 Little Owl distribution in the Arabian Peninsula (after M. Jennings, personal communication).

agriculture in central and northern Arabia in recent decades is likely to have allowed many more to breed than did previously. In the United Arab Emirates the population has been estimated at 300–1000 pairs. This population includes an estimated 50 pairs at Jebel Hafit on the United Arab Emirates/Oman border. Population numbers in pairs are: Kuwait, 50; Oman, 1000; Qatar, 50; Saudi Arabia, 3500; United Arab Emirates, 600; Yemen, 500; Bahrain, not available (M. Jennings, personal communication).

North Africa

The species is widespread to the north of the Sahara from Morocco to Egypt, with an extension to the south along the Red Sea coast to eastern Ethiopia and Somalia. The upper elevational limits in its distribution are at 700 m in the Beni Aros Mountains, 1100–1200 m in the High Atlas, 1500 m at Touggana, 1250 m at Iframe, 2200 m to the west of Bon Iblanc, and above 2000 m near Khenifra (Schönn *et al.* 1991).

The exact southern distribution range remains unclear in several countries of northern Africa. The species was reported to Bojador Cape (Spanish Sahara), Ain Najla, Smara, Tifariti, Tarfaia, Seguiet et Hamra, and Guelta Zammour. The species probably reaches south of Rio de Oro and probably further to Mauretania. It is believed to occur south of the 30th parallel in Morocco. In Algeria, the owl is found even more southerly to Zirara, l'Qued Mya. In Libya, the southern boundary of its range remains unclear around Tripoli and Cyrenaica. In the southern area of northern Africa, the distribution is made up of isolated population clusters, e.g., around Timbuktu (Mali), Ahaggar Mountains, Aïr (Azbine), Serin Tibesti, Ennedi, and Jebel Marra (Sudan) (Schönn *et al.* 1991).

Egypt

In Egypt, the northwestern border of the Little Owl's distribution remains unclear. In eastern Egypt, the distribution area covers the delta of the Nile River up to the Sinai, to the Red Sea coast through Sudan, Ethiopia to Somalia. Following the Nile River, the species breeds further south than the Aswan Dam (Schönn *et al.* 1991).

Libya

While the Little Owl is common in the northern part of the country, the southern limit of its distribution remains unclear but is probably around the border with the Oubari desert. In the east, the species is found from Ajadabia to the Egyptian border. The species is believed to breed around Fezzan. Owls observed in Goddua, Traghen, Hun, and Sebha, confirm the sporadic occurrence and migratory behavior in this region (Schönn *et al.* 1991).

Tunisia

The Little Owl is a sedentary breeding species occurring from the Mediterranean coast to the Sahara Desert. It is considered the most common owl species in Tunisia, occurring in open or scarcely forested areas along fields with walls or rocks (P. Isenmann, personal communication). The species is widespread across the country in both the dry, unfertile desert mountains of the south and the green fertile regions of the north. In central and northern Tunisia, the species breeds in olive plantations (Schönn *et al.* 1991). In 2004, 18 pairs were recorded in a 2 km^2 area, surrounding a northern village with fields, piles of stones, and occupied and abandoned houses (B. A. Hajer, personal communication).

Countries with no currently available population data

The global distribution of the Little Owl indicates a presence in the following countries, but no data on the population were available for this project.

Afghanistan
Andorra
Bahrain
Bhutan
Bosnia & Herzegovina
Chad
China
Djibouti
Egypt
Eritrea
Ethiopia
Iran
Iraq
Jordan
Kazakhstan
Kyrgyzstan
Lebanon
Libya
Liechtenstein
Mali
Mauritania
Monaco
Nepal
Niger
Pakistan
Somalia
Sudan
Syria
Tajikistan
Turkmenistan
Uzbekistan

6
Habitat

6.1 Chapter summary

Habitat is one of the main drivers of species presence, and it needs to be able to support entire populations of Little Owls year round. For Little Owls, habitats will typically include open hunting ground rich in small prey, hunting perches, day-roosts and nest-holes, with benign climate and land management regimes that give reasonable long-term continuity. See Figure 6.1. These habitat aspects can be met within a wide diversity of natural and anthropogenic landscapes. Recent declines in numbers and distributions across much of Europe show that tolerable limits are being exceeded. These results are consistent with the view that agricultural change has influenced birds through changes in food quality or quantity (Benton *et al.* 2002).

In this chapter we review the parameters that are of importance for the species, its prey species, and its predators. The favored habitat for the Little Owl varies from the natural landscapes of steppe and arid deserts to anthropogenic areas. The common features are open areas with low grass, perches and cavities in the ground, rocks, trees, or buildings. The species avoids forests, fallow land, and large parcels of arable land. A mosaic effect seems to be important for the species, due to the use of habitat edges, in particular for the richness of prey found there. The relations between the landscape factors will determine local owl densities and demographics. All quantitative studies available were done on anthropogenic habitats. Of natural habitats, only qualitative descriptions were available. We first discuss natural habitats in general terms, then we give an overview of different types of occupied anthropogenic habitats, followed by the actual preference of the species toward certain habitat parameters. The latter studies entail both occupied and unoccupied habitats, while habitat typology studies consider only occupied habitats.

Understanding the parameters that have an influence on the habitat suitability for Little Owls is not always easy since some studies yield contrasting results. The reason for different outcomes of the studies are mostly related to the different local circumstances per study, the size of the area studied, and the spatial scale at which the research was done. We illustrate contrasting results at different landscape scales. While it is recognized that models are just an approximation of the truth, modeling Little Owl presence with different

Figure 6.1 Nest site of Little Owl, Sint-Lievens-Esse, Herzele, northern Belgium. (Photo Trui Mortier.)

Figure 6.2 Typical primary habitat of Little Owl near Chudshirt, Mongolia. Photo Martin Görner.

techniques can help to interpret and understand the behavior of the species at different spatial scales. Multi-scale habitat assessments enhance our understanding of habitat selection, especially when combined with long-term demographic data. Further, models allow prediction of habitat suitability and help to illuminate conservation priorities. Interpretation and communication of the results is enhanced with transparent and repeatable methodologies. The results of such examinations allow us to view Little Owl habitat use from an array of perspectives, and offer insights into key landscape components and associated management options.

In Figures 6.2–6.13 we offer photographs of natural habitats taken in various locations across the Little Owl's range. We have selected these photographs to reflect important and "typical" habitat conditions that have been demonstrated to be of significance to Little Owl populations. Readers will easily notice the striking differences in ecological aspects of the landscapes used by Little Owls.

In Figures 6.14–6.27 we offer photographs of anthropogenic habitats, and habitat elements taken in various locations within the global distribution of the Little Owl. We have

Figure 6.3 Little Owl habitat in the semi-desert of the southern Gobi-Ajmak, Mongolia. Photo Martin Görner.

selected these photographs to reflect habitat conditions that have been demonstrated to be of significance to Little Owl populations.

6.2 Natural habitats

Natural habitats in moderate and warm European regions, northern Africa, and Asia include ravines, gorges, gullies, walls of river terraces, precipitous cliffs, and dry unwooded mountains, as well as dry hilly steppe, semi-deserts, and sandy or clay deserts.

Figure 6.4 Little Owl habitat in the Kysyklum desert of Uzbekistan. Photo A. V. Grazdankin.

Figure 6.5 Little Owl habitat in Uzbekistan. Photo A. V. Grazdankin.

Figure 6.6 Wadi as typical Little Owl habitat in the desert region of Syria. Photo W. Baumgart.

Figure 6.7 Little Owl habitat in New Zealand. Photo P. Child.

Figure 6.8 Little Owl habitat of East Kazakhstan. Photo Paklina & Van Orden. See Plate 1.

Figure 6.9 East Kazakhstan where Little Owls occupy natural cavities. Photo Paklina & Van Orden. See Plate 2.

The few vertical structures, containing cavities, that are available in these habitats are exploited by the species for nesting. In steppe-like areas the species compensates for the lack of these upstanding structures by occupying cavities in the ground dug by mammals.

In Morocco, north of the High Atlas, the species inhabits open land, wadi edges, matorral, palm groves, coastal cliffs, inland cliffs, and rocky hillsides. At elevations of 2000–2600 m at Jbel Yagour and the western High Atlas, the species occurs on upland meadows with scattered boulders and low herbage. In predesert and desert areas it is found on open spiny bush-covered slopes (*Crassicauletum* with *Euphorbia*), palm and Argan groves, open acacia woodlands, cliffs, and embankments (Thévenot *et al.* 2003).

In Somalia the species breeds in holes in steep dry-river banks. Termite heaps are used as perches and to breed in (Schönn *et al.* 1991).

In the Arabian Peninsula Little Owls are found in rocky or broken country or in open well-wooded areas, spending the days in rock crevices and tree holes. They do not frequent precipitous cliffs, and although they have been found up to 2600 m in Yemen and to 2200 m in the northern Oman highlands they seem more numerous in non-mountainous habitats (M. Jennings, personal communication).

In Afghanistan the species is mostly found in the middle–high mountains and along the valleys in the steppe regions where willows, poplars, tamarisks, and loam walls offer breeding sites (Schönn *et al.* 1991).

In Uzbekistan, isolated pairs occupy old wells in the deserts and nest in the holes of rocks, in clayish and sandy deserts, river valleys, on loess precipices, and in the hollows of trees

Figure 6.10 Kyrgyzstan. Photo Paklina & Van Orden. See Plate 3.

Figure 6.11 Ladakh, India between Stok and Shey. Photo Paklina & Van Orden. See Plate 4.

Figure 6.12 Tibetan Plateau as habitat for Little Owl and Kyangs (wild donkeys). Photo Paklina & Van Orden.

Figure 6.13 Upper limit of the distribution of Little Owls at 2100 m in Great Caucasus. Photo Alexander Abuladze.

Figure 6.14 Little Owl habitat in Kruiskerke, Flanders, northern Belgium. Photo Marc De Schuyter. See Plate 5.

Figure 6.15 Little Owl habitat in Butten, the northern Vosges, France. Photo Jean-Claude Génot. See Plate 6.

Figure 6.16 Little Owl habitat in Brakel, Flanders, northern Belgium. Photo Marc De Schuyter. See Plate 7.

Figure 6.17 Little Owl habitat in Kemmel, Flanders, northern Belgium. Photo Marc De Schuyter. See Plate 8.

Figure 6.18 Little Owl habitat in Slovenia. Photo Milan Vogrin. See Plate 9.

Figure 6.19 Little Owl habitat in the Tolfo mountains, Italy. Photo Duccio Centili. See Plate 10.

Figure 6.20 Little Owl habitat with old Almond trees in Luberon, France. Photo Jean-Claude Génot. See Plate 11.

Figure 6.21 The monastery of Stakna as breeding habitat of Little Owl in Ladakh India. Photo Paklina & Van Orden. See Plate 12.

Figure 6.22 Open holm oak "woodland"/parkland (Cabeça da Serra, Castro Verde, Portugal). Photo Ricardo Tomé. See Plate 13.

Figure 6.23 Steppe-like area featured by scattered stone heaps, Portugal. Photo Ricardo Tomé. See Plate 14.

Figure 6.24 Little Owl habitat in Lozère, France. Photo Jean-Claude Génot. See Plate 15.

Figure 6.25 Little Owl habitat in Toscana, Italy. Photo Marco Mastrorilli. See Plate 16.

Figure 6.26 Little Owl habitat in Lombardia, northern Italy. Photo Marco Mastrorilli. See Plate 17.

Figure 6.27 Little Owl habitat near Milano, northern Italy. Photo Marco Mastrorilli. See Plate 18.

(Sagitov 1990). In the desert they nest in burrows of Great gerbils (*Meriones opinus*) and Thin-toed Ground Souslik (*Spermophilopsis leptodactylus*) or in nests of the Rock Nuthatch (*Sitta neumayer*) (Zarudnyi 1896). The species probably digs burrows independently, as it improves existing burrows at the end of the tunnel, creating an internal room or chamber (Pukinskiy 1977).

On the Kazakh plain the Little Owl inhabits various habitats, occurring equally often in the sandy desert, the clayish desert, and in the semi-desert (Osmolovskaya 1953). The main requirement is the availability of any shelter, such as a burrow or deep hole, in which the birds can roost and breed. Most often nests can be found in the vertical walls of wells, in which they use burrows dug by Rollers (*Coracias garrulus*). In the sands it settles in abandoned burrows of the steppe tortoises, foxes, and badgers (Spangenberg & Feigin 1936). In the desert zone the Little Owl willingly settles in river valleys with precipitous banks and tugai groves. The link with the cultural landscape and human settlements is different in the separate Kazakh subspecies. With rare exceptions, the desert subspecies *A. n. bactriana* clearly avoids nesting in settlements, dwelling houses, and industrial constructions. The south-European *A. n. indigena* and the mountain-Asian *A. n. orientalis* subspecies more frequently occupy cultural landscapes and human dwellings (Gavrin 1962). In the northern steppes the species is absent because the tall vegetation inhibits its ability to hunt on the ground (Schönn *et al.* 1991).

In Crimea the owl is a typical rocky-steppe species and inhabits rocky outcrops in steppes and foothills, shell open mines, and coastal precipices (Kostin 1983). The species inhabits the desert mountains and stony deserts of northwestern Mongolia and southeastern Altai (Rogacheva 1988). In Tajikistan the species occupies bank precipices, and also settles in gullies of foothill areas. Loess precipices are the most favorable nesting habitat. In the mountains, it breeds in niches of rocks on pebble slopes (Abdusalyamov 1971). In the Volga-Kama Territory the species is widely spread in the open steppe or forest-steppe habitats (Kulaeva 1977). On the Kyzylkum sands, the species inhabits practically all wintering sites on the grey-ground plain and in the depth of sands. The species breeds in small numbers in "ostantsakh" or isolated remains of mountains, occupying empty cavities in the ground (Gubin 1998).

In the mountains of central and east Asia the species is found on the treeless and shrubless steppes where the Little Owl is a clear ground dweller. It has adapted to the environment by mainly roosting and breeding in burrows of Himalayan Marmots (*Marmota himalayana*), which seems to be a limiting factor for the owl's distribution (Schönn *et al.* 1991).

In the Asian desert areas, the Saxaul sand deserts are frequently occupied by Little Owls. In the Mongolian northern Gobi Desert large undulating sand planes with Saxaul trees (*Haloxylon ammodendron*) intermingle with steep sand cliffs. Holes created by erosion offer nesting cavities. These holes are complemented by burrows dug by mammals such as Pikas (*Ochotana*), Siesel (*Citellus citellus*), Gerbils (*Rhombomys*) and Voles (*Microtus*) in northern Gobi (Schönn *et al.* 1991). Smaller mammals serve as prey for the Little Owls.

6.3 Anthropogenic habitats

The essential criteria of the habitat for Little Owls can be found in many different combinations of landscape parameters. The types of Little Owl habitat described here reflect only those habitats actually occupied by owls.

In central Europe, pastures and meadows flanked by pollarded trees provide ample nest sites, hunting perches, and year-round short herbage with plenty of invertebrate prey (see Figure 6.28). These components reflect optimal habitat, and are characteristic of conservative small farming economies (Exo 1983; Schönn *et al.* 1991).

Around the Mediterranean Sea, the maquis vegetation community is widespread and widely used by the owls. This vegetation is limited to three meters in height and features different species of Oak (e.g., *Quercus ilex, Q. coccifera, Q. suber*) with holes that are extensively used by the owls. Olive plantations are favored in northern Africa (Schönn *et al.* 1991).

On the outskirts of Ankara (Turkey), the Little Owl inhabits, at very high densities, apartment buildings that are under construction (E. Vaassen, personal communication).

The species occurs in settlements, rocky gorges, on forest edges, inhabits solitary buildings, and isolated trees with hollows, amidst fields in Bulgaria. Modern urban landscapes provide good conditions for the species (Iankov 1983).

The Little Owl occupies a great variety of habitat types from sea-level up to 1000 m in Portugal. Frequented habitat types include a great variety of agricultural fields with hedges of different types, and certain regions of extensive crop fields with some trees or rock outcrops. It is found in different kinds of agro-forest systems, mainly the extensive systems of Cork and Holm Oaks, and of Chestnut woods. In these habitat types it is normally more common in areas with old trees featuring cavities, especially in Cork and Chestnut. It also frequents parks and gardens in cities, small towns, and villages. In Lisbon it is not a rare species, frequenting the great urban parks and some gardens (L. Reino, personal communication).

In Spain, the species shows a preference for open habitats, being found in olive groves, cereal fields, orchards, vineyards, vegetable gardens, wastelands, scattered wooded areas, and populated areas. It avoids excessively wet zones, dense forests, and high mountain habitats (Muntaner *et al.* 1983; Olea 1997).

In Britain, the species is found in agricultural countrysides well endowed with hedgerow trees and farm buildings, old orchards with parkland, drained fenland with lines of pollarded willows (*Salix*), and marginal areas such as industrial waste ground, sand-dunes, moorland edges, old quarries, sea cliffs and inshore islands, treeless rising ground, and settlements (Sharrock 1976). It is found only infrequently and temporarily within major cities (Sharrock 1976).

Old orchards are particularly valuable to Little Owls, and their use is widely recognized (Juillard 1984a; Fuchs 1986; Génot 1990b). Even under intensive management, occupancy of orchards by owls may be continued, provided a nucleus of old trees remains. In areas of West-Flanders densely populated by Little Owls, modern low-stem orchards with integrated fruit growing (reduced pesticide use) offer good habitat when nestboxes are provided.

Figure 6.28 Little Owl habitat featuring pollarded willows (François Génot).

The Little Owl mainly occupies agricultural landscapes with tree lines and farm buildings in New Zealand. When shrubs and bushes are present the species also occupies mountains of middle–high altitude. Farmland and plantations with *Pinus radiata*, poplars, and willows offer hunting grounds with commanding perches and breeding cavities. In cities, the owls can easily forage on the ground in parks due to the short vegetation. Around Alexandra, Central Otago, apricot plantations are characteristic habitats. The plantations are mostly surrounded by pine or poplar trees. Near cities old trees and sand cliffs are occupied when available. Goldmining activities during previous centuries created new cliffs and holes in the landscape (Schönn *et al.* 1991).

Mosaic-like landscapes

Landscape heterogeneity is the only determining factor for the species in some areas of France, highlighting the importance of optimal mosaic structure (Ferrus *et al.* 2002) rather than a minimal occurrence of some landscape elements. The species prefers landscapes in Flanders with intermediate openness while they avoid extremely open (e.g., lakes, dunes, large parcels) and extremely closed (e.g., forests, built-up) landscapes.

Mosaic-like landscape structure was found to be more important than the actual land-cover types. The perimeter of the grassland parcels, or the number of parcels rather than their area, have a positive influence on habitat suitability (Table 6.1) (Van Nieuwenhuyse & Leysen 2001). Landscape heterogeneity (convergence of points from three or four areas having different agricultural land uses) and structuring of grassland were related to population density in Austria (Ille & Grinschgl 2001). The number of such parcels per 25 ha had a positive impact on habitat suitability in Deux-Sèvres, western France (Bretagnolle *et al.* 2001). Spatio-temporal rotation in the management of landscape patterns can offer optimal habitats year-round (see Chapter 10), as the species can very quickly react to changing environments by shifting its hunting ranges, even beyond its territories (Finck 1990).

Plasticity of the species in habitat requirements

"Plasticity" here refers to the capability of the Little Owl populations to remain viable, given the relatively wide range of open and semi-open habitats that it occupies across its 9000 km-wide range.

Some research stresses the heterogeneity of the habitats that the owl occupies (Van Nieuwenhuyse & Leysen 2001). A multitude of habitat typologies of the Little Owl in Europe (Table 6.1) clearly confirms the plasticity of the species in most of the European part of its range. Research results also confirm that the small size of the scale is more important than the actual land-use. All typologies feature the essential criteria for suitable Little Owl habitats, which is met within a wide diversity of natural and anthropogenic landscapes and ecosystems (Génot & Van Nieuwenhuyse 2002). Figures 6.29a–g illustrate the seven main types of habitats occupied by Little Owls in Flanders, Belgium.

Table 6.1 *Overview of clearly distinguished types of occupied Little Owl habitats in western Europe (after Van Nieuwenhuyse & Leysen 2001).*

Van Nieuwenhuyse & Leysen (2001)	Van Nieuwenhuyse & Nollet (1991)	Juillard *et al.* (1992)	Mastrorilli (2001)	Centili (1996)	Barthelemy & Bertrand (1997)	Fajardo *et al.* (1998)	Loske (1986)	Ferrus *et al.* (2002)	Ferrus *et al.* (2002)	Ferrus *et al.* (2002)	Ille (1996)	Blache (2004)
Flanders (North Belgium)	West-Flanders (Belgium)	Causse Méjean (Lozère, France)	Bergamo (Northern Italy)	Tolfa mountain area (Italy)	Garlaban hills (15 km from Marseilles city, southern France)	southern Spain	Germany	Northern Vosges (North-East France)	Scarpe-Escaut Plain (north France	Causse Méjean	Austria	Plaine de Valence; Drôme, France
half open grasslands	humid pasture areas with many pollarded willows	pasturelands grazed by sheep with typical mounds of stones "clapas"	woods and zones with a particularly closed tree cover	open meadows	suburban area with houses and gardens	marshland	orchard meadow	urban-type	pastures (20–40% of the area) and arable land in about the same proportion	arable land and pastures in equal proportions (40–60% of the area)	46% arable land and 17% grassland in Marchfeld	mixture of fields, farms and isolated houses, hedges, isolated trees, and roads
grasslands around farms	large isolated meadows with very few trees	hay meadows and cereal fields with "clapas"	urban area including villages and industrial areas	partially wooded	undisturbed hill area with low shrubs and rocky patches	grassland	grassland with pollarded trees	mixed landscape	arable land, pastures, and woodlands are intermingled	large proportions of pastures (60%-80%) and a clearly lower proportion of arable land (up to 20%)	34% arable land, 35% grassland, and 11% vineyards northern Burgenland	mixture of fields and roads

(cont.)

Table 6.1 *(cont.)*

Van Nieuwen-huyse & Leysen (2001)	Van Nieuwenhuyse & Nollet (1991)	Juillard *et al.* (1992)	Mastrorilli (2001)	Centili (1996)	Barthelemy & Bertrand (1997)	Fajardo *et al.* (1998)	Loske (1986)	Ferrus *et al.* (2002)	Ferrus *et al.* (2002)	Ferrus *et al.* (2002)	Ille (1996)	Blache (2004)
cattle breeding landscape without farm buildings	field habitats at the backyards of the linear built-up areas along the roads	abandoned farms with ruined buildings	residential area with parks	covered meadows and cultivated (grass)land		sunflower fields	borders around villages		arable land and pastures and built-up areas representing the built-up and related areas			small-scale mixture of fields, farms and isolated houses, hedges, isolated trees, roads, tree nurseries, and urban zones
horticultural landscape	fields at slightly higher altitude with many farm-buildings	active farms	gardens at the perimeters of the urban areas	cultivated (grass)land		olive tree plantations	grassland with solitary trees					
rural areas with cattle breeding		disused quarries	arable lands with or without irrigation and cultivated or uncultivated grasslands			orchards	isolated farms					
cereals and orchards far away from farms		linear habitats of coppiced trees (*Fraxinus excelsior*)				urban areas	bridges					
cattle breeding landscape in the vicinity of farm buildings						Eucalyptus plantations	churchyards					
							"Lössgrube"					

Figure 6.29a Seven main types of occupied habitats in Flanders (after Van Nieuwenhuyse & Leysen 2001). Grasslands around farms. B: buildings; C: row crops; F: orchards; G: grass; M: maize; N: fallow land; O: other; S: cereals; T: silviculture. Circle includes 25 ha. See Plate 19.

Figure 6.29b Seven main types of occupied habitats in Flanders (after Van Nieuwenhuyse & Leysen 2001). Urbanized cattle breeding. B: buildings; C: row crops; F: orchards; G: grass; M: maize; N: fallow land; O: other; S: cereals; T: silviculture. Circle includes 25 ha. See Plate 20.

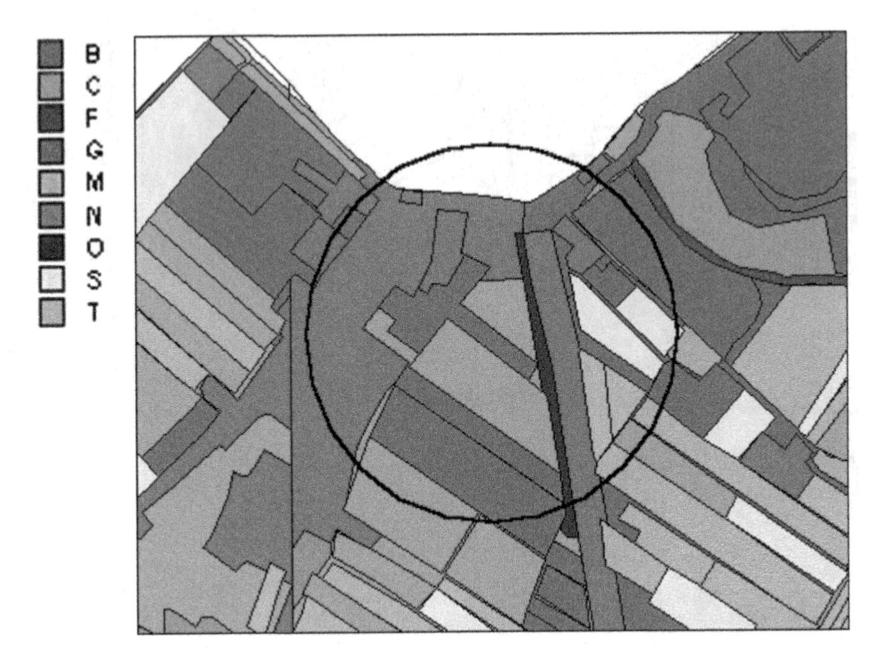

Figure 6.29c Seven main types of occupied habitats in Flanders (after Van Nieuwenhuyse & Leysen 2001). Horticulture. B: buildings; C: row crops; F: orchards; G: grass; M: maize; N: fallow land; O: other; S: cereals; T: silviculture. Circle includes 25 ha. See Plate 21.

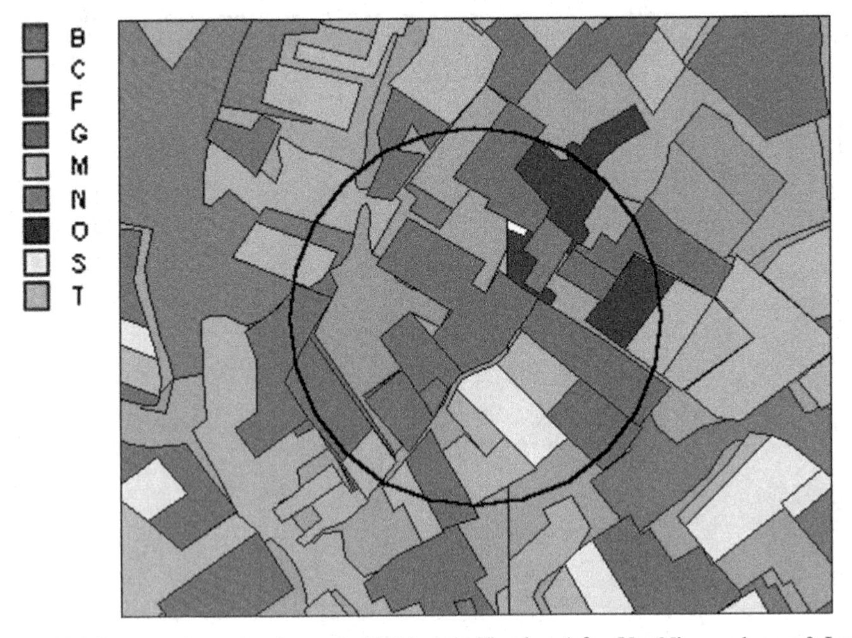

Figure 6.29d Seven main types of occupied habitats in Flanders (after Van Nieuwenhuyse & Leysen 2001). Rural cattle breeding. B: buildings; C: row crops; F: orchards; G: grass; M: maize; N: fallow land; O: other; S: cereals; T: silviculture. Circle includes 25 ha. See Plate 22.

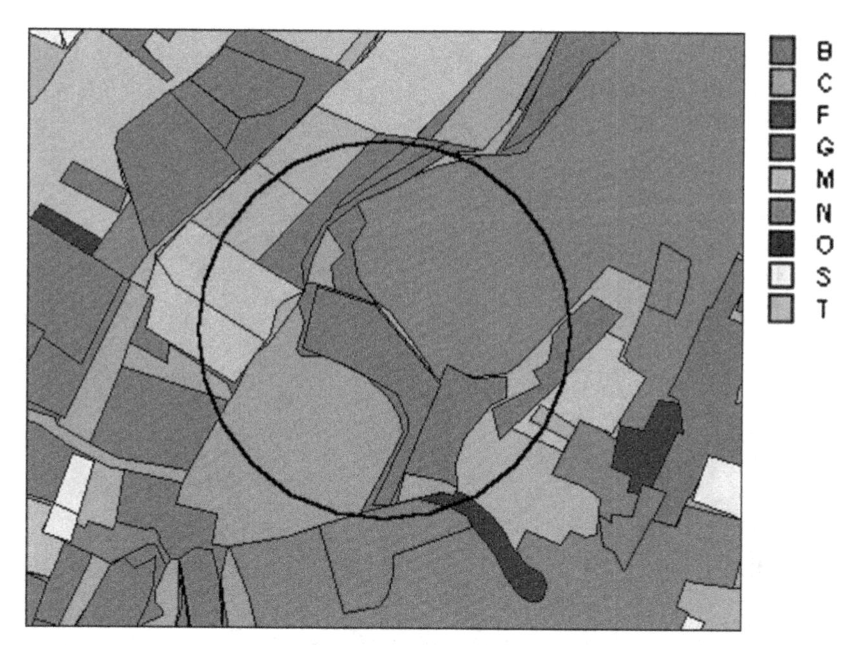

Figure 6.29e Seven main types of occupied habitats in Flanders (after Van Nieuwenhuyse & Leysen 2001). Half open grasslands. B: buildings; C: row crops; F: orchards; G: grass; M: maize; N: fallow land; O: other; S: cereals; T: silviculture. Circle includes 25 ha. See Plate 23.

Figure 6.29f Seven main types of occupied habitats in Flanders (after Van Nieuwenhuyse & Leysen 2001). Remote cereals and orchards. B: buildings; C: row crops; F: orchards; G: grass; M: maize; N: fallow land; O: other; S: cereals; T: silviculture. Circle includes 25 ha. See Plate 24.

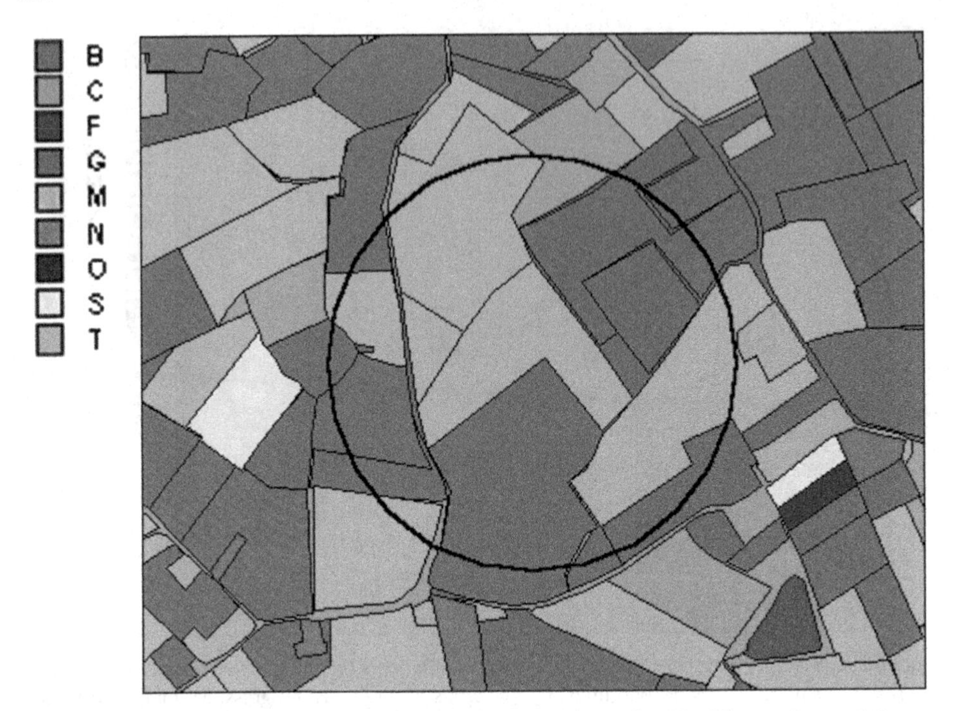

Figure 6.29g Seven main types of occupied habitats in Flanders (after Van Nieuwenhuyse & Leysen 2001). Farmless cattle breeding. B: buildings; C: row crops; F: orchards; G: grass; M: maize; N: fallow land; O: other; S: cereals; T: silviculture. Circle includes 25 ha. See Plate 25.

This heterogeneity illustrates that certain landscape parameters are interchangable. Splitting the habitats into discrete groups or habitat types allows us to identify and quantify the number of different habitat typologies, and furthermore it allows us to select representative sampling points for surveillance of the Little Owl in its environment. In terms of sustainability of inventory and monitoring actions, most of the resources should go to assessing the main habitat types. Monitoring these types allows the tracking of conditions in an objective, representative, and optimal way since both optimal and sub-optimal habitats are included and distinguished. See Figure 6.30.

6.4 Habitat preferences

Essential criteria for suitable Little Owl habitat are year-round prey availability, prey accessibility, vertical landscape structures with cavities, and limited predation pressure. An optimal mosaic of short vegetation and commanding perches for prey observation, tall vegetation for prey reproduction, and availability of secure cavities for roosting and nesting is preferred (Génot & Van Nieuwenhuyse 2002).

Figure 6.30 Little Owl habitat featuring meadows and pollarded willows (François Génot).

Recent studies have clarified the preference of the Little Owl for specific habitat parameters such as areas of land use (e.g., grassland, fields, forest, built-up areas, orchards, surfaces with short/tall vegetation), lengths of linear elements (e.g., roads, tree-lines, fences, field and meadow edges) and absolute number of specific habitat elements (e.g., individual trees, number of cavities, number of perches, number of abandoned buildings) (e.g., Van Nieuwenhuyse & Leysen 2001; Génot & Van Nieuwenhuyse 2002). An array of statistical techniques is used to detect actual habitat preferences from the range of conditions that are available in landscapes. All these techniques try to distinguish occupied from unoccupied landscapes. Hence both occupied and unoccupied parts of the landscape need to be compared. For simplicity, cartographic representations of landscapes are cut into square grid-cells. The landscape within each grid-cell is then quantified and used for analysis (Ferrus *et al.* 2002; Blache 2004; A. Lampe personal communication). The parameter that will be explained is the Little Owl's presence using observed associations between the landscape and the owl's occurrence.

Habitat preference can be studied through a multitude of statistical methods. Most methods use information on occupied and unoccupied habitats in their tests. This way the methods look for parameters that best explain the difference between both types of habitats. Only telemetry uses just the occupied habitats, but analyzes the difference between the used parts of the habitat and those that are not used to distinguish positive and negative impacts of the habitat features.

Figure 6.31a Lubin region study area Poland (G. Grzywaczewski, personal communication).

Telemetry

Telemetry analysis uses observed versus expected use of landscape elements in relation to their availability in the landscape (Finck 1990; Génot & Wilhelm 1993; Orf 2001; G. Grzywaczewski, personal communication). If there were no real preference of the species for a certain vegetation height, then all vegetation heights would be used as proportionally available. A special liking of the species for specific heights of perches can be studied by calculating how many perches are available per height class compared to how many of these are used by the owls.

G. Grzywaczewski (personal communication) measured the vegetation height of an occupied Little Owl territory situated near Lublin. Figures 6.31a,b and 6.32a–d illustrate the habitat, its vegetation, and the use of the habitat by the owl.

Counting the available grid-cells per vegetation height yields the expected use by the owl if no preferential vegetation height exists. Counting the observed use of the grid-cells by the owls yields the observed use. Table 6.2 illustrates that the Little Owl features in a higher

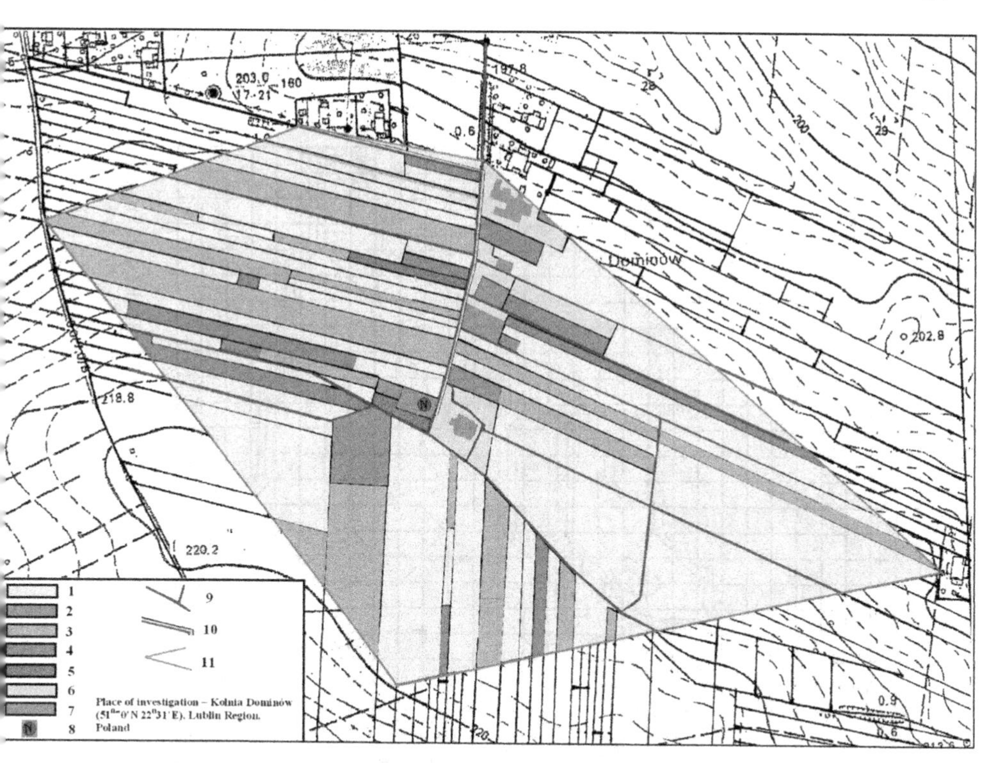

Figure 6.31b Example of Little Owl habitat in Kolnia Dominów (51° 70′ N 22° 31′ E), Lublin Region, Poland in 2001 (G. Grzywaczewski, personal communication).

1. cereals – rye, barley, wheat, oat, cole
2. vegetable garden
3. barren
4. root crops – potatoes, sugar-beet
5. trees
6. grass
7. buildings
8. nest
9. fieldway
10. lighted fieldway
11. border of maximum home range

proportion of used grid-cells with a limited vegetation height, compared to the proportion of this habitat that is available in its territory.

Median test

The median test (Van Nieuwenhuyse & Bekaert 2001; Bekaert 2002) sorts the habitats both occupied and unoccupied, and labels the observations above the median and below the median differently (Table 6.3).

A label is also given to every occupied and unoccupied habitat. A frequency table shows the relative distribution of a parameter by function of its occupation (Table 6.4).

Finally associations between each landscape parameter and its occupation can be derived from the frequency table. When each cell contains a similar amount of observations there

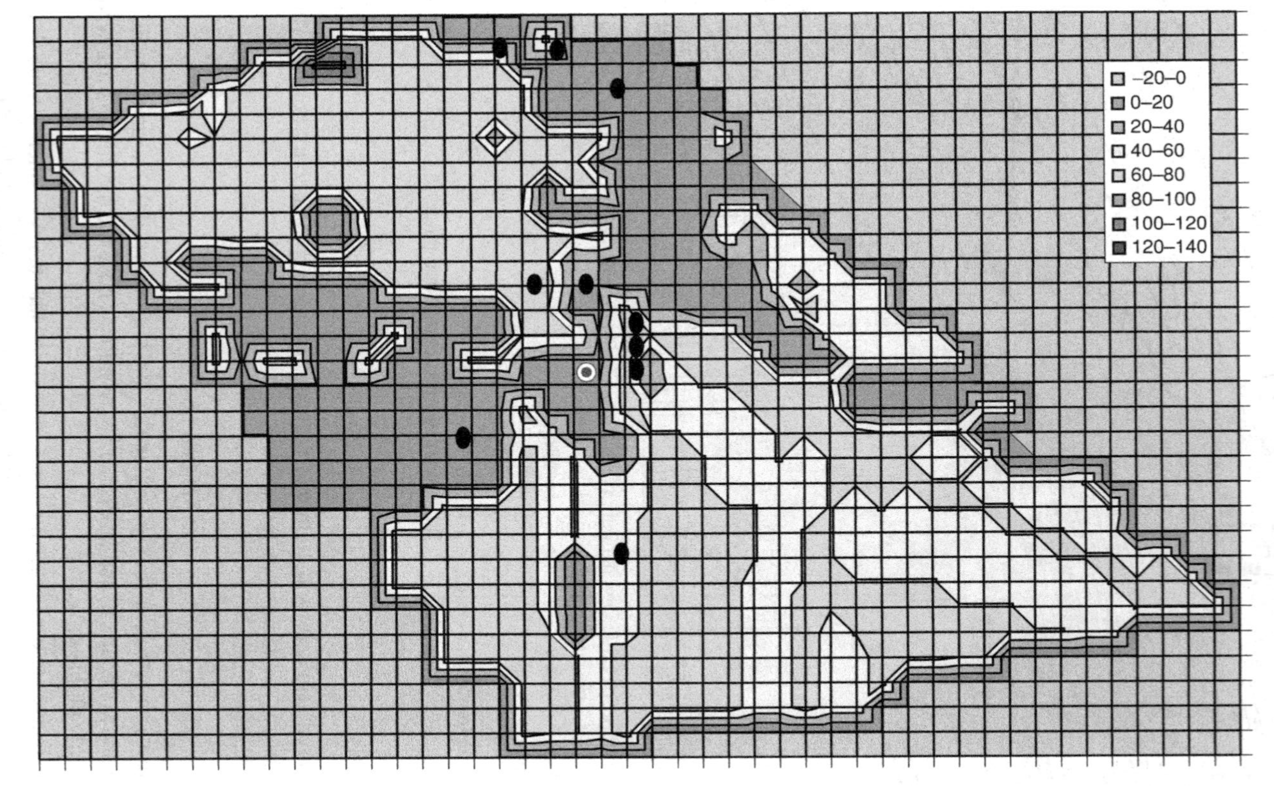

Figure 6.32a Vegetation height (cm) and owl observations at a Little Owl territory in Kolnia Dominów (51° 70′ N 22° 31′ E), Lublin Region, Poland in 2001 (G. Grzywaczewski, personal communication). During the incubation period 21 April–11 May. ◉ Nest site. See Plate 26.

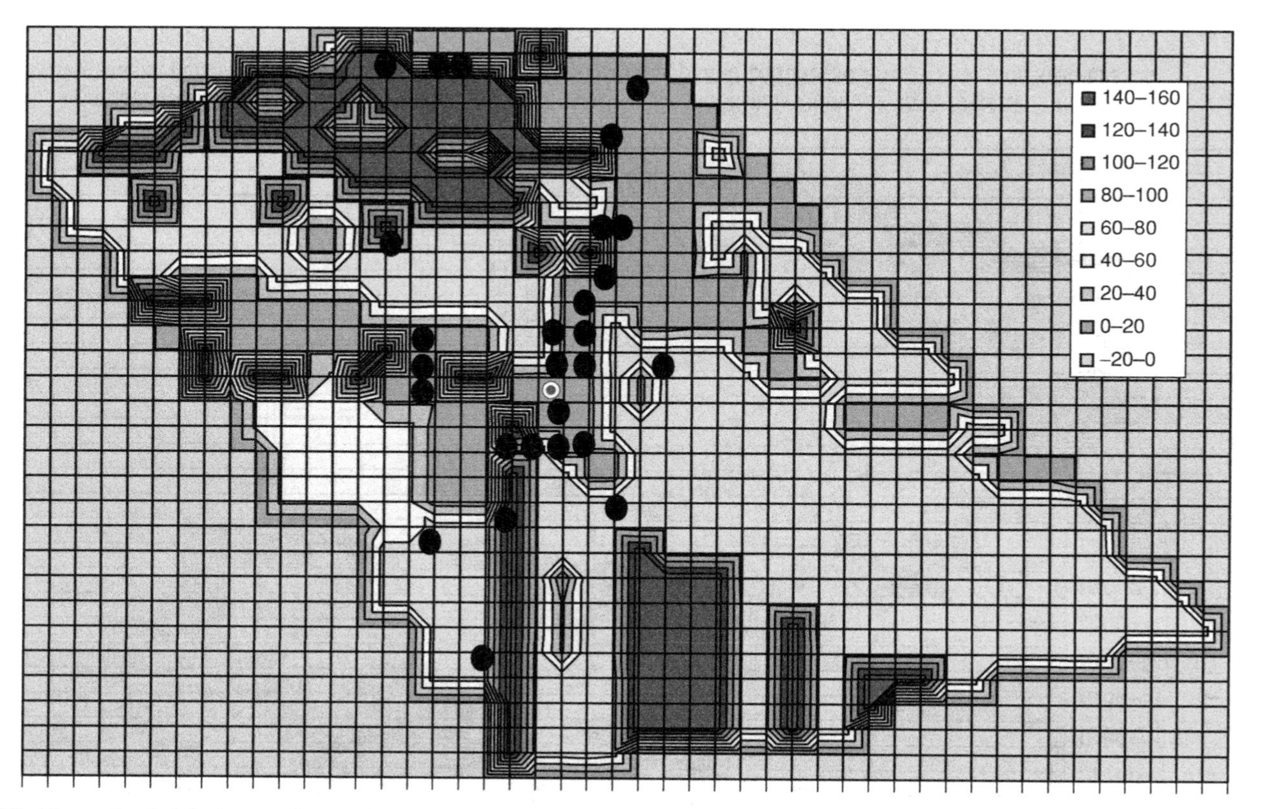

Figure 6.32b Vegetation height (cm) and owl observations at a Little Owl territory in Kolnia Dominów (51° 70′ N 22° 31′ E), Lublin Region, Poland in 2001 (G. Grzywaczewski, personal communication). During the nestling period 12 May–25 June. ◉ Nest site. See Plate 27.

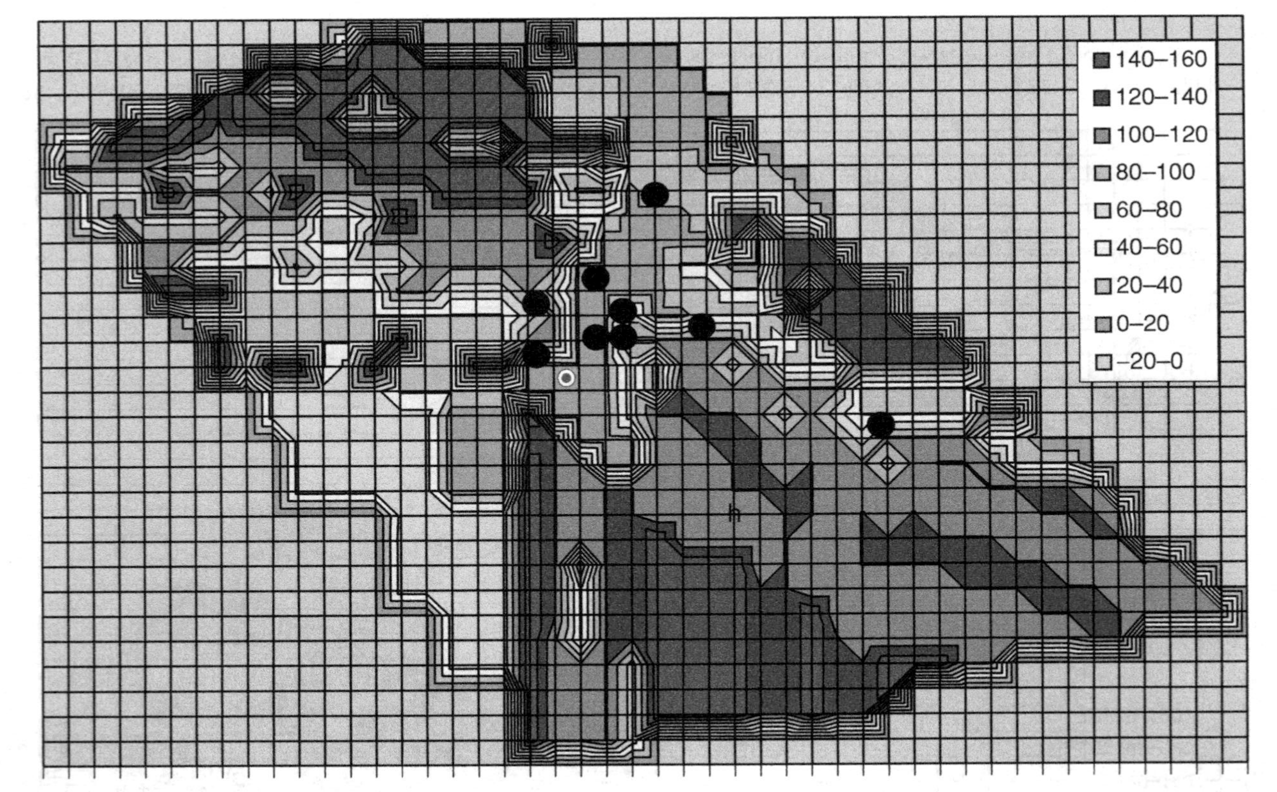

Figure 6.32c Vegetation height (cm) and owl observations at a Little Owl territory in Kolnia Dominów (51° 70′ N 22° 31′ E), Lublin Region, Poland in 2001 (G. Grzywaczewski, personal communication). During the early fledgling period 26 June–30 July. ◉ Nest site. See Plate 28.

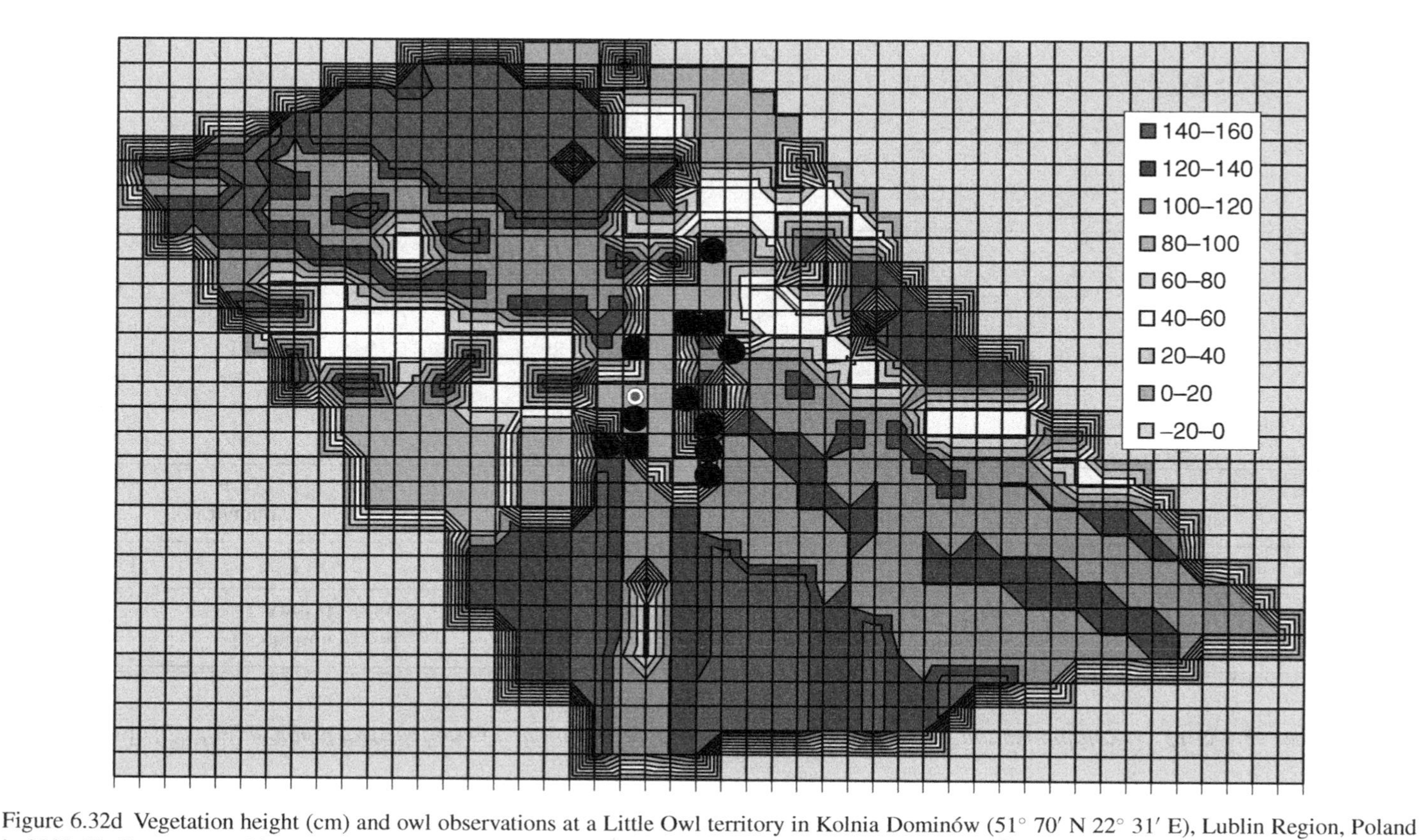

Figure 6.32d Vegetation height (cm) and owl observations at a Little Owl territory in Kolnia Dominów (51° 70′ N 22° 31′ E), Lublin Region, Poland in 2001 (G. Grzywaczewski, personal communication). During the dispersion period 31 July–15 August. ◉ Nest site. See Plate 29.

Table 6.2 *Proportional distribution of habitat use by Little Owls by function of habitat availability (G. Grzywaczewski, personal communication).*

home range (ha) – mean	Incubation period 21 April–11 May		Nestling period 12 May–25 June		Young fledglings 26 June–30 July		Dispersing young 31 July–15 August	
	22.3 [n = 2]		37.5 [n = 3]		17.0 [n = 2]		1.3 [n = 1]	
vegetation height (cm)	availability (%)	used (%)	availability (%)	used (%)	availability (%)	used (%)	availability (%)	used (%)
0–20	44	92	40	96	38	86	47	89
21–40	2	0	0	0	18	0	0	0
41–60	2	2	5	0	3	0	11	0
61–80	53	6	38	3	7	7	0	0
81–100	0	0	0	0	0	0	0	0
101–120	0	0	0	0	18	5	31	11
>120	0	0	17	1	16	2	11	0

Table 6.3 *Determining the median grid-cell for each landscape parameter and labeling of the grid-cells above or below the median (after Bekaert 2002).*

x	y	Pasture	Label pasture	Little Owl presence	Label Little Owls
622	391	71	above median	1	occupied
641	391	195	above median	1	occupied
651	391	254	above median	2	occupied
671	391	260	above median	1	occupied
612	382	263	above median	3	occupied
622	391	295	above median	0	unoccupied
612	391	297	above median	0	unoccupied
611	391	301	below median	1	occupied
661	391	301	below median	2	occupied
642	391	322	below median	1	occupied
611	382	324	below median	0	unoccupied
652	391	345	below median	0	unoccupied
631	391	356	below median	0	unoccupied
621	391	364	below median	0	unoccupied

Table 6.4 *Frequency table indicating how many grid-cells occur with above and below median pasture by function of the occupation by Little Owls (after Bekaert 2002).*

Cell percentage

Row percentage

Column percentage

	Occupied	Unoccupied	Total
above median	58	27	85
	34%	16%	50%
	68%	32%	
	47%	56%	
below median	65	21	86
	38%	12%	50%
	76%	24%	
	53%	44%	
	44%		
total	123	48	171
	72%	28%	100%

Table 6.5 *Determination of associations between Little Owls and a landscape parameter (after Bekaert 2002).*

	Occupied	Unoccupied
A		
above median	=	=
below median	=	=
B		
above median	+	−
below median	−	+
C		
above median	−	+
below median	+	−

=: equal distribution of grid-cells over four quadrants
+: proportionally more grid-cells in this quadrant
−: proportionally fewer grid-cells in this quadrant
A: no associations
B: positive association between landscape element and occupancy
C: negative association.

is no association between Little Owls and the landscape element, given the fact that there are almost as many occupied as unoccupied grid-cells. (Table 6.5)

Prey availability and prey accessibility

A Little Owl habitat needs to have sufficient food for both territory-owners and their offspring. As Little Owls eat live prey (or sometimes recently dead animals), these prey animals also have to be able to live in a sustainable way, maintaining viable populations through time. Beside having prey at their disposal, Little Owls also need to be able to access and catch the prey. As nearly all of the Little Owl's prey are ground-dwelling species, the owl needs to catch its prey in short vegetation. To catch prey in an energy-efficient manner, the owl uses the "perch-and-pounce" hunting strategy. It also uses the terrestrial, walking "on foot", hunting method (see Figure 6.33). While hunting for rodents and insects, the birds tend to be perched. They are perched on a relatively low perch (1–3 m) to obtain a commanding view over their immediate surroundings. Once they discover prey, they pounce on it from the perch. Being positioned above the ground allows the owls to discover their prey visually. Kestrels (Village 1990) and Barn Owls (Shawyer 1998) are able to hunt more in flight, rather than being perched, and have fewer problems catching prey in tall grass. Due to its limited ability to fly above tall grassland, its nocturnal lifestyle, and its more limited hearing capabilities, the Little Owl is more dependent on its sight for hunting.

Figure 6.33 Little Owl running on the ground (François Génot).

In western Europe, earthworms make up an important part of the owl's diet allowing it to hunt successfully on bare ground or in short grasslands. Ideal hunting grounds for the species are to be found at the borders of parcels. These borders function as ecotones with a much higher prey diversity than in the center of the parcels. Ideal habitats are small-scaled pastures with year-round grazing and very small parcel sizes (e.g., <1 ha), and hence with a lot of parcel border for a given area. These small-scaled heterogeneous conditions combine prey presence with prey accessibility. Larger grasslands with tall vegetation favors voles, and hence is more preferred by Kestrels, Harriers, Barn Owls, and Short-eared Owls.

Prey accessibility

Prey accessibility for the species is maximized by short vegetation with commanding perches that allow either "perch-and-pounce" or the "walking" hunting techniques. The Little Owl hunts mainly on the ground for earthworms or insects, and prefers short vegetation or bare ground for this. Short vegetation can be found as grazed grasslands of all kinds, preferably with an average plot size of <0.6 ha (villages with the highest population densities had an average plot size of 0.55 ha) (Dalbeck *et al.* 1999). Preference for pastures over other types of grasslands, especially hayfields, shows the importance of year-round prey accessibility (Dalbeck *et al.* 1999). During the six summer months, utilization of arable land by owls in Germany is disproportionate to its availability. In the summer, territories are reduced in size, with the owls concentrating on allotments that support optimal accessibility of food (Finck 1990; Orf 2001). As the habitat quality increases, active home ranges tend to decrease in size, with the owls using grassland with short vegetation within a radius of 400 m around the nest (Orf 2001). Depending on the relative area of fields and grasslands, both are sustainable habitats as long as there is heterogeneity in time and space. Sometimes even gardens provide suitable foraging habitat (Orf 2001). During some periods of the year, road edges provide a large part of the short or absent vegetation, making hunting owls very vulnerable to vehicle collisions (Hernandez 1988; Clech 2001b). In Germany, newly created vineyards (Weinberg area) were extremely attractive to Little Owls in the first years after creation. The vineyards were characterized by bare ground with poles shorter than 2 m placed at regular intervals (Eick 2003). Vegetation cover in foraging grounds of southern Spain generally does not exceed 2 cm in height; use of areas with cover higher than this was not observed. The most favored hunting grounds consisted of a heterogeneous mosaic of bare ground with short grass patches (Fajardo *et al.* 1998).

Prey availability

Insect prey density and diversity is related to population density of owls in Austria, and the density of small mammals showed a positive association with average reproductive success (Ille & Grinschgl 2001). Grimm (1986) found more and larger prey in a meadow edge beside a brook than in grasslands. The impact of cattle trampling yielded more *Carabid* ground beetles. Tall vegetation is crucial for small mammals, with edges tending to have larger diversities and densities (Dalbeck *et al.* 1999).

Earthworm densities are always higher in grassland than fields because plowing leads to higher oxidation rates and the associated reduction of organic material, and hence less food for earthworms due to reduced humus levels. A higher use of pesticides, and higher predation and mortality are typically associated with plowed fields. Densities in grassland depend on the fertility of the soil (a balanced nutrient influx yields higher densities than in unfertilized grasslands), and higher densities are found with organic fertilizer rather than inorganic, due to direct intake of the former by earthworms (Edwards & Bohlen 1996).

Action to conserve vole numbers has no effect on either productivity or population regulation of Little Owls in the lower Rhine area, Germany (Exo 1987), perhaps because the species predominantly feeds its young on earthworms (*Lumbricidae*) and beetles (*Coleoptera*) (Juillard 1984a) or due to the fact that voles found there are in lower densities because of the regular flooding of the Rhine (M. Exo, personal communication). Research by Génot (2005) in the northern Vosges of France did not reveal any relationship between vole numbers and the Little Owl. Zwölfer *et al.* (1981) showed that hedges and other vegetation gradients have a significant positive impact on prey species in terms of both density and diversity.

Permanent tall vegetation is needed for the breeding grounds for the owl's prey (and thus indirectly for prey availability), but reduces the accessibility of the owl. Hence a combination of short and tall vegetation (with accessibility from commanding perches) yields the highest diversity of prey numbers and species (Van Nieuwenhuyse *et al.* 1999). Edges have a special importance for the owl, in this context edges reflect hedges, walls, and a high proportion of edge structures relative to the area (Dalbeck *et al.* 1999); meadow-edges, orchard-edges

(Van Nieuwenhuyse & Bekaert 2002); field-edges (Van Nieuwenhuyse & Bekaert 2001); tree lines and field-edges (Van Nieuwenhuyse *et al.* 2001a); roads (Hernandez 1988); pasture fences (Loske 1986); hedged open fields (Zuberogoitia & Campos 1997); length of hedges in the Netherlands (Visser 1977); and vertical structuring of grasslands (Ille & Grinschgl 2001). This is probably due to a mix of different parameters – a high diversity of prey breeds in the tall vegetation at high densities, prey becomes visible and reachable in the shorter vegetation nearby, and can be caught easily from fence poles along the edges of two adjacent parcels.

Snow

Most Little Owls have difficulties in surviving more than three weeks of snow cover (Helbig 1981; Schönn 1986). In the former Soviet Union, Gusev (1952) showed a sharp correlation between snow cover depth and hunting success. With a snow depth of 0.5–3 cm, the hunting success of the Little Owl on rodents was higher than in snow-free periods. With shallow snow cover, rodents cannot dig through it and have to move to the surface, where they become clearly detectable on the white background and therefore are easily caught. Friable snow, 7–9 cm deep, sharply decreases the hunting success of Little Owls because small rodents dig tunnels in the snow and seldom emerge at the surface. Snow cover from 12–70 cm further reduces hunting success. However, in deeper snow, the Little Owl might hunt more successfully, for example when melting water refreezes, inhibiting rodents from making tunnels through the snow, and forcing them up to the surface.

Perches

The availability of hunting perches is reported as varying in importance. In central Europe, the species seems to be mainly a ground forager (Exo 1991; Schönn *et al.* 1991) probably because of the relative importance of earthworms in their diet (Exo 1991). Mud-covered entrances to nestboxes in West-Flanders during the nesting period suggests the importance of walking as a hunting technique. Some German studies showed no correlation between the presence of fence poles and the species (Dalbeck *et al.* 1999) while other studies showed a clear positive relationship (Loske 1986). This difference might indicate that other landscape structures, e.g., trees or shrubs, are interchangable for fence poles to serve as perches. In Vosges du Nord (northeastern France) males spend less time on perches at the end of spring and beginning of summer (Génot & Wilhelm 1993). Up to 82% of perch use is on trees, 14% on poles, and 4% on hedges year-round. Fruit trees offer shelter in addition to perches. One radio-tracked female used fence poles significantly more during the nesting than the courtship (prebreeding) period. Modern low-stem orchards with integrated fruit growing (reduced pesticide use) in Flanders (Génot & Van Nieuwenhuyse 2002) and Germany (Orf 2001), and vineyards in Austria (Ille & Grinschgl 2001) and Germany (Eick 2003), offer good hunting grounds due to the short vegetation and the vine stocks or poles that are used as commanding perches. These areas also serve as breeding habitats as long as nest cavities are available.

In Spain, the Little Owl forages mostly from perches (Fajardo *et al.* 1998), probably because of a limited importance of earthworms in its diet. Anti-predator strategies might lead owls to avoid perching on the ground, where they are more vulnerable to predators. Perching sites above 3 m are seldomly used in southern Spain. No change in height of perches was observed during the summer months, when road mortality is at its maximum.

R. Tomé (personal communication) studied the diurnal foraging behavior of Little Owls in two Mediterranean habitats (a pseudo-steppe and a Holm Oak woodland) with different perch characteristics in Portugal. During 1999, 19 observation periods involving Little Owls hunting insects were recorded for each area. He investigated which type of signals Little Owls used for detecting prey, with Little Owls using both visual and acoustic cues while hunting. The use of visual hunting was indicated by the selection of the highest possible perches in the pseudo-steppe, by the positive relationship between perch height or vegetation coverage and giving-up times in the woodland area, and by the positive association between perch height and attack distance in both sites. The use of acoustic hunting was indicated by similar giving-up times in both areas. In spite of major differences in perch height between the areas, owls apparently were able to optimize their foraging behavior in both habitats, maintaining similar detection times and hunting success. Hunting success was probably affected by prey-specific characteristics.

In the district of Modena (Italy), a study of 41 song perches showed that 46.3% were rural buildings, the number of frequently used perches was positively correlated with the extent of the orchard (Selmi *et al.* 2005).

Cavities

Cavities are crucial for the species, as Little Owls use cavities or crevices to breed in and to roost in during the day. Relatively small holes (e.g., 6–7 cm diameter) are preferred for nest cavities. Since the owls do not build their own nests, they are dependent on natural or artificial cavities. Competition for cavities with other bird and mammal species forces the owls to take cavities with small entrance holes. The optimal size of the entrance to the cavity is a balance between the ease of passage of the owls and limiting the access of predators.

Positive correlations have been found between Little Owl densities and the age of trees (rather than the number of trees) in Düren (Northrhine-Westphalia) (Dalbeck *et al.* 1999), with the number of pollarded trees in Northrhine-Westphalia (Loske 1986), and with potential breeding cavities and grasslands (Exo 1983). The Little Owl adapts to nestboxes easily, but this had no influence on population density in Northrhine-Westphalia (Dalbeck *et al.* 1999). The absence of impact might have been related to the availability of natural cavities. No correlation was found between the number of nest sites and density in the Netherlands (Fuchs 1986). In Germany, juvenile Little Owls utilize the crevices in piles of firewood regularly for daytime roosting (Eick 2003).

The addition of nestboxes can have a profound effect on Little Owl populations (Stange 1999; Bultot *et al.* 2001; Génot 2005). The provision of nestboxes can initially appear to increase the local population, as birds without any nest sites readily occupy the boxes (Stange 1999; Bultot *et al.* 2001). Over a period of several years, it appears that the resident owls switch from using natural cavities to nestboxes. Further, the social dynamics of the species (i.e., the apparent desire of the species to form population clusters) is affected; if the availability of natural nest sites is low, owls from outside the local study area will abandon their territories and shift towards the nestbox-rich area. These types of movements were observed by owls in an area of *c.* 200 km^2 in northern France (Génot 2005) over a 20-year period.

Negative habitat relationships

The density of Little Owls is suppressed by factors that decrease the accessibility and availability of prey, decrease the number of nesting cavities, or increase the influence of predator or human actions. The increase in the amount of arable land has had a negative impact in Germany; further, the presence of trees are positively correlated with meadows and negatively correlated with arable land (Loske 1986; Van Nieuwenhuyse & Bekaert 2001). The owl avoids large rape and mustard seed fields in Deux-Sèvres, western France (Bretagnolle *et al.* 2001). It also avoids larger forests, thus the spread of Little Owls in Europe only occurred after the principal deforestation period of the ninth and tenth centuries (Schönn 1986). The cessation of pastoral farming practices (e.g., husbandry of sheep and cattle) results in a reduction in the amount of short vegetation (Juillard *et al.* 1992). The owl avoids larger cities in the UK (Sharrock 1976), larger built-up areas in Flanders (Van Nieuwenhuyse & Bekaert 2002), traffic (Fajardo *et al.* 1998; Clech 2001b), forest (Schönn *et al.* 1991; Van Nieuwenhuyse & Bekaert 2001, 2002), and villages with few but large grassland areas (Dalbeck *et al.* 1999). The only exceptions are when these areas are surrounded by "traditional" farming practices, i.e., environments where different land uses cover small areas, providing heterogeneous habitats within small areas (Ferrus *et al.* 2002). In the southern part of its distribution, the Little Owl is also found in large cities (Rome, Italy; Budapest, Hungary; Bucharest, Romania; Sofia, Bulgaria; Ankara, Turkey). Built-up areas and maize (*Zea mays*) culture in Flanders (Belgium), up to an optimal coverage of 0.21 ha per 25 ha and 0.65 ha per 25 ha, respectively, have had a positive impact on the presence of owls. The probabilities of finding Little Owls in areas larger than this decrease (Van Nieuwenhuyse & Leysen 2001).

Predator avoidance

The structural diversity of the landscape needs to offer enough security for Little Owls to cope with predators. Landscapes with a multitude of cavities limit the chance of predation. The breeding success in piles of stones near human settlements in Portugal might be higher due to some possible predator avoidance (Tome *et al.* 2004). Apparently suitable but unoccupied habitat exists in Biscay (northern Spain) within the distribution area of the Tawny Owl. Attacks on Little Owls by Tawny Owls have been observed on three occasions (Zuberogoitia & Campos 1997a). Mikkola (1983) reported Little Owls in the Tawny Owl diet, and Schönn (1986) reported the absence of overlapping territories between the two species. In Spain, perches are nearly always used in response to possible ground predators (Fajardo *et al.* 1998). This argument is also used to explain variations in daytime activity in southern Spain (Negro *et al.* 1990).

Edge effects and landscape patterns

Many studies have examined the patterns and processes of landscape fragmentation (e.g., Turner & Gardner 1997; Turner *et al.* 2003) and the impact of habitat edges on bird communities (e.g., Hudson 1991; Sparks *et al.* 1996; Fuller *et al.* 1997; Lehman *et al.* 1999; Baker *et al.* 2002; Berg 2002a; Schroth 2004).

Hedges are an important landscape element that can have a favorable impact on species diversity (Hinsley & Bellamy 2000). Increasing the structural complexity of a hedgerow and its associated habitat elements may reduce the incidence of predation on birds. Hedgerows also provide physical shelter and are an important source of winter food supplies, especially berries and other fruits. The Eurasian Pygmy-owl (*Glaucidium passerinum*) and European Eagle Owl (*Bubo bubo*) have been found to hunt disproportionately along the edges between forest and open areas (e.g., Mikkola 1983; Burton 1992; del Hoyo *et al.* 1999; Dalbeck 2003).

The reproductive success in the Loggerhead Shrike (*Lanius ludovicianus*) was lower from nests along fence lines than from nests found free-standing and distributed at random (Yosef 1994) due to greater rates of predation. Where nests of breeding species are concentrated in a linear fashion, it should be most profitable for a predator to search along such corridors. Hole-nesting species were found to be more at risk at the edge of forests than within the forests due to a higher predation risk (Sandstrom 1991). Similarly, increased nest predation was found in neotropical migrant birds at a forest-stream ecotone (Gates & Giffen 1991).

Since Little Owls are both predator and prey they need to seek a profitable trade-off. When predators are limited, breeding in coppiced tree-lines may be extremely positive since both prey abundance and detectability are favorable. To minimize the risk of nest predation on the owl itself, the habitat should offer plenty of linear elements rather than just a few. Hence mosaic-like landscapes probably will feature a maximal amount of edges to reduce the risk of predation on the owl.

Some landscape elements have a different impact on the species in different concentrations. Farm buildings and their surroundings had a positive impact on Little Owls in France (Bretagnolle *et al.* 2001). In the absence of natural cavities in Flanders, the species is attracted to farm buildings during winter and to breed in. In small numbers the presence of farm buildings had a positive impact on Little Owl occurrence, but when this land-cover surpasses 0.21 ha per 25 ha, the impact becomes negative (Van Nieuwenhuyse & Leysen 2001). A differential impact of roads on the presence of species was found, according to traffic loads in Haut-Léon, western France, where Clech (2001b) found that a 2 km-wide area bordering dual-carriageway roads and a 500 m area around a major road were not occupied, while roads carrying limited traffic were not avoided.

6.5 Assessments at three spatial scales: nest site, home range, and landscape

The multi-scale approach to studying habitat selection is based on the conceptual framework suggested by Johnson (1980), whose basic assumption was that animals are capable of making decisions regarding resources at consecutively smaller scales. Consequently, general habitat selection can be regarded as a hierarchical process, e.g., a fitting patch for breeding at a small scale and appropriate areas for foraging at a larger scale (Martínez & Zuberogoitia 2004b). The multi-scale approach may be especially useful in identifying key factors involved in habitat preferences of owls because they have large home ranges usually consisting of different patches for breeding and foraging (Mikkola 1983; Taylor 1994; Scott 1996). Importantly, some of the apparently conflicting results of habitat preference identified in different research areas are potentially due to researchers working at different spatial scales. Analysis of habitat preferences at different spatial scales within the same research area can reveal this. The idea of different scales was introduced by Schönn *et al.* (1991) who identified three aspects of the species' ecology that a habitat must provide for successful survival, i.e., reproduction, foraging, and roosting. The three areas do not necessarily have to be situated in concentric circles. See Figure 6.34.

Multi-scale habitat assessments

Multi-scale habitat assessments have become a common analysis tool in owl conservation and management. Most often, these assessments reflect the enumeration of habitat conditions within circles of different sizes that are centered over a point in the landscape (e.g., the nest site) and determined from vegetation maps, aerial photographs, or satellite imagery. The smallest circle size is often 0.1 ha and it is common practice for researchers to go into the field and physically measure the types and size of vegetation or other structures at this scale. Subsequent larger circles reflect the average home-range size of a pair of the owl species under review, and a circle that reflects a larger landscape size (e.g., to assess the characteristics of watersheds, soils, and climate). Particularly valuable are assessments where comparisons are made between habitat conditions around owl sites to those around random points in the landscape. Sufficient sample sizes are critical in determining the statistical significance of the observed differences.

Not all nesting territories are of equal benefit to the local owl population (see Newton 2002, 1998 for a review of limiting factors in owls and birds, respectively). Researchers have observed that for some species of owls, relatively few nesting territories support most of the reproduction for a given population. The results of multi-scale habitat assessments

Figure 6.34 Small-scaled landscapes as favorable Little Owl habitat, Flanders, northern Belgium. Photo Ludo Goossens. See Plate 30.

benefit significantly by linking long-term reproductive data from specific nest-site locations This type of information is particularly relevant in the development of specific habitat management recommendations for the species of interest.

We offer a few examples of multi-scale habitat assessments on owls:

Barn Owl (*Tyto alba*) – Van Nieuwenhuyse *et al.* 2004
Sooty Owl (*Tyto tenebricosa*) – Lyon *et al.* 2002
Northern Spotted Owl (*Strix occidentalis caurina*) – Ripple *et al.* 1991; Ripple *et al.* 1997; Swindle *et al.* 1999
Great Grey Owl (*Strix nebulosa*) – Stepnisky 1997
Flammulated Owl (*Otus flameolus*) – Wright *et al.* 1997
Powerful Owls (*Ninox strenua*) – Lyon *et al.* 2002; Soderquist *et al.* 2002
Barking Owl (*Ninox connivens*) – Taylor *et al.* 2002a, b
Long-eared Owl (*Asio otus*) – Martínez & Zuberogoitia 2004b
Little Owl (*Athene noctua*) – Blache 2004; Martínez & Zuberogoitia 2004b; Cornulier & Bretagnolle 2006; Zabala *et al.* 2006; Van Nieuwenhuyse (this study)

To date, only a few multi-scale habitat assessments have been undertaken on Little Owls (i.e., Blache 2004; Martínez & Zuberogoitia 2004b; Cornulier & Bretagnolle 2006; Zabala *et al.* 2006 and Van Nieuwenhuyse – this study). The different radii that are used in these studies are given in Table 6. 6.

In Spain, Martínez & Zuberogoitia (2004b) assessed habitat conditions within three plot sizes, centered on 78 Little Owl nests and 55 non-occupied but apparently suitable territories. A territory was considered as suitable for nesting when it contained mature carob trees, derelict or inhabitated country houses, or ephemeral rivers, according to general preferences of the Little Owl in the study area. Habitat selection was studied using generalized linear models using a logit link to identify the difference between occupied and unoccupied habitats. At the nest-site scale, they found significantly more arid plantations with high availability of linear structures in the proximity of villages in occupied habitats than unoccupied ones. At the home-range scale, Little Owls were found more often in habitats with high percentages of arid plantations, linear structures, the presence of neighboring Little Owls, and in the vicinity of cities. At the landscape scale, the model predicts occupancy of habitats in areas with arid plantations. At all spatial scales, arid plantations are the most discriminating parameter between occupied and unoccupied habitats, stressing the importance of traditional land uses.

Blache (2004) used three different methods to study the habitat preferences of Little Owls in Plaine de Valence, France. First he delimited seven types of landscapes in using a multiple correspondence analysis of 817 grid-cells of 250 m by 250 m with a hierarchical clustering of the first two axes. Jacob's Index (Jacob 1974) was used to determine if some landscape types, at the smallest scale, were disproportionately used by the owls. The habitat type that contained the highest number of houses and isolated farms, but also a significant proportion of fields, hedges, and isolated trees, was preferred by Little Owls.

Table 6.6 *Different scales of study in habitat selection of Little Owls. Radius of plots is given in meters.*

	D. Van Niewenhuyse unpublished observations		Martínez & Zuberogoitia 2004b		Blache 2004	
nest-site scale	282 m	25 ha	200 m	10.5 ha	250 m	19.6 ha
home-range scale	564 m	100 ha	309 m	30 ha	500 m	78.54 ha
landscape scale	2000 m	1256 ha	5600 m	10 000 ha	1000 m	314.15 ha

Second, the importance of 15 individual landscape elements was studied using univariate logistic regression of occupied and unoccupied grid-cells. At two spatial scales (250 m by 250 m and 750 m by 750 m grid-cells), isolated farms and mulberry trees were positively associated with Little Owl presence. Intensively managed orchards and forests showed a negative association with Little Owl presence. At the smallest scale, roads, grasslands, the number of isolated houses, and hedges, showed moderate association with Little Owl presence.

Blache (2004) finally studied the importance of the parcel size around occupied (n = 52) and unoccupied farms (n = 23). T-tests were used to compare the mean parcel area between occupied and unoccupied farms at the three spatial scales (Table 6.6). In contrast to most other studies, the habitat diversity around farms was not a discriminating factor between occupied and unoccupied farms. The species prefers farms with only a few large parcels. Farms that had an occupation of seveal years by owls were found to have the smallest diversity of habitat. The species does not show a preference for a fragmented habitat in Valence. In the Netherlands, Van 't Hoff (2001) found that seemingly good habitats that were isolated, were abandoned sooner than clustered habitats of lower quality. Further research on the role of social interactions in this area might explain this interesting behavior of the species.

Zabala *et al.* (2006) modeled the distribution of a Little Owl population in Spain using a geographic information system, extracting data on land use and landscape composition in occupied and unoccupied areas. Their statistical analyses (logistic regression and multiple linear regression) showed that variables such as topography, altitude, road density, and urban areas had a negative effect at the lower scale, while conifer plantations had a negative effect at every spatial scale. The density of predator species, in particular the Tawny Owl linked to the plantations, had an effect only over habitat selection but not over occupancy.

Cornulier and Bretagnolle (2006) used a recently developed method generalizing Ripley's K function for non-homogeneous point patterns to test the aggregation of the nests of Little Owls exhibiting heterogeneous distributions in response to landscape structure. Initially the Little Owl was found to form clusters at some scale, taking spatial heterogeneity into account revealed that territorial Little Owls showed no clustering of territories when habitat availability was considered. When assuming homogeneity, Little Owls maintained some regular spacing at small scales (*c.* 150 m), as expected from a territorial species, whereas territories were clustered at medium scales, between 500 m and 1000 m. Consequently, opposite patterns can be found in a single data set at different scales. This result highlights that it is recommended to account for non-stationarity when testing for aggregation. The authors demonstrated that accounting for large- or small-scale heterogeneity affects the perception of spacing behaviors differently.

We used discriminant analysis on data collected on our East-Flanders (Belgium) study area, Van Nieuwenhuyse examined the habitat relationships between the presence of Little Owls and Barn Owls for an array of landscape variables within three different circular-sized plots (25 ha, 100 ha, 1256 ha) (Table 6.6). He used the number of land parcels, length of the parcel edges, and the parcel areas of row crops, cereal grains, maize, built-up areas, farm buildings, and forests. The length of waterways, tree-lines, and roads was also included in the analysis.

Earlier research at the nest-site scale revealed a positive impact of conspecifics on the probability of finding Little Owls (Van Nieuwenhuyse & Bekaert 2002).

Thus far, only three analyses have been undertaken (Martínez & Zuberogoitia 2004b; Blache 2004; own results) using three different radii around the nest, i.e., nest-site scale, active home-range scale, and landscape scale.

Martínez and Zuberogoitia (2004b) found that at the nest-site scale, Long-eared Owls preferred wooded areas with few paved roads while Little Owls preferred arid plantations. Furthermore, the probability of finding an occupied territory increased with the proximity of another occupied territory in the surroundings. The home-range-scale models mirror the feeding requirements of the owls. Long-eared Owls occupied areas with high percentages of forest, arid plantations, edges between these two land-uses, short distances between nests, with presence of conspecifics and little human disturbance. Little Owls occupied arid plantations with high availability of linear structures and the proximity of villages. At the landscape scale, Long-eared Owls eluded extensive forests, and Little Owls preferred arid plantations.

We examined the relationship between the presence of Little Owls and the landscape in three different sized circular plots (25 ha, 100 ha, 1256 ha) in East-Flanders (northern Belgium) with discriminant analysis. We used the number of parcels, the length of the parcel edges, and the parcel area of row crops, cereals, maize, built-up areas, farm buildings, and forests. The length of waterways, tree-lines, and roads was also included in the analysis.

The predicted probability of presence of the Little Owl dropped with the area covered by farm buildings at the nest-site scale and increased with these areas at the home-range and the landscape scale. Earlier research at the nest-site scale showed a positive association of Little Owls with conspecifics (Van Nieuwenhuyse & Bekaert 2002). The perimeter of the farm buildings associated positively at the nest-site scale but negatively at larger scales. At the three spatial levels, a positive association of Little Owls with maize edges was found, while there was a negative association with the area of maize or number of maize parcels. This indicates that maize parcels do not have a negative influence on the species as long as the parcel size is very limited, confirming the preference by the owl for mosaic-like landscapes.

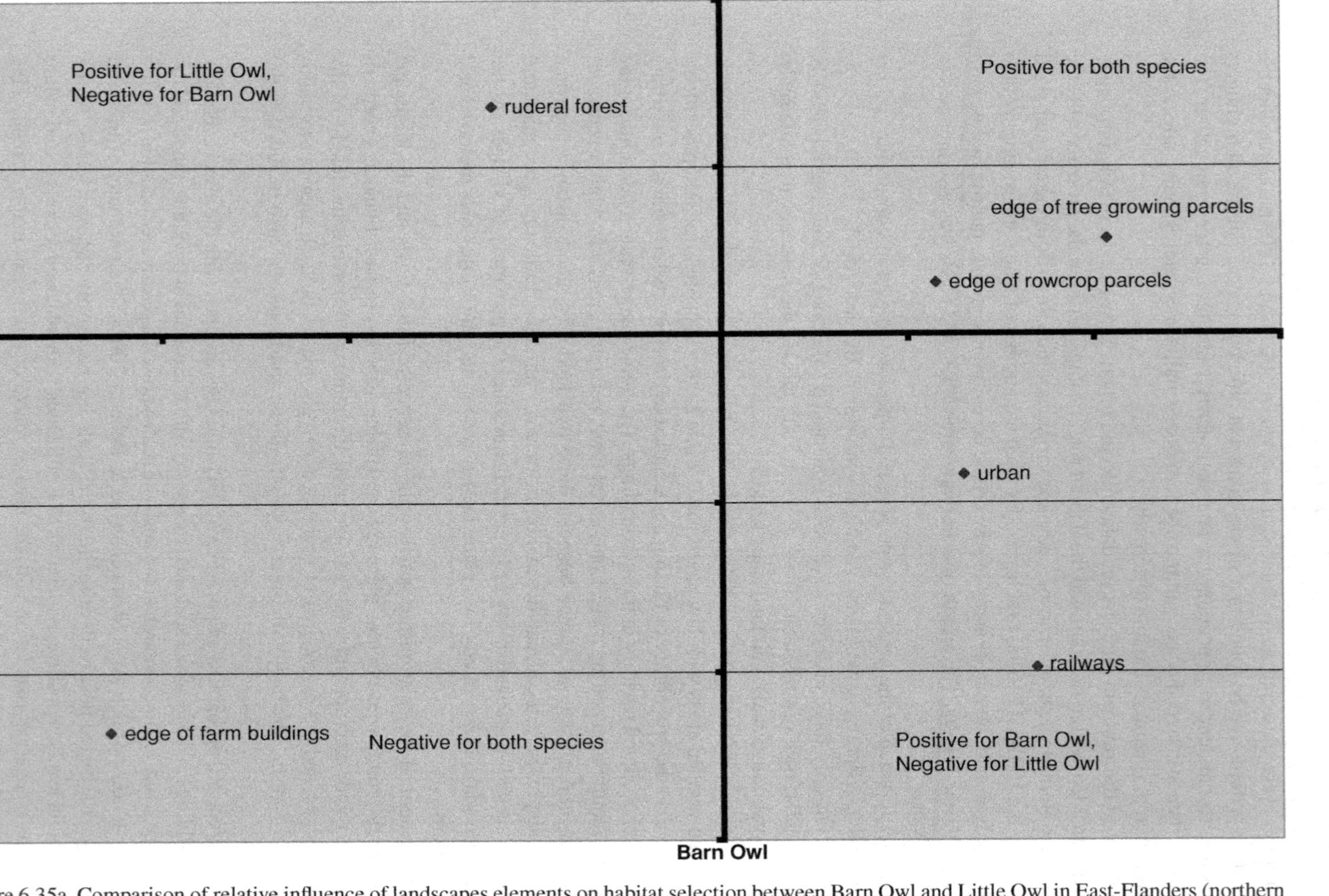

Figure 6.35a Comparison of relative influence of landscapes elements on habitat selection between Barn Owl and Little Owl in East-Flanders (northern Belgium); in 25 ha circular plots; landscape data and analysis as of 2005 (Van Nieuwenhuyse, in prep.).

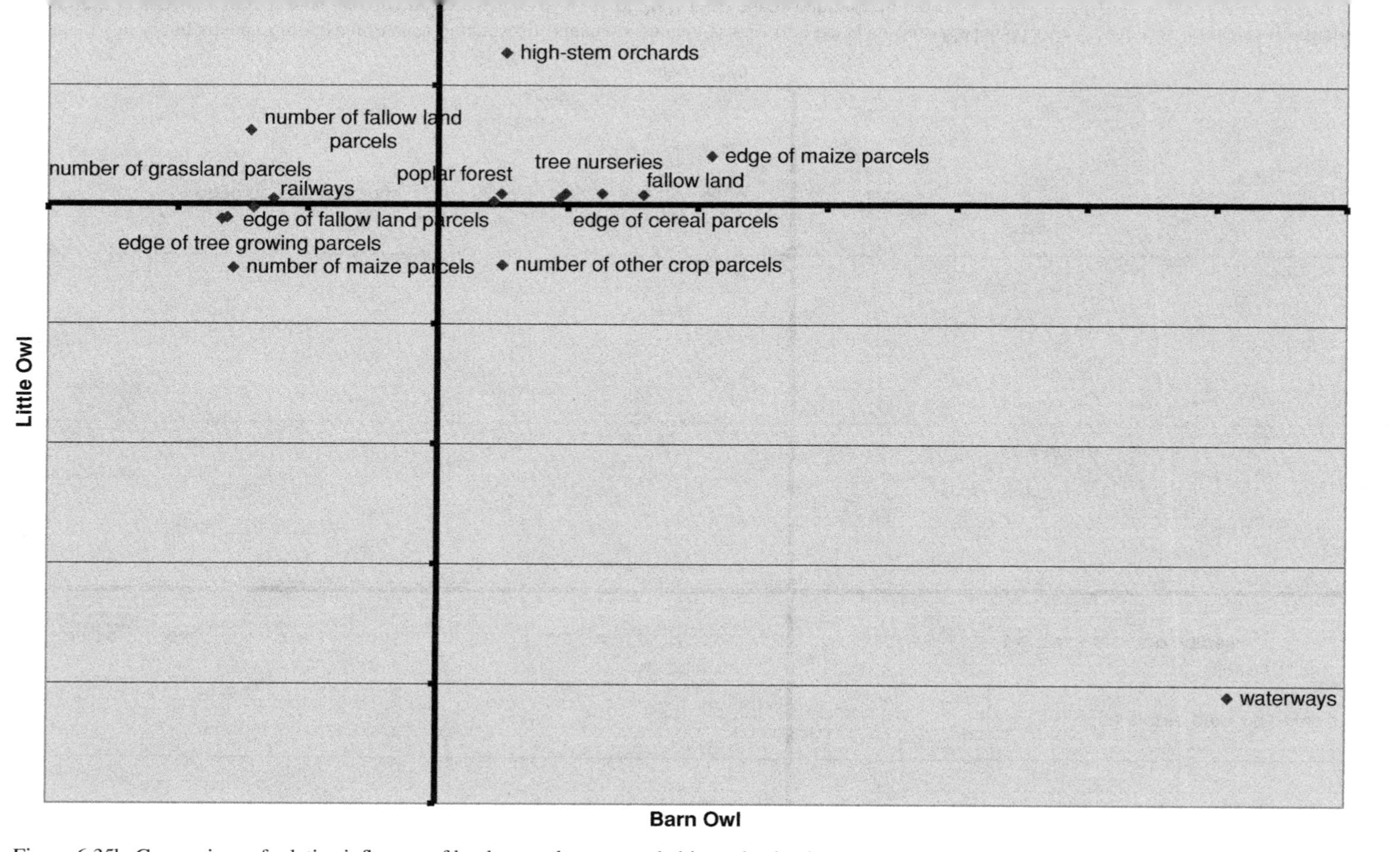

Figure 6.35b Comparison of relative influence of landscape elements on habitat selection between Barn Owl and Little Owl in East-Flanders (northern Belgium); in 100 ha circular plots; landscape data and analysis as of 2005 (Van Nieuwenhuyse, in prep.).

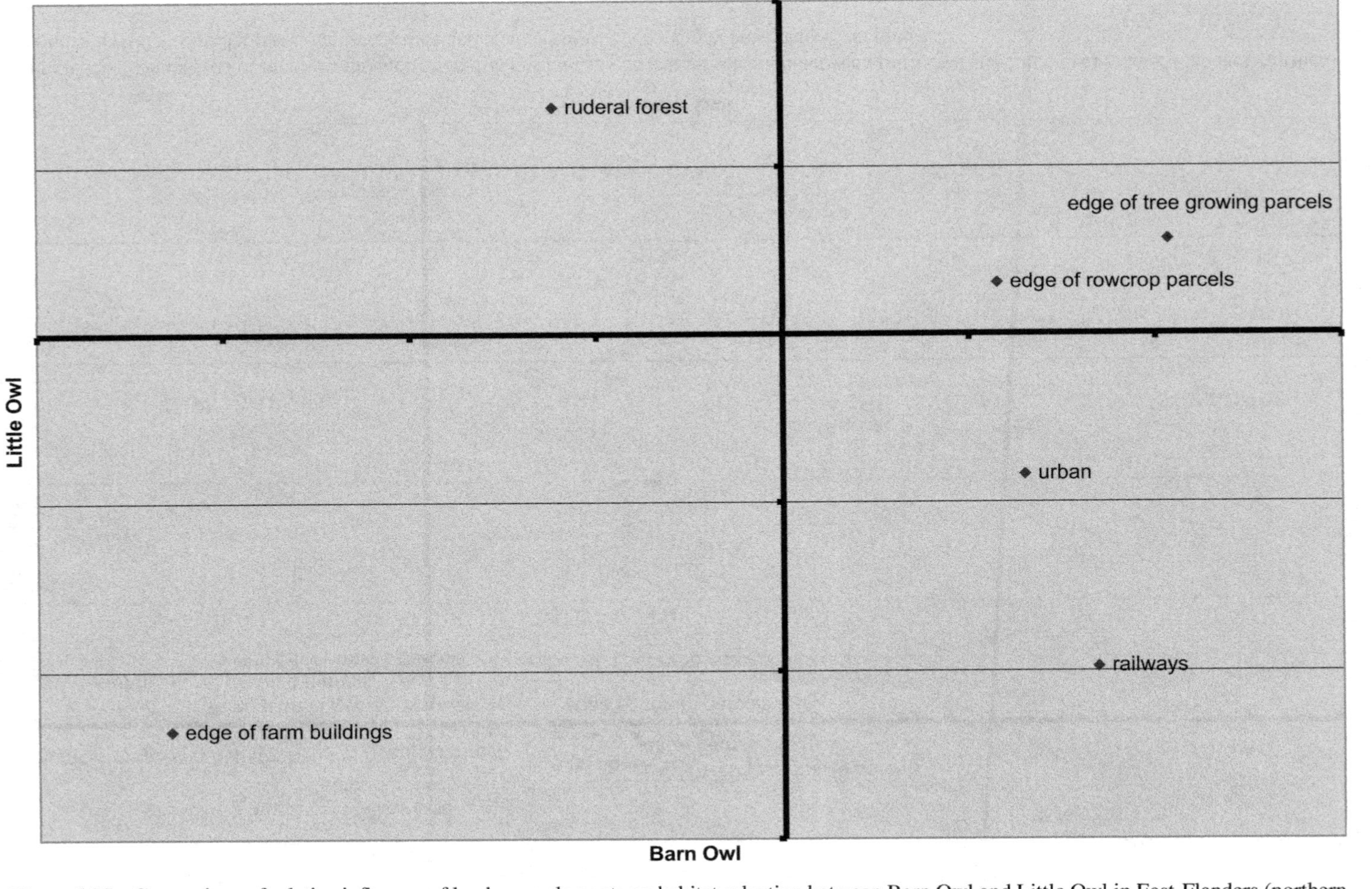

Figure 6.35c Comparison of relative influence of landscape elements on habitat selection between Barn Owl and Little Owl in East-Flanders (northern Belgium); in 1256 ha circular plots; landscape data and analysis as of 2005 (Van Nieuwenhuyse, in prep.).

For a certain period during the growing of maize, the parcels remain mainly bare land, which is favorable for Little Owls. Large maize parcels were negatively associated with Little Owls. Roads and waterways featured a negative association with the species at all levels in contrast to Barn Owls that seem to be attracted by waterways. Little Owls were negatively associated with forests at the nest-site level but positively at larger scales. Railways were positively associated with the owls at the nest-site and home-range scale but negatively at the landscape scale. The species can tolerate short distances of railways. Longer lengths of railways are most associated with urban centers and are hence avoided.

To compare the habitat selection of Little Owls with Barn Owls, we quantified for each landscape element the difference between the coefficients of both discriminant functions (both good and bad habitats). The differences in impact on both species are given in Figures 6.35a–c.

As they are a common landscape feature, railways are associated with both species, as is shrubbery and tree growing parcels at the nest-site scale. Barn Owl presence was mainly associated with the amount of cereal grain fields. Tree-lines discriminate for Little Owls but not for Barn Owls.

At the 100 ha scale, waterways become very distinctive for Barn Owls and not at all for Little Owls. On the other hand, the area of fallow land is distinctive for Little Owls. High-stem orchards are positive for both species.

Finally, at the landscape level, Barn Owls showed a positive association with urban areas and railways, while for Little Owls these landscape elements have a negative association with species presence. The occurrence of elm forests is positively associated with Little Owls and negatively with Barn Owls.

7
Diet

7.1 Chapter summary

The Little Owl has a generalist diet and takes a high diversity of small prey (see Figure 7.1). It catches and eats a wide variety of small-sized prey across its entire distribution area. The diet of the owl varies according to the season and the geographical area. From north to south and from winter to summer, an increase in the numbers of insects in the diet has been observed. Regardless of the numbers of insects, small mammals remain the key prey category by biomass and energetic yield, and they contribute significantly to the ecology and welfare of the species. As with other insect-eating owl species, it is very difficult to get a true picture of its diet by only studying its pellets, because many of the invertebrate prey are difficult to find, or are simply not present in Little Owl pellets (e.g., insects and especially earthworms). In this chapter we look at this prey diversity through time and through space, and we focus on the hunting method of the owl, as well as catching behavior in larders. We offer a thorough review of the owl's diet and individual prey species, with attention given to the importance of voles and other large prey. Pellet contents are first described in detail.

7.2 Pellets

Most of the food studies have examined pellets or prey remains in nests, but early authors also analyzed stomachs (Collinge 1922; Farsky 1928; Madon 1933; Hibbert-Ware 1937/1938; Vachon 1954; Lovari 1974). Current studies are done with infra red cameras (Juillard 1983) or video equipment (Van Zoest & Fuchs 1988; Blache 2001; van Harxen & Stroeken 2003a; H. Keil personal communication).

Pellets are pale gray in color when they contain rodent remains (Thomas 1939), and black (Hibbert-Ware 1937/1938) or red brick when their content is only insects (Festetics 1959). The shiny parts of beetles are clearly visible on the outside edge of the pellets (Thomas 1939).

They are usually rounded at the ends, though occasionally one tip is like a thread (Hibbert-Ware 1937/1938). They look like the pellets of the Kestrel (*Falco tinnunculus*) but those of the Kestrel are narrower than those of the Little Owl and are usually harder and more compact. See Figures 7.2a–e.

Figure 7.1 Little Owl pulling an earthworm out of the ground (François Génot).

Captive Little Owls evacuate one or two small pellets daily, about 11.5 hours after each meal, and the pellet is produced when digestion is still incomplete (Hanson 1973). In nature, the pellets are evacuated after the night's feeding and before daytime roosting (Hibbert-Ware 1937/1938). The time between eating prey and producing pellets depends on the season and the type of prey (Schönn *et al.* 1991). Thus, a pellet may be produced 4 hours after catching grasshoppers or 12 hours after catching vertebrates (Angyal & Konopka 1975); in another case the pellet was produced 5.5 hours after the owl ate a mouse, and 7.5 hours after eating a bird (Oles 1961).

Figure 7.2a Little Owl pellets. Photo Jacques Bultot.

Figure 7.2b Little Owl pellets. Photo Jacques Bultot.

Figure 7.2c Little Owl pellets. Photo Jacques Bultot.

Figure 7.2d Little Owl pellets. Photo Jan Van't Hoff.

Figure 7.2e Little Owl pellets. Photo Jan Van't Hoff.

Normally, Little Owls spread pellets across the fields, except in winter when it is possible to collect up to 40 pellets at the same place because the species often uses the same roost (Génot & Bersuder 1995).

In Austria, Sageder (1990) found an average pellet volume of 5 cm^3 (0.9–20.1 cm^3; n = 433) in Seewinkel (a region with grassland and a greater diversity of prey), and an average pellet volume of 5.91 cm^3 (1.8–14.5 cm^3; n = 165) in Marchfeld (a region with more intensive agriculture and less diversity of prey). According to the author, the size of the pellets was linked to the activity of the owls, e.g., in winter and summer Little Owls produce more, but smaller, pellets due to their limited activity during this period.

The number of prey items per pellet have been reported at 1.2 in Bulgaria (Simeonov 1968) and 10.5 (minimum 1, maximum 60) in Italy (Moschetti & Mancini 1993). In France, Génot and Bersuder (1995) found in 234 winter pellets: no prey (n = 12), one prey (n = 102), two prey (n = 108) and three prey (n = 3). In Moldavia, pellets contained an average of 1.93 prey (0.87 insects and 1.06 vertebrates; n = 634) with the following seasonal composition: 1.59 prey (0.39 insects and 1.19 vertebrates) in autumn–winter and 2.17 prey (1.19 insects and 0.98 vertebrates) in spring–summer (Ganya & Zubkov 1975).

The content of pellets also delivers information about the foraging area of the Little Owl. In this way the analysis of beetles caught by the owl showed that the species hunted along forest edges and in very open woodlands (Kuhn 1995).

Pellets consist mainly of the fur and bones of rodents (vertebrae and ribs, skull fragments, metacarpus and phalanges, femurs, tibias, jaw bones with molars, humerus and radius), bird bones (skull fragments, beaks, scapulas, clavicles, tarsometatarsus, phalanges and claws, vertebrae, and ribs) (Hibbert-Ware 1936), bones of frogs or snakes, fragments of snail shells (Lancum 1925; Berghmans 2001d), and chitinous remains of insects. A German-based project reported insect remains in 352 pellets (Kuhn 1992). Of 9893 insect body parts recorded from these 352 owl pellets, 4027 (41%) were heads, 2329 (23%) were of the left elytra, 2250 (23%) were of the right elytra, and 1287 (13%) were thorax. The average size of beetle parts recovered from the pellets was 1.2 cm, with a range of 0.2–3.5 cm (Kuhn 1992). Earthworms can be identified in pellets thanks to their bristles, but a detailed quantification is not possible. In two German study areas, 57% and 80% of 75 pellets contained earthworms (Koop 1996).

Pellets can also contain grass as vegetal fibers or seeds of weeds (Collinge 1922; Hibbert-Ware 1937/1938; Quadrelli 1985), soil (Hounsome *et al.* 2004), stones (Hulten 1955; Sagitov 1990; Bel'skaya 1992), cherry stones (Misonne 1948), or eggshell (Sagitov 1990). Grass is not eaten as food but facilitates the formation of pellets with the chitinous parts of insects (Madon 1933; Libois 1977; Quadrelli 1985).The same reason is given by Thiollay (1968) for the ingestion of stones. Also Finck (1989) noted that pellets also contain sand, which is linked to the eating of earthworms.

7.3 Hunting method

The typical nocturnal activity pattern of the Little Owl has two peaks of about 1–2 hours each, occurring just after sunset and before sunrise, separated by a stage of less activity around midnight (Exo 1989). The Little Owl's crepuscular and nocturnal feeding habits (Hibbert-Ware 1937/1938; Haverschmidt 1946; Juillard 1984a; Van Zoest & Fuchs 1988; Schönn *et al.* 1991; Fajardo *et al.* 1998; Ryabitsev 2001) are linked to prey activity (Zerunian *et al.* 1982; Fattorini *et al.* 2000). But the owl also hunts by day, especially during the nestling phase (Hibbert-Ware 1937/1938; Van Zoest & Fuchs 1988; Ryabitsev 2001) and during the warm season in the Mediterranean (Zerunian *et al.* 1982; Lo Verde & Massa 1988; Negro *et al.* 1990), and in eastern Palearctic areas (Osmolovskaya 1953).

The method of hunting is related to the type of vegetation in the foraging area (Schönn *et al.* 1991). When the vegetation is low, the Little Owl hunts mainly beetles, earthworms, and rodents on the ground from a perch, dropping onto prey below or nearby (Hibbert-Ware 1937/1938; Van Zoest & Fuchs 1988; Schönn *et al.* 1991; Fajardo *et al.* 1998). In Belgium, Libois (1977) reported that the Little Owl looked for 65% of its prey (n = 1782) on the ground. Hunting from a perch in the Kazakhstan desert is not without risk, due to predators when the perches are without shelter (Osmolovskaya 1953).

When the vegetation is high, it hunts flying insects from a perch, using a twisting and turning flight pattern (Haverschmidt 1946; Beven 1979). In the Netherlands, the type of perch is related to the type of prey chosen. The time spent on a perch is variable and depends on the perch height (Van Zoest & Fuchs 1988).

The owl can catch beetles on the ground by approaching in a short low flight, followed by rapid runs or a few hops (Beven 1979). A Little Owl ran across a lawn in very rapid bursts and at the end of each burst stopped with its head cocked as if listening and occasionally made a grab at something with its beak, than hopped like a thrush and ran rapidly again

Figure 7.3 Little Owl with moth. Photo Ludo Goossens. See Plate 31.

(Cameron-Brown 1975). The Little Owl hunts earthworms by hopping over the ground, stopping suddenly, bending forward, and seizing a worm with its bill and pulling it, often flapping its wings to maintain its balance, and sometimes falling onto its back as the worm comes loose (Haverschmidt 1946). The method of transporting captured prey depends on the prey item. In Switzerland, Juillard (1984a) determined that for 4918 items of prey that were taken to nests, invertebrates were transported almost completely with the beak, 98% of birds with the beak (the other manner was with the foot), and 65% of small mammals were transported with the bill and 35% with the foot. There is almost no significant difference between males and females in the method of transport, but the male brings more prey by the head than the female who prefers carrying them by the neck.

During summer nights, the owls like hunting under lamps, and around well-lit houses, yards, streets, and roads, where plenty of insects are concentrated, especially moths (Nankinov 2002), or grasshoppers on asphalt roads (Adam 1973). See Figure 7.3. Besides fields, the Little Owl hunts on dunghills, in gardens and on lawns by dusk and during the first hours of the night in Denmark (Laursen 1981). The species is able to hunt bats (Reinard 1977).

It raids the nests and nest-holes of birds, such as the Starling (*Sturnus vulgaris*) (Tinbergen & Tinbergen 1932; Elliott 1940; Anderson 1949; Burton 1983), House Sparrow (*Passer domesticus*) (Desmots 1988), and Tree Sparrow (*Passer montanus*) (Steffen 1958). One owl regularly stole worms from a Blackbird (*Turdus merula*) (Tricot 1968). Birds can be taken from the water (Barber 1925) or ground (Ellis 1946). The Little Owl also hovers like a kestrel (Gregory 1944; Tayler 1944; Gyllin 1968; Malmstigen 1970; Wahlsted 1971; Martin & Rollinat 1982).

The Little Owl can combine different types of hunting according to the available prey. Thus, by daylight in southern France, a nervous bird hunted by perching, hovered sometimes as a kestrel, caught beetles on the ground with its bill or claws, changed perches, hovered and walked on the ground picking up beetles (Génot & Lecomte 2002).

In the Cis-Balkhash Lake area (Russia) or in Kazakhstan, the Little Owl is a frequent visitor of Gerbil burrows (Shtegman 1960; Gavrin 1962). At night, they catch Gerbils as they sleep in their nests and eat them immediately. They use certain burrows as larders and can settle in the colonies of underground Gerbils, an interesting parallel with the behavior of the Burrowing Owl (*Athene cunicularia*). This special "underground" behavior of the Little Owl, visiting and living in burrows, is also demonstrated by the wear patterns observed on the feathers of examples in museum collections in Central Asia and Middle Asia. The upper parts of the feathers are missing, especially on the head, back, and shoulders, and the feathers' blades are bare stems jutting like needles. This underground behavior might help explain the mystery regarding the bad condition of the feathers of Little Owl's from the eastern subspecies (Shtegman 1960).

7.4 Larder (prey caching)

Larders (holes or cavities holding excess food) contain the remains of prey stored before feeding it to the nestlings. The term "larder" is only partly correct, as its primary purpose is to serve as a deposit location for surplus food, rather than for formal food storage (Hibbert-Ware 1937/1938, 1938a). The male takes food from the larder to the female in the nest site as she requires it (Owen 1919). The food reserves hidden in holes and niches are often eaten during one day (Gavrin 1962). Certain Little Owls take their food directly to the nest as they probably have no other suitable hole for a larder (Hibbert-Ware 1937/1938).

The larder is used in winter too, e.g., in Switzerland where a larder containing 22 voles was found (Blanc 1958) and where a nestbox containing 95 rodents was found (Juillard 1984). In Russia, a Little Owl built reserves of up to 30 and more animals (Koblik 2001). In Britain, a cache of 167 headless Storm Petrels (*Hydrobates pelagicus*) was recorded by Lockley (1938). The short-term storage of food in larders is a behavior shown by a large number of owl species, especially the Pygmy-owls (*Glaucidium* sp.) (del Hoyo *et al.* 1999:117).

7.5 Catching prey

The ontogenetic development of catching and handling prey in the Little Owl was studied by Ille (1983) in an aviary. The capture of prey by an adult owl entails different steps, i.e., locating the prey, approaching, attacking the prey, and eating or storing the prey, and also non-obligatory acts such as ruffling the feathers, or mantling the prey with its wings. These different steps of catching prey appear to depend on the condition of the owl (age, state of hunger) or/and on environmental factors (size and resistance of prey, competition for food). The ruffling of the feathers and mantling of the prey are examples of threat behavior shown towards prey and towards competitors for food. The beating of wings on large prey that is still moving, serves only for maintaining balance. Every kind of movement of the prey, real or artificial, triggers an act of catching the prey, explaining the aggressive behavior of the Little Owl. The owl reacts to the form of prey according to a process of conditioning. In captivity, the species prefers to seize a mouse by the body and strike it with its beak (Csermely *et al.* 2002).

The complete act of catching prey appears after a maturation process involving several stages. The fledglings go through this process at an average age of 62–76 days. The choice of prey and the foraging method is not instinctive but rather a learned process (Sageder 1990). Although experiments in captivity have shown that Little Owls prefer catching live prey over dead, the species can eat dead prey such as hedgehogs that have been victims of road traffic (Van de Velde & Mannaert 1980).

Males and females differ slightly in the prey they catch and bring to the nest. In the Netherlands, during the nestling phase, Van Zoest and Fuchs (1988) found that males catch more small mammals and earthworms, while females catch more invertebrates. On the other hand R. van Harxen & P. Stroeken (personal communication) showed that males catch more small mammals and cockchafers, while females catch more earthworms and insects, in particular aquatic larvae of *Eristalis tenax* (a diptera). During the nestling phase, Beersma and Beersma (2003) found a dead adult male owl, its stomach completely filled with 210 caterpillars weighing some 13 g. This suggests the potential for transport of small prey in the bird's stomach and for vomiting it back out for the young once back at the nest. In the Netherlands, during the nestling phase, the parents brought more small mammals but fewer earthworms and caterpillars to the nest when the young were just hatched to 9 days old, than when the young were 10–22 days old (R. van Harxen & P. Stroeken, personal communication).

7.6 Prey species

In the western Palearctic, the diversity of prey species can be summarized as follows: 54 mammals, 82 birds, 15 reptiles, 14 amphibians, 2 fishes, 350 insects, and 27 other invertebrate species, totaling 544 different prey species. In Appendix A we offer a complete list of prey items as found in the literature.

A summary of the major western Palearctic diet and prey studies is as follows:

Armenia: Sosnikhina (1950)
Austria: Sageder (1990), Ille (1996)
Britain: Collinge (1922), Hibbert-Ware (1937/1938), Hounsome *et al.* (2004), Hayden (2004)
Belgium: Libois (1977)
Bulgaria: Simeonov (1983)
Denmark: Laursen (1981)
France: Génot (1992a), Génot & Bersuder (1995), Blache (2001)
Germany: Haensel & Walther (1966), Kuhn (1992), Riedel (1996)
Greece: Angelici *et al.* (1997), Goutner & Alivizatos (2003)
Hungary: Marian & Schmidt (1967)

Italy: Zerunian *et al.* (1982), Contoli *et al.* (1988), Lo Verde & Massa (1988), Moschetti & Mancini (1993), Gotta & Pigozzi (1997), Fattorini *et al.* (2000), Manganaro *et al.* (2001), Nappi (2005), Bux & Rizzi (2005)
Kazakhstan: Osmolovskaya (1953), Formozov & Osmolovskaya (1953), Gavrin (1962), Murzov & Berezovikov (2001)
Moldavia: Ganya & Zubkov (1975)
Netherlands: Van Zoest & Fuchs (1988)
Poland: Romanowski (1988), Grzywaczewski *et al.* (2006)
Portugal: R. Tome (personal communication)
Romania: Barbu & Sorescu (1970), Popescu & Blidarescu (1983), Popescu *et al.* (1986), Popescu & Negrea (1987)
Russia: Kiselev & Ovchinnikova (1953), Shtegman (1960), Afanasova & Khokhlov (1995)
Spain: De la Hoz (1982), Manez (1983a,b), Delibes *et al.* (1983)
Switzerland: Juillard (1984a), Gusberti (1998), Schmid (2003)
Tajikistan: Abdusalyamov (1971)
Turkmenistan: Sukhinin *et al.* (1972), Bels'skaya (1992), Zarkhidze & Loskutova (1999)
Ukraine: Bashta (1994)
Uzbekistan: Ishunin (1965), Ishunin & Pavlenko (1966), Sagitov (1990).

Table 7.1 suggests the preponderance of insects (especially beetles) in the diet, mainly in the Mediterranean countries of Italy, Spain, and Portugal. The degree of insects in the diet is still large in the winter, reflecting 93.6% by number, 33.6% by biomass (Delibes *et al.* 1983) and 96.8% by number, 57.8% by weight (R. Tome, personal communication) for Spain and Portugal, respectively.

The three main families of insects found in the diet are beetles, earwigs, and crickets. Sometimes ants can be eaten, as in Portugal. Complete insects are eaten, and while caterpillars or larvae are noticeable in stomach analyses, or are observed with camera equipment, they are not found in pellets. When compared to larger prey, insects are easier for the owls to catch and their means of self-defense are limited.

With the help of camera or video equipment, earthworms have been quantified as being very important in the diet during the feeding of nestlings: 65.5% by number and 57.9% by biomass in Switzerland (Juillard 1984a), 19.2% by number and 20% by biomass in the Netherlands (Van Zoest & Fuchs 1988) and also 10.4% by number and 10.5% by biomass (R. van Harxen & P. Stroeken, personal communication) and 50% by number and 30% by biomass in France (Blache 2001). Earthworms are taken mainly after rainfall.

In Denmark, a large proportion of earwigs (82% by number) were taken by two owls that developed a strategy of hunting within buildings (Laursen 1981).

Mammals, and rodents, are important in number and in biomass, as reported in the diet of Little Owls in central Europe: northeastern France (Génot & Bersuder 1995), Austria (Sageder 1990), Poland where the Little Owl specializes on the Common Vole (Romanowski 1988), Romania (Popescu & Blidarescu 1983), Bulgaria (Simeonov 1983), and Kazakhstan (Murzov & Berezovikov 2001); or just by biomass (59%) as in the Netherlands (Van Zoest & Fuchs 1988).

Table 7.1 *Food composition of the Little Owl* (Athene noctua) *in different countries.*

Country	Study period	Author	Mammals	Birds	Reptiles, amphibians	Insects	Coleoptera	Orthoptera	Dermaptera	Other invertebrates	Total of prey
Denmark	year	Laursen (1981)	**6.8**	0.3	–	**92.9**	16.5	–	76.4	–	8635
Netherlands	summer	Van Zoest and Fuchs (1988)	**8**	9.3	1.9	**63.2**	–	–	–	26.5	1026
Netherlands	summer	Van Harxen and Stroeken unpublished	**5.9**	0.5	2.3	**80.9**	32.4	0.06	–	10.4	3379
Belgium	year	Libois (1977)	**10.49**	0.56	0.05	**86.4**	66.72	–	9.7	2.41	1782
Germany	year	Kuhn (1992)	**16.85**	0.44	–	**82.7**	72.79	–	9.91	–	6568
Switzerland	summer	Juillard (1984a)	**2.9**	0.4	–	**28.2**	6	2	–	65.5	8474
Austria	year	Sageder (1990)	**64.3**	1.5	1.2	**32.9**	30.2	0.4	1.3	1	2813
France	year	Génot and Bersuder (1995)	**32.4**	0.9	0.6	**60.5**	41.3	8.6	10.6	5.4	9181
France	summer	Blache (2001)	**6.51**	0.59	–	**24.61**	2.92	5.59	–	46.94	2768
Italy	year	Zerunian *et al.* (1982)	**2.7**	1.1	0.5	**93.1**	76.9	1.9	12	2.6	3420
Italy	winter	Manganaro *et al.* (2001)	**3.7**	0.6	0.9	**94.4**	65.3	1.1	26.3	–	2558
Spain	year	Manez (1983a)	**4.09**	0.48	0.72	**92.29**	39.63	27.76	14.18	2.41	15993
Spain	winter	Delibes *et al.* (1983)	**5**	0.3	0.2	**93.6**	22	40	13.2	0.9	4118
Portugal	winter	Tome unpublished	**1**	–	–	**96.8**	53	8.2	3.3	32.3	1903
Poland	summer	Romanowski (1988)	**73.6**	0.3	–	**25.4**	22.2	–	2.8	0.3	288
Romania	summer	Popescu and Blidarescu (1983)	**36.5**	1.5	–	**62**	48.2	10.5	0.1	–	1075
Bulgaria	year	Simeonov (1983)	**47.7**	4.3	0.5	**45.5**	43.9	1.5	–	2	3081
Kazakhstan	year	Murzov and Berezovikov (2001)	**62.4**	7.8	6.5	**23.3**	12.9	3.5	–	6.7	1687

While carrion is very seldom used as food (Hibbert-Ware 1937/1938), the Little Owl can be necrophagous in some cases, feeding on dead adult chickens (Blache 2001) or day-old chicks (Berghmans 2001a).

Before legal protection was extended to birds of prey and owls in European countries, the analysis of their diet was done only to classify the Little Owl as a "useful" species (Mercier 1921; Farsky 1928; Sosnikhina 1950; Blagosklonov 1968; Barabash-Nikiforov & Semago 1963). In Britain, after its introduction, a big debate took place about the extent of Little Owl predation on game and poultry, as some individual owls occasionally feed on chicks (Hibbert-Ware 1938b).

Through the diet of the Little Owl, we can study the impact of changes in agricultural techniques. Some prey species are linked to the diversity of habitat and are sensitive to biocides. Before the 1950s, many authors stressed the importance of cockchafers in the May, June, and July diet of the owl when this "extremely abundant" beetle was fed to the young (Farsky 1928; Madon 1933; Hibbert-Ware 1937/1938). The young in the cavities could be found standing on a thick layer of cockchafer remains (D'Hamonville 1895). Nowadays, probably due to the use of biocides in agriculture and gardening, cockchafers have become very rare in nature and likewise in the diet of the Little Owl, except for some local "invasions" as in the Netherlands (van Harxen & Stroeken 2003a). This situation is similar for crickets, which, along with the owls, have disappeared from the very intensively used landscapes of Switzerland (Juillard 1984a).

The higher the diversity of the habitat the wider the trophic niche of the owl. In very intensively managed agricultural landscapes, the habitat diversity is low and very often the Little Owl is more "specialized" as it is limited to one kind of prey, such as earthworms (Juillard 1984; Blache 2001), which must be taken in large quantities. However, the problem is that the earthworm contains less energy value (with 3.01 kcal per gram), and is lower in fat (with 1.49% vs. 19.75% for the Mole Cricket [*Gryllotalpa gryllotalpa*]), and contains less than 40% protein (Juillard 1984). By comparison, a Wood Mouse contains 70% protein and is richer in mineral salts. While earthworms may be abundant, especially after rain, their low weight (less than 4 g) and their weak energetic yield is not as beneficial as a rodent for feeding the young. Thus, the presence of earthworms in the diet may well be more an expression of poor prey diversity in very intensively managed agricultural landscapes rather than the adaptative capacity of the Little Owl in foraging. Moreover, earthworms contain a lot of water and when adults feed the nestlings only on worms, the nest becomes very wet (Luder & Stange 2001). In Flanders there are indications that the abundance of the species is related to parameters that have a positive impact on earthworms (Van Nieuwenhuyse & Leysen 2004). The conservation biology literature contains many examples of serious and long-term problems encountered by "dietary specialists" when their food base undergoes a population collapse.

Food given to the young apparently contains more invertebrates than normally taken by adults, probably because of the urgency to catch them during the critical nestling season. The adult owls must achieve a balance between the energy costs of catching and bringing prey to the nest and the energy yield to themselves and to their developing young. In Switzerland,

Juillard (1984a) showed that the average number of food items supplied during one night ranged from 32 for two young up to 60 for four young. Some pairs brought in 110 prey items and another only 6. But the food supplies depend on the size of the items, their abundance, the size and number of the young, and the climatic conditions.

7.7 Large prey

Among raptors the Little Owl is probably the "bravest hunter", catching prey that weighs as much as the owl itself (Ryabitsev 2001). It catches large prey because of its compactness and strength, and in particular its leg muscles seem unusually powerful. Its thighs are thick, and the bundles of flexors in the toes yield special power. The flexor of the metatarsus is strongly developed and serves to pull and hold the prey (Shtegman 1960).

While the Little Owl is small, it is an energetic and very skillful hunter. Its flight is low above the ground and swift, and the owl uses elements of topographic relief to attack suddenly. These are some of the reasons why its prey can be comparatively large. The owl prefers to hunt small mammals and Zerunian *et al.* (1982) consider that it is limited by the size and the activity of its prey. When the prey is large, it catches nearly exclusively young individuals (91.6% young Siesel [*Citellus citellus*] 8.4% adult; n = 71). When the prey is small, it successfully catches young as well as adult specimens (23.4% young Mid-day Jird Gerbil [*Meriones meridianus*], 76.6% adults; n = 128) (Osmolovskaya 1953). Finally, the species has a wide trophic niche ranging from ants (0.01 g), up to rats (180 g) or young rabbits (200 g). The daily food requirement is 59–75 g at 0°C and 23–30 g at 30°C, and averages 46.7 g for the whole year (Manez 1983b). Despite this wide range of prey sizes, the Little Owl has shown a preference for small rodents of about 10–30 g bodyweight (Romanowski 1988).

7.8 Importance of voles

Even if the diet includes insects at all times of the year, voles, in particular the Common Vole (*Microtus arvalis*), are very important to the Little Owl in central Europe. This has been shown in eastern France, where voles represented 70% of the biomass from 8181 prey items recovered in pellets (Génot & Bersuder 1995), Poland (74% by biomass; n = 288 prey items; Romanowski 1988 and 47.2% by biomass; n = 1953 prey items; Grzywaczewski *et al.* 2006), Germany (with 76% of the number of mammal prey; n = 1805 prey; Lack 1946), Switzerland (51% by biomass; n = 2557 prey items; Schmid 2003), Moldavia (with 51% of the number of vertebrates; Ganya & Zubkov 1975), Romania (with 42% of the number of mammals; Popescu & Negrea 1987), and Caucasia (with 85% of the content of 175 pellets; Afanasova & Khokhlov 1995). Even in southern France, the Common Vole is represented by 6.5% in number and 50% of the biomass (n = 2345 prey). Further, even when it is a new species to the region, the Little Owl quickly incorporates small mammals into its diet (Blache 2001).

The energetic yield of voles is better than that of insects or earthworms. This was shown at a nest in Austria where a brood of three, month-old nestlings were fed only with insects up to 50 times per hour (Ille 1992). In Austria, Ille (1996) observed the lowest reproductive rate in 1993 after a Common Vole population decline. In Britain, Leigh (2001a) reported small mammal populations to be a key factor affecting the breeding performance of Little Owls. In Germany, Illner (1991) showed a correlation between the reproductive rate of the owl and the percentage biomass of the Common Vole (*Microtus arvalis*) and the Mole Cricket (*Gryllotalpa gryllotalpa*) in the owl's diet. In his study, Exo (1987) found no similar relationship. Koop (1996) found voles to be important prey when the owls were laying eggs and rearing young, in particular, their availability was important not only in the meadows but also along habitat edges (hedges, sides of way, and field edges). According to Romanowski (1988), the percentage of the Common Vole in the owl's diet does not depend on the yearly fluctuation of this vole. As with the Long-eared Owl (*Asio otus*), when the Common Vole population is low, the Little Owl may be able to focus its hunting effort on small, isolated colonies of this vole (Goszczynski 1981). In its effort to avoid short-term fluctuation of trophic resources, this situation is supported by the "super territory" foraging behavior shown by the Little Owl (Finck 1990).

In Asia, as in Kazakhstan, voles continue to be key prey for the Little Owl but with other species such as the Afghan Vole (*Microtus afghanus*), gerbils (Sukhinin *et al.* 1972), and the hamster (*Cricetulus triton*) in China (Lei 1995).

7.9 Seasonal variation

In many cases, the Little Owl's diet shows seasonal variations due to the climate, but there is no universal trend and patterns vary according to the studies. In most studies, however, invertebrates are important throughout the year.

In Britain, the seasonal percentage by number of main prey shows the following trends: rodents are between 30% and 50% in winter but stay between 25% and 30% during the other months; birds reach 8–10% in winter but remain under 5% in the other months; and insects reach around 30% every season with a maximum of 34% in spring (Collinge 1922). Hibbert-Ware (1937/1938) confirmed the importance of insects in the diet at all times of the year, with the owls catching the dominant species according to their seasonal abundance, i.e., weevils in spring, cockchafers in summer, earwigs and dor beetles in autumn, and even rove beetles in winter.

In northeastern France, insects are numerically represented more in spring (79.8%), summer (75.1%), and autumn (75.9%) than in winter (28.5%), while rodents follow the opposite trend: 15.9% in summer and 70.4% in winter (Génot & Bersuder 1995). However, in southern France, invertebrates (number of prey and biomass) are numerous year-round, but mammals remain the primary food source by biomass (Génot 1992a).

In the former West Germany, beetles remain important numerically in the diet (73%) all year (Kuhn 1992). In Austria, the number of insects in the diet reaches its highest percentage (67%) in summer but remains low in biomass (2%), and even though the proportion of mammals is reduced in summer, their role in the diet, in terms of biomass, remains high all year (Sageder 1990).

In Switzerland, the seasonal variation of prey composition was not significant, except for an increase of Cantharid larvae during the winter (Schmid 2003).

In southern Italy, Coleoptera are present year-round but are more important prey in autumn (27.9% biomass). Ground and rove beetles are mainly represented in spring and winter, respectively. Scarabid and longhorn beetles are found in summer, while histerids are only recorded in autumn. Mammals and birds represent the main diet by biomass in spring; mammals, birds, and reptiles in summer; and birds, mammals, and Coleoptera in winter. Given the numbers of prey items in the winter diet, earwigs are rather strongly represented (22.3%) (Moschetti & Mancini 1993).

In central Italy, the main seasonal difference in the diet is reflected by changes in the amount of birds, scarabid beetles, and earwigs. In summer, birds are more important (31.9% of the biomass) instead of 10–17% for other seasons. Earwigs are only found in spring, and are scarce in autumn–winter and practically absent in summer. A similar situation exists in the seasonality of scarabid beetles. Finally, vertebrates represent 8.7% of prey numbers and 80% of their biomass in summer, and 2.7% of prey numbers and 60.2% of their biomass in autumn–winter (Zerunian *et al.* 1982).

In Spain, even if rodents represent the main part of the diet in spring, arthropods are important all year round (Manez 1983b).

In Ukraine, in particular the Crimea, rodents and beetles are represented by number: 40% and 39% in April, 38% and 53% in May, 54% and 34% in June, and 11% and 48% in September, respectively (Kostin 1983).

In Turkmenistan, rodents are most significant in winter (91% by number of prey) and then decline in spring (65%) and in summer (56%), while birds, reptiles, and insects, mainly beetles, increase (Sukhinin *et al.* 1972). In the same country, the diet can change during winter and in different years, with invertebrates representing 26.8% by number in 1959 to 41.1% in 1960 (Bel'skaya 1992).

In Uzbekistan, rodents dominate the diet almost all year, and especially in winter. During the other seasons, birds, reptiles, and insects comprise the remainder of the food spectrum (Allanazarova 1988).

Finally, the seasonal change in the diet depends on the density and the availability of certain prey. The preceding information on diet supports a clear pattern, that is, rodents form the base by biomass everywhere, but, depending on the region, insects, mainly beetles, grasshoppers, and earwigs, increase in importance during the summer nestling period and in autumn by number and biomass.

7.10 Geographical variation

Variations in the diet of the Little Owl are linked to overarching localized modifications in habitat conditions. For example, in southern Portugal, R. Tome (personal communication) found more invertebrates (mainly beetles, in particular scarabid) in steppe areas (71.4%

by biomass), and more rodents in the Holm-Oak woodland area (58.8% by biomass). In central Italy, the owl's diet was studied at five sites with insects forming 91.1% of prey items by number; Staphylinidae and Tenebrionidae were mostly taken in urban sites while Scarabaeidae were caught at a rural site (Fattorini *et al.* 2000). Manganaro *et al.* (2001) compared the percentage of prey numbers in the diet of owls at two sites (farmland and urban) in Rome. They found more earwigs, grasshoppers, and beetles, and fewer reptiles, birds, and small mammals at the farmland site compared to the urban site (Manganaro *et al.* 2001).

In Spain, the owl's diet was analyzed at 11 sites located in three main climatic zones: wet climate in the north, "cold" Mediterranean in central Spain, and Mediterranean. The diet was based on invertebrates and locally on reptiles in the Mediterranean area, on small mammals in the wet area, and on a mixture of both in the cold Mediterranean area where small mammals always reflect the larger part of the biomass but not so much as in the wet area (Manez 1983b).

In Kazakhstan, a comparison of the Little Owl's diet was made between the desert and semi-desert environment. More Carabidae, Cerambycidae, Tenebrionidae, and Buprestidae, and fewer Scarabaeidae, were found in the desert. This is related to the occurrence of these beetles in the steppe and the sand habitats. For example, Little Owls caught a lot of Buprestidae, in particular *Julodis variolaris*, in the steppe with bushes of *Tamarix* sp. because the larvae of these beetles develop in the roots of these shrubs (Formozov & Osmolovskaya 1953).

The geographic variation of the diet shows that the species is a generalist from the trophic point of view. The Little Owl can occupy three main trophic niches: small mammals in central Europe, insects in the Mediterranean basin, and both kinds of prey in the middle regions.

Finally, the Little Owl is a non-selective predator, catching prey groups according to their relative abundance and biomass, depending on the region, season, and local habitat conditions. Studies show that the Little Owl is more of a generalist than the Barn Owl, globally recognized as a small mammal specialist (Gotta & Pigozzi 1997; Manganaro *et al.* 2001). Delibes *et al.* (1983) also considered the Little Owl to be more of a generalist than the Long-eared Owl (*Asio otus*) in Spain, while Romanowski (1988) showed that in a suburban area of Poland, the Little Owl and the Long-eared Owl share the same trophic niche breadth with a high degree of specialization. In this study, Romanowski suggested that the actual contribution of insects to the Little Owl's diet is generally overestimated. In such a case, competition could occur between both owls, but the generalist nature of the Little Owl's trophic niche generally avoids an overlap with other predators. There may be exceptions in particular situations, such as in Uzbekistan where competition exists between the Little Owl and snakes (Blunt-nosed Viper [*Vipera lebetina*] and Saw-scaled Viper [*Echis carinatus*]) (Ishunin 1965).

8

Breeding season

8.1 Chapter summary

This chapter will cover the whole of the Little Owl's breeding cycle (see Figure 8.1). The breeding season is obviously a critically important period during which reproduction can be influenced by many different factors, i.e., weather, food, habitat, density, geographical location, and parental experience. The season begins in January or February (Glue & Scott 1980; Exo 1987) with the affirmation of territorial boundaries and onset of courtship. The Little Owl does not have high productivity due to very few replacement clutches, moderate fledging success, and relatively high egg failure. According to the mortality rate of adults and juveniles, each pair should produce between 1.7 and 2.34 fledged young per year to compensate for mortality (Exo 1992) and actually most of the long-term breeding studies across Europe show results ranging between both values. Analysis of consistently organized long-term demographic data is needed to enhance our understanding of Little Owl population dynamics. Further, this demographic data needs to be efficiently linked with specific habitat conditions at the nest-site, home-range, and landscape scales.

In our preparation of this chapter, we became aware that clarification of terminology related to nesting success and reproduction was needed. Clarification of these terms is important to provide an accurate and consistent foundation for the data that will be used to assess the reproductive performance of the owls, as well as in the longer term monitoring of status and trends.

8.2 Courtship and mating

Given sufficient food resources, Little Owls occupy the same territory all year long. Courtship begins simultaneously with displays of territorial defense. In the period January to March the male patrols his home territory emitting a high pitched "gooock" call that serves as both a territorial call, warning off potential male challengers, as well as a contact call to his mate. Once contact has been established, pair bonding behavior occurs; this follows a pattern of flying in pairs and alighting on favorite perches on the same tree, with the female uttering begging calls (a similar call to begging young as noted by Haverschmidt [1946]). The male will respond to the calls by feeding the female with prey he has caught. The prey

Figure 8.1 Litte Owls mating (François Génot).

Figure 8.2 Little Owls mating. Photo Ludo Goossens.

provision stage is very important for the prediction of reproductive output, as it both serves as an indication of prey availability and it conditions the female with extra fat resources for egg formation and body mass for incubation (Leigh 2001a). The vocal activity increases and the paired owls indulge in frequent calling; which Haverschmidt notes "I witnessed in broad daylight a duet, which both birds were sitting on an exposed branch on a tree constantly calling and answering each other" [technically this is not "dueting" but rather call-response vocalizations]. This behavior increases and intensifies until it leads to the more advanced stages of courtship of nest visits and copulation (see Figure 8.2). Nest visiting and showing consists of flying as a pair to the various nest sites, which are preselected by the male within his territory. The visits consist of the female entering the cavity or site to determine its suitability as a nest site and it seems that the ultimate nest-site choice is undertaken by the female. This activity can be carried out either before or during the copulation process. Copulation is a very under-recorded phase of courtship; however, Haverschmidt undertook a very detailed study of the local Little Owl behavior and described the copulation behavior as follows:

"The session begins with the male owl alighting on to his favourite perch and then continues with a questioning call 'hoo,hoo?' which he repeats continuously increasing intensity and volume, building up to a 'hoo-ee, hoo-ee?' The male shifted position a few times on the branch during the calling session. Suddenly a second owl appeared and alighted near to the male, the female in contrast with the male was silent. Without any further uttering the male jumped up onto the female's back, the female leaned forward so her body became horizontal and lifted up her tail a little, to which the male sunk down onto her, soon after copulation occurred with frenzied wing flapping. The female then produces a shrill shrieking sound. Sometimes the male continued to call while sitting on the female's back, and often flew round in a semi-circle to alight into a cavity at the rear of the tree, as if to show the cavity to the female. He then flew back to his favourite perch again.

The time just before the copulation, when the male is on the back of the female, can last 30 seconds (average 27s, n = 10), while the copulation time is shorter with an average of 3.1s (n = 18) (Etienne 2003). The mating includes the "male on the female's back" and the copulation. It lasts on average 32.9s (n = 56) and there are 0–4 copulations per evening, 66% of copulations occur during the evening (n = 64) (Ancelet 2004). The maximum time of mating can reach five minutes (Torregiani 1981). The mating season is from the beginning of February up to the beginning of May. Some copulations were seen on 28th November (Etienne 2003) and on 6th December (Exo 1987) but this could have had a social function, to decrease the aggressiveness between the two partners (Exo 1987). Ancelet (2004) saw 53 matings at the same place, on a particular branch of an apple tree.

8.3 Nest sites

Little Owls will use a variety of nest sites; however, the specific requirements are (i) a chamber with a large enough capacity to hold the eggs and subsequent young as they grow, (ii) the chamber has to be able to protect the contents from wind, rain, and also preferably be dark, (iii) the entrance must be accessible to the owls but prevent access by predators. See Figures 8.3, 8.4 and 8.5.

Little Owls will use any suitable, available site within their territory that is situated in close proximity to their foraging habitat. When available, Little Owls prefer to nest in tree cavities, and splits in trunks or branches, that give the owls excellent opportunities for nesting.

Among many alternative nest sites are crevices in ruins and down wells, adobe buildings in the Sahara Desert, holes in quarries, walls, sand-pits, stick nests (Uspenskiy 1977; Sagitov 1990) and disused rabbit burrows (Bannerman 1955). In the former USSR, breeding was recorded in windmills, mud tombs, granaries, haylofts, haystacks, below overhanging rocks, in burrows of large gerbils, and nests of Rock Nuthatches (*Sitta neumayer*); some nest sites are tunneled by the birds themselves (Dementyev & Gladkov 1951). "Clapas", walls, ruins, rocks, and "chazelles" offer cavities at Causse Méjean, France (Juillard *et al.* 1992). In Haut-Léon, western France, only buildings are used as nesting sites, in 55% of cases they are under the roof (Clech 2001a). Rockpiles are used as nesting in Syria (Figure 8.6). Specific nicknames of owls relate to their nesting places, for example "Chouette des tuiles" (roof owl) in southern France (Barthelemy & Bertrand 1997), "Baumkauz" (tree owl) (Weimann 1965) and "Stockeule" (willow owl) (Schönn *et al.* 1991) in central Europe. Buildings and mounds of stones are used in Tolfa, Italy (Centili 1996). Fruit trees (especially apple), pollarded trees, man-made constructions, and nestboxes are used as nest sites in Betuwe (Netherlands) (See Fuchs 1986 for proportional distribution of the use). Buildings

Figure 8.3 Little Owl in cavity (François Génot).

Figure 8.4 Examples of nest-site cavities (after R. van Harxen & P. Stroeken, personal communication).

used as nest sites include stables, haylofts, and used and disused farm buildings (Schönn *et al.* 1991).

Glue and Scott (1980) studied a selection of 482 nest sites, they found that 24% were in oak, 23% in ash, 18% in fruit trees, and 15% in willows. Of 316 nests in Germany, 54.5% were in trees (27% in fruit trees, 17% in pollard trees 7% Tiliaand Quercus, 4% in other trees), 27.5% in buildings, and 18% in assorted places, e.g., quarries and nestboxes (Schönn 1986). Of 530 nests in France, 18% were in fruit trees, 11% were in pollarded trees, 12% in other trees, 32% in buildings, and 26% in other places such as nestboxes and rock faces. (Génot 1992c). Of 100 nests in buildings in Brittany (western France), 46% were in agricultural buildings, 6% in manors, 9% in inhabited houses, 33% in isolated or abandoned houses, 5% in hangars, and 1% in pigeon houses (Clech 2001a). Of 144 nests in Austria, 17 were in trees, 62 in barns, 53 in wine cellars, 6 in farms, 4 in stacks of straw bales, and 2 in churches (Ille & Grinschgl 2001). In the Mediterranean region, agricultural practices provide nesting opportunities by clearing all large rocks and stones from the land. Farmers pile them up at the side of the land they have cleared. These stone piles provide alternate nest sites in areas that lack suitable nest trees (Juillard *et al.* 1992). Centili (2001a) recorded 39 nests in Italy: 25 were in stone piles, 13 in buildings, and 1 in an iron pole. In Britain at high elevation, tree-lined hedgerows give way to stone walls that are utilized by Little Owls as nest sites. In a study area in Cheshire, England, a pair of owls had

Figure 8.5 Sketches of Little Owl breeding cavities (François Génot).

Figure 8.6 Rockpiles as breeding location of Little Owls in the Daraa region of Syria. (Photo: Yousef Ali Alzaoby.)

nested for several years in a nestbox situated five meters up an oak tree; storms damaged the box just prior to the breeding season one year and the owls vacated the nestbox and nested in a rabbit burrow close to the tree. The following year the nestbox had been replaced and the owls returned to the box. The burrow was inspected and it seemed to be in good condition, showing that the owls preferred the elevated site in this case (Leigh 2001a). The diversity of buildings used as nesting places is shown in the Belarus Republic where Little Owls use barns, windmills, water-towers, silo-towers, building blocks, churches, cemetery groves, old bee-hives, and even unusual places such as haystacks (Nikiforov *et al.* 1989).

In natural biotopes, e.g., deserts and semi-deserts of the former Soviet Union, the species uses as nesting sites: holes in rocks and cliffs in river valleys, holes in loess precipices or in sandy deserts, hollows of trees, burrows of Great Gerbils or Thin-toed Souslik (Sagitov 1990), abandoned burrows of foxes, badgers, and steppe tortoises in hilly sands, burrows of Rollers (*Coracias garrulus*) in the banks of rivers or slopes of ravines, and burrows of the Rock Nuthatch in clefts of rocks (Gavrin 1962). In Moldavia, the Little Owl occupies empty nests of the Magpie (*Pica pica*) far from human settlements. On about 50 ha, out of 100 Magpie nests, 42 were occupied by the Little Owl for breeding, in a sort of "nesting colony" (Uspenskiy 1977). In the Caucasian area, Little Owls also breed in old nests of the Carrion Crow (*Corvus cornix*) (Il'yukh 2002).

Nestboxes are widely used as a conservation tool in areas where nest-site availability is a limiting factor, and in these cases the nestboxes are readily used. A number of different designs are available (see the general specifications, provided in Chapter 11). In areas where natural nest sites are not a limiting factor, it can take a number of years to entice the owls into a nestbox. The nest height will vary according to availability; in a study in France, Génot (1990b) found the range of nest heights of 25 nests in fruit trees to be from 1–4.3 m high. The average nest height is 3.9 m in trees, buildings, and nestboxes in Switzerland (Juillard 1984) (n = 59), and 2.4 m in trees in Germany (Exo 1981) (n = 28). In buildings, the owls mainly use cavities in or under the roof space, often under loose or missing roof tiles. Agricultural buildings offer further nesting opportunities, barns with ledges are readily used, as are hay barns.

8.4 The nest

Nest dimensions recorded in British cavities were on average 20 cm wide (10–50 cm) (n = 74). Ten entrance passage lengths had an average length of 80 cm (50–130 cm) (Glue & Scott 1980). Génot (1990b) undertook a similar exercise in France and found that the average cavity depth of 25 nest sites was 77 cm (32–200 cm). The positions of the entrance holes and the cavities inside trees were analyzed by Exo (1981) and Génot (1990b). When the species uses burrows, the nest can have a diameter of 20–30 cm and be located at 1–3 m from the entrance (Gavrin 1962) or a diameter of 20–25 cm and at 1–1.5 m from the entrance (Sagitov 1990). A nest chamber was recorded with dimensions of 25 cm by 14 cm (Abdusalyamov 1971). When the dimensions of the nest do not provide enough space for the young to move around, they leave the nest early and often suffer increased mortality (Ancelet 2003).

Different types of holes can be distinguished, including horizontal and vertical (Exo 1981; Génot 1990b). In Switzerland, one cavity suitable for the Little Owl was identified from 110 fruit trees (n = 561 including 250 apple trees) (Juillard 1984), in France, one suitable cavity was identified from 60 fruit trees (n = 8839 including 2386 apple trees) (Génot 1990b). Natural cavities, such as those in apple trees, used during a limited number of years, give rise to more frequent changing of nest site (Fuchs 1986). During sunny days, overheating can occur as the temperature inside a nestbox may be 7–9 degrees above the outside temperature, putting the development of the embryos at risk (Beersma & Beersma 2000). This is also the case with nests under roofs. It is advisable to place nestboxes in trees in the shadow of their leaves. Little Owls may not be able to distinguish safe nest locations from those that are likely to overheat, or they may not have a choice of locations (Van Den Burg *et al.* 2003).

In terms of construction no nest building is carried out, although the female will form a scrape in the litter in the chamber (Ullrich 1973). A bed of pellets forms the base for

Table 8.1 *Clutch size and characteristics of eggs of the Little Owl in the Front-Caucasian area (after Il'yukh 2002).*

Indices	N	Range	M ± m
Clutch size	31	–8	4.87 ± 0.22
Length (mm)	89	32.2–36.1	34.32 ± 0.09
Width (mm)	89	26.9–31.1	29.10 ± 0.09
Volume (cm^3)	89	12.5–17.8	14.86 ± 0.12
Index of length (%)	89	78.2–88.5	84.81 ± 0.20

N = number of eggs
M = mean
± m = standard error of mean

the eggs in previously used nest sites. Normally there is no litter in the nest, but only a layer of food remains or pellets. In some nests, stems of dry grass were discovered (Sagitov 1990) and pieces of pine wood, which have a positive effect on the nestlings by providing drier conditions in the nest during wet weather (Robiller 1987). Prior to laying, the female moults her belly feathers creating a brood patch on the underbelly of the owl. The brood patch has increased blood flow directed to it, which increases the heat-exchange efficiency when incubating the eggs. The moult and the appearance of the brood patch depends on clutch size; Illner (personal communication) states that in his experience with small clutch sizes, the brood patch does not develop until the second or third egg is laid.

8.5 Eggs

The eggs are white, with a silky smooth finish. The shape is a slightly elliptical sphere. The newly laid eggs average approximately 14–15 g in weight, and measure 33–40 mm in length by 27–31 mm in breadth. For *Athene noctua noctua*, the mean dimensions are: 34.4 mm × 28.8 mm (n = 140) (Schönwetter 1964; Schönwetter *in* Schönn *et al.* 1991); for *Athene noctua vidalii*: 36 mm × 30 mm (n = 100) (Witherby *et al.* 1938); for *Athene noctua indigena*: 34 mm × 28 mm (n = 38) (Makatsch 1976); and for *Athene noctua bactriana*: 30.9–34.5 mm × 26.1–28.9 mm (Sagitov 1990).

The mean weight of a fresh egg is 15.6 g, and the range is 12.5–19 g (n = 95) (Illner *in* Glutz & Bauer 1980).

Il'yukh (2002) examined Little Owl eggs in the Caucasian area (Table 8.1). With data spanning several years (1999, 2000, 2001) he found that measurements of the size and shape of the eggs were not significantly different. Further, he found that there was a very low coefficient of variation in the dimensions, which is typical for birds that breed in the limited, stable spaces of closed nest sites.

Table 8.2 *Inter-year variability of sizes and form of eggs of the Little Owl in the Front-Caucasian area (after Il'yukh 2002).*

Years	N	Range	M ± m
		Length (mm)	
1999	38	32.2–36.1	34.36 ± 0.16
2000	12	32.5–35.0	33.55 ± 0.23
2001	39	32.8–36.0	34.53 ± 0.10
		Width (mm)	
1999	38	27.5–31.1	29.03 ± 0.14
2000	12	26.9–29.5	28.31 ± 0.23
2001	39	28.1–31.1	29.42 ± 0.11
		Volume (cm^3)	
1999	38	12.5–17.8	14.80 ± 0.21
2000	12	12.7–15.3	13.72 ± 0.22
2001	39	13.5–17.5	15.26 ± 0.15
		Index of length (%)	
1999	38	81.2–87.7	84.51 ± 0.25
2000	12	78.2–87.9	84.44 ± 1.02
2001	39	83.0–88.5	85.22 ± 0.22

N = number of eggs
M = mean
± m = standard error of mean

However, there was significant inter-year variability in some egg measurements (Table 8.2). In 2001, a year with a rainy and cool spring, Little Owl eggs were significantly larger and more round than in the dry hot season of 1999 and the spring of 2000. The difference in Little Owl egg length was significant between 1999 and 2000 ($t = 2.89$; $p < 0.01$) and between 2000 and 2001 ($t = 3.91$; $p < 0.001$). Egg width in 2001 was significantly higher than in 1999 ($t = 2.19$; $p < 0.05$) and in 2000 ($t = 4.35$; $p < 0.001$), and in 1999 it was higher than in 2000 ($t = 2.67$; $p < 0.01$). By volume, eggs laid in 2000 were considerably smaller than those from 1999 ($t = 3.55$; $p < 0.001$) or from 2001 ($t = 5.78$; $p < 0.001$). In 2001 the eggs were more round than in 2000 or 1999 ($t = 2.13$; $p < 0.05$). It is interesting to note that the length of eggs (and the width to a lesser extent) varied more in the cool and rainy seasons, than in those that were hot and dry.

The Little Owl also shows seasonal changes in the size and shape of eggs (Table 8.3). In early clutches, eggs were significantly smaller than those from later clutches in length ($t = 4.36$; $p < 0.001$), width ($t = 2.76$; $p < 0.01$) and volume ($t = 3.41$; $p < 0.01$).

Besides that, there are clear links in quantitative indices of Little Owl eggs and clutch size (Table 8.4). The largest eggs were observed in clutches consisting of six eggs, the

Table 8.3 *Seasonal variability of sizes and form of eggs of the Little Owl in the Front-Caucasian area (after Il'yukh 2002).*

Laid eggs	N	Range	M ± m
		Length (mm)	
Early	55	32.2–35.6	34.04 ± 0.12
Late	34	33.7–36.1	34.78 ± 0.12
		Width (mm)	
Early	55	26.9–30.1	28.91 ± 0.10
Late	34	27.5–31.1	29.43 ± 0.16
		Volume (cm^3)	
Early	55	12.5–16.3	14.53 ± 0.14
Late	34	13.0–17.8	15.39 ± 0.21
		Index of length (%)	
Early	55	78.2–88.5	84.94 ± 0.27
Late	34	81.2–87.8	84.60 ± 0.27

N = number of eggs
M = mean
± m = standard error of mean

smallest eggs were observed in clutches of seven and eight eggs. The lengths of eggs in eight-egg clutches were significantly shorter than those in four-egg ($t = 3.54$; $p < 0.01$), five-egg ($t = 3.17$; $p < 0.01$) and six-egg ($t = 3.75$; $p < 0.001$) clutches. Likewise, the width of eggs in eight-egg clutches were significantly less than in four-egg ($t = 5.76$; $p < 0.001$), five-egg ($t = 3.56$; $p < 0.01$) and six-egg ($t = 6.67$; $p < 0.001$) clutches. Seven-egg clutches contained narrower eggs than four-egg ($t = 2.54$; $p < 0.05$) and six-egg clutches ($t = 2.78$; $p < 0.05$). Finally, by volume, eggs of eight-egg clutches were significantly lower than four-egg clutches ($t = 5.41$; $p < 0.001$), five-egg ($t = 3.70$; $p < 0.001$) and six-egg ($t = 5.62$; $p < 0.001$) clutches; likewise seven-egg clutches consisted of eggs significantly smaller by volume than four-egg ($t = 3.30$; $p < 0.01$), five-egg ($t = 2.31$; $p < 0.05$) and six-egg ($t = 3.60$; $p < 0.01$) clutches.

Finally, in rainy springs, Little Owl eggs were significantly bigger, the eggs hatched later, and clutch sizes were smaller.

8.6 Laying of eggs

In western Europe, egg-laying generally takes place in the last week of April (Knötzsch 1978; Ullrich 1980; Génot 1992b; Lederer & Kämpfer-Lauenstein 1996) or in the second half of April (Bultot *et al.* 2001). In the arid regions of the owl's distribution, laying starts

Table 8.4 *Characteristics of Little Owl eggs in different sized clutches in the Front-Caucasian area (after Il'yukh 2002).*

Egg number in clutch	N	Range	M ± m
		Length (mm)	
4	28	32.8–35.4	34.40 ± 0.12
5	25	32.5–36.1	34.39 ± 0.19
6	18	33.0–36.0	34.61 ± 0.20
7	7	33.0–35.0	33.93 ± 0.32
8	8	32.2–34.4	33.29 ± 0.29
		Width (mm)	
4	28	28.1–31.1	29.34 ± 0.14
5	25	27.5–31.1	29.09 ± 0.20
6	18	28.6–30.3	29.42 ± 0.12
7	7	26.9–29.5	28.31 ± 0.38
8	8	27.6–28.7	28.24 ± 0.13
		Volume (cm^3)	
4	28	13.5–17.5	15.12 ± 0.18
5	25	12.8–17.8	14.88 ± 0.28
6	18	13.8–16.5	15.29 ± 0.21
7	7	12.7–15.3	13.88 ± 0.33
8	8	12.5–14.4	13.54 ± 0.23
		Index of length (%)	
4	28	83.0–88.5	85.31 ± 0.30
5	25	81.2–87.9	84.58 ± 0.34
6	18	83.1–87.7	85.04 ± 0.27
7	7	78.2–87.6	83.53 ± 1.64
8	8	83.0–86.9	84.85 ± 0.50

N = number of eggs
M = mean
± m = standard error of mean

in the second week of March (Uzbekistan; Sagitov 1990) or in the third week of March (Turkmenistan; Ataev 1977). In the Saratov region (Russia), laying occurs in late May to early June (Zav'yalov *et al.* 2000). If we take into account the very early clutches and replacement clutches, egg-laying is extremely extended through the distribution area and can last for three months as in Kazakhstan (from the beginning of April to the end of June; Gavrin 1962) or even five months as in Turkmenistan (from the first week of March to the middle of August; Bel'skaya 1992).

The earliest recorded laying dates are March 16, 1950 in Britain (Glue & Scott 1980); March 19, 1959 in Tajikistan (Abdusalyamov 1971); March 22, 1974 in Turkmenistan

(Ataev 1977); April 4, 1938 in France (Labitte 1951); April 6, in Germany (Schönn 1986) and in France (Blache 2004); April 9, 1977 in Switzerland (Juillard 1984a). Unusually early records for eggs were December 1949 in Hungary (Solymosy 1951), and January in France and Turkey (Pardieu 1927, E. Vaassen, personal communication).

The laying date is influenced by the availability of food, with early laying dates corresponding to peak vole years (Illner 1979; Ullrich 1980; Finck 1989; Sill & Ullrich 2005). Climatic conditions also have a relationship to the laying date. If the duration of snow cover in winter is long, the laying date can be delayed (Illner 1979; Ullrich 1980; Exo 1992). Lederer and Kämpfer-Lauenstein (1996) reported that the laying date was earlier when there were higher amounts of rainfall in spring (March and April). Their explanation is that during wet springs, the availability of earthworms is greater and the female can achieve a better physical state. But Bultot *et al.* (2001) found the contrary; for each 100 mm of extra rainfall egg-laying was delayed an average of six days.

External factors may affect the breeding dynamics greatly. In other owl species when negative factors are prevalent, e.g., when prey cycles are extremely low and the weather factors are not good, the amount of resources is too low and the owls' poor physical condition will not allow them to lay eggs that year. Additional research is needed on the status and physical conditioning of non-breeding Little Owls.

8.7 Clutch size

While the main studies on breeding have focused on western European countries, some data on clutch sizes come from countries of the former Soviet Union. Clutch sizes vary, and most commonly contain between one and seven eggs, but clutches of eight and nine eggs have been recorded in eastern countries (Gavrin 1962; Abdusalyamov 1971; Ataev 1977; Hudec 1983; Sagitov 1990; Hudec *in* Schönn *et al.* 1991; Nemeth *in* Schönn *et al.* 1991; Rakhimov & Pavlov 1999; Il'yukh 2002) and even ten in the Netherlands (Willems *et al.* 2004) or twelve in the desert of Turkmenistan (Sopyiev 1982) (Table 8.5). The high numbers of eggs in these latter cases could also be attributed to two clutches in the same nest. The average clutch size in Europe varied from 2.65 (Leigh 2001a) to 5.24 (Kohl *in* Schönn *et al.* 1991) (Table 8.6). Average clutch sizes were 6.2 (n = 12) in Kazakhstan (Gavrin 1962), 4.4 (n = 9) in Russia, Saratov Region (Zav'yalov *et al.* 2000), 6.67 (n = 15) in Turkmenistan (Ataev 1977) and 6 (n = 6) in Uzbekistan (Bakaev *in* Sagitov 1990).

Clutch size can be influenced by a number of factors such as weather, food availability, timing of laying, population density, age of the female, geographic location, etc. In Europe, the clutch sizes become larger from west to east, and decrease from south (larger clutches) to north (smaller clutches). Changes in clutch size are thought to be related to climatic differences and the owl's preference for semi-arid conditions (Exo 1992). In France, Génot (1992c) reported an increase in clutch size in relation to the climatic shift from oceanic west to the south. Soler and Soler (1992) hypothesized in their study of latitudinal trends that there is a significant positive correlation between egg sizes and clutch sizes, and increasing latitude.

Table 8.5 *Clutch sizes of the Little Owl* (Athene noctua) *recorded across Europe.*

Country	1	2	3	4	5	6	7	8	9	10	Mean	Number of clutches	Author
Britain	1.00	10.00	35.00	39.00	13.00	1.00	1.00	0.00	0.00	0.00	3.6	268	Glue & Scott 1980
Britain	11.50	30.97	33.63	23.89	0.00	0.00	0.00	0.00	0.00	0.00	2.6	115	Leigh 2001a
France	0.00	0.00	23.00	62.00	15.00	0.00	0.00	0.00	0.00	0.00	3.9	80	Labitte 1951
France	6.00	9.00	35.00	46.00	19.00	7.00	1.00	0.00	0.00	0.00	3.67	123	Génot 2005
Switzerland	3.00	11.00	39.00	27.00	15.00	5.00	0.00	0.00	0.00	0.00	3.1	153	Juillard 1984
Netherlands	2.07	7.25	25.48	43.02	17.68	3.94	0.41	0.07	0.07	0.07	3.82	1448	Willems *et al.* 2004
Germany	0.2	2.6	11.8	37.7	32.8	10.7	4.2	0.00	0.00	0.00	4.1	126	Schönn 1986
Germany	5.00	9.00	31.00	41.00	12.00	2.00	0.00	0.00	0.00	0.00	3.5	269	Gassmann & Bäumer 1993
Belgium (Flanders)	3.8	10.3	43.3	33	8	1.3	0	0.00	0.00	0.00	3.36	312	Ph. Smets, R. Huybrechts & S. Cerulis pers. comm.
Belgium (Wallonia)	5.60	18.30	36.80	32.10	6.40	0.60	0.10	0.00	0.00	0.00	3.18	1706	J. Bultot & Groupe Noctua pers. comm.
Netherlands (Achterhoek)	1.0	5.5	20.1	44.7	22.6	5.9	0.2	0.00	0.00	0.00	4.01	477	R. van Harxen & P. Stroeken pers. comm.
Belgium (Dijleland)	0.8	5.50	29.70	40.60	20.30	2.30	0.80	0.00	0.00	0.00	3.85	123	Vanden Wyngaert 2005

pers. comm. = personal communication

Table 8.6 *Average clutch sizes of the Little Owl* (Athene noctua) *recorded across Europe.*

Country	Years	Number of clutches	Average clutch size	Range	Author
Netherlands	1977–2003	1448	3.82	1–10	Willems *et al.* 2004
Britain	1939–77	268	3.59	1–7	Glue & Scott 1980
Britain	1993–2000	115	2.65	1–4	Leigh 2001a
France	1918–50	80	3.91	3–5	Labitte 1951
France	1984–2004	123	3.67	1–7	Génot 2005
Switzerland	1973–80	153	3.12	1–6	Juillard 1984
Switzerland	1984–2000	189	3.70	–	Meisser & Albrecht 2001
Germany	1974–84	96	3.61	2–6	Exo 1987
Germany	1973–88	265	4.42	1–8	Knötzsch *in* Schönn *et al.* 1991
Germany	1960–85	126	4.11	1–7	Schönn 1986
Germany	1971–97	326	3.60	–	Furrington 1998
Germany	1969–88	161	4.13	2–7	Ullrich *in* Schönn *et al.* 1991
Germany	1974–82	154	3.85	2–7	Illner *in* Schönn *et al.* 1991
Germany	1977–85	153	3.65	2–5	Kimmel *in* Schönn *et al.* 1991
Germany	1988–2001	812	3.71	–	Keil unpublished
Germany	1978–92	269	3.51	–	Gassmann & Baümer 1993
Germany	1974–94	486	4.15	–	Kämpfer-Lauenstein & Lederer 1995
Hungary	1910–76	89	5	3–8	Nemeth *in* Schönn *et al.* 1991

Table 8.6 *(cont.)*

Country	Years	Number of clutches	Average clutch size	Range	Author
Romania	1901–75	21	5.24	4–7	Kohl *in* Schönn *et al.* 1991
Russia	1989–2002	31	4.87	3–8	Il' yukh 2002
Czechoslovakia	–	37	4.76	3–9	Hudec 1983
Holland	1986–2003	477	4.01	1–7	Stroeken & van Harxen 2003b
Holland	2003	17	3.33	2–4	Blanke 2003
Belgium (Flanders)	2000–4	312	3.36	1–6	Ph. Smets, R. Huybrechts & S. Cerulis pers. comm.
Belgium (Wallonia)	1999–2004	1706	3.18	1–7	J. Bultot & Groupe Noctua pers. comm.
Belgium (Dijleland)	1999–2004	123	3.85	1–7	Vanden Wyngaert 2005

pers. comm. = personal communication

Clutch size is affected by prey populations, especially small mammal populations, before the laying phase. On average, females will lay 1.0–1.2 more eggs in the peak vole cycle years than in low vole years (Exo 1992). The influence of rodents on clutch size was shown in captivity where Little Owls easily produce large clutches (7–8 eggs) after optimal feeding during the pre-laying period (Robiller & Robiller 1986). Large clutches in the field were found in peak vole years before 1945 (Collinge 1922; Uttendörfer 1939). Pairs that breed in clusters lay fewer eggs than pairs that nest in isolation. Bultot *et al.* (2001) found in their study in Wallonia (Belgium) that clustered pairs laid on average 2.5 eggs whereas isolated pairs laid 3.5 eggs. Also Bultot *et al.* (2001) and Finck (1989) both reported information indicating that larger clutch sizes were associated with earlier laying dates. Gassmann and Bäumer (1993) found a correlation between larger clutch sizes and the amount of precipitation in March, before the egg-laying season.

8.8 Replacement clutches

When environmental and prey resources are favorable, replacement clutches can be laid in cases where the first clutch has been lost through predation or other factors. Génot (2005) recorded 8 replacement clutches out of 123 nest records in France; all replacement clutches were smaller in size than the original clutch. In Germany, Knötzsch (1978) reported that 6 out of 64 clutches were replacements; Furrington (1998) found that 5 out of 287 were replacement clutches; H. Keil (personal communication) reported that in 2004, 2 out of 166 clutches were replacements, and J. Bultot (personal communication) that in 2004, 2 out of 166 clutches were replacements. When all these observations were combined, the clutch replacement rate was found to be 1.7% (25 out of 1446 clutches).

8.9 Incubation

Gavrin (1962), Bakaev (*in* Sagitov 1990), and Glue and Scott (1980) reported that incubation of the eggs began with the first egg laid. Exo (1983) and Ataev (1977) found that incubation started after the laying of the second or third egg. In other cases, incubation took place with the laying of the second-to-last egg (Illner *in* Glütz & Bauer 1980; Robiller & Robiller 1986; Génot & Sturm 2001) or even when the clutch was complete (Labitte 1951; Enehjelm 1969; Zav'yalov *et al.* 2000). Illner (personal communication) believes that this is influenced by the clutch size; in clutch sizes larger than two eggs, the brood patch was not fully formed until the later stages of clutch formation. While the period of incubation has often been stated at 28 days, actual measurements of the period indicate a variation in the time frame. Exo (1983) found the incubation period to be from 27 to 33 days, Juillard (1984) reported 28 days, and Ullrich (1973) recorded an incubation period of 35 days. When Génot and Sturm (2001) studied the breeding cycle of 25 captive Little Owls, incubation ranged between 18–29 days, and averaged 25 days. An average incubation period of 25 days was also found by Robiller and Robiller (1986). The shortest period of incubation was 19.5 days, recorded for an owl in captivity (König 1969). Incubation will begin earlier under good weather or prey conditions, and later under poor weather or prey conditions (Ullrich 1973). It is the female who mainly incubates the eggs, the male provides food to sustain her. She leaves her nest an average of three to four times in a 24-hour period. Abdusalyamov (1971) suggested that both sexes participate in incubation, while Gavrin (1962) noted that participation of the male in incubation was not clear because both birds always stayed near the nest. In years of low prey resources the male struggles to provide enough food to keep the female's body fat at a comfortable level. When this happens she will leave the nest on more occasions, the implications of this being an increase in egg deaths (Leigh 2001a).

8.10 Hatching

Glutz and Bauer (1980) stated that eggs hatched at about one-day intervals. In an aviary Exo (1983) recorded that of 9 clutches (containing 15 eggs) 2 eggs hatched simultaneously, and in 3 cases that 3 eggs hatched during an 18-hour period. Also in an aviary, Génot and Sturm (2001) found that out of 22 clutches, 4 clutches hatched simultaneously and that the eggs in the remaining clutches hatched over an average of 1.2 days (range 1–4 days). Ataev (1977) found that chicks hatched over a three-day period. R. van Harxen and P. Stroeken (personal communication) recorded the hatching of a Little Owl in their study area in eastern Holland, as follows.

> On May 26th 14.45 h we found two owlets and two eggs in one of our monitored nest-sites. The young owls were only a few hours old due to dampness of their down. The weight of the owlets was 10.8 g and 10.6 g. One egg, which was not hatched, was emitting a peeping sound, and the owlet inside was breaking out of the egg with its beak and egg tooth. The weight of the egg was 12.7 g. Whilst we were marking the other two young for individual recognition at our next visit the egg was broken in the middle. The owlet was pushing the eggshell back by his head trying to free itself, resulting in the egg to split into two pieces, first the owlet's head became more visible and the owlet released itself from the eggshell. The whole process took approximately two minutes, after which the owlet was exhausted, and lay down motionless for a few seconds. After which we weighed the newly hatched owlet that weighed 10.1 g, the weight of the eggshell was 1.7 g.

The average weight of newly hatched owlets has been reported as 11 g and 12 g in Mikkola (1983) and Schönn *et al.* (1991) respectively.

Hatching success is defined as the proportion of eggs that hatch divided by the total number of eggs laid. The hatching success from Germany has been reported in several studies. Schönn (1986) reported that 62% (258 of 416) of the eggs recorded in his study nests hatched. Similarly, Exo (*in* Schönn *et al.* 1991) reported hatching success of 67% (180 out of 269 eggs); Knötzsch (*in* Schönn *et al.* 1991) reported hatching success of 79% (925 out of 1171 eggs); and H. Keil (personal communication) noted 86.5% (2609 of 3016 eggs). In Switzerland, Meisser and Albrecht (2001) reported hatching success of 81% (566 out of 700 eggs), and Juillard (1984a) found the rate to be 71% (339 out of 418 eggs). Génot (2005) reported hatching success of 49.3% (223 of 452 eggs) in France. J. Bultot (personal communcation) found a hatching success rate of 83.1% (6265 out of 7536 eggs) in Belgium for 1989 through 2004. Smets *et al.* (personal communcation) found a hatching success rate of 80.1% (836 out of 1044 eggs) in Belgium.

In addition to infertility, loss of or abandonment by the adults, problems with the nest structure, or other complications, eggs are also lost to predators. The list of predators on eggs includes: Magpie (*Pica pica*), Fox (*Vulpes vulpes*), Ermine (*Mustela erminea*), Hedgehog (*Erinaceus europaeus*) (Glue & Scott 1980), Jackdaw (*Corvus monedula*) (Staats Von Wacquant-Geozelles 1890), Brown Rat (*Rattus norvegicus*), Squirrel (*Sciurus vulgaris*) and Dormouse (*Glis glis*) (Juillard 1984); the Stone Marten (*Martes foina*) is a particularly effective predator on the eggs or young of the Little Owl, especially in nestboxes without protection systems (Furrington 1979; Schwarzenberg 1981; Schönn 1986; Génot 1992a).

As shown in Table 8.7, the percentage of nest losses ranges from 17% in Britain (Leigh 2001a) to 72% in France (Génot 2005).

The loss of eggs and young from nests can be attributed to many causes. Juillard (1984) reported losses of 31 complete clutches to the following: 16 were infertile eggs, 8 were lost to predation, 2 were abandoned, and the cause of loss of 5 others was unknown.

Of 48 nest failures reported by Schönn (1986), 15 were lost to predation, 4 were infertile, 3 were abandoned, 2 were lost to flooding, 2 to human disturbance, 2 to the nest falling, 1 due to a fallen nest tree, 1 due to ants, and in 18 others the causes were unknown.

Of 52 nest failures reported by Glue & Scott (1980), 28 were due to humans taking the eggs or young, destroying the nest, or shooting the adults; 8 were lost to predation; and 16 were lost to other causes.

8.11 Owlet development

The weight gain of Little Owl nestlings is shown in Figure 8.7.

One to five days

When Little Owls hatch they are covered in a short white down, their eyes are closed, and they rely on their mother and each other for warmth in a "warmth pyramid" (Schönn *et al.* 1991). The average weight is 10 g (Ataev 1977), 11 g (Bakaev *in* Sagitov 1990) or 12 g (Juillard 1979). The diamond-shaped "egg tooth" on the bill, which breaks the shell of the egg, is visible until the 13th to 15th day. The owlets stay on their abdomen or breast with their angular legs underneath or next to their body; the head is supported by the bill or

Table 8.7 *Hatching success of the Little Owl* (Athene noctua) *across Europe.*

Country	Eggs (clutch number)	Hatched (%)	Fledged (%)	Failure (%)	Author
Britain	477 (156)	269 (56.4)	234 (49)	243 (51)	Glue & Scott 1980
Britain	305 (120)	265 (86.9)	253 (83)	52 (17)	Leigh 2001a
France	452 (123)	223 (49.3)	201 (46.6)	229 (50.7)	Génot 2005
Switzerland	478 (153)	339 (71)	279 (58.4)	199 (41.6)	Juillard 1984a
Switzerland	700 (189)	566 (81)	358 (51)	342 (49)	Meisser & Albrecht 2001
Belgium	984 (257)	689 (77)	585 (65.4)	309 (34.6)	Bultot *et al.* 2001
Germany	1171 (265)	925 (79)	706 (60.3)	465 (39.7)	Knötzsch in Schönn *et al.* 1991
Germany	416 (102)	258 (62)	201 (48.3)	215 (51.7)	Schönn 1986
Germany	3016 (812)	2609 (86.5)	2238 (74.2)	778 (25.8)	Keil unpublished
Netherlands	1886 (471)	1473 (78.1)	1073 (56.9)	813 (43.1)	Stroeken & van Harxen 2003b
Belgium (Flanders)	1044 (312)	836 (80.1)	614 (58.8)	430 (41.2)	Ph. Smets, R. Huybrechts & S. Cerulis pers. comm.
Belgium (Wallonia)	7536 (2346)	6265 (83,1)	5476 (72.6)	2063 (27.4)	J. Bultot & Groupe Noctua pers. comm.
Belgium (Dijleland)	473 (123)	344 (72.7)	273 (58)	200 (42)	Vanden Wyngaert 2005
Germany	269 (76)	180 (66.9)	142 (52.8)	127 (47.2)	Exo in Schönn *et al.* 1991

pers. comm. = personal communication

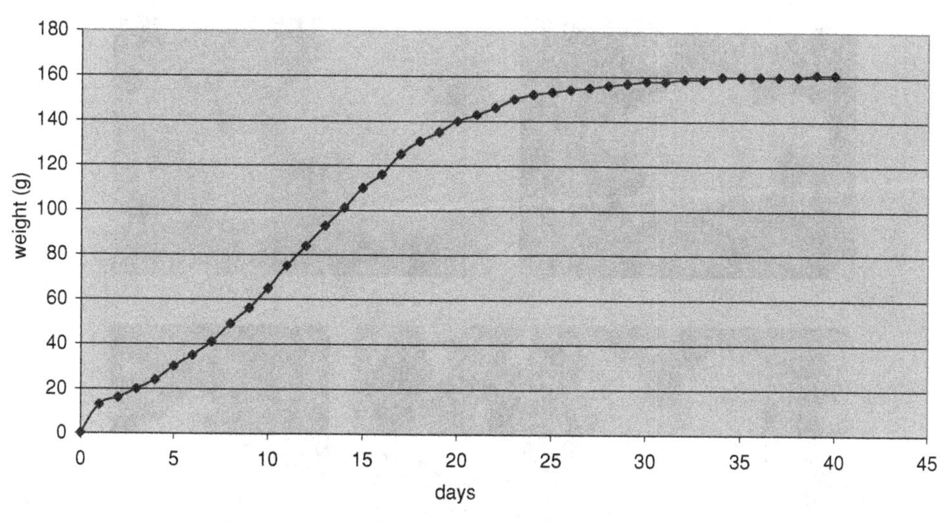

Figure 8.7 Weight gain profile of Little Owl (*Athene noctua*) nestlings while in the nest (data from R. van Harxen & P. Stroeken personal communication).

the chin, the wings are feeble. When they stand on their talons, they can trot a few steps (Schönn *et al.* 1991). See Figure 8.8.

One week

The down on the owlets changes from a white color to a gray and has a more fluffy texture; their primary feather pins are just beginning to show; their legs, feet, and talons are developing. Their cries sound like whistled begging ("chsij") or a shrill defense ("uin", "wuin") (Schönn *et al.* 1991). Between one and ten days, their weight increases from 12 g to 75 g. In captivity, they weigh between 20 g and 45 g at 5 days, between 60 g and 90 g at 10 days, and up to 150 g between 20 and 25 days (Génot & Sturm 2001). Festetics (1952/1955) noted the following weight increase: 20 g (day 1), 56 g (day 5), 112 g (day 10), 120 g (day 15), and 120 g (day 20). See Figure 8.9.

Two weeks

Their eyes are open after eight to ten days (Schönn *et al.* 1991) and sometimes at six to seven days (Bakaev *in* Sagitov 1990). The owlets develop the ability to tear prey provided for them. The down becomes more sparse, and changes color to light gray. They are now able to regulate their own temperature, allowing both parents to forage for food. Some sibling skirmishes begin to occur at this age with the oldest becoming the most dominant in the nest, therefore getting priority over the other siblings when parents bring in food. Short facial feathers begin to develop, the beak seems to grow faster than other parts of the body and dominates the face. The legs and feet grow faster and proportionally larger than

Figure 8.8 Development of Little Owl chick: one to seven days. Photos Rob Hendriks. Top left: one day; top right: three days; bottom left: five days; bottom right: egg tooth. After Beersma *et al.* 2007.

the other body parts. The owlets can move forwards and backwards on their talons with their wings outstretched for balance. They can also climb. They sleep lying or squatting by lowering one wing or leaning against a wall (Schön *et al.* 1991). At 14 days, the owlets begin to be able to focus their eyes on something. The main feathers of the wings and tail appear at 9 days and 12–16 days, respectively.

Three weeks

An accelerated transformation occurs between weeks two and three, with the head, breast, flanks, and nape feathers all developing in one week. See Figure 8.9. For more details on the feathers, see Schönn *et al.* (1991). Their eyes become clearer, their legs and feet lose their fatty appearance, their talons grow in line with the bone structure, and the face pattern becomes characteristic of the Little Owl. The owlets are alone in the nest site because the female leaves them at 16 days. Their behavior also develops, and the owls show more mobility in the nest chamber than before, sometimes running, therefore predator evasion is better. They begin to handle prey; they catch prey at 28 days and are able to kill living prey at 34 days (Ille 1983). The owlets strengthen their begging behavior getting up onto their legs (Schönn *et al.* 1991). Within the nest sibling rivalry tends to be dominated by the oldest

Figure 8.9 Development of Little Owl chick: one week. Photos Rob Hendriks. Top left: seven days; top right: nine days; bottom left: eleven days; bottom right: thirteen days. After Beersma *et al.* 2007.

owlet. The owlets become more adventurous, and climb out of the nest onto the tree trunk climbing up and along the branches using their bill. This provides good exercise, and again minimises the effect of predators on the young in the nest. The third primary feather (P8) measures on average 75 mm (Juillard 1979).

Figure 8.10 Development of Little Owl chick: two weeks. Photos Rob Hendriks. Top left: 15 days; top right: 17 days; bottom left: 19 days; bottom right: 19 days (back). After Beersma *et al.* 2007.

8.12 Fledging period

A fledgling is defined by its capability to attain its first sustained flight. Unlike most birds, the young of many owl species "jump" or climb from their nest before they can fly, and thus become "branchers" for a period of time before they can fly. See Figures 8.11 and 8.12.

The owlets become ready to fledge at about 28–32 days, 35 days according to Haverschmidt (1946), or even 40 days according to Harthan (1948). The owlets leave the nest to explore their surroundings as early as 21 days, depending on the circumstances of the nest site and the pressure of disturbance. However, at this time they cannot fly, as their primary, secondary, and tail feathers have not fully developed. This is sometimes a noisy period, as the parents try to entice the owls to fly from their natal site. Lysaght (1919) noted "When the young are about to quit the nest, the parents are in a very excited state, scolding like

Figure 8.11 Development of Little Owl chick: three weeks. Photos Rob Hendriks. Top left: 21 days; top right: 23 days; bottom left: 25 days; bottom right: 27 days. After Beersma *et al.* 2007.

Figure 8.12 Development of Little Owl chick: post-fledging period. Photos Rob Hendriks. Top left: 29 days; top right: 35 days; bottom left: 39 days; bottom right: 43 days. After Beersma *et al.* 2007.

the Mistle-Thrush, all who approach, and on one occasion they attacked me in the boldest manner." The Little Owl is able to fly well at 42 days (Schönn *et al.* 1991). The longest feathers are completely developed at 40 days and then the owlets know the basics of flight. They also know the alarm cry. According to Schönn *et al.* (1991), they take a threatening posture at 37 days. Once the owlets take flight they do so in a clumsy way, flying in low short flights from perch to perch, landing unsteadily. The activity occurs during both day and night as the owls become accustomed to their surroundings. It normally takes a few days for all members of the brood to undertake their first flights. This activity continues for a number of weeks, during this time the parents continue to feed the owlets, for one month after the fledging time (Haverschmidt 1946). The adults choose perches during the daytime that allow them to watch all of their young (Eick 2003). The owlets fight together with bill and claws and fly high at 45 days (Schönn *et al.* 1991). At 50 days, they are more active at twilight and scarcely active during daytime.

After fledging, nestboxes that were used for nesting are no longer used by the adult or young (Eick 2003).

Post-fledging period

The juveniles leave their parents from the beginning of September to the beginning of November, later than generally indicated in the literature (Eick 2003). During the post-fledging period the owlets begin to develop their life skills, these include foraging for different types of prey, predator evasion, habitat usage, etc. After two months, their voice repertoire is similar to the adults. Some rare observations describe the young taking dust baths and dew baths (Ancelet 2001). The young frequent their natal locality, using their parents' territory to hunt and develop skills. This continues until early autumn when the young disperse from their natal territory, trying to locate an area or territory in which to settle. However, the first year holds many potential risks for the birds, and mortality is high. See Figure 8.13.

8.13 Fledging success

Table 8.8 summarizes the data collected in the major long-term studies of Little Owl breeding success across Europe. Results in the table indicate the highest average number of fledged young per breeding pair is 2.75 in Württemberg near Ludwigsburg, Germany (H. Keil, personal communication) and 2.66 near Lake Constance, Germany (Knötzsch *in* Schönn *et al.* 1991). The lowest average numbers of fledged young per breeding pair were 1.34 in Germany in Voreifel (Zens 2005), 1.78 in Switzerland near Basel (Baur *in* Krischer 1990) and 1.82 in Ajoie (Juillard 1984a), 1.79 in Vosges du Nord, France (Génot 2005), and 1.87 in Leipzig and Dresden (Schönn 1986). The low fledging success obtained near Basel in Switzerland, (Baur *in* Krischer 1990) helps to explain the disappearance of the species in that region (Luder & Stange 2001). It is almost the same situation in Ajoie (Switzerland)

Figure 8.13 Juvenile Little Owl on pole. Photo Ludo Goossens. See Plate 32.

where Juillard (1984) recorded 1.82 fledged young per breeding pair with just a few isolated pairs remaining. In addition to these reports, other data on fledged young per breeding pair include: 2.84 (n = 39) in Czechoslovakia (Danko 1985a, b, 1986, 1989), 2.51 (n = 102) in Mediumrhine, Germany (Ingendahl & Tersteegen 1992), and 2.91 (n = 97) in Kaisersthuhl Germany (C. Stange, personal communication), 1.91 (n = 17) in Austria (Ille 1996), and 1.6 (n = 12) in Russia (Khoklov 1992).

In Germany, the average fledging success seems to be better in nestboxes (2.47 fledged young per breeding pair in 76 breeding attempts) than in natural hollows (2.20 fledged young per breeding pair in 10 breeding attempts).

The number of fledged young per successful breeding pair increases from west to east in Europe (Génot 1992b) and in France too (Génot 1992c). Breeding success, in particular the proportion of fledged young by clutch size, depends on the condition of the female before incubation, and therefore on feeding conditions (Gassmann & Bäumer 1993). The authors found higher fledging success than in other German regions due to the good food conditions in Western Germany. Climatic factors also influence fledging success. Precipitation in March seems to improve food availability (earthworms) and thus the mass of the female in April. While high rainfall in March and April resulted in higher breeding success, high rainfall in May and June, especially when the nestlings are younger than two to three weeks, is correlated with high mortality of the young and lower fledging success (Lederer & Kämpfer-Lauenstein 1996), probably due to the moist environment in the nestboxes (Gassmann & Bäumer 1993). In Belgium too, for each 100 l/m^2 of extra precipitation in

Country	Years	Clutch size		Brood size at hatch				Brood size at fledging				Author
		Eggs/BP		Hatchling/BP		Hatchling/SBP		Fledging/BP		Fledging/SBP		
		$\bar{x}$	n	$\bar{x}$	n	$\bar{x}$	n	$\bar{x}$	n	$\bar{x}$	n	
GERMANY												
Leipzig, Dresden	1960–85	4.11	126	2.35	111	3.48	75	1.87	167	2.84	110	Schönn 1986
North Rhineland, Kleve	1974–84	3.61	96	2.52	81	3.24	63	1.89	89	2.75	61	Exo 1987
North Württemberg	1971–97	3.6	326					2.4	326	3.3	235	Furrington 1998
Württemberg, Ludwigsburg	1988–2001	3.71	812	3.21	812			2.75	812	3.39	659	Keil unpublished
Lake Constance	1973–88	4.42	265	3.49	265			2.66	265			Knötzsch *in* Schönn *et al.* 1991
Württemberg, Göppingen	1969–88	4.12	141	2.78	203	3.78	149	2.35	203	3.34	143	Ullrich *in* Schönn *et al.* 1991
Northrhine-Westfalia	1978–92	3.51	269					2.66	272	3.19	409	Gassmann & Bäumer 1993
Mid-Westfalia	1974–94	4.15	482	2.99	486			2.28	486			Kämpfer-Lauenstein & Lederer 1995
Voreifel	1987–97	3.85	268	3.09	254	3.71	210	1.34	412	2.78	189	Zens 2005
SWITZERLAND												
Ajoie	1973–80	3.12	153	2.22	153	2.78	122	1.82	153			Juillard 1984a
Bale	1981–90	3.75	76	2.98	76	4.72	48	1.78	76	2.83	48	Baur *in* Krischer 1990
Genève	1994–2000	3.70	189	2.99	189	3.50	116	2.22	161	3.09	116	Meisser & Albrecht 2001
BRITAIN												
Nationally	1939–75	3.59	268							2.40	241	Glue & Scott 1980
Cheshire	1993–2000	2.54	120	2.13	120	2.3	115	2.2	120	2.1	115	Leigh 2001a
FRANCE												
Nationally	1957–90	4.24	94	3.62	53	3.70	30	2.84	237	2.96	218	Génot 1992c
Northern Vosges	1984–2004	3.67	123	1.86	123			1.79	185	2.98	111	Génot 2005
NETHERLANDS												
Zuidoost-Achterhoek	1986–2003	4.00	471	3.13	471	3.77	348	2.25	471	3.08	348	Stroeken & van Harxen 2003b
BELGIUM												
Flanders (Vlaams Brabant)	2000–4	3.36	312	2.68	312	3.13	267	1.97	312	2.3	267	Ph. Smets, R. Huybrechts & S. Cerulis pers. comm.
Wallonia	1989–2004	3.2	2346	2.7	2346			2.33	2346			J. Bultot & Groupe Noctua pers. comm.
Flanders (Dijleland)	1999–2004	3.85	123	2.797	123			2.22	123			Vanden Wyngaert 2005

BP = breeding pair; SBP = successful breeding pairs; pers. comm. = personal communication

Table 8.9 *Overestimations in breeding success of Little Owl in Achterhoek (Netherlands) with or without taking mortality at fledging and post-fledging stage into account (after Stroeken & van Harxen 2005).*

		1999	2000	2001	2002	2003	1999–2003
age at ringing (eldest young, days)	average	13	13	15	15	12	13
	s.d.	6.5	6.0	4.4	6.6	4.7	5.8
	median	11	12	13	13	11	11
	number of nests	24	26	24	24	22	120
number of ringed young per nest	average	2.57	2.09	2.72	2.12	2.00	2.29
	s.d.	1.98	1.61	1.92	1.74	1.61	1.8
	number of nests	35	34	32	41	33	175
number of fledged young per nest	average	2.29	1.56	2.53	1.9	1.85	2.02
	s.d.	1.85	1.38	1.9	1.6	1.54	1.7
	number of nests	35	34	32	41	33	175
overestimation of breeding success (%)		12.2	34	7.5	11.6	8.1	13.4

s. d. = standard deviation

May and June, the number of fledglings decreased by 0.6 fledglings per nest (Bultot *et al.* 2001).

Stroeken and van Harxen (2005) studied the difference in "breeding success" by comparing the number of young that reached *ringing age* to the number of owls that reached *fledging age* (24–30 days), minus post-fledging mortality (dead owls found after >30 days). The estimates for the number of owls reaching fledging age was, on average, 13% lower when counting the live birds >30 days old, compared to counts of owls that reached ringing age (Table 8.9). These data have very important implications, as many field studies really reflect the numbers of owls only reaching ringing age, and thus overestimate true breeding success.

To obtain comparable results of breeding success in future studies, it is hence of utmost importance to register the age of the owls at the time of each observation. This will allow improved interpretations of reproductive success later. Post-fledging counts at a subset of nests (i.e., for the number of young >30 days of age) will provide a very important control and a base from which to better evaluate overall breeding success.

8.14 Mortality

In a breeding dynamics study involving 15 nestboxes monitored over eight years, Leigh (2001a) showed that the highest mortality occurred at the egg stage with 13% (40 out of 305) of the eggs not hatching. The suggested causes for this were chilling of eggs, sperm viability, experience of the female, and possibly pesticide pollutants. The mortality rate for owlets was significantly lower at 4.5% (12 out of 265 hatched eggs). Death of the young owls was from chilling, cainism, and starvation. Most of the owlets that died did so within five days of hatching (90%). The high mortality levels in eggs and owlets correlated with years when prey populations were low. In years when both small mammal and invertebrate populations were low the mortality of both eggs and owlets were highest. A simple explanation for this increase in mortality, is that very little prey was available to the parents. Field observations suggest that after hunting earthworms under wet conditions, the adults transfer a lot of moisture into the nest and onto the young, and this cool dampness may increase the risk of mortality. More detail is given in Chapter 10 on the causes of natural and human-related mortality, and the impact of mortality on population regulation. In Germany, Zens (2005) recorded a total mortality rate of 49.9% (n = 412 broods), of these failures 73% concerned owlets and 27% eggs. The mortality due to predators reached 69.3%.

Out of 51 dead young in Germany: 18 were caused by falling from the nest, 12 from predation, 2 from human destruction, 6 from loss of parents, 3 from disease, 2 from flooding, and for 8 the causes were unknown (Schönn 1986). Out of 60 mortality cases in Switzerland: 9 were from predation, 20 from cainism, 6 from human destruction, 17 from the loss of parents, 1 from disease, and for 7 the causes were unknown (Juillard 1984a). Out of 54 mortality cases in France: 7 were from predation, 7 from flooding, 3 from cainism, 1 from falling from the nest, 18 were from causes unknown, and 18 disappeared from the nest probably as victims of predation (Génot unpublished data). Mortality due to the trampling by cattle was noted by Juillard (1984a). All the eggs from 9 out of 37 nests in Flanders did not hatch, while 2 hatched clutches were predated by Stone Martens (Vanden Wyngaert 2005).

8.15 Cainism

Cainism was recorded in France involving 3 out of 57 owlets (Génot 1992b), and in Germany (Ullrich 1973; Knötzsch 1978; Exo 1983) where Zens (2005) recorded 33 out of 332 owlets. Juillard (1984a) noted the losses in a declining population due to cainism in the nests was 33% (20 out of 60 young). Victims of cainism are usually the youngest owlets, which are eaten by the oldest (Ullrich 1973; Exo 1983; Génot; unpublished data). Some young were reported feeding on the carcass of a dead adult in the nest (Mills 1981).

8.16 Life-time reproductive success

One of the most important research priorities in owls is the acquisition of demographic data and the development of good demographic models. To obtain good models the input variables are crucial. Age or stage-specific estimates of survival and fecundity can be used to parameterize population projection matrices with different distinct age or stage-classes and to project future population trajectories (Noon & Franklin 2002). Estimates of survival rates for Little Owls can be based on intensive capture–recapture studies and can yield age pyramids of populations. Fecundity rates can be estimated by determining the reproductive output of birds of known age. First data on life-time reproduction are becoming available. Table 8.10 shows the cumulative reproduction of individual Little Owl females

Table 8.10 *Cumulative reproduction in Little Owls of individual females (R. van Harxen & P. Stroeken, personal communication; J. Bultot, personal communication).*

			Totals			Averages		
Individual female	Age of female	Years	Eggs	Nestlings	Fledglings	Clutch size	Nestlings	Fledglings
123	6	4	9	0	0	2.25	0	0
348	11	9	35	9	9	3.89	1.67	1.00
428	7	6	33	13	13	5.50	4.33	2.17
135	15	14	51	32	28	3.64	2.91	2.55
354	11	9	39	25	25	4.33	4.00	2.78
478	8	7	23	20	20	3.60	3.60	2.86
407	8	7	35	32	32	5.00	4.71	4.57
E142605	7	4	9	7	5	2.25	1.75	1.5
E196139	8	4	10	8	7	2.50	2.00	1.75
E196057	9	4	17	17	12	4.25	4.25	3
E183378	7	5	17	17	14	3.4	3.4	2.8
E183126	8	4	16	16	15	4	4	3.75
E196286	7	4	15	13	8	3.75	3.25	2
E214458	8	5	19	11	11	3.8	2.2	2.2
E229875	7	4	12	10	8	3	2.5	2

(R. van Harxen & P. Stroeken, personal communication; J. Bultot, personal communication). Some females produced increasing clutch sizes with age; others attain a maximum at a certain age, as observed in a female breeding over 12 years in southwest Germany where the average clutch size was reduced by 0.8 eggs per year from the age of 10 years (Sill & Ullrich 2005). In Germany, Zens (2005) noted that 83% of the females were breeding in their first year (n = 30). The average lifetime reproductive success was determined to be 5.1 juvenile owls (range 0–21) per females in the 30 females followed (Zens 2005).

In the Netherlands, Stam (2003) studied an owl population located east of Arnhem and was able to record the specifics on 152 ringed female Little Owls during 1983–2002. He found 54 females that had bred once; 32, twice; 23, three times; 14, four times; 10, five times; 5, six times; 7, seven times; 4, eight times; and 3, nine times. The life expectancy of these breeding Little Owl females averaged 2.8 years (all females bred at least once), while Zens (2005) recorded in Germany a life expectancy of 3.7 years.

Stroeken and van Harxen (2003a) followed a population with the following age structure in 2003: ten females were in their second year, seven in their third year, five in their fourth year, two in their fifth year, three in their sixth year, four in their seventh year, and one in its eleventh year. The life expectancy of these females averaged four years. Stroeken and van Harxen (2003a) also reported the age structure for male owls: three males were in their second year, one in its third year, five in their fourth year, two in their fifth year, and one in its seventh year – indicating a life expectancy averaging 3.8 years. Please note: the ages reported here are minimums, since most birds were ringed when they were found for the first time in a nestbox.

8.17 Extra-pair copulation

Little Owls are monoterritorial, show very high mate fidelity, and the male will undertake aggressive mate guarding principles. However, in some populations where high population densities occur, and the prey resources allow, some extra-pair copulations do sometimes occur. Müller *et al.* (2001) investigated the mating behavior of a population of Little Owls in Westphalia, Germany, by taking blood samples from 53 nestlings produced by 16 breeding pairs. They then applied multi-locus DNA fingerprinting, to establish parentage of the young owls from each nest. The results from their study indicated that no successful extra-pair copulations occurred in this population (Müller *et al.* 2001). Additional studies conducted in high-density populations will provide further insights into the validity and impact of extra-pair copulations.

In a study in Cheshire, England, one record of bigamy was recorded in ten years. This occurrence involved two females each nesting approximately 100 m apart, one in a nestbox and the other in a natural cavity. The timing of the breeding efforts was spread so the male could provide food for both nest sites, and so that both nest sites did not hit their peak food consumption at the same time. R. Leigh (personal communication) also noted that the male was caching food in the nests, which allowed him some rest if the weather prevented foraging for food. Illner (personal communication) observed two females incubating in the

Figure 8.14 Juvenile Little Owl at nest site. Photo Bruno d'Amicis. See Plate 33.

same nest on two occasions during his 25 years of studying over 1000 nest sites. Berghmans (2001c) recorded nine owlets in one nest site being cared for by two females. The instances of bigamy are mainly illustrated by females, which nest near to each other or lay in the same nest site, incubate together, and share parental care for the owlets. However, in these instances the ability to incubate the eggs and brood the young are the main influences on the survival of the eggs or owlets. See Figure 8.14.

9

Behavior

9.1 Chapter summary

Based upon detailed observations derived primarily from owls in captive breeding programs, Schönn *et al.* (1991) reported extensively on the behavior of the Little Owl. In this chapter, we draw heavily from their important work, and offer updates on their major findings. See Figure 9.1.

9.2 Body movements

The Little Owl can easily be observed during the day, especially when perched at exposed locations such as on chimneys and power lines. In the Netherlands, nestboxes designed with special balconies offer ideal roosting opportunities for the owls and allow verification of nestbox occupation (P. Beersma, personal communication). Other owls prefer to roost during the day in cavities. For this purpose, some conservation initiatives install two or three nestboxes per territory to allow males to roost in a different cavity from breeding females. When roosting during the day, most owls lean against one of the walls of the cavity (Scherzinger 1980). While most owls roost relatively low down (<10 m above the ground), radio-marked owls have been found to roost in trees, quite high above the ground.

Remarkable postures of Little Owls, also shown by other owls and raptors, are the cryptic posture to minimize attacks by flying enemies, defensive turning against enemies at close range, and covering behavior to hide prey. The cryptic posture reflects a narrow, rigid, alert upright stance, adopted as soon as the bird detects any kind of alien sound or disturbance. Approaching people or predators are located with short, nervous, horizontal, and vertical circular movements of the owl's head, with the body in an upright position or with bended knees. On arousal, the owls alternate bending their knees with extreme upright stretching of the body (Heinroth & Heinroth 1926; Hubl 1952; Scherzinger 1971a; Pukinskiy 1977; Busse 1983). This behavior gives the impression that the owl is jumping (Daanje 1951). Flicking of the tail, wingtips, and entire wings is frequent among aroused and nervous owls (Scherzinger 1980).

Figure 9.1 Little Owl. (François Génot).

9.3 Locomotion

Little Owls like to stroll but also to run when hunting for earthworms and other invertebrate prey. Before flying off, little Owls regularly step repeatedly ("trample") on the spot. When pursuing voles, owls prefer to walk, jump, and run. Running also occurs when Little Owls are going to catch some prey, or are escaping danger. In contrast to most other owls, the flight of the Little Owl is audible, because the distal radii of the outermost primaries are not serrated (Sick 1937) and because the relationship between body mass and wing surface (i.e., wing loading) is rather high (Schönn *et al.* 1991). The species mostly flies very low (e.g., 1–10 m) above the ground. During the day the Little Owl flies straight for short distances, and crosses longer stretches in a woodpecker-like style. When departing from a perch, the owl first drops to gain speed, flies, and then lands in an upward arc, reflecting a U-shaped flight pattern (Schönn *et al.* 1991). Flights can be interrupted suddenly when the owl enters shelter or cover. Owls hunt from low perches, and then capture prey after a straight flight or pounce (Baumgart 1980; Brands 1980; Scherzinger 1980; Busse 1983; Ille 1983; Mikkola 1983). When hunting, hovering and gliding flight can also be observed.

Aspects of the altitude of the Little Owl's flight has long been underestimated. Eick (2003) observed Little Owls flying above high-tension power lines, crossing steep valleys and urban areas at altitudes ranging from 20–30 m and even higher. While it is easier to see owls flying low above the ground, observations of owls flying higher shed important insights into dispersal, migration, and degree of population isolation. Data suggest that rivers and forest edges do not influence the owl's choice of directional movements. Even relatively broad valleys (e.g., the Neckar and Enz valleys in Germany) are crossed. Telemetry research has confirmed that Little Owls fly above – rather than through – large wooded areas. Large wooded areas and urban areas have not been found to be barriers to the dispersal or movements of the species.

9.4 Preening

When preening, the owls partly or completely close their eyes by lifting their lower eyelids (Heinroth & Heinroth 1926). The head feathers are cleaned as they lay on the head, while the larger body, wing, and tail feathers are cleaned one by one. The back, rump, and base of the tail are reached and preened over the shoulder and slightly lowered wing, under-tail feathers are reached from beneath the wing. The primaries, secondaries, and tail feathers are pulled through the beak one by one, starting at the base. The base of the quill is intensively nibbled. Legs and toes are mostly cleaned by stretching the entire leg while keeping the foot closed. The toes are nibbled individually or pulled through the beak. Little Owls only use their talons in grooming when they scratch the back of the head, the ears, and facial region (Brands 1980; Scherzinger 1980). The head is also cleaned by rubbing with their wing or shoulder. The owls grease their feathers every two to three days (Angyal & Konopka 1975). The beak is cleaned by rubbing it on branches, especially after food intake (Scherzinger 1980). Incubating females tend to rub their beaks against one of the walls of the breeding cavity (Brands 1980).

Mutual preening and cleaning of the feathers of both partners, especially scratching of the facial region, is observed regularly among adults, but is also seen in sibling fledglings (Haverschmidt 1946; Brands 1980; Scherzinger 1980). The act of mutual preening is most often initiated by the cleaning of an owl's own feathers (Schönn *et al.* 1991).

Feathers are shaken or shuffled during or after preening, after producing a pellet, after rain or dust baths, and after swallowing large prey. Body and head feathers are shuffled consecutively, sometimes followed by the shuffling of the wings and the base of the tail (Scherzinger 1980).

Parallel stretching of wing, leg, and tail on one side of the body often preceeds preening activities and is also associated with the intent to fly (Schönn *et al.* 1991) or to leave the daytime roost. Sometimes just one wing is stretched, sometimes both wings are stretched towards the front and upwards, until both wings almost meet above the head. Simultaneous stretching of both legs has also been observed.

Dust and sand baths are taken regularly (Ancelet 2001). The owl squats or lies sideways on its chest, legs stretched backwards, and it tosses sand or dust with its wings towards its body across its back (Schönn *et al.* 1991). After long dry periods, the Little Owl bathes in rain with its wings spread until it is soaking wet (Schönn *et al.* 1991). Bathing in smoke has also been observed on the tops of chimneys (Tubbs 1953; Cleine 1976). After rainy periods or during the nestling phase, owls frequently sunbathe, during which all feathers are fluffed up, wings are slightly open, and the head bent backwards at the neck, with the sunshine on its face (Scherzinger 1980).

Little Owls gape with a completely opened beak and closed eyes. This is frequently observed just before the owl produces a pellet (Schönn *et al.* 1991).

9.5 Intraspecific behavior

Social and sexual behavior

Little Owls form pair bonds, and partners regularly fly together and like to hunt nearby one another. They look for body contact while dozing or sleeping, and scratch each other outside the breeding season too (Exo 1987). The pair formation period during winter is believed to reduce intraspecific aggression between the partners. During the courtship period (January till the end of April in Europe) both birds are frequently found together in the same cavity or roost during the day. Mutual preening is common then (Haverschmidt 1946; Brands 1980; Glue & Scott 1980; Scherzinger 1980; Exo 1987).

Copulation is preceded by individual song, partner attraction, begging, and simultaneous singing. The female flies in and lands close to the male in response to his singing. The male approaches the perched female, circles around her and hovers above her several times (Glue & Scott 1980). The male then lands besides his partner while offering prey. During courtship, prey is only transferred outside the breeding cavity (Scherzinger 1980). Then the partners fly together to another perch. The male perches close to his partner for short periods, while they preen each other's head and beak and eventually touch each other's head (Racz 1917; Haverschmidt 1946; Glue & Scott 1980). The female lifts a wing and lowers it quivering. The male then jumps on the back of the female, while the female bends forward, adopting a horizontal position. The female's tail is lifted up or moved sideways. When the willingness to copulate by the female is very strong, she lands in front of the male and adopts the copulation position (Brands 1980). The male holds the female's back and utters some partner calls while flapping his wings and gently trampling on her back before he bends his knees and copulates (Racz 1917; Haverschmidt 1939; Hosking & Newberry 1945; Haverschmidt 1946; Scherzinger 1980; Exo 1987). During copulation the male grabs the female's neck or stern feathers or touches her beak with his. After copulation, the male flies a short distance, often uttering calls. The female often starts calling too, perched with hanging wings and shuffling her feathers or preening. After copulation, both partners preen each other while begging together with possible bill-snapping. Little Owls always copulate outside their nesting cavity and mostly at favorite perches near the nest (Racz 1917; Haverschmidt 1939; Scherzinger 1980), or sometimes on the ground (Gluz & Bauer 1980). During the peak courtship period, two to five copulations per night are possible (Exo 1987). Copulation activities are often interrupted. After successful mating, the male often shows the female potential nesting cavities. During this nest-showing behavior, the male flies to the cavity and attracts the female to the entrance hole or into the cavity (Haverschmidt 1946; Ullrich 1973; Scherzinger 1980; Exo 1981, 1987). The female mostly follows her partner and inspects the cavity (Haverschmidt 1939). This cavity-showing behavior is limited to the spring courtship period. The same cavities are revisited again in late fall (Scherzinger 1980). In addition to showing potential nest sites, the male also deposits prey in the prospective nest cavity up to six weeks before the start of the incubation (Ullrich 1973; Scherzinger 1980; Exo 1981; Exo 1987). During this period, the female begs for food (similar to fledglings) outside the nest cavity. Before egg-laying, the begging behavior increases in intensity.

After the start of egg-laying, the frequency of copulation decreases (Racz 1917; Haverschmidt 1946). At that time, prey transfers are mainly done in the nesting cavity. When the male brings in food, he calls and stimulates the female to come and take the food at the entrance of the breeding cavity, or she waits on the eggs and receives the food while incubating. After giving the food, the male leaves the nest at once and sometimes even gets chased away by the female. Only when the nestlings are at the age of nine to ten days can the male start feeding them himself. After fledging of the young the pair bond between the adults declines, and partners tend to become aggressive towards each other and seem to be willing to separate (Exo 1987).

Territorial behavior

In large measure, Little Owls occupy their territories for multiple years. The feeding and breeding territory is defended year round against conspecific competitors (Richter 1973; Knötzsch 1978; Exo & Hennes 1980; Glue & Scott 1980; Juillard 1984a; Fuchs 1986). The female lives in the territory of the male. Neighboring territories do not overlap. Territorial defense is at its peak during the spring courtship period (January till April in Europe) until the start of egg-laying. A second peak in defense is observed in October/November during fall dispersal (Scherzinger 1980; Mikkola 1983; Exo 1987). From the end of July/early August, adults produce calls associated with chasing conspecifics, aimed at scaring juveniles or non-territorial "floaters" from their territory. Besides broadcasting acoustic signals, the birds delimit their territory by regular flight patrols. Territory owners chase off intruders by aggressively pursuing them to the boundary of (or briefly just beyond) the territory while uttering alarm calls (Exo 1987). Little Owls were observed attacking the rear of the head of decoy adult owls. The pseudo-face on the rear of the head can be made more pronounced by placing feathers upright, which possibly serves to scare off attacking birds (Scherzinger 1980; Scherzinger 1986).

Year-round territorial defense requires a significant energetic investment for the owls (especially males). Important new insights into how owls minimize energetic output has recently been found by Hardouin *et al.* (2006), who found that Little Owls can discriminate between the hoots of neighbor (known) owls (i.e., owls on neighboring territories) and those of stranger (unknown) owls. Neighbor owls frequently call from known locations (i.e., "usual" locations). Hardouin *et al.* (2006) found that male Little Owls responded significantly less to their neighbors' hoots played back from usual locations than to hoots from unknown (stranger) owls. Responses to playback of a neighbor from an unusual location were similar to responses to the playback of a stranger's hoots from either location.

9.6 Enemies

Little Owls are predated upon by larger owls, large diurnal raptors, corvids, and mammals. To cope with these, Little Owls have developed specific behavioral traits for enemy avoidance. Even without the presence of a threat, the owls move their head to and fro for their own security (Schönn *et al.* 1991), scanning for danger. The owls focus their attention on threatening objects with short, nervous movements from a squatted or upright position. Once seen, the precise position of potential enemies is determined by moving the body quickly up and down, while uttering defensive calls or bill-snapping (Stadler 1932; Runte 1954; Scherzinger 1971b, 1980; Mikkola 1983). To avoid humans in uncovered locations, Little Owls during the day move to a distance of 50 m to 100 m away. In covered places, the owls move shorter distances away (e.g., 10 m or less). When escaping, the Little Owls fly in a straight line just above the ground if seated on it, or pounce from a perch. At longer distances, the owls adopt the cryptic posture to avoid being seen. Hinde (1953) considers this posture as a conflict between staying and escaping (i.e., the "fight or flight" response). Sometimes the owls adopt a sleeping posture with all feathers sticking upright (Scherzinger 1971b; Glutz & Bauer 1980) giving the owl a ball-like shape. In contrast to roosting, the eyes are closed by lowering the upper eye lids (Heinroth & Heinroth 1926). The pseudo-face at the rear of the head is often considered as mimicry of eyes to defend against approaching enemies from the back (Wickler 1965; Karalus & Eckert 1974; Nicolai 1975; Davison 1983). To hide from terrestrial enemies, the Little Owl presses its body flat to the ground or runs towards cover in a crouched position. Incubating females prefer positions in which they can keep the entrance of the nesting cavity within vision (Brands 1980). Many females allow themselves to be taken from the nest by humans without any reaction, only to escape after being put back again (Scherzinger 1980). In Belgium, it was found that some owls could be caught by hand; an owl that was perching during the day on a pollarded willow was caught by somebody slowly moving behind the owl and taking it quietly in the hand (J. Bultot, personal communication).

Defensive turning is displayed by owls that are caught in tight locations such as nesting cavities. When threatened in tight quarters, the owl reacts with defensive calls, turns onto its back, and jabs with its beak and talons (Exo 1987).

Towards the end of incubation, Little Owls are especially aggressive toward enemies. In the vicinity of the nest, Little Owls attack other owls, diurnal raptors, and feral cats (Nash 1925; Haverschmidt 1946; Richter 1973; Glutz & Bauer 1980). Although occurring only rarely, nesting female owls may attack humans at the nest site by scratching or biting the head and hair.

When the adult owls give alarm calls, the juveniles stop begging. Fledged juveniles instantly look for cover, or disappear into the nesting cavity (Scherzinger 1971b, 1980; Exo 1987). In high density areas, alarm calls will cause adults to escape too (Schönn *et al.* 1991). Little Owls are frequent victims of the mobbing behavior of small passerines (Curio 1963). Owls will either ignore the passerines (Scherzinger 1980) or will react to them by looking for cover (Staats Von Wacquant-Geozelles 1890) or rarely by grabbing at mobbing birds with their talons (Haverschmidt 1946). The cryptic posture adopted by the owls helps them to avoid being mobbed (Scherzinger 1971b). Using live owls as bait, the mobbing behavior of passerines towards Little Owls was exploited as a technique to capture passerines in Italy (Schaaf 2005).

9.7 Roosting and sleeping

Sleeping birds are often almost invisible due to their excellent camouflage colors. The feathers are fluffed up, and the head is sunk towards the body, causing all the white features of the bird to be hidden. A leg is often pulled into the belly feathers with closed toes (Scherzinger 1980). The eyes are closed by lifting up the lower eye lids (Heinroth & Heinroth 1926).

Little Owls sleep or rest in places providing immediate cover and with open sightlines and open escape routes, e.g., tree crowns, shrubberies, attics, and holes in walls. During bad weather, Little Owls mostly use holes for resting (Exo 1987). During winter, the owls regularly spend the daytime in stables for domestic livestock, in hay-barns, and in holes in trees. The avoidance of cold temperatures plays a crucial role in their choice of roosts (Schönn 1986). Roosts and sleeping places are situated within the owl's territory and are used by the same bird for many years. These favorite locations are easy to detect, because of the white chalk stripes of their excrement and the large number of pellets beneath them.

10
Population regulation

10.1 Chapter summary

A wide range of factors regulate Little Owl populations. These factors are typically categorized as environmental (external) and demographic (intrinsic) factors (Newton 1998). We have written this chapter in two main sections following this categorization theme. Little Owls have been shown to be directly and indirectly affected by habitat loss, vehicle collisions, limited availability of nest sites, pesticides (i.e., secondary poisoning) and heavy metals, entrapment in anthropogenic structures (i.e., hollow metal power poles and chimneys, and drowning in water troughs), predators, and weather. See Figure 10.1. They have also been shown to be susceptible to parasites, diseases, and injuries. While the Little Owl has co-evolved with a few of these (e.g., weather, predators, diseases, parasites), anthropogenic activities have substantially altered the landscape within which Little Owls exist(ed). When the population grows and owl densities become higher, density-dependent processes take place and serve to stabilize the population. In a meta-population context, as populations become increasingly small, immigration helps to support them, extending the survival time of these population clusters. We summarize evidence along these lines from northern France and the Netherlands.

With the primary backdrop of habitat alteration, other factors (e.g., collisions with vehicles, lack of suitable nest sites, reductions in prey due to pesticides) become additive in reducing owl populations, and at some point in the decline the fragmented populations become susceptible to stochastic events (such as a major storm event) and individual populations become extirpated from a given region. As a population becomes smaller, it fluctuates from year to year and becomes more vulnerable to stochastic events (e.g., several consecutive severe winters or poor nesting years). Unless the population decline is arrested and reversed, the population will be unable to maintain itself over a hundred-year time frame (the accepted time frame for modeling trajectories of most wildlife populations), and is doomed in the "extinction vortex". This is what has happened to Little Owl populations in places like Holland and Germany.

The Little Owl is generally a very sedentary bird and will generally frequent the areas close to its natal area. Dispersal begins when young are from 12–16 weeks of age and lasts up to at least the end of October. Occasionally some owls do embark on long distance

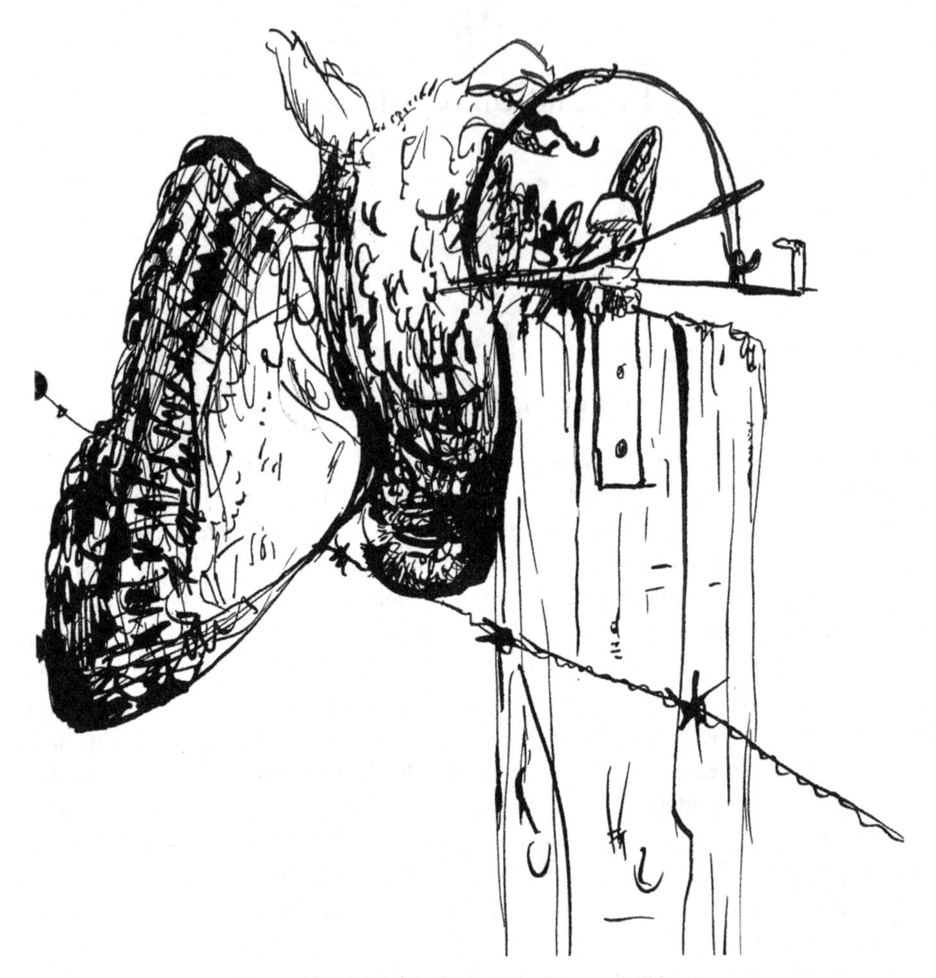

Figure 10.1 Little Owl mortality (François Génot).

movements. Mean home-range sizes are from 14–120 ha. Home-range and territory size depend on the sex, the age of the birds, the population density, the season, and the habitat. Density might influence the reproductive success of the Little Owl, its habitat selection, and its dispersing behavior. More and more Little Owl populations function as meta-populations because of the fragmentation of the countryside.

For Little Owls, there is a critical need for demographic information for juvenile owls (specifically on dispersal and survival data derived from radiotelemetry studies), and for adults (specifically on survival and reproduction associated with long-term mark–recapture data on populations). Ringing data thus far indicates that Little Owls that were ringed as young and then later recovered dispersed an average straight-line distance of 6.2 km (n = 252). The combined data of Gassmann and Bäumer (1993) and Knötzsch (1988) (n = 77 females, n = 72 males) showed that both sexes dispersed on average 4.43 km. The mating system hypothesis, which predicts that the sex that establishes the territory should disperse shorter distances, remains to be studied on larger sample sets. Nearly all of the demographic data we have on Little Owls to date comes from studies conducted in Europe. We strongly encourage and support researchers to undertake projects focused on the demographic aspects of Little Owls in the Middle Eastern and Asian portions of its range.

Introduction to population regulation

In his book on population limitation in birds, Newton (1998) states that to understand what limits bird populations within the habitats they occupy, we have to separate the external (environmental) factors and intrinsic (demographic) features that these external factors affect. External limiting factors include resources (e.g., food supplies, nesting cavities), competing species, and natural enemies (predators, parasites, and competition). Any one of these factors can be considered limiting if it prevents a population from increasing or causes it to decline. Particular populations may be affected by more than one, perhaps all, of these different factors, but often one factor emerges in overriding importance at any one time. It is this factor that must be alleviated if we are to achieve an increase in breeding numbers. Intrinsic (demographic) features include the rates of births and deaths, immigration, and emigration, the net effects of which mediate the influence of external factors on local population trends. Both extrinsic and intrinsic factors can be considered as "causing" population changes, the former as ultimate factors and the latter as proximate ones. Thus, within suitable habitat, a population might be said to decline because of food shortage (the ultimate cause) or because of the resulting mortality (the proximate cause). Some extrinsic factors act in a direct density-dependent manner, affecting a greater proportion of individuals as their numbers rise, while other extrinsic factors act in a density-independent manner, sometimes causing big changes in numbers regardless of density. Severe winter weather might act in this way. A review of factors involved with the population limitation in holarctic owls was given by Newton (2002).

The main challenge in understanding local and global population trends is to isolate and quantify the different limiting factors. This can be done through experimentation by manipulating the available resources (adding food, adding extra breeding cavities) or the impact of predators (using special predator-proof nestboxes), or by collecting various types of circumstantial evidence. In this chapter, we give an overview of some of the factors that limit or influence Little Owl populations. We first consider environmental factors, followed by demographic factors.

10.2 Environmental limiting factors

The main environmental resources on which Little Owls depend are food and nesting sites. Other external factors, which serve to reduce owl numbers, include predators and parasites, weather, and human activity (e.g., habitat alteration, disturbance, and pesticide use).

Figure 10.2 Dead Little Owl (François Génot).

Human influences

In terms of human influences, the primary impacts on Little Owls come from changes in human land use, persecution, application of agrochemicals, vehicle traffic, drowning, and electrocution. Each of these influences serves to limit the breeding output, and limit the health and sustainability of populations. See Figure 10.2.

Van't Hoff (2001) provides a good overview of the situation facing a Little Owl population in response to large-scale landscape changes and loss of nest sites in the Dutch region of Groningen. He found that the population declined markedly, and became much more fragmented. As in other parts of Europe, a strong decline has been noticed in Slovenia in recent years, attributed mainly to loss of suitable habitats (i.e., traditional orchards,) and the use of pesticides in large quantities (Vogrin 2001). Other examples of severe declines come from the Lower Savinja valley where the species was common and abundant about 50 years ago (Dolinar 1951). Today it is extirpated in this region probably due to the increase of intensive agriculture (hop fields) that have radically changed the valley. Changes in agricultural practices and land use are the main factors responsible for the gradual decrease of the species in Spain. Urban development is a real threat, especially in the eastern coastal semi-arid habitats, and in the Basque countryside of the Atlantic coast (northern Spain) (Martínez & Zuberogoitia 2004a). The reduction of dry-farmed tree crops, in particular the carob tree (*Ceratonia siliqua*), has caused severe population declines in the eastern coast. Pesticides diminish prey availability, land redistribution reduces suitable sites for nesting or hunting, and the reduction of hedges and trees along roads increases the risk of collisions

Figure 10.3 Traffic as a major cause of Little Owl mortality. Photo Vildaphoto Ludo Goossens.

with vehicles (Olea 1997; Garcia & Muñoz, 2003). Likewise, the enlargement of forested areas, especially in Pre-Pyrenees has led to the reduction of suitable habitats.

Effects of traffic

The Little Owl suffers heavy losses to vehicle collisions on roads, which is perceived to be the most important cause of human-induced mortality in Europe (Knötzsch 1978; Exo & Hennes 1980). See Figure 10.3. Among 15 bird species, the Little Owl was the second most common victim of road traffic ($n = 150$) among birds brought to wildlife hospitals in Belgium between 1987 and 1997 (Rodts 1994). Illner (1992) suggested that Little Owl deaths due to vehicle collisions increase when roads transect villages. This correlates with the owl's relationship with man, as they routinely nest in or near human dwellings in rural landscapes. Baudvin (1997) found that the rate of vehicle collisions with Barn Owls and Long-eared Owls in France was based on the physical structure of the roadbed (i.e., many fewer owls were killed where the roadway surface was set below that of the surrounding terrain) and the structure of roadside vegetation (there were fewer owls in areas with taller natural or shrubby vegetation that limited access to voles).

It is difficult to know the exact proportion of deaths caused by traffic, because deaths from most of the natural causes (e.g., starvation, predation) go unobserved. Road casualties may predominate because these are the owls that are more readily detected by people. In a Belgium sample, collisions with vehicles was the cause of 34% of the total Little Owl deaths (Bultot *et al.* 2001). Vehicle collisions represented 40% of the known mortality causes in eastern France ($n = 28$) (Génot 1995) and 36% ($n = 53$) in western France (Clech 2001a). Clech (2001b) showed that 68% of breeding sites were situated more than one kilometer from high-traffic roads and speculated that this was because all owls living closer had been killed. Of 89 Little Owl casualties in Belgium, 65% were birds less than one-year old and 35% were adults (Bultot *et al.* 2001). These data are consistent with the mortality of Barn Owls in Britain, where Newton *et al.* (1997) found that 45% of 1101 owls died from collisions with vehicles, and that 76% and 24% of these owls were juveniles and adults, respectively.

The important mortality on roads can be explained as follows. Little Owls are vulnerable to vehicle deaths due to their habit of flying low from post to post or hedge to hedge in pursuit of prey. Little Owls spend a great deal of time foraging for food on the ground, which also makes them vulnerable to vehicles. When the owls forage on the ground, they become dazzled by the headlights of vehicles, tend to freeze, and then cannot escape. The Little Owl can sometimes be found feeding on the remains of an animal killed by traffic (Van de Velde & Mannaert 1980).

Increasing road traffic, faster cars, the higher density of major road networks, and perhaps the physical structure of the roadbed and adjacent vegetation all serve to increase the vehicle-related mortality in the owl population. This perspective is supported by comparisons of the data from Britain, Germany, and France. A study in Britain by Glue (1971) showed that 12% (24 out of 198) of the owls died from vehicle collisions; in Germany, Schönn (1986) found that road traffic was the cause of 23% (19 out of 82) of the mortalities there. Génot (1991) analyzed 751 Little Owls recovered in France for the cause of death; 52% of these owls were road traffic victims. A report commissioned by the Dutch Department of Public Works on road traffic mortality among birds reviewed all records of road traffic deaths and ringing recoveries for the period 1961–90 (Van den Tempel 1993). The Little Owl topped the list of bird mortalities with 382 cases reported; 181 (47.4%) of these were ringed birds. The importance of this finding was that, based on ringing data, nearly half of the owls in this study were killed by vehicle collisions.

This increase in road-related mortality through time is consistent with that found for Barn Owls in Britain (Newton *et al.* 1997). Barn Owl deaths attributed to road traffic increased from 6% in 1910–54 and 15% in 1955–69 (Glue 1971, based on ring recoveries), to 35% in 1963–70 and 50% in 1991–6 (Newton *et al.* 1997).

Mortality due to railways is less well described in the literature, presumably because the tracks carry less traffic than roads and are less frequented by people able to pick up carcasses. Spencer (1965) showed that of 116 avian casualties due to railways, 41 were owls and only one was a Little Owl. Génot (1991) found that of 751 dead Little Owls in France, only one case of railway casualty and three cases of plane casualties were recorded. In former East Germany, three cases of railway casualty were recorded out of 98 mortalities (Schönn 1986).

Most studies showed that road deaths occurred mainly in July to August, which suggests a link with post-fledging dispersal. For example, an analysis of 418 deaths over a three-year period in Spain shows the road mortality peak to occur in August (Hernandez 1988).

Effects of poisoning

Little Owls take a relatively wide range of small insect and rodent prey, which makes them vulnerable to the effects of a wider variety of poisons.

Rodenticides, anthelmintics, and organochlorines Rodenticides are readily available and are used to poison and control rodents that inhabit human dwellings, or that eat and contaminate foodstuffs that are destined for human consumption. Little Owls rarely eat carrion; therefore the first-generation rodenticides that kill the rodents outright rarely affect the owls directly. However, most current rodenticides (which inhibit blood coagulation) are ingested by rodents and are carried for several days prior to death. In terms of LD_{50} values (the lethal dose for 50% of a sample, expressed as mg/kg bodyweight), the new chemicals are roughly 100–1000 times more toxic than the first-generation (i.e., warfarin) rodenticides. It is the combination of greater toxicity and greater persistence that gives the potential for secondary poisoning of rodent predators (Newton *et al.* 1997). While the rodents are alive (but sick) they are available to the owls (perhaps more so than healthy rodents), and when caught and eaten the poison within the rodents is transferred to the owls. Beersma and Beersma (2001) reported that owls affected by rodenticide poisoning suffer internal bleeding from organs and blood vessels. This becomes visible in the acute stages when joints bleed, and the owls seem emaciated even during times of good prey populations. Eventually the owls die, the speed of which is denoted by the amount of rodenticide accumulated from the prey.

Veterinary anthelmintics are used widely in the countryside to protect horses, cattle, and sheep against helminth fauna. The most used vermifuge (anthelmintic medicine) is ivermectine, which accumulates in the food chain through sustained bolus treatment (medicine given in a pill or capsule) to grazing cattle and is toxic to dung beetles (Lumaret 1993). Cows are typically given these medicines through their feed, and dung beetles pick up this insecticide through ingestion of the cow dung. Little Owls consume the target and non-target species that these pesticides are aimed at, ingesting an accumulation of the pesticide from the prey. The effects on Little Owl biology are marked; the ability to use its senses for foraging is affected and pesticide accumulation affects the breeding performance of the owls.

While eggshell thinning is a result of an accumulation of organochlorines and polychlorinated biphenyls (PCBs) in the metabolism of the owls, little information is available from the national monitoring programs on the effect of organochlorines and PCBs on Little Owls. In France, Génot *et al.* (1995) undertook a study of organochlorines found in 12 dead Little Owls and 40 sterile eggs. When they compared their results to earlier studies in Europe, concentrations were found to be lower than they had been 15 years earlier, and much lower than they were 30 years ago. More specifically, dieldrin levels in Britain were reported at 6.5 ppm (Cramp 1963), and 0.1 ppm ($n = 7$) (Walker *et al.* 1967), while levels in France were found to be 0.03 ppm ($n = 4$) (see Tables 10.1, 10.2, and 10.3; Génot *et al.* 1995). The concentrations found in Little Owls were still significantly higher than in forest owls (Ravussin *et al.* 1990).

Table 10.1 *Concentration of organochlorines and PCBs in Little Owls in France expressed in ppm (mg/kg fresh weight). Data from Génot* et al. *1995.*

Origin of sampled owls		HCB	HCH total	HE	D	DDE	TDE	DDT	PCB
Moselle	Rahling	–	0.002	0.001	–	0.005	–	–	0.075
Bas-Rhin	Ormersviller	<0.01	<0.01	<0.01	–	<0.01	<0.01	<0.01	<1
Bas-Rhin	Weyer	<0.01	<0.01	<0.01	–	<0.01	<0.01	<0.01	<1
Bas-Rhin	Weyer	<0.01	<0.01	<0.01	–	<0.01	<0.01	<0.01	<1
Nord	Wallers	–	–	0.021	0.003	0.343	0.022	0.02	0.05
Nord	Petite-Forêt	–	0.002	0.059	0.002	0.012	–	–	0.122
Nord	Rumegies	0.029	0.002	0.016	0.002	0.028	–	–	0.028
Lozère	Drigas	–	–	–	–	0.005	–	–	0.027
Lozère	Drigas	–	0.002	–	–	0.01	–	–	0.026
Lozère	Drigas	–	–	–	–	0.005	–	–	0.022
Lozère	Aven Armand	–	–	0.01	0.11	0.4	–	–	–
Lozère	Aven Armand	–	–	–	–	0.004	–	–	0.011

HCB = hexachlorobenzene; HCH = hexachlorocyclohexane; HE = heptachlor epoxide; D = dieldrin; DDE = dichlorophenyldichloroethylene; TDE = trichlorophenyldichloroethane; DDT = dichlorodiphenyltrichloroethane; PCB = polychlorinated biphenyls.

Table 10.2 *Concentration of organochlorines, PCBs, and lead in Little Owl eggs in France expressed in ppm (mg/kg fresh weight). Data from Génot* et al. *1995.*

Year	Number of eggs	HCB	HCH total	HE	DDE	TDE	DDT	PCB	Pb
1987	4	0.038	0.018	0.003	0.059	<0.001	<0.001	0.200	–
1988	4	0.013	0.003	0.008	0.080	<0.001	<0.001	0.420	0.043
–	6	0.012	0.018	0.019	0.062	<0.001	<0.001	0.400	0.056
1989	2	<0.001	<0.001	<0.001	<0.001	0.48	<0.001	1.28	–
–	2	<0.001	<0.001	<0.001	<0.001	<0.001	<0.001	–	3.209
–	5	<0.001	<0.001	<0.001	<0.001	<0.001	<0.001	–	0.958
–	5	–	<0.001	<0.001	–	–	–	–	7.116
1991	7	<0.01	<0.01	<0.01	<0.01	<0.01	<0.01	<0.01	5.101
1992	5	<0.01	<0.005	<0.01	0.214	–	0.214	<0.05	5.284

HCB = hexachlorobenzene; HCH = hexachlorocyclohexane;
HE = heptachlor epoxide; DDE = dichlorophenyldichloroethylene;
TDE = trichlorophenyldichloroethane; DDT = dichlorodiphenyltrichloroethane;
PCB = polychlorinated biphenyls; Pb = lead.

Table 10.3 *Comparison of the levels of organochlorines and PCB in Little Owls across Europe. Mean and range of concentrations are expressed in ppm (mg/kg) in comparison to fresh weight.*

Country	Author	Years	Number of eggs	HCB	DDE	HE	PCB
Belgium	Joiris & Delbecke 1981	1969–74	4	0.01	–	0.11	0.7
The Netherlands	Fuchs & Thissen 1981	1972–7	102	0.02–0.27	3–10	0.02–0.06	0.4–0.9
Switzerland	Juillard 1984	1976–81	112	0.12	0.80	0.04	1.45
Germany	Bednarek *et al.* 1975	1972–3	–	0.56	0.26	0.09	1.2
	Hahn 1984	1981–2	23	0.049	0.344	0.026	0.647
France	Génot 1995	1987–91	35	0.01	0.03	0.005	0.46

HCB = hexachlorobenzene; DDE = dichlorophenyldichloroethylene; HE = heptachlor epoxide; PCB = polychlorinated biphenyls.

In Table 10.3, we offer results from a number of studies that have examined organochlorine and PCB levels in Little Owls across Europe. In the Netherlands, Van der Brink *et al.* (2003) showed that PCBs as well as dichlorophenyldichloroethylene (DDE) bioaccumulate from soil to worms and from worms to owls. Mean concentrations and ranges in uropygial oil, expressed in ppm, were 2 (0.8– 6.2) for PCBs and 1.9 (0.8–17.9) for DDE. PCB levels in a Dutch contaminated area were elevated, and PCB patterns indicated hepatic enzyme induction in owls. The levels of PCBs, possibly combined with other dioxin-like compounds, may pose a risk to Little Owls in the Rhine floodplain.

Recently, brominated diphenylethers (BDEs), organochlorinated pesticides (OCPs), and PCBs were measured in 40 Little Owl eggs in Belgium (Jaspers *et al.* 2005). The major organochlorine pollutants detected were PCBs (median 2.6 ppm lipid, range 0.786–23.204 ppm lipid). Brominated diphenylethers were measurable in all eggs, although concentrations were low compared to levels of persistent organic pollutants. The results revealed that concentrations of most persistent organic pollutants were significantly higher in eggs from deserted nests in comparison to addled eggs, even though there was no difference in eggshell thickness between deserted and addled eggs (Jaspers *et al.* 2005).

Heavy metals The effect of industrial contamination of watercourses, which carry the contaminants and deposit them on flood plains, have been studied in detail in the Netherlands by Groen *et al.* (2000). The Little Owl was an excellent study species, due to its widely varying diet, allowing different poison absorption rates to be analyzed at different trophic levels (see Figure 10.4). The study examined the breeding biology of Little Owls in two areas (one contaminated area and an uncontaminated control) to compare breeding output.

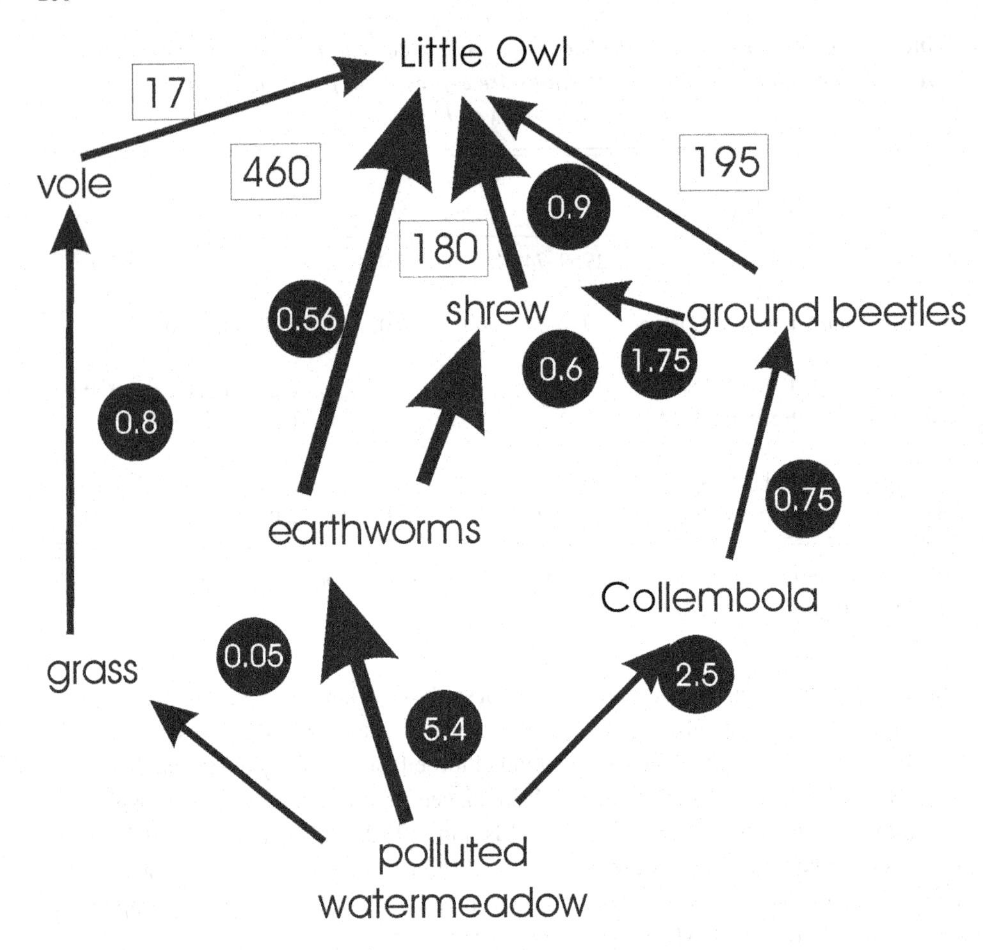

Figure 10.4 Accumulation rates of heavy metal poisons (cadmium) absorbed into the food chain consumed by Little Owls (after Groen *et al.* 2000). Bioaccumulation is indicated in circles in µg Cd/day. Squares indicate the exposure to µg Cd/day.

The results were noteworthy, in that the output of fledged birds from the contaminated area was dramatically lower than from the uncontaminated control.

Jongbloed *et al.* (1996) found that the Little Owl was particularly susceptible to the uptake of cadmium, through the following food chain: soil to worm, and worm to owl. This perspective was also confirmed by Van der Brink *et al.* (2003) who considered that cadmium posed a risk when the owl fed only on earthworms over a long period of time.

In eastern France, the mean concentration of lead in Little Owl eggs was 2.75 ppm (n = 6) (Génot *et al.* 1995). This level was more than ten times higher than the level of 0.2 ppm

($n = 18$) reported by Juillard *et al.* (1978) for eggs from Switzerland. As the Little Owl is linked with prey in more open habitats, it appears much more likely to pick up lead pollutants than forest owls. For example, Ravussin *et al.* (1990) found the mean concentration of lead levels in eggs of the Tengmalm's Owl to be 0.07 ppm ($n = 25$). Except for some samples, the contamination of eggs with heavy metals was low in France (Génot *et al.* 1995) but a monitoring program should continue to track these important pollutants in the future.

In southeastern Spain, the exposure of raptors to lead (Buzzard, Kestrel, Eagle Owl, and Little Owl) was recorded through samples of liver, kidney, brain, blood, and bone. Relationships were found between the size and age of the birds, the proximity to areas of human activity, and lead concentrations in tissues. The larger the bird, the higher the lead concentration in all tissues. Adults accumulated more lead than nestlings and immature birds except for lead in the bones, which was lower for adults possibly due to the removal of lead from the circulation and deposition in ossifying bone. The lead distribution pattern revealed that bone is the principal part of the body for accumulation (900–6000 ppm, dry weight), followed by the kidney (46–375 ppm, dry weight), the liver (35–315 ppm, dry weight) and the brain (14–149 ppm, dry weight). The correlation between lead concentrations in soft tissues and in the blood were high (Garcia-Fernandez *et al.* 1997). See Table 10.4.

Other anthropogenic causes

Drowning Most drowning cases are of soiled females who have endured a breeding season without bathing, and young birds that are leaving their natal territory. These owls become immersed in cattle water troughs, which have steep slippery sides, their feathers become waterlogged, they cannot recover and drown. In a French inquiry about mortality causes between 1980 and 1990, Génot (1991) found that 21 out of 751 (2.8%) owl deaths were due to drowning. In the French region of Ile-de-France, drowning represented 20% of the mortality among young owls (Lecomte 1995). Out of 53 cases of mortality in western France, four (7.5%) were due to drowning (Clech 2001a). In Belgium, drowning represented 9.6% out of 89 mortalities (Bultot *et al.* 2001).

Electrocution One of the most common views that people have of Little Owls is of them sitting on top of a telephone or electricity pole. Reports of deaths resulting from electrocution are surprisingly rare. However, a report from a local birdwatcher in Cyprus reported a small owl hanging by its feet from overhead electricity supply wires; it was assumed that the owl had come into contact with two or more wires as it perched on the wires.

In an analysis of ringing recovery data on Little Owls in France from 1927–90, P. Raevel (personal communication) noted that 1.9% of the recoveries were of owls recovered as a result of electrocution by power lines. This low rate of electrocution-related recoveries places the species in "class 5" – reflecting a species that is not sensitive to the risk of power-line mortality.

Table 10.4 *Comparison of the levels of heavy metals in Little Owls across Europe. Mean of concentration expressed in ppm (mg/kg dry or fresh weight according to the study).*

Country	Author	Result	Sample	Lead (Pb)				Cadmium (Cd)				Mercury (Hg)		Chromium (Cr)	
				Liver	Muscle	Brain	Feathers	Liver	Muscle	Brain	Feathers	Liver	Muscle	Liver	Brain
France	Génot *et al.* 1995	fresh weight	n = 8	0.46	0.057	–	–	0.382	0.045	–	–	0.139	0.029	–	–
Italy	Battaglia *et al.* 2004	dry weight	n = 38	0.35	0.10	–	2	0.05	0.003	–	0.05	–	–	–	–
Italy	Zaccaroni *et al.* 2003	fresh weight	n = 41	0.20	–	0.057	–	0.110	–	0.054	–	–	–	0.288	0.057
			n = 11	0.232	–	0.070	–	0.105	–	0.061	–	–	–	0.297	0.076
Netherlands	Van den Brink *et al.* 2003	dry weight	n = 4	–	–	–	<40	–	–	–	<2.3	17	–	27	–

Shooting and other persecution Like most birds of prey, the Little Owl has been the subject of a high degree of persecution, particularly in the late eighteenth century and early to mid nineteenth century. The Little Owl's diet was not fully researched, and the suspicion was that the main component of its diet was game birds. Many methods of persecution were used, including shooting, pole or tunnel trapping (Hewson 1972), poisoning, and egg pricking, to name a few. Hibbert-Ware (1936, 1937/1938) undertook a survey of the Little Owl's diet and disproved the gamekeepers' and hunting fraternity's perspective of the owl. In England and Wales it was noticeable that during World War I and II, when most gamekeepers were serving their country, the population exploded and colonized new areas, unhindered by the controls of persecution.

Even though laws have protected the Little Owl in the UK since the 1950s, persecution continues today in some places. The diurnal activities make the owl an easy target for unscrupulous shooters. Likewise, the species is still killed in southern Italy where 24 Little Owls were reported shot in 1974–5, and 9 reported in 1981 and 1983 (Falcone 1987). Similarly, in North Caucasia the Little Owl is shot by local inhabitants in all seasons and especially in the breeding period (Afanasova & Khokhlov 1995). In the Ural region of Russia, shepherds rob the nests of Little Owls that breed in farms and summer cattle camps (Karyakin 1998). It seems likely that part of this indiscriminate killing may have some roots in cultural beliefs (see Chapter 2).

Trapping in hollow metal telephone poles In France, hollow metal telephone poles throughout the countryside constitute a trap for many birds that are looking for cavities to nest or roost in (see Figure 10.5). The hollow poles are structured such that once the birds enter the hollow posts (which are open at the top) they are unable to escape. In western France, eight Little Owls were found dead in 32 telephone poles (Clech 1993). In the region of Ile-de-France, 13 Little Owls were found dead in 305 telephone poles surveyed (Savry, personal communication). In the beginning of the 1970s some 3 500 000 such telephone poles were put up in France (Clech 1993). From the analysis of local records, we can imagine that these poles could very likely cause the local extirpation of populations already suppressed due to other different factors.

Trapping in chimneys As Little Owls are cavity nesters, and since many live in proximity to villages and towns, many Little Owls visit chimneys. Once inside, as with the hollow metal power poles, they are unable to fly or climb the steep walls, and they become trapped and die. Zvaral (2002) believes that deaths due to chimneys and shafts are the most important mortality factors in the Czech Republic. Of the overall mortality rates, this cause of death has been documented in a number of areas: 13% (7 out of 53) in western France (Clech 2001a), 7% (2 out of 31) in eastern France (Génot 2005), 19.5% (19 out of 98) in eastern Germany (Schönn 1986), and 4% in Britain (Glue 1971).

Land-use changes The loss of habitat that has taken place from the 1960s until now is mainly responsible for the decline of Little Owl populations in all of western Europe.

Figure 10.5 Electricity poles as traps for Little Owls. Little Owls become trapped and die within the open-topped poles. Photo Jean-Claude Génot.

The land-use changes reflect the conversion of meadows into crop fields, drainage of wet meadows, cutting of isolated trees (e.g., willows and oaks) or hedges, transformation of traditional orchards into "fruit tree crops" (also called "low-stem" orchards), and the general shift to industrial farming with its associated large fields without perches, nesting sites or hunting ground, and with the use of pesticides. In addition, areas that were previously managed in a pastoral or agricultural context have been encroached upon with the development of big cities and intermediate-sized towns, highways, railways, roads, airports, and of industrial and business areas. The loss of habitat was measured in a number of study areas in Germany. In Schleswig-Holstein, the area of pasturelands was reduced by 20% between 1960 and 1980 (Ziesemer 1981). These former pasturelands were transformed into arable lands, thus becoming less suitable for the Little Owl. In Northrhine-Westphalia, on a study area of 4316 ha, the proportion of the landscape in grassland decreased by 34% between 1919 and 1991 (Kämpfer-Lauenstein & Lederer 1995). In Baden-Württemberg, between 1970 and 1973, 16 693 fruit trees (mainly apple and pear, which offer the best holes for the Little Owl) were cut down (Ullrich 1975). In Westphalia, Illner (1981) reported that the decrease in Little Owl numbers was due to the loss of nest sites and food resources due to the conversion of a mosaic landscape composed of meadows, pastures, hedges, orchards, and field edges.

In Switzerland, Juillard (1989) explained that the decrease in the Little Owl population between 1950 and 1985 was mainly due to the destruction of favorable biotopes, in particular, the cutting of several million fruit trees. This cutting came from organized campaigns to restructure the orchards in order to obtain fruit tree crops.

A smaller scale study was conducted in Northrhine-Westphalia (Germany) where Loske and Loske (1981) noted the following transformation of an area about 1 km^2: 7% in crops as of 1970, and 62% in 1980; 545 pollarded willows in 1970, and 271 in 1980. Survey efforts conducted during this time found four calling Little Owl males in 1970 and only one in 1980.

Urban development is also a threat to Little Owl populations. An evaluation of land-use plans in three municipalities in Northrhine-Westphalia where 115 breeding pairs of Little Owl were present, predicted the loss of 8% to 50% of the breeding pairs (Breuer 1998). The importance of "green systems" including mown meadows and orchards for the species has been illustrated through the size of the home range of the Little Owl, i.e., 20.3 ha in the "green system" and 74.8 ha in the "arable system" (Orf 2001).

In France, the transformation of the rural landscape has been substantial. Four areas (totaling approximately 26 000 km^2) in western France lost 174 000 km of hedges between 1962 and 1990. For all of France, there were some 48 million apple trees in 1929, and 11 million in 1990. A substantial portion (20%) of this loss in apple orchards occurred during the period 1981–90. Also, significantly, 25% of the area in meadows was converted to other land uses between 1970 and 1995 (Pointereau & Bazile 1995).

In eastern France, the last available habitats for the Little Owl remain around villages in "green belts", which include orchards with old fruit trees, meadows, pastures, and gardens. Between villages, the former landscape has been modified by industrial agriculture, reallotment of land during 1970–90 (when people exchanged land to obtain larger contiguous parcels mainly for larger-scale agricultural purposes), and the recent conversion of meadows into arable land resulting from European agricultural policies. So outside of the green belts containing habitat close to the villages, the landscape has become unfavorable for Little Owls. Pairs have become more and more isolated with the fragmentation of the former habitat and local populations are weak and very susceptible to stochastic environmental disturbances, such as the December 26th, 1999 storm when several thousand fruit trees were felled (Génot, personal observation).

In Austria changes in the landscape, notably the increase in area of arable fields, decrease of grasslands, and the reduction in habitat heterogeneity, explained the large decline in the Little Owl. The patch size of agricultural areas and the insect prey density and diversity are the important factors in maintaining populations (Ille & Grinschgl 2001). The enlargement of the size of parcels to more than 20 ha was followed by a decrease in both the number of beetle species and number of individuals (Deveaux *in* Lefeuvre 1981).

In the steppe region of the former Soviet Union, the ongoing desertification process, as seen in Dagestan, is a threat to the Little Owl (Vilkov 2003).

Finally, the decline of the Little Owl is often due to a synergy between different factors. In Northern Vosges (France) the combination of adverse climate (winters of 1985–7), predation, road traffic, loss of orchards and grasslands, and fragmentation of good habitat, were the primary drivers behind the reduction in owl numbers (Génot 1992a). Similarly, in the Basel region of Switzerland, the decline was due to predation by Stone Martens, climatic conditions (in particular the rainfall during the feeding of nestlings), lack of prey in hay lands, road traffic, and the lack of connectivity to other surrounding populations (which could offer demographic support) (Luder & Stange 2001).

While the majority of the data in the above discussion on habitat loss and modification has come from Europe, it is reasonable to expect that similar types of (and impacts from) alterations in land uses are occurring in other portions of the Little Owl's range.

Food

The prey resources available to owls vary from year to year. Small mammal prey (e.g., voles) typically follow cycles with peaks of abundance every three to five years. In years of scarcity, the breeding performance and survival rates of both adults and first-year birds may be low, causing a population decline (Taylor 1994; Van Veen & Kirk 2000). Resident owls respond to prey cycles in two ways, first by switching to other prey, and second by producing fewer eggs, or in some cases none at all (Newton 1998, 2002). On average, female Little Owls lay 1.0–1.2 more eggs in good vole years than in poor ones (Exo 1992).

The Little Owl's ability to forage for a wide variety of small prey aids it when one of its prey species becomes scare, enabling it to compensate to some extent. However, in years when small mammal populations crash, the alternative prey, invertebrates or birds, often provide less energy reward per unit of hunting effort.

Juvenile mortality tends to rise as the young gain independence from their parents, and dispersal from natal areas tends to be later in years when prey is scarce than years where prey is abundant.

Nest and roost sites

The lack of natural cavities has been shown to limit Little Owl populations (Exo 1983), as no new sites are available for birds to take up as first-time breeders or to relocate to from damaged nests. The Little Owl has high nest-site fidelity, making the problem worse should any damage occur to their nest site.

The provision of additional nest sites provides support for the Little Owls and allows for other cavity nesting species (which compete with Little Owls for the same cavities). We much prefer the planting and long-term protection of native tree species that cost-effectively provide natural cavities for owls, as well as offering other ecological benefits. As a short-term remedy for the shortage of these cavities, nestboxes are used. In the chapter on conservation, we give an overview of nestbox campaigns that have enhanced Little Owl populations. See Figure 10.6.

In the steppe area of Portugal, the number of alternative suitable cavities emerged as the main variable explaining nest-site selection, with the owls showing a preference for stone piles containing larger sized stones (Tomé *et al.* 2004).

Predation

The predation rates on Little Owl nests occurring in nestboxes with and without anti-predator devices were monitored from 2000 through 2004 by J. Bultot (personal communication). The overall predation rate on 1004 active nestboxes without anti-predator devices was 9.45%. No predation occurred on 509 nests (predation rate of zero) equipped with anti-predator devices. Thus, with the addition of anti-predator devices, the overall rate of predation went down to 6.3% on 1513 active nests. Blanke (2005) reported for five research areas in the Netherlands, a predation rate of 10% on 20 active nests in Ruinen, 80% on 5 active nests in Geleen, 3% on 305 active nests in Zuidoost-Achterhoek, 5% on 433 active nests in Doesburg, and 2% on 69 active nests in Raalte. Vanden Wyngaert (2005) reported the predation rate by Stone Martens to be 5% on 37 active nests in Dijleland, Belgium. Eick (2003) radio-tracked ten juvenile Little Owls of which five (50%) were killed by predators (i.e., Goshawk, Fox, Tawny Owl or Goshawk, unknown mammal, and Sparrowhawk). These owls were tracked for 1, 1, 2, 48, and 101 days before they were predated.

Mammals The small size of the Little Owl means that it has many possible predators among mammals and birds of prey. For the mammals, the Stone Marten (*Martes foina*) is recognized as an important predator on the Little Owl, as stressed by many authors (Knötzsch 1978; Ullrich 1980; Hahn 1984; Schönn 1986; Veit 1988; Krischer 1990;

Figure 10.6 Little Owl conservation activities by Ph. Smets, Flanders, northern Belgium. (Photo Maarten Bekaert.)

Haase 1993; Peter 1999; Luder & Stange 2001). Starting in the 1970s, ornithologists have been building nestboxes designed to protect Little Owls from the Stone Marten (Schwarzenberg 1970; Furrington 1979; Marié & Leysen 2001). The impact of Stone Marten predation on the eggs of nesting Little Owls has been quantified in Germany and in France. In Germany, Illner (1979) found a predation rate on eggs to be from 10–20%. In France, Génot (unpublished data) found a predation rate of 28% (61 out of 218) on unhatched eggs. The Stone Marten is a generalist predator, and will eat the eggs as well as the young or adult owls. Martens are an important predator on Little Owls as they are agile climbers, will search all of the tree holes and other cavities, and eat whatever small prey they find within. Visitation to holes and nestboxes by martens is revealed by their fecal droppings found inside the boxes. Like others in the Mustelidae family, martens are territorial, and by marking the places they visit, they also scent-mark the cavity against other martens. Its predation probably does not depend on Little Owl densities, because during foraging the Stone Marten visits the same cavities again and again even after a clutch has been found. Little Owls faithful to their nest even after a failure might become easy victims to marten predation. The Stone Marten benefits from human activities, and is expanding in numbers in the cities as well as in the countryside (Van Den Berghe 1998). Even if the Little Owl is not the main prey of the Stone Marten in a simple ecosystem, it could be considered as an important food source (Bergerud 1984) for Stone Marten.

In Portugal, the presence of two predators (the Stone Marten and Common Genet, *Genetta genetta*), emerged as the main factor linked to nest-site selection by Little Owls in woodland areas (Tomé *et al.* 2004). In the Volga Region of Russia, Frolov *et al.* (2001) considered that the Little Owl suffered from predation from feral cats and Stone Martens, whose numbers had increased in anthropogenic landscapes. While predation by Stone Martens may threaten some isolated and small Little Owl populations, as Illner (1988) said, in the long term, predation has probably contributed little, if anything, to the population decrease in comparison to the reduction in food supply.

Other mammals can destroy the eggs or chicks of Little Owls, including the Stoat (*Mustela erminea*), Hedgehog (*Erinaceus europaeus*) (Glue & Scott 1980), Brown Rat (*Rattus norvegicus*) (Glue & Scott 1980; Coppée *et al.* 1995); or can kill the adults, such as the feral cat (Glue & Scott 1980; Haase 1993), Fox (*Vulpes vulpes*) (Knötzsch 1978), and domestic dog (Glue 1971; Juillard 1984a). We have even witnessed trampling by cows as a cause of mortality among young that had fallen out of (or jumped from) the nest. During a reinforcement experiment in France, out of 35 Little Owls born in captivity and equipped with radio-transmitters at release, 22 were found dead within one day. In some cases the reason was known: fox (2), cat (1), bird of prey (2), stoat or weasel (1), unknown carnivorous mammal (2) (Génot & Sturm 2003). Through the radio-tracking of Little Owls, Eick (2003) found one juvenile Little Owl caught by a fox, and one by an unidentified mammal.

Birds Little Owls are also found in the diet of birds of prey, as shown by pellet analyses. Among the birds of prey, the Eagle Owl (*Bubo bubo*) seems to be the main predator (Blondel & Badan 1976; Herrera & Hiraldo 1976; Mikkola 1976; Orsini 1985; Simeonov 1988;

Martinez *et al.* 1992; Jay 1993; Karyakin & Kozlov 1999), followed by the Tawny Owl (*Strix aluco*) (Schnurre 1940; Staton 1947; Gunston 1948; März & Weglau 1957; Mikkola 1976; Baumgart 1980; Schönn 1980; Ullrich 1980; Zuberogoitia & Campos 1997a; Eick 2003). A peaceful cohabitation is possible during the breeding season (Rusch 1988) and might be due to the fact that the female does not hunt during that period (I. Newton, personal communication). When Little Owl calls are broadcast and a Tawny Owl answers, Little Owls remain silent (Schönn *et al.* 1991).

While observed much less frequently, other birds of prey can kill Little Owls: Barn Owl (*Tyto alba*) (Castellucci & Zavalloni 1989; Bonvicini & Maino 1993; Sgorlon 2005), Long-eared Owl (*Asio otus*) (Glue 1972; Mikkola 1976), Peregrine (*Falco peregrinus*) (Formon 1969; Mikkola 1976; Schönn *et al.* 1991), Buzzard (*Buteo buteo*) (Christie 1931; Mikkola 1983), Goshawk (*Accipiter gentilis*) (Mikkola 1976; Nore 1977; Schönn 1986; Eick 2003), Sparrowhawk (*Accipiter nisus*) (Nore 1979; Mikkola 1983), Red Kite (*Milvus milvus*) (Mikkola 1983; Schönn *et al.* 1991; Eick 2003), Tawny Eagle (*Aquila rapax*), Rough-legged Buzzard (*Buteo lagopus*), Booted Eagle (*Hieraaetus pennatus*), Lanner (*Falco biarmicus*) (Mikkola 1983), Marsh Harrier (*Circus aeruginosus*), Black Kite (*Milvus migrans*), Imperial Eagle (*Aquila heliaca*), and Long-legged Buzzard (*Buteo rufinus*) (Gavrin 1962). In the desert habitat of Kazakhstan without shelters, the Little Owl can become locally extirpated in winter by Long-legged Buzzards and large falcons (Osmolovskaya 1953).

Besides birds of prey, a case of egg predation by a Magpie (*Pica pica*) was recorded (Glue & Scott 1980). Magpies may also mob the owls (Ritzel & Wulf 1978).

Parasites and disease

A complete list of 67 species of parasites of the Little Owl has been provided by Juillard (1984) and Schönn *et al.* (1991). Additional insights into the parasites has been offered in the works of Borgsteede *et al.* (2003), Sanmartin *et al.* (2004), and Ferrer *et al.* (2004). The latter authors showed that the Little Owl had the highest level of infestation by parasitic helminths in the digestive tract of six species of owls in Spain, likely due to its generalist diet. The main types of parasites include representatives from the following groups: Protozoa (*Falgellata*), Sporozoa (*Coccidia* and *Haemosporidia*), Metazoa (Plathelminthes, *Trematoda*, and *Cestoda*), Nemathelminthes (*Nematoda* and *Acanthocephala*), Arachnida (*Acarina*), Insecta (*Mallophaga*, *Diptera*, and *Siphonaptera*). Another acarine species, *Ornithogastia ariadnae*, found in Ukraine can be added (Gushcha 1982).

Tomé *et al.* (2005) studied the relationships between the occurrence of blood parasites and host traits in wild Little Owls in the Mediterranean habitats of southern Portugal. Out of 39 owls captured between February and August 1999, 16 (41.0%) were infected with *Leucocytozoon ziemanni*. One individual was infected with *Trypanosoma* sp. and another with a microfilaria. Age was the only variable influencing the prevalence of *L. ziemanni* in a multivariate logistic regression model comparing infected and non-infected owls. Prevalence was much higher in adults (82.4%; $n = 17$) than among juveniles (9.1%; $n = 22$). This difference was probably due to the scarcity of appropriate vectors (black flies; Diptera;

Simuliidae) in the area at the time of fledging. Univariate analysis also suggested a trend for higher prevalence among females than males and a difference between bill lengths of infected (shorter billed) and uninfected (longer billed) adult owls. Korpimäki *et al.* (2002), Hakkarainen *et al.* (1998), Korpimäki *et al.* (1995), and Korpimäki *et al.* (1993) studied the important relationships between blood parasite loads, nest defense, and reproductive performance in the Tengmalm's Owl and Eurasian Kestrel. Because reproduction and immune defenses are both thought to be energetically costly, the birds may face a trade-off between allocation of resources to reproduction and immunity. The authors found that defense against blood parasites is indeed costly to the owls and kestrels (e.g., reduced clutch size, less vigorous defense of offspring by male owls), and that costs may vary with the sex of the host species and with prevailing environmental conditions. Little Owls are subjected to a very wide range of parasites (both internal and external). The implications from the studies described above are that blood parasite levels do have a very real impact on raptors, and through additional investigations, new insights into the breeding ecology of Little Owls can be ascertained.

During his study, Juillard (1984a) found seven species of ectoparasites, mainly on nestlings. Depending on the nature of the ectoparasites, they are typically found around the eyes and the bill, on the eyelids, on the body, and in the feathers. One of the ectoparasites found by Juillard in Switzerland, the dipteran *Carnus hemapterus*, was also identified in Germany as a parasite on nestlings of many other birds, including the Peregrine (*Falco peregrinus*), Barn Owl (*Tyto alba*), Kestrel (*Falco tinnunculus*), Starling (*Sturnus vulgaris*), Great Tit (*Parus major*), Stock Dove (*Columba oenas*), Jackdaw (*Corvus monedula*), and Carrion Crow (*Corvus corone*) (Walter & Hudde 1987). Walter and Hudde showed that this ectoparasite did not reduce breeding success. Two endoparasites have been found in the intestine and in the liver of the Little Owl (Juillard 1984a). Beside the parasites, two commensal species of Lepidoptera, *Trichophagata petzella* and *Monopis laevigella*, have been found in the pellets of a Little Owl in France (Courtois 1988).

Regarding diseases, a list of infections is given in Schönn *et al.* (1991). Among the main non-specific diseases affecting Little Owls are: (1) Newcastle-infection (Schoop *et al.* 1955), a viral infection that appears when the birds have limited food supplies; (2) Hepatosplenitis Infectiosa Strigum that can cause intestinal inflammation in Little Owls (Schönn 1986); (3) bacterial infections such as Salmonellosis, Tuberculosis, Pseudomonasis; and (4) fungal diseases such as Aspergillosis. Finally, *Streptococcus* and *Staphylococcus* infections affect Little Owls in captivity (Schönn *et al.* 1991). Bultot *et al.* (2001) note that Belgian bird-care centers have recorded 23% of incoming Little Owls as being victims of disease.

In addition to parasites and diseases, which are much more prevalent, a few anomalies or anatomical malformations have been reported. Juillard (1984a) found a nestling with a crossed bill, which was probably lethal because the young bird could not feed itself; another with a necrotic leg, which was due to a thorn under the ring of the bird (the bird died despite the removal of the thorn). A third anomaly is very rare and is called "iris colobome" (Reisinger 1926a; Juillard 1981). It is a tear of the iris caused either during embryonic development or due to a collision in a fully fledged bird. During field studies, two other cases of injuries to the iris were recorded in, one in France (Génot, unpublished data) and another in Belgium (J. Bultot, personal communication).

Weather

Rain

Weather has been shown to influence the breeding performance of Little Owls (Knötzsch 1988). For example, in the period leading up to egg-laying, rainy weather conditions can hinder and sometimes prevent hunting, affecting the breeding condition of adult birds, and delaying the start of breeding. Bultot *et al.* (2001) highlighted the constraints that weather puts on later breeding performance by showing that a higher rate of owlet mortality was related to the amount of rainfall during May and June.

At Lake Constance, Germany, Knötzsch (1988) considered that if the number of cavities was sufficient, the only limiting factor for a population was the climate; this also implied that food supplies were influenced by the prevailing weather. Larger amounts of precipitation in March seem to result in improved food availability (earthworms) and thus higher weights of adult owls. In contrast, higher rainfall in May and June, especially when the nestlings were younger than two to three weeks old, reduced fledging success, as a result of the moist environment in cavities or nestboxes (Knötzsch 1978; Exo 1983; Gassmann *et al.* 1994; Lederer & Kämpfer-Lauenstein 1996). However, the reduced fledging success could also be due to reduced hunting success of the adults at this time. Fledging success in Belgium (Bultot *et al.* 2001) showed a negative association with cumulative precipitation in May and June. For every 100 L/m^2 extra precipitation in this period, 0.6 fewer young were fledged per active nest.

Snow

Cold and snowy winters cause substantial losses in Little Owls, as was seen in Switzerland in 1951 (Büchi 1952), in Denmark in 1956 (Poulsen 1957), in England in 1962–3 (Dobinson & Richards 1964), and more recently in France in 1985–7 (Génot 1990a), and Germany (Kämpfer-Lauenstein & Lederer 1995). The annual mortality patterns of adult and first-year Little Owls in Germany indicate a peak in mortality in January to February (Exo & Hennes 1980). Little Owls have more fat reserves (15.1% of bodyweight) than Barn Owls (Piechocki 1960), and hence more capacity to cope with periods of cold weather (Helbig 1981; Schönn 1986). During severe or prolonged cold weather or snow events, the owls cannot feed themselves and die of starvation or exposure (Poulsen 1957; Géroudet 1964). In France, Génot (personal observation) found a Little Owl dead in January 1997 after three weeks of snow; the owl weighed 110 g compared to a normal weight of 160–250 g.

After cold winters, as in 1939–42 in Germany and 1894–95 in Switzerland, Peitzmeier (1952) and Luginbühl (1908), respectively, saw that Little Owl populations declined and took about ten years to recover. In the Kaliningrad Region, the number of Little Owls decreased considerably after a severe winter (Grishanov 1994). In France, after two severe winters in 1985 and 1986 with 25–26 days with 10–15 cm of snow, the population decreased almost by 30% (Génot 1992a). In Germany, after the winters of 1978–9 and 1985–6, 30% and 38% of the occupied sites were abandoned, respectively (Schönn 1986).

The Tibetan Little Owls, (*Athene noctua ludlowi*) use the burrows of Himalayan Marmots (*Marmota himalayana*). An unusual winter, with exceptional amounts of snow, flooded the burrows in the spring, killing the majority of the marmots. The owls survived this event by moving to lower elevations. Upon their return to the area, the Little Owls did not breed in the marmot burrows, but rather looked for shelter in them, and appeared to use the burrows that were occupied by marmots. One could speculate that perhaps this was because the occupied burrows were warmer, or because the burrows were not so full of small rocks and sediment caused by the flooding (van Orden & Paklina 2003).

Hard winters, which inhibit foraging activity, are a major cause of both first-year and adult mortality. This is reflected across the species' range by the higher winter mortality in countries that tend to have lower temperatures and longer periods of snow cover. Snow cover presents problems for ground-feeding predatory birds, which cannot gain access to their prey below the snow (Schröpfer 1996).

Interspecific relationships

Aggressive interactions exist between the Little Owl and some other species. As shown by Zuberogoitia *et al.* (2005) among European owls, most attacks were aimed at Little Owls and the main aggressors were Barn Owl and Tawny Owl. While the Barn Owl may have been noted as being a potential predator on Little Owls (Lord & Ainsworth 1945), in captivity the species can get along peacefully (Festetics 1952/1955), and in nature the two species can live in the same building (Reynaud, personal communication) and even breed as in Slovenia (Denac *et al.* 2002). Magpies (*Pica pica*), mob the Little Owl (Ritzel & Wulf 1978), and the Eurasian Kestrel (*Falco tinninculus*) competes for nest sites (Bonin & Strenna 1986) or prey (Moreau 1947). As noted in the chapter on history and traditions, the mobbing behavior of small birds on Little Owls (and the human desire to catch these small birds) was one of the main reasons that young owls were taken from nests and raised in captivity for hunting for nearly 2000 years. In Italy, the mobbing of small birds was concentrated within the breeding season (e.g., Swallow, *Hirundo rustica*, and House Martin, *Delichon urbica*), although some species displayed mobbing behavior in autumn and in winter (e.g., Tree Sparrow, *Passer montanus*, Italian Sparrow, *Passer italiae*, Great Tit, *Parus major*, Blue Tit, *Parus caeruleus*, Greenfinch, *Carduelis chloris*, Goldfinch *Carduelis carduelis*, Serin, *Serinus serinus*, Stonechat, *Saxicola torquata*, and Meadow Pipit, *Anthus pratensis*) (Casalini 2005). Due to either alternative and readily available food sources, or fear of injury, some tolerance is observed between the Little Owl and some animals that could be its predators or competitors. Thus, Little Owls can breed close (less than 50 m) to Tawny Owls (Rusch 1988), or in the same nestbox as a squirrel, *Sciurus vulgaris*, (Palmer 1989) or a Jackdaw, *Corvus monedula* (Berghmans 2001b). Little Owls are able to breed in a big nestbox already occupied by Starlings, *Sturnus vulgaris* (Henrioux 1980). These events do not need to have biological significance. Nest-site sharing may be a symptom of the shortage of suitable sites. A final example of living together is given in Italy where a Little Owl shared the daytime use of a Long-eared Owl's (*Asio otus*) winter roost (Mastrorilli 1999b).

10.3 Demographic limiting factors

The following discussion reflects a review of studies that contain the type of demographic data and analysis techniques that are needed to help answer fundamental questions of population dynamics in Little Owls. Detailed long-term studies from different geographical and climatic regions are required to resolve whether the observed population decrease in Little Owls across Europe is due to poor reproduction or enhanced mortality (Exo 1992).

Demographic factors reflect the number of births and deaths within a given local population, the immigration of owls coming in from other areas, and the emigration of owls out from the local population. We first look at the factors that regulate births and deaths within the local population and then zoom in on dispersal and movements of the birds. Some density-dependent influences might be occurring (e.g., smaller clutch sizes from owls within higher densities, later breeding from owls in lower densities). Therefore, we describe the densities that are observed throughout the distribution area and then focus on the effects that this might have on local populations. The connectivity of local small population clusters has been found to be increasingly important in the conservation biology of wildlife species. In relation to this, we further discuss meta-populations of Little Owls and the impact this type of structure has on the species.

Births and deaths

A local population that is considered isolated from the outside can be seen as the sum of births and deaths. To keep things simple we first consider that the local population is "closed", i.e., there is no immigration into or emigration from the local population.

In stable populations, reproduction balances mortality. Mortality studies on birds involve some kind of marking (usually ringing) and subsequent recapture or re-sighting. As owls are mostly nocturnal and reclusive, recovery rates of ringed owls are invariably low, as few people re-sight, recapture, or recover dead ringed owls. From the Finnish Bird Ringing Centre data, Saurola (2002: 43) reported that of 198 887 individuals of ten species of owls ringed between 1913–99, some 25 492 (12.8%) were recovered. This level of ringing and recovery represents a truly significant amount of volunteer and professional effort. For Little Owls, the recovery rate was 2.1% for 755 ringed owls (Furrington 1998), 2.8% for 541 ringed owls (Juillard 1984a), 3.9% for 616 ringed owls (Knötzsch 1985), and 2.5% for 435 ringed owls (Génot 2005).

Plate 1 Little Owl habitat of East Kazakhstan. Photo Paklina & Van Orden. See Figure 6.8.

Plates 1-39 are available for download in colour from www.cambridge.org/9780521714204

Plate 2 East Kazakhstan where Little Owls occupy natural cavities. Photo Paklina & Van Orden. See Figure 6.9.

Plate 3 Kyrgyzstan. Photo Paklina & Van Orden. See Figure 6.10.

Plate 4 Ladakh, India between Stok and Shey. Photo Paklina & Van Orden. See Figure 6.11.

Plate 5 Little Owl habitat in Kruiskerke, Flanders, northern Belgium. Photo Marc De Schuyter. See Figure 6.14.

Plate 6 Little Owl habitat in Butten, the northern Vosges, France. Photo Jean-Claude Génot. See Figure 6.15.

Plate 7 Little Owl habitat in Brakel, Flanders, northern Belgium. Photo Marc De Schuyter. See Figure 6.16.

Plate 8 Little Owl habitat in Kemmel, Flanders, northern Belgium. Photo Marc De Schuyter. See Figure 6.17.

Plate 9 Little Owl habitat in Slovenia. Photo Milan Vogrin. See Figure 6.18.

Plate 10 Little Owl habitat in the Tolfo mountains, Italy. Photo Duccio Centili. See Figure 6.19.

Plate 11 Little Owl habitat with old Almond trees in Luberon, France. Photo Jean-Claude Génot. See Figure 6.20.

Plate 12 The monastery of Stakna as breeding habitat of Little Owl in Ladakh, India. Photo Paklina & Van Orden. See Figure 6.21.

Plate 13 Open holm oak "woodland"/parkland (Cabeça da Serra, Castro Verde, Portugal). Photo Ricardo Tomé. See Figure 6.22.

Plate 14 Steppe-like area featured by scattered stone heaps Portugal. Photo Ricardo Tomé. See Figure 6.23.

Plate 15 Little Owl habitat in Lozère, France. Photo Jean-Claude Génot. See Figure 6.24.

Plate 16 Little Owl habitat in Toscana, Italy. Photo Marco Mastrorilli. See Figure 6.25.

Plate 17 Little Owl habitat in Lombardia, northern Italy. Photo Marco Mastrorilli. See Figure 6.26.

Plate 18 Little Owl habitat near Milano, northern Italy. Photo Marco Mastrorilli. See Figure 6.27.

Plate 19 Seven main types of occupied habitats in Flanders (after Van Nieuwenhuyse & Leysen 2001). Grasslands around farms. B: buildings; C: row crops; F: orchards; G: grass; M: maize; N: fallow land; O: other; S: cereals; T: silvaculture. Circle includes 25 ha. See Figure 6.29a.

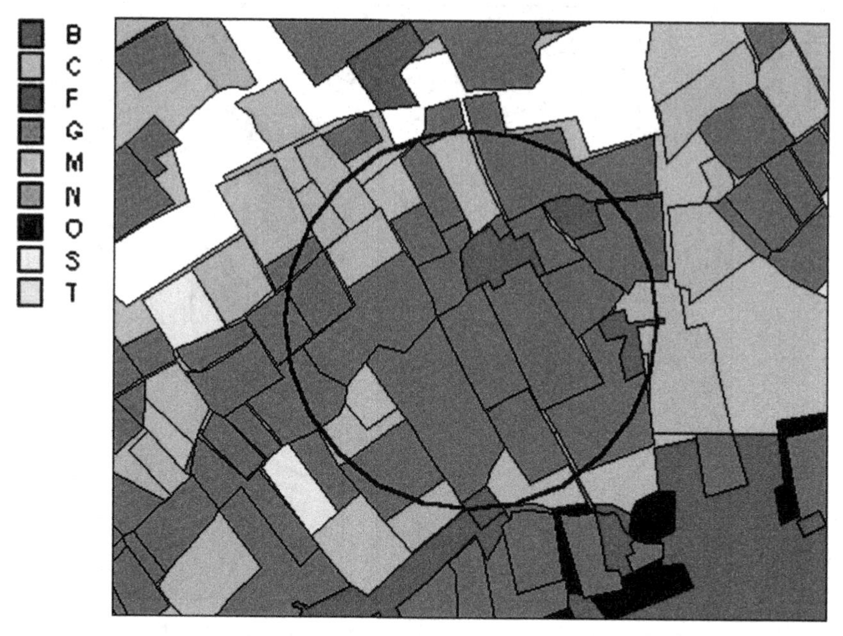

Plate 20 Seven main types of occupied habitats in Flanders (after Van Nieuwenhuyse & Leysen 2001). Urbanized cattle breeding. B: buildings; C: row crops; F: orchards; G: grass; M: maize; N: fallow land; O: other; S: cereals; T: silvaculture. Circle includes 25 ha. See Figure 6.29b.

Plate 21 Seven main types of occupied habitats in Flanders (after Van Nieuwenhuyse & Leysen 2001). Horticulture. B: buildings; C: row crops; F: orchards; G: grass; M: maize; N: fallow land; O: other; S: cereals; T: silvaculture. Circle includes 25 ha. See Figure 6.29c.

Plate 22 Seven main types of occupied habitats in Flanders (after Van Nieuwenhuyse & Leysen 2001). Rural cattle breeding. B: buildings; C: row crops; F: orchards; G: grass; M: maize; N: fallow land; O: other; S: cereals; T: silvaculture. Circle includes 25 ha. See Figure 6.29d.

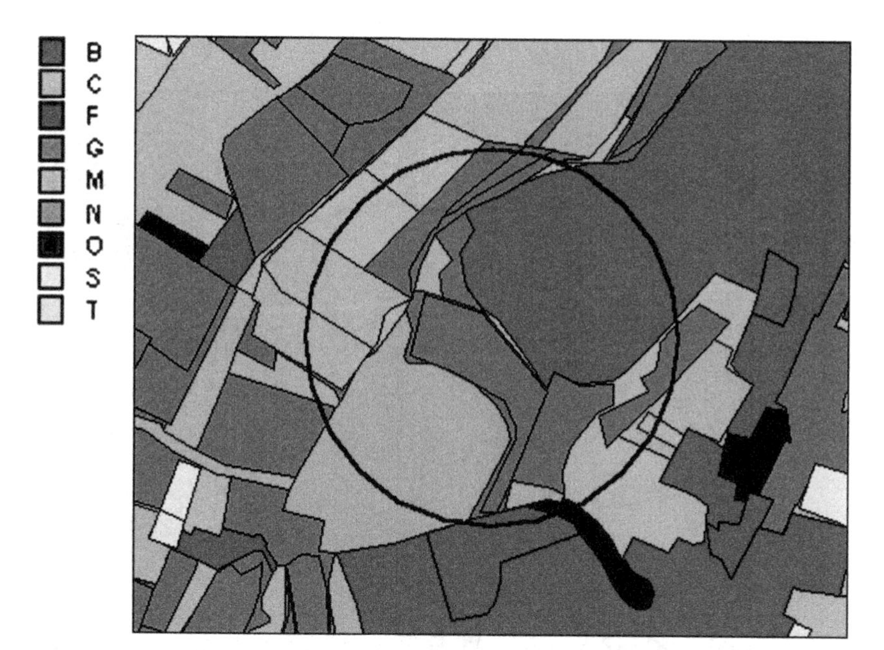

Plate 23 Seven main types of occupied habitats in Flanders (after Van Nieuwenhuyse & Leysen 2001). Half-open grasslands. B: buildings; C: row crops; F: orchards; G: grass; M: maize; N: fallow land; O: other; S: cereals; T: silvaculture. Circle includes 25 ha. See Figure 6.29e.

Plate 24 Seven main types of occupied habitats in Flanders (after Van Nieuwenhuyse & Leysen 2001). Remote cereals and orchards. B: buildings; C: row crops; F: orchards; G: grass; M: maize; N: fallow land; O: other; S: cereals; T: silvaculture. Circle includes 25 ha. See Figure 6.29f.

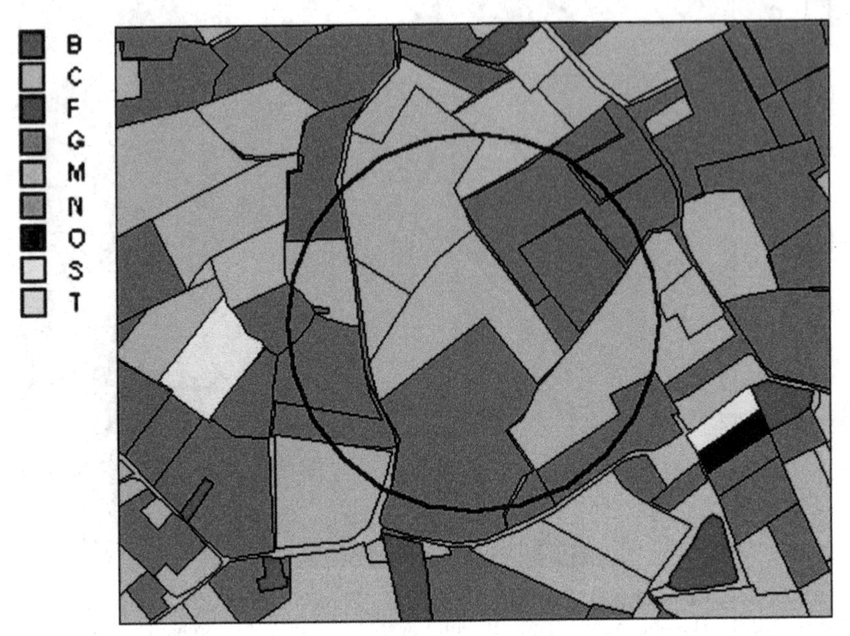

Plate 25 Seven main types of occupied habitats in Flanders (after Van Nieuwenhuyse & Leysen 2001). Farmless cattle breeding. B: buildings; C: row crops; F: orchards; G: grass; M: maize; N: fallow land; O: other; S: cereals; T: silvaculture. Circle includes 25 ha. See Figure 6.29g.

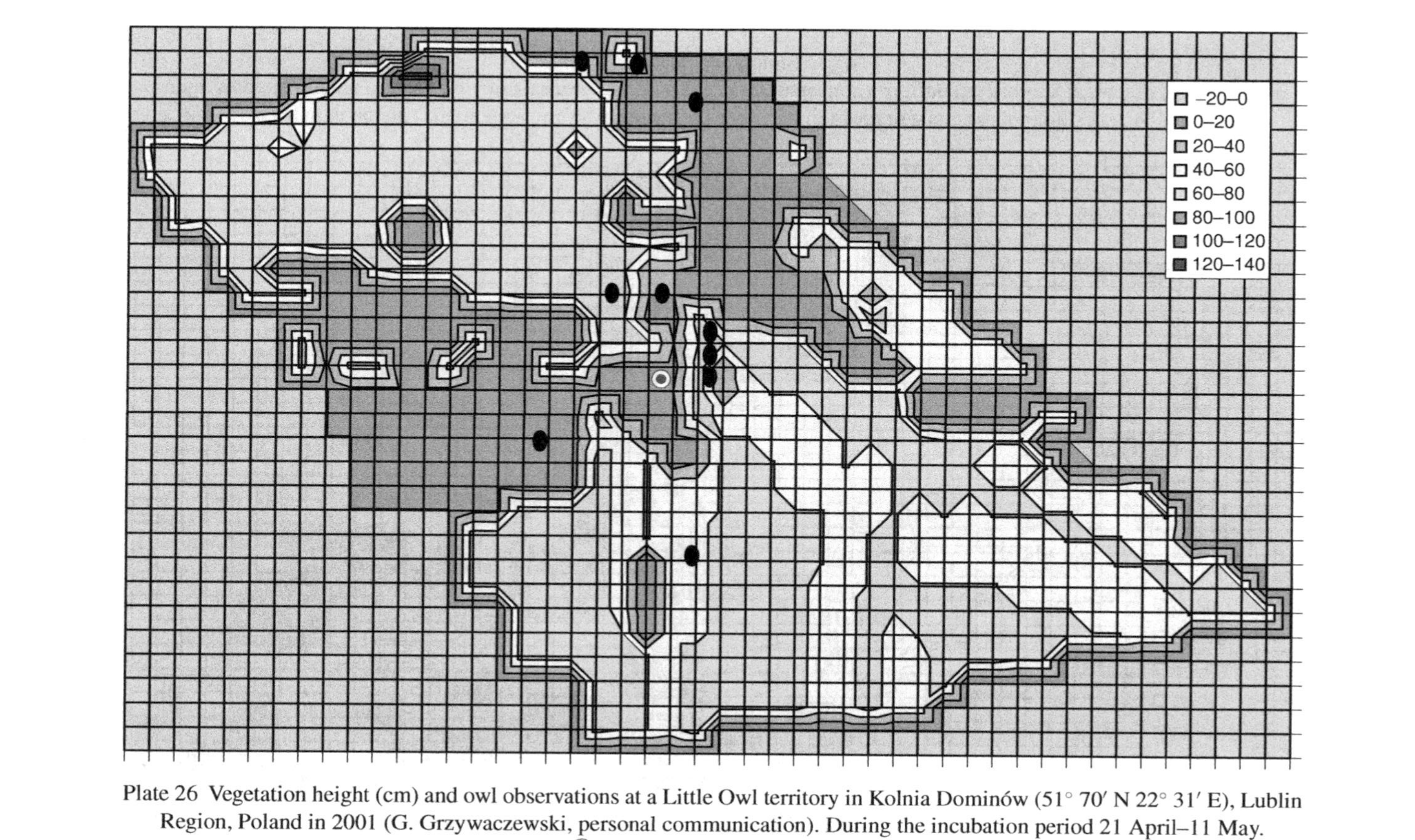

Plate 26 Vegetation height (cm) and owl observations at a Little Owl territory in Kolnia Dominów (51° 70′ N 22° 31′ E), Lublin Region, Poland in 2001 (G. Grzywaczewski, personal communication). During the incubation period 21 April–11 May. ◉ Nest site. See Figure 6.32a.

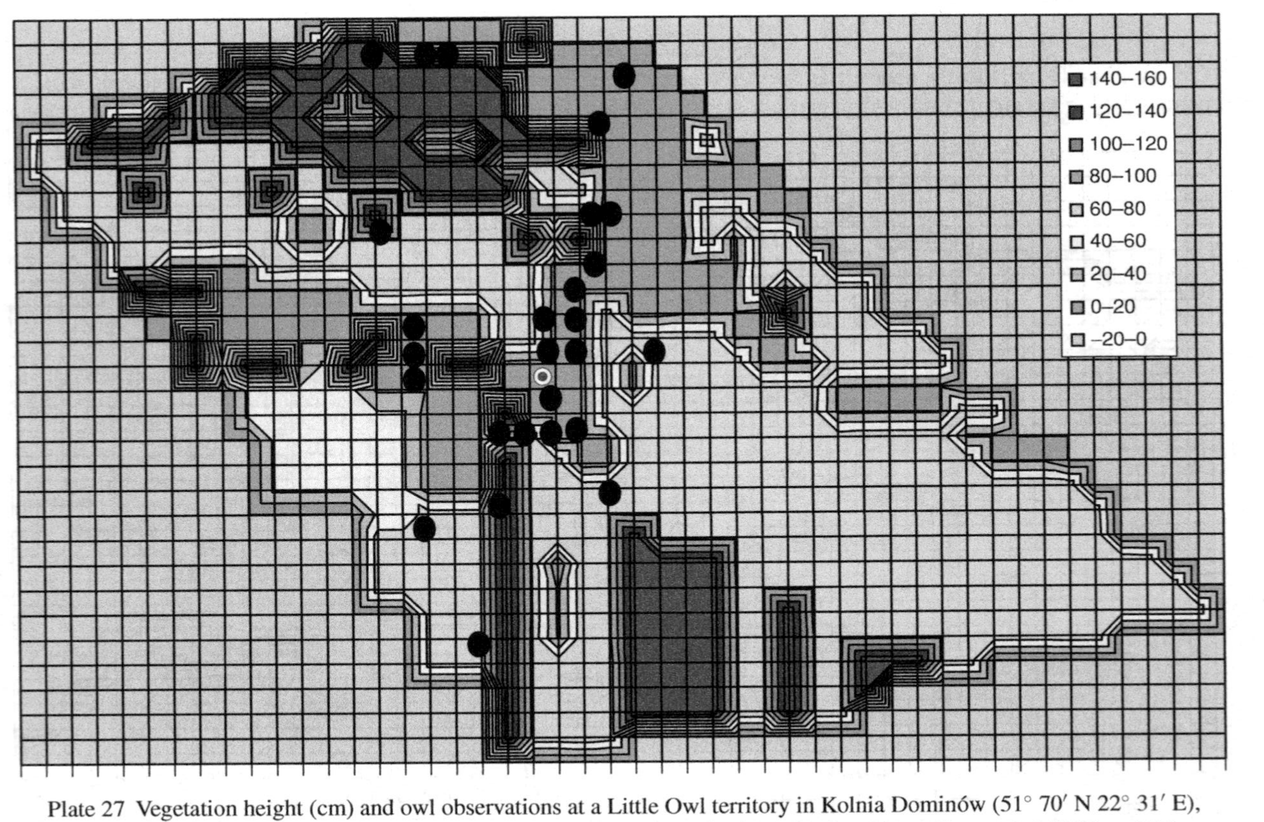

Plate 27 Vegetation height (cm) and owl observations at a Little Owl territory in Kolnia Dominów (51° 70′ N 22° 31′ E), Lublin Region, Poland in 2001 (G. Grzywaczewski, personal communication). During the nestling period 12 May–25 June. ◉ Nest site. See Figure 6.32b.

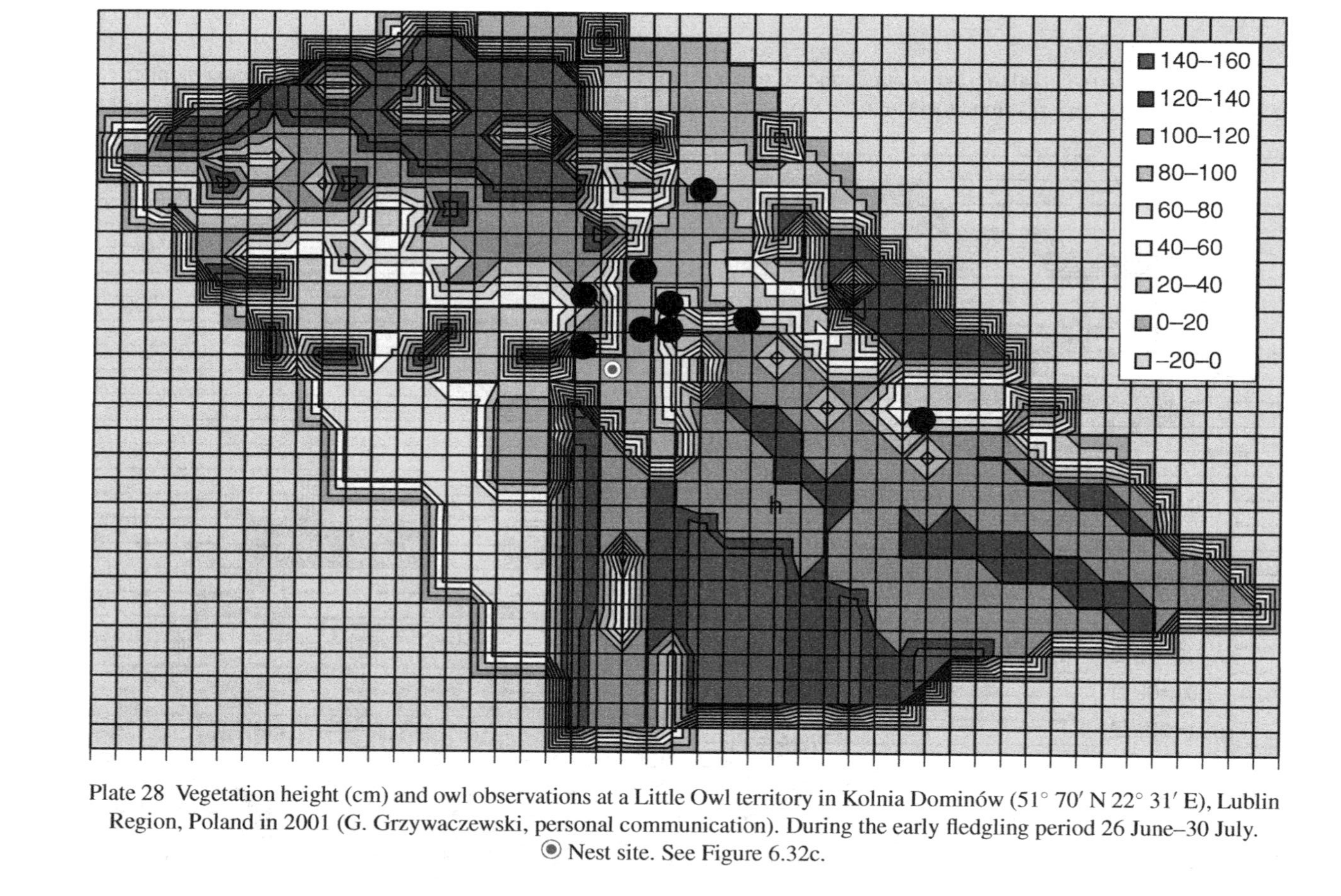

Plate 28 Vegetation height (cm) and owl observations at a Little Owl territory in Kolnia Dominów (51° 70′ N 22° 31′ E), Lublin Region, Poland in 2001 (G. Grzywaczewski, personal communication). During the early fledgling period 26 June–30 July. ◉ Nest site. See Figure 6.32c.

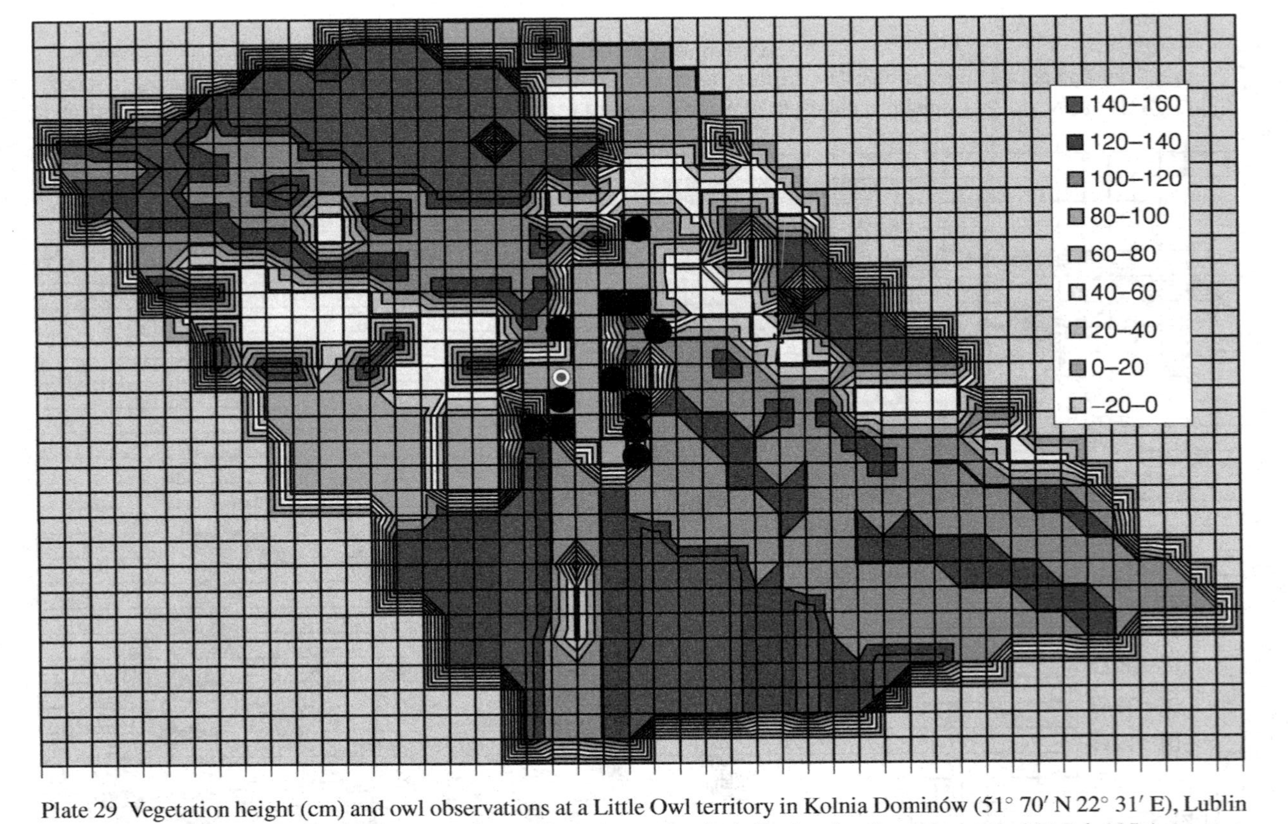

Plate 29 Vegetation height (cm) and owl observations at a Little Owl territory in Kolnia Dominów (51° 70′ N 22° 31′ E), Lublin Region, Poland in 2001 (G. Grzywaczewski, personal communication). During the dispersion period 31 July–15 August. ◉ Nest site. See Figure 6.32d.

Plate 30 Small-scaled landscapes as favorable Little Owl habitat Flanders, northern Belgium. Photo Ludo Goossens. See Figure 6.34.

Plate 31 Little Owl with moth. Photo Ludo Goossens. See Figure 7.3.

Plate 32 Juvenile Little Owl on pole. Photo Ludo Goossens. See Figure 8.13.

Plate 33 Juvenile Little Owl at nest site. Photo Bruno d'Amicis. See Figure 8.14.

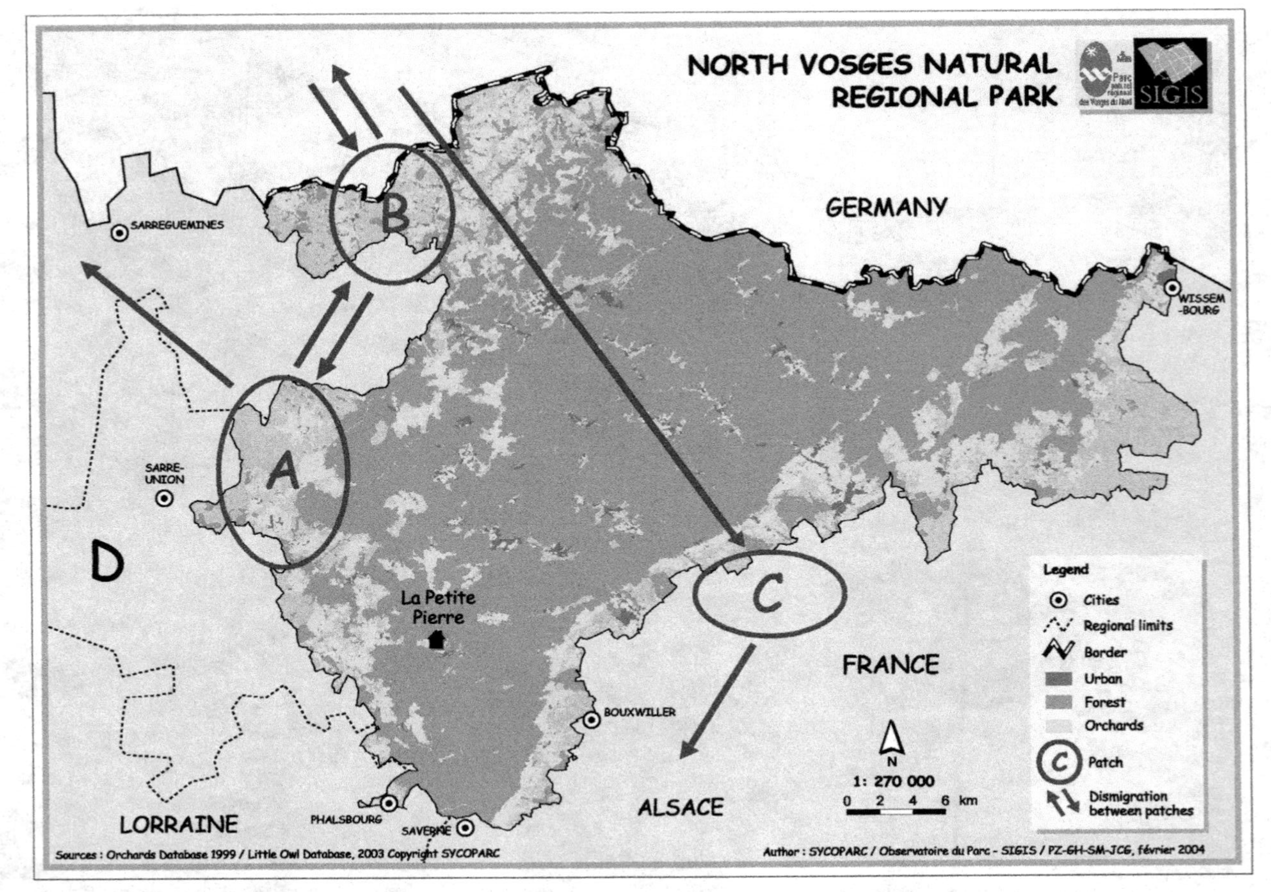

Plate 34 Subpopulations and immigration of Little Owls in Northern Vosges, France (after Génot 2001). See Figure 10.9.

Plate 35 Little Owl habitat in Flanders near Halle, northern Belgium. Photo Marc de Schuyter. See Figure 11.5.

Plate 36 Little Owl habitat in Flanders near Wakken, northern Belgium. Photo Marc de Schuyter. See Figure 11.6.

Plate 37 Adult Little Owl at nest site. Photo Bruno d'Amicis. See Figure 12.2.

Plate 38 Little Owl adult. Photo Bruno d'Amicis. See Figure 13.3.

Plate 39 Little Owl on a fence pole. Photo Ludo Goossens. See Figure 14.8.

Based on their work in western Germany and the Netherlands, Exo and Hennes (1980) found mortality rates in adult owls to be 35% and for young owls it was 70%. From this data, they estimated that each pair of Little Owls must produce 2.34 fledgings each year for reproduction to balance mortality. Figures derived from ring recoveries can over- or underestimate the reproduction rate, for example, if the annual mortality rate for adult and juvenile owls was only 5% lower (30% and 65%, respectively) then each pair must produce only 1.7 fledged young per year to compensate for mortality (Exo 1992). In Germany, the mortality rate for owls in their second year of life was 30% (Mohr 1990). The mortality rate of male owls between 12 and 24 months of age was 27%, and was 30% for the females; for owls between their second and sixth year of life, the mortality rate was 26% for males and 27% for females (Knötzsch 1988). Exo (1992) reported that extraordinarily high mortality occurs in severe snowy winters. In France, the annual mortality rate was 85% for owls in their first year of life and 36% for adults (Letty *et al.* 2001). In Switzerland, the first-year mortality rate was 74%, and the second-year was 64% (Glutz & Bauer 1980). In Bönnigheim (Landkreis Ludwigsburg, Germany) 11 juveniles were followed by telemetry after dispersal and had suffered a mortality rate of 63% (Eick 2003).

Glue (1972) reported two main periods of seasonal mortality of Little Owls in England: July–August for the young and May–June for adults. In France, Génot (1991) found seasonal mortality peaks of July–September for the young and April–June for adults. In Germany, the peak mortality period was June–July for adults, with a third peak in adult mortality occurring just after fledging, probably because of the energetic bottleneck of rearing young and moulting (Exo 1992). In Britain, Glue (2002) analyzed ringing recoveries data and showed that most juvenile birds die from starvation in June–July, which correlates to the time when the young owls gain independence from their parents.

The following data reflect longevity records for ringed Little "Methuselah" Owls. The oldest Dutch bird was seen until its fifteenth year (Stroeken & van Harxen 2003b). The oldest Belgian owl was found to live twelve years and seven months (Vercauteren 1989). The oldest bird still alive and breeding in Wallonia (Belgium) was in its thirteenth year (J. Bultot, personal communication). The oldest ringed owl from Germany was stated to be fifteen years and seven months of age (Rydzewski 1978). Peter (1999) reports a German owl that lived to be eleven years old and Sill and Ullrich (2005) a female of twelve years old. Another old German owl was nine years and six months old (Furrington 1988), and the oldest French owl was also nine years and six months old (Génot 2005).

Juvenile dispersal and adult movements

Earlier in this chapter, we described a "closed" population as one in which there is no immigration or emigration. In wild populations of Little Owls, this simplified situation is less frequently the case, and the dispersal of juveniles and movements of adults play a significant role in the demographic (and genetic) connectivity between/among populations. Dispersal is the term that applies to a young owl as it leaves its natal territory to mature, and search out and settle into a territory of its own. A "successful" dispersing owl is one who moves from his or her natal territory, disperses, and lives long enough to become part of a mated pair. Movements reflect the shifts that adults make, for example, a female may breed in one location one year, and may move to a different territory for the next breeding season.

The distance moved by dispersing male and female owls has provided key information on aspects of population dynamics and meta-population analyses (for examples of juvenile dispersal studies, see Ganey *et al.* 1998; Forsman *et al.* 2002; Ganey & Block 2005). Further, the juvenile survival *rate* (simply the inverse of the mortality rate) has been shown to be a key driver in the dynamics of other owl populations such as the California Spotted Owl (*Strix occidentalis occidentalis*) (Franklin *et al.* 2004), Northern Spotted Owl (*S. o. caurina*) (Forsman *et al.* 1996), and Mexican Spotted Owl (*S. o. lucida*) (Ganey *et al.* 2004).

In some owl species (e.g., Northern Spotted Owls [Forsman *et al.* 2002], Tengmalm's Owls [Korpimäki 1988], Lanyu Scops Owls *Otus elegans* [Severinghaus 2002], and as will be shown for Little Owls below) females disperse nearly twice as far as males. The ecological basis for this appears to be the different breeding strategies of the sexes. In the context of perpetuating the species, in order to be successful, males need to have a territory, and so will select one of the first vacant territories they come upon, with even a low-quality territory being better than no territory at all. In order for the female to be successful, she must produce young, so she will "shop around" for a territory that contains a mate, food, and nest resources that would allow her to be successful. The mating system hypothesis, which predicts that the sex that establishes the territory should disperse shorter distances, remains tenable as an explanation for female-biased dispersal in Little Owls.

Unlike the more northerly *Asio* and *Surnia* owls that embark on long-distance migrations or movements, Little Owls are considered relatively sedentary, and will generally settle in areas close to their natal origin. This was noticed when the Little Owl colonized Britain in the late 1800s to early 1900s. It did so in small incremental movements establishing new clusters of nests, and then moving out from the core densities (Marchant *et al.* 1990).

Life is a balance, and within the context of the Temperate, Mediterranean and related interior climates, the combination of their knowledge of local foraging habitat combined with their varying prey capture techniques allows them to survive within a limited range, and they do not have to migrate to search out areas with higher prey densities. The home range of territorial Little Owls is pretty much fixed in size throughout the year; however, in some cases the birds will range a little further in the winter months due to the hardships of finding adequate prey. In some portions of their range, Little Owls are less sedentary than elsewhere. For example, in Kazakhstan the Little Owl is considered nomadic in winter, leaving the central part of the country and migrating southwards (Gavrin 1962).

Juvenile dispersal characteristics

We examined data from 12 European-based dispersal studies (see Table 10.5), and offer a brief summary of each study in this section. Eleven of these studies involved the recoveries of ringed birds and one study involved the use of radiotelemetry. Dispersal begins when the owls are 12–16 weeks of age (Eick 2003) and lasts through to at least the end of October. Most birds do not settle at once. The earliest settlement was observed at the end of September. Young birds disperse in random directions, and most settle within 20 km of their natal area. While a small proportion of ringing recoveries were from birds dispersing over 50 km, some studies indicate that between 4–9% of juvenile owls disperse over 100 km, and distances of 182 km, 190 km, 220 km, 230 km, 270 km, 297 km, and 600 km have been recorded.

In typical ringing studies, researchers capture and ring the juveniles and (ideally) all of the adult owls within a given area. Ringing continues in subsequent years, with the researchers ringing the new offspring and recapturing the owls that remain as members of breeding pairs, or the juveniles that have since been recruited into the study sites' breeding population. Unless the study area is large, few of the owls that move away from the study area are recovered (as there are simply fewer people looking for owls in those locations), thus establishing a recognized bias in the recovery rates of ringed birds. With ringing programs, it is particularly challenging to determine what proportion of the owls emigrate (move out of the population), become non-territorial "floaters", or simply die, from the recovery of ringed owls. Radiotelemetry studies such as that of Eick (2003) are badly needed for better determining the dispersal distances and survival rates of juvenile Little Owls.

Given very large sample sizes (e.g., Saurola 2002), ringing studies do offer some information about owl dispersal and movements (i.e., straight-line distance moved, and longevity) and are particularly valuable for investigating within-population dynamics (e.g., reproduction, and survival rates for known adults). In the paragraphs that follow, we provide short summaries of the juvenile dispersal and adult movement studies that have been conducted thus far on Little Owls.

Glue (2002) examined the British ringing data on 57 owls that had been ringed as juveniles and later recovered. He found that the median straight-line dispersal distance of these owls was 7 km (range 0–92 km, $n = 57$). In 1962, one bird ringed as a nestling in Northumberland, northern England, was found dead later that year in Musselburgh, Lothian, Scotland, a distance of approximately 35 km.

Birds from the introduced British population disperse on average further than the sympatric Tawny Owl. The sample sizes and distance dispersed (by distance category) for 174 owls ringed as nestlings or May–July juveniles and recovered from August onwards were: 111 dispersed 0–10 km, 30 dispersed 11–20 km, 21 dispersed 21–50 km, 7 dispersed 51–100 km, and 5 dispersed over 100 km (maximum distance was 150 km) (Glue 2002).

A juvenile ringed in September was recovered two years later 175 km north-northeast of its ringing site. A full-grown bird ringed in December was recovered 16 months later (April) some 110 km northeast of its ringing site. Most of those ringed as adults were recovered

Table 10.5 *Juvenile dispersal studies – straight-line distances between natal site and recovery during the following breeding season.*

Author	Total young	Males	Females	Results
Glue 2002	57	unk	unk	Glue found that the median straight-line dispersal distance of these owls was 7 km. However, he found that there was a regional difference between the owls originating in northern and southern England. Median recovery distances for birds ringed (pulli and adults) in the northwest (15 km, $n = 4$) and northeast (7.5 km, P5–P95 0–92 km, $n = 6$) exceeded those of the southwest (2 km, $n = 111$) and southeast (1 km, $n = 171$) regions.
Exo & Hennes 1980	48	unk	unk	From November–March (inclusive) of their first winter, the median dispersal distance for 30 owls in Germany was 15 km, and for 18 owls from the Netherlands, the median recovery distance category was 10–19 km. Overall, about 55% of these juvenile owls settled within 10 km of their birthplace and only 9% dispersed distances over 100 km.
Ullrich 1980	21	16	4	For 21 owls ringed at their nests as nestlings, 18 were recovered 0–16 km (average 6.6 km) from their birthplace, and 1 each at 22 km, 55 km, and 190 km. The 20 owls that dispersed less than 55 km (cases at 0–55 km) involved 16 males and 4 females, suggesting females more likely to move away (though sex-ratio at ringing unknown).
Kämpfer & Lederer 1988	130			In their Central Westphalia study area, juveniles settled on average 4.1 km away from their birthplace.
Gassmann & Bäumer 1993	29	9	13	In a 15-year population study in Northrhine-Westphalia, the distance of settlement of first-year owls was 5.7 km ($n = 29$), females (6.4 km, $n = 13$) moved further than males (3.9 km, $n = 9$).
Veit 1988	11	unk	unk	In Hesse, young moved an average of 19 km, ranging from 2 km to 48 km.
Juillard 1984	20	unk	unk	In Switzerland young moved 0.5–16 km (average 5.8 km; $n = 20$).

Table 10.5 (*cont.*)

Bultot *et al.* 2001	26	unk	unk	In Wallonia (Belgium), the distance of settlement of first-year owls in two research areas with plenty of vacant nestboxes averaged 2.4 km in Ransart, min. = 1.2 km, max. = 4.4 km, SD = 1.2 km, $n = 9$; and 1.2 km in Neufville, min. = 0 km, max. = 3.2 km, SD = 0.9 km, $n = 17$).
Génot 2005	33	12	11	In Vosges du Nord (east France) juveniles dispersed on average 6.5 km ($n = 33$) from their birthplace, max. 25 km. Males 5.6 km ($n = 12$) females 9 km ($n = 11$).
Stastny *et al.* 1996	unk	unk	unk	50% of birds found not further than 10 km from the place of ringing, only 4% found further than 100 km.
Knötzsch 1988	127	68	59	Lake Constance study area between 1976 and 1987, the dispersal of first-year owls averaged 4.5 km for males ($n = 68$) and 4 km for females ($n = 59$).
Eick 2003 (short-term radio-telemetry study)	11	unk	unk	In Germany (Bönnigheim, Landkreis Ludwigsburg) 11 juveniles were followed by telemetry during their dispersal during 31 nights. The birds moved on average 9.8 km and maximally 41 km per night; nearly all juveniles came back to their birthplace after exploratory movements in all directions after a few days.

unk = unknown

within 10 km (usually locally), but several were found up to 45 km away. One adult female was ringed in May in Dorset, and was recovered 13 months later in Hereford, having moved an impressive 182 km to the north-northwest. Stragglers (none ringed) have crossed the Irish Sea to the Isle of Man and eastern Ireland (Glue 2002). Continental birds also disperse relatively short distances. In their review of German and Dutch ringing data, Exo and Hennes (1980) found that juvenile dispersal began in August. From November–March (inclusive) of their first winter, the median dispersal distance for 30 owls in Germany was 15 km, and 18 owls from the Netherlands were recovered in the distance category of 10–19 km. Overall, about 55% of these juvenile owls settled within 10 km of their birthplace and only 9% dispersed distances over 100 km.

Owls ringed after attaining breeding age (i.e., April onwards) were recovered at a median distance of 7.5 km ($n = 53$ owls) and in the distance category of 0–9 km ($n = 25$ owls) in Germany and the Netherlands, respectively. Of these adult owls, 74% were recovered within 10 km of their original ringing site (Exo & Hennes 1980).

In a population study conducted in North-Württemberg, Germany, Ullrich (1980) found first-year birds to disperse 0.5–220 km. Most of the owls dispersed less than 40 km, and there were three cases in which juveniles that dispersed in autumn later returned to near their natal area, one coming from 36 km away.

As found in other owl species, recoveries of siblings show different within-brood dispersal characteristics (i.e., siblings disperse different directions and distances). Ullrich (1980) reported on dispersal aspects for pairs of siblings from three nests: (1) 3 km west, 36 km northeast; (2) 3 km west-southwest, 38 km north; and (3) 3 km northeast, 190 km southwest. The different young from the same clutch dispersed in random directions.

For 21 owls ringed as nestlings, 18 were recovered 0–16 km (average 6.6 km) from their natal area, and one each at 22 km, 55 km, and 190 km. Here, the 20 owls that dispersed less than 55 km were 16 males and 4 females. The different young of the same clutch showed random dispersal direction (Ullrich 1980).

In a population study in Central Westphalia, juveniles settled an average of 4.1 km away from their natal area ($n = 130$). Some 75% of the first-year owls trapped after November 1 were recaptured later within the breeding area (Kämpfer & Lederer 1988).

In a population study around Lake Constance between 1976 and 1987, the dispersal of first-year owls averaged 4.5 km for males ($n = 68$) and 4 km for females ($n = 59$) (Knötzsch 1988).

In a 15-year population study in Northrhine-Westphalia, the dispersal distance and settlement of first-year owls was 5.7 km ($n = 29$); with females (6.4 km, $n = 13$) moving further than males (3.9 km, $n = 9$). Out of two young from six nests (siblings), one bird was found in a different direction and at a significantly greater distance than the other (Gassmann & Bäumer 1993). In Hesse, young moved an average of 19 km ($n = 11$) with a range from 2–48 km (Veit 1988).

In an 11-year population study in Northrhine-Westphalia, the dispersal distance and settlement of first-year owls was 5.17 km ($n = 88$); with females (6.8 km, $n = 34$) moving further than males (3.8 km, $n = 51$). Out of 153 owls born, only 19 will breed in the study area (Zens 2005).

In Switzerland, young owls dispersed an average of 5.8 km (range 0.5–16 km, $n = 20$) (Juillard 1984a).

In a population study in Wallonia (Belgium) (Bultot *et al.* 2001), the dispersal distance and settlement of first-year owls in two research areas with plenty of vacant nestboxes averaged 2.4 km in Ransart (min. = 1.2 km, max. = 4.4 km, $n = 9$) and averaged 1.2 km in Neufville (min. = 0 km, max. = 3.2 km, $n = 17$). Here, owls appeared to show a preference for dispersing to a south and east direction. No relationship was found between the density and the emigration distance, possibly because of the lack of recaptures outside the research area.

In Vosges du Nord (eastern France) juveniles dispersed on average 6.5 km (max. 25 km, $n = 33$) from their natal area, with females (9 km, $n = 11$) moving further than the males (5.6 km, $n = 12$) (Génot 2005).

Table 10.6 *Exploratory excursions of radio-marked juvenile Little Owls in Germany (after Eick 2003).*

Excursion duration	Total distance (km)	Average distance per night (km)	Maximum distance per night (km)
3 nights	15	7.5	7.5
3 nights	20	7	8.5
5 nights	35	7	12.5
5 nights	76	15.2	41
>2 nights	>25	–	25
2 nights	4	2	2.5
5 nights	>80	16	25
3 nights	>25	8.3	12
2 nights	25	12.5	19
31 nights	>30.5	9.8	41

In the Czech Republic, 50% of the ringed birds were found within 10 km of the place of ringing, and only 4% were found further than 100 km away (Stastny *et al.* 1996).

Occasionally some owls do embark on long-distance movements. The longest recoveries and direction of travel of ringed birds are as follows: Mittel-Franken, Germany (June nestling) to Bas-Rhin, France (January), 230 km west-southwest; Hessen, Germany (October) to Halle, Germany (April), 270 km northeast; Dresden, Germany (July, full-grown) to Austria (February), 297 km south; two North-Württemberg nestlings (June) were found in October in Switzerland (220 km southwest) and Zielona Gura, Poland (600 km northeast) (Glutz & Bauer 1980; Furrington 1998). Such erratic movements as these are consistent with aspects of vagrancy to Helgoland and southern Fenno-Scandia.

Recently, Eick (2003) conducted a radiotelemetry study on juvenile Little Owls during and after dispersal in Germany (Bönnigheim, Landkreis Ludwigsburg). He tracked 11 juveniles with telemetry during 31 nights (Eick 2003). On average, the birds moved 9.8 km and maximally 41 km per night (see Table 10.6). The movements showed no systematic direction. Such random dispersing behavior is typical for this rather sedentary owl species (Bauer 1987; Berthold 2000). The juveniles tended to revisit their natal site at regular intervals between exploratory excursions of three to five days (Eick 2003). This behavior of exploratory excursions followed by a return to the nest was also observed in radio-marked Northern Spotted Owls prior to their formal dispersal from the natal territory (Miller 1989).

Zens (2005) followed 11 juvenile Little Owls with radio-transmitters in Germany (Mechernicher Voreifel). In July and August, juveniles flew around their nest site up to 300 m. In September, they flew further during the night up to 1.25 km from their birthplace. During one night they moved from 560–690 m to 980–3100 m. At this time 54.5% of the juveniles were already dead or missing. At the end of September or beginning of October, the juveniles left their birthplace.

Radio-telemetry has been used in juvenile survival studies on only small numbers of Little Owls. Because of the high mortality rate of juveniles, one has to equip sufficient numbers of birds with radio-transmitters to be able to track a sufficient sample of them from dispersal through to ultimate pairing and nesting. The critical advantage provided by radio-telemetry studies over regular ringing efforts is that the "outcomes" of individual owls become known.

Adult movements

Relatively few data exist to show the movements in birds of breeding age. We examined data from eight studies and a few anecdotal notes regarding the movements of adults, and offer a summary of those here. Available data confirm field observations that the great majority stay within their territories through the autumn and winter, but that travel up to 45 km and longer movements away from their previous nest site rarely occur. Some owls return to their previous nests the following year. Again, we caution readers in recognizing that these studies involved ringed owls, and of the inherent recovery bias that ringing studies are subject to.

Exo and Hennes (1980) recorded the movements of owls ringed after attaining breeding age (i.e., April onwards). These owls were recovered a median distance of 7.5 km ($n = 53$ owls) and in the median distance category of 0–9 km ($n = 25$ owls) from their natal sites in Germany and the Netherlands, respectively. Of these adult owls, 74% were recovered within 10 km of their original ringing site.

In their Central Westphalia study area, adult movements away from their last nest sites differed according to sex, with males moving an average of 2.3 km and females 6.3 km. Among adults, 10.2% of males and 1.8% of females showed no site fidelity; they settled 1.7 km away from their former sites (Kämpfer & Lederer 1988).

In a Northrhine-Westphalia study area, adult males also moved shorter distances (1.37 km, $n = 11$) than females (2.2 km, $n = 19$) (Zens 2005).

In Switzerland, adults moved an average of 1.6 km (range 0.4–4.7 km, $n = 12$) (Juillard 1984a). In Vosges du Nord (eastern France), two adults were recorded as moving 12 km and 32 km (Génot 2001). Ullrich (1980), Fuchs (1986), and Eick (2003) also reported recaptures of birds that had successfully bred somewhere, then moved up to 36 km to breed elsewhere and subsequently returned to the original breeding place the year after.

Nightly movements of adults of 1–1.5 km up to 3–3.5 km were reported by Exo (1989) especially in the fertile period. This might indicate the willingness of adults to solicit extra-pair copulations.

A median dispersal distance of just 2 km for Little Owls ringed and recovered, including all age classes, provided factual evidence for a strongly sedentary species, generally site-faithful but undertaking limited post-natal movements spanning autumn and winter (Glue 2002).

Movements of adult owls are done in random directions, sometimes at higher elevations and do not appear to follow linear landscape structures, e.g., canals, power lines. Eick (2003) even observed Little Owls crossing larger forests. The final settlement of these owls is relatively late and might be related to finding a partner. Their settlement is a dynamic process and is driven by two factors, i.e., birds want to settle in close vicinity to existing population clusters as much as possible and adult birds want to push conspecifics out of their active territory to avoid competition (Génot & Van Nieuwenhuyse 2002). This dynamic process is confirmed by the juvenile displacement behavior. It should be stressed that presumably isolated populations might be less isolated than previously assumed. Little Owls disperse at flying altitudes of at least 20–30 m and sometimes even higher avoiding physical barriers such as forests and cities (Eick 2003).

Home range and territory

A bird's home range is defined as "the area that embraces all activities of a bird or pair over a given time period." Generally a yearly home range is used, but sometimes a seasonal area (e.g., breeding season area) is given. A territory is defined as "the area around the nest that is defended" according to Newton (1979). Finck (1990) on the other hand considers a territory as "the defended area for food and nesting." No matter which definition is used, the best way to get home-range values is by attaching a small radio-transmitter to a bird, allowing the movements and activity of the Little Owl to be tracked. Generally, the transmitter is put in place through a backpack harness and weighs 6–9 g, that is to say 2.5–4.5% of the bodyweight of the bird. Typically, the main reason for assessing home-range sizes is to define the area within which the species finds selected habitats and resources. While the backpack transmitter may seem an inconvenience for the Little Owl, it has been shown that the transmitter does not influence the birds' breeding success. Exo (1987) and Zens (1992), respectively, reported the following breeding success data: 2.5 fledglings/pair with transmitter ($n = 4$ pairs) and 1.33 fledglings/pair without transmitter ($n = 15$ pairs); 1.5 fledglings/pair with transmitter ($n = 4$ pairs) and 1.29 fledglings/pair without transmitter ($n = 31$pairs).

To accurately determine home-range sizes, a minimum of 30 locations per bird are needed (Kenward 1987) over a specified time frame (e.g., 12 months).

Home-range sizes are often quantified by the use of the minimum convex polygon method. This is the area encircled by a line that connects the outermost observations (see Figure 10.7a). The minimum convex polygon and adapative kernel are methods to assess the home-range dimensions of owls (and other species). In Table 10.7, we summarize home-range sizes based on the use of the minimum convex polygon method. In addition to the data presented in Table 10.6, Eick (2003) monitored three adults and four juveniles during the dispersal time from July to November (48–120 days of monitoring). The available data on home-range sizes reflect efforts that were either data-intensive for a given season (e.g., breeding season), or reflect a relatively low number of locations gathered over a longer period. In order for annual home-range sizes to be calculated, additional rigor in data collection is

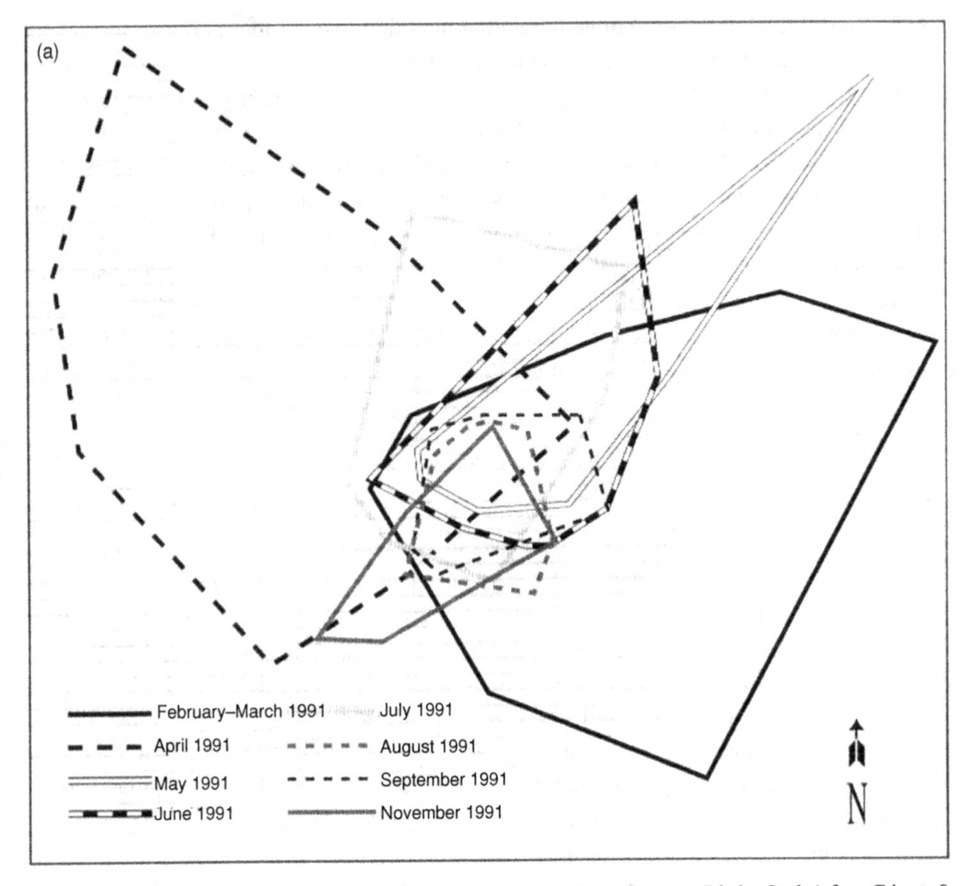

Figure 10.7a Examples of monthly home ranges: calculations for one Little Owl (after Génot & Wilhelm 1993).

warranted. For most owl species, it is recognized that an asymptote in radio-locations occurs with increasing home-range size. That is to say, after about 100 or so radio-locations, the home-range sizes either do not increase in area, or do so only minimally. It is important to note that for conservation applications, radio-locations gathered over a full year (or more), for both male and female members of a pair, are the most useful, as the area involved most accurately represents the area actually used by a pair of owls. Thus, ideally, researchers should gather information locations of radio-marked owls at least twice a week (during the night when the owls are active) for 12 months. Gathering data at only one location per night would minimize problems associated with auto-correlation and the independence of data. Finck (1990) was the only author to study the size of the Little Owl territory using a decoy and a loudspeaker. The territory size of 19 tracked males ranged from 1–68 ha (mean 12.3 ha). Finck showed that Little Owls defended their territories all year round, with the aggressiveness of the owls changing according to the season.

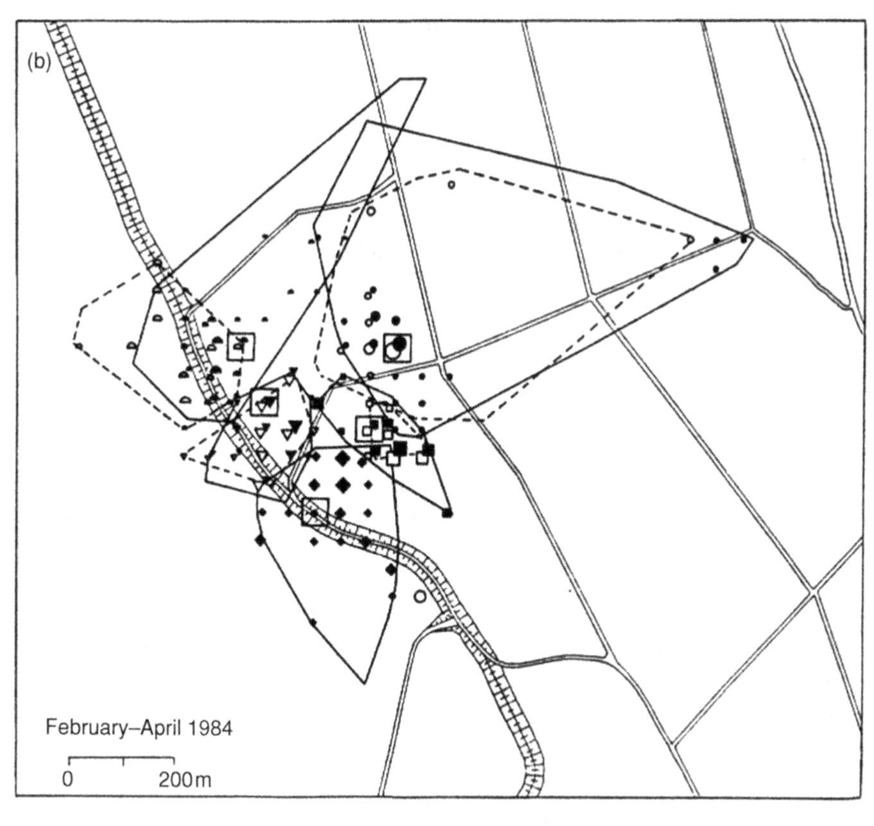

○ Pair 1
□ Pair 2
◇ Pair 3
▽ Pair 4
△ Pair 5
⌓ Pair 6

♂
♀
□ Tree with nesting cavity

Percentage of time spent in grid cell of 50 m × 50 m

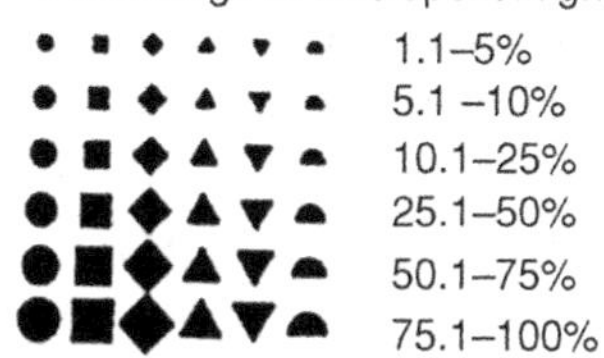

1.1–5%
5.1 –10%
10.1–25%
25.1–50%
50.1–75%
75.1–100%

Figure 10.7b Examples of home ranges for six pairs of Little Owls from February to April 1984 (after Exo 1989).

Table 10.7 *Summary of Little Owl home-range sizes in ha, from different studies as calculated using the minimum convex polygon or MCP method. The home-range values reflect the size of area within the MCP during the time the owl was followed.*

Number of Little Owls monitored	Home range, annual mean	Home range, minimum	Home range, maximum	Density, pairs/km^2	Study area, km^2	Author	Country
12	14.5	1	50	1.7	35	Exo 1987	Germany
19	14.6	2	107	1.6	29	Finck 1990	Germany
4	27.4	1	150	0.15	260	Zens 1992	Germany
8	31	5	107	0.09	437	Génot & Wilhelm 1993	France
15	51.3	6.5	137	0.5	–	Orf 2001	Germany

Use of the home range

As with many other owl and raptor species, breeding Little Owls have been found to concentrate their activities in a portion of their overall home ranges. For example, Génot and Wilhelm (1993) found that owls they studied in France had a mean home-range size of 31 ha, and spent 80% of their time in an area of 3.5–6 ha. Zens (1992) found that owls concentrated their activities in an area of between 20% and 80% of the home ranges; Exo (1987) found that owls concentrated their time in an area of between 5% and 50% of the home ranges.

Eick (2003) studied some adults and dispersing juveniles during the breeding season. The breeding area used by the adults was 20–30 ha. Pre-dispersal juveniles used different-sized areas depending on their age; during the first week after leaving the nest: 1.5 ha ($n = 5$ individuals); 9 weeks after: 15.8 ha ($n = 6$ individuals); and 12 weeks after: 31 ha ($n = 3$ individuals). Please note that the juveniles did not have a "home range" but rather, the areas given here reflect their pre-dispersal excursions. Exo (1987) showed that 50% of the Little Owl vocal activity is observed in a radius of less than 50 m around the nest or roost. Orf (2001) found a mean activity radius (distance from the nest or roost) of 78 m during the courtship, 219 m during the breeding season, and 181 m in winter. For Zens (1992), the activity radius ranged 60–120 m in June, the radius increased to 170–300 m in September–November, and to 420–530 m up to 1.5 km in winter. In France, the mean activity radius was 430 m, depending on the sex and the season; it was longer in winter but some distances were also long in July (up to 2.6 km).

Exo (1989) determined the daily activity pattern. Activity peaks of about one to two hours just after sunset and before sunrise, when light conditions of less than one lux, were observed. Less activity was recorded around midnight. Males and females showed the same activity patterns except during the breeding time (April/May–July). Their activity lasted

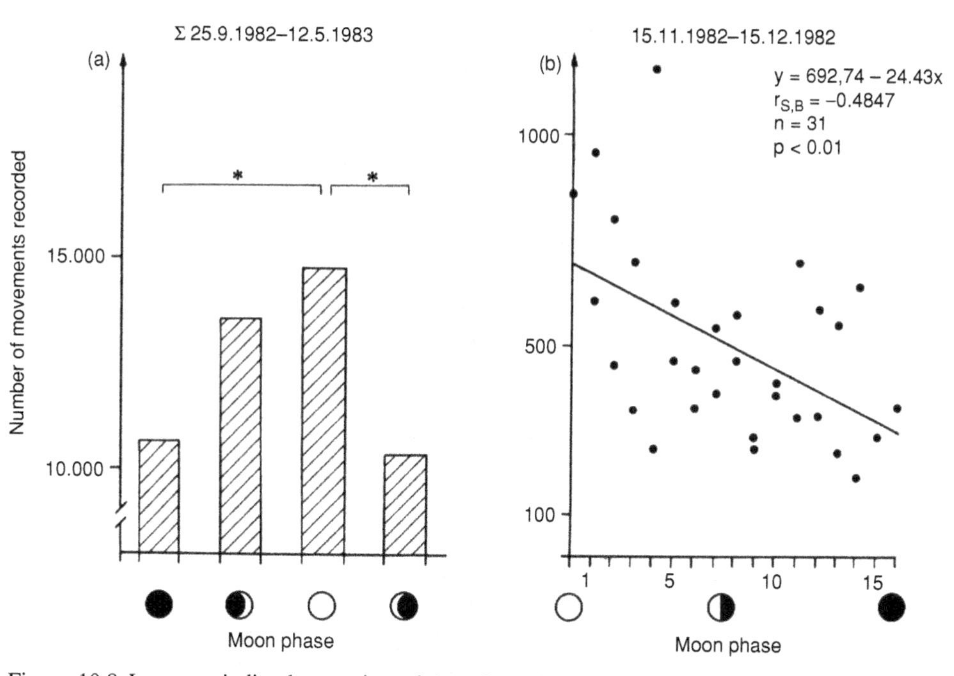

Figure 10.8 Lunar periodic changes in activity of a pair of Little Owls kept under natural light conditions (after Exo 1989). (a) Daily activity in relation to moon phase, total from seven lunar cycles (* = $p<0.05$). (b) Example of variation in the daily activity total during one lunar cycle.

between 5.5 hours and 9 hours per night, while during the breeding period the daytime activity was longer than nighttime activity.

Little Owls are more active during the full moon phase, but they move less when it is raining and the wind speed is stronger than three on the Beaufort scale. See Figure 10.8.

In France, hunting perches used by Little Owls were in trees (82.2%), on fence poles (13.7%), in hedges (3.9%), and in buildings (0.2%) (Génot & Wilhelm 1993). Orf (2001) showed that Little Owls spent the daytime mainly in nestboxes (44.3%), hedges or trees (38.5%), natural cavities (15.5%), and in other sites such as buildings (1.7%). Out of 39 dispersing juvenile owls, Eick (2003) found that 56% of the day roosts were in wood piles, 23% were in branches of trees, 8% were in nestboxes, 5% were in tree holes, and 8% were in other places.

Variation of home range and territory size

Home range and territory size depend on different factors, e.g., the sex and the age of the birds, the population density, the season and the structure of the habitat, and the density and availability of prey.

Sex and age of the Little Owls Exo (1987) monitored pairs of owls over the longest time period (Table 10.8) (one pair over one year and three pairs over ten months), and

Table 10.8 *Summary of Little Owl home-range sizes in ha, from different studies in Europe.*

Author	Study location	Sex of owl	Age of owl	Time followed in months	Number of locations	Home range (MCP) in ha
Orf (2001)	Main-Taunus (Germany)	male	adult	5	102	137.1
		male	juvenile	6	74	44.7
		male	adult	6	127	6.5
		male	adult	6	251	43.1
		male	adult	6	123	56
		female	adult	6	163	32
		male	juvenile	4.5	105	41.5
		male	adult	3.5	97	128.9
		female	adult	4.5	43	39.1
		?	adult	2	25	11.2
Génot & Wilhelm (1993)	Vosges du Nord (France)	male	adult	9	180	50.2
		male	adult	2	65	13.1
		male	adult	3	50	15.6
		male	adult	1	30	46.5
		female	adult	8	190	84.1
		female	adult	3	95	26.2
		female	adult	3	55	28.4
		female	adult	4	110	107.4
Zens (1992)	Voreifel (Germany)	male	juvenile	10	25	148.9
		female	adult	6	14	17.8
		male	adult	7	18	51.6
		female	adult	9	16	73.8
Exo (1987)	Niederrhein Niederrhein	male	adult	14	45	41.5
		female	adult	10	30	38.4
		male	adult	14	29	10.1
		female	juvenile	10	17	9.6
		male	adult	10	14	4.6
		female	adult	12	20	3.1
		male	adult	12	26	4.9
		female	adult	12	15	2.5
		male	adult	8	19	17.8
		female	adult	8	14	20.7

all authors found Little Owl males to have home ranges larger than females, except for certain males during the end of summer and beginning of autumn (Exo 1987; Génot & Wilhelm 1993). The overlap of the home ranges between males and females can match perfectly (Exo 1987; Zens 1992) or can overlap from 38–94%, depending on the season (Génot & Wilhelm 1993). During the breeding season, Eick (2003) found the range of adult females was within the range of their partner for two pairs (males on average 31.7 ha, females 21.9 ha). After independence of the young, both adults used different plots. The territories of neighboring males did not overlap throughout the year (Exo 1992). Orf (2001) found no significant relationship between the age of the Little Owl and the size of the home range. Finck (1993) showed that new settlers defended larger territories (15 ha, $n = 5$) than well-established Little Owls or long-term settlers (8.3 ha, $n = 6$). While established owls are more knowledgeable of their territory because of their longer occupation, other possibilities for the differences in home-range sizes are possible (e.g., quality of sites, food availability).

Density No differences in territory size were detected between population densities of 0.75 calling males/km^2 and 2.25 calling males/km^2 (Finck 1990). We examined the density data and average home range size (as seen in Table 10.8), and found no significant linear relationship between them.

Season With the exception of data from Exo (1987) and Orf (2001), home range sizes are larger in winter than during the other periods (Table 10.9). Home ranges are substantially larger in winter than they are during the courtship period, as behaviorally the attachment to the breeding cavity is looser than during the courtship and breeding periods. The seasonal variation in home range size is likely also linked to the changing availability of prey. Orf (2001) found males to use larger areas in May and June, during the feeding of the female and nestlings, and considered this a result of the owl's need to move further afield within a poor-quality hunting range.

Structure of the habitat The structure of the habitat has a large influence on the size of the home range and territory. In terms of habitat structure for Little Owls, we often think of agricultural land uses linked to the seasonal development and height of the vegetation. Finck (1990) considered four types of land use having an influence on Little Owl foraging:

- grassland (meadow and pasture) with free access for hunting year round
- barley and wheat fields, which are unsuitable (due to their height) from early May to mid July and from mid May to mid August, respectively
- corn fields, which are unsuitable for hunting from July to September/October
- sugar beet fields, which are unsuitable from September to November.

Many authors have stressed the importance of grasslands in the home range of the Little Owl. The species can survive in only 0.5–3 ha of low-cut grassland (Exo 1987; Zens 1992; Eick 2003). Finck (1990) observed that throughout the year, grassland within the owls' home ranges was used proportionally more than its availability. This is because pastures

Table 10.9 *Comparison of the median areas (ha) used by Little Owls during different seasons.*

Area used in winter Nov–Feb	Area used in courtship Mar–Apr	Area used in breeding May–Jun	Area used in summer Jul–Oct	Author
4.2	4.7	3.4	7.2	Exo (1987)
26.7	15.2	21	9	Finck (1989)
48.2	–	12.9	15.6	Zens(1992)
14.9	11.9	4.6	4.5	Génot & Wilhelm (1993)
21.8	12.3	31.5	–	Orf (2001)

and meadows offer a continuous food supply almost year round. Further, in a study of the Little Owl's prey in France, H. Dewulf (personal communication) found that the timing of grassland mowing had an impact on the availability of beetle prey: the diversity of beetles was greater if the grass was mowed early, and smaller if mowed later in the season. This observation was also supported by Grimm (1986) who recommended that grass be mowed early (end of May) rather than grazed, to obtain the highest diversity of prey.

Access to ground-based prey is negatively correlated with vegetation height. Uncut meadows could be a "reserve" of food that is simply not available to the owls because of the vegetations' height. Exo (1987) showed that all of the Little Owls studied used crop fields. For example, they spent 20–40% of their time in the crop fields from August to February/March. Several unpaired birds had larger areas of crop fields than grasslands in their home ranges. Sometimes, grasslands have no perches for the owls to hunt from, which might be a limiting factor for them. Eick (2003) recorded 769 locations of juvenile Little Owls, and found 45% of their locations to be in mowed grassland and 26% in vineyards. Little Owls can use intensive vineyards, preferring fallows and areas with short grass. For hunting, out of 698 locations of foraging juveniles, 62% were found to be in orchards, 20% in vineyards, 15% in gardens, and 3% in grassland with high grass.

The preponderance of evidence suggests that Little Owls get their food supplies from different areas within a large diversity of natural and pastoral landscapes having meadows, pastures with variable grazing intensity, a limited area of crops (because of pesticides issues), and plenty of edges between grasslands and other types of land uses.

Density dependence

Some processes might be driven by local population density. The higher the density, the fiercer the competition might be and the higher the impact on the breeding success will be (Newton 1998). In Chapter 5, we gave an overview of the variability in observed densities through the number of pairs of calling individuals per km^2, and nearest neighbor distances.

Where habitats are not disturbed, most bird species remain relatively stable in numbers over longer periods. Their breeding numbers fluctuate moderately from one year to another, mostly between limits that are smaller than their birth and death rates would allow (Lack 1954).

It seems that relatively few Little Owl populations live in habitats not subjected to anthropogenic activities. Exo (1992) looked at the role of nest-site availability in limiting population densities of Little Owls, to investigate how the provision of nestboxes could partially compensate for the loss of natural sites. The results illustrated that the population density increased until the carrying capacity was met, then other controlling factors become more prevalent, and after three to five years the growth rate diminished. Similarly, evidence for a strong increase in population numbers after the introduction of additional artificial nesting cavities was illustrated in other locations (Van Nieuwenhuyse & Nollet 1991; Bultot *et al.* 2001; C. Stange personal communication). On the other hand, in some populations that are more severely impacted by anthropogenic factors, such as the Little Owl population in the Northern Vosges (France), the addition of nestboxes has served to slow the rate of population decline, and now helps to support the few small population clusters that remain (Génot 2001).

More detailed observations on density dependence were recorded by Bultot *et al.* (2001), as part of the nestbox provisioning project in Wallonia, Belgium. By comparing isolated pairs and clustered pairs, it was found that owls that nested as isolated pairs produced an average of one more egg per breeding season than owl pairs that nested in clusters. However, for owl pairs that nested as part of clusters, density dependence was found to affect the timing of egg-laying – the higher the owl density the earlier the laying date became. This might indicate that social aspects become important at higher densities, reducing the number of eggs and/or changing the timing of egg-laying. Density dependence in any demographic parameter, whether births or deaths, immigration or emigration, can be caused by competition for resources such as food, nest sites, or territorial space. Natural enemies can also cause density dependence if predators kill an increasing proportion of prey individuals as the prey density rises, or if parasites infect a greater proportion of host individuals as host density rises. In any event, the lines of evidence from the provision of nestboxes has helped to illuminate the onset of density-dependent regulatory affects.

Knowing that densities of the Little Owl population may itself influence the Little Owl in its reproductive success, habitat selection and dispersal behavior need closer attention. In Flanders (Belgium), the vicinity of conspecifics in surrounding grid-cells proved to be a better predictor of a given grid-cell than the landscape composition (Van Nieuwenhuyse & Bekaert 2002). These results suggest that correlations of habitat and demographics be examined and interpreted in light of this density-dependent perspective. Van Horne (1983) identified social interactions within wildlife populations as a potential habitat classification issue. In the structure of some populations, dominant breeding animals exclude more numerous submissive or non-breeding animals from the highest quality habitat. Habitat classification based solely on aural surveys may result in a model that identified sub-optimal habitat as critical, and excluded optimal habitat (Van Horne 1983). Protecting only sub-optimal habitats would negatively influence the breeding success and overall stability of the population (Christie & Woudenberg 1997). This issue can be remedied by acquiring some demographic information (e.g., nest success) alongside basic owl surveys. During a 16-year study on the Flammulated Owl (*Otus flammeolus*) in Colorado (USA), Linkhart and Reynolds (1997) illuminated the demographic relationships for owls on 14 territories within their 452 ha study area. Their data suggests that territory occupancy may be an indicator of habitat quality for Flammulated Owls. The old Ponderosa Pine/Douglas Fir forest type (200–400 years old) was found to be the best breeding habitat for the owls. Territories most consistently occupied by breeding pairs had habitat circles (of 212 m radius centered on the nest tree) containing the highest amount of this vegetation type. A similar correlation between territory occupancy and territory quality has also been found in the Tengmalm's Owl (*Aegolius funereus*) (Korpimäki 1988; Laaksonen *et al.* 2004). Consistent with this is the information from a long-term study on the Snowy Owl (*Bubo scandiacus*) on Wrangel Island (Menyushina 1997). Menyushina found that even when owl numbers were low, the majority of breeding birds were concentrated within the most favorable nesting habitats, whereas the spatial patterns of non-breeding owls were more opportunistic.

Meta-population

A meta-population is a spatial distribution of distinct subpopulations, separated by large distances or barriers and connected by dispersal movements. This patchy population pattern changes over time (Opdam 1991; Hanski 1999). At high densities, the spatial distribution pattern of Little Owls remains clustered, with the "saturation level" of owls determined by the territorial behavior of the species, consisting of the regularly spaced calling males (Bretagnolles *et al.* 2001; Van Nieuwenhuyse *et al.* 2001b). At this point, the variance in territoriality and subsequent increase in population density (Fuchs 1986) appears to stress the importance of social aspects over landscape conditions.

More and more, Little Owl populations in western Europe are functioning like meta-populations (rather than a consistently interactive regional population) as the fragmentation of the habitat results in localized population declines and areas with very low owl densities. In the Northern Vosges (France) Génot studied a meta-population from 1984 (Génot 2001). It was composed of three clusters of owls (Figure 10.9), with distances of 18 km between clusters A and B, 33 km between A and C, and 36 km between B and C. These three subpopulations contained about 10–15, 5–10, and 5–10 pairs of Little Owls, respectively. A semi-barrier of deep forest separated the cluster C from A and B. Arrows in Figure 10.9 show the movements of ringed birds between the different clusters. The mean distance of young dispersal was 6 km ($n = 15$). Cluster D was occupied in 1984–7 but has since disappeared. A population viability analysis applied to this meta-population revealed a high extinction risk for population sizes below five breeding pairs (Letty *et al.* 2001). A

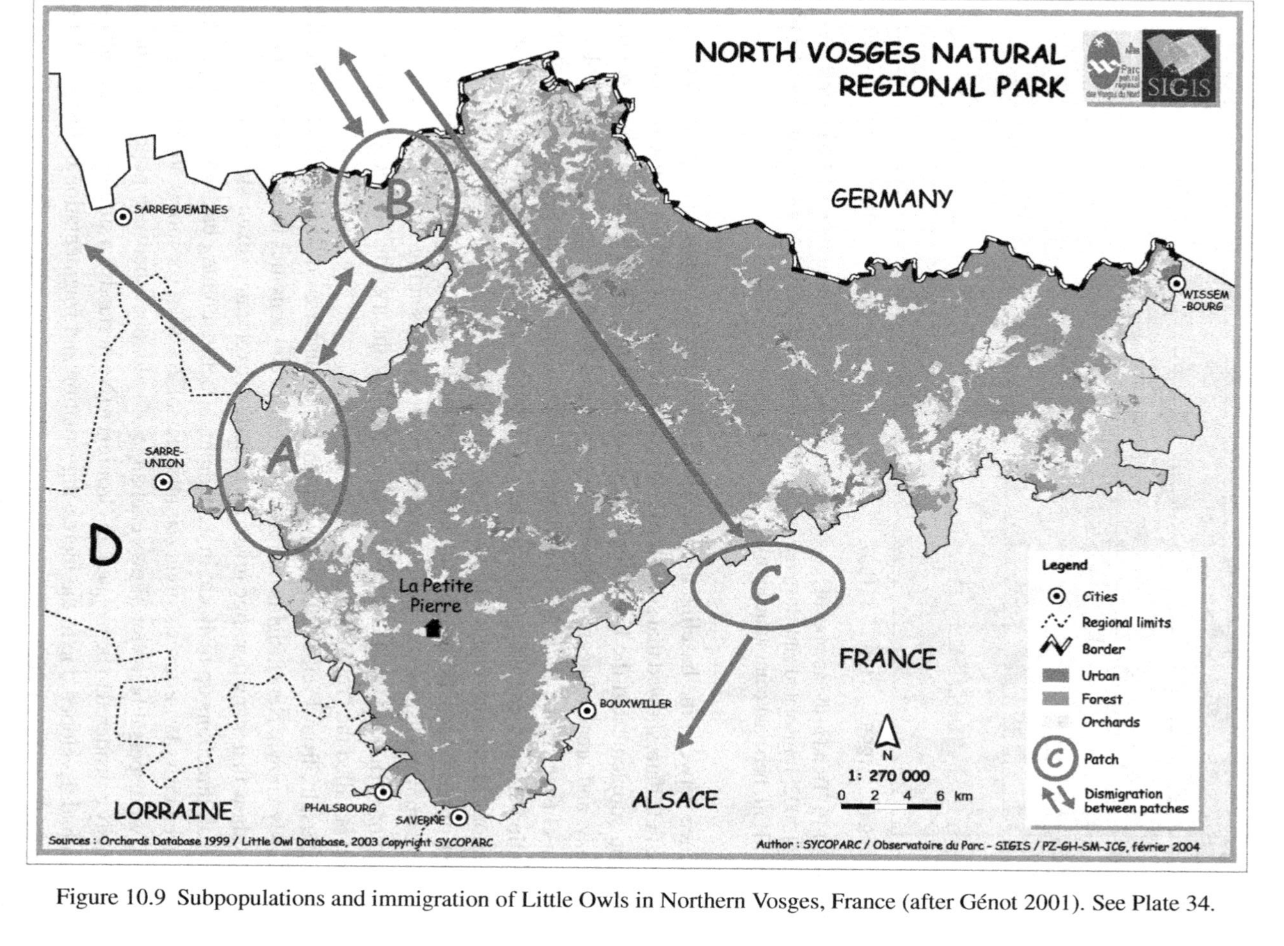

Figure 10.9 Subpopulations and immigration of Little Owls in Northern Vosges, France (after Génot 2001). See Plate 34.

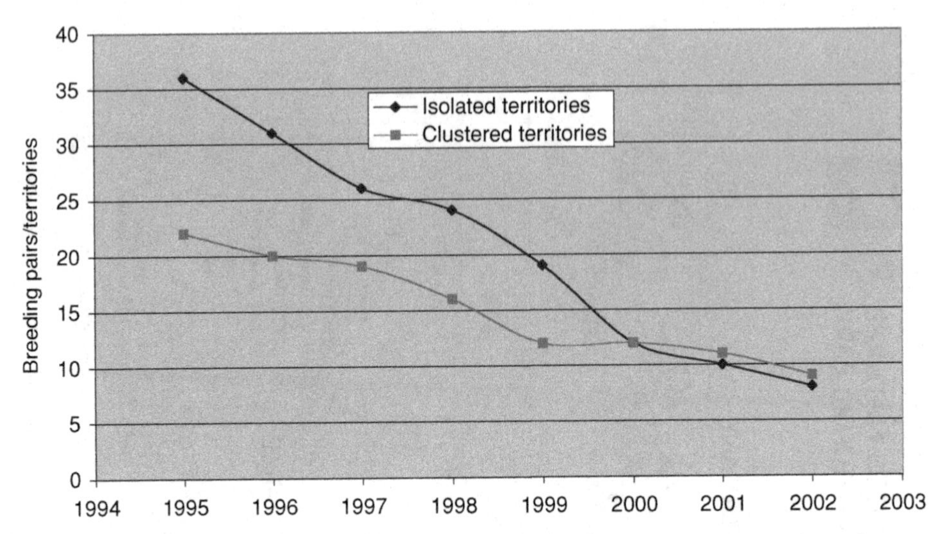

Figure 10.10 Trend in the number of breeding pairs or territories of Little Owls in Groningen, between 1995 and 2002, for isolated and clustered pairs in Oosterwijtwerd and Oldehove (Netherlands) (after J. Van 't Hoff, personal communication).

difference between the baseline model, which indicated a continuous population decline, and the field data, showed that some birds from outside these population clusters immigrated into and supplemented the local stock. In particular Little Owls from the Sarre region of Germany, and areas adjacent to the Northern Vosges, flew across the area to supplement owls in cluster C (see Eick 2003) and also clusters A and B. The sustainability of these population clusters can therefore only be explained by a strong connection with German populations, demonstrating some sink-source dynamics. Both small local population sizes and low observed genetic structure can only be compensated by a considerable immigration rate (>2 immigrants per year per population cluster, representing 17% of the current local population) (Bouchy 2004). In conservation biology theory, this degree of immigration support is called the "rescue effect".

The results from Eick (2003) show much larger dispersal distances of juveniles than previously known. New insights into Little Owl dispersal activities through radio-telemetry studies show that many dynamic juveniles undertake exploratory excursions of three to five days in all directions up to 40 km in a single night before coming back to their birthplace. They even cross large forests, flying at altitudes of 20–30 m (or more) above the ground. Little Owl dispersal behavior allows exchange of individuals between population clusters previously considered to be isolated. In Groningen (Netherlands), a similar meta-population pattern had developed due to the increasing isolation and fragmentation of the population (Van't Hoff 2001). In 2000, three population clusters remained in Groningen: Oosterwijtwerd in the east with six pairs; Oldehove in the west with five pairs; and Peize (north-Drenthe) with 11 pairs (J. Van't Hoff, personal communication). The distances between the different clusters are: Oosterwijtwerd to Oldehove, 27 km; Oosterwijtwerd to Peize, 28 km; and Oldehove to Peizenorth, 13 km. The average nearest neighbor distance between owl pairs within each cluster was 1494 m (Oldehove), 1140 m (Oosterwijtwerd), and 763 m (Peize).

In Groningen, isolated pairs declined much faster than those in remaining population clusters (Figure 10.10). The vicinity of the last population clusters in Groningen seemed more attractive to owls than the actual quality of the habitats. Most recently abandoned (and isolated) territories were situated in apparently optimal habitat while the population clusters remained in sub-optimal habitat. A key value of these vacant "habitat isolates" lies in their role for a future recovering population.

Finally, besides stochastic and deterministic factors, the spatial extent and the isolation of a local population may have an impact on the contribution of the different demographic components. Using long-term demographic data, Schaub *et al.* (2006) performed retrospective population analyses of four Little Owl populations with differential spatial extent and degree of isolation, to assess the contribution of demographic rates to the variation of the growth rate of each local population and to the differences between the growth rates among populations. The relative importance of immigration to the growth rate tended to decrease with increasing spatial extent and isolation of local populations.

11
Conservation

11.1 Chapter summary

We open this chapter with a brief overview of the status of and threats to the Little Owl. See Figure 11.1. We then offer a conservation strategy for the owl that involves five main components: knowledge, limiting factors, landscape conditions, legislation and policies, and people. Thereafter, we describe four main drivers to implement this strategy, focused on monitoring the owl and its habitat, standardizing methodologies, data management, and measuring success – assessing the strategy. While we offer aspects of conservation in this chapter, Chapter 13 is dedicated to the topic of a monitoring plan for the Little Owl.

The long-term conservation of the Little Owl is complicated as the species is largely linked to an agriculturally dominated landscape. This landscape condition can change rapidly and significantly due to changes in policies and management. The conservation strategy described in this chapter requires a multi-scale, multi-disciplinary approach, with collaboration between different stakeholders (conservationists, scientists, different authorities, farmers), and additional research into the ecology of the species. This strategy must be applied at different levels: local, regional, national, and international. We encourage people involved in this conservation strategy to work broadly, openly, and to freely co-ordinate on issues, data, and management efforts that will benefit the broader array of species and environments of which the Little Owl is a part.

In closing, as nestboxes are currently a primary conservation tool in supporting Little Owls in some regions, we offer some examples of nestboxes and anti-marten protection devices.

11.2 Conservation status and threats

In Europe, the Little Owl is considered as a Species of European Conservation Concern (SPEC) category three with a European Threat Status that is "Declining" (Manez 1994). This means that the species is not of global conservation concern, that it has an unfavorable conservation status (declining), and that its populations are not concentrated in Europe. For the arable land and improved grassland habitats, the species obtains a priority class B because of its SPEC 3 status and because more than 75% of the European population

Figure 11.1 Little Owl fledgling (François Génot).

occurs at any stage of the annual cycle in this habitat (Tucker & Evans 1997). For the steppe habitats, the species has a priority class C, that is, less than 25% of its population occurs in this habitat.

The threats to the Little Owl are mainly due to changes in human land-use practices. Some threats are becoming more prominent across the European range of the Little Owl, such as the reduction in the amount of tree-lines, the deterioration of high-stem orchards, and the increase in the area of subsidized maize. The Little Owl is also negatively impacted by collisions with vehicles on roadways, and by pesticides that reduce the amount of invertebrate prey available (Manez 1994).

Populations can disappear because of low density, lack of immigration, and fragmentation of habitat and populations (Génot 1992c; Van't Hoff 2001). More and more Little Owls near villages in central Europe are threatened by building activities without consideration of nature conservation in land-use plans (Breuer 1998). Throughout the rest of the breeding range, little is known about the current population levels, threats, or long-term trends.

11.3 Conservation strategy for the Little Owl

We propose a conservation model featuring five components: knowledge, limiting factors, landscape conditions, legislation and policies, and people. We first characterize the components, and follow with a review of the drivers that are essential in implementing the components.

Five components of the conservation strategy

Knowledge

One of the first steps in developing a practical and cost-effective conservation program is to identify the current status of knowledge on the species and define the gaps in the key scientific aspects (Tucker & Evans 1997). Several authors have given an overview of what is currently known on the species for several areas: the Netherlands (van Dijk & Ottens 2001), Italy (Mastrorilli 2005), and Belgium (Van Nieuwenhuyse, Leysen & Leysen 2001d).

To assist in examining the current state of knowledge (and knowledge gaps) on the species, we offer a summary of the Little Owl literature, through time, by geography, and by subject area. Importantly, we include a Little Owl bibliography (Appendix B) in this book. The timeline of the bibliography is for publications from 1769 through early 2005. This bibliography represents all of the Little Owl literature of which we are currently aware; we welcome any additions to this literature base, and apologize for any missed citations.

Most of the references appear after the 1960s. The relative number of publications since this time is 1544, and 416 of these come from the period 1980–2000 (Figure 11.2). Most of the publications from 1880–1940 come from Britain, and are associated with the status of the Little Owl after its introduction there at the end of the nineteenth century.

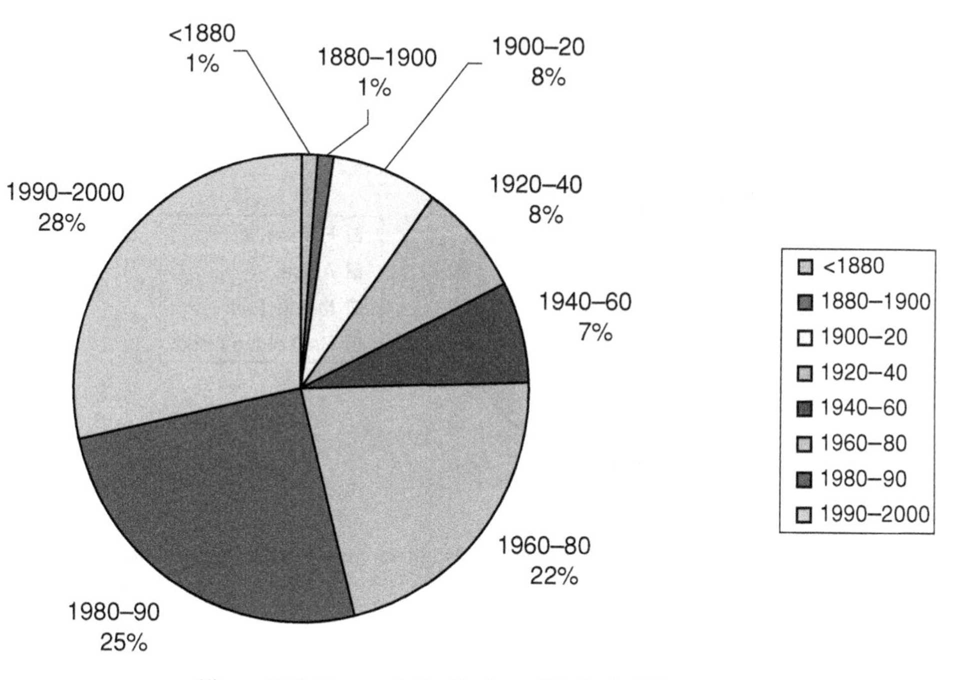

Figure 11.2 Temporal distribution of Little Owl literature.

Regarding the geographic distribution of the Little Owl literature, a total of 1512 titles were assigned to geographic areas (Figure 11.3); with 1452 of the references concerning Europe. This can be explained partly by the distribution area of the species but mostly by a readily accessible European literature base. Western Europe (France, Germany, Belgium, Netherlands, Britain, Switzerland) accounts for 1115 references, Mediterranean Europe (Spain, Italy, Greece, Crete, Cyprus) 118, with eastern Europe (former Soviet Union, Poland, Czech Republic, Hungary, Romania) reflecting 56 references (Figure 11.4). Even if France is partly considered as a Mediterranean region, the majority of French references do not refer to this climatic area.

France and Germany account for 685 of the European references. If France shows the highest number of publications ($n = 359$), this is partly due to the nationality of one of the authors who had easy access to the French bibliography. But Germany (326 references) is the country where the Little Owl has been the most well studied and is the source for particularly important knowledge on the owl.

In Switzerland, 26 out of 67 publications were written by one particularly prolific author, Michel Juillard.

The majority of publications come from countries where the species is declining (France, Germany, Switzerland, Belgium, Netherlands, Austria, Luxembourg). This is confirmed by the chronology of studies whose 1188 references appear after 1960, which is the beginning of the Little Owl population decline across Europe.

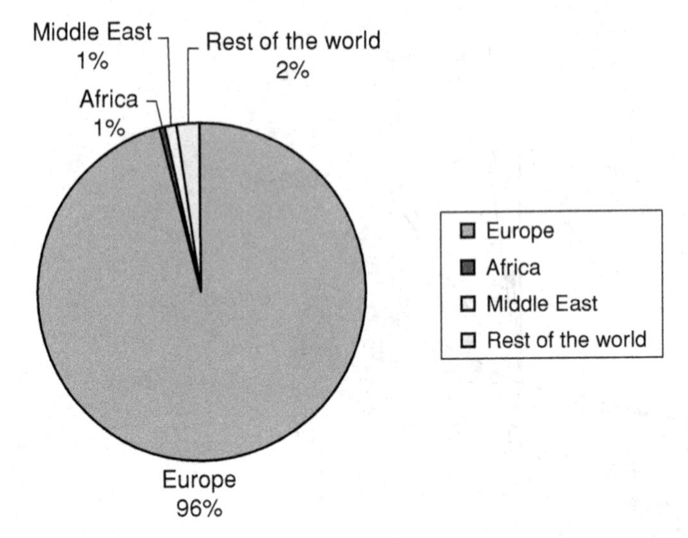

Figure 11.3 Geographic distribution of Little Owl literature across the range of the species.

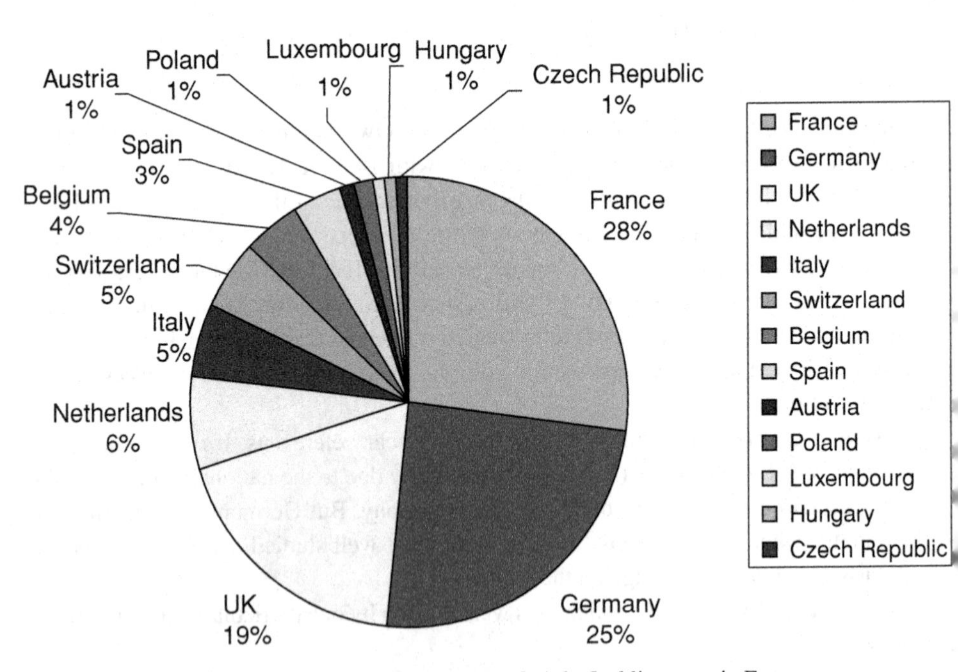

Figure 11.4 Geographic distribution of Little Owl literature in Europe.

Table 11.1 *Types of Little Owl publications.*

Type of document	Number	Percentage
Ph.D. thesis	14	0.7
Red list book	5	0.3
Monographs	7	0.4
Proceedings	104	5.5
Atlas	63	3.3
Report	106	5.6
Book	214	11.3
Journal	1387	73.0
Total	**1900**	**100**

Subject area of the literature Each reference has several keywords. The most studied subjects are the following:

Distribution: 407
Food: 291
Habitat: 243
Breeding: 200
Density: 175
Behavior: 135
Mortality: 130 with the subtopics of predation (29), road (30), or pesticide (39)
Population, abundance: 114
Nestboxes: 115
Protection: 116
Generalities: 86
Population dynamics: 60
Taxonomy: 56
Vocalization: 25

Certain topics remain poorly studied, such as physiology (14), landscape (7), territory (8), or genetics (7).

The majority of the references (1387) come from journals (Table 11.1). Other types of documents include: books (214), reports (106), atlases (63), proceedings (44), monographs (7), red list books (5), and eight masters theses and six Ph.D. theses.

The six Ph.D. theses focused on the Little Owl are those of: Juillard (1984b) in Switzerland, Exo (1987) and Finck (1989) in Germany, Sageder (1990) in Austria, and Génot (1992a) and Bouchy (2004) in France.

Among the monographs, three titles are very important: Cramp (1985) in English, Juillard (1984a) in French, and Schönn *et al.* (1991) in German, which is the most complete monograph. An update of the *Birds of the Western Palearctic* was written by Génot and Van Nieuwenhuyse (2002).

To assess the missing critical aspects of knowledge, we distinguish the following key topic areas: (1) taxonomy, (2) habitat use, and (3) demographics.

Taxonomy: Additional knowledge on the taxonomic diversity of the species based on DNA analysis (e.g., mitochondrial and nuclear) is needed. An example of an important taxonomic clarification was recently made in China, with the determination of a new subspecies, *Athene noctua impasta*, (Qu *et al.* 2002) based on DNA evidence.

Habitat use: The general knowledge of the Little Owl stems mainly from research carried out in eastern European agriculturally dominated landscapes. Important habitat work needs to take place in areas more in the core of the species range. The Little Owl has been characterized as having an important plasticity in habitat use (Van Nieuwenhuyse & Leysen 2001). The species displays an ability to cope with local circumstances in some situations but not in others. When combined with specific projects that monitor the demographic performance of Little Owls, a habitat quality assessment that takes into consideration the heterogeneity of different habitat types occupied by Little Owls will provide a significant advance for cost-efficient monitoring and the possible distinction of population source and sink areas (Pulliam 1988).

Demographics: Only a few Little Owl populations have been monitored (in the countries of Belgium, France, Netherlands, Switzerland, and Germany) for more than ten years – a time period long enough to begin to yield meaningful results on population performance (see: Knötzsch *in* Schönn *et al.* 1991; Ullrich *in* Schönn *et al.* 1991; Gassmann & Bäumer 1993; Kämpfer-Lauenstein & Lederer 1995; Furrington 1998; Bloem *et al.* 2001; Génot 2001; Meisser & Albrecht 2001; and Van Nieuwenhuyse *et al.* 2001a). In general, more knowledge is needed on important demographic parameters (e.g., immigration and emigration, juvenile and adult survival, age structure of the breeding population) and population regulating forces.

As part of demographic work on the Little Owl, insights are needed as to the behavioral aspects of the species, especially the clustering behavior of the species. While plenty of ringing data exist among active ringers, it has not been rigorously analyzed or published. To obtain a quick start on this topic, we suggest a maximal exploitation of existing ringing data on a short time frame. A similar exploration of Little Owl data from Wallonia is presented in Bultot *et al.* (2001) and shows that even non-experimental data can yield important knowledge of the species. This knowledge is crucial to assess the local context of breeding performance and health of the population.

Other topics: In addition, we suggest an extension of the tasks of ringers towards the focused sampling of additional data, e.g., the acquisition of feathers for the analysis of genetics and presence of heavy metals, eggs for analysis of pollutants, and dead birds for the study of mortality causes. Due to the simplicity of sampling, we highly recommend opportunities to examine dead animals, e.g., road casualties and unhatched eggs to analyze the impact of pesticides (Vogrin 2001) and rodenticides (Beersma & Beersma 2001).

Owls released from bird-care centers also provide unique opportunities to gather more crucial knowledge and information on the owl's ecology. For example, these owls could be released as experimental samples using telemetry to study local movements and settlement patterns in regions currently vacant of owls. The supply of such birds is large – some 156 Little Owls were released from bird-care centers in Flanders in 1995.

Limiting factors

Essential criteria for suitable Little Owl habitat are year-round prey availability, prey accessibility, vertical landscape structures with cavities (for nesting and roosting), and limited predation pressure (see Chapter 6). We recommend management activities that increase the availability and long-term replacement of nesting cavities, provide better availability and accessibility to prey, and provide other habitat conditions that serve to limit the impact of predators. Finally, before we can get a clear picture of the relative importance of the different factors that influence the population dynamics of the species (Exo 1992) in a given region, we need reference data on these key resources and aspects.

Food availability Improving the amount of prey for the Little Owl can be done through different management activities that increase habitat patch heterogeneity by reducing the average parcel size (ha). Grimm (1986) found larger and more numerous prey in a meadow edge beside a brook than in grasslands. The trampling of vegetation by cattle yielded more Carabidae beetles, especially near parcel edges (but trampling by cattle has other potentially habitat-damaging consequences, such as the reduction in growth of cavity-bearing trees). Tall vegetation is crucial for small mammals, with edges tending to have larger diversities and densities (Dalbeck *et al.* 1999). Planting shrubs, hedges, and tree-lines serves to promote the abundance of insects and other prey.

If we want to study food availability and its possible impact on owl population numbers, we need more information on the prey choice of the owls, as well as the variance of those prey in space and time. An international database on small-mammal distribution and especially densities is needed, as is better data on the relative abundance of earthworms (Lumbricidae) and key beetle species. See Figure 11.5.

Food accessibility Being principally a perch-and-pounce hunter in southern regions (Fajardo *et al.* 1998; R. Tomé, personal communication), the availability of perches is fundamental to Little Owls. A substantial portion of perches are provided in anthropogenic landscapes by fences and fence poles. Temporal-spatial rotation in mowing activities can create gradients between tall and short vegetation, and offers favorable conditions for the owl's access to prey. We propose a hypothesis for testing: that local habitat conditions can be optimized by a combination of artificial perches and phased mowing to increase prey accessibility (similar to shrikes) (see Yosef 1993b; Van Nieuwenhuyse *et al.* 1995). See Figure 11.6.

Nesting cavities The provision of nestboxes or other cavities, while labor intensive, has been shown to be a successful short-term solution to the lack of nesting and roosting cavities in different regions (Schwarzenberg 1970; Knötsch 1978; Loske 1978; Furrington 1979; Schönn 1980; Schönn 1986; Schönn *et al.* 1991; Exo 1992; Haase 1993;

Figure 11.5 Little Owl habitat in Flanders near Halle, northern Belgium. Photo Marc de Schuyter. See Plate 35.

Figure 11.6 Little Owl habitat in Flanders near Wakken, northern Belgium. Photo Marc de Schuyter. See Plate 36.

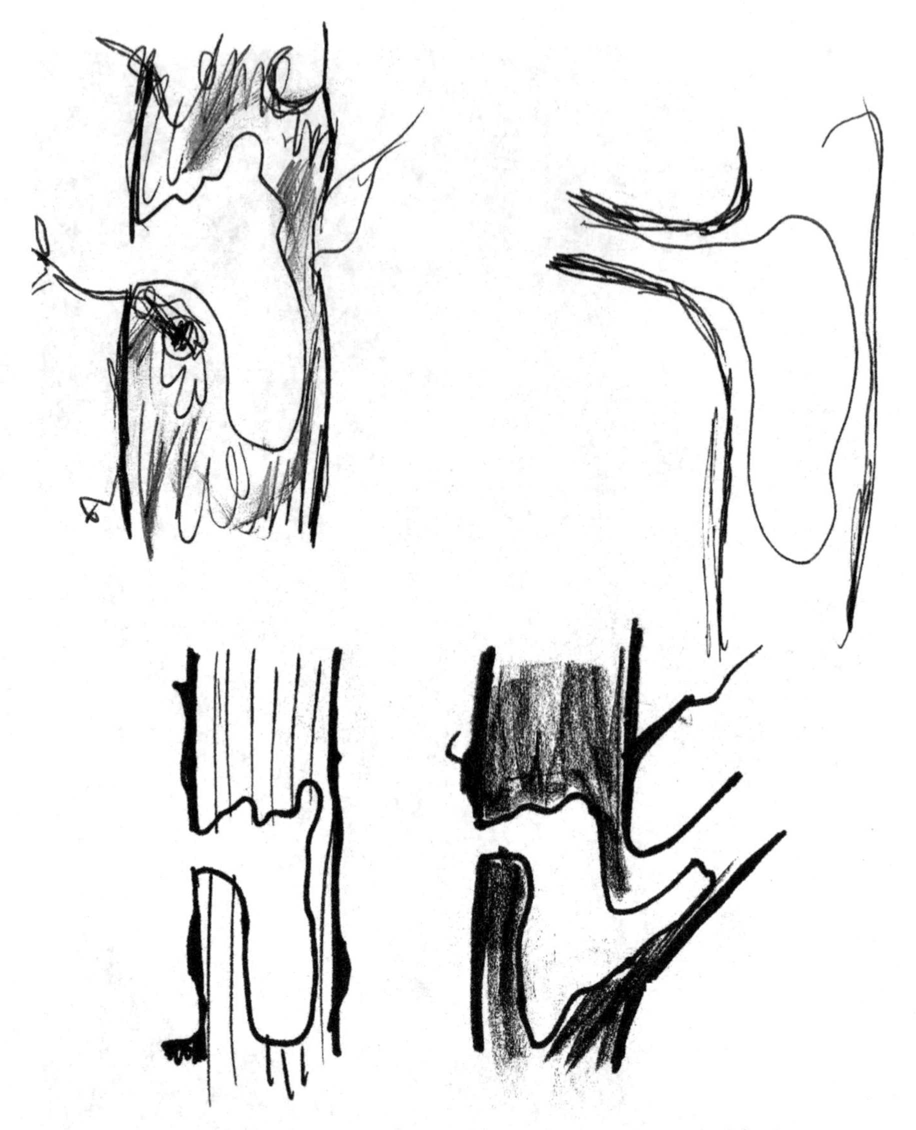

Figure 11.7 Sketches of Little Owl breeding cavities (François Génot).

Bultot *et al.* 2001; Stange, personal communication). Nestboxes have been readily accepted, even when a minimal owl population remains in a given area. In some cases, researchers have found that nestboxes that were newly installed at the beginning of April were occupied after only two weeks, indicating that nest cavities were a limiting factor for the floater owls that were otherwise capable of nesting (Bultot *et al.* 2001). Nesting and roosting sites can be created for the short term by enhancing existing previously unsuitable cavities, by modifying buildings to provide holes and crevices for owls, or by hanging nestboxes. In the longer term, cavities can be offered through planting new trees and managing existing trees (fruit trees, pollarded willows). See Figure 11.7.

Natural cavities More suitable sites can be created by the enhancement of natural cavities. Cyclical management of coppiced willows by volunteers offers the best guarantee of a durable number of suitable sites. This method is preferred since the trees also offer opportunities for prey species, and can act as hunting perches too. Importantly, provision of these types of nesting structures is the most natural, and cost-efficient.

Buildings Across large parts of its breeding range, the Little Owl occupies buildings. Hence, modifying barns, chapels, and dilapidated structures by adding suitable holes in the walls or under roofs is easy and sustainable. Furthermore, it also offers good opportunities for interacting positively with people to help the species due to its association with humans. See Figure 11.8.

Nestboxes We wish to stress that the last resort to increase the number of nest sites is through the installation of nestboxes. Nestboxes should be avoided when possible since predators such as Stone Martens recognize the boxes easily and remember their locations during their hunting rounds. This is especially the case when larger numbers of a standardized type of nestbox are used (Kirchberger 1988). Importantly, nestboxes are often not available for Little Owls due to use by other animals such as the Starling (*Sturnus vulgaris*), Tit (*Parus* sp.), Tree Sparrow (*Passer montanus*), Dormouse (*Glis glis*), Garden Doormouse (*Eliomys quercinus*), hornet, bee and wasp, bumble-bee, and House Mouse (*Mus musculus*) (Pitzer 1995). Thus, before installing any nestboxes, the initial population should be determined through a standardized survey method. This will provide specific insights into what extent the owl population may actually need newly offered nesting places. Furthermore, a clear view of all available nestboxes in the study area should be obtained and communicated during this inventory. Nestboxes that are not known to researchers prior to efforts can influence interpretation of the results. When installing nestboxes we advise this is done in both seemingly suitable and unsuitable habitats and that differences are studied in the owls in both habitats afterwards.

In Figures 11.9a–e we offer differing designs of nestboxes to guarantee a certain degree of heterogeneity. Installing nestboxes should be seen as creating excellent opportunities for research. Sampling of the species is made very easy through the use of nestboxes. We consider the follow-up monitoring of the response of the owls to the addition of nestboxes as mandatory. It is important for researchers to understand that population trends produced by nestbox studies alone are not always reliable (Illner 1990).

Predators The acrobatic skills of Stone Martens were illustrated by Marié and Leysen (2001) and possible measures to avoid predation by Martens on Little Owls are suggested (see Figures 11.10a,b and Figures 11.11a,b,c). To get a better assessment of the impact of predators, we are very supportive of co-ordination with existing Stone Marten (*Martes foina*) monitoring projects. These types of investigations/co-ordination efforts are important given the population increase of the Marten species in Europe (Van Den Berghe 1998) and its possible impact on areas with low populations of Little Owls.

Figure 11.8 Barn improvement to help Little Owls. Photo Rene Krekels.

Figure 11.9a An example of a Little Owl nestbox. Photo Marc De Schuyter.

Figure 11.9b An example of a Little Owl nestbox. Photo Jacques Bultot.

Figure 11.9c An example of a Little Owl nestbox. Photo Rottraut Ille.

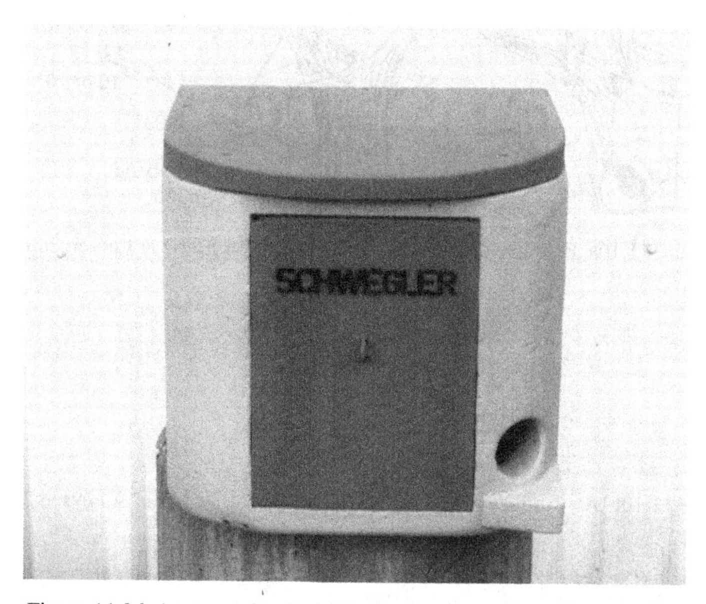

Figure 11.9d An example of a Little Owl nestbox. Photo Rottraut Ille.

Figure 11.9e An example of a Little Owl nestbox. Photo Rottraut Ille.

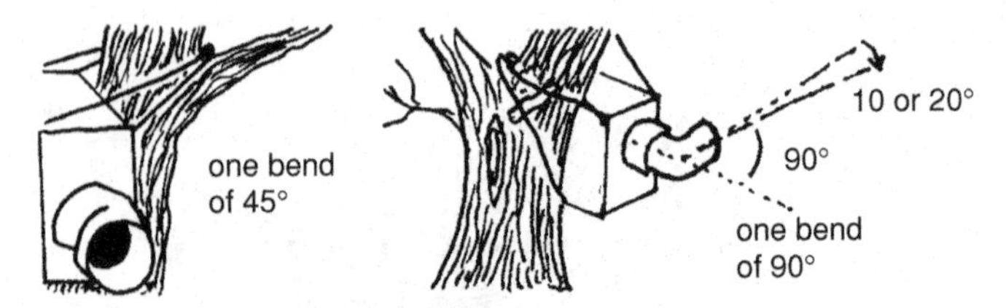

Figure 11.10a Example of anti-Marten devices (after Marié & Leysen 2001).

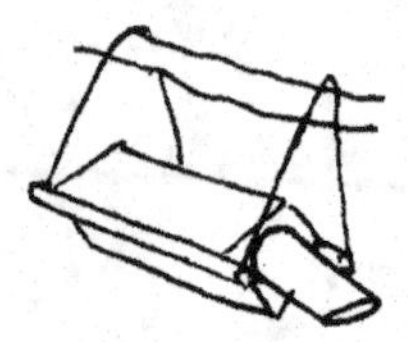

Figure 11.10b An example of an anti-Marten device (after Marié & Leysen 2001).

Figure 11.11a An example of an anti-Marten device. Photo Jacques Bultot.

Figure 11.11b An example of an anti-Marten device. Photo Luc Van Den Wijngaert.

Figure 11.11c An example of an anti-Marten device. Photo P. Winterink.

Landscape conditions

Across its global range, the Little Owl is closely associated with the agricultural landscape (Tucker & Evans 1997). The impact of changes to what was likely a largely pastoral-like landscape is unclear due to a lack of data from a historical baseline condition. However, as illustrated by Van't Hoff (2001) in the Netherlands, this kind of data is crucial in the study of the causes of population trends. The Flemish Little Owl Project (Van Nieuwenhuyse, Leysen & Leysen 2001d) paid much attention to the set-up of a referential database of soil data, land-use data, biological quality data, roads, rivers, and related spatial landscape conditions. This reference database will be used to follow changes in landscape in a digital and exploitable way, and made available in the planned European Little Owl Data Warehouse via the internet. Contacts and co-operation between conservation groups and official organizations, e.g., Intra Eco Network Europe (IENE) need to be established to jointly and more efficiently assess the impact of habitat fragmentation on birds by roads and other societal infrastructure.

To date all habitat selection studies have been carried out in anthropogenic landscapes. New research on habitats are needed in natural areas such as deserts (Northern Africa), steppe (Kazakhstan), and at higher altitudes (Tibet, India).

Legislation and policies

An excellent overview of the opportunities for conserving the wider environment, and indirectly the Little Owl, through international and European legislation is given in Tucker and Evans (1997). An important challenge that remains is to analyze these different instruments in more detail and to translate the opportunities into a broader international context for the conservation of the Little Owl and its habitat.

For Flanders, Belgium legislation on nature conservation has evolved from a species protection scheme for "useful" species towards a more territory directed conservation strategy, e.g., Natura 2000 (http://europa.eu.int/comm/environment/nature/home.htm). In recent years, more attention has been directed at species protection again due to the formal adoption of the "Communication on a European Biodiversity Strategy", adopted on February 4, 1998 (http://europa.eu.int/comm/environment/docum/9842sm.htm). With this strategy, the EU reinforced its role in finding solutions for biodiversity within the framework of the United Nations' Convention on Biological Diversity (also called the Biodiversity Treaty, see http://www.ciesin.org/TG/PI/TREATY/bio.html).

In Flanders, the Little Owl is a protected species through the European Bird Guideline 79/409 (Royal decree of September 9, 1981 changed by decree of the Flemish Government of November 20, 1985). Since 1998, the Decree on Nature Conservation (Belgisch Staatsblad January 10, 1998) offers different opportunities for Little Owl conservation. The fundamentals of the decree are species protection, territory-directed conservation, general basic ecological quality, and a focused tackling of species groups of interest. The implementation of Little Owl conservation measures can be done through management contracts as mentioned in the European Union ordinance 2078/92. With these contracts, farmers can get money to create or maintain certain habitat elements in the landscape (e.g., wooded banks,

Figure 11.12 Little Owl front view. (François Génot).

pollarded willows, ponds, field edges). These contracts are essential for safeguarding viable breeding places for the Little Owl. Despite their availability, there is no systematic use of management contracts. Additional instruments in Flanders include financial subsidies for the planting and maintenance of pollarded willows, high-stem orchards, wooded banks, and other small landscape elements in most communities through the GNOPs (Gemeentelijk Natuur Ontwikkelings Programma, or communal nature development plans). Despite some good and pragmatic initiatives, there remains a huge gap in the use of economic and policy mechanisms and their conservation impact on the species (Tucker & Evans 1997) in most parts of Europe, let alone the rest of the Little Owl's distributional range.

In Luxembourg, the Règlement Grand-Ducal of March 22, 2002 (Règlement de biodiversité) regulating a set of measures for the conservation of biodiversity, was made official on April 4, 2002. This law provides funds and the opportunity to take measures for the protection of rare and/or endangered species. The costs of actions especially helpful for the Little Owl, such as providing nestboxes or planting fruit trees around villages, are 90% covered by the Ministry of the Environment. See Figure 11.12.

People

People are another key factor in any proposed Little Owl conservation strategy. Up to 75% of the European Little Owl population occurs in agricultural and grassland habitats (Tucker & Evans 1997). Especially for this species, environmental protection and biodiversity conservation should be an integral part of all uses of the environment, and of all policies of all socioeconomic sectors (agriculture, forestry, tourism). In the longer term, sustainable use of the environment will require substantial changes in society's use of natural resources, energy, and transport. It is therefore essential that the public is aware of the implications of the current and increasing intensity of its uses of the environment, and of the fact that everyone has a role to play in the conservation of biodiversity in the wider environment (Tucker & Evans 1997). The Little Owl might offer an excellent model to implement this in Europe because of its positive emotional impact on humans. We propose the species become a flagship for well-managed agricultural landscapes in Europe.

Specific action plans should be made to educate and influence policy makers, landowners, land-users, conservationists, and children. Below, we give an overview of possible initiatives for different age-classes and target groups.

Educational initiatives and materials for Little Owl programs

General

- Research material, e.g., survey maps, old radio-transmitters, examples of nestboxes, study skins, photographic keys.
- Habitat management working days for volunteers (see Bultot *et al.* 2001).
- Video "Histoire de chouettes chevêches". A 28-minute film by Marcel Thonnon. Ministère de la Région Wallonne. Direction générale des Ressources naturelles et de l'Environnement, Service de documentation et communication. Avenue Prince de Liège, 15, B-5100 Jambes.
- "Uilen, mannen van de nacht" (De Laet J., as part of the VUBPRESS series of "Vogels rondom ons", [Birds around us]).
- Excursions for the general public during the "Night of the Little Owl" (France, Italy).
- Identify specific target groups of adults (e.g., farmers, landowners, communities, schools, conservationists).

Farmers

- Active promotion of subsidized management contracts among farmers through specific leaflets.
- Creation of a clear link with organic farming; try to promote the Little Owl as their label.
- Publication of articles in agricultural magazines, and magazines of other land-users.

Conservationists

- Offer scientific consultancy to volunteers, i.e., international workshops on monitoring in Belgium 2004, Dutch Little Owl research manual (e.g., Bloem *et al.* 2001).

- Publication of theme issues of *Oriolus* (Van Nieuwenhuyse, Leysen & Leysen 2001d), *Ciconia* (Génot *et al.* 2001).

Primary school

- Uilenboekje (Van't Hoff 2003) (in Dutch) dealing with owl ecology in very simple words for school-children.

Secondary school

- Provide assistance to students participating in nature study competitions (e.g., Prize for Biology Jacques Kets, Zoo Antwerp and ECYS).
- Maarten Bekaert's study on Little Owls (Bekaert 2002) was selected as the best Flemish biology study for the 13th European Union Contest for Young Scientists (ECYS).
- He also received the "Prijs voor Biologie Jacques Kets", a prestigious award given by the "Koninklijke Maatschappij voor Dierkunde van Antwerpen".

Websites

Currently there are several Little Owl websites.

http://www.diomedea.org/ILOWG/ILOWG.index.htm
http://owlpages.com/species/Athene/noctua
http://votquenne.multimania.com/
In Belgium: http://noctua.en-action.org/
In the Netherlands:

http://www.steenuil.nl
http://www.steenuilgroningen.nl

In France: http://cheveche.free.fr
In Switzerland: http://www.orniplan.ch/seiten/th_steinkauz_sh.htm
In Italy: http://www.progettoathena.it
In Germany:

http://www.nabu-rlp.de/html/projekte/steinkauz/steinkauz.html
http://www.ageulen.de

Global:

www.owlpages.com (general website on owls of the world)
www.globalowlproject.com (a more scientific focus on owls of the world)

Implementing the strategy: four drivers

The components described above reflect the overarching initiatives that embody a Little Owl conservation strategy. The four drivers that we propose in this section are specific measures to implement the initiatives: monitoring the owl and its habitat, standardizing methodologies, data management, and measuring success. See Figure 11.13.

Figure 11.13 Little Owl *Athene noctua lilith* with prey. Prey species is the House Gecko (*Hemidactylus turcicus*). Photo Amir Ezer.

Monitoring the Little Owl and its habitat

The overall goal of natural resource monitoring is to develop scientifically sound information on the current status and long-term trends in the composition, structure, and function of ecosystems, and to determine how well current management practices are sustaining those ecosystems. While we offer a simple summary here, Chapter 13 describes a detailed monitoring plan for the Little Owl. The purpose of the Little Owl monitoring plan is to assess trends in owl populations and habitat. Monitoring data will be used to evaluate the success of the various conservation measures in arresting the downward trends in some owl populations and in assessing habitat conditions necessary to support viable owl populations throughout the range of the owl.

Monitoring the owl: demographic study areas In our monitoring plan, we propose the development of a network of some thirty vital sign demographic monitoring areas where detailed mark–recapture studies are employed. The primary objective of demographic studies is to detect trends in the vital rates of the species. More specifically, demographic analyses on owls employ mark–recapture studies, typically on large identified areas, where repeated surveys are conducted to locate, mark, and re-observe or recapture pairs of owls and their offspring. Many of the demographic areas we are recommending are in operation now, and for these we suggest the formalization of a more detailed and rigorous study design. For some areas of the range of the species, this would mean the identification of new demographic study areas. Rather than having a lot of study areas, we propose having fewer, but more intensively investigated, areas, within which some 50–100 pairs of owls (or habitat for same) are monitored routinely. Adult survival rates, fecundity, and population and meta-population trend analyses would be based upon the work conducted by Franklin *et al.* (2004), Forsman *et al.* (1996), and Ganey *et al.* (2004). For countries without vital sign demographic study areas, we recommend distributional surveys and ecological studies.

Monitoring the habitat Van't Hoff (2001) offered interesting insights from the historical data on landscape and Little Owl distributions in Groningen, the Netherlands. Unfortunately few historical landscape data are available in digital formats. Acquiring a digital version of the landscape information (e.g., satellite images) at regular intervals of time (say, every ten years) is extremely important because we can then monitor the changes in landscape variables important to Little Owl populations. Van Nieuwenhuyse and Leysen (2001) described some key principles for a long-term monitoring program. In relationship to Little Owl ecology, they identified which landscape elements cause habitat heterogeneity (i.e., dependent and independent variables), which habitat types should be monitored, and determined the relevant sample sizes to be acquired.

To obtain better insights into habitat selection, a multi-scale approach is suggested to more readily detect differing patterns that occur at different scales (May 1994; Swindle *et al.* 1999). The multi-scale approach to the study of habitat selection is based on the conceptual framework suggested by Johnson (1980). Consequently, general habitat selection can be regarded as a hierarchical process, with owls determining what is an appropriate place for breeding at a small scale, with foraging areas determined at a broader scale (see Laaksonen *et al.* 2004). Knowledge should become available on the macro-scale of countries and regions, on the meso-scale (a population cluster), at the individual home-range scale, and at the fine-scale (components within individual territories). The first projects following a scaled approach were done in Flanders (northern Belgium) (Van Nieuwenhuyse *et al.* 2001c), in France (Ferrus *et al.* 2002), and in Spain (Martínez & Zuberogoitia 2004b). A detailed monitoring scheme was documented for the Netherlands (SOVON 2002) with special emphasis on the selection of a stratified sampling scheme.

Standardizing methodologies

Standardizing the methodologies for Little Owl studies involves using consistent terminology, survey methods, demographic data, and habitat analyses. Certainly, we are not urging researchers and managers to suppress their innovative talents, but rather, feel we must standardize core aspects of our work in order for us to maximize the benefits from our collective efforts. Part of the standardization process starts with the use of common terminology; in support of this, we offer definitions for some frequently used terms in the glossary.

Survey method Survey protocols are detailed study plans that explain how data are to be collected, managed, and reported, and are a key component of quality assurance for natural resource monitoring programs. There are typically three primary objectives in surveying for owls: (1) documentation of presence/absence, (2) enumeration (density, absolute or relative abundance), or (3) population dynamics (size, distribution, trends in abundance over time). In our survey protocol, we focus on surveying to determine population dynamics. This is the most rigorous of the three objectives, and documentation of owl presence and enumeration of owls can also be determined using this protocol. While a census is defined as a complete count of every individual in a given population (very rarely possible in wild populations), a survey is an estimate of the number of animals in a given area (e.g., number of territorial pairs) or of the relative frequency of encounters of animals (e.g., number per unit transect length or time period). Repeated surveys can be used to estimate the trend of a population. Surveys provide an efficient way of collecting information from a large number of sites and can also help determine the overall distributional range of a species in different seasons. The structure of surveys is intended to reduce observer bias, and surveys are best conducted in a standardized way to ensure reliability and validity.

Thus, for owls, as it is truly unlikely that a complete count of all individuals could realistically be determined for any wild population (due to methods, budgets, or detectability of the species), a survey is a more appropriate approach to estimating numbers, frequencies, and trends (see also Thompson *et al.* 1998).

Surveys efforts are ultimately aimed at determining the total number of territorial owls (paired and unpaired) present on a given area. Not all territorial owls readily respond to playback, either because they are not within hearing range when playback surveys are conducted, or because they are engaged in other more pressing behaviors (e.g., hunting). It requires energy from the owls to respond to territorial intruders, and the owl must balance its energy expenditure in territorial defense against its other nightly and seasonal activities.

The broadcasting of pre-recorded conspecific vocalizations using a cassette tape player (or CD) and some type of speaker is a common technique used to detect secretive and vocally active raptors (Fuller & Mosher 1981). More simply called "playback", this technique is widely used to solicit responses from territorial owls, whose elusive and nocturnal habits typically prevent visual detection (Bibby *et al.* 1992). Playback is the most common method for surveying Little Owls.

An important aspect of surveying using playback is quantifying the playback efficiency (i.e., the response rate). In determining the response rate, we are trying to determine how many of the owls holding territories that can hear the playback, actually respond to it. In order to determine the response rate, one needs to know the locations of the existing owls (prior to playback solicitation), as determined through radio-telemetry, and then to broadcast the playback and record the number of responding owls. Thus, the response rate reflects the proportion of owls that, when within hearing range of the broadcasted playback calls, actually respond to the playback. For example, if five out of ten owls that were within the broadcast range of playback recordings actually responded to the playback, this is a response rate of 50%. However, unless the owls' locations are first determined with radio-telemetry, observers employing playback have no way of knowing whether surveys without responses reflect the absence of owls, whether the territory holders were outside of hearing range, or whether owls were present and did not reply. The detection rate is a summary of the number of owls that were recorded across a given area, e.g., (number of individuals heard)/(number of playback sessions performed) (Galeotti 1989; Sará & Zanca 1989; Centili 2001b). While easier to calculate, these summaries invariably yield rather low rates of detection, as they cannot separate owl absence from owl silence and incorporate surveys conducted in areas of unsuitable habitat (no owls present to respond). The calculated detection rate is inversely proportional to the number of playback stations surveyed with no owls responding.

An alternative way of determining the response rate is to establish survey call points along a transect, and conduct multiple (e.g., six to ten) visits across the survey route, conducting playback surveys at the survey points over a given time frame (e.g., the breeding season). During these surveys, the observer records the responding owls and maps their locations. During each visit across the route, it is probable that some "new" owls (those that did not respond, or were not within hearing range of the broadcast during previous visits) would be detected. Additional survey visits would be conducted until no new owls were detected along the transect. Analysis of the data would indicate the number of visits required until 90–95% of the territorial owls had been detected. As an example, for the Northern Spotted Owl, a species considered to be a relatively good responder, some six visits across the survey routes are required to effectively detect the resident owls. Field tests of detectability for the Powerful Owl (Australia) have indicated that it may require up to 18 visits to survey stations involving listening and call–playback to detect 90% of the territorial owls within a given area (Wintle *et al.* 2005). Similarly, eight visits to survey stations involving listening and call–playback were required to detect 90% of the Sooty Owls (*Tyto tenebricosa*) (Wintle *et al.* 2005).

Further, the sex of the owl whose vocalizations are being broadcast may make a difference. For example, if you were broadcasting the vocalizations of a male owl (this is most appropriate, since males are the primary defenders of territorial boundaries), why would the resident female owl respond aggressively to another male owl (and potential mate)? Further, if you were to simultaneously broadcast the vocalizations of a pair of owls (i.e., a male and female), a resident owl who was unpaired, (i.e. a resident single) may well be too intimidated to respond to two intruders. However, the broadcasting of female calls (soley) may be a good way to efficiently locate nests (as females often respond aggressively to other female owls). Field surveyors should be aware that sex, behavioral, and seasonal aspects all have a role to play in the response rates of owls.

In developing a standardized survey protocol, we examined the array of methods currently being used. In Table 11.2, we give an overview of the factors that influence the efficiency of the broadcasting methods. Temperature does not appear to have much of an effect; however, the weather does and is therefore used as a criterion for broadcasting. Most researchers do not perform playback in conditions of rain or strong wind, as reduced owl vocalizations have been observed during these times.

Centili (2001b) examined the response rate of Little Owls to playback surveys at known occupied sites. The overall response rate between February and July was 49.6%; three playback visits to each site were necessary to locate 87% of the resident owls. The response rate did not change significantly during these months, consistent with results of Zuberogoitia & Campos (1997b) and Mastrorilli (1997) (see Table 11.3). There was no significant

Table 11.2 *Overview of the factors that influence the efficiency of Little Owl broadcasting methods.*

Reference	Efficiency	Method	Weather	Time of the year	Temperature	Distance	Day/Night	Census period
Exo & Hennes (1978)	80–90%	percentage of known owls answering	influence	influence	no influence	–	–	end February till mid April
Centili (2001b)	50%	percentage of known owls answering	–	limited influence	–	no influence	–	February till July
Zuberogoitia & Campos (1997b)	>100%	number of individuals heard / number of playback sessions performed	no influence	no influence	–	–	–	January till December
Mastrorilli (1997)	28%	number of individuals heard / number of playback sessions performed	influence	limited influence	–	–	–	September till May
Galeotti & Morimando (1991)	25%	number of individuals heard / number of playback sessions performed	–	–	–	–	–	unknown
Pirovano & Galeotti (1999)	55%	number of individuals heard / number of playback sessions performed	–	–	–	–	–	unknown

Table 11.3 *Overview of the response probability to broadcasting by month of the year.*

Reference		Jan	Feb	Mar	Apr	May	Jun	Jul	Aug	Sep	Oct	Nov	Dec
Centili (2001b)	percentage of known owls	–	47	52	69	40	53	38	–	–	–	–	–
	stimuli	–	17	21	13	20	31	21	–	–	–	–	–
Zuberogoitia &	percentage of successful stimuli	21	37	45	52	46	19	49	19	18	32	10	7
Campos (1997b)	stimuli	28	46	71	52	69	68	65	47	40	53	41	30
Mastrorilli (1997)	percentage of successful stimuli	23	29	22	29	28	28	–	–	31	33	35	24
	stimuli	56	24	22	54	57	37.4	–	–	29	33	17	45

relationship between the response rate and the distance from playback station to occupied sites that were within broadcast reach. This suggests that playback is as effective in eliciting an owl's vocal reaction inside a territory as it is outside of the territory. Thirty-six percent of the replies came from occupied sites over 450 m away from playback stations (max. distance 1100 m). Since a commonly used detection radius around playback stations is 400 m, beyond which owl calls are considered not audible, it is suggested that caution be used in incorporating detection radii in survey schemes in order to minimize the risk of overestimating owl numbers.

To better deal with the issue of estimating owl numbers, we recommend the use of triangulating on responding owls, to better discern and accurately map their locations. In triangulation, an observer acquires two (or three) compass bearings on the responding owl (e.g., from points that are 100 m apart), and then plots both the observers' locations (on a detailed map or aerial photograph) as well as the compass bearings to the located owl from those locations. This will serve to pinpoint the location of the responding owl. Observers will need to be attentive, to insure that the owl does not move between the acquisition of the two to three compass bearings; if it does, then a new series of bearings will be required to map its location.

Table 11.4 gives an overview of published survey methods displaying an array of the call types used, time of survey activities, playback sequences, number of visits across the survey routes, inter-station distance, and suggested periods for experiments. Important differences exist between the current methods; some groups use rather stringent responses (i.e., only "ghuck"-calls) to accept a calling Little Owl as a territorial bird (Bloem *et al.* 2001), while other methods accept all the different calls of the owls as responses (Verwaerde *et al.* 1999). Hardouin (2002) showed that only the call of the male has a territorial function, suggesting the use of the "ghuck" call during broadcasting efforts. These slightly different survey programs (Table 11.4) lead to differences in the data collected, frustrate interpretations, and limit the utility of the acquired data. Clearly, a shared goal of our programs is to assess and monitor the status and trends of Little Owl populations. A key step in achieving our shared goal will be based on standardizing the survey methods and associated terminology (see the glossary).

Field observations during Little Owl surveys have indicated that there appears to be a population density-based response involved with response rates (Exo & Hennes 1978; Pirovano & Galeotti 1999; Centili 2001b). That is, owls residing in high population densities appear to respond more readily to playback. This could be because of the additional commitment needed by resident owls to continually affirm their territorial boundaries in the face of increased competition for space and resources. More detailed field trials are urgently needed to determine response rates of Little Owls occurring at low, moderate, and high population densities.

Demographic data Standardization is needed for the description and measurement of demographic data across the breeding range. We consider the work by Bloem *et al.* (2001) an important step in this undertaking. We strongly encourage readers to examine Forsman *et al.* (1996), Franklin *et al.* (2004), and Ganey *et al.* (2004) for their particularly valuable

Reference	Calls	Timing	Sequence	Visits across survey routes	Locations	Inter-station distance	Seasonal period	Complementary methods
Exo & Hennes (1978)	8 territorial calls / 30 s.	30 min after sunset till midnight or 2 a.m. till sunrise	15 s. Broadcasting / 1 min. Silence / 30 s. Broadcasting / 1 min. Silence / 1 min. Broadcasting / 2–3 min. Silence.	1	potential habitats	300–400 m	end February till mid April	–
Bloem *et al*. (2001)	territorial calls	till 2 hours after sunset or 2 hours before sunrise	10 ghuck calls / 1 min. Silence. 10 ghuck calls / 1 min. Silence. 10 ghuck calls / 2–3 min. Silence. if no response 10 ghuck calls / 1 min. Silence. 10 ghuck calls / 1 min. Silence. 10 ghuck calls / 2–3 min. Silence. if no response 10 ghuck calls / 1 min. Silence.	3	potential habitats	250 m	mid February till mid April	visual observations at day
Verwaerde *et al*. (1999)	alarm / territorial / kiew calls	sunset till midnight or two hours before sunrise	78 s. Broadcasting / 1 min. Silence / 78 s. Broadcasting / 1 min. Silence / 78 s. Broadcasting / 5 min. Silence.	1	centre of 500 m UTM squares	500 m	mid February till mid March	–
Mastrorilli (1997) Mastrorilli (2001)	territorial alarm calls	one hour after sunset	3 min. silence / 3–4 min. Broadcasting / 3 min. Silence / 2–3 min. Broadcasting / 3 min. Silence.	5	a route selected with spot comprised in the study area, with spot inserted in function of the habitat	in function of available sites	January till December	visual observations at day

(*cont.*)

Table 11.4 *(cont.)*

Reference	Calls	Timing	Sequence	Visits across survey routes	Locations	Inter-station distance	Seasonal period	Complementary methods
Bretagnolles *et al.* (2001)	territorial calls	at sunset till midnight	1 min. Silence / 1 min. Broadcasting / 1 min. Silence.	2	potential habitats around villages	min. 500 m; preferably 750–1000 m	March and June	–
Van't Hoff (2001)	territorial calls	unknown	unknown	unknown	potential habitats	in function of available sites	unknown	visual observations at day
Lorgé (pers. comm.)	territorial calls	sunset till midnight	15 s ghuuk call, 30 s silence, 35 s ghuuk call, 60 s silence, 40 ghuuk call by pair and then waiting 5 min. for any response	1 – max 3 (if no reply)	potential habitats	300–400 m	mid February till mid April	yearly control of nestboxes
J. Navarro (pers. comm.)	territorial calls (of male owl)	2 h before sunset; 2 h after sunset; results indicated that nearly twice as many owls responded after sunset as before	Two broadcasting sequences of: 2 min. tape & 1 min. silence, and then an additional 2 min. tape & 1 min. silence.	5 visits (calling was done before and after sunset on these 5 occasions)	potential habitats. A total of 14 survey stations were used in this study.	500 m	19 April to 17 May 2002 – peak of vocal activity in SE Spain	surveys at sunset would be helpful when looking for nests, as owls could be observed responding to playback
Survey protocol for the Little Owl (Johnson *et al.* 2008) proposed standard	territorial calls	sunset till midnight and/ or two hours before sunrise	1 min. Silence / 2 min. Broadcasting / 1 min. Silence 1 min. Silence / 2 min. Broadcasting / 1 min. Silence 1 min. Silence / 2 min. Broadcasting / 3 min. Silence.	3 for general surveys; 4–6 for demographic monitoring (number needs more testing), 4 for density estimates of territorial pairs and singles	centre of 500 m UTM squares	500 m	February–April for Western Europe; March–May for Eastern Europe; March–May for Middle East & Asia	visual observations at day in low-density areas

contributions to the scientific assessment and analysis of demographic parameters in the three subspecies of spotted owls. In particular, the focus of demographic analyses, as based upon mark–recapture methods, was directed at adult and juvenile survival, fecundity, and population trends. The consistent collection and reporting of ringing and radio-telemetry data on juveniles and adults, especially in regards to dispersal, will be of particularly significant value in understanding this key aspect of population dynamics. For owl dispersal data, we strongly urge readers to examine the monograph by Eric Forsman *et al.* (2002), to better understand the compilation and analysis of substantial ringing and radio-telemetry data. Based upon the work of Van Nieuwenhuyse & Leysen (2001) the international standardization is also suggested for the analysis of habitat data (see Chapter 12) as this issue is closely associated with demographic performance.

Nestboxes will likely remain one of the main venues for gaining demographic data on Little Owls. Following the thoughtful work of Paul Marié in Soignies (Bultot *et al.* 2001), an experimental design should be applied when placing nestboxes in research areas. Marié installed nestboxes at regular intervals, thus minimizing aspects of habitat quality on the resulting aspects of box use. Further, this method allowed additional types of analyses in subsequent years. Even with volunteer-based activities, an experimental design of actions should be undertaken. We urge the development of standardized guidelines for the installation of nestboxes to aid in this effort.

Other aspects Further research is also needed into the better use of birds released from care centers as experimental samples to further improve the guidelines for release.

It should be stressed that there is a strong need for making data collection methods straightforward and clear. Simple but useful methods can help to narrow the "gap" between scientists and volunteers. Furthermore, simplified methods allow us to let larger groups of interested people and non-professional bird-watchers contribute to the conservation of the Little Owl. Optimal exploitation and publication of existing volunteer data should be undertaken along with statements that describe some of the methodological shortcomings. This will serve to make more apparent volunteer-based data, and to validate the important conservation efforts of these caring people.

Data management

The importance of good data management cannot be overstated. Future research and international co-ordination should emphasize the use of modern information technology (IT) methodologies, including storing and exploiting data using data warehousing techniques. To avoid the inefficient use of funds and data, and subsequent delays in critical conservation efforts, we recommend that all future Little Owl projects have mandatory standards for meta-data, data collection and storage, and a data exploitation steering committee. To use all information available throughout the complete distribution range, more literature overviews like those provided by Génot (2001), Mastrorilli (Italy), and J. Shergalin (former USSR) should be made and published on paper or via the Internet. An international Little Owl data warehouse is in the developmental stages. This should serve to support a complete overview of literature, data, and conservation insights, and provide easy access and distribution of the available knowledge. We believe that within-country data sets (especially in developing countries) are needed, or need to be upgraded. With careful and reasonable design, these upgraded data sets can support international efforts such as:

- *Global Biodiversity Information Facility* (GBIF): http://www.gbif.org/. The mission of the GBIF is to facilitate digitization and global dissemination of primary biodiversity data, so that people from all countries can benefit from the use of the information.
- *The Species Analyst Project* (Peterson *et al.* 2003): http://speciesanalyst.net/. The Species Analyst Project is a research project developing standards and software tools for access to the world's natural history collection and observation databases. The project has given rise to, or provides the technical infrastructure for, a number of other projects including: FishNet, MaNIS, HerpNET, Canadian Biodiversity Information Facility, and the Global Biodiversity Information Facility.

Measuring success: assessing the strategy

We need to determine if the overall Little Owl conservation strategy is being effective. Is it being implemented as we intended? Is it having the effects we desire for Little Owls? What changes should be made to make the program more effective? It is not our intention to create additional workloads for already overworked biologists and managers; however, conservation, like any serious profession, needs to highlight its products and demonstrate its accountability.

We propose the establishment of a Little Owl balanced scorecard (following Kaplan & Norton 1996) that contains the key performance indicators needed to obtain a holistic view of the status of the strategy. The accomplishments of work on Little Owls should be provided in annual summary reports. We recommend that a panel of science and management personnel be assigned the tasks of (1) reviewing the annual monitoring summary reports, and (2) preparing the interpretive scorecard report assessing progress in meeting the goals, objectives, and expected values for the Little Owl under the array of applicable legislation and national policies. Importantly, the scorecard report will profile the state of Little Owl populations and their habitat, evaluate the effectiveness of current conservation measures in arresting and reversing the decline in owl habitat and populations and maintaining the viability of owl populations, point out areas of progress and concern, and, as necessary, make recommendations on changes in management practices. This scorecard report will provide decision-makers with a scientifically credible evaluation of the state of Little Owl populations and habitat, and would be a reference document for decision-makers during periodic land-use planning reviews.

To fill key gaps in our collective knowledge, we would monitor the number of topics that are and will become available, the number of people working on them, and follow the number realized versus those needed. For breeding data we suggest keeping track of the number of owl sites (with each site having detailed latitude and longitude co-ordinates), tracking the actual associated breeding data, and that we offer comparisons of demographic and other results from the different study areas.

More specifically, components of a Little Owl balanced scorecard would reflect aspects of: knowledge, limiting factors, landscape conditions, legislation and policies, and people.

For example, concurrent with demographic data collection, range-wide landscape *conditions* would be monitored to track changes in conditions by using owl habitat maps derived from the national and international vegetation maps and imagery. This analysis would reflect analysis at several spatial scales (i.e., nest-site, home-range, landscape). Landscape components, such as area of pastureland, would be monitored on the basis of critical structural and management activities that reflect ecological conditions for the owl.

An evaluation of policies and legislation relevant to land-use practices that affect Little Owls should include: (1) country-specific evaluations, and (2) an international evaluation. Compliance with existing policies and legislation is part of this review process.

For people, part of the scorecard evaluation could be a status report on the current and anticipated infrastructure of people and programs working on Little Owl core studies and conservation projects. Implementation of the monitoring plan, outlined in Chapter 13, will require a co-ordinated international effort. The key to successfully implementing the monitoring program is a co-ordinated network of personnel and co-operators who will implement individual elements of the monitoring strategy. The annual population surveys, periodic habitat assessments, cumulative data analyses, and integrated syntheses of the individual monitoring efforts will implement the monitoring strategy as a whole.

Steps to accomplish the strategy are assigning specific tasks to an administrative body, gathering and analyzing the data through standardized methods (some of which await development), and implementing the monitoring program on schedule. We recommend that a Little Owl monitoring leader be assigned. This individual will work with international counterparts (such as representatives of the International Little Owl Working Group [ILOWG], BirdLife International, World Wide Fund for Nature [WWF], Global Owl Project, etc.) and oversee the organization and implementation of the monitoring plan. Fortuitously, the ILOWG encompasses personnel and co-operators enlisted to conduct the demographic studies, assess habitat conditions, and develop predictive models. The monitoring leader and associated program staff will require agency or non-governmental organization (NGO) support, with an estimated equivalent of two or three permanent, full-time positions among the participating agencies. In addition to agency support of co-ordination activities, support will be needed for permanent, full-time agency personnel engaged in implementing the plan. Assuming that demographic studies are continued under the current co-operative working relations and habitat assessments are done in-house, these positions would involve research scientists, research wildlife biologists, statisticians, and geographical information systems (GIS) specialists.

As part of this review process, the status of the implementing components of: monitoring the owl and its habitat, standardizing methodologies, data management, and measuring success: assessing the strategy will be reported. For example, is the monitoring plan (or a subsequent version) being adopted across the range of the owl? Is a standard protocol for demographic surveys and habitat assessments in place? Have within-country and range-wide results been presented? A meta-analysis of owl population data from the demographic study areas could be completed every five years beginning in 2010. Range-wide habitat maps would be recompiled every five to ten years in line with monitoring plan schedules.

11.4 Conservation projects

To address the alarming situation of declining Little Owl populations in some areas, conservation plans have been set up at an international level (Leigh 2001b), at national levels in Austria (Kirchberger 1988), France (Lecomte *et al.* 2001), the Netherlands (Plantinga 1999), Switzerland (Meisser in press), and at regional levels in Flanders (Van Nieuwenhuyse *et al.* 2001b), in Genève (Meisser, personal communication) and in Ajoie (Collectif d'associations & OEPN 2003), and in Groningen, the Netherlands (Van't Hoff 1999). In many countries, actions are underway to protect and manage Little Owl habitat, such as through the maintenance of pollarded willows (Loske 1978; Bultot 1996; Bultot *et al.* 2001), oak hedges (Meisser 1998; Meisser & Albrecht 2001), and orchards (Harbodt & Pauritsch 1987; Grimm 1989; Juillard 1997). Since 1970 (Schwarzenberg 1970), many nestboxes have been established partly to compensate for the lack of natural cavities and also for scientific study. To reduce collisions with vehicles on roads, special actions on perches along roads have been found to decrease mortality (Hernandez 1988). Optimization of existing territories by improvement of prey abundance and accessibility through the rotation of grassland management, and the placement of additional perches (similar to those for shrikes, *Laniidae*; Van Nieuwenhuyse *et al.* 1999) are suggested to help conserve remaining owl populations. To improve post-fledging survival, concrete or PVC tubes might be laid on the ground below or near breeding cavities. These tubes are easily adopted by fledglings as shelter and prevent them from getting trampled by cows (van Harxen & Stroeken 2003b).

Types of nestboxes Schwarzenberg (1970) developed the original artificial nestbox, which is basically a wooden pipe (Figure 11.14). Its use has spread widely in German-speaking countries. Such nestboxes are made of narrow planks, held together by glue. The length of the pipe is 80–100 cm, it has an inner diameter of about 18 cm, and the diameter of the entry-hole is 6.5–8.0 cm. As a rule, the upper part of the pipe somewhat juts out ahead, creating small naves over the entry-hole. In the rear part of the pipe a hole for checking and cleaning the nestbox is present. It is possible to make an analogous four-angle nestbox of boards, or to use a piece of an empty tree trunk. Usually nest-pipes are protected against penetration by Martens by installing two front walls at a distance of 7–8 cm with entry holes in different sides. On the side opposite to the entry-hole, ventilation holes with diameters of 1.5–2 cm are made. Predators are supposed to be unable to do two turns under the direct angle in such a narrow space, while the Little Owl has no entry problems. Pipes are attached above or below horizontal branches of trees or under the roofs of buildings. In the Czech Republic the species occupied plastic boxes shaped like cubes with sides being 25–30 cm

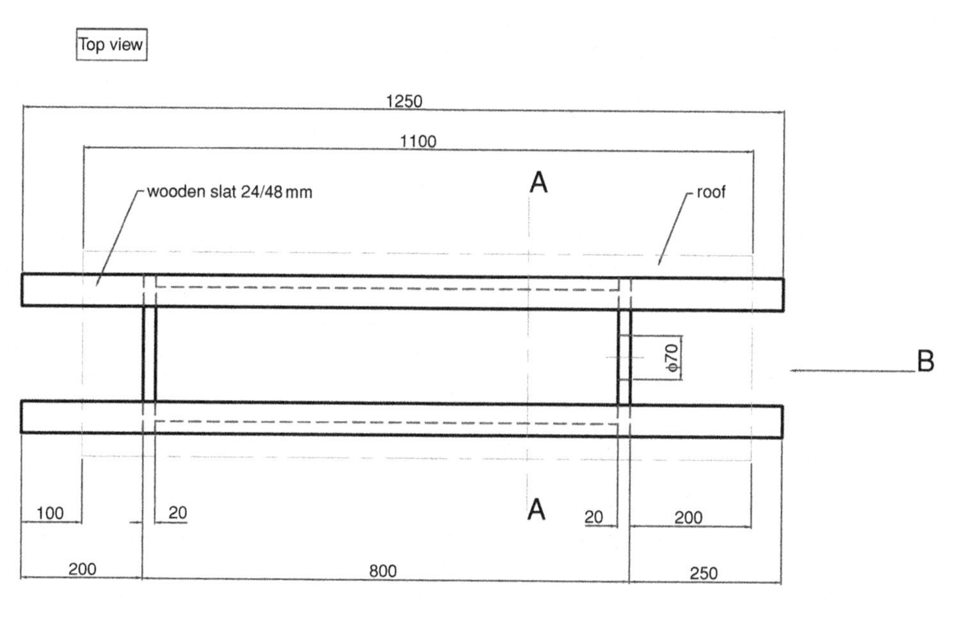

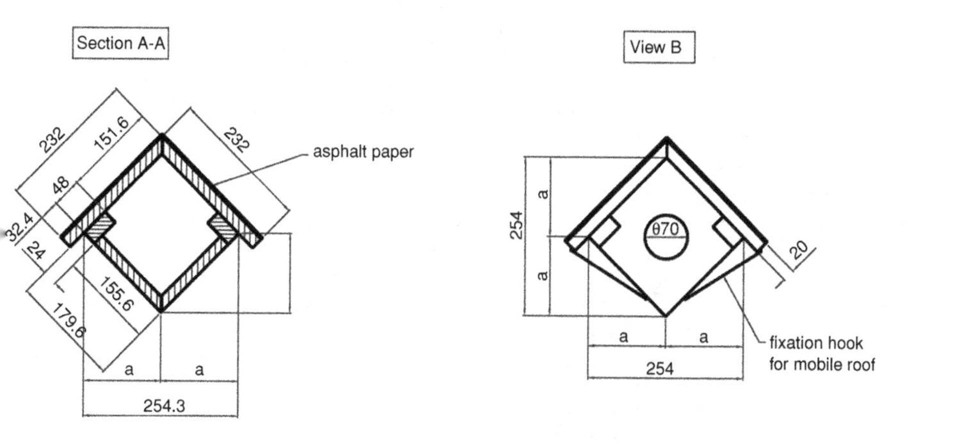

Figure 11.14 Schwarzenberg nestbox modified by Juillard (1984a).

(L. Oplustil, personal communication). Bultot (1990) developed nestboxes using recycled 12-bottle wine cases (Figure 11.15).

More and more nestboxes are equipped with either video or photographic tools to automatically monitor the visiting frequency and the prey brought in by the parent owls during the breeding season (Figure 11.16) (Juillard 1984a; Blache 2001; H. Keil, personal

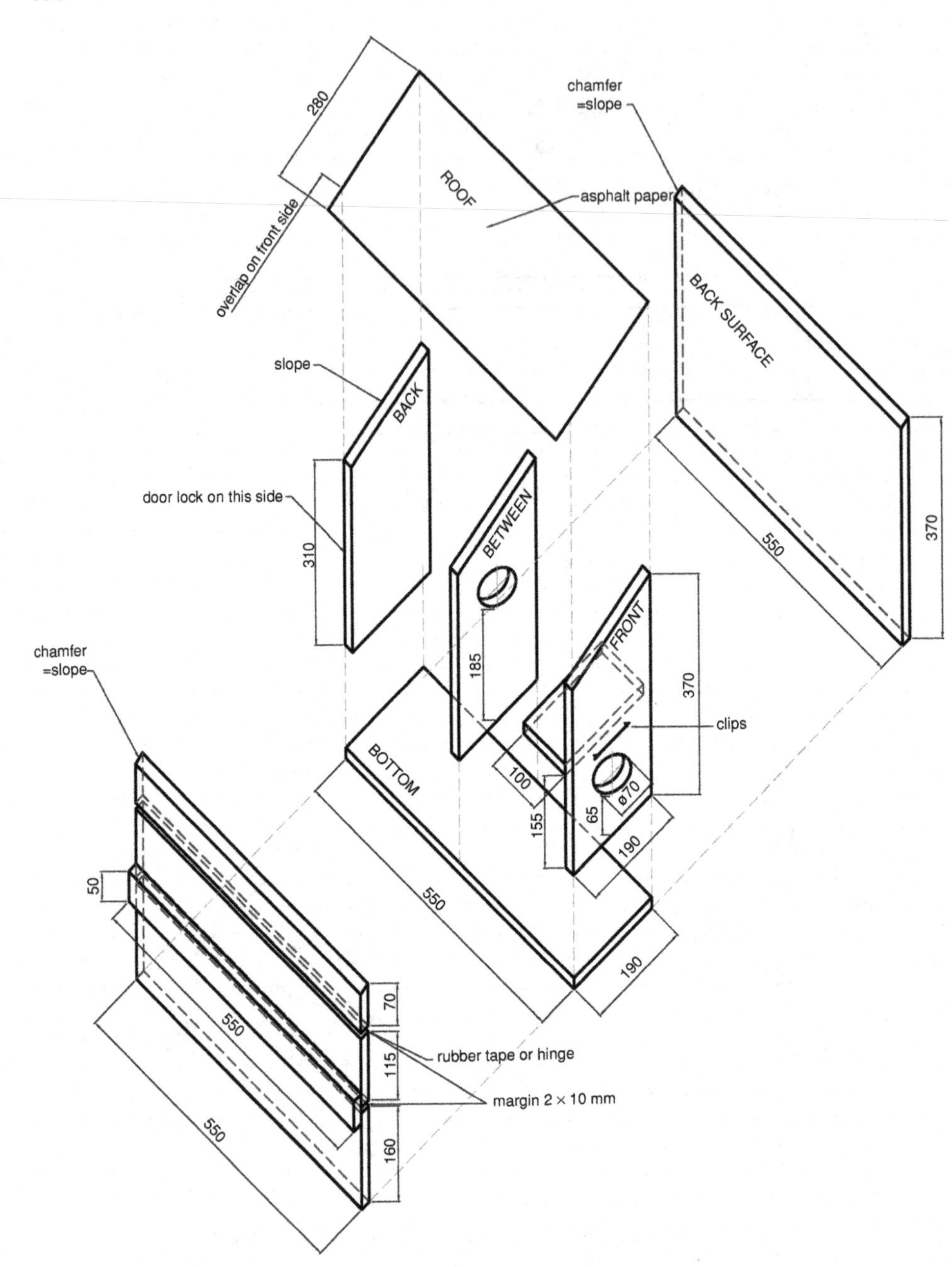

Figure 11.15 Bultot nestbox modified by Smets (Smets, personal communication).

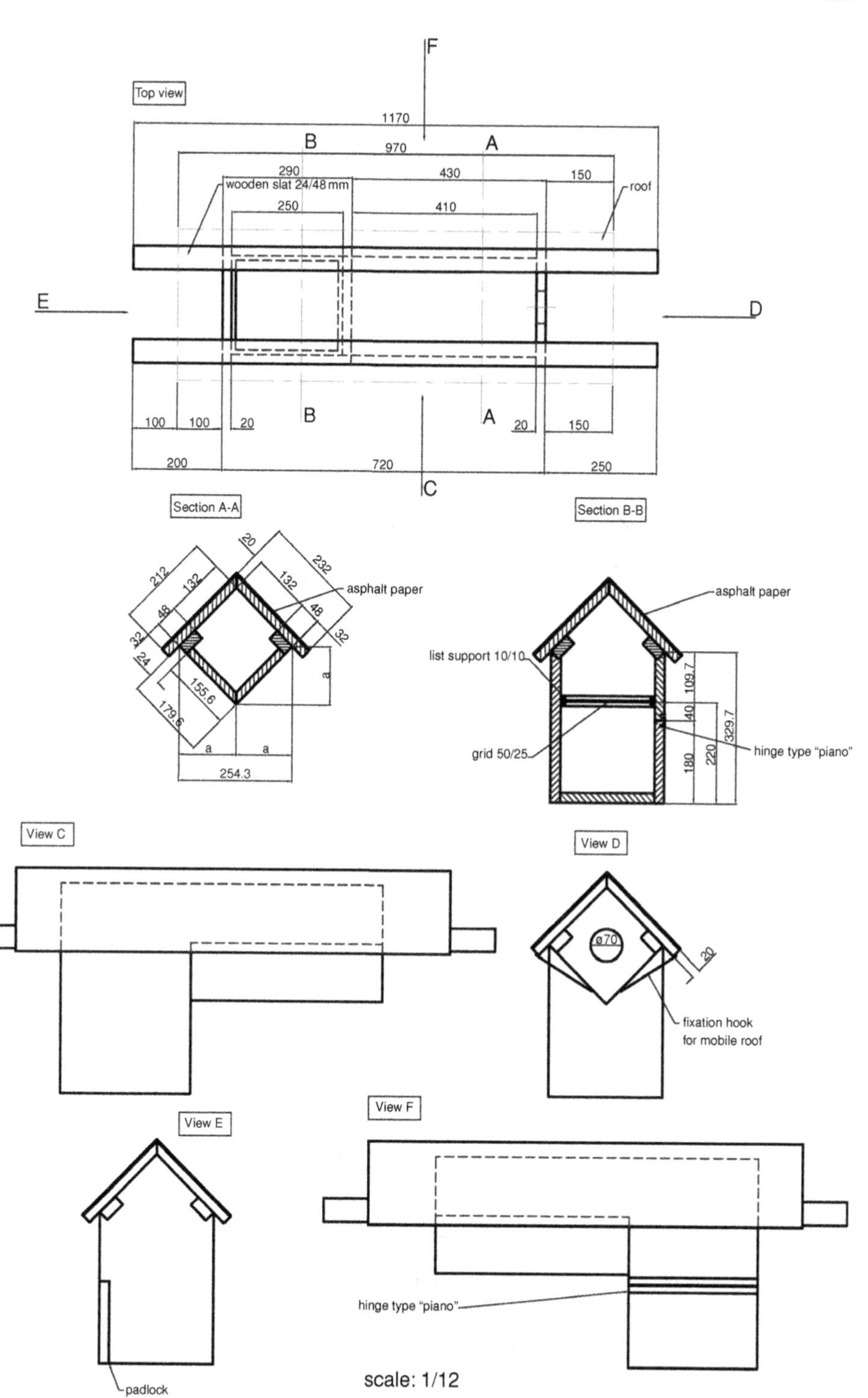

Figure 11.16 Juillard nestbox L-type by Juillard (1984a).

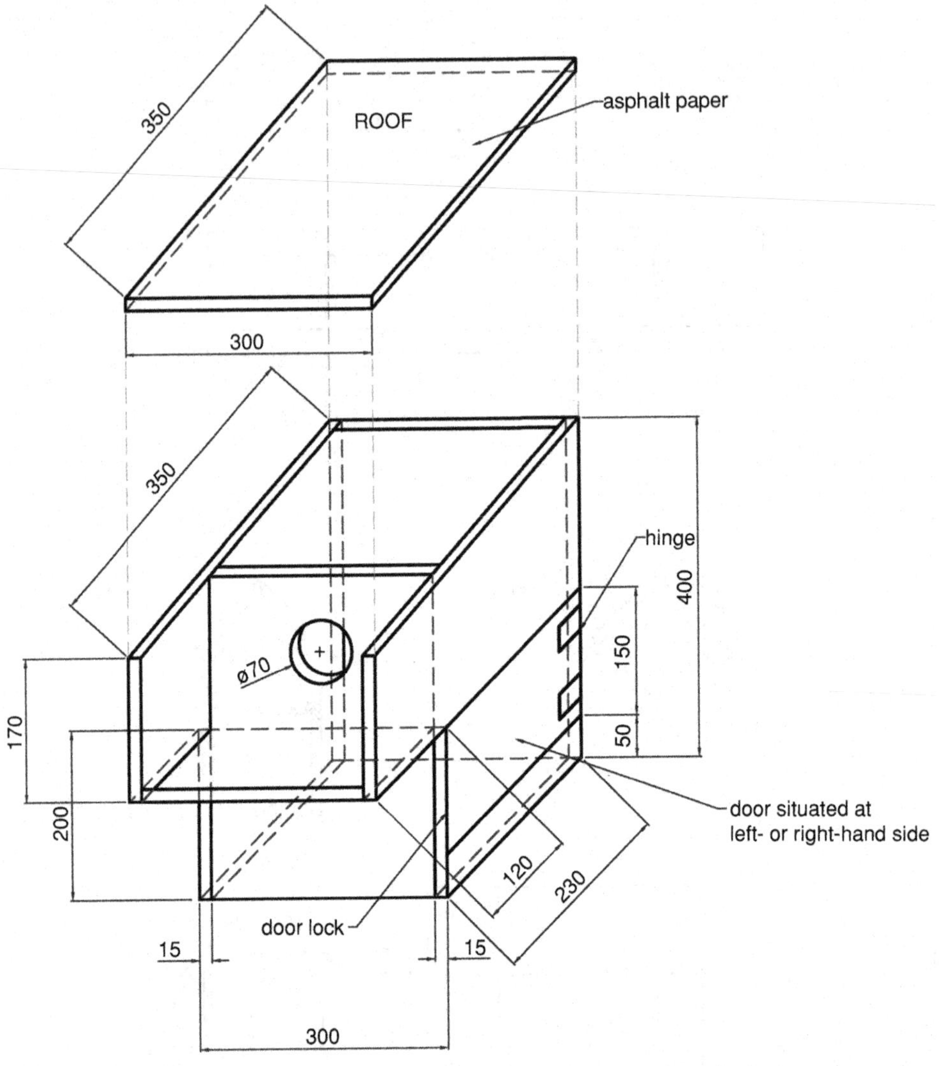

Figure 11.17 Meisser nestbox (C. Meisser, personal communication).

communication). The observed variability in the choice of nesting sites gives the possibility of development of the more diverse variants of artificial nestboxes (Figures 11.17, 11.18).

11.5 Reintroduction and supplementation

As the species has been disappearing or declining in many European regions, some experiences with reintroduction (where the species had been extirpated and owls were released) or supplementation (where the species was still present but extra owls were released to increase the population) have been documented (Table 11.5). These efforts took place in

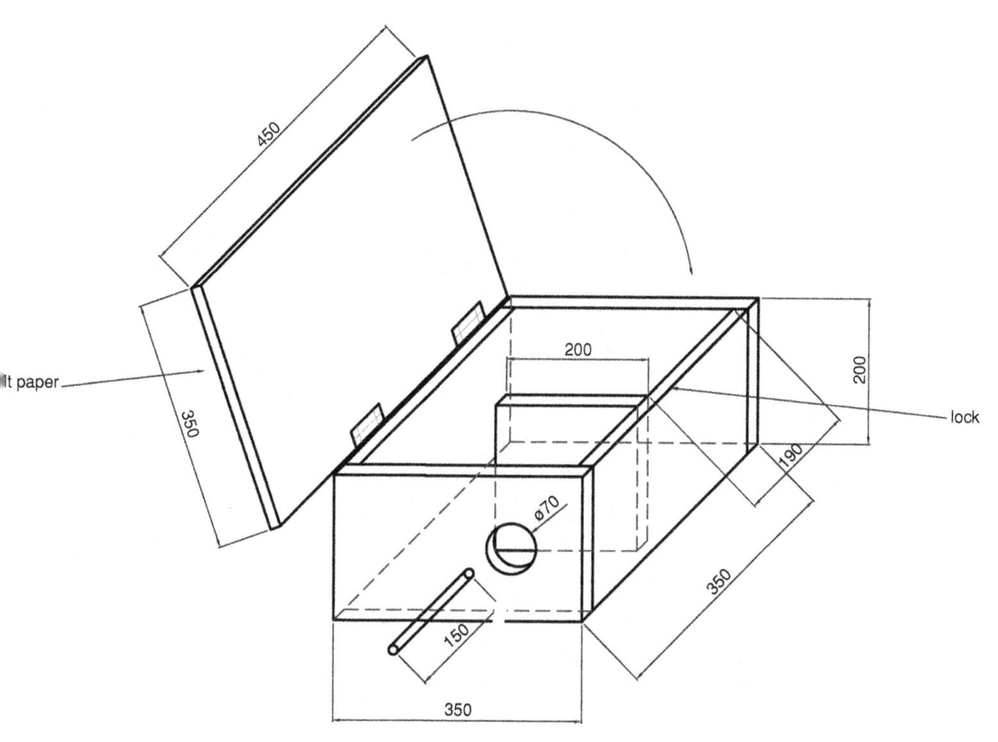

Figure 11.18 English nestbox by Burton, Glue and Johnson.

Germany in Hesse (Stahl 1982), in Jura Souabe (Mohr 1989), in the Coburg region of Bavaria (Leicht 1992), in Thuringen (Robiller 1993), in Lower-Saxony (Möller 1993), in Mecklenburg (Bönsel 2001), in south Lower-Saxony, particularly in the Göttingen region (Illner & Kartheuser, personal communication), in Brandeburg (C. Haase, personal communication), in Switzerland in the Bern region (Robin, personal communication) and in France in Alsace (Génot *et al.* 2000). All of these efforts required aviaries to produce the nestlings and to raise the juveniles in captivity. The attempts in Switzerland and in Göttingen failed.

The project in Jura Souabe (Mohr 1989) was started in 1990, with 15 Little Owls hatched and raised in captivity and then released into the wild. This experiment was a supplementation of a wild population that increased from 2 pairs in 1982 to 25 pairs in 1990 thanks to the use of nestboxes (Mohr, personal communication).

In Hesse the project was a reintroduction (Stahl 1982). The captively raised owls were released during September 1980 and 1981. They were placed in nestboxes with a metal balcony, the balcony helped the young birds to stay inside and acclimatize to their environment. In 1982, two released pairs bred in the wild, producing nine fledged young.

In Lower-Saxony (Möller 1993), a reintroduction project was started in 1989; five Little Owls were released in the first year and ten birds released every year thereafter. The young were released from an aviary containing young and adults from which only the young could

Table 11.5 *Different projects of Little Owl reinforcement and reintroduction in Europe. Breeding pair in nature means a pair containing at least one released owl.*

Number of Little Owls released	Year	Type of experience	Final result	Region	Country	Author
79	1983–2002	reintroduction	0 BP in nature	Harz	Germany	E. Kartheuser (pers. comm.)
102	1983–2002	reintroduction	–	Göttingen	Germany	F.-K. Schöttelndreier (pers. comm.)
266	1987–2002	reinforcement	4 BP in nature	Brandenburg	Germany	P. Haase (pers. comm.)
109	1978–90	reintroduction	0 BP in nature	Coburg	Germany	Leicht (1992)
15	1990	reinforcement	–	Jura Souabe	Germany	Mohr (1989)
–	1980–81	reintroduction	2 BP in nature	Hesse (Limburg)	Germany	Stahl (1982)
31	1987–90	reintroduction	–	Hesse (Fulda)	Germany	K. Burbach (pers. comm.)
25	1989–92	reintroduction	7 BP in nature	Lower Saxony	Germany	Möller (1993)
–	1988–99	reintroduction	3 BP in nature	Thüringen	Germany	F. & F. C. Robiller (pers. comm.)
171	1996–99	reintroduction	4 BP in nature	Mecklenburg	Germany	Bönsel (2001)
92	1982–86	reinforcement	O BP in nature	Bern-Freiburg	Switzerland	K. Robin (pers. comm.)
85	1993–2002	reinforcement	1 BP in nature	Alsace	France	Génot (2005)

BP = breeding pair; pers. comm. = personal communication

fly out. In 1992, seven pairs bred in the wild but with the low breeding success of 1.57 fledgings per breeding pair.

In Mecklenburg, the project was also a reintroduction (Bönsel 2001). A total of 42 Little Owls were released in 1996, 49 in 1997, 49 in 1998, and 31 in 1999. In 1999, seven territories were surveyed and four breeding pairs were found.

In the Coburg region where the Little Owl was extirpated (Leicht 1992), 109 young owls reared in captivity were released between 1978 and 1990 by differing methods. Some owls were released directly in the spring and in the autumn, others were reared through the adoption by wild owls of young born in an aviary, and were kept in contact with the parents in the aviary. But, ultimately, no breeding pairs were recorded in the region.

In Thüringen, in particular the Weimar region (Robiller & Robiller 1986, 1992) a reintroduction experiment was conducted from a complex of 16 aviaries built as veterinary equipment and to study Little Owl behavior. The species disappeared around 1970 from the Weimar area. Thanks to the few adults released from captivity and to the wild birds probably attracted by the birds in the aviary, two pairs bred in the wild and produced 15 fledglings between 1988 and 1992.

In the Göttingen region, from 1999–2003, 79 young Little Owls were ringed and released in an area of about 200–300 km^2 (E. Kartheuser, personal communication). In 2001, there were still no wild Little Owls forming a population, although some single birds stayed for some months in that region. Illner (personal communication) gives several reasons for these results:

- The habitat for Little Owls at the release area was too isolated from the densely populated areas in Westfalia (more than 100 km apart), and a wooded hill of some 600 m wide presented a potential barrier to immigrating owls.
- The climate of this area is cooler and there is on average much more snow than in Westfalia.
- The number of owls released was too low.
- The optimal habitats were too isolated from each other and it is not clear whether the available food and nesting sites were sufficient.

In Brandenburg, and in Haveland in particular, an experiment began in 1987 to supplement a declining population (Haase 1993). Between 1987 and 2002, a total of 266 birds were released from captivity: 90 in spring (March–April) and 176 in the autumn (August–September). The survival of the owls after release was better in autumn than in the spring. During this period, 9 birds were recorded as breeding in the wild and 15 died (of which 12 were killed and one was wounded by collisions with vehicles). Different methods for release were used:

- a mobile aviary was put in place and young owls stayed inside it for seven to ten days prior to their release; the owls were fed with live prey before the aviary was opened at twilight and was kept open for several days
- a small mobile cage was put in place for release and the young owls stayed inside it for 14–26 hours and were also fed, the cage was opened at twilight

Table 11.6 *Detail of the Little Owls released in Alsace (France).*

Number of released birds	1993	1994	1995	1996	1997	1998	1999	2000	2001	2002	Total
Nestboxes with balcony	5	–	6	4	2	–	–	–	–	–	17
Nestboxes	–	–	–	–	–	12	17	8	13	9	59
Complete wild nest	3	–	4	2	–	–	–	–	–	–	9
Total	**8**	**0**	**10**	**6**	**2**	**12**	**17**	**8**	**13**	**9**	**85**

- young owls born in captivity were put in wild clutches of the same age and not exceeding four to five young, and thereafter were raised by the wild adults. But this method was limited due to the low numbers of Little Owls in the wild.

In Sarre, 40 Little Owls were released between 2001 and 2003 and two females bred in 2003 (W. Stelzl, personal communication).

Most of these experiments had no real monitoring; we encourage observers involved with experimental studies to routinely provide status reports on these efforts, so that we can collectively gain from their specific insights.

Since 1993 a supplementation experiment has been undertaken in the Northern Vosges Biosphere Reserve (France). This supplementation took place in a poplation that was isolated due to habitat fragmentation, had weak breeding success, and in which the demographic viability had been analyzed (Letty *et al.* 2001). The aim was to avoid extinction due to inbreeding (Frankel & Soule 1981). This project was part of a program of global activities aimed at protecting the Little Owl, in the framework of the national conservation plan of the French Ministry of the Environment (Lecomte *et al.* 2001).

Genetic fingerprinting was carried out for several reasons:

- to analyze the existence or absence of genetic links between owls of different geographic regions in order to find out whether these birds could be cross-bred in captivity
- to establish the degree of relationship between the birds.

The analysis of owls from different geographic origins revealed that genetic variability was not unique for each region. Two unrelated birds from the same region may be more distant (genetically speaking) than two birds from different regions. Even if the sampling of nine individuals is restricted, the marked similarity coefficients coming from the owls of different localities translate the existence of an important genetic proximity between the different individuals studied. Such values thus reflect both the absence of great genetic diversity locally and the homogeneity of the similarity coefficients obtained.

Young born in captivity were released either into nestboxes, or placed in wild broods in order to complement them. A total of 85 young owls were released from 1993–2002 (Table 11.6) in an area where a population of about ten pairs was living. Of 35 birds fitted with transmitters from 1995–2001: 22 were found dead shortly after being released, 12 disappeared from the release site and have not been traced, and one was followed for five months before disappearing.

Young owls placed in wild broods were readily adopted by the parent birds. The birds born in captivity were of an equivalent age to those born in the wild. Extra prey was supplied to facilitate feeding. One young, born in captivity in 1996 and placed in a wild brood of only one chick to complement it, survived the very snowy winter of 1997. This male paired with a wild female, together they produced two young in 1997 and four in 1998 from a nestbox situated 600 m from the release site.

Despite staying for three to four weeks in balcony nestboxes, the Little Owls left them once the metal balcony was opened and did not come back later. This is the reason why birds were ultimately released from nestboxes having no balcony.

Among the known causes of death, the Little Owls found dead after being released in the wild were either victims of predators or were crushed by tractors. The mortality rate among the young born in captivity and released was high because the birds spend a lot of time on the ground.

Due to the difficulties in producing young in captivity and the logistic conditions of such experiments, the decision was taken to stop the supplementation and to wait for the results of a genetic study to find out more details about the structure of the population and to establish a demographic and genetic viability model (Génot & Sturm 2003).

Some movement experiments were carried out in the Netherlands by Jacobs (2003). Manipulation of eggs, young, and entire broods were undertaken with varying degrees of success. The manipulation of eggs (addition or replacement of eggs) yielded the most positive results, so long as the nesting place remained the same. Moving an entire clutch including the brooding female did not work. The success of adding one or more nestlings to another nest (after death of one of their parents) was most successful when the difference between the nestlings of both clutches was not too large and the nestlings were at least one week of age. Artificial feeding with dead mice, sparrows, and day-old chicks helped the birds. Displacement of an entire clutch of nestlings is too risky if the birds are less than ten days old.

Four young owls were released directly into nestboxes in North Holland in the summer of 2001. While this method probably results in higher mortality than aviary-raised owls, this method was chosen, and one pair and two males were released. The female died during the winter (killed on a road), but the three males stayed around the release site for one year. Finally the three males disappeared too, having dispersed or died (J. Van't Hoff, personal communication).

A supplementation or reintroduction program can only succeed if some key criteria are fulfilled:

- an analysis of the factors explaining the decline or the disappearance of the species in the study area should first be undertaken
- no wild owls should be used – only birds reared in captivity or taken from care centers (in Flanders up to 150 Little Owls are cared for and released per year [Rodts 1994])

- good habitats in the release area should be identified through a survey
- in the case of supplementation, that there is the ability of Little Owls in the release area to establish a connection to the existing population
- landowners, farmers, and other inhabitants living in the release area should be informed
- there should be good logistic conditions for rearing in captivity; for dimensions of aviaries see Stahl (1982), Robiller and Robiller (1992), Möller (1993), and Génot *et al.* (2000)
- radio-transmitters are used to monitor some owls after release
- enough owls are released per year (because of the high mortality rate)
- a conservation management plan for the habitat is carried out at the same time as the reintroduction or supplementation programs
- DNA fingerprinting analysis of the wild and released owls is carried out to know the genetic diversity of the former and the new population and its viability. The problem, however, it that nobody knows exactly what is the minimum viable population for Little Owls.

For rearing in captivity one needs to guarantee:

- a complete separation of the breeding pairs
- not to use owls that have been living in captivity for a long time due to the possibility of abnormal behavior
- to have an outside aviary – to allow the training of the Little Owl for catching prey
- to give living prey and different items (birds, rodents, earthworms, insects)

The release must happen after the owls are 80 days old, either during their normal dispersal period (August–September) or the following spring. There are no comparative studies from which to choose a method, but for the Tawny Owl the owls become less suited to release after longer periods of captivity compared to younger owls (Meyer-Holzapfel & Räber 1975; Meyer-Holzapfel & Räber *in* Schönn *et al.* 1991). The three main methods of release are as follows.

1. Young born in captivity are placed in wild broods (if their ages are the same); or sterile eggs in the wild are exchanged for fertile eggs from an aviary.
2. The young are placed with their parents in a special aviary, the same one used for rearing, and placed at the release site. The aviary is opened until the young leave the parents.
3. The young are released directly by placing them in nestboxes located in suitable habitats.

The released owls must be ringed. There is not enough knowledge to establish the optimal number of owls to release per year to be successful. But because of the high mortality among the young, we consider that a minimum of 20–30 individuals per year over a ten-year period is necessary with the latter method of release.

12
Research priorities

12.1 Chapter summary

The framework of this chapter was taken from Rich *et al.* (2004) who proposed a large-scale conservation plan for American landbirds. We complement this framework with some ideas of Noon and Franklin (2002) who have given an excellent overview of the current and required knowledge of the ecology of the Spotted Owl, *Strix occidentalis*. This chapter applies the given approaches and insights to the context of the Little Owl. See Figure 12.1.

Until now Little Owl research has been mostly descriptive. Research has proceeded in a logical, sequential fashion, beginning with descriptive studies of life history, food, behavior, and habitat ecology. Initial studies focused on characterizing the geographic distribution of the species and its general habitat associations. That work was followed by detailed investigations into patterns of distribution within its geographic range, specific habitat relationships within these areas, estimates of minimum area requirements, and the extent of geographic variation in these patterns. However, the mechanisms underlying the observed patterns of variation are mostly unknown. Causal models to better understand relationships and to seek mechanistic explanations for the patterns that have been observed (similar to Franklin *et al.* 2000 for the Northern Spotted Owl) need to be developed. New research should be applied, and should move away from descriptive, correlative and short-term work in small geographic areas, to large-scale replicated studies, controlled experiments, and long-term studies of demography (Donovan *et al.* 2002). Results that lead to concrete recommendations for habitat management are needed.

Many of the priority information needs that should be addressed by research are too specific to local or regional circumstances to be summarized at the level of the entire distribution range. These needs should be included in regional conservation plans and made available in a searchable database (similar to www.partnersinflight.org/pifneeds/searchform.cfm). There are, however, some common themes reflected by these regional research priorities, many of which are inter-related.

Identifying critical habitat components

General habitat associations of Little Owls are well known. However, we lack important information on specific structural features, landscape configurations, and amounts of habitat

Figure 12.1 Little Owl with prey in tree (François Génot).

that are required. Such information is critical for guiding the development of effective management strategies for meeting large-scale population objectives. For the Little Owl, this is an important challenge since different landscape elements might serve similar purposes for the species and some landscape features might have a different impact on habitat suitability at different concentrations. The insights that we obtain might be misleading since current densities might not be revealing real habitat quality (Van Horne 1983). This has been shown in the Netherlands where isolated high-quality habitats are becoming less favorable due to the degree of isolation, while clustered sub-optimal habitats remain occupied (Van't Hoff 2001). In France, Blache (2004) found the majority of occupied territories in exceptionally homogeneous habitats rather than mosaic-like landscapes, suggesting the occupation of sub-optimal habitat rather than optimal. Hence we will need to link habitat selection to demographic performance (i.e., survival and reproduction) to be sure that we are modeling for habitat quality and not purely for habitat occupation (Rosenzweig 1991). More research is also needed at multiple landscape scales. Finally we also need more insight into the selection process of natural habitats. The use of comparable satellite imagery should prove to be crucial for this.

Demographics

Measurement of demographic parameters (nest success, productivity, survival, dispersal) is needed to identify factors limiting populations and to contribute to understanding of meta-population dynamics (e.g., gene flow, source vs. sink populations). Until now, minimum viable population and meta-population studies have only been undertaken for a small population in the Northern Vosges (France) (Letty *et al.* 2001). Measurement of survival, particularly for juveniles as they move from their natal sites to eventually acquire their own territories and mates, is needed to both assess when Little Owls are most at risk and to identify sources of mortality. In the near term, there is a need to analyze existing multi-year data sets that are available among the many Little Owl volunteer working groups (e.g., R. van Harxen & P. Stroeken in the Netherlands; J. Bultot & Groupe Noctua in Wallonia, Belgium; Ph. Smets, R. Huybrechts & S. Cerulis in Flanders, Belgium). A first example has already been given by Bultot *et al.* (2001) where fruitful co-operation between volunteers and researchers yielded important insights. Age-specific estimates of survival and fecundity can be used to parameterize population metrics to project future population trajectories assuming constant vital rates (similar to Blakesley *et al.* 2001). Estimates of survival rates could be based on existing capture–recapture data, and fecundity rates could be estimated by determining the reproductive output of known-aged birds. There are examples of female owls whose lifetime reproductive success is now becoming available (Belgium: J. Bultot, personal communication; the Netherlands: R. van Harxen & P. Stroeken, personal communication). Since Hardouin (2002) showed that individual recognition of individuals is feasible through vocal recognition, new opportunities have arisen to better identify individual birds on a large scale without catching them. This is very promising for high-density regions where crucial factors that determine the distribution of the species might be further researched, e.g., spatial auto-correlation, social interactions, occurrence of floaters, and exchange between populations. There is currently enough historical data available to study the auto-correlative processes of the disappearance and expansion of population clusters.

Examining responses of Little Owl populations to land uses

There is a critical need to determine the effects of various types of land use on Little Owl populations in order to devise effective measures for minimizing the negative consequences of such land use. Land uses affecting owl populations include livestock grazing, silviculture, recreation, fire management, oil and gas development, mining, water control and development, agriculture, roads, suburbanization, communication towers, and wind-power development. Only by understanding the responses and tolerances of owls to land use and management regimes can effective mitigation actions be developed. Research should involve stakeholders from the beginning so that solutions will be accepted and used by those who control the land.

GIS and landscape modeling

More geographical information system (GIS) data and landscape modeling is needed to identify geographic focus areas for habitat protection, restoration, and management. We need to test whether the currently available digital maps contain enough variance to distinguish different habitat qualities in the longer term, without a need for additional and hence time-consuming data collection of independent variables. To make conservation of Little Owls sustainable, we need to focus on parsimonious models that explain the maximum amount of variance with a minimal amount of information. Identification of this limited number of crucial landscape variables to monitor is a priority. Storage of landscape descriptions in a GIS by national administrations allows us to take a snapshot of the actual historical status for later analysis without any considerable work for conservationists. To be sure we do not lose too much time collecting our own independent data, we need to acquire specific and easily available landscape data (Safriel 1995). This would go a long way in supporting long-term monitoring and conservation activities. A good example of the historical importance of regular measurements of the landscape was illustrated by Van't Hoff (2001). However, further analysis will be needed to determine if there is a need for additional information, and which variables should be included. Comparable data sources throughout Europe (e.g., the Corine land-cover database from the European Environment Agency, see www.eea.europa.eu/documents) will allow studies on an international scale within Europe and hopefully beyond the European borders in the future.

Examining the effects of abiotic environmental factors

More research is needed on the importance of abiotic factors on owl population regulation, including climate change, drought, and contaminants (acid deposition, pesticides). The special case for Little Owls is the abundant predation on earthworms and hence a possible accumulation of heavy metals or polychlorinated biphenyls (PCBs). Previous results are limited to habitats like the Rhine flood plains (Groen *et al.* 2000) that are not always representative of larger areas. Special attention needs to go to vermicide contamination of Little Owls.

Adoption of the information-theoretic approach

Being a plastic species, statistical analysis of Little Owl data might yield unexpected results through null-hypothesis testing, which could sometimes be considered as a "fishing expedition" hoping to hook a significant P-value (Noon & Franklin 2002). An information-theoretic approach avoids the numerous problems with using null-hypothesis testing and instead focuses on posing a set of multiple a-priori research hypotheses for a particular data set, and then ranking and weighting those hypotheses to find the most plausible hypothesis, given the data, using an objective criterion. Additional features of this approach include multi-model inference because it is unlikely that a single model will be uniquely supported by the data; the possibility of averaging effects across competing models also exists (Burnham & Anderson 1998). The link between statistical methods and the biology of the organism occurs through a set of a-priori hypotheses. These begin as verbal ideas of what factors may influence the observed system, which are then translated into statistical models (Franklin *et al.* 2001). See Figure 12.2.

Focus on process variation

There is a clear need for a mutual dependence between model formulation and prior biological knowledge (Noon & Franklin 2002). White (2000) emphasized three sources of variation in the dynamics of wild populations as important sources of variation effecting life-history traits: (1) temporal, (2) spatial, and (3) individual sources. Factors can be explicitly incorporated as covariates into a set of a-priori statistical models that seek to explain variation in demographic rates of possible causal factors in order to address the variation of population processes. This requires partitioning the variance of the dependent variable into its sampling and process components (White 2000). By removing variation attributable to the sampling process, subsequent statistical modeling focuses exclusively on the process component of variation. The presence of spatial auto-correlation has already been shown by Van Nieuwenhuyse and Bekaert (2002) in the Little Owl. Elimination of this auto-correlation on Little Owl presence in habitat selection studies, should yield statistically more relevant results in terms of population processes. Spatial auto-correlation in the independent variables furthermore needs to be studied to assess its impact on Little Owl models.

Search for mechanisms

Morrison (2001) criticized studies on wildlife habitat use and selection for being mostly descriptive in nature and concentrating solely on identifying patterns of use, rather than the causes of those patterns. Mechanistic explanations for habitat use in Little Owls is still

Figure 12.2 Adult Little Owl at nest site. Photo Bruno d'Amicis. See Plate 37.

missing, making it difficult to understand why certain components must be managed in a given manner. In addition, habitat is rarely considered in terms of quality (Van Horne 1983) where variation in demographic traits, such as survival and reproduction, may be explained by different habitat components. Little Owl research needs to strive for mechanistic explanations of why certain habitat components separate high-quality from low-quality habitats by combining habitat selection studies with demographic studies. Establishing real causes can either be done through experimentation or through structural equation modeling (Pugesek 2003).

The Little Owl is a perfect species for experimental research due to its extraordinary ease of manipulation, its relative commonness in human-dominated landscapes, and its availability of individuals from bird-care centers (e.g., in Italy, M. Mastrorilli, personal communication; in Belgium, own observations). Manipulation of occupied habitats by adding or removing perches, by adopting rotation mowing of haylands, and adding additional daytime roosts, might influence the hunting process of the owls. The influence of the manipulations on the home ranges and the reproductive process could be examined in a process similar to shrikes (Yosef 1993b). Artificial feeding experiments could allow the calibration of the energetic balance for specific prey items similar to the Red-backed Shrike (Carlson 1985). Density-dependence studies could use experiments by adding new nesting cavities and even by adding some extra birds to the population under study. Hardouin (2002) was one of the few to undertake manipulative research using playback of recorded Little Owls at other locations than normal to test for the individual recognition of birds and for territorial aggression. Structural equation modeling can be used to test causality (after Pugesek 2003).

Long-term observational studies

Observational studies serve three useful purposes: (1) providing clear and explicit hypotheses that can be further examined with experiments, either as large-scale manipulative experiments or as smaller-scale experiments that test key predictions posed by a larger question, and (2) estimating variation in demographic parameters, such as climate, that may influence that variation, and (3) long-term observational studies are required to capture sufficient environmental variation and examine hypotheses on the large-scale effects of that variation on Little Owl populations (e.g., Noon & Franklin 2002).

Cumulative effects

Because owl populations are affected by multiple factors, understanding the cumulative effects of these factors is critical to all future management strategies. Design of experiment (DOE) techniques and statistical interaction effects hence will need to be used to cancel out or better quantify specific combinations.

Investigating interactions between Little Owls and other flora and fauna

We need to understand the relative importance of disease, predation, nest parasitism, and introduced species. Effects may be magnified by land use and abiotic factors, so these should not be studied in isolation. Also, there is significant conservation value to be achieved in co-ordinating Little Owl research efforts with those of other agro-pastoral species. There can be larger implications of this work, with connections being made to efforts such as the Millennium Ecosystem Assessment (Millennium Ecosystem Assessment Board 2005).

Combining research and management

Owl conservation plans are built upon information about the ecological and environmental factors affecting bird populations that is inadequate for many species. Research should be combined with ongoing management to evaluate assumptions and contribute new information for revision and improvement of those plans (adaptive management). Combining research and management also is fundamental to testing the effects of management action on bird population response. Experimental research within agricultural landscapes is fairly easy for Little Owls and should be expanded.

Improving monitoring

There is need for research on monitoring methods, dispersal assessments, and analysis procedures. In particular, there is a pressing need for the detailed and rigorous testing of the response rates of Little Owls surveyed through the use of tape playback. This is especially needed in areas of low, moderate, and high population densities. This information, once obtained, should quickly be incorporated into subsequent peer-reviewed and published versions of the Survey Protocol for the Little Owl (Johnson, Van Nieuwenhuyse & Génot 2008). We strongly urge additional research on juvenile dispersal – from natal grounds to ultimate paired territory-holding status. This will involve radio-telemetry on relatively large numbers of owls, but this will be the primary way that this crucial information on dispersal, survival, and inter-population connectivity will be acquired. Data about demographic trends in survival, reproduction, and annual rate of population change will need to be summarized for each of the vital sign demographic study areas (see Chapter 13). A comprehensive meta-analysis of all data sets (for example, Franklin *et al.* 2004) will need to be undertaken, with specific analyses identifying estimates of adult survival, fecundity, juvenile survival, and rates of population change. This will require a new level of data organization and co-ordination for researchers and managers working on the Little Owl. The development of the first range-wide habitat map is technically within reach, but still no easy task. Detailing the specifics of this satellite imagery-based mapping process will help to ensure that future updates will be accomplished every five to ten years in synchrony with monitoring plan schedules. Finally, formal connections to major policies and legislation need to be developed, with Little Owl demographic and habitat data providing a key set of measurement tools (*metrics*) for evaluating and supporting such agreements as the Convention on Biological Diversity (http://www.sdinfo.gc.ca/docs/en/biodiversity/Default.cfm).

Expansion of investigated geographic range

Few publications deal with Little Owl populations occurring in natural habitats. While western Europe will probably remain the scientific nucleus for the species, substantial efforts will need to be undertaken in the Middle East and Asian portions of the owl's range. For some areas, basic distributional and ecological studies are still needed for the owl. An asset to help more rapidly bring information and insights to these areas can be offered through the enlargement of the International Little Owl Working Group (ILOWG). The first steps towards this is shown by active participation of members from the former Soviet Union and China in the ILOWG. There is a need for additional DNA studies to clarify some subspecies-level classification of taxonomy of the Little Owl; this would involve some relatively low-cost projects in the Middle East and Asian portions of the owl's range. Funding and institutional support are, of course, the foundation for ensuring that the necessary research is undertaken. Providing adequate resources will require co-operation and collaboration among management agencies, research facilities, industry, and non-governmental organizations, all of which have a role to play in support of Little Owl research. See Figure 12.3.

Figure 12.3 Little Owl adopting upright position. Photo Vildaphoto Ludo Goossens.

13

Monitoring plan for the Little Owl

13.1 Chapter summary

In this chapter, we offer an outline and specifics for the range-wide monitoring plan for the Little Owl. See Figure 13.1. The Little Owl is distributed within 84 countries from the UK to China. We propose a network of 30 vital sign demographic monitoring areas where mark–recapture studies are employed. These "Little 30" demographic study areas are located within 20 countries across the range of the Little Owl. Importantly, we recommend that these study areas overlap and be integrated with monitoring efforts in places with other natural resource values (e.g., Important Bird Areas, UNESCO Biosphere Reserves, National Parks, Natura 2000 sites, etc.). Within this framework, we describe a program for monitoring the long-term status and trends of Little Owls to evaluate the success of various plans to arrest downward population trends, and in maintaining and restoring the habitat conditions necessary to support viable owl populations. We describe how population and habitat data from the demographic studies would be integrated in individual population and meta-population analysis. Finally, we present a process to report status and trend results as a reference document for decision-makers during their periodic land-use plans and policy reviews. For countries without vital sign demographic study areas, we recommend baseline distributional surveys and ecological studies. This monitoring program is designed to produce results directly applicable to the articles and details of the Convention on Biological Diversity (http://www.sdinfo.gc.ca/docs/en/biodiversity/Default.cfm).

13.2 Background

The overall goal of natural resource monitoring is to develop scientifically sound information on the current status and long-term trends in the composition, structure, and function of ecosystems, and to determine how well current management practices are sustaining those ecosystems. Use of vital sign monitoring information will increase confidence in managers' decisions and improve their ability to manage resources, and will allow managers to confront and mitigate threats to those natural resources and operate more effectively in legal and political arenas. To be effective, the monitoring program must be relevant to current management issues as well as anticipate future issues based on current and potential threats

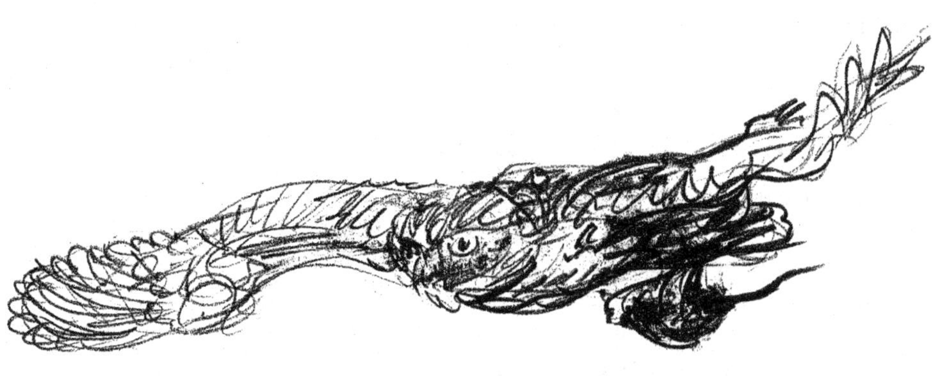

Figure 13.1 Little Owl in flight with prey (François Génot).

to the resources. The program must be scientifically credible, produce data of known quality that are accessible to managers and researchers in a timely manner, and be linked explicitly to management decision-making processes (e.g., Lint *et al.* 1999; Mulder *et al.* 1999; Greenwood & Carter 2003; Noon 2003).

A survey (or inventory) is an extensive point-in-time effort to determine the location or condition of a resource, including the presence, class, distribution, and status of plants, animals, and abiotic components such as water, soils, landforms, and climate. Surveys provide information on the distribution, abundance, habitat associations, and primary ecological aspects of species – information required for developing sound conservation strategies. Monitoring differs from surveys or inventories in adding the dimension of time, and the general purpose of monitoring programs is to illuminate population trends, identify specific aspects needed for conservation action, and evaluate the effectiveness of current management programs. Survey protocols and monitoring programs are necessary to ensure that changes detected by fieldworkers actually are occurring in nature and not simply a result of measurements taken by different people or in slightly different ways.

Elzinga *et al.* (1998) defined monitoring as "The collection and analysis of repeated observations or measurements to evaluate changes in condition and progress toward meeting a management objective." Monitoring objectives should include six components to be complete (Elzinga *et al.* 1998:41): the indicator or "vital sign" to be monitored, the location or geographical area, the attribute of the indicator to be measured (e.g., population size, density, percentage cover), the intended management action (increase, decrease, maintain), the measurable state or degree of change for the attribute, and the time frame. Monitoring data are most useful when the same methods are used to collect data at the same locations over a long time period (e.g., more than 10–12 years). It is important to note that cause-and-effect relationships usually cannot be demonstrated with monitoring data, but monitoring data might suggest a cause-and-effect relationship that can then be investigated with a research study. The key points in the definition of monitoring are that: (1) the same methods are used to take measurements over time; (2) monitoring is done for a specific purpose, usually to determine progress towards a management objective; and (3) some action will be taken based on the results, even if the action is to maintain the current management.

13.3 Overview of monitoring programs and demographic studies on owls

Long-term demographic monitoring programs have been established for a relatively few owls. While some of these monitoring programs are explicitly tied to recovery plans or conservation strategies, others are not. Importantly, though, these studies have advanced the science of conservation biology, have been influential in affecting the land-use changes, and have provided some fundamental information on the ecology and conservation of owls and their habitats. Further, these studies have provided an important springboard in furthering the science of wildlife species, and have offered important insights into the broader health of the ecosystems these owls inhabit. A listing of monitoring programs for owls around the world is given in Table 13.1. For our purposes here, we list only those programs that are directly involved with long-term (e.g., longer than five years) detailed demographic studies on owl populations.

The primary objective of demographic studies is to detect trends in the vital rates of the species. More specifically, demographic analyses employ mark–recapture studies, typically on large identified areas, where repeated surveys are conducted to locate, mark, and re-observe or recapture pairs of owls and their offspring. Some studies also include radio-telemetry to examine dispersal of juvenile and adult owls. As is frequently the case, demographic studies often begin through projects that are designed to illuminate the basic ecology of the species, such as estimating owl numbers and densities, and identifying habitat relationships. With demographic studies, the emphasis of such projects shifts to estimating trends in reproduction and survival. The types of questions that drive demographic studies include: “How many owls are needed to maintain viable populations?”, “Are the populations really declining because of habitat loss?”, “Are individual populations isolated from other populations?”, and “How is the population (or meta-population) responding to our management efforts?” Demographic information and estimates of habitat loss are often central issues in litigation and procedural arguments levied by competing advocacy groups (e.g., environmental and industry groups). In some situations, analyses of demographic data have been used to demonstrate that proposed (or current) conservation plans are insufficient to provide for the long-term welfare of the species. Finally, the science behind species and ecosystem management, of which demographic analyses are an important part, has become a biopolitical cornerstone in national and international conservation policy. See Figure 13.2.

Alongside demographic data come a large number of mathematical population models that have been designed to investigate the hypothetical responses of owl populations to different kinds of landscape management. The distinction between population modeling and population analysis is that the former is an abstraction and the latter is an objective assessment of empirical data. Population models are constructed by depicting,

Table 13.1 *A sample of monitoring programs that have a primary focus on gathering demographic data for owls around the world.*

Owl(s)	Location	Source for program information
Northern Spotted Owl (*Strix occidentalis caurina*)	western United States (Washington, Oregon, California)	Northern Spotted Owl population data from 14 demographic study areas within Washington, Oregon, and California. Focus: survival, reproduction, color ringing. See http://www.reo.gov/monitoring/nso/index.htm. See Forsman *et al*. 1996. These 1996 findings are updated in a report at: www.reo.gov/monitoring/trends/Final%20Demographic%20report%20Q%20and%20A%20.htm
California Spotted Owl (*Strix occidentalis occidentalis*)	western United States (California)	Demographic study areas; survival, reproduction, color ringing. See Franklin *et al*. 2004. For a technical assessment see Verner *et al*. 1992.
Mexican Spotted Owl (*Strix occidentalis lucida*)	southwestern United States	Demographic study areas; survival, reproduction, color ringing. See Ganey *et al*. 2004. http://ifw2es.fws.gov/Documents/R2ES/MSO_Recovery_Plan.pdf
Ten species of owls: Eurasian Eagle Owl, Snowy Owl, Hawk-owl, Eurasian Pygmy-owl, Tawny Owl, Ural Owl, Great Gray Owl, Long-eared Owl, Short-eared Owl, Tengmalm's Owl.	Finland	Ringing effort conducted by hundreds of volunteers, data co-ordinated by the Finnish ringing center. See Saurola 2002, 2006.
Ural Owl and Tawny Owl	Finland	Pertti Saurola has been studying the demographics of Ural and Tawny Owls since 1965. Ringing; mark–recapture. See Saurola 1989.
Barn Owl (*Tyto alba*)	UK	Nest monitoring. British Trust for Ornithology, see the Barn Owl monitoring program at: http://www.bto.org/survey/bomp/ index.htm Also see the integrated population monitoring program at: http://www.bto.org/survey/ipm.htm

(*cont.*)

Table 13.1 *(cont.)*

Owl(s)	Location	Source for program information
Barn Owl (*Tyto alba*)	Scotland	20+ years of nest monitoring; ringing. See Taylor 1994, 2002.
Great Horned Owl (*Bubo virginianus*)	Yukon, Canada	Telemetry, playback. Research study conducted from 1988–93; see Rohner 1997. See also Houston & Francis 1995 (ringing recoveries).
Christmas Island Hawk-Owl (*Ninox natalis*)	Christmas Island (off of Australia)	See Hill & Lill 1998. See also the (Australian) national recovery plan for the Christmas Island Hawk-Owl, *Ninox natalis*. Nest monitoring. http://www.deh.gov.au/biodiversity/threatened/publications/recovery/n-natalis/recovery.html
Great Gray Owl (*Strix nebulosa*)	Manitoba, Canada	25 years of ringing; mark–recapture; radio-telemetry. See Duncan & Hayward 1995.
Lanyu Scops Owl (*Otus elegans botelensis*)	Lanyu Island (a 46 km^2 island southeast of Taiwan)	Studies began in 1986; since 1990 owl population has been surveyed by census and playback counts; color ringing. See Severinghaus 2002.
Tengmalm's Owl (*Aegolius funerus*)	Finland	Long-term research studies (ringing) conducted by Erkki Korpimäki, students, and associates. See Korpimäki 1981. See also http://users.utu.fi/ekorpi/index.htm
Multiple owl species	Germany	The working group for the conservation of threatened owls ("AG Eulen") is an association of German-speaking owl experts. Monitoring program based on ringing and nestbox programs. See http://www.ageulen.de/
Multiple owl and raptor species	Europe (although mostly Germany based)	Monitoring program that follows breeding pairs through time. The database contains records and reproduction figures for breeding pairs of 17 species. See Mammen & Stubbe 2001. See also http://www.greifvogelmonitoring.uni-halle.de/

Figure 13.2 Telemetry (François Génot).

nathematically, the characteristics of a population and then examining the hypothetical pop-
ılation behavior under a variety of assumptions. They can be simple or complex depending
on the number of parameters used and their purpose. On the other hand, population analyses are based on objective evaluation of the life-history information of the owl derived by
capturing, marking, and re-sighting the same birds over long periods of time. It is this latter
ıspect of data for population analyses that our Little Owl monitoring plan is focused on
ıcquiring.

In addition to these formalized demographic studies, there are a number of additional
ecovery plans, conservation plans, and conservation assessments for other owl species. We

offer a sample of these, as population data on these owls have been instrumental in advancing conservation and land management issues. The Powerful, Sooty, Masked and Barking Owls have all been extremely influential in affecting the land-use changes in Australia over the past ten or more years. For example, in New South Wales alone, between 1995 and 2003, more than 1.1 million hectares of state forest were transferred to national park status – and habitat models developed for these owls played a big part in the process (R. Kavanagh personal communication).

Barn Owl (*Tyto alba*) – draft recovery plan – Ontario, Canada: http://www.bsc-eoc.org/download/baowplan.pdf
Burrowing Owl (*Speotyto cunicularia*) – Canadian recovery plan: http://www. speciesatrisk.gc.ca/publications/plans/Burrowing_Owl_e.cfm

Burrowing Owl (*Speotyto cunicularia*) – conservation assessment, US-based, 2004: http://www.fs.fed.us/r2/projects/scp/assessments/burrowingowl.pdf

Cactus Ferruginous Pygmy-owl (*Glaucidium brasilianum cactorum*) – US-based, 2003. Draft Recovery Plan 2003. US Fish and Wildlife Service. 2003. Cactus Ferruginous Pygmy-owl (*Glaucidium brasilianum cactorum*) Draft recovery plan. Albuquerque, NM. 164 pp. plus appendices: http://arizonaes.fws.gov/Documents/DocumentsBySpecies/ CactusFerruginousPygmyOwl/DRAFT%20CFPO%20RECOVERY% 20PLAN%20January%202003.pdf
http://arizonaes.fws.gov/Documents/DocumentsBySpecies/ CactusFerruginousPygmyOwl/Appendices.pdf

Barking Owl (*Ninox connivens*) – draft recovery plan, New South Wales (Australia): http://nationalparks.nsw.gov.au/PDFs/recoveryplan_draft_barking_owl.pdf

Southern Barking Owl (*Ninox connivens connivens*) – summary: http://www.deh.gov.au/biodiversity/threatened/action/birds2000/pubs/ barking-owl.pdf

Powerful Owl (*Ninox strenua*) – summary: http://www.deh.gov.au/biodiversity/threatened/action/birds2000/pubs/powerful-owl.pdf

Masked Owl – summaries

Northern Masked Owl (*Tyto novaehollandiae kimberli*): http://www.deh.gov.au/biodiversity/threatened/action/birds2000/pubs/ masked-owl-n.pdf

Tasmanian Masked Owl (*Tyto novaehollandiae castanops*): http://www.deh.gov.au/biodiversity/threatened/action/birds2000/pubs/masked-owl-tas.pdf

Northern Spotted Owl (*Strix occidentalis caurina*) – British Columbia, Canada. Population assessment of the Northern Spotted Owl in British Columbia 1992–2001: http://wlapwww.gov.bc.ca/wld/documents/spowtrend_1992_2001.pdf

13.4 Goals of the Little Owl monitoring plan

The purpose of the Little Owl monitoring plan is to assess trends in owl populations and habitat. Monitoring data will be used to evaluate the success of the various conservation measures in arresting the downward trends in some owl populations and in assessing habitat conditions necessary to support viable owl populations throughout the range of the owl. See Figure 13.3.

Figure 13.3 Little Owl adult. Photo Bruno d'Amicis. See Plate 38.

Specific goals of the Little Owl monitoring plan are to:

- identify status and trends in the health of the Little Owl population and its key habitat components
- provide early warning of situations that require intervention
- suggest remedial treatments and frame research hypotheses
- determine compliance with laws and regulations
- provide reference data for comparison with more disturbed sites
- provide scientifically rigorous data to enable managers to make better informed management and policy decisions.

Specific population-related and habitat monitoring questions that direct the plan are as follows.

Population questions

Is implementation of the various conservation measures and policies effective in reversing the declining population trend and maintaining the historical geographic distribution of the Little Owl?

a. What is the trend in rates of demographic performance (adult survival, reproduction, turnover, and the annual rate of change of owl populations)? Do these trends support a conclusion that the conservation measures are working to achieve a stable or increasing population?
b. Can the status and trends in Little Owl abundance and demographic performance be inferred from the distribution and abundance of habitat?
   (i) Can the relation between owl occurrence and demographic performance be reliably predicted given a set of habitat characteristics at the landscape scale?
   (ii) How well do habitat-based models predict occurrence and demographic performance of owls in different land allocations (that is, parks, reserves, and matrix lands)?

Habitat questions

The habitat monitoring portion of the monitoring plan is focused on the following question.

Is Little Owl habitat being maintained and restored as prescribed under the various conventions, plans, and policies? This general question has two key components:

a. What is the trend in amount and changes in distribution of habitat? Questions relevant to specific parameters and how they are changing through time, include:
   (i) What is the structure and composition of habitat at a variety of spatial scales (nest site, home-range, landscape, provincial population)?
   (ii) What proportion of the total landscape is owl habitat?
   (iii) What is the distribution of sizes of habitat patches?
   (iv) What is the distribution of distances (connectivity) among habitat patches?
   (v) What are the primary factors leading to loss and fragmentation of habitat?
b. What is the trend in the habitat that provides dispersal connectivity between populations? Questions relevant to specific parameters and how they are changing through time include:
   (i) What is the structure and composition of dispersal habitat at the home-range and landscape scales?
   (ii) What proportion of the landscape represents dispersal habitat? How is it changing through time?
   (iii) What are the primary factors leading to dispersal habitat changes at the home-range, population, and landscape scales?

A number of Little Owl study areas currently exist, and some of these are suitable to become part of a network of formal demographic study areas (see Sampling design below). Most of the current study areas are in European countries; study areas in other portions of the owl's range would need to be developed. In this plan, the ongoing population demographic monitoring would continue in selected areas. Data about the abundance and demographic performance of owls would be combined with habitat data from the demographic study areas to develop models to predict owl occurrence and demographic performance. Sampling the population and habitat as described in this plan will ensure efficient use of resources and consistency of the information needed to conserve owl populations.

Concurrent with demographic data collection, range-wide habitat conditions would be monitored to track changes in conditions by using owl habitat maps derived from the national and international vegetation maps and imagery. This analysis would reflect analysis at several spatial scales (i.e., nest site, home range, landscape). Landscape components, such as area of pastureland, would be monitored on the basis of critical structural and management activities that reflect ecological conditions for the owl. The assumption is that the larger ecosystem will retain its ecological integrity to the extent that key habitat elements and physical processes are sustained. Population data from the demographic studies and habitat information from companion studies on vegetative characteristics at these sites will provide the data needed to "model" patterns of Little Owl occurrence and demographic performance to home-range and landscape characteristics of the vegetation. Provided the models are field tested and shown to be predictable at a level of uncertainty (associated risk) mutually acceptable to scientists and decision-makers, emphasis could shift from mark–recapture studies to increased reliance on monitoring of owl habitat and use of predictive models to indirectly estimate the occurrence and demographic performance of owls.

13.5 Development of conceptual models at the home range and landscape scales

An initial step in the monitoring task is to select indicators that reflect the underlying processes governing the dynamics of a species' populations. Indicator selection is facilitated by first developing a conceptual model (Lint *et al.* 1999; Noon 2003). The conceptual model outlines the interconnections among ecosystem processes (key system components), the structural and compositional attributes (resources), and how the condition of the resources affects the owl population (see Van Nieuwenhuyse *et al.* 2001c). The model should also indicate the pathways by which populations accommodate natural disturbances and how they demonstrate resilience to such disturbances. A framework to guide model development is to link ecosystem process and function to measurable components of the resources (that is, structural and compositional attributes). Changes in resource value, in turn, can be used to make predictions of expected biological response. Measurements of biological systems are affected by the scale of observation. To determine the appropriate scale(s) of indicator measurement, the temporal and spatial scales at which processes operate and populations respond must be estimated (at least to a first approximation) and clearly identified in the conceptual model. The indicators selected for measurement reflect known or suspected cause–effect relationships to population dynamics. A satisfactory model provides a justification, in terms of current ecological principles and theory, for the indicator(s) selected for monitoring and how knowledge of the status and trend of the indicator reflects underlying process, function, and population response.

On the basis of other demographic work done on owls (Forsman *et al.* 1996; Lint *et al.* 1999; Mulder *et al.* 1999; Franklin *et al.* 2004; Ganey *et al.* 2004), and on the conservation biology science upon which that work was based (e.g., Gaines *et al.* 1999; Noon 2003), selected indicators should:

- reflect fundamental population processes and changes in stressor levels (that is, habitat change)
- be representative of the state of the population
- be measurable and quantifiable

Further refinement identified indicators with the following properties:

- their dynamics would parallel the dynamics of the species' populations
- they would reflect a rapid and persistent response to change in the state of the environment
- an accurate and precise estimation is possible (that is, a high signal-to-noise ratio)
- there is a high likelihood of detecting a change in the magnitude of the indicators given a change in the state of the ecosystem
- they would demonstrate a low level of natural variability, or additive variation, and changes in their values can readily be distinguished from background variation
- measurement costs are not prohibitive.

The conceptual models developed for the Little Owl focus on habitat change (both loss of area and fragmentation) as the prime determinants of the species' population dynamics at this time. We developed models at two spatial scales. One model focuses on habitat change (stressor) processes relevant at the home range scale, a meaningful scale in the context of the dynamics of individual pairs of owls (Figure 13.4). The other model focuses on changes in habitat pattern at the landscape scale (Figure 13.5). At this scale, habitat change affects the size, and thus likelihood of persistence of local populations, and the connectivity among populations.

For both conceptual models, possible indicators are in two broad categories – habitat and populations. This was done (1) to document the well-established relation between habitat change, at a variety of spatial scales, and the dynamics of owl populations; and (2) to suggest the list of habitat variables that may serve as surrogate indicators once habitat-based models are developed.

13.6 Selection of vital sign indicators

Population and meta-population studies on owls have indicated that the adult survival rate is the primary indicator driving the population dynamics. Franklin *et al.* (2004:34) stated that in the California Spotted Owl, while the adult survival rate provided the baseline indicator for the population trajectory, fecundity and recruitment provided the variability around that trajectory. For fecundity, demographic analyses focus on the number of female young produced per adult female.

At the home range scale (Figure 13.4), habitat stressors affect the birth and death rates of individual owls. The consequences of habitat change at this scale can be assessed directly by measuring various demographic rates, or indirectly by measuring habitat attributes

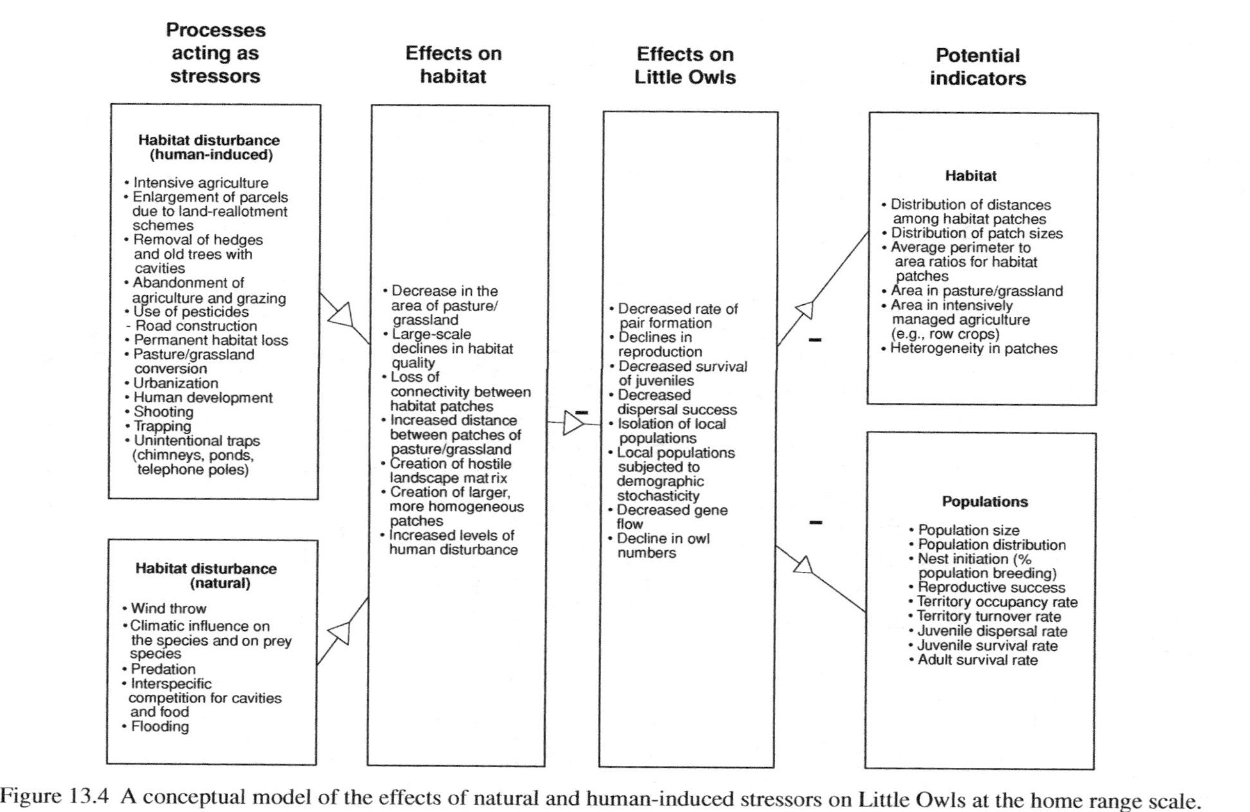

Figure 13.4 A conceptual model of the effects of natural and human-induced stressors on Little Owls at the home range scale.

correlated with variation in these rates. Both approaches are appropriate, but direct measures of population attributes are clearly less ambiguous. For the near term, we recommend a continued focus on direct measures of population attributes. Specifically, we propose that monitoring efforts focus on precise and accurate estimates of adult survival rate by using capture–recapture methods (see details in Forsman *et al.* 1996; Franklin *et al.* 2004; Ganey *et al.* 2004). We emphasize adult survival rate as the most relevant indicator of population status because (1) variation in this rate most affects changes in population growth rate, and (2) it is responsive to habitat change at the home-range scale.

The second conceptual model addressed the landscape scale, focusing on stressors affecting local populations and the interactions among these populations (Figure 13.5). Relevant population-level indicators at this scale are population size and distribution. At this scale, habitat change affects not only dynamics of local pairs of owls but also connectivity among local populations. Attributes such as habitat patch size, shape, number, and distribution determine the connectivity and stability of local populations. The population consequences of habitat change at the landscape scale are ultimately the same as at the home range scale, even though the mechanisms driving the change are different (compare Figure 13.4 with Figure 13.5). In practice, though, landscape scale habitat variables will probably be of insufficient resolution to predict population response at the home range scale.

The conceptual models document the relations between habitat dynamics and population response at two spatial scales. Until reliable habitat-based predictive models are developed, we recommend monitoring at the population level by refining current European-based studies and developing new demographic studies in Middle Eastern and Asian regions. At the same time, we propose an active research effort to develop predictive models relating population dynamics to habitat variation at the home range scale and the landscape scale. This parallel research effort is necessary to link habitat conditions to Little Owl population dynamics. For the most part, landscape scale models will be based on habitat attributes estimated from remotely sensed data. Home range-scale habitat models will rely on low-level aerial photography and ground plots to estimate the relevant variables. Habitat models applicable at the landscape scale will be more extensive but will predict population status with less resolution than models applicable at the home range scale. Because landscape scale models make simplifying assumptions about habitat quality at the home range scale, both types of predictive models are necessary.

13.7 Sampling design

Population monitoring

The capability of a monitoring method to provide data about vital rates for individual owls was considered key to whether a method would meet the indicator monitoring needs identified in the conceptual models. Three alternative methods for monitoring owl populations have been evaluated. The first, based on mark–recapture methods, would provide detailed

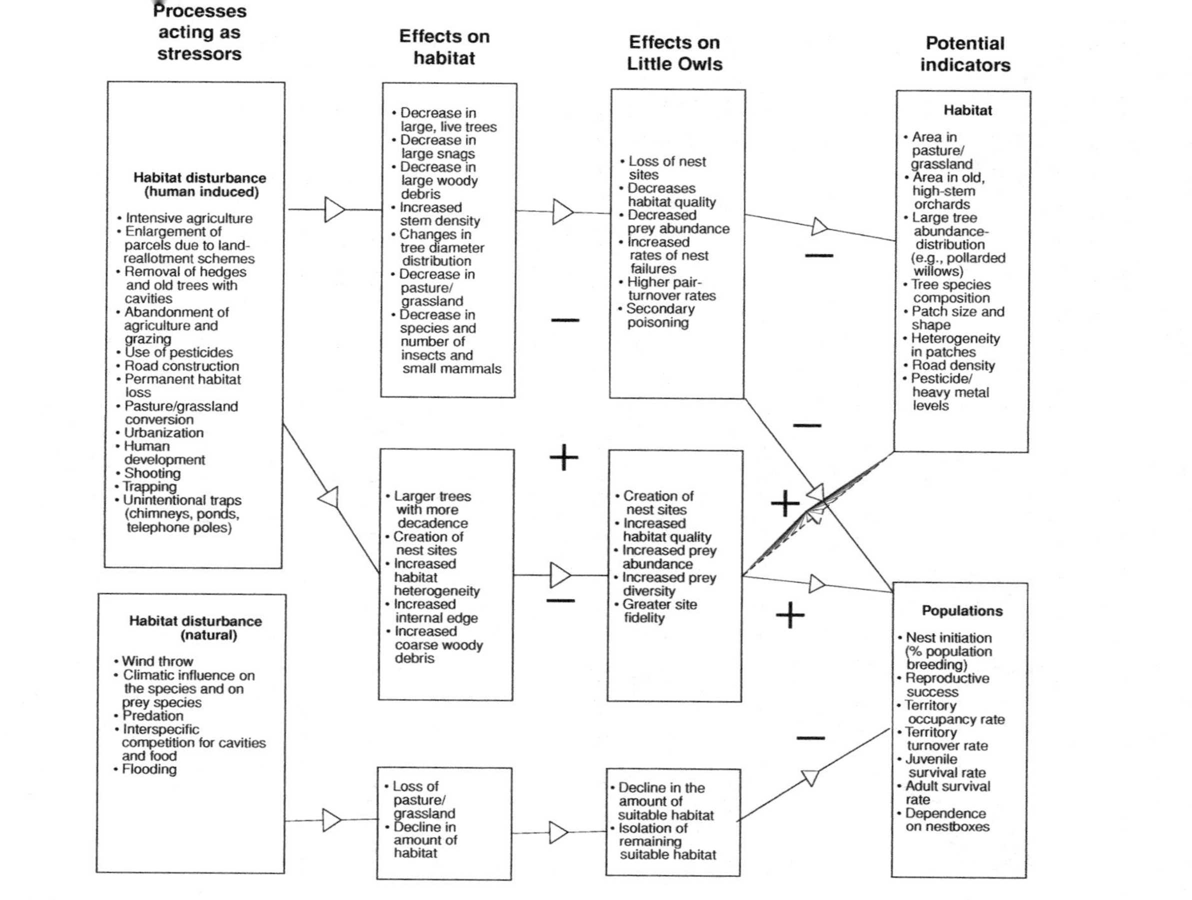

Figure 13.5 A conceptual model of the effects of natural and human-induced stressors on Little Owls at the landscape scale.

Figure 13.6 Systematic capturing of both adults in Flanders, northern Belgium by Ph. Smets. Photo Maarten Bekaert.

estimates of demographic rates and the annual rate of population change (see Franklin *et al.* 2004; Forsman *et al.* 1996 for methods and rationale). The second alternative method is based on repeated counts in randomly selected census plots and would provide estimates of rates of population change and rates of change in occupied habitat. The third method, density study areas, would require a total count of individual territorial owls in a multiple-kilometer area and would provide estimates of the rate of population change and change in occupied habitat.

Mark–recapture methods are currently being used to monitor territorial owl populations in western USA (see Forsman *et al.* 1996; Franklin *et al.* 2004; Ganey *et al.* 2004) and elsewhere in the world. The studies provide estimates of rates of survival, reproduction, and turnover, as well as detailed data about location of nests, roosts, and habitat conditions at nest sites. Data about survival and reproduction are used to estimate the annual rate of population change and to investigate trends in reproduction.

Bart and Robson (1995) suggest that a random survey using the playback technique (for Northern Spotted Owls) could be done for considerably less money over broader areas than the intensive mark–recapture approach. The results can be used to estimate rates of population change, but not to estimate rates of survival, reproduction, or turnover.

The density-estimate study area approach has been used in parallel with mark–recapture projects for several years. The density-estimate method assumes a 100-percent survey of a land unit of about 20–100 km^2. In each density study area, observers record the number and location of all owls encountered in repeated surveys during a single season. Density areas are resurveyed annually. The results can be used to identify substantial changes in the number of owls, but do not estimate rates of survival, reproduction, or turnover.

Based on previous analyses on spotted owls (e.g., Forsman *et al.* 1996; Lint *et al.* 1999), only the mark–recapture method within demographic study areas will satisfy the array of indicator data needs. It also is the only method that will provide the option to pursue predictive model development, which is a key option within the overall monitoring strategy. Given the declining nature of Little Owl populations, we consider estimates of the demographic rates to be key elements of a monitoring plan for the owl at this time. We therefore propose that the primary sampling method for monitoring should be mark–recapture studies in the demographic study areas.

An additional consideration for population monitoring is the use of an independent estimate of population trend for comparison with the results from the demographic studies. An additional, scientifically credible method would provide an independent, empirical assessment of the estimated population trend to compare with observed demographic parameters. For example, declines in survival and reproductive rates discerned from demographic studies should be reflected in declines in number of owls encountered during repeated surveys. We recognized both the random census plot and density study area methods as having the potential to provide the independent estimate of population trend. Considerable uncertainty exists, however, about the sample size of survey units and number of years of survey required to detect small changes in population trend. If the rate of population change is small (say 1% per year), and owl detection rates differ among years, many years would be required to detect a <5% rate of change in the population. If the rate of change is large (>5% per year), however, the method would probably detect a significant downward trend in only a few years. Because of uncertainty about the statistical power of the census plot approach, we recommend additional estimates of the statistical power of the method and its cost effectiveness. Likewise, further analysis should be conducted on the efficacy of the density study areas to provide an independent estimate of population trend. Once these additional analyses are completed, the adoption of a method to obtain an independent estimate of population should be given further consideration.

Habitat monitoring

The dynamic changes in vegetation structure and composition of grassland ecosystems in the Little Owl's range reflect the underlying biotic and physical driving forces, including intensive and pastoral agricultural practices and other human-caused sources of disturbance. For this document, we assumed that knowledge of vegetation structure and composition (amount and distribution) does not yet allow reliable prediction of Little Owl occurrence or the demographic performance of the owls.

Habitat monitoring for Little Owls will consist of two separate, but related, initiatives. One is tied directly to each of the demographic study areas. Patch-specific vegetation classifications and habitat evaluations will be completed in each study area (e.g., see Van Nieuwenhuyse *et al.* 2001c). Habitat assessments of demographic study areas will be based on plot data, vegetation description information, and standard aerial photography to develop structure and composition attributes for owl habitat. These data will be used with the population data to assess the relations among varying demographic responses and varying habitat conditions at the nest-site, home range, and landscape scales. These relations will form the basis of the (future) predictive models. Accurate and timely monitoring of changes in the status and trends of vegetation (habitat) should provide a reliable early warning system to eventual changes in the population viability of Little Owls.

The second habitat assessment will estimate baseline, range-wide Little Owl habitat conditions, and track change in habitat conditions over time. The range-wide coverage also will provide unique capabilities for spatial analysis, portrayal of geographic distribution of habitat, and analysis of patch statistics. The range-wide vegetation map will be developed from satellite imagery. This is a case where the current analytical tools permit the assessment of the complete set of data, thus eliminating the need to stratify and sample the data, which would require statistical analysis to describe the precision of the estimates and would be more expensive. The baseline for range-wide habitat condition will be established and periodically reassessed to describe habitat condition and trend. This will allow tracking of the amount of nesting and foraging habitat at a variety of spatial scales. It also will provide information on habitat conditions in population connectivity areas.

Habitat-based aspects of the monitoring plan offer several advantages:

- Habitat monitoring can build from existing inventory programs.
- Estimating the trends in vegetative structure and composition represent a prospective, as opposed to a retrospective, approach to ecological monitoring.
- Monitoring vegetation change is more cost-effective than directly monitoring populations of all the possible species for which management agencies are responsible.
- The baseline and subsequently vegetation maps will be applicable to other species, and supportive of the broader array of conservation assessments and biodiversity management.

Some limitations to a strictly habitat-based approach to monitoring include:

- Some unknown proportion of the variation in species' population dynamics may not be driven by changes in habitat amount and distribution (for example, population fluctuations due to behavioral and prey-related influences).
- Changes in habitat may not predict population responses to other stressors (for example, environmental toxins, changing environmental conditions, and competitive interactions).
- On the basis of the above two limitations, a strictly habitat-based monitoring program may have limited ability to predict changes in species viability and distribution.

The following variables are a minimum set of attributes thought to influence presence and abundance of Little Owls at the landscape scale and to be monitored range wide:

- distribution and area of the pastureland/grassland corresponding to owl nesting and foraging habitat
- frequency distributions, by area size-class, of habitat patches
- frequency distributions of distances between habitat patches.

These parameters may be further summarized at a variety of spatial scales including land allocation, administrative unit, physiographic province, and range of the Little Owl.

Predictive modeling

At some future point, to make the transition from population-based monitoring to a completely habitat-based program will require an intensive period of habitat model development. We recognize this task to be primarily one for research. We envision that a completely habitat-based monitoring program must identify those aspects of vegetation structure and composition that have the greatest power and precision to predict the number, distribution, and demographic performance of owls at the landscape scale, as well as to explain the observed variation in demographic rates at a home range scale. This task will require characterizing the vegetation at a variety of spatial scales in the existing demographic study areas. The combination of spatially referenced data from both the owl demographic studies and mapped vegetation attributes provides the fundamental data for the model-building phase. The degree to which these models explain the observed variation in owl distribution and demographic performance will estimate the certainty with which habitat variation predicts population performance and stability. Explained variation is thus a direct measure of the confidence we have in habitat as an appropriate monitoring surrogate for population performance. In addition, validating model predictions by independent field surveys is essential; that is, the models will be used to predict owl population response, which must then be verified by direct field measurement of owls from several landscapes with different population levels. Validation testing will use the range-wide habitat map derived from the regional or country-based vegetation maps. Once reliable models are developed, existing habitat conditions across extensive landscapes can be assessed and the expected owl occupancy, distribution, or demographic performance predicted depending on which predictive level of monitoring is implemented. We have a very long way to go before we will be able to shift to a habitat-only monitoring program.

13.8 Development of a network of Vital Sign monitoring areas

Vital Signs are key elements that indicate the health of the Little Owl population and the key habitat components it depends on. Vital Signs can be any measurable feature of the population or environment that provides insights into the state of the species. The term is synonymous with "ecological indicator", but use of the term and the analogy to an individual's health will be helpful in explaining the need for monitoring to managers, politicians, and the public.

We propose the development of a network of some 30 monitoring areas across the range of the Little Owl. The scientific framework and conceptual approach for this work is based upon Mulder *et al.* (1999), Lint *et al.* (1999), the National Park Monitoring Program for the USA (http://science.nature.nps.gov/im/monitor/vsmTG.htm), Gaines *et al.* (1999), the European Biodiversity Monitoring and Indicator Framework (http://www.strategyguide.org/ebmf.html), and the Framework for Monitoring Biodiversity Change in Canada (www.eman-rese.ca/eman/reports/publications/framework/context.html).

Our proposed network of monitoring areas reflects an array of study areas spread across the major ecological regions (biomes) within the range of the Little Owl. We offer a conceptual map of this monitoring network across the range of the Little Owl in Figure 13.7. Each monitoring area should contain from 50–100 pairs of Little Owls, or equivalent habitat in areas where population restoration is planned. A network of such monitoring areas is recommended because it is currently unrealistic to expect that all countries within the range of the Little Owl will be able to adequately fund and consistently support monitoring at the optimum scale and intensity desired. A monitoring program requires professional-level staff who can analyze data, interpret data, prepare reports, and provide the information in a usable format to managers, scientists, and other interested parties. Thus, the compromise position is to identify a base network of monitoring areas, and to develop a core program for the Little Owl. Thereafter, the participating parties in the network can identify what specialists are needed, and allow each network component to then leverage their core resources with other resources and partnerships to build a monitoring program that can also be integrated with monitoring efforts for other natural resources (e.g., Important Bird Areas, UNESCO Biosphere Reserves, National Parks, Natura 2000 sites, etc.).

The experimental design of the demographic study areas is crucial for estimating owl birth and survival rates. This is important because the individual vital rate estimates provide data about population indicators. These data will be used in the population meta-analysis and as input to predictive models. While we are not selecting demographic study areas randomly (in the statistical sense) we expect the non-random design to provide data adequate to meet the monitoring needs. In addition, the large, quality data sets associated with existing Little Owl study areas provide a foundation for the monitoring program not present in a random design. Based on this reasoning, we selected the pattern of 30 study areas as the entry-level option for population monitoring. This conclusion was supported by the knowledge that the 30 areas had sufficient pairs of owls to assure continued low standard errors of the vital rate data.

We expect there to be uncertainty about the adequacy of the coverage of the Little Owl range with a design of less than 30 study areas and the capability to expand the results of a meta-analysis to the range of the owl. Although the study areas do not cover all provinces or ecological conditions across the range of the owl, we believe they represent a sample of sites that spans a wide range of habitat conditions and represent the range of

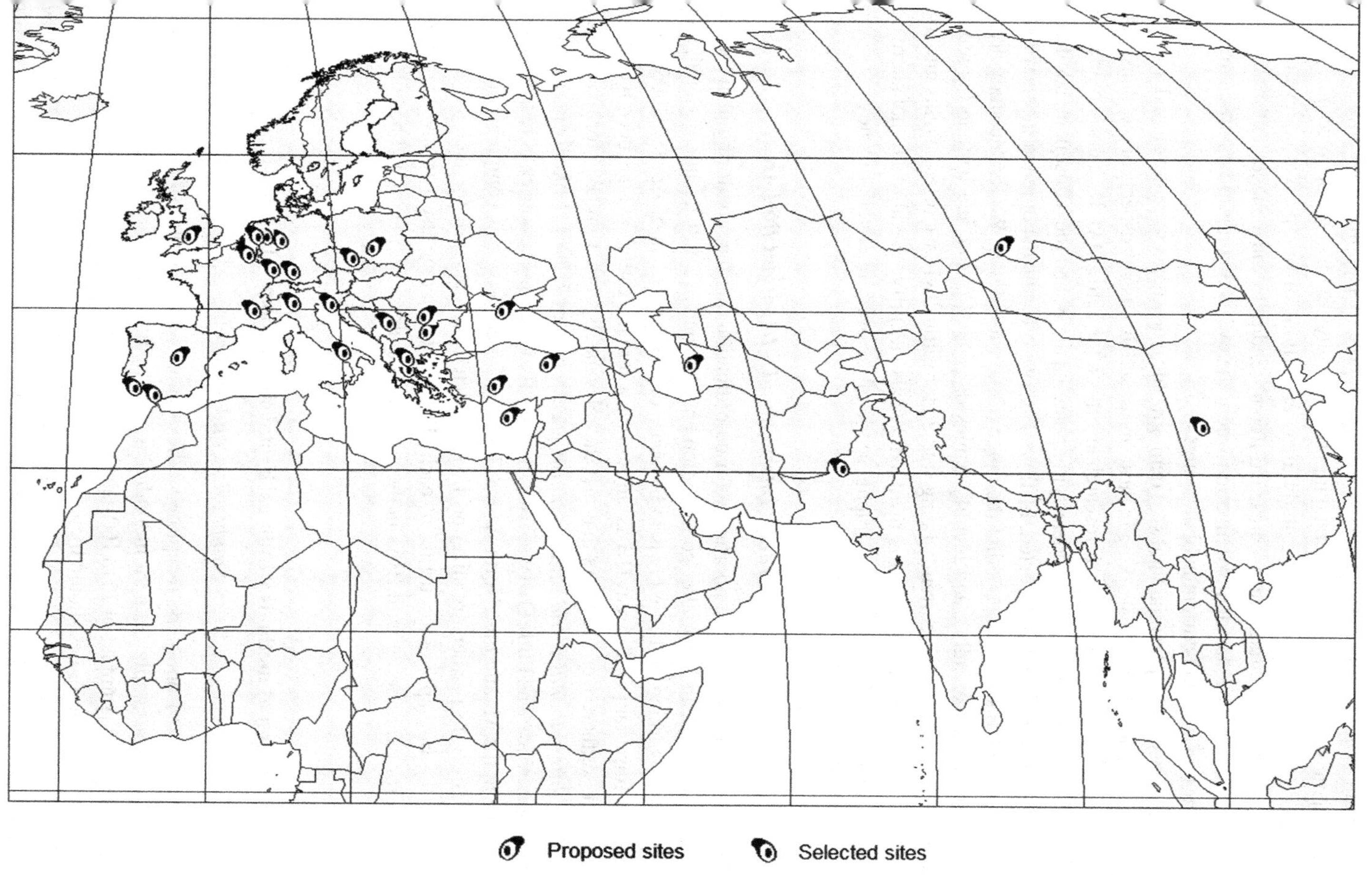

Figure 13.7 Conceptual map of the proposed Vital Sign demographic monitoring area network across the range of the Little Owl.

variation in demographic rates. As such, we are not able to support an option with fewer demographic study areas and still defend it as scientifically credible. We are recommending a stronger network of demographic study areas in portions of Europe. We believe this is justified because those study areas represent different conditions and management strategies, and will allow a more detailed examination where the owl is thought to be particularly at risk because of past and current habitat changes.

The majority of the potential vital sign demographic study areas were developed by contacting researchers currently working in these areas, and discussing aspects of owl densities, distribution, logistics, and previous research efforts on the Little Owl. Readers should note that this list was our best effort at identifying potential areas, and the adjustment, addition, or deletion of study areas is expected. In Table 13.2, we offer the list of recommended vital sign demographic study areas within the range of the Little Owl). We list the study areas by country (alphabetically), and identify names of general places and latitude-longitude co-ordinates. If the area is currently being studied for Little Owls, we offer insights into the size (km^2) and numbers of owl pairs involved. We also identify several additional alternate study areas, to complement the geographic scope of the overall network of study.

Integration with BirdLife International – Important Bird Areas

We recognize the important opportunity for co-ordination that the monitoring plan for the Little Owl represents with other conservation entities such as BirdLife International, and urge people working on Little Owls to be aware of this aspect. As of April 7, 2005, there were 278 Important Bird Areas (IBAs) in Europe where the Little Owl has been recorded. Within these, there were 15 IBAs that met the relevant IBA criterion (B2) for the species (*BirdLife International World Bird Database*, accessed April 7, 2005, Mike Evans personal communication). The Little Owl was listed in many IBAs in some countries, at few in others, and none in the Middle East/North Africa. This was based upon how comprehensively the national BirdLife partner organization had populated the database for their country, in terms of the bird species occurring at each IBA. We encourage users to investigate the potential for overlapping with IBAs in their monitoring areas for the mutual benefit it can provide.

13.9 Design of methods within each vital sign demographic monitoring area

Sample sizes and protocols

Precise estimates from mark–recapture studies require large samples of marked owls. We recommend that sample sizes for individual demographic studies be 50 pairs of owls (100 individuals), with samples of 75 to 100 pairs preferable. Demographic parameters monitored in mark–recapture studies should include survival and reproductive rates by age-group and sex. From these data, the annual rate of population change can be estimated. Occupancy and turnover rates at historical owl sites in the demographic study areas should also be

Table 13.2 *List of recommended Vital Sign demographic study areas within the range of the Little Owl.*

Country	Study sites recommended	Comments
Afghanistan	–	Recommend baseline ecological surveys.
Albania	–	Recommend additional baseline ecological surveys.
Algeria	–	Recommend additional baseline ecological surveys.
Armenia	–	Recommend additional baseline ecological surveys.
Austria	–	Very common species in Austria until the 1970s, declined to *c.* 60 pairs.
Azerbaijan	–	Recommend baseline ecological surveys.
Bahrain	–	No data available to make any recommendation.
Belarus	–	Recommend additional baseline ecological surveys; the Little Owl is a rare resident breeding species throughout all of Belarus.
Belgium	2	Recommend two monitoring areas in Belgium, one in Flanders and one in Wallonia. These study areas reflect long-term efforts already established and ongoing. The Wallonia study area is at: N 50°20′, E 4°30′, is approximately 90 km^2 in size, and involves some 40 pairs of owls (J. Bultot personal communication). This project started in 1989. The Flanders study area is at: N 50°45′, E 4°45′ and is 225 km^2 in size, and involves some 64 pairs of Little Owls (P. Smets personal communication). This project started in 1993. Smets captures all adult females and males (males are captured in a special box attached in front of the nestbox with a light indicating if the male has entered and been caught).
Bhutan	–	Recommend baseline ecological surveys.
Bosnia & Herzegovina	–	Recommend baseline ecological surveys.
Bulgaria	1	Recommend one study area, specific location undefined.
Chad	–	Recommend baseline ecological surveys.
China	1	Recommend one study in Qishan County, Shaanxi Province, location: N 34°27′, E 107°38′ (Lei Fu-Min personal communication).
Croatia	–	Recommend additional baseline ecological surveys.
Cyprus	1	Recommend one area; location to be determined; considered a common but decreasing species.
Czech Republic	1	The Little Owl was the most abundant and widespread owl in Bohemia in the 1940s; opportunity to study population restoration. A recommended monitoring area in South Moravia is a current Little Owl study area. The center of the study area is at N 49°95′, E 16°75′. The study area is about 2300 km^2. In 2004, 11 calling males and 4 breeding pairs in nestboxes were found in the area (Libor Oplustil, personal communication).

(*cont.*)

Table 13.2 *(cont.)*

Country	Study sites recommended	Comments
Denmark	–	Recommend studies associated with restoration actions.
Djibouti	–	Recommend baseline ecological surveys.
Egypt	–	Recommend additional baseline ecological and distributional surveys.
Eritrea	–	Recommend baseline ecological surveys.
Estonia	–	The Little Owl is only a vagrant to Estonia.
Ethiopia	–	Recommend baseline ecological surveys.
France	2	Recommend two monitoring areas in France. These study areas reflect long-term efforts already established and ongoing. These study areas are: In Vosges du Nord, N 48°50′, E 7°20′, 177 km^2 and 30 pairs (J-C Génot, personal communication), monitoring began in 1986. In Plaine de Valence, N 44°55′, E 5°0′, 48 km^2 and 40 pairs (S. Blache, pers. comm.), monitoring began in 1997.
Georgia	–	Recommend additional baseline ecological and distributional surveys.
Germany	3	Recommend three monitoring areas in Germany: one in Goeppingen, southern Germany studied since 1971; one in Werl, NW Germany studied since 1974; and one in Lippstadt, NW Germany, studied since 1980. These study areas reflect long-term efforts already established and ongoing. The Goeppingen study area is at: N 48°40′, E 9°38′, is approximately 40 km^2 in size, and involves some 21 pairs of owls (B. Ullrich, personal communication). The Werl study area is at: N 51°33′, E 7°55′ and is 127 km^2 in size, and involves some 50 pairs of owls (H. Illner, personal communication). The Lippstadt study area is at: N 51°42′, E 8°30′, and is approximately 127 km^2 in size, and involves some 65 pairs of owls (A. Kaempfer-Lauenstein & W. Lederer, personal communication).
Greece	2	Recommend two study areas in Greece: one in the Thessaly region of central Greece (N 39°30′, E 22°00′); and the other in northern Greece (approx. N 40°30′, E 22°00′ in the general area of Olimpos or Ptolemaida) (Christos Vlachos, personal communication).
Hungary	–	The population has declined mainly due to agricultural intensification, recommend studies associated with restoration actions.
India	–	Species limited to far northwestern India, mainly Ladakh and North Sikkim.
Iran	–	Recommend baseline ecological surveys.
Iraq	–	Recommend baseline ecological surveys.
Israel	–	Recommend additional baseline ecological surveys.

(cont.)

Table 13.2 *(cont.)*

Country	Study sites recommended	Comments
Italy	2	Recommend two study areas: one study area to be located in Lombardia, Province of Bergamo; N 45°42′, E 9°39.′ This study area is 55 km^2 in size, and contains about 60–70 pairs of Little Owls (M. Mastrorilli, personal communication). The location for the other area has yet to be defined.
Jordan	–	Recommend additional baseline ecological surveys.
Kazakhstan	–	Recommend additional baseline ecological surveys.
Korea	–	Reportedly very rare in North and South Korea.
Kuwait	–	Very small population (est. 50 pairs).
Kyrgyzstan	–	Recommend baseline ecological surveys.
Latvia	–	The Little Owl is very rare in Latvia.
Lebanon	–	Recommend baseline ecological surveys.
Libya	–	The Little Owl is common in the northern part of the country; southern limit of its distribution remains unclear; recommend additional baseline ecological and distributional surveys.
Liechtenstein	–	Recommend baseline ecological surveys.
Lithuania	–	The Little Owl is a rare resident breeder in Lithuania.
Luxembourg	–	Species has undergone significant declines, and is largely extirpated.
Macedonia	–	Recommend baseline ecological surveys.
Mali	–	Recommend baseline ecological surveys.
Malta	–	Recommend baseline ecological surveys. Recommend immediate halt to illegal hunting, and development of a conservation program for all raptors and owls.
Mauritania	–	Recommend baseline ecological surveys.
Moldova	–	Recommend additional baseline ecological surveys.
Monaco	–	Recommend baseline ecological surveys.
Mongolia	–	The Little Owl is considered rare everywhere in Mongolia.
Morocco	–	Recommend additional baseline ecological surveys.
Nepal	–	Recommend additional baseline ecological surveys.
Netherlands	2	Recommend two study areas. One study site is Meddo-Huppel in Achterhoek-Liemers; 12 km^2: 30–35 pairs; N 52°00′, E 6°44′ (Ronald van Harxen & Pascal Stroeken, personal communication). The other is the Midden-Betuwe in the River Area; 55 km^2: 74 pairs; N 51°58′, E 5°35′ (Frans Jacobs, personal communication).
New Zealand	–	Introduced population in New Zealand.
Niger	–	Recommend baseline ecological surveys.
Oman	–	Very small population (est. 100 pairs).

(*cont.*)

Table 13.2 *(cont.)*

Country	Study sites recommended	Comments
Pakistan	1	Recommend one study area for Pakistan. The Little Owl exists in Ladakh in the North West Frontier Province, and on the western border region of Pakistan around Ziarat (30°22′N and 67°44′E) and Zhob (31°30′ N and 69°30′E) (M. Mahmood-ul-Hassan, personal communication). Given logistic issues, a study area in either the Ziarat or Zhob area is offered for consideration.
Poland	1	Recommend one study area for Poland; location yet to be defined (G. Grzywaczewski, personal communication). The species is scarce in most areas, with marked declines after severe winters; substantial population decline since the 1960s.
Portugal	2	Two study areas are recommended for southern Portugal, and two additional study areas have been identified as alternates (Ricardo Tome, personal communication): 1. "Beja" district, in the "Baixo Alentejo" region. Pseudo-steppe study area (named "S. Marcos da Atabueira", near the village of "Castro Verde"); 37°42′N, 7°50′W; study area is 16.8 km² in size, and contains 42 pairs of Little Owls. 2. "Beja" district, in the "Baixo Alentejo" region. Open Holm oak woodland study area (named "Cabeça da Serra", between the villages of "Castro Verde", "Ourique" and "Rosário"); 37°37′N, 8°09′W; study area is 6.1 km²', and has 43 pairs of Little Owls. Two additional areas in southern Portugal:
	& 2 Alt	3. "Beja" district, in the "Baixo Alentejo" region. Olive groves of "Sobral da Adiça" area, near the village of "Moura" and the border with Spain. 38°03′N, 7°17′W; more than 30 km², where the densities of Little Owls are very high (more than 50 pairs). 4. "Faro" district, in the Algarve region (southernmost region of Portugal). This is a very large area, occupied with many different habitats. Near the southern coast, Little Owls are very common and occupy mostly open woodland with almond and ficus trees, near the villages of "Lagos" and "Vila do Bispo" (and others); 37°11′N, 8°44′W. Easy to identify an area with more than 50 pairs (high densities). It would also be possible to map other areas with 50+ pairs in central and northern Portugal. However, since densities are much lower in those regions, the areas would probably extend to some hundred square kilometers or so.
Qatar	–	Very small population (est. 50 pairs).

(cont.)

Table 13.2 *(cont.)*

Country	Study sites recommended	Comments
Republic of Bashkortostan	–	Recommend additional baseline ecological surveys.
Republic of Dagestan	–	Recommend additional baseline ecological surveys.
Republic of Ossetia	–	Recommend additional baseline ecological surveys.
Republic of Tatarstan	–	Recommend additional baseline ecological surveys
Romania	1	Recommend one study area, specific location to be defined.
Russian Federation	1 Alt	Recommend additional baseline ecological surveys. Possibility for one alternate study area to be in the southern Tuva region of eastern Russia.
Saudi Arabia	–	Recommend additional baseline ecological surveys.
Serbia and Montenegro	1	Recommend one study area, reflecting 50 owl pairs within 6–7 UTM grids, centered on Čačak, central Serbia (N 43°50′, E 20°20′): DP46 Cacak; DP45 Parmenac i okolna sela; DP56 Gornja Trepca; DP55 Atenica, Trnava, Lipnica, Viljusa; DP65 Mrcajevci; DP64 Samaila, Milocaj (vise u Kraljevu, ali ne mari); DP66 (mozda) sela ka Knicskom (Gruzanskom) jezeru (Milan Ružic, personal communication).
Slovakia	–	Small populations.
Slovenia	1	Recommend one study area in SW Slovenia: the IBA site Karst and the Vipava valley (N 45°40′, E 13°55′); around 650 km^2; estimated population of 100 breeding pairs. Although the Little Owl population has declined greatly in the NE and central Slovenia, it is still relatively numerous in SW Slovenia. (Al Vrezec', personal communication).
Somalia	–	Recommend baseline ecological surveys.
Spain	2	Recommend two study areas; specific locations yet to be determined. Connection with Project NOCTUA sites and IBAs are recommended. Potential study sites include: 1. Seville (Southern Spain). Olea. 2. Madrid (Central Spain). Alonso and Orejas (Brinzal- Owl rehabilitation center). 3. Biscay (Northern Spain). Zuberogoitia (SEAR- society for the study of birds of prey).
Sudan	–	Recommend baseline ecological surveys.
Switzerland	–	Species present in very low numbers in three small regions. Conservation programs under development or running (Christian Meisser personal communication).

(cont.)

Table 13.2 *(cont.)*

Country	Study sites recommended	Comments
Syria	–	Recommend additional baseline ecological surveys.
Tajikistan	–	Recommend additional baseline ecological surveys.
Tunisia	–	Recommend baseline ecological surveys.
Turkey	2	Recommend two study areas, specific locations undefined as yet. Organizational aspects of surveys will be through Doga Dernegi (Onder Cirik and Lale Aktay, personal communication).
Turkmenistan	1	Recommend one study area, specific location undefined; suggest consideration of west or northwestern Turkmenistan.
Tyva (formerly Tuva)	–	The Little Owl is a rare, probable breeding, and rare wintering species in Tyva.
Ukraine	1	Recommend one study area, specific location undefined; suggest the Crimea region.
United Arab Emirates	–	Small population (est. 600 pairs).
United Kingdom	1 Alt	Introduced population in the UK; recognize the value in having an alternate demographic study area in the UK near Manchester (Roy Leigh personal communication).
Uzbekistan	–	Recommend additional baseline ecological surveys.
Yemen	–	Small population (est. 500 pairs).

monitored because these variables may provide key information about the response of owls to variation in home-range and landscape-scale features. We anticipate that mark–recapture studies will employ ringing (or color ringing) as the primary means of tracking individual owl histories. Protocols for assessing survival, reproduction, occupancy rates, and turnover rates of owls will need to be consistently followed across all study areas (for examples, see Forsman *et al.* 1996; Franklin *et al.* 2004).

A peer-reviewed survey protocol has been developed to provide consistent and cost-effective methods for monitoring demographic aspects of Little Owl populations (Johnson *et al.* 2008).

Demographic area habitat assessment

Habitat classification and mapping will be completed for each of the demographic study areas. To date, the habitat characterization has been conducted independently for each of the demographic study areas. As a result, there are differences in the measured attributes, classification standards, and resulting map products. Consistency of the vegetation classifications is important to the modeling that will follow, thus there is a need to resolve differences between the methods. The resolution of these differences may involve some re-sampling and reclassification of variables in some of the demographic study areas. The responsibility for completing this will be assigned to the leaders of the respective demographic studies. All reassessment and new classification efforts will follow the consensus standards developed during the resolution process.

The set of vegetation attributes that best characterize Little Owl habitat will differ by physiographic province and study area. Investigators will therefore first identify the key attributes in each demographic study area to be used to define habitat depending on presumed habitat relations in that province. The landscape-scale habitat maps will be based on attributes acquired from the regional vegetation map or specified for inclusion in subsequent versions of the vegetation map. This approach is essential so that more general predictive models can be built that apply outside the demographic study area boundaries.

Range-wide habitat assessment

The basic information needed for range-wide monitoring of owl habitat is a set of map layers that collectively characterize Little Owl nesting and foraging habitat. An overlay of map layers will allow the development of a GIS-compatible database used to describe amount and distribution of habitat in relation to land allocations or other geographic areas of interest. Once developed, the map would be updated periodically to track habitat change. Periodic updates of the map layers in the near term will allow the estimation of changes in amount and distribution of habitat over time resulting primarily from agricultural intensification and human development. In the initial years of monitoring, detecting biologically significant changes in habitat condition will require periodic inventory at ten-year intervals. The map would be recompiled and habitat conditions reassessed in synchrony with the schedule for monitoring map product updates.

13.10 Data management and methods of individual population and meta-population

o that inferences about population trend from all demographic study areas are comparable, data for annual survival, reproduction, and owl turnover rate at sites in each area will be stimated by standardized methods and protocols. These methods reflect the best available cience and provide consistency and rigor to the data collection and analysis process. For xample, consistency will be provided for field surveys, the estimation of reproductive effort, or capture, ringing, sex and age identification, and for pre-analysis data screening. Estimates of survival rates will be based on Jolly-Seber open-population models (for example, Program MARK), as described by Franklin *et al.* (2004). Estimates of reproduction will be based on mpirical counts of numbers of young produced by each female in the sample. Similar rigor will be applied to estimating adult survival, fecundity, juvenile survival, and estimating rates f population change.

Data about demographic trends in survival, reproduction, and annual rate of population change will be summarized annually for each demographic study area. A more comprehensive meta-analysis of all data sets (for example, Franklin *et al.* 2004) will be conducted every five years, starting no later than December 2010. Specific analyses will identify estimates of adult survival, fecundity, juvenile survival, and rates of population change. Interpretation of results will continue to address uncertainties about the significance of adult emigration, possible biases in estimating fecundity, the effects of aging, and differential detectability of nesting and non-nesting pairs. Reports will include an annual summary for each demographic study area and a more detailed report for the meta-analyses.

For additional details on data management and methods for population and meta-population analyses, please see Franklin *et al.* (2004), Ganey *et al.* (2004), and Forsman *et al.* (1996).

Data tables and field headings for Little Owl databases

We recommend that six relational data tables, as detailed in Table 13.3, be used to track the specific data aspects of Little Owls within the vital sign study areas. The field headings and definitions reflect those used in the Spotted Owl demographic analyses programs in the USA, and have been adapted to fit the "Little 30" network. The six tables reflect a many-to-one relationship. The "Fecundity Table" tracks reproductive output with analysis focused on the total number of young owls produced, and not on the specific pairs of owls that actually nested. The "Lambda Table" reflects a new method that uses only the number of territorial owls in the study without including young of the year. The "Reproduction Table" has a record for all of the pairs that are encountered and the nest status, and the reproductive output of each. The "Survival Table" is the basis for calculating survival; juveniles are included in the estimate. In the "Visit Table", one record represents each visit to an owl site. The "Owl Table" tracks the data for all owls encountered. Data sheets and an associated field manual are being developed to reflect these data components.

13.11 Habitat condition and trend

Habitat condition and trend information will be estimated every five to ten years after the baseline habitat map is developed. Monitoring over time will allow for estimates of change in amount and distribution of owl habitat and for relating such changes to implementation of conservation measures. Habitat trend reports will tabulate information about hectares of habitat by land allocation and their percentage of change over time. It also may be possible (and useful) to estimate habitat status and trend at the home range scale. Such measurements would be based on the use of extensive inventory plot data. Here the appropriate habitat indicators would be attributes associated with home ranges (e.g., nest trees, nest rock piles, road density, area of pastureland and old orchards) and how their number and distribution are changing through time. See Figure 13.8.

Table 13.3 *Proposed data model for Little Owl monitoring database.*

FECUNDITY TABLE

MSNO = Master Site Number: a unique number used to identify the site in a particular country.

SITE = Name given to the location of the owl site/territory for identity purposes.

YR = YEAR: Year the data pertains.

AM = Age of the male

AF = Age of the female

NYF = Number of young fledged, must have left the nest site. Note: the number of young fledged is important; and the recording of the number ringed is not sufficient.

AREA = Name of Demographic Study area to distinguish from other study areas across the range of the Little Owl.

MRING = ring (band) number on the male.

FRING = ring (band) number on the female.

OC = Occupancy status code for the site. Pair, Single Male, Single Female, Unknown, Extra bird at the site, X (as a one time response, does not constitute a single).

B was used for birds with radio-transmitters to exclude them from the calculations.

COMMENTS: short text about owl site aspects, if needed.

VISITS: Number of visits to the territory that met the survey protocol. 2R means two reproductive surveys and that there was no judgment call on the data. 1N1R meant one nesting visit, one reproductive visit.

LAMBDA TABLE

COUNTRY = Country of study

AREA = Name of Demographic Study area to distinguish from other study areas across the range of the Little Owl. Note: all demographic study areas must use same survey protocol.

RING = ring (band) number for the individual.

SEX = Sex of the individual (M, F, U). Unknown sex individuals are excluded from the analysis.

AGE = Little Owls are in three age classes: A, SY, and J (Adult, Second-year, and Juvenile). Note: Juveniles are not included in Lambda calculations using RJS method.

Y?? = Year XXXX through XXXX year of the study. Binomial field, 0 for not re-sighted, 1 for re-sighted during the survey season interval.

M = Binomial field, 1 for male, 0 for not male.

F = Binomial field, 1 for female, 0 for not female.

C = semicolon (;) which is in the file for ease of creating a text file that is ready for survival analysis. Program.

MARK only needs the YXX data fields, M,F,C in that order.

TX = type of radio transmitter: N or blank for none, T for tail-mount, B for backpack

COMMENTS

OWL TABLE

AREA = Name of Demography Study area.

MSNO = Master Site Number: a unique number used to identify the site in a particular country.

SITE = Name given to the location of the owl site/territory for identity purposes.

DY = Day of the month data was collected (two numbers).

(*cont.*)

Table 13.3 *(cont.)*

MO = Month data was collected (two numbers).
YR = YEAR: Year the data pertains.
SPEC = Species code, we use first two letters of genus, followed by first two letters of species. ATNO is Little Owl.
TIME = Time of first detection for the survey.
Sex = Sex of individual detected.
OT = Observation Type (AN = Auditory only, VN = visual only; MC = male captured; RA: first rings attached; RR = rings re-sighted, etc.)
AGE = Age (A, SY, J, U). Put in the known age where possible (e.g., six for an owl that is in its sixth year of life.
WT = Net weight of individual in grams.
UTMX, UTMY = Co-ordinates in Universal Transverse Mercator. X is the Easting and Y is the Northing.
QUALITY = Degree of certainty of the assigned co-ordinates. Use the reading from the GPS that indicates it is 3D and within a few meters (3D009) is three-dimensional on the GPS unit and +/− 9 meters. Otherwise estimate to within 50 or 100 meters. For many species, a high degree of accuracy is necessary.
KMT = Kilometers traveled from the last known site. Pertinent only to the owls that have changed sites This is not really a necessary field and can be generated in summaries.
AZTR = Azimuth traveled from last known site. Pertinent only to the owls that have changed sites. This is not really a necessary field and can be generated in summaries.
COMMENT = Y, N, blank for comments entered. Not necessary as comments should be visible; helpful if comments get accidentally deleted to know if there were comments.
ENTDATE = Entry date of data. In some programs it can be generated.
TX= Type of radio transmitter: N or blank for none, T for tail-mount, B for backpack.

REPRODUCTION TABLE
MSNO = Master Site Number: a unique number used to identify the site in a particular country.
SITE = Name given to the location of the owl site/territory for identity purposes.
LAND = Primary landowner division (e.g., federal, private).
YR = Year the data pertains.
OC = Occupancy status of owl site (P = Pair, M = Single Male, F = Single Female, U = Unknown, E = Extra bird at the site, MXT = male with transmitter; FXT = female with transmitter, N = no owls present, Z = no visits to site).
AM = Age of the male.
AF = Age of the female.
NV = Number of Night visits conducted to the site that met protocol.
DV = Number of Day visits conducted to the site that met protocol.
VERNS = Date format of the survey that verification of the nesting status occurred.
NS = nesting Status for the survey season. Y= nesting, N = not nesting, U = unknown. Note: this is different for the visit (see Visit Table).
CNS = Y for those records that are to be used in the Calculation of Nesting parameters such as the proportion of pairs nesting.

(cont.)

Table 13.3 *(cont.)*

VERNYF = Date format of the survey where reproductive surveys were completed.

NSUCC = Nesting success. (NN= not nesting, FY= fledged young, UN= unknown, FN = nesting documented, but failed to fledge young.)

NYF = Number of young fledged, must have left the nest site. Note: the number of young fledged is important; and the recording of the number ringed is not sufficient. Do not include the number of young seen or heard that did not actually fledge.

CNYF = Y for those records to be used in calculating Reproductive parameters such as reproductive success and fecundity. Enter a "Y" only if a pair or female was present and reproductive status was determined.

BP = Brood patch present or not. Used in determining non-nesting status in some cases.

FEC = Fecundity. Number of young produced divided by two (to give the number of female young produced).

MRING = ring (band) number on the male.

FRING = ring (band) number on the female.

AREA = Name of Demographic Study area to distinguish from other study areas across the range of the Little Owl.

UTMX, UTMY = Co-ordinates in Universal Transverse Mercator of nest site or activity center. X is the Easting and Y is the Northing.

SOURCE = Source map of the UTM location(s); showing UTM zone and year of map publication.

The following are used in special analyses.

ALTNAME = Alternate name. Due to habitat reduction, some sites that used to be two are now one; you can combine the records since the same individuals were clearly occupying both sites.

ACATF = Age category of female. Used in SPSS to distinguish S1females, Unknown, and Adult females from no data or males.

NSTAT = Coding for SPSS. 1 is nesting, 0 is not nesting, 9 is missing data, blank is for no data. SPSS needs it to be in 1 or 0 categories.

SUCRATE = 1 for successful fledging sites, 0 for sites that did not successfully fledge young, and 9 for those sites that did not attempt nesting.

SURVIVAL TABLE

MSNO = Master Site Number: a unique number used to identify the site in a particular country. Use the most recent location.

SITE: This is the most recent location for the owl.

SITE1: This is the next most recent location for the owl.

SITE2: This is the second most recent location for the owl.

YR1: The first year the owl was located on the study, this should agree with the first record in the capture matrix.

YREND: When an owl has moved (within or out of study), the first year the owl was located at the most recent site. If the owl emigrated out, only use the first year it was located off the study area.

AREA: Name of study area.

(cont.)

Table 13.3 *(cont.)*

JM: The age at original ringing. When juveniles are recaptured, enter the correct sex field. For analysis, non-recaptured juveniles will be assigned to one of the sexes keeping each cohort at a 50:50 ratio. An entry into the JSX field will be made for "sex assigned" individuals. There are two choices, 1 = the owl was a juvenile male, 0 = the owl was not this age and sex.

JF: The age at original ringing. When juveniles are recaptured, enter the correct sex field. For analysis, non-recaptured juveniles will be assigned to one of the sexes keeping each cohort at a 50:50 ratio. An entry into the JSX field will be made for "sex assigned" individuals. There are two choices, 1 = the owl was a juvenile female, 0 = the owl was not this age and sex.

JSX: This is used to identify juveniles that have been ringed but not recaptured and are temporarily assigned a sex. Y = Yes the sex was assigned. N = No the sex has been confirmed.

SYM: The age and sex at original ringing. There are two choices, 1 = the owl was a one-year-old male, 0 = the owl was not this age and sex.

SYF: The age and sex at original ringing. There are two choices, 1 = the owl was a one-year-old female, 0 = the owl was not this age and sex.

AM: The age and sex at original ringing. There are two choices, 1 = the owl was an adult male, 0 = the owl was not this age and sex.

AF: The age and sex at original ringing. There are two choices, 1 = the owl was an adult female, 0 = the owl was not this age and sex.

NOTE: An entry of −1 in any of the 8 fields above represents a removal from analysis after the last 1 in the capture matrix. This is usually used for emigrated individuals.

PRESITE: If an owl has moved from the original ringing site, then enter the year, age class, and name of any site the owl has been located at, including the current site (text).

SEX: As determined at first capture. M = Male, F = Female, U = Unknown. For juveniles that have their sex determined later, use a second letter code, UM = Unknown when first ringed, but determined a male on recapture, UF = determined a female.

AGE: The age of owl when it first entered the study area, usually the same age as at first capture but immigrating individuals would be the age when they immigrated in. 0 = young of the year, 1 = one year old, and A = Adult.

Q: The reliability of the age determination. E = Exact, a juvenile recapture. A = Approximate, within one year using subadult characteristics. C = Confirmed age. M = Minimum, used for all owls initially captured as adults.

RING: Ring (band) number.

UTMX: UTMX for most current location.

UTMY: UTMY for most current location.

DTRING: The date of the original ringing.

KMT: Only used for owls that have been recaptured. The kilometers an owl has traveled from its previous site to the most recent site.

AZTR: Only used for owls that have been recaptured. The azimuth an owl has traveled from its previous site to the most recent site.

COMMENTS: If an owl has moved from the original ringing site, then enter the year, age class, and name of any site the owl has been located at, including the current site. Also, any other brief comments.

(cont.

Table 13.3 *(cont.)*

VISIT TABLE

DN = Day, night, or visit that overlaps before and after dark (D,N,B).

VT = Visit type, as to whether the visit was a nesting visit, occupation visit, reproductive survey visit, etc.)

BEGT, ENDT, TOTT = In 24 hour clock: beginning survey time, ending survey time, total survey time in minutes.

W, C, P = weather codes for wind, cloud cover, and precipitation level.

SM, SM2= survey method and secondary survey method.

OBSERVER1, OBSERVER2 = Names of observers conducting surveys. If more than two, list the primary names, put other names in comments.

RT = Response type: auditory, visual, none, etc.

NM, NF, NU, NJ, NFG = Number of males, females unknown, juveniles, fledglings respectively encountered in the survey.

NS = Nest status for the visit: incubating, not nesting, nestlings present, fledglings observed, etc.

NSS = Nest status support code for how nest status was determined: direct observation, inferred from roosting female, etc.

NL = Nest site determination: located (L), approximate (A), or center of pair activity but no nest located (C); leave blank for no data.

TS = Type of nest site (tree cavity, nestbox, building, rock pile, etc).

13.12 Quality assurance

Assurance of the quality of data collected and the methods used to summarize, analyze, and interpret the data will be applied to both aspects of the monitoring plan: population survey and habitat assessment. For population surveys, survey timing in demographic study areas, determination of sex and age of individual owls, and capture and marking methods will be conducted according to the forthcoming survey protocols for the Little Owl. Summarization and analyses of the data from the demographic study areas will be subject to these protocols. Survival and reproduction data along with estimates of population trend will be analyzed by using the procedures described in Franklin *et al.* (2004) and Forsman *et al.* (1996). Consistency in the demographic data is paramount, and participants contributing data to the meta-analysis will agree to adhere to a rigid and formal protocol for analytical sessions (e.g., Anderson *et al.* 1999).

For landscape-scale habitat assessment we will rely on the quality assurance protocols for the production of the range-wide vegetation map. Quality assurance for the derived Little Owl habitat maps will rely on the knowledge of province-specific experts as they define the habitat attributes at both the demographic study area scale and province scale. An uncertainty estimate is essential for decision-makers to assess risks associated with their decisions and for scientists to assess the efficacy of the data and associated models.

Figure 13.8 Little Owl monitoring in the Netherlands. Photo René Krekels.

13.13 Organizational infrastructure

The monitoring program outlined here will require a co-ordinated international effort. The key to successfully implementing the monitoring program is a co-ordinated network of personnel and co-operators who will implement individual elements of the monitoring strategy. The annual population surveys, periodic habitat assessments, cumulative data analyses, and integrated syntheses of the individual monitoring efforts implement the monitoring strategy as a whole. Steps to accomplish the strategy are assigning specific tasks to an administrative body, gathering and analyzing the data through standardized methods (some of which await development), and implementing the monitoring program on schedule.

We recommend that a Little Owl monitoring lead be assigned. This individual will work with international counterparts (such as representatives of the International Little Owl Working Group [ILOWG], BirdLife International, World Wide Fund for Nature [WWF], Global Owl Project, etc.) and oversee the organization and implementation of the monitoring plan. Fortuitously, the ILOWG reflects personnel and co-operators enlisted to conduct the demographic studies, assess habitat conditions, and develop predictive models. The monitoring lead and ILOWG can co-ordinate the participation of land managers and key resource management personnel to assure adequate funding and that survey results are integrated into the annual summary reports. The ILOWG also should establish contacts with other scientists outside the European community who are conducting Little Owl monitoring and research.

The monitoring lead and associated program staff will require agency or NGO support, with an estimated equivalent of two to three permanent, full-time positions among the participating agencies. In addition to agency support of co-ordination activities, support will be needed for permanent, full-time agency personnel engaged in implementing the plan. Assuming that demographic studies are continued under the current co-operative working relations and habitat assessments are done in-house, these positions would involve research scientists, research wildlife biologists, statisticians, and GIS specialists.

13.14 Reporting of results and linkage to decision-making

The accomplishments of monitoring for Little Owls will be provided in annual summary reports. A meta-analysis of owl population data from the demographic study areas would be completed every five years beginning in 2010. Range-wide habitat maps would be recompiled every five to ten years in synchrony with monitoring plan schedules.

An interpretive report of monitoring for Little Owls will be completed every five years beginning in December 2010. We recommend that a panel of science and management personnel be assigned the tasks of (1) reviewing the annual monitoring summary reports, and (2) preparing an interpretive report assessing progress in meeting the monitoring goals, objectives, and expected values for the Little Owl under the array of applicable legislation and national policies. Importantly, this report will profile the state of Little Owl populations and their habitat, evaluate the effectiveness of current conservation measures in arresting and reversing the decline in owl habitat and populations and maintaining the viability of owl populations, point out areas of progress and concern, and, as necessary, make recommendations for changes in management practices. This report will provide decision-makers with a scientifically credible evaluation of the state of Little Owl populations and habitat, and would be a reference document for decision-makers during periodic land-use planning reviews.

The status and trend of owl habitat and projected population responses will provide managers with feedback about existing conditions and allow comparison with future expected conditions. The results of these comparisons will provide information for review of the adequacy of management direction. If the trend, or rate of improvement, in habitat conditions is significantly below expectations, then a change in management practices may be required. These changes may involve land-use allocations or management standards and guidelines.

There will be no sirens or alarms wired to these elements to sound a signal for change. We will be responsible for interpreting the monitoring results, and if needed, signaling ourselves of the need for change. In the end, this monitoring plan will provide data only about owl habitat and populations. Knowing how much habitat there is, the survival rates of owls, and how many young they produce are important indicators, but we will be required to assess the indicators and decide whether conservation is proceeding as planned and yielding the results we expected.

14

Citizen conservation and volunteer work on Little Owls: the past, present, and future

Roy S. Leigh

This chapter looks at the role of citizen conservationists, individuals, and groups of voluntary Little Owl enthusiasts who actively engage in the work of conserving the long-term future of the Little Owl in their local areas.

Owls are very evocative birds, which, because of their nocturnal habits, vocal activities, their part in human folklore, and dependence on man's activities, provoke an enormous amount of human interest across the whole wildlife family. The Little Owl is no exception, and the interest in this owl has grown tremendously over the last decade, as communication and conservation initiatives have developed. See Figure 14.1.

14.1 International network

The International Little Owl Working Group (ILOWG) first emerged in 1999, with the aims of increasing communication on Little Owl research, conservation, and education across Europe. A great deal of work had been undertaken by individuals, many of whom began to develop nestbox schemes, tree management initiatives, and research and monitoring programs. A great deal of information was generated by these programs, and all that was needed was a mechanism for sharing the information, hence the conception of the ILOWG. The group began by publishing *Athene News* with the objectives of providing lightweight articles on conservation and monitoring. An internet-based group developed to readily support communications to outposted owl workers so that they were able to tap into a mine of experience from the ILOWG community. The success of this truly international family reflects a membership of over 460 people from 11 countries. This is truly remarkable, and cannot be matched by any other single-species study group! In 2000 the first of three symposia was hosted by French Little Owl workers at Champ-sur-Marne, near Paris. This took the level of exchange of information and co-operative fieldwork up to a new level. This symposium was followed by a second symposium in Flanders (Belgium), which involved national and international presentations, and exchange of discussions regarding standardizing methodologies. These symposia generated great interest from across Europe, and owl workers returned home full of ideas, and energized about developing more initiatives on Little Owls in their regions. Importantly, the participants were eager to contribute to the

Figure 14.1 Little Owl with prey (François Génot).

larger program, and to understand and share in the conservation priorities of Little Owls across Europe.

14.2 National initiatives

A great deal of work has been undertaken by individual ornithologists monitoring local populations and undertaking conservation initiatives, to promote an understanding of Little Owl ecology within their own regions. See Figure 14.2.

Figure 14.2 Little Owl on a fence pole (François Génot).

Belgium

)ver the last few years a number of groups have developed that bridge the interests of rnithologists and the local community. A model for this is Groupe Noctua operating vithin the region of Wallonia, Belgium. See Figure 14.3. This group is led by a pure Little)wl enthusiast, Jacques Bultot, who noticed that the population of Little Owls began to lecline rapidly with the onset of mechanized farming and the introduction of farming quotas,

Figure 14.3 Logo of Groupe Noctua (Belgium).

which increased the pressure on the farmers to intensify their land use. This meant removal of hedgerows, trees, and field headland, which hosted a high percentage of the food for Little Owls. Groupe Noctua undertook a large-scale nestbox installation program, to compensate for the loss of natural tree cavities. All boxes are monitored to understand the breeding ecology of the owls and to monitor the success of the initiatives. The owls are ringed by licensed ringers, which provides further information on the survival rates, mortality causes, and dispersal of owls within the region. This work has yielded interesting information on the causes of the continued decline. Groupe Noctua has operated for some 15 years, and has inspected 1853 nests between 1988–2002. This group has provided a massive input into the understanding of the ecology and causes of the decline of the owls.

It is important that the group continues to recruit new members, particularly children, the generation who shape the future. To do this the group undertakes a series of project-based

Figure 14.4a Conservation activities in Belgium (Groupe Noctua). Photo Jacques Bultot.

events, both in the field and social events, and tasks such as nature trails, managing willow trees, and planting new trees, are undertaken, after which the event is completed by a wood fire barbecue picnic (See Figures 14.4a and b). The results so far are given in Table 14.1.

The development of communications enables groups to discuss issues and to learn from each other. Each population of Little Owls is faced by negative issues, these voluntary groups supported by national ornithological organizations work to develop and deliver conservation programs for Little Owls.

Some important conservation actions are also found within the framework of Natuurpunt, the largest nature conservation group in Flanders. A large-scale research program was carried out by over 400 volunteers (Van Nieuwenhuyse *et al.* 2001d). Local ringing and monitoring initiatives are run by volunteers e.g., Philippe Smets, Ronny Huybrechts, and Stanny Cerulis, or by communities like Overijsse (Luc Vanden Wyngaert, personal communication). See Figure 14.5.

Luxembourg

The decline of the Little Owl across Europe has been documented elsewhere within this volume. In Luxembourg the Little Owl is known under its older Luxembourgish name of *Doudevull* (the bird of death) as it had a habit of nesting around cemeteries at village edges, where its calls were heard.

Figure 14.4b Conservation activities in Belgium (Groupe Noctua). Photo Jacques Bultot.

Table 14.1 *Overview of management activities carried out by Groupe Noctua, Wallonia, Belgium, and the results obtained.*

		Pollarded willows					
		Existing trees		New trees		Participants	
	Number of management sessions	Management of trees >10 years	Management of trees <10 years	Planted trees	Managed new trees	Number	Hours
Total	134	328	138	1870	244	1075	599
Average per year	14	33	14	187	24	107	60

In the mid-1980s, some enthusiastic volunteer owl workers noticed that the owl was becoming scarcer so they began to erect nestboxes around the edges of villages. The situation today is that all Little Owls breeding in the region nest in the nestboxes erected for them. This success is a testimony to the volunteers and their foresight. This has now been formalized into a national Little Owl conservation plan, developed by ornithologists under the leadership

Figure 14.5 Logo of Steenuilenproject Dijleland (Belgium).

of Patrick Lorgé (Centrale Ornithologique LNVL), the monitoring continues to be carried out by voluntary fieldworkers.

The Netherlands

In 1974, Piet Fuchs started scientific research on the Little Owl in the river forelands near his hometown Wageningen. Piet was working for the Agricultural University of Wageningen and his research then already showed a decline in the number of Little Owls in this region. At that time, the decline in the owl was due mainly to the disappearing high-stemmed apple tree orchards and the subsequent disappearance of nesting opportunities in these orchards, plus the scaling-up of the agricultural lands, which made them less attractive to Little Owls. When, in the 1970s and 1980s, Johan de Jong, in succession of Sjoerd Braaksma, picked up

Figure 14.6 Logo of Steenuilen Overleg Nederland (STONE) working group (Netherlands).

the protection work of the Barn Owls, interest was also drawn to the delicate and precarious situation of the Little Owl. A lot of people and bird-watching groups started drawing up inventories of the Little Owls in their parishes and started taking conservation actions, such as putting up nestboxes, and managing and planting wood banks to restore the small-scaled landscape. Further, breeding biology research was started. Among the researchers were Ronald van Harxen and Pascal Stroeken, who started their research in 1986. Hein Bloem started his co-ordination work on the Little Owl in 1993 and networked with volunteers all over the country and abroad. In 1997, STONE (Steenuilen Overleg Nederland) was founded by Ronald van Harxen, Pascal Stroeken, and Hein Bloem and had a total of 120 members at start-up. In every province throughout Holland, STONE has a regional contact. The regional contacts mostly run their own research and conservation schemes as members of the local Little Owl working groups. STONE regularly publishes its newsletter *Athene* and dispatches Little Owl questions through its website www.steenuil.nl on which it publishes reports from its members, gives a platform to queries, tips and tricks, and provides advice from its members. See Figure 14.6.

Since its establishment STONE has not only triggered Little Owl conservation in the Netherlands but is also supplying data to SOVON (Dutch Center for Field Ornithology), influences decision-making levels through its board, discusses nationwide conservation schemes, and takes part in establishing a new Little Owl conservation plan in conjunction with the Ministry of Agriculture and Vogel Bescherming Nederland. At the beginning of 2005, there were about 240 members. The development of this network was aided by the publishing of a monitoring handbook, which provided the regional observers with standardized methodology for undertaking population monitoring and conservation initiatives.

The development of Little Owl conservation in the Netherlands has continued on a more regional basis, as the understanding of the decline has developed, the more pressing issues have been addressed. In Groningen, northern Netherlands, Jan Van't Hoff undertook a very detailed study on the very fragmented population of Little Owls (Van't Hoff 2001), and developed a detailed conservation program of nestbox provision, pollarded willow management, and replanting of traditional orchards. In addition he has taken the Little Owl to the people, through the opening of a Little Owl information center at Oldehove, and the production of a "little" owl book for children. See Figure 14.7.

STEENUILwerkgroep Groningen

Athene noctua

Figure 14.7 Logo of Steenuilwerkgroep Groningen (Netherlands).

France

In France, a Little Owl Working Group has existed for more than ten years and was created by Patrick Lecomte, Yvan Tariel, and Jean-Claude Génot under the umbrella of the Fonds d'Intervention pour les Rapaces (FIR), which became a part of Ligue pour la Protection des Oiseaux (LPO). They regularly publish a newsletter, *Chevêche Infos*, and organize an annual meeting. Their aim is to implement the national action plan for the conservation of the Little Owl.

The group of LPO Alsace is carrying out an international project for the Little Owl conservation with Germany and Switzerland, including the conservation of orchards.

Since 1986, a Little Owl working group has existed for the Regional Natural Parks and is co-ordinated by Jean-Claude Génot, in the frame of his job as nature protection officer in the Northern Vosges Regional Natural Park. This working group gathers people from ten parks in different regions that survey and protect the Little Owl. They started the first monitoring program in France, and survey for calling males every four years in the same areas. This group has also created the biannual "Night of the Owl", which first took place in 1997. This event is organized by the LPO and the French Federation of the Regional Natural Parks. The initiative is a great success with several hundred events taking place across the entire country, and it brings together several thousand people. The Little Owl working group in the Regional Natural Parks is also at the origin of the national action plan for the Little Owl, which was approved by the French Ministry of Environment – but was not really applied in the field due to the lack of funding, will, and co-ordination at the ministry. This plan ought to be carried out in each region with all of the partners concerned. Only some of the actions of this plan are implemented now, such as the monitoring program of the Little Owl or the "Night of the Owl". The LPO published a technical document about the conservation measures and study of the Little Owl for the wildlife managers within habitats suitable for the species and for conservationists who need information and data.

Germany

The working group for the conservation of threatened owls ("AG Eulen") is an association of German-speaking owl experts and is honorarily managed. In 1974, a Little Owl group was founded in Northrhine-Westphalia (NRW). It was transformed into the "AG Eulen" working group in 1979, which is active in the whole of Germany. Among the actual 550 members are 30 local and regional working groups and a lot of individual members, who are engaged in the conservation of the Little Owl. The conservation of habitats, for example the managing of fruit trees and orchards, the cutting of pollarded willows, or the installation and monitoring of nestboxes (thousands of them in NRW), are the most important activities. Their journal *Eulen-Rundblick* contains a lot of information about research and conservation projects of owls, as does the website (www.ageulen.de). At the annual conferences many of the contributions deal with Little Owls.

Most people engaged in the conservation of Little Owls are from NRW, the home of 80% of the German Little Owl population. This is the reason why NRW has a particular responsibility for the conservation of owls in Germany, and why voluntary work on the Little Owl is concentrated there. However, the government of NRW does not meet its responsibility (i.e., Little Owl habitats such as orchards are not protected by laws and rules, and other measures for Little Owls ordered by the European Union are not applied).

In the spring of 2004, a conference on the conservation of the Little Owl in NRW took place in Kleve with 200 participants by "AG Eulen", "Nordrheinwestfälische Ornithologen-Gesellschaft", "Naturschutzbund Deutschland (NABU) Landesverband NRW", and "Landesanstalt für Ökologie, Bodenordnung und Forsten". The aim of this conference was to give an overview of the actual situation and development of the Little Owl population in Germany, especially in NRW. Furthermore, there were contributions and discussions on monitoring programs, monitoring methods, aspects of habitat use (e.g., the importance of natural cavities and nest sites in buildings), and measures of conservation. The main conclusion of the conference was that the voluntary work on Little Owls is not enough to ensure the long-term conservation of the population in Germany. Furthermore, systematic monitoring, a better application of existing European and German regulations, and the preservation of extensive meadows through agricultural subsidizing policies are absolutely necessary. See Figure 14.8.

Switzerland

A national action plan for Little Owls is currently being implemented by the Swiss Agency for the Environment, Forests and Landscape (government), SVS/BirdLife Switzerland, and the Swiss Ornithological Institute. The final plan is due in September 2008. All active Little Owl conservation groups are co-operating under the chairmanship of Christian Meisser.

At the regional level, four main projects are running. In the Basel region, an international project was initiated in 2000 between Switzerland, Germany, and France, and is trying to restore historical links between neighboring populations (leaders: Hansruedi

Figure 14.8 Little Owl on a fence pole. Photo Ludo Goossens. See Plate 39.

Schudel, SVS/BirdLife Switzerland; Christian Braun, LPO/BirdLife France; Christian Stange, NABU/BirdLife Germany). The main initiatives are:

- census of the calling males every spring in several areas
- installation, maintenance, and control of over 500 nestboxes and holes
- evaluation of breeding success
- restoration of traditional landscapes by planting new fruit trees, and development of local products made with the fruit of traditional orchards.

In the Geneva region, a local working group lead by Christian Meisser, Patrick Albrecht, and Christian Fosserat (society "Nos Oiseaux"), has concentrated on a population of around 35 pairs since 1982, with the following initiatives:

- census of the calling males every spring (about 50 km^2)
- maintenance and control of over one hundred nestboxes and natural cavities
- evaluation of breeding success
- ringing of the birds (over 800) to analyze the survival rates and the movements of young and adults.

An orchards conservation program (implemented by Pro Natura Geneva) has been running since 1992 and has yielded over 800 new trees with thousands of trimmed trees. The

Agro-Ecological Net ("réseau agro-écologique"), based on the voluntary participation of farmers with professional co-ordination financed by the state, also sponsored specific Little Owl initiatives.

An action plan was made in 2005 by "Nos Oiseaux", with the support of the regional administration.

In the Ajoie region (canton du Jura), a local working group, led by Arnaud Brahier and Damien Crelier, has concentrated on a population of around 15–20 pairs since 1997 (following a break after the work of Michel Juillard during the 1980s), with the following initiatives:

- census of the calling males every spring (about 50 km^2)
- maintenance and control of over 50 nestboxes and holes
- evaluation of breeding success.

The orchards conservation program and the agro-ecological measures have been introduced here recently too.

An action plan has been implemented in Ajoie since 2003. The co-ordination is done by SVS/BirdLife Switzerland, with four other groups among which "Nos Oiseaux" and "Pro Natura Jura" are the most well known. The project is mainly federally financed, but the regional administration and the nature conservation groups also provide sponsorship.

In Tessin Canton, a plan is being developed (by the society "Ficedula" and the regional administration) for the plain of Magadino, in conjunction with an action plan for the Hoopoe (*Upupa epops*).

Czech republic

Libor Oplustil (personal communication) studies and conserves a local Little Owl population within the framework of the Czech Union of Nature Conservation, through nestboxes in an area of 2000 km^2 near the town of Brno, South Moravia Region, Czech Republic.

Italy

In Italy, exciting times are occurring, with the development of GIC ("Gruppo Italiano Civetta" – the Italian Little Owl group) by Marco Mastrorilli and his colleagues; they offer new impetus to the work currently underway in the northern European countries. The regional make-up of the group in Italy crosses many temporal terrains and habitats, which will provide some excellent opportunities to study the adaptability of the Little Owl in different landscapes. The first meeting of 102 Italian Little Owl enthusiasts took place on March 21, 2004 in Osio near Bergamo, northern Italy. A newsletter is edited regularly and the first book on the Italian Little Owl was published in 2005 (Mastrorilli 2005). For the near future, a national standardized survey is planned. Also, "the Night of the Little Owl" (in Italian: *La Notte della Civetta*) was started in Italy in 2005, and the initiative has been

Figure 14.9 Logo of the Italian Little Owl Working Group.

a great success with many events taking place across the entire country (M. Mastrorilli, personal communication). See Figure 14.9.

The future of the Little Owl is in our hands; within the remit of fiscal agri-ecology schemes we need to develop methods to inform farmers and landowners how best to manage the land to the benefit of the Little Owl.

We need to understand the populations and the causes of declines better, so we can be more prescriptive with the conservation measures we apply.

The population data will serve to add strength to lobby for the Little Owl's inclusion as a priority agri-environment target species. This will help promote the needs of the owl, and lead to the development of national action plans.

The role of the voluntary Little Owl fieldworker has never been more important. Individually we need to continue to develop our conservation programs, advise farmers and landowners, monitor nest sites, manage trees, plant more trees, educate, and take the owl to the people.

On a collective basis, the future looks strong with the established groups in northern Europe joined by an Italian group. However, collectively, the need to further the understanding will move on to a more informed level.

This can be achieved if we:

- continue to share experiences
- develop best practice methods
- develop workshops on key monitoring skills
- hold regular symposiums.

The key to achieving these points is communication, sharing information, which is the foundation stone of the ILOWG.

This volume is a testimony to all voluntary fieldworkers and conservationists who have freely given their time and energy; their reward is to see the owl that they love and cherish continue to survive and flourish.

Glossary

The terms found in this glossary relating to the body characteristics of birds are from the US Geological Service website:
http://www.mbr-wrc.usgs.gov/id/framlst/Glossary/glossary.html. Other terms were derived from a variety of other sources. See Figure G.1.

We draw on the definitions provided by Steenhof (1987) for descriptions of some of the demographic terms below.

The different parts of the owl are first given. The foot consists of toes with claws and the tarsus or tarso-metatarsus. The expression "foot" is used when no specific part of it is referred to. Owls have four toes: three at the front and one hind toe. Of the front toes the outer one may be turned backwards (as owls often do when they are roosting).

An illustration and list of terms concerning the wing and tail are also given. The primaries are counted from the middle of the wing outwards to the tip, and the secondaries from the middle of the wing inwards to the body. When wing length is measured, the wing has to be closed and the distance is taken from the wrist or bend to the longest primary feather. The tail length is the distance from the base of the central tail feather to the tip. Tail feathers are counted from the center outwards. Each feather has a shaft and an outer and inner web. See Figures G.2–G.4.

abdomen Ventral part of the bird. Synonym: belly. In Figure G.2 it is referred to as belly.
alula Three feathers springing from the base of the primaries. Synonym: alular quills.
alular quill coverts Feathers overlying the bases of alula.
alular quills Three feathers springing from the base of the primaries. Synonym: alula. In Figures G.2 and G.3 it is referred to as alula.
anthelmintics Quinoline-derived, organophosphorous compound drugs used to kill parasites, including roundworms, whipworms, hookworms, pinworms, trichinella (trichinosis), and other less common organisms. Also indexed as: Albendazole, Albenza®, Antiminth®, Biltricide®, Diethylcarbamazine, Hetrazan®, Ivermectin, Mebendazole, Mintezol®, Oxamniquine, Pin-Rid®, Praziquantel, Pyrantel, Stromectol®, Thiabendazole, Vansil®, and Vermox®.
auricular Area around ear opening. Synonym: ear patch.
axillary Ventral area between the body and the wing. Synonym: wingpit.
back Dorsal part of the bird.

Figure G.1 Little Owl looking sideways (François Génot).

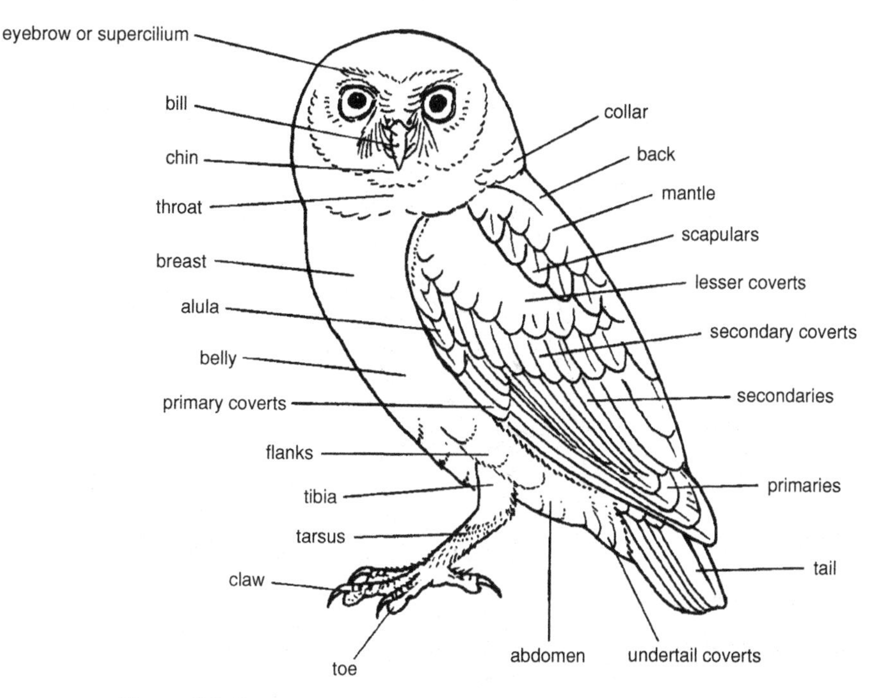

Figure G.2 Topology of a typical *Athene* owl (König *et al.* 1999).

belly Ventral part of the bird. Synonym: abdomen.

bill Beak.

body Main mass of the bird as distinguished from its appendages.

breast Front part of the chest.

breast band Stripe across the breast.

breast spot Small, differently colored area on the breast.

breeding pair A pair of owls that lay at least one egg. A breeding attempt is confirmed by observing an incubating adult, eggs, young, or any field sign that indicated eggs were laid.

brood size Number of nestlings per nest that survive after hatching for a specified period (e.g., to feeding stage).

cap Top of the crown.

census The complete count of every individual in a given population (very rarely possible in wild populations).

cere Fleshy area between the beak and face.

cheek Area bounded by lore, eye, auricular, and lower mandible.

chest Front part of the body.

chin Part of the face below the bill.

clutch size Number of eggs in a nest immediately following the laying of the last egg during a particular reproductive cycle.

collar Rear portion of crown. Synonym: nape, hindneck.

comb Colored area over eye found in males.

commissure Base of the bill where the mandibles join. Synonym: gape, rictus.

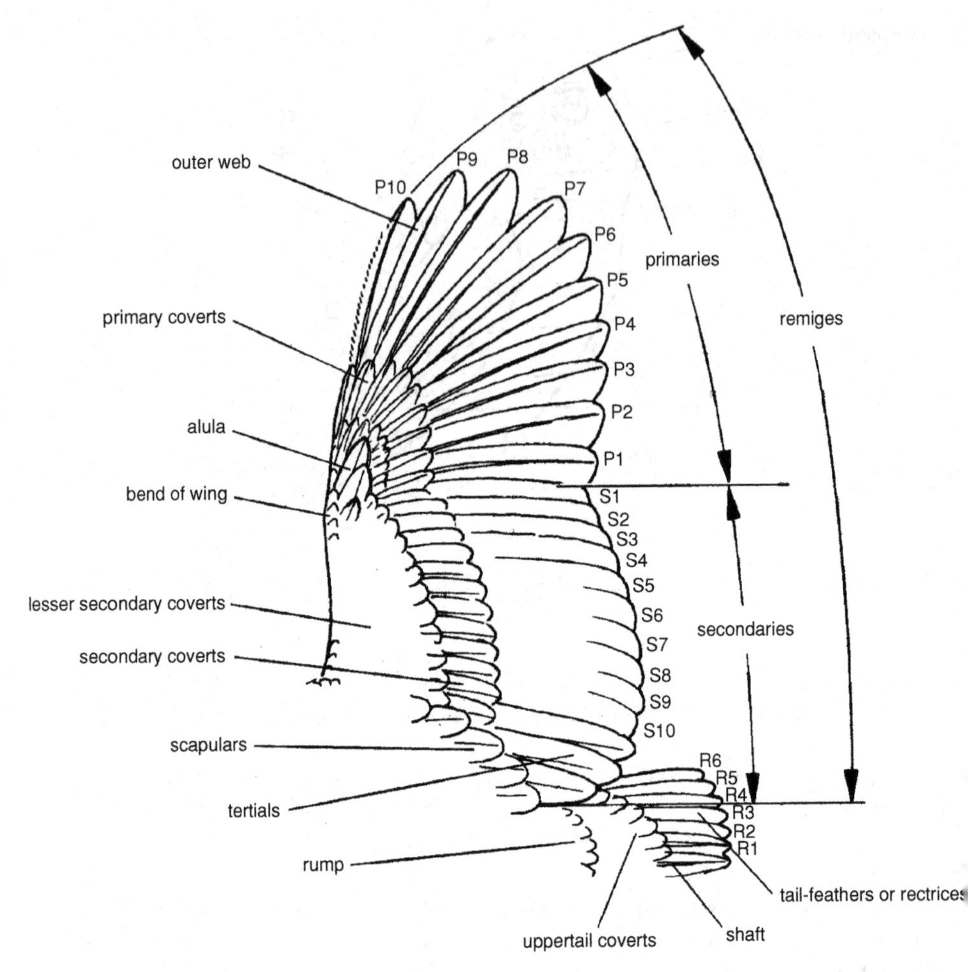

Figure G.3 Topology of wing and tail of a typical *Athene* owl (König *et al.* 1999).

crest Tuft on the head.

crissum Feathers covering underside of base of tail. Synonym: undertail coverts. In Figure G.2 they are referred to as undertail coverts.

crown Top of the head.

culmen Upper ridge on bill.

detection rate A summary of the number of owls that were located across a given area, e.g., (number of individuals heard)/(number of playback sessions performed) (Galeotti 1989; Sará & Zanca 1989; Centili 2001b). Detection rates are rather low, as observers cannot separate owl absence from owl silence and surveys conducted in areas of unsuitable habitat (no owls present to respond) are incorporated into the summary. The calculated detection rate is inversely proportional to the number of playback stations surveyed with no owls responding.

dihedral Wings of a flying bird held at an angle appearing to form a "V".

dispersal The term that applies to a young owl as it leaves its natal territory to mature and search out and settle into a territory of its own.

Figure G.4 Different bill measurements as illustrated by a schematic drawing of a Bald Eagle head showing (A) culmen length with cere, (B) beak depth, (C) culmen length.

Breeding dispersal (demographic context): reflects the shifts that adults owls make, for example, a female may breed in one location one year, and may move to a different territory for the next breeding season.

Successful dispersal: an owl is one who moves from his/her natal territory, disperses, and lives long enough to become part of a mated pair.

ear patch Area around ear opening. Synonym: auricular.

ears Rounded, earlike areas on the face. Synonym: facial discs.

ecoregion A relatively large unit of land or water containing a characteristic set of natural communities that share a large majority of their species, dynamics, and environmental conditions.

eye Organ of sight.

eye line Line of feathers in front of and behind the eye.

eye ring Pale-colored feathers encircling the eye.

eyebrow Line of feathers above the eye. Synonym: supercilium, superciliary line.

eyelid Skin fold covering the eye.

face Front part of the head.

facial discs Rounded, earlike areas on the face. Synonym: ears.

feet Terminal parts of the legs.

flank Area between the belly and the wings, more posterior.

flank stripe Band on the flanks.

fledging rate Number of young that fledge the nest divided by the brood size.

flight feathers Primaries and secondaries.

floater A non-territorial owl, who discretely exists within the population. In territorial owls, adults (typically the male) vocalize to affirm their territorial boundaries. Meanwhile floaters remain silent (or mostly silent), not wanting to be found as they are often aggressively chased by the territory holder, who views the floater as a potential competitor for the available resources (food or mate). The function of a floater is to quickly fill territories as they become vacant through the death or movement of the previous territory holder. For Little Owls, we urge future research conducted in high-density areas to examine the context of floaters, as anecdotal observations to date suggest that floater owls may, in fact, be more vocal and aggressive rather than being discrete.

forehead Part of the face above the eyes.

foreneck Front part of the neck. Synonym: throat, jugulum, throat patch.

frontal shield Extension of the bill onto the forehead.

fruit tree crops: Actively managed, low-stem orchard (fruit trees of low height). These fruit trees are planted very closely in rows to allow easy treatment with chemicals and other recurring management activities.

gape Base of the bill where the mandibles join. Synonym: commissure, rictus.

gonys Lowermost ridge on lower mandible.

greater secondary coverts Feathers overlying bases of secondaries.

gular region Between the chin and the foreneck.

hatching rate (or hatching success, or hatch rate) Number of eggs per nest that successfully hatch into live young divided by the clutch size.

head Upper part of the body.

head stripes Bold lines on the head.

hindhead Rear portion of crown. Synonym: occiput.

hindneck Back of the neck. Synonym: nape, collar.

home range The area that embraces all activities of a bird or pair over a given time period. For owls this is often expressed over the time period of one year.

horns Paired contour feathers arising from head.

inner primaries Group of primaries closest to the body.

inner secondaries Group of secondaries closest to the body.

inner wing Shoulder, secondaries, and secondary coverts.

iris Colored part of eye.

jugulum Front part of the neck. Synonym: foreneck, throat, throat patch.

knee Joint in the middle part of the leg.

leading edge of wing Front edge of the wing in flight.

leg Limb used for supporting the bird.

lesser secondary coverts Feathers overlying bases of median secondary coverts. Synonym: marginal coverts, shoulder.

leucistic Plumage aberration of genetic origin, typically of faded or washed-out coloration; not to be confused with albinism.

lore Area between the eye and the bill.

lower mandible Lower part of the bill.

lower mandibular tomia Cutting edges of lower mandible.

malar streak Area at the sides of the chin. Synonym: whisker, moustache.

mandibular ramus Prong-like, posterior projection from bill.

mantle Upper surface of the wings and the back.

marginal coverts Feathers overlying bases of median secondary coverts. Synonym: lesser secondary coverts, shoulder. In Figure G.3 they are referred to as lesser secondary coverts.

median line Stripe through the crown.

median secondary coverts Feathers overlying bases of greater secondary coverts.

meta-population A spatial distribution of distinct subpopulations, separated by large distances or barriers and connected by dispersal movements.

moustache Area at the sides of the chin. Synonym: whisker, malar streak.

mouth Cavity bounded by the bill.

nape Back of the neck. Synonym: hindneck, collar. In Figure G.2 it is referred to as collar.

nasal canthus Anterior corner of eye.

nasal fossa Depression in which nostril is located.

neck Part connecting the head to the main part of the body.

neck patch Inflatable sac on neck used by males in courtship display.

nest disturbance A nest disturbance is considered to be any occasion when a nest was entered or when adult owls react to the presence of researchers or other human activity in the vicinity of a nest.

nictitating membrane Translucent, vertical fold under the eyelid.

nostril External naris.

occiput Rear portion of crown. Synonym: hindhead.

outer primaries Group of primaries farthest from the body.

outer secondaries Group of secondaries farthest from the body.

outer tail feathers Part of the tail farthest from the center.

outer wing Alula and primaries.

patagial mark Dark patch on leading edge of underside of inner wing.

pelagic Living on the open ocean rather than coastal or inland bodies of water.

pileum Top of the head extending from the base of the bill to the nape.

pinnae Projecting feathers.

plumes Large, conspicuous, showy feathers.

post-fledging check Inspection of nestbox and surroundings after the young have reached more than 30 days.

primaries Flight feathers attached to the "hand".

primary coverts Feathers protecting and covering the primaries.

primary numbering System for assigning a number to each primary.

pupil Contractile aperture in iris.

rectrices Conspicuous feathers forming posterior margin of tail.

remiges Flight feathers of the wing, which include the primaries, secondaries and tertiaries.

response rate Used in the context of survey techniques, the response rate of live, wild owls to the broadcasted calls of their own species. Response rate reflects the proportion of owls that, when within hearing range of the broadcasted playback calls, actually respond to the playback. For example, if five out of ten owls that were within the broadcast range of playback recordings actually responded to the playback, this is a response rate of 50%. However, unless the owls' locations are first determined with radio-telemetry, observers employing playback have no way of knowing whether surveys without responses reflect the absence of owls, whether the territory holders were outside of hearing range, or whether owls were present and did not reply.

rictal bristles Stiffened feathers near bill.

rictus Base of the bill where the mandibles join. Synonym: gape, commissure.

ringing age of young The age of young when their legs are formed and of sufficient strength and size to hold a ring, typically about three weeks of age. Ringing is typically conducted about one week before the young "jump" from the nest. Because young have achieved ringing age does not imply that they have successfully fledged.

ruffs Fringe of feathers growing on the neck.

rump Area between the uppertail coverts and the back.

scapulars Area of feathers between the back and the wings.

secondaries Flight feathers attached to the "elbow".

secondary coverts Feathers protecting and covering the secondaries.

shoulder Feathers overlying bases of median secondary coverts. Synonym: lesser secondary coverts, marginal coverts. In Figure G.3 it is referred to as lesser secondary coverts.

side Area between the belly and the wing.

side of neck Area of neck between foreneck and hindneck.

spectacle Eye ring and supraloral line together.

suborbital ring Eyelids.

subterminal band Stripe before tip of tail.

successful brood (also successful breeding attempt) A pair of owls having one or more young that successfully fledge.

superciliary line Line of feathers above the eye. Synonym: supercilium, eyebrow.

supercilium Line of feathers above the eye. Synonym: eyebrow, superciliary line.

supraloral line Line of feathers above the lore.

survey An estimate of the number of animals in a given area (e.g., number of territorial pairs) or of the relative frequency of encounters of animals (e.g., number per unit transect length or time period). Repeated surveys can be used to estimate the trend of a population. Surveys provide an efficient way of collecting information from a large number of sites and can also help determine the overall distributional range of a species in different seasons. The structure of surveys is intended to reduce observer bias, and surveys are best conducted in a standardized way to ensure reliability and validity.

tail Feathers extending from the rear of the bird.

tail coverts Under- and uppertail coverts.

tail numbering System for assigning a number to each tail feather.

tape playback Survey method using broadcasting of Little Owl call to provoke reaction of birds present.

tarsus Part of the leg between the knee and the foot.

temporal canthus Posterior corner of eye.

terminal band Stripe at tip of tail.

territory The area around the nest that is defended (Newton 1979).

tertiaries Feathers adjoining the secondaries.

throat Front part of the neck. Synonym: foreneck, jugulum, throat patch.

throat patch Front part of the neck. Synonym: foreneck, throat, jugulum.

tibia Part of the leg above the knee.

toe Digit attached to the feet.

trailing edge of wing Rear edge of the wing in flight.

underparts Belly, undertail coverts, chest, flanks, and foreneck.

undertail coverts Feathers covering underside of base of tail. Synonym: crissum.

underwing Underside of wing.

unsolicited calling Spontaneaous calling of Little Owl without stimulation by tape playback.

upper mandible Upper part of the bill.
upper mandibular tomia Cutting edges of upper mandible.
upperparts Back, rump, hindneck, wings, and crown.
uppertail coverts Feathers covering upperside of base of tail.
upperwing Upperside of wing.
whisker Area at the sides of the chin. Synonym: moustache, malar streak.
wing Moveable feathered appendage.
wing bars Pale tips of greater and median secondary coverts.
wing coverts Primary and secondary coverts.
wing lining Median, lesser, and marginal coverts on underwing.
wing stripe Paler area at base of flight feathers.
wingpit Ventral area between the body and the wing. Synonym: axillary.
wrist Area at base of the primaries.

References

Abdusalyamov, I. A. 1971. *Fauna Tadzhikskoi SSR [Fauna of Tajik SSR]*. Vol. XIX. Part 1. Ptitsy [Birds]. Dushanbe, Donish. Little Owl (*A. n. bactriana* Hutton), pp. 332–336. (In Russian).

Adam, G. 1973. Steinkauz (*Athene noctua*) jagt zu Fuss nach Heuschrecken. *Orn. Mitt.* **25**: 249. (In German).

Afanasova, L. V. & Khokhlov, A. N. 1995. Pitanie domovogosycha v period razmnojeniya na zapade Stavropol'skogo kraya [The Little Owl feeding during breeding period in the west of Stavropol Territory]. Khishchnye ptitsy i sovy Severnogo Kavkaza [Birds of prey and owls of the North Caucasia]. *Trudy Teberdinskogo gosudarstvennogo zapovednika [Proceedings of Teberda State Nature Reserve]* **14**: 187–190. (In Russian).

Agelet, A. 1979. Nota sobre la alimentacion del Mochuelo comun *Athene noctua* (*Aves, Strigidae*). *Misc. Zool.* **5**: 186–188. (In Spanish).

Al-Melhim, W. N., Amr, Z. S., Disi, A. M. & Katbeh-Bader, A. 1997. On the diet of the Little Owl, *Athene noctua*, in the Safawi area, eastern Jordan. *Zoology in the Middle East* **15**: 19–28.

Alivizatos, H., Goutner, V. & Zogaris, S. 2005. Contribution to the study of the diet of four owl species (Aves, Strigiformes) from mainland and island areas of Greece. *Belgium Journal of Zoology* **135**: 109–118.

Allanazarova, N. A. 1988. Nekotorye dannye po pitaniyudomovogo sycha v yugo-zapadnom Uzbekistane [Some data on the Little Owl feeding in Uzbekistan]. Ekologiya, okhrana i ratsional'noeispolízovanie ptits Uzbekistana [Ecology, conservation and wise management of birds of Uzbekistan]. Materialy II Respublikanskoi ornitologicheskoi konferentsii [Materials of the 2nd Republican Orn. Conf.]. Tashkent, November 23–25, 1988. Tashkent, Fan Press. pp. 6–7. (In Russian).

Ametov, M. 1981. *Ptitsy Karakalpakii i ikh okhrana [Birds of Karakalpakia and their Conservation]*. Nukus. (In Russian).

Ancelet, C. 2001. Prises de bains par une jeune Chouette chevêche *Athene noctua*. *Héron* **34**: 105–106. (In French).

Ancelet, C. 2003. Exiguïté de la cavité de reproduction et mortalité juvénile chez la Chouette chevêche *Athene noctua*. *Héron* **36**: 160–165. (In French).

Ancelet, C. 2004. Parade et accouplement chez la Chevêche d'Athéna *Athene noctua*. *Alauda* **72**: 211–219. (In French).

Figure R.1 Little Owl attacking (François Génot).

Anderson, D. R., Burnham, K. P., Franklin, A. B. *et al.* 1999. A protocol for conflict resolution in analyzing empirical data related to natural resource controversies. *Wildlife Society Bulletin* **27**: 1050–1058.
Anderson, S. I. 1949. Little Owl raiding starlings' nest. *New Zealand Bird Notes* **3**: 110.
Andreotti, A. 1989. Civetta *Athene noctua. In* Spanò, S., Truffi, G. & Burlando, B. (eds). *Atlante degli uccelli nidificanti in Liguria Regione Liguria*. (In Italian).
Andrews, I. 1995. *The Birds of the Hashemite Kingdom of Jordan*. Mussleburgh.
Andrews, P. 1990. *Owls, Caves, and Fossils*. Chicago, Illinois: University of Chicago Press.

Angelici, F. M., Latella, L., Luiselli, L. & Riga, F. 1997. The summer diet of the Little Owl (*Athene noctua*) on the island of Astipalaia (Dodecanese, Greece). *Journal of Raptor Research* **31**: 280–282.

Angyal, L. & Konopka, H. P. 1975. Beobachtungen an einem zahmen Steinkauz. *Gef. Welt* **99**: 212–217. (In German).

Antczak, J., Kotlarz, B. & Pędziwiatr, R. 1995. Occurrence of Owls Strigiformes in selected areas of the Middle-Pomeranian Region. XVI Meeting of the Polish Zoological Society, Łodź, 14–16 Sept 1995: 6. (In Polish).

Anthonioz, J., Bavoux, C. & Seguin, N. 2000. Prédation du Pélobate cultripède *Pelobates cultripes* par la Chevêche d'Athéna *Athene noctua* dans l'île d'Oléron (Charente-Maritime). *Annales de la Société des Sciences Naturelles de la Charente-Maritime* **8**: 1097–1100. (In French).

Ataev, K. 1977. Novye dannye po ekologii domovogo sycha [New data on the Little Owl ecology]. *VII Vsesoyuznaya ornitologicheskaya konferentsiya [VIIth All-Union Orn. Conference]*. Kishinev, Naukova Dumka Press. 1: 196–197. (In Russian).

Audenaert, T. 2003. Voorkomen en verspreiding van de Steenuil *Athene noctua* te Sint-Pauwels tijdens het broedseizoen 2003. *Groene Waasland* **132**: 10–13. (In Dutch).

Augst, U. & Manka, G. 1997. Vorkommen, Verbreitung und Bestandsentwicklung von Steinkauz (*Athene noctua*), Sperlingskauz (*Glaucidium passerinum*) und Rauhfusskauz (*Aegolius funereus*) in der Sächsischen Schweiz. Beiträge der bundesweiten Vortragstagung "Rauhfusskauz und Sperlingskauz in Deutschland-Vorkommen, Reproduktionsbiologie und Schutz" am 16. und 17. September 1995 in Bad Blankenburg, Freistaat Thüringen. *Naturschutzreport* **13**: 122–131. (In German).

Austin, O. L. 1948. The birds of Korea. *Bulletin of the Museum of Comparative Zoology of Harvard College* **101**: 1–301.

Bakaev, S. 1974. K biologii razmnozheniya domovogo sycha [*A. noctua bactriana* Hutton] v nizov'yakh reki Zerafshan [On breeding biology of the Little Owl (*A. noctua bactriana* Hutton) in Lower Zerafshan River]. *Trudy Samarkandskogo universiteta [Proc. of Samarkand University]* **211**: 35–39. (In Russian).

Baker, J., French, K. & Whelan, R. J. 2002. The edge effect and ecotonal species: bird communities across a natural edge in southeastern Australia. *Ecology* **83**(11): 3048–3059.

Baker, S. 1926. *Athene noctua ludlowi*. *Bulletin of the British Ornithologists' Club* **47**: 58.

Bameul, F. 1982. *Hydrophilus piceus* L. captured by an owl. *Balfour Browne Club Newsletter* **25**: 14.

Bangs, O. & Peters, J. L. 1928. *Athene noctua impasta*. *Bulletin of the Museum of Comparative Zoology of Harvard University*, **68**: 330.

Bannermann, D. A. 1955. *The Birds of the British Isles*. 4. Edinburgh, London: Oliver and Boyd, pp. 198–207.

Barabash-Nikiforov, I. I. & Semago, L. L. 1963. *Ptitsy Yugo Vostoka chernozemnogo tsentra [Birds of Southeast of Black-Earthcentre]*. Voronezh: Voronezh University Press. (In Russian).

Barber, W. E. 1925. Increase of Little Owl in north Lancashire. *British Birds* **19**: 26.

Barbu, P. 1978. Contribution à l'écologie des petits mammifères du Sud de la R. S. Roumanie. *An. Univ. Buc. Biol.* **27**: 101–104. (In French).

Barbu, P. & Sorescu, C. 1970. Contribution concernant la nourriture de la Chouette (*Athene noctua noctua* Scop.). *An. Univ. Buc. Biol.* **19**: 67–72. (In French).

Bart, J. & Robson, D. S. 1995. Design of a monitoring program for northern spotted owls. *In* Ralph, C. J., Sauer, J. R. & Droege, S. (eds). *Monitoring Bird Populations by Point Counts*. General Technical Report PSW-GTR-149. Albany, CA: US Department of Agriculture, Pacific Southwest Research Station, pp. 75–81.

Barthelemy, E. 2000. L'avifaune du massif du Garlaban et de sa périphérie. Suivie de notes sur les mammifères, les reptiles et les amphibiens. *Faune de Provence (C.E.E.P.)* **20**: 29–65. (In French).

Barthelemy, E. & Bertrand, P. 1997. Recensement de la Chevêche d'Athéna *Athene noctua* dans le massif du Garlaban (Bouches-du-Rhône). *Faune de Provence (C.E.E.P.)*, **18**: 61–66. (In French).

Bashta, T. V. 1994. Do ekologii zhyvlennya khatnjogo sycha [On the Little Owl feeding ecology]. Materialy 1-j konferentsii molodykhoritologiv Ukrainy [Lutsk, 4–6 bereznya 1994). [Materials of the 1st Conference of Young Ornithologists of Ukraine, March 4, 1994, Lutsk]. Chernivtsi. pp. 126–127. (In Ukrainian).

Battaglia, A., Ghidini, S., Campanini, G. & Spaggiari, R. 2004. Heavy metal contamination in little owl (*Athene noctua*) and common buzzard (*Buteo buteo*) from northern Italy. *Ecotoxicology and Environmental Safety* **60**: 61–66.

Baudvin, H. 1974. Le surmulot (*Rattus norvegicus*) proie de la Chevêche (*Athene noctua*). *Jean-le-Blanc* **13**: 25. (In French).

Baudvin, H. 1997. Barn owl (*Tyto alba*) and Long-eared Owl (*Asio otus*) mortality along motorways in Bourgogne-Champagne: report and suggestions. *In* Duncan, J. R., Johnson, D. H., & Nichols, T. H. (eds). *Biology and Conservation of Owls of the Northern Hemisphere*: 2nd International Symposium; February 5–9, 1997; Winnipeg, Manitoba. US Forest Service General Technical Report NC-190. St. Paul, MN, pp. 58–61.

Bauer, H. G. 1987. Geburtsortstreue und Streuungsverhalten junger Singvögel. *Vogelwarte* **34**: 15–32. (In German).

Bauer, H.-G., Berthold, P., Boye, P. *et al.* 2002. Rote Liste der Brutvögel Deutschlands. 3., überarbeitete Fassung, 8.5.2002. *Ber. Vogelschutz* **39**: 13–60. (In German).

Bauer, S. & Thielcke, G. 1982. Gefährdete Brutvogelarten in der Bundesrepublik Deutschland und im Land Berlin: Bestandsentwicklung, Gefährdungsursachen und Schutzmassnahmen. *Vogelwarte* **31**: 265–267. (In German).

Baumanis, J. & Blums, P. 1969. *Latvijas putni.* Riga: Liesma. (In Latvian).

Baumgart, W. 1980. Wodurch ist der Steinkauz bedroht? *Falke* **27**: 228–229. (In German).

Bavoux, C. & Burneleau, G. 1983. Statut des rapaces nocturnes à l'Ile d'Oléron. *Bull. Groupe Ornithologique Aunis-Saintonge* **13**: 64–68. (In French).

Bavoux, C., Burneleau, G. & Seguin, N. 2000. Consommation de crabes par la Chevêche d'Athéna *Athene noctua* dans l'île d'Oléron (Charente-Maritime). *Alauda* **68**: 329–330. (In French).

Bayle, P. & Ziano, M.-T. 1989. Les limaciens, proies potentielles des rapaces nocturnes en Provence et dans les Alpes du sud. *Faune de Provence (C.E.E.P.)* **10**: 23–29. (In French).

Bednarek, W., Hausdorf, W., Jörissen, U., Schulte, E. & Wegener, H. 1975. Über die Auswirkung der chemischen Unwelt belastung aut Greifvögel in zwei Probeflächen Westfalen. *Journal für Ornithologie* **116**: 184–185. (In German).

Bednarz, J., Kupczyk, M., Kuźniak, S. & Winiecki, A. 2000. *Birds of the Poznań Region.* Faunistic Monograph. Poznań: Bogucki Scient. Publ. (In Polish).

Beersma, P. & Beersma, W. 2000. Temperatuur metingen in nestkasten voor Steenuilen *Athene noctua*. *Het Vogeljaar* **48**: 269–271. (In Dutch).
Beersma, P. & Beersma, W. 2001. Little Owls (*Athene noctua*) and biocides: reasons for concern? *In* Van Nieuwenhuyse, D., Leysen, M. & Leysen, K. (eds). The Little Owl in Flanders in its international context. Proceedings of the Second International Little Owl Symposium, 16–18 April 2001, Geraardsbergen, Belgium. *Oriolus*, **67**(2): 94–99.
Beersma, P. & Beersma, W. 2003. Vliegen met elke kleine prooi? *Athene nieuwsbrief STONE* **8**: 16. (In Dutch).
Beersma, P., Beersma, W. & van den Brug, A. 2007. Steenuilen. Roodbont Uitgeverij, Zutphen, Nederland. (In Dutch).
Bekaert, M. 2002. Habitatvoorkeur en populatiedichtheden van de Steenuil *Athene noctua* in Vlaanderen. *Oriolus* **68**(4): 181–190. (In Dutch).
Bekhuis, J., Bijlsma, R., Van Dijk, A. *et al.* (eds) 1987. *Atlas van de Nederlandse vogels*. SOVON Nederland. (In Dutch).
Bel'skaya, G. S. 1992. *Osobennosti biologii ptits v aridnykhusloviyakh (Peculiarities of Bird Biology in Arid Conditions)*. Ashgabat: Ylym Press. (In Russian).
Benton, T. G., Bryant, D. M., Cole, L. & Crick, H. Q. P. 2002. Linking agricultural practice to insect and bird populations: a historical study over three decades. *Journal of Applied Ecology* **39**: 673–687.
Berg, A. 2002a. Breeding birds in short-rotation coppices on farmland in central Sweden – the importance of salix height and adjacent habitats. *Agriculture Ecosystems & Environment* **90**(3): 265–276.
Berg, A. 2002b. Composition and diversity of bird communities in Swedish farmland-forest mosaic landscapes. *Bird Study* **49**: 153–165.
Bergerud, A. 1984. Changement de proie dans un écosystème simple. *Pour la Science* **77**: 76–85. (In French).
Berghmans, H. 2001a. Little Owl *Athene noctua* preying on dead day-old chickens. *Natuur. Oriolus* **68**: 191. (In Dutch).
Berghmans, H. 2001b. Adult Little Owl *Athene noctua* sharing nestbox with three Jackdaw *Corvus monedula* chicks. *Natuur. Oriolus* **68**: 191. (In Dutch).
Berghmans, H. 2001c. Nestbox with two incubating female Little Owls *Athene noctua* and nine chicks: a case of bigamy? *Natuur. Oriolus* **68**: 192–193. (In Dutch).
Berghmans, H. 2001d. Shell of snail *Cepaea nemoralis* in pellet of Little Owl *Athene noctua*. *Natuur. Oriolus* **68**: 194. (In Dutch).
Berndt, R. K., Koop, B. & Struwe-Juhl, B. 2002. *Vogelwelt Schleswig-Holsteins*. Band 5, Brutvogelatlas. Ornithologische Arbeitsgemeinschaft für Schleswig-Holstein und Hamburg e.V., Wachholtz Verlag, Neumünster. (In German).
Bernini, F., Dinetti, M., Gariboldi, A., Matessi, G. & Rognoni, G. 1998. *Atlante degli uccelli nidificanti a Pavia*. Comune di Pavia. LIPU. (In Italian).
Bernis, F. 1974. Algunos datos de alimentacion y depredacion de falconiformes y estrigiformes Ibericas. *Ardeola* **19**: 225–248. (In Spanish).
Berthold, P. 2000. Vogelzug: eine kurze, aktuelle Gesamtübersicht. Wiss. Buchges. Darmstadt. (In German).
Bettmann, H. 1951. Waldkauz oder Steinkauz. *Orn. Mitt.* **3**: 132–133. (In German).
Beven, G. 1979. Little Owl's method of catching cockchafers. *British Birds* **72**: 594–595.
Biagioni, M., Coppo, S., Dinetti, M. & Rossi, E. 1996. La conservazione della biodiversità nel Comune della Spezia. *Comune della Spezia*. (In Italian).

Bibby, C. J., Burgess, N. D. & Hill, D. A. 1992. *Bird Census Techniques*. B.T.O. R.S.P.B., London: Academic Press.

BirdLife International. 2004. *Birds in Europe: Population Estimates, Trends and Conservation Status*. Cambridge, UK: BirdLife International. BirdLife Conservation Series No. 12. p. 163.

Bisseling, G. A. L. 1933. Een partieel albinistische Steenuil, *Athene noctua vidalii* A. E. Brehm. *Orgaan Club Ned. Vogelkd.* **5**: 170. (In Dutch).

Bisseling, G. A. L. 1936. Nachtzwaluwen, *Caprimulgus eu. europaeus* L., als prooi van Steenuil, *Athene noctua vidalii* A. E. Brehm. *Orgaan Club Ned. Vogelkd.* **9**: 139–141. (In Dutch).

Blache, S. 2001. Etude du régime alimentaire de la Chevêche d'Athéna (*Athene noctua* Scop.) en période de reproduction en zone agricole intensive dans le sud-est de la France. *In* Génot, J.-C., Lapios, J.-M., Lecomte, P. & Leigh, R. S. (eds). Chouette chevêche et territoires. Actes du Colloque International de Champ-sur-Marne, 25 et 26 Novembre 2000. ILOWG. 2001. *Ciconia* **25**: 77–94. (In French).

Blache, S. 2003. Chevêche d'Athéna *Athene noctua*. *In* CORA (eds). *Atlas des Oiseaux Nicheurs de Rhône-Alpes*. (In French).

Blache, S. 2004. La Chevêche *(Athene noctua)* en zone d'agriculture intensive (plaine de Valence; Drôme): habitat, alimentation, reproduction. Mémoire. Diplôme Ecole Pratique des Hautes Etudes. Montpellier. (In French).

Blagosklonov, C. 1968. *Guide de la Protection des Oiseaux Utiles*. MIR Moscou. p. 246. (In French).

Blakesley, J. A., Noon, B. R. & Shaw, D. W. H. 2001 Demography of the California Spotted Owl in Northeastern California. *Condor* **103**: 667–677.

Blanc, T. 1958. Au garde-manger hivernal de la Chevêche. *Nos Oiseaux* **24**: 321. (In French).

Blanke, E. 2003. Steenuilen in de gemeente Raalte in 2003. *Athene Nieuwsbrief STONE* **8**: 43–44. (In Dutch).

Blanke, E. 2005. Steenuilen en steenmarters: een probleem? Oproep! *Athene Nieuwsbrief STONE* **10**: 7–13. (In Dutch).

Bloem, H., Boer, K., Groen, N. M., van Harxen, R. & Stroeken, P. 2001. De Steenuil in Nederland. Handleiding voor onderzoek en bescherming. Stichting Steenuilenoverleg Nederland (STONE). (In Dutch).

Blondel, J. & Badan, O. 1976. La biologie du Hibou grand-duc en Provence. *Nos Oiseaux* **33**: 212–213. (In French).

Blyth, E. 1847. *Athene noctua bactriana*. *In* Hutton, *Journal of the Asiatic Society of Bengal*, **16**(2): 776–777.

Bogdanov, A. N. 1956. [Birds of the Zerafshan River basin]. Part 1. *Proc. of Inst. of Zool. and Parasith. of Ac. Sc. of the Uzbek SSR*. **5**: 107–163. (In Russian).

Boie, H. 1822. *Athene*. Isis oder Encyclopaedische Zeitung, von Oken 1817–1848 Jena. pp. 10–11; col. 549. (In German).

Bonin, B. & Strenna, L. 1986. Sur la biologie du Faucon crécerelle, *Falco tinnunculus*, en Auxois. *Alauda* **54**: 251. (In French).

Bönsel, A. 2001. Erste Ergebnisse zum Wiederansiedlungsprojekt des Steinkauzes (*Athene noctua*) in Mecklenburg-Vorpommern. Landesverband Eulen-Schutz im Schleswig-Holstein, pp. 29–31. (In German).

Bonvicini, P. & Maino, M. 1993. Predazione di barbagianni *Tyto alba* su Civetta *Athene noctua* in provincia di Arezzo. *Picus* **19**: 69. (In Italian).

Bordignon, L. 1997. *Atlante Degli Uccelli Nidificanti a Cossato. Anno 1989-Anno 1995*. Quaderni di Educazione Ambientale. Comune di Cossato (VC). (In Italian).

Bordignon, L. 1999. *Gli uccelli della Città di Biella*. Comune di Biella. (In Italian).

Borgsteede, H. M., Okulewicz, A., Zoun, P. E. F. & Okulewicz, J. 2003. The helminth fauna of birds of prey (*Accipitriformes*, *Falconiformes* and *Stringiformes*) in the Netherlands. *Acta Parasitologica* **48**: 200–207.

Bouchy, P. 2004. Analyse demo-génétique de viabilité dans le cas de petites populations fragmentées: application à la Chouette chevêche (*Athene noctua*). Ph.D. Thesis, Université Paris 6 Pierre et Marie Curie. (In French).

Boyd, A. W. 1940. Little Owl preying on birds. *British Birds* **34**: 18.

Božič, L. & Vrezec, A. 2000. [Owls of the Pohorje Mountains] Sove Pohorja. *Acrocephalus* **21**(98/99): 47–53. (In Slovene).

Brands, R. 1980. Beiträge zur Brutbiologie des Steinkauzes *(Athene noctua).* Staatsexamensarb [Masters Thesis]. Univ. Köln. (In German).

Brehm, A. E. 1857. *Athene noctua vidalii.* Allgemeine Deutsche Naturhistorische Zeitung Isis, n.f., 3, Heft 11, p. 440. (In German).

Brehm, C. L. 1855. *Athene noctua indigena.* Der Vollstandige Vogelfang, p. 37. (In German).

Bretagnolles, V., Bavoux, C., Burneleau, G. & Van Nieuwenhuyse, D. 2001. Abondance et distribution des Chevêches d'Athéna: approche méthodologique pour des enquêtes à grande échelle en plaine céréalière. *In* Génot, J.-C., Lapios, J.-M., Lecomte, P. & Leigh, R. S. (eds). Chouette chevêche et territoires. Actes du Colloque International de Champ-sur-Marne, 25 et 26 Novembre 2000. ILOWG. *Ciconia* **25**: 173–184. (In French).

Breuer, W. 1998. Käuze, Klüngel, kommunale Nachsicht. Vom Wohl und Wehe der Steinkäuze in Flächenutzungsplänen. *Eulen-Rundblick* **47**: 3–10. (In German).

Brichetti, P. 1997. L'avifauna nidificante. *In* Brichetti, P. & Gariboldi, A. (eds). *Manuale pratico di Ornitologia*. Edagricole, Bologna, pp. 259–267. (In Italian).

Brodkorp, P. 1971. Catalogue of Fossil Birds, Part 4 (Columbiformes through Piciformes). *Bulletin of the Florida State Museum, Biological Sciences.* **15**(4): 163–266.

Broun, L. V. 1986. Osobennosti stroeniya zheludochno-kishechnogo traktaplotoyadnykh ptits v svyazi s pishchevoi spetsializatsiei [Peculiarities of the anatomy of digestive tract of bird-myophages in connection with feeding specialization]. Izuchenie ptits SSSR, ikh okhrana i ratsional'noe ispol'zovanie [The study of birds of the USSR, their protection and wise management]. Materials of the 1st Meeting of All-Union Orn. Soc. and 9th All-Union Orn. Conf. Leningrad. Part 1. pp. 99–100. (In Russian).

Büchi, O. 1952. Forte mortalité de rapaces nocturnes dans le canton de Fribourg. *Nos Oiseaux* **21**: 224. (In French).

Buker, J. B., De Wit, J. N. & Van Zuilen, W. A. 1984. Forse prooien van Steenuil *Athene noctua. Limosa* **57**: 118. (In Dutch).

Bultot, J. 1990. Construction d'un nouveau type de nichoir pour chouette chevêche (*Athene noctua*) et premiers résultats. *Bulletin à l'Usage du Baguer Ornithologique:* 56–59. (In French).

Bultot, J. 1996. Opération Chevêche. *L'Homme et l'Oiseau* **34**: 101–107. (In French).

Bultot, J., Marié, P. & Van Nieuwenhuyse, D. 2001. Population dynamics of Little Owl *Athene noctua* in Wallonia and its driving forces. Evidence for density-dependence. *In* Van Nieuwenhuyse, D., Leysen, M. & Leysen, K. (eds). Little Owl in Flanders in its international context. Proceedings of the Second International Symposium, March 16–18, 2001, Geraardsbergen, Belgium. *Oriolus* **67**: 110–125.

Burbach, K. 1997. Der Steinkauz – *Athene noctua. In* Hesssische Gesellschaft für Ornithologie und Naturschutz e.V. (ed.). *Avifauna von Hessen* (1993–2000). Echzell: Eigenverlag. (In German).

Burnham, K. P. & Anderson, D. R. 1998. *Model Selection and Inference: A Practical Information-Theoretic Approach*. New York: Springer-Verlag.

Burton, J. A. 1992. *Owls of the World*. London: Peter Lowe, Eurobook Ltd.

Burton, P. J. K. 1983. Little Owl raiding Starling's nest. *British Birds* **76**: 314–315.

Busse, H. 1983. Der Steinkauz. *Falke* **30**: 34–35. (In German).

Bux, M. & Rizzi, V. 2005. Dieta della Civetta, *Athene noctua*, in una salina dell'Italia Meridionale. *In* Mastrorilli, M., Nappi, A. & Barattieri, M. (eds). Atti I Convegno italiano sulla Civetta. Osio Sotto (BG), 21 Marzo 2004. Gruppo Italiano Civette: 50–52. (In Italian).

Buxton, E. J. M. 1947. Little Owl taking grass-snake. *British Birds* **40**: 55.

Calderon, J. 1977. El papel de la Perdiz roja (*Alectoris rufa*) en la dieta de los predatores ibericos. *Doñana Act. Vert*. **4**: 77–78. (In Spanish).

Calvi, G., De Carli, E., Buvoli, L. & Fornasari, L. 2005. La Civetta nel progetto MITO2000 dal 2000 al 2002. *In* Mastrorilli, M., Nappi, A. & Barattieri, M. (eds). Atti I Convegno italiano sulla Civetta. Osio Sotto (BG), 21 Marzo 2004. Gruppo Italiano Civette: 13–17. (In Italian).

Cameron-Brown, F. A. 1975. Little Owl behaviour. *J. Gloucs. Nat. Soc*. **26**: 83–84.

Capanna, E., Vittoria, M. & Martinico, E. 1987. I cromosomi de gli uccelli. Citotassonomia de evoluzione cariotipica. *Avocetta* **11**: 101–143. (In Italian).

Carlson, A. 1985. Central place foraging in the Red-backed Shrike (*Lanius collurio*): allocation of prey between foraging and sedentary consumer. *Anim. Behav*. **33**: 664–666.

Casalini, R. 2005. Attivita di *mobbing* nei confronti di *Athene noctua* (Scopoli, 1769) in un ambiente mediterraneo. *In* Mastrorilli, M., Nappi, A. & Barattieri, M. (eds). Atti I Convegno italiano sulla Civetta. Osio Sotto (BG), 21 Marzo 2004. Gruppo Italiano Civette: 41–44. (In Italian).

Casini, L. 1993. *In* Meschini E. & Frugis, S. (eds). Atlante degli uccelli nidificanti in Italia. *Suppl. Ric. Biol. Selvaggina* **20**: 1–344. (In Italian).

Castellucci, M. & Zavalloni, D. 1989. Predazione di barbagianni, *Tyto alba*, su civetta, *Athene noctua*. *Riv. Ital. Orn*. **59**: 282–283. (In Italian).

Centili, D. 1995. Dati preliminari sulla Civetta *Athene noctua* in un'area dei Monti della Tolfa (Roma). *Avocetta* **19**: 113. (In Italian).

Centili, D. 1996. Censimento, distribuzione e habitat della Civetta *Athene noctua* in un'area dei monti della Tolfa. Masters Thesis, Universita di Roma. (In Italian).

Centili, D. 2001a. A Little Owl population and its relationships with humans in central Italy. *In* Génot, J.-C., Lapios, J.-M., Lecomte, P. & Leigh, R. S. (eds). Chouette chevêche et territoires. Actes du Colloque International de Champ-sur-Marne, November 25–26, 2000. ILOWG. *Ciconia* **25**: 153–158.

Centili, D. 2001b. Playback and Little Owls *Athene noctua*: preliminary results and considerations. *In* Van Nieuwenhuyse, D., Leysen, M. & Leysen, K. (eds). The Little Owl in Flanders in its international context. Proceedings of the Second International Little Owl Symposium, April 16–18, 2001, Geraardsbergen, Belgium. *Oriolus* **67**(2): 88–93.

Cesaris, C. 1988. Popolazioni di Allocco *Strix aluco* e di Civetta *Athene noctua* in un'area del Parco Lombardo della Valle del Ticino. *Avocetta* **12**: 115–118. (In Italian).

Charter, M., Leshem, Y., Izhaki, I., Guershon, M. & Kiat, Y. 2006. The diet of the Little Owl, *Athene noctua*, in Israel. *Zoology in the Middle East* **39**: 31–40.

Chauvet, J.–M., Brunel Deschamps, E. & Hillaive, C. 1996. *Dawn of Art: The Chauvet Cave. The Oldest Known Paintings in the World*. Harry N. Abrams Inc.

Cheylan, G. 1986. Inventaire ornithologique préliminaire des îles de Marseille. *Faune de Provence. Bull. C.E.E.P.* **7**: 30–38. (In French).

Christie, D. & Van Woudenberg, A. 1997. Modeling critical habitat for flammulated owls (*Otus flammeolus*). *In* Duncan, J. R., Johnson, D. H. & Nicholls, T. H. (eds). *Biology and Conservation of Owls on the Northern Hemisphere*: 2nd International Symposium; February 5–9, 1997; Winnipeg, Gen. Tech. Rep. NC-190. St. Paul, MN: US Department of Agriculture, Forest Service, North Central Forest Experiment Station, pp. 97–106.

Christie, H. H. V. 1931. Buzzard killing Little Owl. *British Birds* **24**: 367.

Ciani, C. 1988. Predazione di Civetta, *Athene noctua*, su Rondone, *Apus apus*. *Riv. Ital. Orn.* **58**: 87–88. (In Italian).

Cignini, B. & Zapparoli, M. 1996. *Atlante degli Uccelli Nidificanti a Roma*. Rome: Fratelli Palombi Editori. p. 37. (In Italian).

Clech, D. 1993. La Chouette chevêche *Athene noctua* en Bretagne. *Ar Vran* **4**: 5–34. (In French).

Clech, D. 2001a. Etude d'une population de Chevêche d'Athéna dans le Haut-Léon (Bretagne–France). *In* Génot, J.-C., Lapios, J.-M., Lecomte, P. & Leigh, R. S. (eds). Chouette chevêche et territoires. Actes du Colloque International de Champ-sur-Marne, November 25–26, 2000. ILOWG. *Ciconia* **25**: 199–128. (In French).

Clech, D. 2001b. Impact de la circulation routière sur la Chevêche d'Athéna *Athene noctua*, par l'étude de la localisation de ses sites de reproduction. *Alauda* **69**: 255–260. (In French).

Cleine, H. 1976. Onbekend gedrag van Steenuil. *Vogeljaar* **24**: 326. (In Dutch).

Collectif d'associations "Chevêche-Ajoie" & Office des Eaux et de la Protection de la Nature. 2003. Plan d'action pour la Chevêche d'Athéna en Ajoie (JU). Rapport. (In French).

Collinge, W. E. 1922. The food and feeding habits of the Little Owl. *J. Min. Agric.* **28**: 1022–1031 and 1133–1140.

Contoli, L., Aloise, G. & Filippucci, M. G. 1988. Sulla diversificazione trofica di Barbagianni *Tyto alba* e Civetta *Athene noctua* in rapporto al livello diagnostico delle preda. *Avocetta* **12**: 21–30. (In Italian).

Coppée, J.-L., Bultot, J. & Hanus, B. 1995. Population et reproduction de la Chouette chevêche (*Athene noctua*) en Hainaut. Protection et restauration de ses habitats. *Aves* **32**: 73–99. (In French).

Cornulier, T. & Bretagnolle, V. 2006. Assessing the influence of environmental heterogeneity on bird spacing patterns: a case study with two raptors. *Ecography* **29**: 240–250.

Costaglia, A., Lepore, G., De Falco, G. & Cecio, A. 1981. Neuroistologia comparata dell'esofago in uccelli granivori onnivorie carnivori. *Atti Soc. Ital. Sc. Vet.* **35**: 323–324. (In Italian).

Courtois, J.-M. 1988. Deux lépidoptères commensaux de la Chouette Chevêche (*Athene noctua*). *Ciconia* **12**: 177. (In French).

Cramp, S. 1963. Toxic chemicals and birds of prey. *British Birds* **56**: 129–130.

Cramp, S. (ed.) 1985. *The Birds of the Western Palearctic. Vol. 4. Terns to Woodpeckers.* Oxford, New York: Oxford University Press. pp. 514–525.

Crespon, J. 1840. *Ornithologie du Gard*. p. 54. (In French).

Csermely, D., Casagrande, S. & Sponza, S. 2002. Adaptive details in the comparison of predatory behaviour of four owl species. *Ital. J. Zool.* **69**: 239–243.

Curio, E. 1963. Probleme des Feinderkennens bei Vögeln. Proc. 13. Intern. Orn. Congr. Ithaca: 206–239. (In German).

Daanje, A. 1951. On locomotory movements in birds and the intention-movements derived from them. *Behaviour* **3**: 48–98.

Dalbeck, L. 2003. Der Uhu *Bubo bubo* (L.) in *Deutschland – autökologische Analysen an einer wieder angesiedelten Population – Resumee eines Artenschutzprojekts.* Aachen: Shaker Verlag. (In German).

Dalbeck, L., Bergerhausen, W. & Hachtel, M. 1999. Habitatpräferenzen des Steinkauzes (*Athene noctua*) im ortsnahen Grünland. *Charadrius* **35**: 100–115. (In German).

Danko, S. 1985a. Report on activities of the group for research and protection of birds of prey and owls in Czechoslovakia in 1982. *Bull. World Working Group Birds of Prey* **2**: 48.

Danko, S. 1985b. Report on activities of the group for research and protection of birds of prey and owls in Czechoslovakia in 1983. *Bull. World Working Group Birds of Prey* **2**: 50.

Danko, S. 1986. Report on activities of the group for research and protection of birds of prey and owls in Czechoslovakia in 1986. *Buteo* **1**: 22.

Danko, S. 1989. Report on activities of the group for research and protection of birds of prey and owls in Czechoslovakia in 1989. *Buteo* **4**: 1–28.

Danko, S., Diviš, T., Dvorská, J. *et al.* 1994. [The state of knowledge of breeding numbers of birds of prey (Falconiformes) and owls (*Strigiformes*) in the Czech and Slovak Republics as of 1990 and their population trends in 1970–1990]. *Buteo* **6**: 1–89.

Danko, S., Darolova, A. & Kristin, A. 2002. *Bird Distribution in Slovakia.* Bratislava: Veda, vydavatelstv SAV.

Davison, G. 1983. The eyes have it: Ocelli in a rain forest pheasant. *Anim Behav.* **31**: 1037–1042.

Dawson-Smith, F. 1913. British owls. *Bird Notes and News* **4**: 169.

De La Hoz, M. 1982. Algunos datos sobre la alimentacion del Mochuelo comun (*Athene noctua*) en Asturias. *Bol. C. Nat. I.D.E.A.* **29**: 113–120. (In Spanish).

De Smet, A. 2002. Onderzoek naar het voorkomen van de Steenuil in West-Zeeuws-Vlaanderen 2000–2001. Natuurbeschermingsvereniging 't Duumpje. (In Dutch).

del Hoyo, J. E., Elliott, A. & Sargatal, J. 1999. *Handbook of the Birds of the World.* Vol 5. *Barn Owls to Humming Birds*. Barcelona, Spain: Lynx Editions.

Delamain, J. 1938. *Portraits d'Oiseaux.* Paris: Stock. pp. 59–62. (In French).

Delibes, M., Brunet-Lecomte, P. & Manez, M. 1983. Datos sobre la alimentacion de la Lechuza comun (*Tyto alba*), el Buho Chico (*Asio otus*) y el Mochuelo (*Athene noctua*) en una misma localidad de Castilla la Vieja. *Ardeola* **30**: 57–63. (In Spanish).

Delmée, E. 1988. *Athene noctua. In* Devillers, P., Roggeman, W., Tricot, J. *et al.* (eds). *Atlas van de Belgische broedvogels*. Brussels: Koninklijk Belgisch Instituut voor Natuurwetenschappen. (In Dutch).

Dementyev, G. & Gladkov, N. A. 1951. *Ptitsy Sovetskogo Soyusa. [Birds of the Soviet Union].* 1. Moskva. pp. 440–447. (In Russian).

Denac, D., Marčič, M., Radolič, P. & Tomažič, A. 2002. Owls in churches, castles and other buildings in the Vipava valley and the Karst (SW Slovenia) /Sove v cerkvah, gradovih in drugih objektih na območju Vipavske doline). *Acrocephalus* **23**(112): 91–95. (In Slovene).

Desmond, M. J. 1997. Evolutionary history of the genus *Speotyto:* a genetic and morphological perspective. PhD dissertation. Lincoln: University of Nebraska.
Desmots, D. 1988. Prédation de Moineaux domestiques (*Passer domesticus*) par une Chouette chevêche (*Athene noctua*). *Biotopes* **53**(6): 44–45. (In French).
Díaz, M., Asensio, B. & Tellería, J. L. 1996. *Aves Ibéricas. I. No Paserifiomes*. J. M. Reyero Editor, Madrid. (In Spanish).
Dinetti, M. 1994. *Atlante degli uccelli nidificanti a Livorno*. Quaderni dell'ambiente, n. 5, Comune di Livorno e Coop. Ardea. (In Italian).
Dinetti, M. & Ascani, P. 1990. *Atlante degli uccelli nidificanti nel Comune di Firenze*. Studio GE9, Firenze. (In Italian).
Dobinson, H. M. & Richards, A. J. 1964. The effects of the severe winter of 1962/63 on birds in Britain. *British Birds* **57**: 373–433.
Dolinar, I. 1951. Ptiči v območju Savinjske doline pred 45 leti in danes. *Lovec* **34**: 8–12. (In Slovene).
Dombrowski, A., Fronczak, J., Kowalski, M. & Lippoman, T. 1991. Population density and habitat preferences of owls Strigiformes on agricultural areas of Mazowsze Lowland (Central Poland). *Acta Orn.* **26**: 39–53. (In Polish).
Donovan, T. M., Beardmore, C. J., Bonter, D. N. *et al.* 2002. Priority research needs for the conservation of Neotropical migrant landbirds. *Journal of Field Ornithology* **73**: 329–450.
Duncan, J. R. & Hayward, P. H. 1995. Review of technical knowledge: Great Gray Owls. *In* Hayward, G. D. & Verner, J. (eds). *Flammulated, Boreal, and Great Gray Owls in the United States: A Technical Conservation Assessment.* USDA Forest Service General Technical Report RM-253. pp. 154–175 (Chapter 14)
Dupond, C. 1943. *Les Oiseaux de la Belgique*. Patrimoine du Musée Royal d'Histoire naturelle de Belgique. Bruxelles. pp. 71–72. (In French).
Dyrcz, A., Grabinski, W., Stawarczyk, T. & Witkowski, J. 1991. *Birds of the Silesian Region – Faunistic Monograph*. Warsaw University: Department of Bird Ecology, Wroclaw (In Polish).
Eakle, W. L. 1994. A raptor survey in western Turkey and eastern Greece. *Journal of Raptor Research* **28**(3): 186–191.
Edwards C. A. & Bohlen, P. J. 1996. *Biology and Ecology of Earthworms*, 3rd edition. London: Chapman and Hall.
Eick, M. 2003. Habitatnutzung und Dismigration des Steinkauzes *Athene noctua*. Eine Telemetriestudie in Zusammenarbeit mit der Forschungsgemeinschaft zur Erhaltung einheimischer Eulen e.V. (FOGE), dem Staatlichen Museum für Naturkunde Stuttgart und der Max-Planck Forschungsstelle für Ornithologie, Vogelwarte Radolfzell. Diplomarbeit. Universität Hohenheim. (In German).
Elliott, J. S. 1940. Little Owl preying on birds. *British Birds* **33**: 312–313.
Ellis, J. C. S. 1946. Little Owl's method of taking pheasant chick by daylight. *British Birds* **39**: 217.
Elzinga, C. L., Salzer, D. W. & Willoughby, J. W. 1998. Measuring and Monitoring Plant Populations. US Department of the Interior. Burea of Land Management, Technical Reference 1730–1. Denver, Colorado, USA.
Enehjelm af, C. 1969. The breeding of the Little Owl (*Athene noctua*). *Avic. Mag.* **63**: 44–45.
Enriquez, P. L. & Mikkola, H. 1997. Comparative study of general public owl knowledge in Costa Rica, Central America and Malawi, Africa. *In* Duncan, J. R., Johnson, D. H & Nichols, T. H. (eds). *Biology and Conservation of Owls of the Northern*

Hemisphere: 2nd International Symposium; February 5–9, 1997; Winnipeg, Manitoba, Canada. US Forest Service General Technical Report NC-190. St. Paul, MN. pp. 160–166.

Erkert, H. 1967. Beleuchtungsabhängige Aktivitätsoptima bei Eulen und circadiane Regel. *Naturwissenschaften* **54**: 231–232. (In German).

Erlanger von, C. F. 1898. Beiträge zur Vogelfauna Tunesiens. l. Teil. *Journal für Ornithologie* **46**: 377–497. (In German).

Estoppey, F. 1992. Une densité elevée de Chouettes chevêches, *Athene noctua*, dans la plaine du Po en Italie. *Nos Oiseaux* **41**: 315–319. (In French).

Etienne, P. 2003. La reproduction de la chouette chevêche *Athene noctua*: parades et occupation de l'espace. *L'Avocette (numéro spécial)*: 113–116. (In French).

Exo, K.-M. 1981. Zur Nistökologie des Steinkauzes (*Athene noctua*). *Vogelwelt* **102**: 161–180. (In German).

Exo, K-M. 1983. Habitat, Siedlungsdichte und Brutbiologie einer niederrheinischen Steinkauzpopulation (*Athene noctua*). *Ökol. Vögel* **5**: 1–40. (In German).

Exo, K-M. 1984. Die akustistiche Unterscheidung von Steinkauzmännchen und -weibchen (*Athene noctua*). *J. Orn.* **125**: 94–97. (In German).

Exo, K.-M. 1987. Das Territorialverhalten des Steinkauzes (*Athene noctua*). Eine verhaltensökologie Studie mit Hilfe der Telemetrie. Inaugural-Dissertation zur Erlangung des Doktorgrades der Mathematisch-Naturwissenschaftlichen Fakultät der Universität zu Köln. (In German).

Exo, K-M. 1988. Jahreszeitliche ökologische Anpassungen des Steinkauzes (*Athene noctua*). *J. Orn.* **129**: 393–415. (In German).

Exo, K-M. 1989. Tagesperiodische Aktivitätsmuster des Steinkauzes (*Athene noctua*). *Vogelwarte* **35**: 94–114. (In German).

Exo, K-M. 1990. Geographische Variation des Reviergesangs beim Steinkauz (*Athene noctua*) – ein Vergleich des Gesangs nordwestdeutscher und ostenglisher Vögel. *Vogelwarte* **35**: 279–286. (In German).

Exo, K-M. 1991. Der Untere Niederrhein – ein Verbreitungsschwerpunkt des Steinkauzes (*Athene noctua*) in Mitteleuropa. *Natur und Landschaft* **66**: 156–159. (In German).

Exo, K-M. 1992. Population ecology of Little Owls *Athene noctua* in Central Europe: a review. *In* Galbraith, C A., Taylor, I R. & Percival, S. (eds). *The Ecology and Conservation of European Owls*. Joint Nature Conservation Committee. UK Nature Conservation, No. 5. Petersborough. pp. 64–75.

Exo, K.-M. & Hennes, R. 1978. Empfehlungen zur Methodik von Siedlungsdichte Untersuchungen vom Steinkauz (*Athene noctua*). *Vogelwelt* **99**: 137–141. (In German).

Exo, K.-M. & Hennes, R. 1980. Beitrag zur Populationsökologie des Steinkauzes (*Athene noctua*) – eine Analyse deutscher und niederländischer Ringfunde. *Vogelwarte* **30**: 162–179. (In German).

Exo, K-M. & Scherzinger, W. 1989. Stimme und Lautrepertoire des Steinkauzes (*Athene noctua*): Beschreibung, Kontext und Lebensraumanpassung. *Ökol. Vögel* **11**: 149–187. (In German).

Faggio, G. 2005. La Civetta *Athene noctua* in Corsica. *In* Mastrorilli, M., Nappi, A. & Barattieri, M. (eds). Atti I Convegno italiano sulla Civetta. Osio Sotto (BG), March 21, 2004. Gruppo Italiano Civette: 35–36. (In Italian).

Fajardo, I., Pividal, V., Trigo, M. & Jimenez, M. 1998. Habitat selection, activity peaks and strategies to avoid road mortality by the Little Owl *Athene noctua*. A new methodology on owls research. *Alauda* **66**: 49–60.

Falcone, S. 1987. Rapaci abbattuti illegalmente in Sicilia. Rapaci Mediterranei III. Atti del IV Colloquio, Sant'Antioco, Cagliari, October 11–13, 1984. *Suppl. Ric. Biol. Selvaggina* **XII**: 93–95. (In Italian).

Farsky, O. 1928. De l'utilité de quelques oiseaux de proie et corvidés déterminée par l'examen de leurs aliments. Thèse de Docteur en Sciences. Nancy. pp. 99–103. (In French).

Fattorini, S., Manganaro, A., Ranazzi, L., Cento, M. & Salvati, L. 2000. Insect predation by Little Owl, *Athene noctua*, in different habitats of Central Italy. *Riv. Ital. Orn.* **70**: 139–142.

Feiler, M. & Litzbarski, H. 1987. Steinkauz – *Athene noctua. In* Rutschke, E. (ed.). *Die Vogelwelt Brandenburgs*, 2nd edition. Jena. pp. 247–248. (In German).

Ferrer, D., Molina, R., Castella, J. & Kinsella, J. M. 2004. Parasitic helminths in the digestive tract of six species of owls (*Strigiformes*) in Spain. *The Veterinary Journal* **167**: 181–185.

Ferrus, L., Génot, J-C., Topin, F., Baudry, J. & Giraudoux, P. 2002. Répartition de la Chevêche d'Athéna (*Athene noctua* Scop.) et variation d'échelle d'analyse des paysages. *Terre et Vie* **57**: 39–51. (In French).

Festetics, A. 1952/1955. Observations on the Barn-Owl's and Little-Owl's Life-Habits. *Aquila* **59–62**: 452–453.

Festetics, A. 1959. Gewölluntersuchungen an Steinkauzen der Camargue. *Terre et Vie* **106**: 121–127. (In German).

Finck, P. 1989. Variabilität des Territorialverhaltens beim Steinkauz (*Athene noctua*). Inaugural-Dissertation zur Erlangung des Doktorgrades der Mathematisch-Naturwissenschftlichen Fakultät der Universität zu Köln. (In German).

Finck, P. 1990. Seasonal variation of territory size with the Little Owl (*Athene noctua*). *Oecologia* **83**: 68–75.

Finck, P. 1993. Territoriengrösse beim Steinkauz (*Athene noctua*): Einfluss der Dauer der Territorienbesetzung. *J. Orn.* **134**: 35–42. (In German).

Formon, A. 1969. Contribution à l'étude d'une population de Faucons pèlerins *Falco peregrinus* dans l'est de la France. *Nos Oiseaux* **30**: 136. (In French).

Formozov, A. N. & Osmolovskaya, V. I. 1953. Chislennost nekotorykhlandshaftnykh zhivotnykh tsentral'nogo Kazakhstana po dannym analizadobychi, vylovlennoi khishchnymi ptitsami [Number of some landscape animals of the central Kazakhstan on the base of analysis of prey, caught by birds of prey]. Trudy Instituta geografii (Proc. of Inst. of Geography). *AN. SSSR (Ac. of Sc. of the U.S.S.R.)* **54**: 329–350. (In Russian).

Forsman, E. D., Meslow, E. C. & Wight, H. M. 1984. Distribution and biology of the Spotted Owl in Oregon. *Wildlife Monographs* **87**: 1–64.

Forsman, E. D., De Stefano, S., Raphael, M. G. & Gutiérrez, R. J. (eds). 1996. *Demography of the Northern Spotted Owl*. Studies in Avian Biology No. 17. Cooper Ornithological Society. California, USA: Camarillo.

Forsman, E. D., Anthony, R. G., Reid, J. A. *et al.* 2002. Natal and breeding dispersal of Northern Spotted Owls. *Wildlife Monographs* **149**: 1–35.

Fraczek, T. & Szewczyk, P. 2000. Occurrence of owls *Strigiformes* in Kraœnik. 4th Review of the activity of the Scientific Circles of Naturalists, October 20–22, 2000, Slupsk (In Polish).

Fraissinet, M. & Milone, M. 1995. Migrazione ed inanellamento degli uccelli in Campania. Monografia AISOM, Napoli, **2**: 1–166. (In Italian).

Frankel, O. H. & Soule, M. E. 1981. *Conservation and Evolution*. Cambridge, UK: Cambridge University Press.

Franklin, A. B., Anderson, D. R., Gutiérrez, R. J. & Burnham, K. P. 2000. Climate, habitat quality, and fitness in northern spotted owl populations in northwestern California. *Ecological Monographs* **70**: 539–590.

Franklin, A. B., Shenk, T. M., Anderson, D. R. & Burnham, K. P. 2001. Statistical model selection: an alternative to null hypothesis testing. *In* Shenk, T. M. & Franklin, A. B. (eds). *Modeling in Natural Resource Management: Development, Interpretation, and Application*. Washington, DC: Island Press. pp. 75–90.

Franklin, A. B., Gutierrez, R. J., Nichols, J. D. *et al.* 2004. *Population Dynamics of the California Spotted Owl* (Strix occidentalis occidentalis): *a Meta-analysis*. Ornithological Monographs No. 54. Washington, DC: American Ornithologists' Union.

Freethy, R. 1976. Little Owl flying at Dunlin. *British Birds* **69**: 272.

Frolov, V. V., Korkina, S. A., Frolov, A. V. *et al.* 2001. Analiz sostoyaniya fauny nevorob'inykhptits yuga lesostepnoi zony pravoberezhnogo Povolzh'ya v XX veke [Analysis of the state of fauna of non-Passerine birds in south of the forest-steppe zone of the right-bank Volga region in the twentieth century]. *Berkut* **10**(2): 156–183. (In Russian).

Fronczak, J. & Dombrowski, A. 1991. Owls *Strigiformes* in the agricultural and forest landscapes of South Białystok Lowlands. *Acta Orn.* **26**: 55–61 (In Polish).

Fuchs, P. 1986. Structure and functioning of a Little Owl *Athene noctua* population. *Ann. Rep. Res. Inst. Nat. Management* **1985**: 113–126.

Fuchs, P. & Thissen, J. B. M. 1981. Die Pestizid – und PCB – Belastung bei Greifvögeln und Eulen in den Niederlanden nach gesetzlich verordneten Einschränkungen im Gebraut der Chlorierten Kohlenwasserstoffpestizide. *Ökologie der Vögel* **3**: 189–195. (In German).

Fuller, M. R. & Mosher, J. A. 1981. Methods of detecting and counting raptors: a review. *In* C. J. Ralph & J. M. Scott (eds), *Estimating Numbers of Terrestrial Birds. Studies in Avian Biology* 6. pp. 235–246.

Fuller, R. J., Trevelyan, R. J. & Hudson, R. W. 1997. Landscape composition models for breeding bird populations in lowland English farmland over a 20 year period. *Ecography* **20**(3): 295–307.

Furrington, H. 1979. Eine Röhre schütz vor dem Marder. *Wir und die Vögel* **11**: 20–22. (In German).

Furrington, H. 1998. 27 Jahre Steinkauz-Schutz im Stadt-und Landkreis-Heilbronn mit Randgebiet, auf einer Kontrollfläche von ca. 750 km^2. *Orn. Schnellmitt. Bad.-Württ. N. F.* **59**: 14–16. (In German).

Gaffrey, M. F. & Hodos, W. 2003. The visual acuity and refractive state of the American Kestrel (*Falco sparverius*). *Vision Res.* **43**: 2053–2059.

Gaines, W. L., Harrod, R. J. & Lehmkuhl, J. F. 1999. *Monitoring Biodiversity: Quantification and Interpretation*. Gen. Tech. Rep. PNW-GTR-443. Portland, OR: US Department of Agriculture, Forest Service, Pacific Northwest Research Station.

Galeotti, P. 1989. Tavola rotonda: metodi di censimento per gli strigiformi. *In* Fasola, M. (ed.), Atti II Seminario Italiano Censimenti Faunistici dei Vertebrati. *Suppl. Ric. Biol. Selvaggina* **16**: 437–445. (In Italian).

Galeotti, P. & Morimando, F. 1991. Dati preliminari sul censimento della Civetta in ambiente urbano. *Suppl. Ric. Biol. Selvaggina* **16**: 349–351. (In Italian).

Galeotti, P. & Sacchi, R. 1996. Owl census project in the Lombardy Region: preliminary data on the Tawny owl (*Strix aluco*), the Little Owl (*Athene noctua*), and the Long-eared owl (*Asio otus*) populations. Abstracts from the 2nd International Conference on Raptors. Raptor Research Foundation and University of Urbino.

Gallego, A., Baron, M. & Gayoso, M. 1975. Organization of the outer plexiform layer of the diurnal and nocturnal bird retinae. *Vision Res.* **15**: 1027–1028.

Ganey, J. L. & Block, W. M. 2005. *Winter Movements and Range Use of Radio-marked Mexican Spotted Owls: An Evaluation of Current Management Recommendations.* General Technical Report RMRS-GTR-148-WWW. Fort Collins, CO: US Department of Agriculture, Forest Service, Rocky Mountain Research Station.

Ganey, J. L., Block, W. M., Dwyer, J. K., Strohmeyer, B. E. & Jenness, J. S. 1998. Dispersal movements and survival rates of juvenile Mexican Spotted Owls in northern Arizona. *Wilson Bulletin* **110**(2): 206–217.

Ganey, J. L., White, G. C., Bowden, D. C. & Franklin, A. B. 2004. Evaluating methods for monitoring populations of Mexican Spotted Owls: a case study. *In* Thompson, W. L. (ed.). *Sampling Rare or Elusive Species: Concepts, Designs, and Techniques for Estimating Population Parameters.* Island Press. pp. 337–385.

Ganya, I. M. & Zubkov, N. I. 1975. Pitanie domovogo sycha (*Athene noctua* Scop.) in srednei chasti Moldavii [The Little Owl (*Athene noctua* Scop.) feeding at the middle part of Moldavia]. *Ecologiya ptits immlekopitayushchikh Moldavii [Ecology of birds and mammals of Moldavia].* Kishinev: Shiintsa Press. pp. 62–72. (In Russian).

Garcia, J. B. & Muñoz, R. 2003. Mochuelo Europeo. *In* Martí, R. & Moral, J. C. (eds). *Atlas de las Aves Reproductoras de España.* Madrid: Dirección General de Conservación de la Naturaleza-Sociedad Española de Ornitología. pp. 318–319. (In Spanish).

Garcia-Fernandez, A. J., Motas-Guzman, M., Navas, I. *et al.* 1997. Environmental exposure and distribution of lead in four species of raptors in Southeastern Spain. *Arch. Environ. Contam. Toxicol.* **33**: 76–82.

Gassmann, H. & Bäumer, B. 1993. Zur Populationsökologie des Steinkauzes (*Athene noctua*) in der westlichen Jülicher Börde. Erste Ergebnisse einer 15 jährigen Studie. *Vogelwarte* **37**: 130–143. (In German).

Gassmann, H., Bäumer, B. & Glasner, W. 1994. Faktoren der Steuerung des Bruterfoges beim Steinkauz *Athene noctua. Vogelwelt* **115**: 5–13. (In German).

Gates, J. E. & Giffen, N. R. 1991. Neotropical migrant birds and edge effects at a forest-stream ecotone. *Wilson Bulletin* **103**(2): 204–217.

Gavrilov, E. I. 1999. *Fauna i rasprostranenie ptits Kazakhstana [Fauna and Distribution of the Birds of Kazakhstan].* Almaty. (In Russian).

Gavrin, V. F. 1962. *Otryad Sovy Striges [Order Strigiformes]. Ptitsy Kazakhstana [Birds of Kazakhstan].* Vol. 2. Alma-Ata, Publishing House of Academy of Sciences of the Kazakh SSR. pp. 708–779. (In Russian).

Geister, I. 1995. *Ornitološki Atlas Slovenije.* Ljubljana: DZS. (In Slovene).

Geister, I. 1998. *Are the Birds Truly Disappearing? (Ali ptice res izginjajo?).* Tehniška založba Slovenije, Ljubljana (In Slovene).

Génot, J.-C. 1990a. Régression de la Chouette chevêche, *Athene noctua* SCOP., en bordure des Vosges du Nord. *Ciconia* **14**: 65–84. (In French).

Génot, J.-C. 1990b. Habitat et sites de nidification de la Chouette chevêche, *Athene noctua* SCOP., en bordure des Vosges du Nord. *Ciconia* **14**: 85–116. (In French).

Génot, J-C. 1991. Mortalité de la Chouette chevêche, *Athene noctua*, en France. *In* Juillard, M., Bassin, P., Baudvin, H. *et al.* (eds). Rapaces Nocturnes Actes du 30e colloque interrégional d'ornithologie Porrentruy (Suisse): November, 2–4, 1990. *Nos Oiseaux*: 139–148. (In French).

Génot, J.-C. 1992a. Contribution à l'écologie de la Chouette chevêche, *Athene noctua* (Scop.), en France. PhD Thesis. Université de Dijon. (In French).

Génot, J.-C. 1992b. Biologie de reproduction de la Chouette chevêche *Athene noctua* SCOP., en bordure des Vosges du Nord. *Ciconia* **16**: 1–18. (In French).

Génot, J.-C. 1992c. Biologie de reproduction de la Chouette chevêche *Athene noctua*, en France. *L'Oiseau et R.F.O.* **62**: 309–319. (In French).

Génot, J.-C. 1994. Chouette chevêche ou Chevêche d'Athéna. *In* Yeatman-Berthelot, D. & Jarry, G. (eds). *Nouvel Atlas des Oiseaux Nicheurs de France. 1985–1989*. Paris: Société Ornithologique de France. pp. 398–401. (In French).

Génot, J.-C. 1995. Données complémentaires sur la population de Chouettes chevêches, *Athene noctua*, en déclin en bordure des Vosges du Nord. *Ciconia* **19**: 145–157. (In French).

Génot, J.-C. 2001. Etat des connaissances sur la Chevêche d'Athéna, *Athene noctua*, en bordure des Vosges du Nord (Nord-Est de la France) de 1984 à 2000. *In* Génot, J.-C., Lapios, J.-M., Lecomte, P. & Leigh, R. S. (eds). Chouette chevêche et territoires. Actes du Colloque International de Champ-sur-Marne, November 25–26 2000. ILOWG. *Ciconia* **25**: 109–118. (In French).

Génot, J.-C. 2005. La Chevêche d'Athéna, *Athene noctua*, dans la Réserve de la Biosphère des Vosges du Nord de 1984 à 2004. *Ciconia* **29**: 1–272. (In French).

Génot, J.-C. & Bersuder, D. 1995. Le régime alimentaire de la Chouette chevêche, *Athene noctua*, en Alsace-Lorraine. *Ciconia* **19**: 35–51. (In French).

Génot, J.-C. & Lecomte, P. 1998. Essai de synthèse sur la population de Chevêche d'Athéna *Athene noctua* en France. *Ornithos* **5**: 124–131. (In French).

Génot, J.-C. & Lecomte, P. 2002. *La Chevêche d'Athéna*. Paris: Delachaux Niestlé. (In French).

Génot, J.-C. & Sturm, F. 2001. Biologie de reproduction de la Chevêche d'Athéna (*Athene noctua*) en captivité. *Ciconia* **25**: 219–230. (In French).

Génot, J.-C. & Sturm, F. 2003. Bilan de l'expérience de renforcement des populations de Chevêche d'Athéna (*Athene noctua* Scop.) dans le Parc naturel régional des Vosges du Nord. *Alauda* **73**: 175–178. (In French).

Génot, J.-C. & Van Nieuwenhuyse, D. 2002. *Athene noctua* Little Owl. *BWP Update* **4**: 35–63.

Génot, J.-C. & Wilhelm, J.-L. 1993. Occupation et utilisation de l'espace par la Chouette chevêche, *Athene noctua*, en bordure des Vosges du Nord. *Alauda* **61**: 181–194. (In French).

Génot, J.-C., Lecci, D., Bonnet, J., Keck, G. & Venant, A. 1995. Quelques données sur la contamination chimique de la Chouette chevêche, *Athene noctua*, et de ses oeufs en France. *Alauda* **63**: 105–110. (In French).

Génot, J.-C., Juillard, M. & Van Nieuwenhuyse, D. 1997. Little Owl. *In* Hagemeijer, W. J. M. & Blair, M. J. *The EBCC Altas of European Breeding Birds: Their Distribution and Abundance*. London: T. & A.D. Poyser. pp. 408–409.

Génot, J.-C., Sturm, F., Pfitzinger, H. *et al.* 2000. Expérience de renforcement des populations de Chevêche d'Athéna (*Athene noctua* Scop.) dans la Réserve de Biosphère des Vosges du Nord. *Ann. Sci. Rés. Bios. Trans. Vosges du Nord-Pfälzerwald* **8**: 31–51. (In French).

Génot, J.-C., Lapios, J.-M., Lecomte, P. & Leigh, R. S. (eds) 2001. Choutte chevêche et territoiries. Actes du Colloque International de Champs-sur-Marne, 26 et 27 Novembre 2000. ILOWG. *Ciconia* **25**:

Geroudet, P. 1964. L'hiver rigoureux de 1962–1963 en Suisse romande. *Nos Oiseaux* **27**: 209–223. (In French).

Giannotti, M., Balestrieri, R., Gori, F. *et al.* 2005. Dati considerazioni sulla distribuzione della Civetta *Athene noctua* nei Campi Flegrei (Na). *In* Mastrorilli, M., Nappi, A. & Barattieri, M. (eds). Atti I Convegno italiano sulla Civetta. Osio Sotto (BG), March 21, 2004. Gruppo Italiano Civette: 27–28. (In Italian).

Gibbons, D. W., Reid, J. B. & Chapman, R. A. 1993. *The New Atlas of Breeding Birds in Britain and Ireland 1988–1991*. London: T. & A. D. Poyser.

Gimbutas, M. 1989. *The Language of the Goddess*. New York, USA: Harper & Row.

Giuseppe, A. 2005. La Civetta *Athene noctua* nella provincial di Cosenza (Calabria). *In* Mastrorilli, M., Nappi, A. & Barattieri, M. Atti I Convegno italiano sulla Civetta. Osio Sotto (BG), March 21, 2004. Gruppo Italiano Civette: 37–39. (In Italian).

Glue, D. E. 1971. Ringing recovery circumstances of small birds of prey. *Bird Study* **18**: 137–146.

Glue, D. E. 1972. Bird prey taken by British owls. *Bird Study* **19**: 91–95.

Glue, D. E. 2002. Little Owl *Athene noctua. In* Wernham, C. V., Toms, M. P., Marchant, J. H. *et al.* (eds). *The Migration Atlas: Movements of the Birds of Britain and Ireland* pp. 429–431. London, T. & A. D. Poyser.

Glue, D. E. & Morgan, R. 1977. Recovery of bird rings in pellets and other prey traces of owls, hawks and falcons. *Bird Study* **24**: 111–113.

Glue, D. E. & Scott, D. 1980. Breeding biology of the Little Owl. *British Birds* **73**: 167–180.

Glutz von Blotzheim, U. N. 1962. *Die Brutvögel der Schweiz*. Verlag Aargauer Tagblatt A. G. Aarau. pp. 319–321. (In German).

Glutz von Blotzheim, U. N. & Bauer, K. M. 1980. *Handbuch der Vögel Mitteleuropas*. Bd 9 Akad. Verlagsges. Wiesbaden, pp. 501–530. (In German).

Godard, A. 1917. *Les oiseaux nécessaires àll'agriculture, à la sylviculture, à la viticulture, à l'arboriculture et à l'hygiène publique*. Librairie Académique Perrin. (In French).

Golubeva, T. B., Chernyl, A. G. & Il'ichev, V. D. 1970. Summary responses of the auditory nerve in relation to acoustic signal parameters in the owls *Asio otus* and *Athene noctua. J. Evol. Biochem.* **6**: 169–175.

Gooch, G. B. 1940. The bill-snapping of a Little Owl. *British Birds* **33**: 316.

Gorman, G. 1995. The status of owls (*Strigiformes*) in Hungary. *Buteo* **7**: 95–108.

Goszcynski, J. 1981. Comparative analysis of food of owls in agrocenoses. *Ekol. Pol.* **29**: 431–439.

Gotta, A. & Pigozzi, G. 1997. Trophic niche of the Barn Owl and Little Owl in a rice field habitat in northen Italy. *Ital. J. Zool.* **64**: 55–59.

Goutner, V. & Alivizatos, H. 2003. Diet of the Barn Owl (*Tyto alba*) and Little Owl (*Athene noctua*) in wetlands of northeastern Greece. *Belg. J. Zool.* **13**: 15–22.

Grant-Ives, D. M. 1936. Observation on the Little Owl. *Avic. Mag.* **7**: 198.

Graubitz, G. 1983. Domovyi sych (Little Owl). *In* Viksne, J. (ed.). 1983. *Ptitsy Latvii. Territorial'noe razmeshchenie i chislennost' [Birds of Latvia. Territorial Distribution and Numbers*]. Riga: Zinatne Press. pp. 122–123. (In Russian).

Greenwood, J. J. D. & Carter, N. 2003. Organisation eines nationalen Vogelmonitorings durch den British Trust for Ornithology – Erfahrungsbericht aus Großbritannien. [Organizing a national bird monitoring by the BTO experiences from Britain.] *In*

Berichte des Landesamtes für umweltschutz Sachsen-Anhalt. Sonderheft 1/2003 14–26. (In German).

Greenwood, J. J. D., Crick, H. Q. P. & Bainbridge, I. P. 2003. Numbers and international importance of raptors and owls in Britain and Ireland. *In* Thompson, D. B. A., Redpath, S. M., Fielding, A. H., Marquiss, M. & Galbraith, C. A. (eds). *Birds of Prey in a Changing Environment.* Edinburgh: TSO Scotland. pp. 25–49.

Gregory, T. C. 1944. Little Owl hovering. *British Birds* **38**: 55–56.

Grell, M. B. 1998. *Fuglenes Danmark.* Kϕbenhavn, Denmark: GEC. GAD & Dansk Ornitologisk Forening. (In Danish).

Grimm, H. 1986. Zur Strukturierung zweier Graslandhabitate und deren potentielles Nahrungsangebot für den Steinkauz (*Athene noctua*) in Thüringer Becken. *Landschftspflege und Naturschutz in Thüringen* **22**: 94–104. (In German).

Grimm, H. 1989. Die Erhaltung und Pflege von Streuobstwiesen unter dem Aspekt des Steinkauzes (*Athene noctua*). *Abh. Ber. Mus. Nat. Gotha* **15**: 103–107. (In German).

Grimmett, R., Inskipp, C. & Inskipp, T. 1999. *Birds of India, Pakistan, Nepal, Bangladesh, Bhutan, Sri Lanka, and the Maldives.* Princeton, NJ: Princeton University Press.

Grishanov, G. V. 1994. Gnezdyashchiesya ptitsy Kaliningradskoi oblasti: territorialnoe razmeshchenie i dinamikachislennosti v XIX-XX vv. I. Non-Passeriformes [Breeding birds of the Kaliningrad Region: territorial distribution and dynamics of number in the nineteenth to twentieth centuries. I. Non-Passeriformes]. *Russk. ort. zhurnal.* **3**: 83–116. (In Russian).

Groen, N. M., Boudewijn, T. J. & de Jonge, J. 2000. De effecten van overstroming van de uiterwaarden op de Steenuil. *De Levende Natuur* **5**: 143–148. (In Dutch).

Groen, N. M., van Harxen, R. & Stroeken, P. 2002. Steenuil *Athene noctua. In* SOVON. *Atlas van de Nederlands broedvogels 1998–2000. Nederlandse Fauna 5.* Vogelonderzoek Nederland, Nationaal Natuurhistorisch Museum Naturalis, KNNV-Uitgeverij & European Invertebrate Survey-Nederland, Leiden. pp. 276–277. (In Dutch).

Groppali, R. 1994. *Gli uccelli nidificanti e svernanti nella Città di Cremona (1990–1993).* Azienda Energetica Municipalizzata e Museo Civico di Storia Naturale, Cremona. (In Italian).

Grosse, A. & Transehe, N. 1929. *Austrumbaltijas mugurkaulaino saraktis.* Riga. (In Latvian).

Grzywaczewski, G. & Kitowski, I. 2000. Owls (*Strigiformes*) in cities of the Lublin Region (SE Poland). Conference Birds of Prey and Owls, Czech Society for Ornithology. Mikulov, November 24–26, 2000. pp. 22–23.

Grzywaczewski, G., Kitowski, I. & Scibior, R. 2006. Diet of Little Owl *Athene noctua* during breeding in the central part of Lublin region (SE Poland). *Acta Zoologica Sinica* **52**: 1155–1161.

Gubin, B. M. 1998. Gnezdyashchiesya ptitsy vostochnoi kromki peskov Kyzylkum [Breeding birds of eastern edge of Kyzylkum sands]. *Russkiy ornit. Jurnal (Russ. J. of Ornithology)* **55**: 3–23. (In Russian).

Guermeur, Y. & Monnat, J. Y. 1980. *Histoire et géographie des oiseaux nicheurs de Bretagne.* SEPNB Centrale Ornithologique Ar Vran. Ministère de l'Environnement et du cadre de Vie. Direction de la Protection de la Nature. ARPEGE Clermont-Ferrand. pp. 103–104. (In French).

Gunston, D. 1948. Little Owl as prey of Tawny Owl. *British Birds* **41**: 388.

Gusberti, V. 1998. Evaluation des facteurs limitants et des exigences minimales pour la Chevêche d'Athéna (*Athene noctua*, Scop. 1769) au Tessin: le cas de la population du Mendrisiotto. Université de Neuchâtel. Institut de Zoologie. (In French).

Gusev, V. M. 1952. O znachenii glubiny snezhnogo pokrovadlya ptits, pitayushchikhsya myshevidnymi gryzunami [On the significance of the snow cover depth for birds feeding on mouse-like rodents]. *Zool. Zhurnal* **31**: 471–473. (In Russian).

Gushcha, G. I. 1982. *Ornithogastia* Vercammen-Grandjean, 1960, Stat. N. (Acariformes, Trombiculidae) sopisaniem novogo vida Ukrainy [*Ornithogastia* Vercammen-Grandjean, 1960, Stat. N. (Acariformes, Trombiculidae) with description of new species of Ukraine]. *Vestnik Zoologii (Bulletin of Zoology)* **4**: 20–25. (In Russian).

Guyot, H. & Lecomte, P. 1996. La Chouette chevêche et les insectes. *Insectes* **102**: 22–24. (In French).

Gyllin, R. 1968. Ryttling som jaktmethod hos minervauggla (*Athene noctua*). *Var Fagelvärld* **27**: 172–173. (In Danish).

Haase, P. 1993. Zur Situation und Brutbiologie des Steinkauzes *Athene n. noctua* SCOP., 1769 im Westhaveland. *Naturschutz und Landschatspflege in Brandenburg* **2**: 29–37. (In German).

Haase, P. 2001. Steinkauz – *Athene noctua*. *In* Arbeitsgemeinschaft Berlin-Brandenburgischer Ornithologen. 2001. *Die Vogelwelt von Brandenburg und Berlin*. Ringsdorf. pp. 382–385. (In German).

Haensel, J. & Walter, H. J. 1966. Beitrag zur Ernährung der Eulen im Nordharz-Vorland unter besonderer Berücksichtigung der Insektennahrung. *Beitr. Vogelkd.* **11**: 345–358. (In German).

Hahn, E. 1984. Welche der mitteleuropäischen Eulenarten eignet sich als Biomonitor? Diplomarbeit Universität des Saarlandes. Saarbrücken. (In German).

Hakkarainen, H., Ilmonen, P., Koivunen, V. & Korpimäki, E. 1998. Blood parasites and nest defence behaviour of Tengmalm's owls. *Oecologia* **114**: 574–577.

Hamonville d', L. 1895. Les oiseaux de la Lorraine. *Mém. Soc. Zool. France* **8**: 257. (In French).

Handrinos, G. & Akriotis, T. 1997. *The Birds of Greece*. London, UK: Helm.

Hanski, I. 1999. *Metapopulation Ecology*. Oxford, UK: Oxford University Press.

Hanson, D. E. 1973. X-ray photographs of Little Owl pellets. *British Birds* **66**: 33.

Harbodt, A. & Pauritsch, G. 1987. Lebensraum Streuobstwiese. Programme und gesetzliche Schutzmöglichkeiten. *In* Keil, W. (ed.). *Staatliche Vogelschutzwarte für Hessen, Rheinland-Pfalz und Saarland. 1937–1987*. Festschrift. Institut für Angewandte Vogelkunde, pp. 81–91. (In German).

Hardouin, L. 2002. Communication acoustique et reconnaissance individuelle chez la Chevêche d'Athéna CEPE. CNRS. DEA. Strasbourg: Université Louis Pasteur.

Hardouin, L. 2006. Communication acoustique et territorialité chez les rapaces nocturnes. PhD Thesis. Strasbourg: Université Louis Pasteur. (In French).

Hardouin, L., Tabel, P. & Bretagnolle, V. 2006. Neighbour–stranger discrimination in the little owl, *Athene noctua*. *Animal Behaviour* **72**: 105–112.

Hartert, E. 1913. *Athene noctua lilith*. Die Vogel der paläarktischen Fauna, Systematisch Übersicht der in Europa, Nord Asien under der Mittelmeerregion vorkommende *Vögel* **2**: 1006. (In German).

Harthan, A. J. 1948. Fledgling period of Little Owl. *British Birds* **41**: 22.

Hartmann-Müller, B. 1973. Beiträge zur Mauser des Steinkauzes, *Athene noctua* (Scop.). I Der Handflügel. Philippia. *Abh. Ber. Naturkde. Mus. Kassel* **1**: 286–295. (In German).

Hartmann-Müller, B. 1974. Beiträge zur Mauser des Steinkauzes, *Athene noctua* (Scop.). II Armflügel und Schwanzfedern. Philippia. *Abh. Ber. Naturkde. Mus. Kassel* **2**: 182–184. (In German).

Haverschmidt, F. 1939. Beobachtungen über die Bagattung beim Steinkauz (*Athene noctua*). *Beitr. Fortpfl. Vögel* **15**: 236–239. (In German).

Haverschmidt, F. 1940. Prooiresten bij Steenuilenbroedsels. *Ardea* **29**: 56–57. (In Dutch).

Haverschmidt, F. 1946. Observations on the breeding habits of the Little Owl. *Ardea* **34**: 214–246.

Hayden, J. 2004. The Diet of the Little Owl on Skomer Island NNR 1998–2003. CCW Contract Science Report 673. Bangor, Wales: Countryside Council for Wales.

Heckenroth, H. & Laske, V. 1997. Atlas der Brutvögel Niedersachsens 1981–1995. *Naturschutz Landschaftspfl. Niedersachs.* **37**: 173. (In German).

Hegger, H. L. 1977. Steinkauz, Waldkauz und Waldohreule als Brutvögel in Kempener Land. *Heimatbuch des Kreises Viersen* **1977**: 58–63. (In German).

Heidrich, P. & Wink, M. 1994. Tawny Owl (*Strix aluco*) and Hume's Tawny Owl (*Strix butleri*) are distinct species. Evidence from nucleotide sequences of the cytochrome b gene. *Z. Naturforsch.* **49c**: 230–234.

Heidrich, P. & Wink, M. 1998. Phylogenetic relationships in holarctic owls (order *Strigiformes*): Evidence from nucleotide sequences of the mitochondrial cytochrome b gene. *In* Chancellor, R. D., Meyburg, B.-U. & Ferrero, J. J. (eds). *Holarctic Birds of Prey.* Adenex & WWGBP. pp. 73–87.

Heidrich, P., König, C. & Wink, M. 1995a. Molecular phylogeny of the South American Screech Owls of the *Otus atricapillus* complex (Aves, Strigidae) inferred from nucleotide sequences of the mitochondrial cytochrome b gene. *Z. Naturforsch.* **50c**: 294–302.

Heidrich, P., König, C. & Wink, M. 1995b. Bioakustik, Taxonomie und molekulare Systematik amerikanischer Sperlingskäuze (Strigidae: *Glaucidium sp*). *Stuttgarter Beiträge zur Naturkunde A* **534**: 1–47. (In German).

Heilig, D. & Stahlheber, K.-H. 2004. Die Entwicklung des Steinkauzbestandes (*Athene noctua*) 1987–2004 einer Untersuchungsfläche in der Südpfalz (Rheinland-Pfalz). *Fauna Flora Rheinland-Pfalz* **10**: 425–437. (In German).

Heinroth, O. & Heinroth, M. 1926. *Die Vögel Mitteleuropas.* Vol. 2. Berlin: Hugo Bermühler Verlag. (In German).

Helbig, A. 1981. Auswirkungen des strengen Winters 1978–79 auf die Vogelwelt in Westphalia. *Charadrius* **17**: 82–103. (In German).

Hellebrekers, W. P. J. 1928. Aantal eieren bij *Asio otus* (L.), *Athene noctua* A. E. Brehm en *Riparia riparia* (L.). *Ardea* **17**: 165–166. (In Dutch).

Henrioux, P. 1980. Nidification simultanée de l'Etourneau et de la Chouette chevêche dans le même nichoir. *Nos Oiseaux* **35**: 343. (In French).

Hens, P. A. 1954. Een Wezel (*Mustela nivalis*) als prooi van de Steenuil (*Athene noctua*). *Limosa* **27**: 63. (In Dutch).

Hernandez, M. 1988. Road mortality of the Little Owl – *Athene noctua* – in Spain. *J. Raptor Res.* **22**: 81–84.

Herrera, C. M. & Hiraldo, F. 1976. Food-niche and trophic relationships among European owls. *Ornis Scand.* **7**: 29–41.

Heuglin von, M. T. 1863. Beiträge zur ornithologie Nord-Ost Afrika's. *J. Orn.* **13**(1): 3–29. (In German).

Heuglin von, M. T. 1869. *Athene noctua spilogastra.* Ornithologie Nordost-Afrika's, der Nil-quellen-und kusten-gebiete des Rothen Meeres under des nordlichen Somal-lande s. 1, p. 119, pl. 4

Hewson, R. 1972. Changes in the number of stoats, rats and Little Owl in Yorkshire as shown by tunnel trapping. *J. Zool.* **168**: 427–429.

Heydt, J. G. 1968. Las rapaces y ostras *Aves* de la Sierra de Gata. *Ardeola* **14**: 120. (In Spanish).

Hibbert-Ware, A. 1936. Report of an investigation of the food of captive Little Owls. *British Birds* **29**: 302–305.

Hibbert-Ware, A. 1937/1938. Report of the Little Owl food inquiry. 1936–37. *British Birds* **31**: 162–187, 205–229, 249–264.

Hibbert-Ware, A. 1938a. The food habits of the Little Owl (*Carine noctua vidalii*). *Ann. Appl. Biol.* **25**: 218–220.

Hibbert-Ware, A. 1938b. The Little Owl. Is it injurious? *School Nature Study*: 1–5.

Hill, F. A. R. & Lill, A. 1998. Density and total population estimates for the threatened Christmas Island Hawk-Owl *Ninox natalis*. *Emu* **98**: 209–220.

Hinde, R. 1953. The conflict between drives in the courtship and copulation in the chaffinch. *Behaviour* **5**: 1–31.

Hinsley, S. A. & Bellamy, P. E. 2000. The influence of hedge structure, management and landscape context on the value of hedgerows to birds: a review. *Journal of Environmental Management* **60** (1): 33–49.

Holmgren, V. C. 1988. *Owls in Folklore and Natural History*. Santa Barbara, CA, USA: Capra Press.

Hölzinger, J. 2001. *Athene noctua*-Steinkauz. *In* Hölzinger, J. & Mahler, U. (eds). *Die Vögel Baden-Württembergs*. Vol. 2.3. Non-Passeriformes-Nicht-Singvögel (Teil 3). Stuttgart. pp. 195–211. (In German).

Hordowski, J. 1999. *Birds of the Polish Eastern Carpathian Mountains and Sub-Carpathian Regions*. Faunistic Monograph – vol. I Pterodidiformes-Passeriformes. Brzemyoel: Mercator Publishers (In Polish).

Hosking, E. J. & Newberry, C. W. 1945. *Birds of the Night*. London, UK.

Hounsome, T., O'Mahony, D. & Delahay, R. 2004. The diet of Little Owls *Athene noctua* in Gloucestershire, England. *Bird Study* **51**: 282–284.

Houston, C. S. & Francis, C. M. 1995. Survival of Great Horned Owls in relation to the Snowshoe Hare cycle. *Auk* **112**: 44–59.

Hubl, H. 1952. Beiträge zur Kenntnis der Verhaltensweisen junger Eulenvögel in Gefangenschaft. *Z. Tierpsychol.* **9**: 102–119. (In German).

Hudec, K. (ed.) 1983. *Fauna CSSR, Ptaci – Aves*, BD. 3/1. Prague. (In Czech).

Hudson, W. E. (ed.) 1991. *Landscape Linkages and Biodiversity*. Washington, DC, USA: Island Press.

Hulten, M. 1955. Gewöllenuntersuchungen. *Regulus* **35**: 45. (In German).

Hulten, M. & Wassenich, V. 1961. Die Vogelfauna Luxemburgs. 2. Teil. *Bull. Soc. Naturalistes Luxembourgeois* **66**: 446–447. (In German).

Iankov, P. 1983. The ornithofauna of Sofia, its structure and formation characteristics. Ph.D. thesis. Minsk. (In Russian)

Il'yukh, M. P. 2002. Gnezdovaya biologiya domogovo sycha Predkavkazie [Breeding biology of the Little Owl in Front-Caucasian area]. Ptitsy Yuzhnoi Rossii [Birds of south Russia]. (Materials of Intern. Conf. "Results and prospects of ornithology development on the Northern Caucasus in the 21st century" devoted to 20-year anniversary of activity of the North Caucasian Ornithological Group. October 24–27, 2002). Rostov-on-Don. *Proc. of Teberda State Natural Biospher Reserve* **31**: 113–118. (In Russian).

Ille, R. 1983. Ontogenese des Beutefangverhaltens beim Steinkauz (*Athene noctua*). *J. Orn.* **124**: 133–146. (In German).

Ille, R. 1992. Zur Biologie und Ökologie des Steinkauzes (*Athene noctua*) im Marchfeld: Aktuelle Situation und mögliche Schutzmassnahmen. *Egretta* **35**: 49–57. (In German).

Ille, R. 1995. Ergebnisse einer Bestandserhebung beim Steinkauz (*Athene noctua*) im Nordburgenland zwischen 1992–1994. *BFB-Ber.* **83**: 23–29. (In German).

Ille, R. 1996. Zur Biologie und Ökologie zweier Steinkauzpopulationen in Österreich. *Abh. Zool. Ges.Österreich* **129**: 17–31. (In German).

Ille, R. & Grinschgl, F. 2001. Little Owl (*Athene noctua*) in Austria. Habitat characteristics and population density. *In* Génot, J.-C., Lapios, J.-M., Lecomte, P. & Leigh, R. S. (eds). Chouette chevêche et territoires. Actes du Colloque International de Champ-sur-Marne, November 25–26, 2000. ILOWG. *Ciconia* **25**: 129–140.

Illner, H. 1979. Erfahrungsbericht über Steinkauzbrutenin Niströhren. D. B.V. *AG zum Schutz bedrohter Eulen Informationsblatt* **9**: 6. (In German).

Illner, H. 1981. Popultionsentwicklung der Eulen (Strigiformes) auf einer Probefläche Mittelwestphalens 1974–1979 und bestandsbeeinflussende Faktoren, insbesondere anthropogener Art. *Ökologie der Vögel* **3**: 301–310. (In German).

Illner, H. 1988. Langfristiger Rückgang von Schleiereule *Tyto alba*, Waldohreule *Asio otus*, Steinkauz *Athene noctua* und Waldkauz *Strix aluco* in der Agrarlandschaft Mittelwestphalens 1974–1986. *Vogelwelt* **109**: 145–151. (In German).

Illner, H. 1990. Sind durch Nistkasten-Untersuchungen verlässliche Populationstrends zu ermitteln? Eine Fallstudie am Steinkauz (*Athene noctua*). *Vogel und Umwelt* **6**: 47–57. (In German).

Illner, H. 1991. Influence d'un apport de nourriture supplémentaire sur la biologie de reproduction de la Chouette chevêche, *Athene noctua*. *In* Juillard, M., Bassin, P., Baudvin, H. *et al.* (eds). Rapaces nocturnes. Actes du 30e Colloque interrégional d'Ornithologie Porrentruy (Suisse): November 2–4, 1990. *Nos Oiseaux*: 153–157. (In French).

Illner, H. 1992. Road deaths of Westphalian owls: methodological problems, influence of road type and possible effects on population levels. *In* Galbraith, C. A., Taylor, I. R. & Percival, S. (eds). The ecology and conservation of European owls. Peterborough, Joint Nature Conservation Committee. UK. *Nature Conservation* **5**: 94–100.

Ingendahl, D. & Tersteegen, H.-J. 1992. Bestand von Steinkauz und Schleiereule im Sudkreis Kleve (Unterer Mittelrhein). Ein Bericht der Eulenschutzgruppe Issum. *AG zum Schutz bedrohter EulenInformationsblatt* **38**: 11–13. (In German).

Isenmann, P. & Moali, A. 2000. *Oiseaux d'Algérie*. Paris, France: Société d'Etudes Ornithologiques de France. (In French).

Ishtiaq, F., Rahmani, A. R. & Rasmussen, P. C. 2002. Ecology and behaviour of the Forest Owlet (*Athene blewitti*). *In* Newton, I., Kavanagh, R. P., Olsen, J. & Taylor, I. R. (eds). *Ecology and Conservation of Owls*. Melbourne: CSIRO Publishing. pp. 80–88.

Ishunin, G. I. 1965. Pitanie domovogo sycha v zapovednike Aral-Paigambar (Uzbekistan). [The Little Owl feeding in Aral-Paigambar Nature Reserve (Uzbekistan)]. *Ornithologiya* **7**: 471–472. (In Russian).

Ishunin, G. I. & Pavlenko, T. A. 1966. Materialy poekologii zhivotnykh pastbishch Kyzylkuma. [Materials on animal ecology of the pastures of Kyzylkum]. Pozvonochnye zhivotnye Srednei Azii. [Vertebrates of Middle Asia]. Tashkent. pp. 28–66. (In Russian).

Ivanov, A. I. 1940. *Ptitsy Tajikistana* [*Birds of Tajikistan*]. Leningrad–Moscow: Publishing House of the Academy of Science of the USSR.

Ivanov, A. I. 1969. *Ptitsy Pamiro-Alaya [Birds of Pamiro-Alai]. L.*, "Nauka", Leningrad branch. (In Russian).

Ivanov, A. I. 1976. *Katalog ptits SSSR (Catalogue of birds of the USSR).* L., "Nauka". (In Russian).

Jacob, J. 1974. Quantitative measurement of food selection. *Oecologia* **14:** 413–417.

Jacob, J. & Hoerschelmann, H. 1984. Chemotaxonomische Untersuchungen an Eulen (*Strigiformes*). *Funkt. Biol. Med.* **3**: 56–61. (In German).

Jacobs, F. 2003. Succes van het verplaatsen van eieren of jongen van Steenuilen. *Athene nieuwsbrief STONE* **8**: 20–23. (In Dutch).

Jančar, T. 1997. *Ornithological Atlas of Breeding Birds in Triglav National Park, 1991–1996 [Ornitološki atlas gnezdilk Triglavskega narodnega parka, 1991–1996].* Društvo za opazovanje in proučevanje ptic Slovenije, Ljubljana. (In Slovene).

Jančar, T., Gregori, J., Mihelič, T. *et al.* 2001. Red Data List of Breeding Birds in Slovenia, DOPPS 2001 [Rdeči seznam gnezdilk Slovenije, DOPPS 2001]. Ljubljana: DOPPS-BirdLife Slovenia. (In Slovene).

Jaspers, V., Covaci, A., Maervoet, J. *et al.* 2005. Brominated flame retardants and organochlorine pollutants in eggs of little owls (*Athene noctua*) from Belgium. *Environmental Pollution* **136**: 81–88.

Jathar, G. A. & Rahmani, A. R. 2004. Ecological studies of the Forest Spotted Owlet *Athene (Heteroglaux) blewitti*. Final Report. Mumbai, India: Bombay Natural History Society.

Jay, M. 1993. *Athene noctua. In* Centre Ornithologique du Gard. *Oiseaux Nicheurs du Gard. Atlas Biogéographique 1985–1993*. Gard, France: Centre Ornithologique du Gard. (In French).

Jermaczek, A., Czwalga, T. & Stanko, R. 1990. Population numbers and distribution of owls in the landscape of the Gorzów Wielkopolski Region. *Nat. Review* **3**: 41–50 (In Polish).

Jermaczek, A., Czwałga, T., Jermaczek, D. *et al.* 1995. *Birds of the Gorzów Wielkopolski Region – Faunistic Monograph*. Świebodzin: Publishers of Naturalists of the Gorzów Wielkopolski Region. (In Polish).

Jöbges, M. 2004. Der Steinkauz (*Athene noctua*). *In* Gedeon, K., Mitschke, A. & Sudfeldt, C. (eds). *Brutvögel in Deutschland*. Hohenstein-Ernsthal. pp 22–23.

Johnson, D. H. 1980. The comparison of usage and availability measurements for evaluating resource preference. *Ecology* **61**: 6–71.

Johnson, D. H., Van Nieuwenhuyse, D. & Génot, J.-C. 2008. Survey Protocol for the Little Owl (*Athene noctua*). *In* Johnson, D. H., Van Nieuwenhuyse, D. & Duncan, J. R. (eds). Proceedings of the 4th World Owl Conference, Groningen, Netherlands 2008. *Ardea* **96**.

Joiris, C. & Delbecke, K. 1981. Rückstände chlororganischer Pestizide und PCBs in belgischen Greifvögeln. *Ökologie der Vögel* **3**: 173–180. (In German).

Jongbloed, R. H., Traas, T. P. & Luttik, R. 1996. A probabilistic model for deriving soil quality criteria based on secondary poisoning of top predators. II. Calculations for Dichlorodiphenyltrichloroethane (DDT) and Cadmium, *Ecotoxicity and Environmental Safety* **34**: 279–306.

Juillard, M. 1979. La croissance des jeunes Chouettes Chevêches, *Athene noctua*, pendant leur séjour au nid. *Nos Oiseaux* **35**: 113–124. (In French).

Juillard, M. 1980. Répartition, biotopes et sites de nidification de la Chouette chevêche *Athene noctua* en Suisse. *Nos Oiseaux* **35**: 309–337. (In French).

Juillard, M. 1981. Trois malformations anatomiques apparentes chez la Chouette chevêche, *Athene noctua*. *Nos Oiseaux* **36**: 121–125. (In French).

Juillard, M. 1983. La photographie sur pellicule infrarouge: une méthode pour l'étude du régime alimentaire des oiseaux cavicoles. *Terre et Vie* **37**: 267–285. (In French).
Juillard, M. 1984a. *La Chouette Chevêche. Nos Oiseaux*. Prangins. (In French).
Juillard, M. 1984b. Contribution à la connaissance éco-éthologique de la Chouette Chevêche, *Athene noctua* (Scop.), en Suisse. Ph.D. thesis. Université de Neuchâtel. (In French).
Juillard, M. 1989. The decline of the Little Owl *Athene noctua* in Switzerland. *In* Meyburg, B.-U. & Chancelor, R. D. (eds). *Raptors in the Modern World. Proceedings of the III World Conference on Birds of Prey and Owls*. Eilat, Israel, March 22–27, 1987. Berlin, London & Paris: WWGBP. pp. 435–439.
Juillard, M. 1997. Les vergers de la Chouette. *Pro Natura* **5**: 6–8. (In French).
Juillard, M., Praz, J.-C., Etournaud, A. & Beaud, P. 1978. Données sur la contamination des rapaces de Suisse romande et de leurs oeufs par les biocides organochlorés, les PCB et les métaux lourds. *Nos Oiseaux* **34**: 189–206. (In French).
Juillard, M., Baudvin, H., Bonnet, J., Génot, J.-C. & Teyssier, G. 1990. Sur la nidification en altitude de la Chouette chevêche, *Athene noctua*. Observations dans le Massif central (France). *Nos Oiseaux* **40**: 267–276. (In French).
Juillard, M., Baudvin, H., Bonnet, J. & Génot, J-C. 1992. Habitat et sites de nidification de la Chouette chevêche (*Athene noctua*) sur le Causse Méjean (Lozère, France). *Nos Oiseaux* **41**: 415–440. (In French).
Kämpfer, A. & Lederer, W. 1988. Dismigration des Steinkauzes *Athene noctua* in Mittelwestphalen. *Vogelwelt* **109**: 155–164. (In German).
Kämpfer-Lauenstein, A. & Lederer, W. 1995. Bestandsentwicklung einer Steinkauzpopulation in Mittelwestphalen. *Vogelwelt* **109**: 155–164. (In German).
Kaplan, R. & Norton, D. 1996. *The Balanced Scorecard: Translating Strategy into Action*. Boston, Massachusetts: Harvard Business School Press.
Karalus, K. & Ekkert, A. 1974. *The Owls of North America*. New York, USA: Garden City.
Karyakin, I. V. 1998. *Pernatye khishchniki Ural'skogo regiona. [Birds of Prey and Owls of the Ural Region]*. Perm: Centre of Field Researches of Ural Animal Conservation Union. (In Russian).
Karyakin, I. V. & Kozlov, A. A. 1999. *Predvaritelínyikasdastr ptits Chelyabinskoi oblasti [Preliminary Cadastre of Birds of the Chelyabinsk Region (Oblast)]*. Novosibirsk: Manuscript Press. (In Russian).
Kasparek, M. 1992. *Die Vögel der Türkei*. Heidelberg: Kasparek Verlag. (In German).
Kasparek, M. & Bilgin, C. C. 1996. Aves. *In* Kence, A. & Bilgin, C. C. (eds). 1996. *Türkiye Omurgalilar Tür Listesi*. Ankara: DPT Yayinlari. (In Turkish).
Kasprzykowski, Z. & Golawski, A., in press. Habitat use of the Barn Owl *Tyto alba* and Little Owl *Athene noctua* in east-central Poland. *Buteo*.
Kavanagh, R. P. 2002. Conservation and management of large forest owls in southeastern Australia. *In* Newton, I., Kavanagh, R., Olsen, J. & Taylor I. (eds). *Ecology and Conservation of Owls*. Australia: CSIRO. pp. 201–219.
Keil, H. 2001. Artenschutzprojekt Steinkauz im Kreis Ludwigsburg. Forschungsgemeinschaft zur Erhaltung einheimischer Eulen e.V. (eds): Eulen als Kulturfolger – Probleme und Chancen. Tagungsband zum 1. Treffen der Eulenfachleute in Baden-Württemberg 2001. (In German).
Kenward, R. 1987. *Wildlife Radio Tagging. Equipment, Field Techniques and Data Analysis*. London: Academic Press.
Keve, A., Kohn, I., Matousek, F., Mosansky, A. & Rucner-Kroneisl, R. 1960. Über die taxonomische Stellung der Südesteuropäischen Steinkäuze, *Athene noctua* (Scop.). *Larus* **14**: 26–74. (In German).

Khokhlov, A. N. 1992. Osobennosti ekologii sov vantropogennykh landshaftakh Tsentral'nogo Predkavkz'ya [Ecological characters of owls in the anthropogenic landscapes of the central Pre-Caucasus]. Sovremennatya ornitologiya 1991 [Modern ornithology 1991]. Nauka. pp. 85–95. (In Russian).

Khokhlov, A. N. 1995. *Ornitologicheskie nablyudeniya vZapadnoi Turkmenii [Ornithological Observations in Western Turkmenia].* Stavropol: Stavropol State Pedag. University. (In Russian).

Khokhlov, A. N., Il'yukh, M. P., Emel'yanov, S. A. *et al.* 1998. K letnei ornitofaune nizoviy reki Kumy i prilezhashchikh territoriy [On the summer ornithofauna of lower parts of Kuma River and adjoining territories]. *Kavkazskiy ornitologicheskiy vestnik [Caucasian Ornithological Bulletin]* **10**: 135–143. (In Russian).

Kinsky, F. C. 1973. The subspecific status of the New Zealand population of the Little Owl, *Athene noctua* (Scopoli, 1769). *Notornis* **20**: 9–13.

Kirchberger, K. 1988. Artenschutzmöglichkeiten beim Steinkauz und Schwarzmilan. *Vogelschutz im Österreich* **2**: 52–55. (In German).

Kiselev, F. A. & Ovchinnikova, E. N. 1953. Domovoi sych kakistrebitel myshevidnykh gryzunov v stepnom Krymu [The Little Owl as extirpator of mouse-like rodents in the steppe Crimea]. *Trudy Krymskogo filiala AN SSSR [Proc. of the Crimean Branch of Ac. of Sc. of the USSR]* **3**: 51. (In Russian).

Kitowski, I. & Grzywaczewski, G. 2003. The monitoring of Little Owl *Athene noctua* in Chelm (SE Poland) in 1998–2000. *Ornis Hungarica* **12–13**: 1–2.

Kiyko, A. O. & Yakubenia, O. I. 1995. Vydovyi sklad, poshyrennya ta zhyvlenya sov u Lvivs'kiy oblasti [Species structure, spreading and feeding of owls in Lviv Region]. Problemy vyvchennya ta okhorony ptakhiv [Problems of study and protection of birds]. Materials of the VI conference of Western Ukraine Ornithologists (Drohobych, Fenruary 1–3, 1995). pp. 61–63. (In Ukrainian).

Kleinschmidt, O. 1909. Beschreibung neuer Formen. *Falco* **5**: 19–20. (In German).

Knötzsch, G. 1978. Ansiedlungsversuche und Notizen zur Biologie des Steinkauzes (*Athene noctua*). *Vogelwelt* **99**: 41–54. (In German).

Knötzsch, G. 1985. Zwölfjährige Populationsuntersuchungenan einem isolierten Vorkommen des Steinkauzes im Bodenseegebiet. Bundesweiter *Tagung der A. G. zum Schutz bedrohter Eulen am 2/3 März 1985 in Grävenwiesbach. D.B.V. Zusammenfassung der Vorträge*: 3–4. (In German).

Knötzsch, G. 1988. Bestandsentwicklung einer Nistkasten-Population des Steinkauzes *Athene noctua* am Bodensee. *Vogelwelt* **109**: 164–171. (In German).

Koblik, E. A. 2001. *Raznoobrazie ptits (po materialam ekspozitsii Zoologicheskogo Muzeya MGU) [Birds diversity (on materials of exposition of the Zoological Museum of Moscow State University)].* Part 3. Moscow Univ. Press. (In Russian).

Komarov, Y. E. 1998. Ptitsy sel'skikh naselennykh punktov respubliki Severnaya Ossetia-Alania [Country side settlement birds of the Republic of Northern Ossetia-Alania]. *Kavkazskiy ornitologicheskiy vestnik [Caucasian Ornithological Bulletin]* **10**: 65–74. (In Russian).

König, C. 1969. Extrem kurze Brutdauer beim Steinkauz (*Athene noctua*). *Vogelwelt* **90**: 66–67. (In German).

König, C., Weick, F. & Becking, J.-H. 1999. *Owls. A Guide to the Owls of the World.* Sussex, UK: Pica Press.

Koop, M.-H. 1996. Freilandökologischer Vergleich der Kleinsäuger-, Laufkäfer-und Regenwurm-Abundanz mit deren Nutzung als Beute durch den Steinkauz *Athene noctua* (Scopoli, 1769) in drei Steinkauzrevieren der Mechernicher Voreifel.

Diplom-Arbeit am Institut für Angewandte Zoologie der Rheinischen Friedrich-Wilhelms Universität Bonn. (In German).

Kopystynska, K. 1962. Investigations on the vision of infra-red in animals. Part III. Preliminary experiments on the Little Owl *Athene noctua* (Scop.). *Prace Zoologiczne* **7**: 95–107.

Korpimäki, E. 1981. On the ecology and biology of Tengmalm's Owl (*Aegolius funereus*) in Southern Ostrobothnia and Suomenselkä western Finland. Acta Universitatis Ouluensis. Series A. *Scientia Rerum Naturalium* **118**: 1–84.

Korpimäki, E. 1988. Effects of territory quality on occupancy, breeding performance, and breeding dispersal in Tengmalm's owl. *Journal of Animal Ecology* **57**: 97–108.

Korpimäki, E., Hakkarainen, H. & Bennett, G. F. 1993. Blood parasites and reproductive success of Tengmalm's Owls: detrimental effects on females but not males? *Functional Ecology* **7**: 420–426.

Korpimäki, E., Tolonen, P. & Bennett, G. F. 1995. Blood parasites, sexual selection and reproductive success of European kestrels. *Ecoscience* **2**: 335–343.

Korpimäki, E., Hakkarainen, H., Ilmonen P. & Wiehn, J. 2002. Detrimental effects of blood parasites on parental effort and reproductive success of Tengmalm's owls (*Aegolius funereus*) and Eurasian Kestrels (*Falco tinnunculus*). *In* Newton, I., Kavanagh, R., Olsen, J. & Taylor, I. (eds). *Ecology and Conservation of Owls.* Australia: CSIRO Publishing, pp. 68–73.

Kostadinova, I. (ed.) 1997. *Important Bird Areas in Bulgaria.* BSPB Conservation Series. Book 1. Sofia: BSPB. (In Bulgarian).

Kostin, Y. V. 1983. *Ptitsy Kryma [Birds of the Crimea].* Moscow: Nauka. (In Russian).

Kostrzewa, A., Ferrer-Lerin, F. & Kostrzewa, R. 1986. Abundance, status and vulnerability of raptors and owls in parts of the Spanish Pyrenees. *Birds of Prey Bull.* **3**: 182–190.

Kowalski, M., Lippoman, T. & Oglecki, P. 1991. Population numbers of owls *Strigiformes* in the eastern part of Kampinos Primeval Forest. *Acta Orn.* **26**: 23–29 (In Polish).

Krischer, O. 1990. Rückgang der Steinkauzpopulation in der Region Basel. *Mitglieder-informationen AG zum Schutzbedrohter Eulen* **35**: 3–4. (In German).

Krzanowski, A. 1973. Numerical comparison of *Vespertilionidae* and *Rhinolophidae* (Chiroptera: *Mammalia*) in the owl pellets. *Acta Zool. Cracov.* **18**: 133–140.

Kuhn, W. 1992. Analyse von Steinkauzgewöllen mit besonderer Berücksichtigung der Koleopteren. Dilpomarbeit des Fachbereichs Forstwirtschaft der Ludwig-Maximilian-Universität. 116 p. (In German).

Kuhn, W. 1995. Struktur und jahreszeitliche Verteilung von Käfern in Steinkauzgewöllen. *Eulen Rundblick* **42**: 12–15.

Kulaeva, T. M. 1977. *Sovoobraznye [Strigiformes]. Ptitsy Volzhosko-Kamskogo kraya [Birds of the Volga-Kama Territory].* Moscow: Nauka Press. pp. 239–257. (In Russian).

Kulczycki, A. 1964. Badania na skladem pokarmu sow zBeskidu Niskiego [Study of the make-up of the diet of owls from the Niski Beskid Mts]. *Acta Zool. Cracov.* **9**: 535–558. (In Polish).

Kumerloeve, H. 1955. Spalax und Skorpione als Steinkauz-Nahrung. *Vogelwelt* **76**: 110. (In German).

Laaksonen, T., Hakkarainen, H. & Korpimäki, E. 2004. Lifetime reproduction of a forest-dwelling owl increases with age and area of forests. *Proc. R. Soc. Lond. B (Suppl.)* **271**: 461–464.

Labes, R. & Patzer, J. 1987. Steinkauz – *Athene noctua*. *In* Klafs, G. & Stübs J. (eds). *Die Vogelwelt Mecklenburgs*, 3rd edition. Jena. pp. 247–248. (In German).

Labitte, A. 1951. Notes biologiques sur la Chouette chevêche, *Carine noctua vidalii* A. E. Brehm, 1857. *L'Oiseau et RFO* **21**: 120–126. (In French).

Lack, D. 1946. Competition for food by birds of prey. *J. Anim. Ecol.* **15**: 128.

Lack, D. 1954. *The Natural Regulation of Animal Numbers*. Oxford, UK: Oxford University Press.

Lancum, F. H. 1925. Further observations on the food of the Little Owl. *J. Min. Agric.* **32**: 170–173.

Laursen, J. T. 1981. Kirkeuglens, *Athene noctua*, fodevalg i Ostjylland. *Dansk Orn. Foren.Tidsskr.* **75**: 105–110. (In Danish).

Lecomte, P. 1995. Le statut de la Chouette chevêche *Athene noctua* en Ile-de-France. Evolution et perspectives. *Alauda* **63**: 43–50. (In French).

Lecomte, P., Lapios, J.-M. & Génot, J.-C. 2001. Plan de restauration des populations de Chevêches d'Athéna en France. *In* Génot, J.-C., Lapios, J.-M., Lecomte, P. & Leigh, R. S. (eds). Chouette chevêche et territoires. Actes du Colloque International de Champ-sur-Marne, November 25–26, 2000. ILOWG. *Ciconia* **25**: 159–171. (In French).

Lederer, W. & Kämpfer-Lauenstein, A. 1996. Einfluss der Witterung auf die Brutbiologie einer Steinkauzpopulation (*Athene noctua*) in Mittelwestphalen. *Populationsökologie Greifvögel und Eulenarten* **3**: 353–360. (In German).

Lefeuvre, J.-C. 1981. Les études scientifiques, un préalable indispensable à la restructuration foncière et à l'aménagement des zones agricoles bocagères. Ecologie et Développement, journées scientifiques des 19–20 Septembre 1979. Editions du CNRS. (In French).

Lehman, L., Wisniewski, J. & Rochella, J. A. (eds). 1999. *Forest Fragmentation: Wildlife and Management Implications*. Boston, Massachusetts, USA: BRILL Academic Publishers.

Lei, F.-M. 1995. A study on diet of the Little Owl (*Athene noctua plumipes*) in Qishan, Shaanxii Province, China. *Wuyi Science Journal* **12**: 136–142.

Lei, F. M., Cheng, T. H. & Yin, Z. H. 1997. On distribution, habitat and the clinal variations of the subspecies of the Little Owl *Athene noctua* in China (*Strigiformes*: Strigidae). *Acta Zootaxonomical Sinica* **22**(3): 327–334.

Leicht, U. 1992. Erfahrungen mit der Steinkauzzucht und der Auswilderung. *Naturschutzzentrum Wasserschloss Mitwitz* **2**: 35. (In German).

Leigh, R. 2001a. The breeding dynamics of Little Owls (*Athene noctua*) in North West England. *In* Génot, J.-C., Lapios, J.-M., Lecomte, P. & Leigh, R. S. (eds). Chouette chevêche et territoires. Actes du Colloque International de Champ-sur-Marne, November 25–26, 2000. ILOWG. *Ciconia* **25**: 67–76.

Leigh, R. 2001b. Working structure of the European conservation plan for the Little Owl *Athene noctua*: the way forward. *In* Van Nieuwenhuyse D., Leysen, M. & Leysen, K (eds). The Little Owl in Flanders in its international context. Proceedings of the Second International Little Owl Symposium, April 16–18, 2001, Geraardsbergen, Belgium. *Oriolus* **67**(2): 142–146.

Letty, J., Génot, J.-C. & Sarrazin, F. 2001. Viabilité de la population de Chevêche d'Athéna *Athene noctua* dans le Parc naturel régional des Vosges du Nord. *Alauda* **69**: 359–372. (In French).

Leveque, G. 1997. Reproduction et régime alimentaire de la population de Chouette chevêche (*Athene noctua*) du Causse Méjean. Rapport Maîtrise Sciences et Techniques. Parc national des Cévennes. (In French).

Libois, R. 1977. Contribution à l'étude du régime alimentaire de la Chouette chevêche (*Athene noctua*) en Belgique. *Aves* **14**: 165–177. (In French).

Lilleleht, V. & Leibak, E. 1992. Rarities in Estonia till 1989: report of the Estonian Rarities Committee (3). *Hirundo* **10**: 3–20.

Linkhart, B. D. & Reynolds, R. T. 1997. Territories of Flammulated Owls (*Otus flammeolus*): is occupancy a measure of habitat quality? *In* Duncan, J. R., Johnson, D. H. & Nichols, T. H. (eds). *Biology and Conservation of Owls of the Northern Hemisphere*: 2nd International Symposium; February 5–9, 1997; Winnipeg, Manitoba. US Forest Service Gen. Tech. Rep. NC-190. St. Paul, MN. pp. 250–254.

Lint, J., Noon, B., Anthony, R. *et al.* 1999. Northern Spotted Owl effectiveness monitoring plan for the Northwest Forest Plan. General Technical Report PNW-GTR-440. Portland, Oregon: US Department of Agriculture, Forest Service, Pacific Northwest Research Station.

Lipej, L. & Gjerkes, M. 1992. Bats in the diet of owls in NW Istra. *Myotis* **30**: 133–138.

Lippens, L. & Wille, H. 1972. *Atlas van de vogels van België*. Tielt, Lannoo. (In Dutch).

Lipsberg, Y. K. 1985. Domovyi sych (Little Owl). *Sarkana gramata Latvijas PSR [Red Data Book of the Latvian SSR]*. Riga: "Zinatne" Publ. House. pp. 452–453. (In Russian and Latvian).

Lo Verde, G. & Massa, B. 1988. Abitudini alimentari del la Civetta (*Athene noctua*) in Sicilia. *Naturalista Sicil.* **12**: 145–149. (In Italian).

Lockley, R. M. 1938. The Little Owl inquiry and the Skokholm-Petrels. *British Birds* **31**: 278–279.

Lord, J. & Ainsworth, G. H. 1945. Fight between Little Owls and Barn Owls. *British Birds* **38**: 275.

Lorgé, P. 2006. Gehört der Steinkauz *Athene noctua* in Luxemburg bald zum alten Eisen? *Regulus Wissenschaftliche Berichte* **21**: 54–58. (In German).

Loske, K-H. 1978. Pflege, Erhaltung und Neuanlage von Kopfbäumen. *Natur und Landschaft* **53**: 279–281. (In German).

Loske, K-H. 1986. Zum Habitat des Steinkauzes (*Athene noctua*) in der Bundesrepublik Deutschland. *Vogelwelt* **107**: 81–101. (In German).

Loske, K.-H. & Loske, R. 1981. Quantitative Erfassung von Biotopverlusten dargestellt am Beispiel des Langeneicker Bruches (Kreis Soast). *Natur u. Landschaft Westf*. **17**: 79–82. (In German).

Loudon, H. 1909. Vorlaufiges Verzeichnis der Vogel der russischen Ostzeeprovinzen Estland, Livland und Kurland. *Annu. Musee Zool. Acad. Imp. Sci. St.-Petersbourg* **1**: 192–222. (In German).

Lovari, S. 1974. The feeding habits of four raptors in central Italy. *Raptor Research* **8**: 45–57.

Lucas, P. 1996. Comparaison entre deux populations de Chouette chevêche (*Athene noctua*) en province de Liège. *Aves* **33**: 85–92. (In French).

Luder, R. & Stange, C. 2001. Entwicklung einer Population des Steinkauzes *Athene noctua* bei Basel 1978–1993. *Orn. Beob.* **98**: 237–248. (In German).

Luginbühl, J. 1908. Etwas vom Steinkauz. *Orn. Beob.* **6**: 166–167. (In German).

Lumaret, J. P. 1993. Insectes coprophages et médicaments vétérinaires: une menace à prendre au sérieux. *Insectes* **91**: 2–3. (In French).

Lyaister, A. F. & Sosnin, G. V. 1942. Materialy po ornitofaune Armyanskoi SSR (*Ornis Armenica*) [Materials on ornithofauna of the Armenian SSR (*Ornis Armenica*)]. Yerevan. (on the cover 1944). 402 /XV/ p. *AN SSSR Arm. filial. Biol. in-t (Ac. of Sc. of the USSR Biol. Inst.)*: 5–10. (In Russian).

Lyon, R. H., McNabb, E. G., Volodina, L. & Willig, R. 2002. Modelling distributions of large forest owls as a conservation tool in forest management: a case study from Victoria, southeastern Australia. *In* Newton, I., Kavanagh, R., Olsen, J. & Taylor, I. (eds). *Ecology and Conservation of Owls*. Australia: CSIRO. pp. 242–254.

Lysaght, W. R. 1919. Little Owl breeding in Monmouthshire. *British Birds* **12**: 237–238.

Madon, P. 1933. Les rapaces d'Europe, leur régime, leurs relations avec l'agriculture et la chasse. Toulon. pp. 97–113. (In French).

Mädlow, W. & Model, N. 2000. Vorkommen und Bestand Seltener Brutvogelarten in Deutchland 1995/96. *Vogenwelt*. **121**: 189–205. (In German).

Makatsch, W. 1976. *Die Eier der Vögel Europas*. Band 2. Verlag J. Neumann. Neudamn. Melsungen. Berlin. Basel. Wien. pp. 43–45. (In German).

Malchevskiy, A. S. & Pukinskiy, Y. B. 1983. *Ptitsy Leningradskoi Oblasti Isopredelnykh Territoriy [Birds of the Leningrad Region and Adjoining Territories]*. Leningrad: Leningrad University Press. Vol. 1. (In Russian).

Malmstigen, J. E. 1970. Ryttling hos minervauggla, *Athene noctua*. *Var Fagelvärld* **29**: 231. (In Danish).

Mambetjumeaev, A. M. 1998. Pitanie khishchnykh ptits i sovv nijnem i srednem techenii Amudarii na Ustyurte i v Kyzylkumakh [Diet of birds of prey and owls in Lower and Middle Amu Darya, Ustyurt and KyzylKum]. *Russk. orn. zhurnal (Russ. J. of Ornithology)* **47**: 6–16. (In Russian).

Mammen, U. & Stubbe, M. 2001. Jahresbericht 2000 zum Monitoring Greifvögel Eulen Europas. Jahresber. *Monitoring Greifvögel Eulen Europas*, **13**: 1–99.

Manez, M. 1983a. Espectro alimentacio del Mochuelo comun (*Athene noctua*) en Espana. *Alytes* **1**: 275–290. (In Spanish).

Manez, M. 1983b. Variaciones geograficas y estacionales en la dieta del Mochuelo comun (*Athene noctua*) en Espana. *XV Congr. Int. Fauna Cinegetica y Silvestre. Trujillo*, 1981: 617–634. (In Spanish).

Manez, M. 1994. Little Owl *Athene noctua*. *In* Tucker, G. M. & Heath, M. F. (eds). *Birds in Europe: their Conservation Status*. Cambridge: UK BirdLife International. BirdLife Conservation Series No. 3. pp. 330–331.

Manganaro, A., Ranazzi, L. & Salvati, L. 2001. Diet overlap of Barn Owl (*Tyto alba*) and Little Owl (*Athene noctua*) in a Mediterranean urban area. *Buteo* **12**: 67–70.

Marchant, J. H., Hudson, R., Carter, S. P. & Whittington, P. 1990. Population trends in British breeding birds. Nature Conservancy Council, BTO. pp. 114–116.

Marcot, B. G. & Johnson, D. H. 2003. Owls in mythology and culture. *In* Duncan, J. R. (ed.). *Owls of the World: Their Lives, Behavior and Survival*. Toronto ON, Canada: Key Porter Books. pp. 89–105

Marian, M. & Schmidt, E. 1967. Adatok a kurvik (*Athene noctua* (Scop.)) gerinces taplalenak ismeretehez Magyarorszagon. Daten zur Kenntnis der Nahrung an Wirbeltieren der *Athene noctua* (Scop.) in Ungarn. *Mora F. Muzeum Evkonyve*: 271–275. (In Hungarian).

Marié, P. & Leysen, M. 2001. Contribution to the design of an anti-marten *Martes foina* system to limit predation in Little Owl *Athene noctua* nestboxes. *In* Van Nieuwenhuyse, D., Leysen, M. & Leysen, K. (eds). Little Owl in Flanders in its international context. Proceedings of the Second International Symposium, March 16–18, 2001, Geraardsbergen, Belgium. *Oriolus* **67**: 126–131.

Marinatos, S. 1968. Die Eulengottin van Pylos. *Mitteilungen des Deutschen Archaologischen Instituts, Athenische Abteilung* (Berlin) **83**: 167–174.

Martí, R. & Moral, J. C. (eds) 2003. *Atlas de las Aves Reproductoras de España*. Madrid: Dirección General de Conservación de la Naturaleza-Sociedad Española de Ornitología. (In Spanish).

Martin, R. & Rollinat, R. 1914. *Description et Moeurs des Mammifères, Oiseaux, Reptiles, Batraciens et Poissons de la France Centrale*. Paris: Paul Lechevalier. pp. 97–98. (In French).

Martin, R. & Rollinat, R. 1982. *Les oiseaux du département de l'Indre à la fin du 19ème siècle*. Texte intégral du chapitre ornithologique de l'ouvrage "Les vertébrés sauvages du département de l'Indre". Groupe d'Etude de l'Avifaune de l'Indre. pp. 17–18. (In French).

Martínez, J. A. & Zuberogoitia, I. 2003. *Las Aves Rapaces en los Mosaicos Agroforestales de Alicante y Valencia*. Sura Eds Alicante. (In Spanish).

Martínez, J. A. & Zuberogoitia, I. 2004a. Effects of habitat loss on perceived and actual abundance of the Little Owl *Athene noctua* in Eastern Spain. *Ardeola* **51**(1): 215–219.

Martínez, J. A. & Zuberogoitia, I. 2004b. Habitat preferences for Long-eared Owls *Asio otus* and Little Owls *Athene noctua* in semi-arid environments at three spatial scales. *Bird Study* **51**: 163–169.

Martínez, J. A., Sanchez, M. A., Carmona, D. *et al.* 1992. The ecology and conservation of the Eagle Owl *Bubo bubo* in Murcia, southeast Spain. *In* Galbraith, C. A., Taylor, I. R. & Percival, S. (eds). The ecology and conservation of European owls. Peterborough, UK: Joint Nature Conservation Committee. *Nature Conservation* **5**: 84–88.

Martínez, J. A., Zuberogoitia, I. & Alonso, R. 2002. *Rapaces Nocturnas. Guia para la Determinacion de la Edad y el Sexo en las Estrigiformes Ibericas*. Madrid: Monticola Ed. (In Spanish).

März, R. & Weglau, I. 1957. Rupfungs-und Gewöllaufsammlung bei Darfel/Westphalia. *Vogelwelt* **78**: 110–112. (In German).

Maslov, N. M. 1947. *Ptitsy Bukharskoi oblasti [Birds of the Bukhara region]*. Sb. trudov Bukharskogo Gos. ped. in-ta "Estestvoznanie" [Collection of papers of the Bukhara State Ped. Inst., "Natural Sciences"]. Bukhara. pp. 45–70. (In Russian).

Mastrorilli, M. 1997. Popolazioni di Civetta (*Athene noctua*) e selezioni dell'habitat in un'area di pianura della provincia di Bergamo. *Riv. Mus. Civ. St. Nat. "E. Caffi" Bergamo* **19**: 15–19. (In Italian).

Mastrorilli, M. 1999a. Caratteristiche dei ricoveri e dei siti riproduttivi di Civetta *Athene noctua* nelle province di Bergamo e Cremona. *Avocetta* **23**: 163. (In Italian).

Mastrorilli, M. 1999b. Ripetuta ed inusuale osservazione di Civetta *Athene noctua* in un dormitorio invernale di Gufo comune *Asio otus*. *Picus* **25**: 47. (In Italian).

Mastrorilli, M. 2000. Seconda nidificazione di Civetta *Athene noctua*, in Corsica. *Riv. Ital. Orn.* **70**: 89–91. (In Italian).

Mastrorilli, M. 2001. Little Owl *Athene noctua* status and habitat selection in the town of Bergamo (Lombardy, Northern Italy). *In* Van Nieuwenhuyse D., Leysen, M. & Leysen, K. (eds). The Little Owl in Flanders in its international context. Proceedings of the Second International Little Owl Symposium, March 16–18, 2001. Geraardsbergen, Belgium. *Oriolus* **67**(2–3): 136–141.

Mastrorilli, M. 2003. La comunità ornitica nidificante della città di Crema (CR) nel biennio 2000–2001. *Riv. Ital. Orn.* **72**(2): 261–268. (In Italian).

Mastrorilli, M. 2005. *La Civetta in Italia*. Araspix Editrice, Brescia. (In Italian).

Mastrorilli, M., Barattieri, M., Calvi, G. & Fasano, D. 2005. Le popiolazioni di Civetta *Athene noctua* nei comprensori urbani d'Italia: proposta di standardizzazione del metodo di raccolta. *In* Mastrorilli, M., Nappi, A. & Barattieri, M. (eds). Atti I Convegno italiano sulla Civetta. Osio Sotto (BG), 21 Marzo 2004. Gruppo Italiano Civette: 20–23. (In Italian).

Matvejev, S. D. 1950. La distribution et la vie des oiseaux en Serbie (Ornithogeographia serbica). Academie Serbe des Sciences. Monographies – T. Clxi. Institut d'ecologie et de Biogéographie No. 3. Belgrade (In French).

May, R. M. 1994. The effects of spatial scale on ecological questions and answers. *In* Edwards, P. J., May, R. M. & Webb, N. R. (eds). *Large-scale Ecology and Conservation Biology.* (35th Symposium of the British Ecological Society, with the Society for Conservation Biology). London, UK: Blackwell Scientific Publications. pp. 1–17.

Medlin, F. 1967. *Centuries of Owls in Art and the Written Word.* Norwalk, Connecticut: Silvermine Publishers.

Meisser, C. 1998. Suivi et protection de la Chouette chevêche *(Athene noctua)* dans le canton de Genève, Suisse. Aperçu de la période d'étude 1984–1997. *Nos Oiseaux* **46**, 1–4. (In French).

Meisser, C. In press. Plan d'action national Chevêche d'Athéna. Programme de Conservation des Oiseaux en Suisse. (In French).

Meisser, C. & Albrecht, P. 2001. Suivi et protection de la Chevêche d'Athena *Athene noctua* dans le canton de Genève, Suisse (période 1984–2000). *In* Génot, J.-C., Lapios, J.-M., Lecomte, P. & Leigh, R. S. (eds). Chouette chevêche et territoires. Actes du Colloque International de Champ-sur-Marne, 25 et 26 Novembre 2000. ILOWG. *Ciconia* **25**: 191–197. (In French).

Meisser, C. & Juillard, M. 1998. *Athene noctua. In* Schmid, H., Luder, R., Naef-Daenzer, B., Graf, R. & Zbinden, N. (eds). *Atlas des Oiseaux Nicheurs de Suisse et du Liechtenstein. Distribution des Oiseaux Nicheurs en Suisse et au Liechtenstein en 1993–1996.* Station ornithologogique suisse. Sempach. (In French).

Meklenburtsev, R. N. 1936. Materialy po mlekopitayushchim i ptitsam Pamira [Materials on mammals and birds of Pamir]. Tashkent. ill. Ser. 8a. Zool. Issue 22. Rezul'taty Pamir. ekspeditsii Sredne-Az. gos. un-ta [Results of Pamir expedition of Middle-Asian State Univ.]. *Trudy Sredne-Az. un-ta [Proc. of the Middle-Asian Univ.].* Issue **4**: 39–40. (In Russian).

Meklenburtsev, R. N. 1946. Zimuyushchie i proletnye ptitsy Vostochnogo Pamira [Wintering and passage birds of Eastern Pamir]. *Byull. MOIP., Otd. biol.* [*Bulletin of Moscow Naturalist's Society, Dept. Biology*] **54**(1): 87–110. (In Russian).

Meklenburtsev, R. N. 1958. Materialy po nazemnym pozvonochnym basseina r. Kashkadar'i [Materials on terrestrial verbebrates of Kashkadarya River basin]. Proc. of Central-Asian State University, new series, issue 130, biol, sciences, book 30. Tashkent. pp. 1–140. (In Russian).

Melchior, E., Mentgen, E., Peltzer, R., Schmitt, R. & Weiss, J. 1987. *Atlas of the Breeding Birds in Luxembourg.* LNVL. (In English, French and German).

Melendro, J. & Gisbert, J. 1978. Notas ornithologicas breves. *Ardeola* **24**: 261. (In Spanish).

Menyushina, I. E. 1997. Snowy Owl (*Nyctea scandiaca*) reproduction in relation to Lemming population cycles on Wrangel Island. *In* Duncan, J. R., Johnson, D. H. & Nichols, T. H. (eds). *Biology and Conservation of Owls of the Northern Hemisphere:* 2nd International Symposium; February 5–9, 1997; Winnipeg, Manitoba. US Forest Service Gen. Tech. Rep. NC-190. St. Paul, MN. pp. 572–582.

Mercier, A. 1921. Valeur économique de quelques rapaces. *Gerfaut* **11**: 65. (In French).

Meyer, B. 1815. *Kurze Beschreibung der Vögel Liv- und Esthlands*. Nürnberg. (In German).

Meyer-Holzapfel, M. & Räber, H. 1975. Verhaltensstörunges beim Beuteerwerb handaufgezogener Waldkäuze (*Strix aluco*) und deren experimentelle Abklärung. *Orn. Beob. Bern* **72**: 18–22. (In German).

Meyknecht, J. 1941. Farbensehen und Helligkeitsunterscheidung beim Steinkauz (*Athene noctua vidalii* A. E. Brehm). *Ardea* **30**: 129–170. (In German).

Mienis, K. H. 1979. Predation on *Testudo graeca* by the Little Owl in Israel. *Salamandra* **15**: 107–108.

Mikkola, H. 1976. Owls killing and killed by other owls and raptors in Europe. *British, Birds* **69**: 144–154.

Mikkola, H. 1983. *Owls of Europe*. Calton: Poyser.

Millennium Ecosystem Assessment Board. 2005. Living beyond our means: natural assets and human well-being. United Nations. March 2005. http://www.millenniumassessment.org/ en/Products.BoardStatement.aspx

Miller, G. S. 1989. Dispersal of juvenile Northern Spotted Owls in western Oregon. M.S. thesis. Corvallis, Oregon: Oregon State University.

Mills, D. G. H. 1981. Cannibalism among Little Owls. *British Birds* **74**: 354.

Mishchenko, A. 2004. Otsenka chislennosti i eyo dinamiki dlya ptits Evropeiskoi chasti Rossii (Ptitsy Evropy –II) [Estimation of numbers and trends for birds of the European part of Russia (Birds in Europe-II]). Moscow: Russian Bird Conservation Union. pp. 10–14. (In Russian).

Misonne, X. 1948. Le régime de la Chouette chevêche *Athene noctua vidalii* A. E. Brehm. *Gerfaut* **38**: 89. (In French).

Mitschke, A. & Baumung, S. 2001. Brutvogel-Atlas Hamburg. *Hamb. Avifaunist Beitr.* **31**: 1–343. (In German).

Mlikovsky, J. & Piechocki, R. 1983. Biometrische Untersuchungen, zum Geschlechtsdimorphismus einiger mitteleuropäische Eulen. *Beitr. Vogelkd.* **29**: 1–11. (In German).

Mohr, H. 1989. Steinkäuze brüten wieder in Schwaben. *Gef. Welt* **113**: 308–309. (In German).

Mohr, R. 1990. Bemerkenswertes aus dem Leben hessischer Steinkäuze (*Athene noctua*). *Vogel und Umwelt* **6**: 63–69. (In German).

Möller, B. 1993. Erste Ergebnisse zur Wiedereinbürgerung des Steinkauzes (*Athene noctua*) in den Landkreisen Hildesheim und Peine. *Beitr. Naturk. Niedersachsens* **46**: 72–81. (In German).

Moreau, W. M. 1947. Kestrel robbing Little Owl. *British Birds* **40**: 216–217.

Morrison, M. L. 2001. A proposed research emphasis to overcome the limits of wildlife-habitat relationship studies. *Journal of Wildlife Management* **65**: 613–623.

Moschetti, G. & Mancini, D. 1993. Dieta della Civetta *Athene noctua* (Scopoli) e sue variazioni stagionali in un parco urbano in ambiente mediterraneo. *Gli Uccelli Italia* **28**: 3–12. (In Italian).

Moschetti, G. & Mancini, D. 1995. Variazioni stagionalinella dieta della Civetta *Athene noctua* (Scopoli) in un parco urbano in ambiente mediterraneo. *Suppl. Ric. Biol. Selvaggina* **22**: 261–263. (In Italian).

Mostini, L. 1988. *Athene noctua*. *In* Mingozzi, T., Boano, G. & Pulcher, C. (eds). *Atlante Degli Uccelli Nidificanti in Piemonte e Valle d'Aosta*. Monograf. VIII. Mus. Region. Sc. Nat., Torino. (In Italian).

Mulder, B. S., Noon, B. R., Spies, T. A. *et al.* 1999. *The Strategy and Design of the Effectiveness Monitoring Program for the Northwest Forest Plan*. General Technical Report PNW-GTR-437. Portland, OR: US Department of Agriculture, Forest Service, Pacific Northwest Research Station.

Müller, W. 1999. Untersuchungen zum Auftreten alternativa Fort planzungsstratagien und Geschlechterverhältnis beim Steinkauz (*Athene noctua*). Masters thesis. Universität Bonn. (In German).

Müller, W., Eppler, J. T. & Lubjuhn, T. 2001. Genetic paternity analysis in Little Owls (*Athene noctua*): does the high rate of paternal care select against extra-pair young? *J. Ornithol.* **142**: 195–203.

Muntaner, J., Ferrer, X. & Martínez-Vilalta, A. 1983. *Atlas del Ocells Nidificants de Catalunya i Andorra*. Barcelona: Ketres Ed. (In Catalan).

Murzov, V. N. & Berezovikov, N. N. 2001. Pitanie domovogosycha *Athene noctua* v pustynnykh landshaftakh Yuzhnogo Pribalkashya (Yugo-Vostochnyi Kazakhstan) [The food of the Little Owl *Athene noctua* in the deserts of southern part of Balkhash region (Southeastern Kazakhstan)]. *Russk. orn. zhurnal (Russ. J. of Ornithology)* **134**: 176–178. (In Russian).

NABU Altkreis Lingen e.V. (ed.) 2003. Steinkauz. Brochure. (In German).

Nankinov, D. N. 2002. Sovremennoe sostoyanie populyatsiy sov Bolgarii (Present situation of population of owls in Bulgaria). *Berkut*. **11**: 48–60. (In Russian).

Nappi, A. 2005. Dati sull'alimentazione della Civetta in una risaia vercellese. *In* Mastrorilli, M., Nappi, A. & Barattieri, M. Atti I Convegno italiano sulla Civetta. Osio Sotto (BG), 21 Marzo 2004. Gruppo Italiano Civette: 53–54. (In Italian).

Nappi, A. & Mastrorilli, M. 2003. The Little Owl *Athene noctua* diet in Italy. Third International Symposium on Wild Fauna. Ischia 24–28 May 2003. Luigi Esposito & Bianca Gasparrini. pp. 403–406.

Nash, E. G. 1925. Little Owls attacking cat. *British Birds* **19**: 25–26.

Negro, J.-J., De La Riva, M. J. & Hiraldo, F. 1990. Daytime activity of Little Owls (*Athene noctua*) in southwestern Spain. *J. Raptor Res.* **24**: 72–74.

Newton, I. 1979. *Population Ecology of Raptors*. London, UK: T & AD Poyser.

Newton, I. 1998. *Population Limitations in Birds*. London, UK: Academic Press.

Newton, I. 2002. Population limitation in Holarctic owls. *In* Newton, I., Kavanagh, R., Olsen, J. & Taylor, I. (eds). *Ecology and Conservation of Owls*. Australia: CSIRO Publishing. pp. 3–29.

Newton, I., Wyllie, I. & Dale, L. 1997. Mortality causes in British Barn Owls (*Tyto alba*), based on 1101 carcasses examined during 1963–1996. *In* Duncan, J. R., Johnson, D. H. & Nichols, T. H. (eds). *Biology and Conservation of Owls of the Northern Hemisphere*: 2nd International Symposium; February 5–9, 1997; Winnipeg, Manitoba. US Forest Service General Technical Report NC-190. St. Paul, MN. pp. 299–307.

Nicolai, B. 2000. Bestandsentwicklung und Situation des Steinkauzes (*Athene noctua*) in Sachsen-Anhalt. *Apus* **10**: 55–64. (In German).

Nicolai, J. 1975. *Vogelleben*. Stuttgart: Belser Verlag. (In German).

Nikiforov, M. E., Yaminskiy, B. V. & Shklyarov, L. P. 1989. *Belorussii [Birds of Belarus]. Spravochni-kopredelitel gnedz i yaits [Manual-guide of Nests and Eggs].* Minsk: Vysheishaya Shkola Press. (In Russian).

Nikiforov, M. E., Kozulin, A. V., Grichik, V. V. & Tishechkin, A. K. 1997. *Ptitsy Belarusi na rubezhe XXI veka [Birds of Belarus at the Edge of Twenty-first Century].* Minsk: N. A. Korolev Producer. (In Russian).

Nitsche, G. & Plachter, H., 1987. *Atlas der Brutvögel Bayerns 1979–1983.* Ornithologische Gesellschaft in Bayern, Bayerisches Landesamt für Umweltschutz. München. p. 126. (In German).

Noon, B. R. 2003. Conceptual issues in monitoring ecological resources. *In* Busch, D. E. & Trexler, J. C. (eds). *Monitoring Ecosystems: Interdisciplinary Approaches for Evaluating Ecoregional Initiatives.* Washington, DC: Island Press. pp. 27–72.

Noon, B. R. & Franklin, A. B. 2002. Scientific research and the Spotted Owl (*Strix occidentalis*): opportunities for major contributions to avian population ecology. *The Auk* **119** (2): 311–320.

Norberg, R. A. 2002. Independent evolution of outer ear asymmetry among five owl lineages; morphology, function, and selection. *In* Newton, I., Kavanagh, R., Olsen, J. & Taylor, I. (eds). *Ecology and Conservation of Owls.* Australia: CSIRO. pp. 329–342.

Nore, T. 1977. L'autour et l'épervier en Limousin. *Bulletin de la Société pour l'Etude et la Protection des Oiseaux en Limousin* **7**: 40. (In French).

Nore, T. 1979. Rapaces diurnes communs en Limousin pendant la période de nidification (II: autour, épervier et faucon crecerelle). *Alauda* **47**: 265. (In French).

Obuch, J. & Kristin, A. 2004. Prey composition of the little owl *Athene noctua* in an arid zone (Egypt, Syria, Iran). *Folia Zool.* **53**: 65–79.

Olea, P. P. 1997. Mochuelo comun. *Athene noctua. In* Purroy, J. (ed.). *Atlas de las Aves de Espana (1975–1995).* Barcelona, Spain: Lynx Editions. pp. 260–261. (In Spanish).

Oles, T. 1961. Obserwacje nad obyczajami pokarmowymipojdzki. [Observations on the food habits in Little Owl]. *Przeglad Zoologiczny* **4**: 377–378. (In Polish).

Olsen, J., Wink, M., Sauer-Gürth, H. & Trost, S. 2002. A new *Ninox* owl from Sumba, Indonesia. *Emu* **102**: 223–232.

Opdam, P. 1991. Metapopulation theory and habitat fragmentation: a review of holarctic breeding bird studies. *Landscape Ecology* **5**: 93–106.

Orf, M. 2001. Habitatnutzung und Aktionsraumgrösse des Steinkäuze *Athene noctua* im Main-Taunus-Kreis. Diplomarbeit im Fachbereich Biologie der Johann Wolgang Goethe-Universität, Frankfurt am Main. (In German).

Orsini, P. 1985. Le régime alimentaire du Hibou grand-duc *Bubo bubo* en Provence. *Alauda* **53**: 11–28. (In French).

Osieck, E. R. & Hustings, F. 1994. Rode lijst van bedreigde en kwetsbare vogelsoorten in Nederland, Technisch rapport Vogelbescherming Nederland 12. Vogelbescherming Nederland Zeist. (In Dutch).

Osmolovskaya, V. I. 1953. Geograficheskoe raspredeleniekhishchnykh ptits ravninnogo Kazakhstana i ikh znachenie v istrebleniivreditelei [Geographical distribution of the birds of prey of the plain Kazakhstan and their importance in extermination of pests]. Trudy Instituta Geografii (Proc. of Inst. of Geography). *AN SSSR* **54**: 219–307. (In Russian).

Otto, A. & Ullrich, T. 2000. Schutz des Steinkauzes (*Athene noctua*) in der südlichen Ortenau und in angrenzenden Gebieten. *Naturschutz südl. Oberrhein* **3**: 49–54. (In German).

Owen, J. H. 1919. On the procuring of food by the male for the female among birds of prey. *British Birds* **13**: 29.

Owen, J. H. 1939. Little Owl taking House-Martin. *British Birds* **33**: 111.

Pailley, M., Pailley, P. & Preau, L-M. 1991. Prédation de la Chouette chevêche (*Athene noctua*) sur des populations nicheuses de Sternes pierregarins (*Sterna hirundo*) et de Sternes naines (*S. albifrons*) en Maine-et-Loire. *L'Oiseau et RFO* **61**: 337–338. (In French).

Palmer, P. 1989. Little Owl and grey squirrel in same nest-box. *British Birds* **82**: 221.

Pardieu, M. 1927. Nidification précoce d'une Chevêche. *Revue Fr. Orn.* **11**: 118. (In French).

Parigi-Bini, P. 1948. Strano reperto nel contenuto gastricodi una Civetta. *Riv. Ital. Orn.* **18**: 146–147. (In Italian).

Paris, P. 1909. Anomalies observées chez quelques oiseaux de la Côte d'Or. *Revue Fr. Orn.* **1**: 104. (In French).

Park, K. J., Calladine, J. R., Graham, K. E., Stephenson, C. M. & Wernham, C. V. 2005. *The impacts of predatory birds on waders, gamebirds, songbirds and fisheries interests.* A report to Scotland's Moorland Forum by the Centre for Conservation Science. University of Stirling and the British Trust for Ornithology, Scotland. Edinburgh: Scottish Natural Heritage. www.moorlandforum.org.uk/documents.php

Parslow, J. L. F. 1967. Changes in status among breeding birds in Britain and Ireland. *British Birds* **60**: 190–200.

Patten, C. J. 1930. The Little Owl and little rabbits. *Ir. Nat. J.* **3**: 92–94.

Peitzmeier, J. 1952. Langsamer Ausgleich der Winterluste beim Steinkauz. *Vogelwelt* **73**: 136. (In German).

Peter, W. 1999. Schutzmassnahmen für den Steinkauz (*Athene noctua*) und dessen Bestandsentwicklung im Main-Kinzig-Kreis. *Eulen-Rundblick* **48/49**: 24–25. (In German).

Peterson, A. T., Vieglais, D. A., Navarro Sigüenza, A. G. & Silva, M. 2003. A global distributed biodiversity information network: building the world museum. *Bulletin of the British Ornithologists' Club* **123A**: 186–196.

Petzold, H. & Raus, T. 1973. Steinkauz (*Athene noctua*) Bestandsaufnahmen in Mittelwestphalen. *Anthus* **10**: 25–38. (In German).

Pfister, O. 1999. Owls in Ladakh. *Oriental Bird Club Bulletin* **29**: 22–28.

Piechocki, R. 1960. Über die Winterverluste der Schleiereule (*Tyto alba*). *Vogelwarte* **29**: 277–280. (In German).

Piechocki, R. 1968. Die Grossfieder-Mauser des Steinkauzes (*Athene noctua*). *J. Orn.* **109**: 30–36. (In German).

Pirovano, A. & Galeotti, P. 1999. Territorialismo intra- e interspecifico della Civetta *Athene noctua* in provincia di Pavia. *Avocetta* **23**: 139. (In Italian).

Pitzer, W. 1995. Schleiereulen-und Steinkauzvorkommen am südlichen Rand des Ruhrgebiets. *Der Falke* **4**: 121–124. (In German).

Plantinga, J-E. 1999. Plan van aanpak Steenuil. Actierapport Vogelbescherming Nederland 14, Zeist. (In Dutch).

Pointereau, P. & Bazile, D. 1995. *Arbres des Champs, Haies, Alignements, Prés Vergers.* Toulouse, France: Solagro. (In French).

Polak, S. 2000. *Important Bird Areas (IBA) in Slovenia [Mednarodno pomembna območja za ptice v Sloveniji].* Ljubljana: DOPPS-BirdLife Slovenia (In Slovene).

Ponebšek, J. 1917. *Our Raptors, Part I. Owls [Naše ujede, I. del: Sove]*. Ljubljana: Muzejsko društvo za Kranjsko. (In Slovene).

Popescu, A. & Blidarescu, S. 1983. Date referitoare lahrana devara a cucuvelei (*Athene noctua noctua* Scop.). *An. Univ. Buc. Biol.* **32**: 77–82. (In Romanian).

Popescu, A. & Negrea, I. 1987. Date privind hrana unorpasari rapitoare de noapte (*Athene noctua noctua* Scop.: *Strigiformes*) din judetul timis. *An. Univ. Buc. Biol.* **36**: 67–70. (In Romanian).

Popescu, A., Nitu, E. & Negrea, I. 1986. Insects, an important component part of the food of the Little Owl (*Athene noctua noctua* Scop.) in summer time. *An. Univ. Buc. Biol.* **35**: 17–20.

Popov, A. V. 1959. *Ptitsy Gissaro-Karategina: Ekologo-geograficheskiy ocherk [Birds of the Gissaro-Karategin: Ecology-geographical sketch]*. Stalinabad: Publ. House of Ac. of Sc. of the Tadjik SSR. (In Russian).

Popov, V. V. 1991. K raspredeleniyu sov v yugo-zapadnoi Tuve [On owl distribution in southwestern Tuva (Tyva, Russia)] Ornithologicheskie problemy Sibiri [Ornithological problems of Siberia]. Barnaul: Altai University Press. pp. 152–154. (In Russian).

Porciatti V., Fontanesi G., Raffaelli A. & Bagnoli P. 1990. Binocularity in the Little Owl, *Athene noctua* II. Properties of visually evoked potentials from the wulst in response to monocular and binocular stimulation with sine wave gratings. *Brain Behav. Evol.* **35**: 40–48.

Poulsen, C. M. 1957. Massedodsfald blandt kirkeugler (*Athene noctua* (Scop.)). *Dansk Orn. Foren.Tidsskr*. **51**: 40–41. (In Danish).

Pugesek, B. H. 2003. Concepts of structural equation modeling in biological research. *In* Pugesek, B. H., Tomer, A. & Von Eye, A. (eds). *Structural Equation Modeling*, Cambridge, UK: Cambridge University Press. pp. 40–57.

Pukinskiy, Y. B. 1977. *The Life of Owls*. Leningrad: Leningrad State University Press (In Russian).

Pulliam, H. R. 1988. Sources, sinks, and population regulation. *Am. Nat.* **132**: 652–661.

Pykal, J., Krafka, Z., Klimes, Z. *et al.* 1994. Populacni hustota sycka obecneho (*Athene noctua*) ve vybranych oblastech jiznich Cech. [Population density of Little Owl (*Athene noctua*) in selected regions of Southern Bohemia (Czech Republic)]. *Sylvia* **30**: 59–63. (In Czech).

Qu, Y.-H., Lei, F.-M. & Yin, Z.-H. 2002. A study on genetic polymorphisms within *Athene noctua*. *In* Lian, Z. M., Xi, G. S., Huang, Y. *et al.* (eds). *Animal Science*. Shaanxi: Xian, Normal University Press. pp. 189–195.

Quadrelli, G. 1985. Pesenza di fibre vegetali nelle borredi civetta *Athene noctua* (Scopoli). *Picus* **11**: 69–71. (In Italian).

acz, B. 1917. Winterpaarung von *Glaucidium noctuum* retz. *Aquila* **24**: 286. (In German).

Raevel, P. 1986. Concentration ponctuelle de rapaces nocturnes sur un lieu de passage de crapauds communs (*Bufo bufo*). *Le Héron* **4**: 52. (In French).

Rakhimov, I. I. 1995. *Sych domovyi [Little Owl]. Krasnaya kniga Respubliki Tatarstan [Red Data Book of Republic of Tatarstan]*. Kazan: Publishing House "Priroda / Nature/", Publisher TOO "Star" pp. 78–79. (In Russian).

Rakhimov, I. I. & Pavlov., Y. I. 1999. *Sych domovyi [Little Owl]. Khishchnye ptitsy i sovy Tatarstana [Birds of Prey and Owls of Tatarstan]*. Kazan: "Tatpoligraf" Press, pp. 76–78. (In Russian).

andi, E., Lorenzini, R. & Fusco, G. 1991. Biochemical variability in four species of Strigiformes. *Biochemical Systematics and Ecology* **19**: 13–16.

Ravussin, P.-A., De Alencastro, L.-F., Humbert, B., Rossel, D. & Tarradellas, J. 1990. Contamination des oeufs de Chouette de Tengmalm, *Aegolius funereus*, du Jura vaudois (Suisse) par les métaux lourds et les organochlorés. *Nos Oiseaux* **40**: 257–266. (In French).

Reichenow, A. 1905. *Athene noctua somaliensis. Die Vogel Afrikas*, Vol. 3. German: Neudamm, p. 822.

Reinard, M. 1977. Steinkauz zieht seine Jungen in einer Erdhöhle gross. *Regulus* **12**: 234–236. (In German).

Reiser, O. 1925. *Die Vögel von Margurg an der Drau*. Graz, Austria. (In German).

Reisinger, L. 1926a. Ein interessanter Fehler am Auge eines Steinkauzes. *Biol. Zentralbl. Leipzig* **46**: 632. (In German).

Reisinger, L. 1926b. Hypnose des Steinkauzes. *Biol. Zentralbl. Leipzig* **46**: 630–631. (In German).

Rey, E. 1910. Mageinhalt einiger Vögel. *Orn. Mschr*. **35**: 227. (In German).

Rich, T. D., Beardmore, C. J., Berlanga, H. *et al.* 2004. *Partners in Flight North American Landbird Conservation Plan*. Ithaca, NY: Cornell Lab of Ornithology.

Richetti, F., Vittoria, A., Costagliola, A., De Falco, G. & Lepore, G. 1980. Prime osservazioni istomorfologiche sull'apparatodigerente di alcuni uccelli rapaci diurni (*Buteo buteo*) enoturni (*Athene noctua*). *Atti Soc. Ital. Sc. Vet.* **34**: 125. (In Italian).

Richter, B. 1973. Der Steinkauz (*Athene noctua*) im Hamburger Raum. *Hamb. Avifaun. Beitr*. **11**: 1–16. (In German).

Riedel, W. 1996. Gewölleuntersuchung von Steinkäuzen (*Athene noctua*) aus dem mittleren Neckarraum. *Orn. Jh. Bad.-Württ.* **12**: 313–318. (In German).

Ripple, W. J., Johnson, D. H., Hershey, K. T. & Meslow, E. C. 1991. Old-growth and mature forests near Spotted Owl nest-sites in western Oregon. *Journal of Wildlife Management* **55**: 316–318.

Ripple, W. J., Lattin, P. D., Hershey, K. T., Wagner, F. F. & Meslow, E. C. 1997. Landscape composition and pattern around Northern Spotted Owl nest-sites in southwestern Oregon. *Journal of Wildlife Management* **61**: 151–158.

Risenthal von, D. 1879. Verkannte und Missachtete. Das Steinkäuzchen. *Orn. Mschr*. **4**: 119–124. (In German).

Ritzel, L. & Wulf, J. 1978. Elstern (*Pica pica*) attackiert Steinkauz (*Athene noctua*). *Orn. Mitt*. **30**: 255. (In German).

Robiller, F. 1987. Über den Bodenbelag in Bruthöhlen des Steinkauzes (*Athene noctua*). *Acta Ornithoecol*. **1**: 299–301. (In German).

Robiller, F. 1993. Zur Lokalen Bestandsentwicklung des Steinkauzes, *Athene noctua*, am Stadtrand von Weimar. *Beiträge zur Vogelkunde* **39**: 1–3.

Robiller, F. & Robiller, M. 1986. Ein Beitrag zur Brutbiologie des Steinkauzes (*Athene noctua*). *Beitr. Vogelkd*. **32**: 161–174. (In German).

Robiller, F. & Robiller, F. C. 1992. Gelungene Ansiedlung des Steinkauzes (*Athene noctua*) am Stadtrand von Weimar. *Beitr. Vogelkd.* **38**: 1–5. (In German).

Rochon-Duvigneaud, A. 1934. Les yeux de la Chouette chevêche (*Athene noctua*). *Bull. Soc. Zool. Fr*. **59**: 224–226. (In French).

Rockenbauch, D. 1976. Ergänzungen zur Nahrungsbiologie einiger Eulenarten. *Anz. Orn. Ges. Bayern* **15**: 78–84. (In German).

Rodts, J. 1994. La faune et le trafic routier. *L'Homme et l'Oiseau* **2**: 109–118. (In French).

Rogacheva, E. V. 1988. *Ptitsy Srednei Sibiri [Birds of Middle Siberia]*. Moscow: Nauka. (In Russian).

Rohner, C. 1997. Non-territorial floaters in Great Horned Owls (*Bubo virginianus*). *In* Duncan, J. R., Johnson, D. H. & Nichols, T. H. (eds). *Biology and Conservation of Owls of the Northern Hemisphere*: 2nd International Symposium; February 5–9, 1997; Winnipeg, MB. US Forest Service General Technical Report NC-190. St. Paul, MN. pp. 347–362.

Romanowski, J. 1988. Trophic ecology of *Asio otus* (L.) and *Athene noctua* (Scop.) in the suburbs of Warsaw. *Pol. Ecol. Stud.* **14**: 223–234.

Rosenzweig, M. L. 1991. Habitat selection and population interactions: the search for mechanism. *Amer. Natur.* **137**: 5–28.

Rufino, R. (ed). 1989. *Atlas das Aves que Nidificam em Portugal Continental*. Lisboa: SNPRCN. (In Portuguese).

Rule, M. 1977. Diet of nesting Little Owl. *Notornis* **24**: 40.

Runte, P. 1951. Zur "kuwitt Frage" (Steinkauz oder Waldkauz?). *Orn. Mitt.* **3**: 133–136. (In German).

Runte, P. 1954. Instrumentallaute bei Singvögeln und einigen anderen Vogelgrüppen. *Orn. Mitt.* **6**: 28–31. (In German).

Rusch, W. 1988. Zwei Brutnachweise des Steinkauzes (*Athene noctua*) in der Nähe eines Waldkauz-Brutplatzes (*Strix aluco*). *Kiebitz* **8**: 145. (In German).

Russow, V. 1880. Die Ornis Ehst-, Liv- und Curlad's mit besonderer Berucksichtingung der Zug- und Brutverhaltnisse. *In* Pleske, T. (ed.). Archiv f. d. Naturkude Liv-, Ehst- und Kurlands. Dorpat. Volume 9. (In German).

Ryabitsev, V. K. 2001. *Ptitsy Urala, Priural'ya i Zapadnoi Sibiri [Birds of the Ural, Ural Area and West Siberia]*. Manual-guide Ekaterinburg, Uralian University Press. (In Russian).

Rydzewski, W. 1978. The longevity of ringed birds. *Ring* **96/97**: 239.

Ryslavy, T. 2002. Zur Bestandssituation ausgewählter Vogelarten in Brandenburg-Jahresbericht 2000. *Naturschutz und Landschaftspflege in Brandenburg* **11**(3): 183–197. (In German).

Safriel, U. N. 1995. What's special about Shrikes? Conclusions and recommendations. *In* Yosef, R. & Lohrer, F. E. (eds). Shrikes (Laniidae) of the world: biology and conservation. *Proc. West Found. Vert. Zool.* **6**: 299–308.

Sageder, G. 1990. Nahrungsspektrum und Mechanismen der Entstehung der Beutewahl beim Steinkauz: eine vergleichende Freiland-und Laboruntersuchung. Dissertation zur Erlangung des Doktorgrades an der Formal-und Naturwissenschftalichen Fakultät der Universität Wien. (In German).

Sagitov, A. K. 1990. *Otryad Sovy Striges [Order Strigiformes]. Ptitsy Uzbekistana [Birds of Uzbekistan]*. Vol. 2. Tashkent: Fan Press. pp. 225–244. (In Russian).

Šálek, M. 2004. Vývoj populace sýčka obecného (*Athene noctua*) na Českobudějovicku a Písecku [Population development of the Little Owl (*Athene noctua*) in Ceské Budejovice and Pisek Region]. *Sluka Holysov* **1**: 87–88. (In Czech)

Salikhbaev, K. S., Kashkarov, D. Y. & Sharipov, A. 1970. *Ptitsy* [*Birds*]. *Ekologiya Pozvonochnykh Zhivotnykh Khrebta Nuratau* [*Ecology of Vertebrates of the Nuratau Ridge*]. Tashkent: Fan Publishers. pp. 42–99. (In Russian).

Sandstrom, U. 1991. Enhanced predation rates on cavity birds nests at deciduous forest edges – an experimental study. *Ornis Fennica* **68**(3): 93–98.

Sanmartin, M. L., Alvarez, F., Barreiro, G. & Leiro, J. 2004. Helminth fauna of Falconiform and Stringiform birds of prey in Galicia, Northwest Spain. *Parasitology Research* **92**: 255–263.

ará, M. & Zanca, L. 1989. Considerazioni sul censimento degli strigiformi. *Rivista Italiana di Ornitologia* **59**(1–2): 3–16. (In Italian).

Saurola, P. 1989. Ural Owl. *In* I. Newton (ed). *Lifetime Reproduction in Birds*. London, Great Britain: Academic Press. pp. 327–345.
Saurola, P. 2002. Natal dispersal distances of Finnish owls: results from ringing. *In* Newton, I., Kavanagh, R., Olsen, J. & Taylor, I. (eds). *Ecology and Conservation of Owls*. Australia: CSIRO Publishing. pp. 42–55.
Saurola, P. 2006. Monitoring "common" birds of prey in Finland in 1982–2005. *In* Koskimies, P. & Lapshin, N. V. (eds). *Status of Raptor Populations in Eastern Fennoscandia*. Petrozavodsk. pp. 133–145.
Savigny, 1809. *Athene noctua glaux*. Description d'Egypte; ou Recueil des observations et des recherches qui ont été faites en Egypte pendant l'expédition de l'armée française, livr. 1, p 105. (In French).
Sawitzky, W. 1899. Beitrage zur Kenntnis der Baltischen Ornis. Die Vogelwelt der Stadt Riga und Umgegend. *Korr.-Bl. Naturforscher-Vereins zu Riga* **62**: 191–218. (In German).
Schaaf, R. 2005. Der Steinkauz von Crespina. *Kauzbrief* **17**: 14–24. (In German).
Schaub, M., Ullrich, B., Knötzsch, G., Albrecht, P. & Meisser, C. 2006. Local population dynamics and the impact of scale and isolation: a study on different little owl populations. *OIKOS* **115**: 389–400.
Scherzinger, W. 1971a. Beobachtungen zur Jugendentwicklung einiger Eulen (*Strigidae*). *Z. Tierpsychol.* **28**: 494–504. (In German).
Scherzinger, W. 1971b. Zum Feindverhalten einiger Eulen (Strigidae). *Z. Tierpsychol.* **29**: 165–174. (In German).
Scherzinger, W. 1980. Verhalten und Stimme des Steinkauzes. *In* Glutz von Blotzheim & Bauer, K. M. (eds). *Handbuch der Vögel Mitteleuropas*. Vol. 9. Wiesbaden, Germany: AULA Verlag. (In German).
Scherzinger, W. 1986. Kontrastzeichnungen im Kopfgefieder der Eulen (Strigidae) – als visuelle Kommunikationsmittel. *Ann. Naturhist. Mus. Wien* **88/89**: 37–56. (In German).
Scherzinger, W. 1988. Vergleichende Betrachtung der Lautrepertoires innerhalb der Gattung *Athene* (Strigiformes). Proc. Int. 100. Do-G Meeting, Current Topics in Avian Biol. Bonn, pp. 89–96. (In German).
Schmid, P. 2003. Gewöllenanalyse bei einer Population des Steinkauzes *Athene noctua* im Grossen Moos, einer intensive genutzen Agrarlandschaft des schweizerischen Mittellandes. *Orn. Beob.* **100**: 117–126. (In German).
Schmidt, E. & Szabo, L. V. 1981. Data to the small mammal fauna of the Hortobagy based on owl pellet examinations. *Natural Hist. Ntn. Pks. Hung.* **1**: 409–411.
Schnurre, O. 1940. Drei Jahre aus dem Leben eines Berliner Waldkauzes. *Beitr. z. Fortpflanzungsbiol. Vögel* **16**: 84. (In German).
Schönn, S. 1980. Käuze als Feinde anderer Kauzarten und Nisthilfen für höhlenbrütende Eulen. *Falke* **27**: 294–299. (In German).
Schönn, S. 1986. Zu Status, Biologie, Ökologie und Schutz des Steinkauzes (*Athene noctua*) in der DDR. *Acta Ornithoecol.* **1**: 103–133. (In German).
Schönn, S., Scherzinger, W., Exo, K.-M. & Ille, R. 1991. *Der Steinkauz.* Die Neue Brehm-Bücherei. Wittenberg Lutherstadt: A. Ziemsen Verlag. (In German).
Schönwetter, M. 1964. *Handbuch der Oologie*, BD.I. Berlin. In German.
Schoop, G., Siegert, R., Galassi, D. & Klöppel, G. 1955. Newcastle-Infektionen beim Steinkauz (*Athene noctua*), Hornraben (*Bucorvus* sp.), Seeadler (*Haliaetusalbicilla*) und Rieseneisvogel (*Dacelo gigas*). *Mh. Prakt. Tierheilk.* **7**: 223–235. (In German).
Schröpfer, L.1996. The Little Owl (*Athene noctua*) in the Czech Republic – abundance and distribution in the years 1993–95. *Buteo* **8**: 23–38 (In Czech).

Schröpfer, L. 2000. The Little Owl (*Athene noctua*) in the Czech Republic – abundance and distribution in the years 1998–99. *Buteo* **11**: 161–174 (In Czech).

Schroth, G. (ed). 2004. *Agroforestry and Biodiversity Conservation in Tropical Landscapes*. Washington, DC: Island Press.

Schwarzenberg, L. 1970. Hilfe unserem Steinkauz. *DBV Jahresheft*: 20–23. (In German).

Schwarzenberg, L. 1981. Nisthilfen für Steinkauz, Schleiereule und Turmfalken. *Ökol. Vögel* **3**: 349–353. (In German).

Scopoli, G. A. 1769. *Athene noctua*. Historico-Naturalis Annus I Hist.-Nat. Annus I-V: p 22. (In Italian).

Scott, D. 1980. Owls allopreening in the wild. *British Birds* **73**: 436–439.

Scott, D. 1996. *The Long-eared Owl*. Taunton, UK: The Hawk and Owl Trust.

Selmi, E., Ronchetti, G., Zoboli, A. & Conventi, L. 2005. Posatoi di Civetta *Athene noctua* in un'area rurale della Provincia di Modena. *In* Mastrorilli, M., Nappi, A. & Barattieri, M. Atti I Convegno italiano sulla Civetta. Osio Sotto (BG), 21 Marzo 2004. Gruppo Italiano Civette: 33–36. (In Italian).

SEO/BirdLife 2002. *Seguimiento de Aves Nocturnas en España*. Informe 2001. Madrid: SEO/BirdLife. (In Spanish).

Serrano, J., De Los Santos, A. & Manez, M. 1987. Loscaraboidea de Donana y zonas adyacentes (Coleoptera). *Graellsia* **43**: 39–48. (In Spanish).

Severinghaus, L. L. 2002. Home range, movement and dispersal of Lanyu Scops Owls (*Otus elegans*). *In* Newton, I., Kavanagh, R., Olsen, J. & Taylor, I. (eds). *Ecology and Conservation of Owls*. Australia: CSIRO Publishing. pp. 58–67.

Severtzov, N. A. 1873. *Athene noctua orientalis*. Izvestia Imperskogo Obschestva Liubiteley Estestvoznaniya, Antropologii i Etnografikii, Moskva, 8 Dec. 1872, pt. 2, p. 115. (In Russian).

Sgorlon, G. 2003. Densità e spaziatura dei siti di nidificazione di Civetta *Athene noctua* in un ambiente urbano del Veneto Orientale. *Avocetta* **27**: 88. (In Italian).

Sgorlon, G. 2005. Exitus di Civetta *Athene noctua* dopo interazione con il Barbagianni *Tyto alba*. *In* Mastrorilli, M., Nappi, A. & Barattieri, M. (eds). *Atti I Convegno italiano sulla Civetta*. Osio Sotto (BG), 21 Marzo 2004. Gruppo Italiano Civette: 49. (In Italian).

Sharrock, J. T. R. 1976. *The Atlas of Breeding Birds in Britain and Ireland*. Berkhamsted: Poyser. pp. 254–255.

Shawyer, C. 1998. *The Barn Owl*. Chelmsford, Essex, UK: Arlequin Press.

Shirihai, H. 1996. *The Birds of Israel*. London, UK: Academic Press.

Shtegman, B. K. 1960. K biologii domovogo sycha v yuzhnom Pribalkhashie [On Little Owl biology in southern Pribalkhash'ye (Cis-Balkhash Lake area)]. *Ornitologiya* **3**: 315–319. (In Russian).

Sibley, C. G. & Monroe, E. B. L. 1991. *Distribution and Taxonomy of Birds of the World*. New Haven: Yale University Press.

Sick, H. 1937. Morphologisch-funktionelle Untersuchungen über die Feinstruktur der Vogelfeder. *J. Orn.* **85**: 206–372. (In German).

Signoret, F. 2002. La Chouette chevêche *Athene noctua* en marais Breton Vendéen: inventaire 1999–2000. *La Gorgebleue* **17/18**: 25–30. (In French).

Sill, K. & Ullrich, B. 2005. Reproduktive Leistung eines über zwölf Jahre brütend kontrollierten Steinkauzweibchens *Athene noctua*. *Vogelwarte* **43**: 43–45. (In German).

Simakin, L. V. 2000. Zimnyaya fauna i naselenie ptits Badkhyza (Yugo-vostochnaya Turkmenia) [Winter fauna and bird population of Badkhyz]. *Ornitologiya* **29**: 87–92. (In Russian).

Simeonov, S. 1968. Materialen über die Nahrung des Steinkauzes (*Athene noctua* Scopoli) in Bulgarien. *Fragmenta Balcanica Mus. Maced. Sci. Nat.* **6**: 157–165. (In German).

Simeonov, S. 1983. Novy materially verkhu khranata na kukumyavkata (*Athene noctua* (Scop.)) v Bulgariya. [New data on the food of the Little Owl (*Athene noctua* (Scopoli)) in Bulgaria]. *Ekologiya Sofiya* **11**: 53–60. (In Bulgarian).

Simeonov, S. 1988. A study of the nutritive spectrum of the Eagle Owl (*Bubo bubo* L.) in Bulgaria. *Ekologiya Sofiya* **21**: 47–56.

Simeonov, S., Michev, T. & Nankinov, D. 1989. *The Fauna of Bulgaria*. Vol. 20. *Aves*. Part I. Sofia: BAS. (In Bulgarian).

Simmons, W. S. 1971. *Eyes of the Night – Witchcraft among a Senegalese people*. Boston: Little, Brown and Company.

Simon, L. 1982. Ergebnisse der Bestandserfassungen an Steinkauz (*Athene noctua*) und Schleiereule (*Tyto alba*) in Rheinland-Pfalz-Erklärung der Bestandssituation. Kurzfassungen der Vorträge (November 7, 1981, Gießen), Anhang zum Informationsblatt Nr. 15 der AG zum Schutz bedrohter Eulen: 10. (In German).

Soderquist, T. R., Lowe, K. W., Lyon, R. H. & Price, R. 2002. Habitat quality in Powerful Owl (*Ninox trenua*) territories in the Box-Ironbark forest of Victoria, Australia. *In* Newton, I., Kavanagh, R., Olsen, J. & Taylor, I. (eds). *Ecology and Conservation of Owls*. Australia: CSIRO. pp. 91–99.

Sokolov, V. Y. & Syroechkovsky, Y. Y. (eds). 1990. *State Nature Reserves of The Caucasus*. Moscow: Mysl.

Sokolowski, J. 1953. *Owls*. Warsaw: National School Publishers. (In Polish).

Soler, M. & Soler, J. J. 1992. Latitudinal trends in clutch size in single brooded hole nesting bird species: a new hypothesis. *Ardea* **80**: 293–300.

Solymosy, L. 1951. A kuvik téli költése. *Aquila* **55–58**: 242. (In Hungarian).

Sopyiev, O. 1982. On some adaptations of birds due to the anthropogenic effects on the Kara-Kum desert. *Abstracts XVIII Cong. Int. Orn. Moscow*: 1060–1061.

Sosnikhina, T. M. 1950. Khozyaistvennoe znachenie domovogo sycha vusloviyakh polupustyni yuga Armyanskoi SSR [Practical importance of the Little Owl in conditions of semidesert of the south of Armenian SSR]. Izvestiya Akademii nauk Armyanskoi SSR [Proc. of Ac. of Sc. of the Armenian SSR]. *Biol. i sel'skokhoz.nauki (Biol. & Agricultural Sc.)* **3**: 95–100. (In Russian).

SOVON 1987. *Atlas van de Nederlandse Vogels*. Arnhem: SOVON. pp. 318–319. (In Dutch).

SOVON 2002. Naar een betere monitoring van de Steenuil in Nederland. SOVON-informatierapport 2002/06. (In Dutch).

Spangenberg, E. P. & Feigin, G. A. 1936. Ptitsy nizhnei Syr-Dari i prilegayushchikh raionov [Birds of Lower Syr-Daria ad adjacent regions]. *Sbornik Tr. Zool, muzeya Moskovskogo Universiteta [Proceedings of Zool. Museu of Moscow State University]* **3**: 41–184. (In Russian).

Sparks, T. H., Parish, T. & Hinsley, S. A. 1996. Breeding birds in field boundaries in an agricultural landscape. *Agriculture Ecosystems & Environment* **60** (1): 1–8.

Spencer, K. G. 1965. Avian casualties on railways. *Bird Study* **12**: 257.

Staats Von Wacquant-Geozelles. 1890. Zur Lebensweise des Steinkauzes (*Athene noctua* Retz). *Orn. Mschr.* **15**: 194–202. (In German).

Stadler, H. 1932. La voix des chouettes de l'Europe moyenne. *Alauda* **4**: 271–283. (In French).

Stahl, D. 1982. Zucht und Auswilderung des Steinkauzes (*Athene noctua*). *Die Voliere* **5**: 161–200. (In German).

Stall, F. E. 1904. Ornithologische Notizen. *Korrespondenzblatt des Naturforscher-Vereins zu Riga* **47**: 77–107. (In German).

Stalling, T. 1997. Zur Nahrung des Steinkauzes (*Athene noctua*). *Orn. Schnellmitt. Bad-Württ. N.F.* **55/56**: 30. (In German).

Stam, F. 2003. De overlevingskans en levensverwachting van Steenuilen in het werkgebied van de vogelwerkgroep Stad en Ambt Doesborgh. *Athene nieuwsbrief STONE* **8**: 9–11. (In Dutch).

Stange, C. 1999. Steinkauz – Artenschutztagung in Niedersachsen vom 16–18 April 1999. *Eulen Rundblick* **48/49**: 45–49.

Stankovics, B. 1997. Kukumavka (*Athene noctua*) u urbanoj sredini Jagodine [Little Owls (*Athene noctua*) living in an urban area of Jagodina town]. *Ciconia* **6**: 91–92. (In Serbian).

Stastny, K., Bejcek, V. & Hudec, K. 1996. *The Atlas of Breeding Birds in the Czech Republic 1985–1989*. H&H. Jinocany. (In Czech)

Staton, J. 1947. Little Owl as prey of Tawny Owl. *British Birds* **40**: 279.

Steenhot, K. 1987. Assessing raptor reproductive success and productivity. *In* Giron, B. A., Pendleton, B. A., Cline, K. W. & Bird, D. M. (eds). *Raptor Management Techniques Manual*. Washington, DC: National Wildlife Federation. pp. 157–170.

Steffen, J. 1958. Chevêches et Friquets. *Nos Oiseaux* **24**: 321. (In French).

Stepanyan, L. S. 1990. *Konspekt Ornitologicheskoi Fauny SSSR [Conspectus of the Ornithological Fauna of the USSR]*. Moskow: Nauka. (In Russian).

Stepnisky, D. P. 1997. Landscape features and characteristics of Great Gray Owls (*Strix nebulosa*) nests in fragmented landscapes of central Alberta. *In* Duncan, J. R., Johnson, D. H. & Nicholls, T. H. (eds). *Biology and Conservation of Owls of the Northern Hemisphere*: 2nd International Symposium; February 5–9, 1997; Winnipeg, Manitoba. General Technical Report NC-190. US Department of Agriculture, Forest Service, North Central Research Station, St. Paul, MN. pp. 601–607.

Stevanovic, V. & Vasic, V. (eds). 1995. Biodiverzitet Jugoslavije sa pregledom vrsta od medunarodnog znaćaja. Biološki fakultet i Ecolibri, Beograd. [Biodiversity of Yugoslavia with review of the species of international significance. Faculty of Biology & Ecolibri, Belgrade]. (In Yugoslavian).

Stival, E. 1990. *Avifauna e Ambienti Naturali del Comune di Marcon* (VE). Club Marcon, Marcon (VE). (In Italian).

Stój, M. & Dyczkowski, J. 2002. *Birds of Jaslo – Population Numbers, Distribution and Protection*. Poznan: Bogucki Scientific Publishers. (In Polish).

Stresemann, E. & Stresemann, V. 1966. Die Mauser der Vögel. *J. Orn.* **107**: 357–362. (In German).

Stroeken, P. & van Harxen, R. 2003a. Steenuil bereikt leeftijd van 15 kalenderjaren. *Athene nieuwsbrief STONE* **8**: 12–16. (In Dutch).

Stroeken, P. & van Harxen, R. 2003b. Verslag van het broedbiologisch onderzoek in de Zuidoost-Achterhoek in 2003. *Athene nieuwsbrief STONE* **8**: 29–31. (In Dutch).

Stroeken, P. & van Harxen, R. 2005. Overschatting broedsucces Steenuil – het effect van controles na het ringbezoek op de berekening van het broedsucces. *Athene nieuwsbrief STONE* **10**: 38–43. (In Dutch).

Stroeken, P., Bloem, H., van Harxen, R., Groen, N. & Boer, K. 2001. The Little Owl in the Netherlands: distribution, breeding densities, threats and protection measures. *In*

Génot, J.-C., Lapios, J.-M., Lecomte, P. & Leigh, R. S. (eds). Chouette chevêche et territoires. Actes du Colloque International de Champ-sur-Marne, 25 et 26 Novembre 2000. ILOWG. *Ciconia* **25**: 185–190.

Stuart, J. 1977. *The Magic of Owls*. New York, NY: Walker and Company.

Sukhinin, A. N., Bel'skaya, G. S. & Zhernov, I. V. 1972. Pitanie domovogo sycha v Turkmenii [The Little Owl feeding in Turkmenia]. *Ornitologiya* **10**: 216–227. (In Russian).

Sushkin, P. P. 1938. Ptitsy Sovetskogo Altaya I prilezhashchikh chastei severo-zapadnoi Mongolii [Birds of the Soviet Altai and adjoining parts of northwestern Mongolia]. *M. L.* **1**: 1–317. (In Russian).

Swindle, K. A., Ripple, W. J., Meslow, E. C. & Schafer, D. 1999. Old-forest distribution around Spotted Owl nests in the central Cascade Mountains, Oregon. *Journal of Wildlife Management* **63**: 1212–1221.

Swinhoe, R. 1870. *Athene noctua plumipes*. Proceedings of the Zoological Society of London, Pt. 2, p. 448.

Taczanowski, W. 1882. *Polish Birds,* Vols. I-II, Cracow, Poland (In Polish).

Tarr, C. L. & Fleischer, R. C. 1993. Mitochondrial DNA variation and evolutionary relationships in the amakihi complex. *Auk* **110**: 825–831.

Tarres, M. A. G., Baron, M. & Gallego, A. 1986. The horizontal cells in the retina of the owl, *Tyto alba*, and owlet, *Carine noctua. Exp. Eye Res.* **42**: 315–321.

Taurinsh, E. A. & Vilks, K. A. 1949. Ornitofauna Latvijskoi SSR. *Okhrana prirody [Conservation of Nature].* (M.) **9**: 52–73. (In Russian).

Tayler, A. G. 1944. Little Owl hovering. *British Birds* **37**: 178.

Taylor, I. 1994. *Barn Owls. Predator–Prey Relationships and Conservation*. Cambridge, UK: Cambridge University Press.

Taylor, I. 2002. Occupancy in relation to site quality in Barn Owls (*Tyto alba*) in south Scotland. *In* Newton, I., Kavanagh, R., Olsen, J. & Taylor, I. (eds). *Ecology and Conservation of Owls*. Australia: CSIRO. pp. 30–41.

Taylor, I., Kirsten, I. & Peake, P. 2002a. Distribution and habitat of Barking Owls (*Ninox connivens*) in central Victoria. *In* Newton, I., Kavanagh, R., Olsen, J. & Taylor, I. (eds). *Ecology and Conservation of Owls*. Australia: CSIRO. pp. 107–115.

Taylor, I., Kirsten, I. & Peake, P. 2002b. Habitat, breeding and conservation of the Barking Owl *Ninox connivens* in northeastern Victoria, Australia. *In* Newton, I., Kavanagh, R., Olsen, J. & Taylor, I. (eds) *Ecology and Conservation of Owls*. Australia: CSIRO. pp. 116–124.

Teixeira, R. M. 1979. *Atlas van de Nederlandse Broedvogels*.Vereniging tot Behoud van Natuurmonumenten in Nederland. S Graveland, the Netherlands. (In Dutch).

Tella, J.-L. & Blanco, G. 1993. Possible predation by Little Owl *Athene noctua* on nestling Red-billed Choughs *Pyrrhocorax pyrrhocorax. Bull. GCA* **10**: 55–57.

Thévenot, M., Vernon, R. & Bergier, P. 2003. *The Birds of Morocco*. BOU Checklist No. 22. Peterborough, UK: British Ornithological Union.

Thibault, J.-C. 1983. *Les oiseaux de la Corse. Histoire et répartition aux XIXe et XXe siècles*. Parc Naturel Régional de Corse. Le Gerfaut., Paris. pp. 148–149. (In French)

Thibault, J. & Bonaccorsi, G. 1999. *The Birds of Corsica*. BOU Checklist. No.: 17.

Thiollay, J.-M. 1966. Note sur le régime de *Tyto alba* et *Athene noctua* en Corse. *L'Oiseau et RFO* **36**: 282–283. (In French).

Thiollay, J.-M. 1968. Le régime alimentaire de nos rapaces: quelques analyses françaises. *Nos Oiseaux* **29**: 249–269. (In French).

Thom, V. M. 1986. *Birds in Scotland*. Calton: T. & A. D. Poyser.

Thomas, J. F. 1939. The food of the Little Owl. Derived from pellets, February 1936 to January 1937. *Discovery New Series* **2**: 94–99.

Thompson, W. L., White, G. C. & Gowan, C. 1998. *Monitoring Vertebrate Populations.* New York: Academic Press.

Thorpe, W. H. & Griffin, D. R. 1962. The lack of ultrasonic components in the flight noise of owls compared with birds. *Ibis* **104**: 256–257.

Tinbergen, L. & Tinbergen, N. 1932. Ueber die Ernährungeiner Steinkauzbrut (*Athene noctua vidalii* A. E. Brehm). *Beitr. z. Fortpflanzungsbiol. Vögel* **8**: 11–14. (In German).

Tinbergen, N. 1932. Eine Beobachtung über die Ernährung des Steinkauzes (*Athene noctua vidalii* A. E. Brehm). *Ardea* **21**: 74–75. (In German).

Tischler, F. 1941. *Die Vögel Ostpreussens und seiner Nachgebiete.* Königsberg, Berlin. pp. 553–555. (In German).

Toffoli, R. & Beraudo, P. 2005. Considerazioni sulla densita della Civetta *Athene noctua* in Provincia di Cuneo. *In* Mastrorilli, M., Nappi, A. & Barattieri, M. (eds). Atti I Convegno italiano sulla Civetta. Osio Sotto (BG), 21 Marzo 2004. Gruppo Italiano Civette: 24–26. (In Italian).

Tomassi, R., Piattella, E., Manganaro, A. *et al.* 1999. Primi dati su dieta e densità della Civetta *Athene noctua* nella Tenuta Presidenziale di Castelporziano (Roma). *Avocetta* **23**: 159. (In Italian).

Tome, D. 1996a. Vertical distribution of owls in Slovenia [Višinska razširjenost sov v Sloveniji]. *Acrocephalus* **17**(74): 2–3. (In Slovene).

Tome, D. 1996b. Owls in Slovenia – present status and distribution. *In* Stubbe, M. & Stubbe, A. (eds). *Populationsökologie Greifvogel-und Eulenarten*, 3. Wittenberg, Germany: Martin-Luther Universität. pp. 343–351.

Tomé, R., Bloise, C. & Korpimäki, E. 2004. Nest-site selection and nesting success of Little Owls (*Athene noctua*) in Mediterranean woodland and open habitats. *J. Raptor Res.* **38**: 35–46.

Tomé, R., Santos, N., Cardia, P., Ferrand, N. & Korpimäki, E. 2005. Factors affecting the prevalence of blood parasites of Little Owls *Athene noctua* in southern Portugal. *Ornis Fennica* **82**: 63–72.

Tomialojc, L. 1972. *Polish Birds – Index of Species and Distribution.* Warsaw: State Scientific Publishers. (In Polish).

Tomialojc, L. 1976. *Birds of Poland: Ptaki Polski.* Warsaw: State Scientific Publishers. p. 123. 253 p.

Tomialojc, L. 1990. *Polish Birds – Distribution and Population Numbers.* Warsaw: State Scientific Publishers. (In Polish).

Toms, M. P., Crick, H. Q. P. & Shawyer, C. R. 2000. *Project Barn Owl Final Report.* British Trust for Ornithology, Research Report 197. Thetford, Norfolk: British Trust for Ornithology.

Torregiani, F. 1981. Osservazione dell'accopiamento della civetta. *Gli Uccelli d'Italia* **6**: 178–179. (In Italian).

Toschki, A. 1999. Bestand und Verbreitung des Steinkauzes (*Athene noctua*) in Aachen. *Eulen-Rundblick* **48/49**: 16–20. (In German).

Transehe, N. & Sinats, R. 1936. *Latvijas putni.* Riga, Latvia: Mežu dep. Izdevums (In Latvian).

Tricot, J. 1968. A propos de la capture des lombrics par la Chouette chevêche (*Athene noctua*). *Aves* **5**: 11. (In French).

Trimnell, H. C. 1945. Little Owl feeding young on newts. *British Birds* **38**: 174.

Trontelj, P. 1994. Ptice kot indikator ekološkega pomena Ljubljanskega barja (Slovenia). *Scopolia* **32**: 1–61. (In Slovene).

Tubbs, C. R. 1953. Little Owl "smoke-bathing". *British Birds* **46**: 377.

Tucker, G. M. & Evans, M. I. 1997. *Habitats for Birds in Europe: A Conservation Strategy for the Wider Environment.* BirdLife Conservation Series No. 6. Cambridge, UK: BirdLife International.

Turner, M. & Gardner, R. H. (eds). 1997. *Quantitative Methods in Landscape Ecology: The Analysis and Interpretation of Landscape Heterogeneity*. New York: Springer Verlag.

Turner, M., Gardner, R. H. & O'Neill, R. V. 2003. *Landscape Ecology in Theory and Practice: Pattern and Process*. New York: Springer Verlag.

Ullrich, B. 1970. Ersatzbrut und Mauserbeginn beim Steinkauz (*Athene noctua*). *Vogelwelt* **91**: 28–29. (In German).

Ullrich, B. 1973. Beobachtungen zur Biologie des Steinkauzes (*Athene noctua*). *Anz. Orn. Ges. Bayern* **12**: 163–175. (In German).

Ullrich, B. 1975. Bestandsgefährdung von Vogelarten im Ökosystem "Streuobstwiese" unter besonderer Berücksichtigung von Steinkauz und den einheimischen Würgerarten der Gattung Lanius. *Beih. Veröff. Naturschutz u. Landschaftspflege Baden-Württemberg* **7**: 90–110. (In German).

Ullrich, B. 1980. Zur Populationsdynamik des Steinkauzes (*Athene noctua*). *Vogelwarte* **30**: 179–198. (In German).

Urios, V., Escobar, J. V., Pardo, R. & Gómez, J. A. (eds). 1991. *Atlas de las Aves Nidificantes de la Comunidad Valenciana.* Valencia: Generalitat Valenciana. (In Spanish).

Uspenskiy, G. A. 1977. O gnezdovanii pustel'gi obyknovennoi, kobchika i domovogo sycha v gnezdakh vranovykh ptits [On nesting of the Common Kestrel, Red-footed Falcon and the Little Owl in the nests of Corvids] Tez. dokl. VII Vsesoyuz. orn. konf. [Abstracts of the VIIth All-Union Orn. Conf.]. Kiev: Naukova Dumka. Part 1. pp. 330–331. (In Russian).

Uttendörfer, O. 1930. Studien zur Ernährung unserer Tagraubvögel und Eulen. *Abhandlungen der Naturforschenden Geslleschaftzu Görlitz* **31**: 136–141. (In German).

Uttendörfer, O. 1939. *Die Ernährung der deutschen Raubvögel und Eulen und ihre Bedeutung in der heimischen Natur*. Verlag J. Neumann. Neudamm. Berlin. pp. 257, 266, 376–377. (In German).

Uttendörfer, O. 1952. *Neue Ergebnisse über die Ernährung der Greifvögel und Eulen.* Eugen. Ulmer. Stuttgart. pp. 123–128. (In German).

Vaassen, E. W. A. M. 2000. Some notes on urban raptors and their habitats in Turkey. *International Hawkwatcher* **2**: 34–38.

Vachon, M. 1954. Remarques sur les ennemis des scorpions. A propos de la présence de restes de scorpions dans l'estomac de la Chouette *Athene noctua. L'Oiseau et RFO* **24**: 171–174. (In French).

Van de Velde, E. & Mannaert, P. 1980. Steenuil (*Athene noctua*) eet wegslachtoffers. *Veldorn. Tijdsch.* **3**: 137–139. (In Dutch).

Van den Berghe, K. 1998. Marterachtigen in Vlaanderen. *De Levende Natuur* **99**: 169–170. (In Dutch).

Van den Brink, N. W., Groen, N. M., de Jonge, J. & Bosveld, A. T. C. 2003. Ecotoxicological suitability of floodplain habitats in the Netherlands for the Little owl (*Athene noctua vidalli*). *Environmental Pollution* **122**: 127–134.

Van den Burg, A., Beersma, P. & Beersma, W. 2003. De temperatuur van nest-en roestplaasten van de Steenuil *Athene noctua*. *Het Vogeljaar* **51**: 147–152. (In Dutch).

Van den Tempel, R. 1993. Vogelslachtoffers in het wegverkeer. Technisch Rapport 11: Vogelbescherming Nederland, Zeist. (In Dutch).

Vanden Wyngaert, L. 2005. Steenuilproject Dijleland, Verslag 2004. (In Dutch).

Van Dijk, A. J. & Ottens, H.-J. 2001. Actuele verspreiding van de Steenuil en van Steenuilonderzoekers in Nederland. SOVON-informatierapport 2001/15 uitgevoerd in opdracht van Vogelbescherming Nederland. (In Dutch).

van Harxen, R. & Stroeken, P. 2003a. Prooiaanvoer bij een steenuilenbroedpaar. *Athene nieuwsbrief STONE* **7**: 17–28. (In Dutch).

van Harxen, R. & Stroeken, P. 2003b. Bijzondere schuilplaats van uitgevlogen Steenuiljongen. *Athene nieuwsbrief STONE* **8**: 45–46. (In Dutch).

Van Horne, B. 1983. Density as a misleading indicator of habitat quality. *Journal of Wildlife Management* **47**: 893–901.

Van Nieuwenhuyse, D. 2004. Steenuil. *In* Vermeersch, G., Anselin, A., Devos, K. *et al.* (eds). *Atlas van de Vlaamse Broedvogels 2000–2002*. Mededelingen van het Instituut voor Natuurbehoud 23, Brussel, pp. 256–257.

Van Nieuwenhuyse, D. & Bekaert, M. 2001. Study of Little Owl *Athene noctua* habitat preference in Herzele (East-Flanders, Northern Belgium) using the median test. *In* Van Nieuwenhuyse, D., Leysen, M. & Leysen, K. (eds). The Little Owl in Flanders in its international context. Proceedings of the Second International Little Owl Symposium, March 16–18, 2001, Geraardsbergen, Belgium. *Oriolus* **67**(2–3): 62–71.

Van Nieuwenhuyse, D. & Bekaert, M. 2002. An (auto)logistic regression model for prediction of Little Owl (*Athene noctua*)-suitability of landscapes in East-Flanders (northern Belgium). Evidence for socially induced distribution patterns of Little Owl. *In* Yosef, R., Miller, M. L. & Pepler, D. (eds). *Raptors in the New Millennium*. Eilat, Israel: Int. Birding & Res. Centre. pp. 80–90.

Van Nieuwenhuyse, D. & Leysen, M. 2001. Habitat typologies of Little Owls *Athene noctua* in Flanders (northern Belgium). Focusing on what really matters through Principal Component Analysis and Cluster Analysis. *In* Van Nieuwenhuyse, D., Leysen, M. & Leysen, K. (eds). The Little Owl in Flanders in its international context. Proceedings of the Second International Little Owl Symposium, March 16–18, 2001, Geraardsbergen, Belgium. *Oriolus* **67**(2–3): 72–83.

Van Nieuwenhuyse, D. & Leysen, M. 2004. Distribution of Little Owl (*Athene noctua*) in Flanders (northern Belgium), in relation to environment: spatial modeling through GIS data and logistic regression. *In* Rodríguez, R. E. & Bojórquez Tapia, L. (eds). *Spatial Analysis in Raptor Ecology and Conservation*. Centro de Investigaciones Biologicas del Noroeste/Comision Nacional para el Conocimiento y Uso de la Biodiversidad. pp. 75–109.

Van Nieuwenhuyse, D. & Nollet, F. 1991. Biotoopstudie van de Steenuil *Athene noctua* met behulp van het clusteringsprogramma TWINSPAN. *Oriolus* **57**: 57–61. (In Dutch).

Van Nieuwenhuyse, D., Coussens, P. & Nollet, F. 1995. Digital method for recording and analyzing the territory use and activity budget of the Red-backed Shrike *Lanius collurio*. *In* Yosef, R. & Lohrer, F. E. Shrikes (Laniidae) of the world: biology and conservation. *Proc. West Found. Vert. Zool.* **6**: 268–275.

Van Nieuwenhuyse, D., Nollet, F. & Evans, A. 1999. The ecology and conservation of the Red-backed Shrike *Lanius collurio* breeding in Europe. *Aves* **36**: 179–192.

Van Nieuwenhuyse, D., Bekaert, M., Steenhoudt, K. & Nollet, F. 2001a. Longitudinal analysis of habitat selection and distribution patterns in Little Owls *Athene noctua* in Meulebeke (Belgium). *In* Van Nieuwenhuyse, D., Leysen, M. & Leysen, K. (eds). The Little Owl in Flanders in its international context. Proceedings of the Second International Little Owl Symposium, March 16–18, 2001, Geraardsbergen, Belgium. *Oriolus* **67**(2): 52–61.

Van Nieuwenhuyse, D., Leysen, M., De Leenheer, I. & Bracquene, J. 2001b. Towards a conservation strategy for Little Owl *Athene noctua* in Flanders. *In* Van Nieuwenhuyse, D., Leysen, M. & Leysen, K. (eds). The Little Owl in Flanders in its international context. Proceedings of the Second International Little Owl Symposium, March 16–18, 2001, Geraardsbergen, Belgium. *Oriolus* **67**(2): 21.

Van Nieuwenhuyse D., Leysen, M. & Steenhoudt, K. 2001c. Analysis and spatial prediction of Little Owl *Athene noctua* distribution in relation to its living environment in Flanders. Modelling spatial distribution through logistic regression. *In* Van Nieuwenhuyse, D., Leysen, M. & Leysen, K. (eds). The Little Owl in Flanders in its international context. Proceedings of the Second International Little Owl Symposium, March 16–18, 2001, Geraardsbergen, Belgium. *Oriolus* **67**(2): 32–51.

Van Nieuwenhuyse, D., Leysen, M. & Leysen, K. (eds). 2001d. Little Owl in Flanders in its international context. Proceedings of the Second International Symposium, March 16–18, 2001. Geraardsburgen, Belgium. *Oriolus* **67**.

Van Nieuwenhuyse D., Lefebvre, J. & Leysen, M. 2004. Kwaliteitsbepaling en – voorspelling van Kerkuil *Tyto alba*-habitat in Oost-Vlaanderen op verschillende ruimtelijke niveaus. *Natuurpunt. Oriolus* **70**: 11–20. (In Dutch).

Van't Hoff, J. 1999. Nu het nog kan . . . Een soortenbeschermingsplan voor de Steenuil in Groningen. Steenuilwerkgroep Groningen. (In Dutch).

Van't Hoff, J. 2001. Balancing on the edge. The critical situation of the Little Owl (*Athene noctua*) in an intensive agricultural landscape. *In* Van Nieuwenhuyse D., Leysen, M. & Leysen, K. (eds). The Little Owl in Flanders in its international context. Proceedings of the Second International Little Owl Symposium, April 16–18, 2001, Geraardsbergen, Belgium. *Oriolus* **67**(2): 100–109.

Van't Hoff, J. 2003. Uilenboekje over Nederlandse uilen. Steenuilenwerkgroep Groningen. (In Dutch).

Van Orden, C. & Paklina, N. V. 2003. An association between Tibetan Owlets *Athene noctua ludlowi* and Himalayan Marmots *Marmota himalayana* in West-Tibet and Southeast Ladakh. *De Takkeling* **11**: 80–85.

Van Veen, J. C. & Kirk, D. A. 2000. Dietary shifts and fledging success in breeding Tawny Owls *Strix aluco*. *Ökol. Vögel* **22**: 237–281.

Van Zoest, J. G. A. & Fuchs, P. 1988. Jaaggedrag en prooiaanvoer van een Steenuil *Athene noctua* broedpaar. *Limosa* **61**: 105–112. (In Dutch).

Vanderlee, B. 1984. Poging tot vergelijkend voedselonderzoek aan de hand van braakballen van enkele roofvogels. *Wielewaal* **50**: 232–240. (In Dutch).

Vaurie, C. 1960. Systematic notes on palearctic birds. No. 42 *Strigidae*: The genus *Athene*. *American Museum Novitates* **2015**: 1–21.

Vaurie, C. 1965. *The Birds of the Palearctic Fauna. A Systematic Reference. Non Passeriformes.* London: Witherby Limited. pp. 607–613.

Veit, W. 1988. Die Bestandsentwicklung zweier Steinkauzpopulationen in den Kreis Limburg-Weilburg, Lahn-Dill und Giessenvon 1978–1987. *Vogel und Umwelt* **5**: 87–91. (In German).

Vercauteren, P. 1989. Steenuil *Athene noctua*. Vogels in Vlaanderen, Voorkomen en verspreiding. I.M.P. pp. 248–249. (In Dutch).

Verner, J., McKelvey, K. S., Noon, B. R., Gutiérrez, R. J. & Beck, T. W. (eds). 1992. *The California Spotted Owl: A Technical Assessment of its Current Status.* General Technical Report PSW-GTR-133. Albany, CA: Pacific Southwest Research Station, Forest Service, US Department of Agriculture; http://www.fs.fed.us/psw/rsl/projects/wild/gtr_133/gtr133_index.html

Verwaerde, J., Van Nieuwenhuyse, D., Nollet, F. & Bracque, J. 1999. Standaardprotocol voor het inventariseren van Steenuilen *Athene noctua* Scop. (Strigidae) in de West-Europese Laagvlakte. *Oriolus* **65**: 109–116. (In Dutch).

Vezinet, P. 2003. A propos du comportement alimentaire de la Chevêche d'Athéna dans notre région. *La chouette d'Eoures, Bull. Association La Chevêche* **45**: 4. (In French).

Vilkov, E. V. 2003. Status, dinamika chislennosti i osobennosti ekologii sov Dagestana [Status, number dynamics and peculiarities of ecology of owls of Dagestan]. (In Russian).

Village, A. 1990. *The Kestrel*. London: Poyser.

Visser, D. 1977. De Steenuil in het Rijk van Nijmegen. *De Mourik* **3**: 13–27. (In Dutch).

Vogrin, M. 1997. Little Owl (*Athene noctua*): a highly endangered species in NE Slovenia. *Buteo* **9**: 99–102.

Vogrin, M. 2001. Little Owl *Athene noctua* in Slovenia: *In* Van Nieuwenhuyse D., Leysen, M. & Leysen, K. (eds). *Little Owl in Flanders in its international context.* Proceedings of the Second International Symposium, March 16–18, 2001, Geraardsbergen, Belgium. *Oriolus* **67**: 132–135.

Wahlsted, J. 1971. Jaktmethod hos minervauggla *Athene noctua. Var Fagelvärld* **30**: 46–47. (In Danish).

Walker, C. H., Hamilton, G. A. & Harrisson, R. B. 1967. Organochlorine insecticide residues in wild birds. *J. Sci. Fd. Agric.* **18**: 123–129.

Walter, G. & Hudde, H. 1987. Die Gefiederfliege *Carnus hemapterus* (Milichidae, Diptera), ein Ektoparasit der Nestlinge. *J. Orn.* **128**: 251–255. (In German).

Weimann, R. 1965. Die Vögel des Kreises Paderborn. *Schr. R. Paderborn, Heimatver.* **3**: 1–87. (In German).

Weinstein, K. 1989. *The Owl in Art, Myth, and Legend.* London: Savitri Books Ltd.

Wemer, P. 1910. Etwas vom Steinkauz (*Athene noctua*). *Zool. Beob.* **51**: 137–141. (In German).

White, G. C. 2000. Population viability analysis: data requirements and essential analyses. *In* Boitani, L. & Fuller, T. K. (eds). *Research Techniques in Animal Ecology: Controversies and Consequences.* New York: Columbia University Press. pp. 288–331.

Wickler, W. 1965. Mimicry and the evolution of animal communication. *Nature* **208**: 519–521.

Willems, F., van Harxen, R., Stroeken, P. & Majoor, F. 2004. Reproductie van de Steenuil in Nederland in de periode 1977–2003. SOVON-onderzoeksrapport 2004/04. (In Dutch)

Williams, R. 1945. Little Owl attacking and carrying off Jackdaw. *British Birds* **38**: 194–195.

Wilson, A. C., Ochman, H. & Prager, E. M. 1987. Molecular timescale for evolution. *Trends Genetics* **3**: 241–247.

Wink, M. 2000. Advances in DNA studies of diurnal and nocturnal raptors. *In* Chancellor, R. D. & Meyburg, B.-U. (eds). *Raptors at Risk.* WWGBP/Hancock House. pp. 831–844.

Wink, M. & Heidrich, P. 1999. Molecular evolution and systematics of the Owls (Strigiformes). *In* König, C., Weick, F. & Becking, J. H. (eds). *Owls. A Guide to the Owls of the World*. Sussex: Pica Press. pp. 39–57.

Wink, M. & Heidrich, P. 2000. Molecular systematics of owls (Strigiformes) based on DNA sequences of the mitochondrial cytochrome b gene. *In* Chancellor, R. D. & Meyburg, B.-U. (eds). *Raptors at Risk*. London: WWGBP/Hancock House. pp. 819–828.

Wink, M., Sauer-Gürth, H. & Fuchs, M. 2004. Phylogenetic relationships in owls based on nucleotide sequences of mitochondrial and nuclear marker genes. *In* Chancellor, R. D. & Meyburg, B.-U. (eds). *Raptors Worldwide*. Proceedings of the World Working Group on Birds of Prey and Owls (WWGBP) – Conference in Budapest, Hungary, May 18–23, 2003. pp. 517–526.

Wintle, B. A., Kavanagh, R. P., McCarthy, M. A. & Burgman, M. A. 2005. Estimating and dealing with detectability in occupancy surveys for forest owls and arboreal marsupials. *Journal of Wildlife Management* **69**: 905–917.

Witherby, H. F., Jourdain, F. C. R., Ticehurst, N. F. & Tucker, B. W. 1938. *The Handbook of British Birds*. 2. London: Witherby Ltd. pp. 322–327.

Wright, V., Hejl, S. J. & Hutto, R. L. 1997. Conservation implications of a multi-scale study of Flammulated Owl (*Otus flammeolus*) habitat use in the northern Rocky Mountains, USA. *In* Duncan, J. R., Johnson, D. H. & Nicholls, T. H. (eds). *Biology and Conservation of Owls of the Northern Hemisphere*: 2nd International Symposium; Feb 5–9, 1997; Winnipeg, Manitoba. General Technical Report NC-190. US Department of Agriculture, Forest Service, North Central Research Station, St. Paul, MN. pp. 506–516.

Wuczyński, A. 1994. Population numbers and distribution of owls Strigiformes in the agricultural landscape of the Dzierżonów Dale. *Birds of the Silesian Region* **10**: 118–121. (In Polish).

Wüst, W. 1986. *Avifauna Bavariae. Die Vogelwelt Bayems im Wandel der Zeit*. Vol. II. München: Ornithologische Gesellschaft in Bayern. (In German).

Yeatman, L. 1976. *Atlas des Oiseaux Nicheurs de France*. Paris: Soc. Orn. de France. pp. 128–129. (In French).

Yew, D. T. 1980. Neuronal cells and types of contacts in the owl's retina. *Anat. Anz.* **147**: 255–259.

Yosef, R. 1993a. Effects of Little Owl predation on Northern Shrike postfledging success. *Auk* **110**: 396–398.

Yosef, R. 1993b. Influence of observation posts on territory size of Northern Shrikes. *Wilson Bull.* **105**: 180–183.

Yosef, R. 1994. The effects of fencelines on the reproductive success of Loggerhead Shrikes. *Conservation Biology* **8**: 281–285.

Zabala, J., Zuberogoita, I., Martinez-Climent, J. A. *et al.* 2006. Occupancy and abundance of Little Owl *Athene noctua* in an intensively managed forest area in Biscay. *Ornis Fennica* **83**: 97–107.

Zaccaroni, A., Amorena, M., Naso, B. *et al.* 2003. Cadmium, chromium and lead contamination of *Athene noctua*, the little owl, of Bologna and Parma, Italy. *Chemosphere* **52**: 1251–1258.

Zarkhidze, V. A. & Loskutova, E. A. 1999. Dinamika troficheskikh svyazeikhishchnykh ptits i melkikh mlekopitayushchikh v severozapadnoi Tirkmenii [Relationships between the birds of prey and small mammals in northwestern Turkmenistan and their dynamics]. *Russk. orn. zhurnal* [*Russ. J. of Ornithology*] **82**: 3–17. (In Russian).

Zarudnyi, N. A. 1896. Ornitologicheskaya fauna Zakaspijskogo kraya (Severnoi Persii, Zakaspijskoi oblasti, Khivinskogo oazisa i ravninnoi Bukhary) [Ornithologica fauna of Trans-Caspian territory (Northern Persia, Trans-Caspian Region, Khiva oasis and the Plain Bukhara)]. Mat-ly k poznaniyu fauny i flory Ros. Imperii [Materials on the study of fauna and flora of the Russian Empire]. *Otd. Zool. (Dept. of Zoology)*. **2**: 1–555. (In Russian).

Zav'yalov, E. V., Tabachishin, V. G., Shlyakhtin, G. V., Yakushev, N. N. & Kochetova, I. B. 2000. Sovy Saratovskoi oblasti [Owls of the Saratov Region]. *Berkut*. **9**: 74–81. (In Russian).

Zens, K.-W. 1992. Ökologische Studien an einer Population des Steinkauzes (*Athene noctua* SCOP. 1769) in der Mechernicher Voreifel unter Einbeziehung der radiotelemetrischen Methode. Diplom-Arbeit am Institut für Angewandte Zoologie der Rheinischen Friedrich-Wilhelms-Universität Bonn. (In German).

Zens, K.-W. 2005. Langstudie (1987–1997) zur Biologie, Ökologie und Dynamik einer Steinkauzpopulation (*Athene noctua* SCOP. 1769) im Lebensraum der Mechenicher Voreifel. Dissertation zur Erlangung des Doktorgrades der Mathematisch-Naturwissenschaftlichen Fakultätc der Rheinischen Friedrich-Wilhelm-Universität Bonn. (In German).

Zerunian, S., Franzini, G. & Sciscione, L. 1982. Little Owls and their prey in a Mediterranean habitat. *Boll. Zool.* **49**: 195–206.

Zhalakevichius, M., Paltanavichius, S., Shvazhas, S. & Stanevichius, V. 1995. Lietuvos Paukshchiai [Birds of Lithuania]. *Acta Ornithologica Lituanica* **11**: 59. (In Lithuanian).

Ziesemer, F. 1981. Zur Situation der Eulen (Strigiformes) in Schleswig-Holstein. *Ökol. Vögel* **3**: 311–316. (In German).

Zmihorski, M., Altenburg-Bacia, D., Romanoswki, J., Kowalski, M. & Osojca, G. 2006. Long-term decline of the Little Owl (*Athene noctua* Scop., 1769) in Central Poland. *Polish Journal of Ecology* **54**: 321–324.

Zuberogoitia, I. 2002. Ecoetologia de las rapaces nocturnas de Bizkaia. Ph.D. thesis. Universidad del Pais Vasco. (In Spanish).

Zuberogoitia, I. & Campos, L. F. 1997a. Intensive census of nocturnal raptors in Biscay. *Munibe* **49**: 117–127.

Zuberogoitia, I. & Campos, L. F. 1997b. Censusing owls in large areas: a comparison between methods. *Ardeola* **45**(1): 47–53.

Zuberogoitia, I. & Martínez, J. A. 2001. The Little Owl in "Proyecto *Noctua*". *In* Génot, J.-C., Lapios, J.-M., Lecomte, P. & Leigh, R. S. (eds). Chouette chevêche et territoires. Actes du Colloque International de Champ-sur-Marne, 25 et 26 Novembre 2000. ILOWG. *Ciconia*, **25**(2): 13–108.

Zuberogoitia, I., Martínez, J. A., Zabala, J. & Martinez, J. E. 2005. Interspecific aggression and nest-site competition in a European owl community. *J. Raptor Res.* **39**: 156–159.

Zvářal, K. 2002. Can "architectural traps" be the cause of the critical decrease of Little Owl (*Athene noctua*)? *Crex* **18**: 94–99. (In Czech).

Zwölfer, H., Bauer, G. & Heusinger, G. 1981. Ökologische Funktionsanalyse Von Feldhecken. Tierökologische Untersuchungen Über Struktur Und Funktion Biozönotischer Komplexe. Schlussber. Lehrstuhl Tierökol. Univ. Bayreuth; München Bayer. Landesamt F. Umweltschutz. (In German).

Appendix A List of prey

The food of the Little Owl is mainly composed of small mammals and birds, reptiles, amphibians, beetles (Coleoptera), crickets and grasshoppers (Orthoptera), earwigs (Dermaptera), and earthworms (Lumbricidae).

Western Palearctic

The following prey were recorded in the Western Palearctic. The diversity of prey species in the Western Palearctic was: 54 mammals, 82 birds, 15 reptiles, 14 amphibians, 2 fishes, 377 and invertebrates, giving a total of 544 prey species.

Mammal

Western Hedgehog, *Erinaceus europaeus*
Pygmy Shrew, *Sorex minutus*
Common Shrew, *Sorex araneus*
Sorex alpinus (Popescu & Negrea 1987)
Water Shrew, *Neomys fodiens*
Southern Water Shrew, *Neomys anomalus*
Pygmy White-toothed Shrew, *Suncus etruscus*
Bicolored Shrew, *Crocidura leucodon*
Lesser White-toothed Shrew, *Crocidura suaveolens*
Greater White-toothed Shrew, *Crocidura russula*
Northern Mole, *Talpa europaea*
Talpa caeca
Talpa romana (Zerunian *et al.* 1982)
Noctule Bat, *Nyctalus noctula* (Krzanowski 1973; Ganya & Zubkov 1975)
Greater Mouse-eared Bat, *Myotis myotis* (Kulczycki 1964; Krzanowski 1973)
Lesser Mouse-eared Bat, *Myotis blythi* (Lipej & Gjerkes 1992)
Common Pipistrelle, *Pipistrellus pipistrellus* (Simeonov 1983; Moschetti & Mancini 1993; Gusberti 1998)
Nathusisus's Pipistrelle, *Pipistrellus nathusii* (Lipej & Gjerkes 1992)
Savi's Pipistrelle, *Pipistrellus savii* (Lipej & Gjerkes 1992)
Horseshoe Bat, *Rhinolophus* sp. (Manez 1983a)
Parti-coloured Bat, *Vespertilio murinus* (Farsky 1928; Ille 1992)
Long-eared Bat, *Plecotus* sp. (Barbu & Sorescu 1970; Popescu & Blidarescu 1983)

Rabbit, *Oryctolagus cuniculus* (Patten 1930; Tinbergen & Tinbergen 1932; Hibbert-Ware 1937/1938; Uttendörfer 1952; Libois 1977; Manez 1983a)
Spotted Souslik, *Citellus suslicus* (Cramp 1985)
Citellus citellus (Marian & Schmidt 1967)
Garden Dormouse, *Eliomys quercinus* (Manez 1983a)
Hazel Dormouse, *Muscardinus avellanarius* (Simeonov 1968; Thiollay 1968; Zerunian *et al.* 1982; Bashta 1994; Nappi 2005)
Common Hamster, *Cricetus cricetus* (Haensel & Walter 1966; Barbu 1978; Ille 1992)
Migratory Hamster, *Cricetulus migratorius* (Cramp 1985)
Spalax leucodon (Schönn *et al.* 1991)
Bank Vole, *Clethrionomys glareolus*
Northern Water Vole, *Arvicola terrestris*
Arvicola sapidus (Manez 1983a)
Pitymys subterraneus
Pitymys duodecimcostatus
Microtus savii (Lovari 1974; Contoli *et al.* 1988; Moschetti & Mancini 1993; Gotta & Pigozzi 1997)
Microtus lusitanicus (De la Hoz 1982; Manez 1983a)
Common Vole, *Microtus arvalis*
Field Vole, *Microtus agrestis*
Snow Vole, *Microtus nivalis*
Root Vole, *Microtus oeconomus* (Schönn *et al.* 1991)
Alpine Pine Vole, *Microtus multiplex* (Gotta & Pigozzi 1997)
East European Vole, *Microtus rossiaemeridionalis* (Alivizatos *et al.* 2005)
Harvest Mouse, *Micromys minutus*
Striped Field Mouse, *Apodemus agrarius*
Yellow-necked Mouse, *Apodemus flavicollis*
Wood Mouse, *Apodemus sylvaticus*
Black Rat, *Rattus rattus* (Zerunian *et al.* 1982; Manez 1983a; Vanderlee 1984; Contoli *et al.* 1988; Génot & Bersuder 1995; Barthelemy 2000)
Brown Rat, *Rattus norvegicus* (Collinge 1922; Thiollay 1968; Baudvin 1974; Rule 1977; Juillard 1984a; Génot & Bersuder 1995; Gotta & Pigozzi 1997; Blache 2001)
House Mouse, *Mus musculus*
Algerian Mouse, *Mus spretus* (Barthelemy 2000)
Northern Birch Mouse, *Sicista betulina* (Bashta 1994)
Southern Birch Mouse, *Sicista subtilis* (Marian & Schmidt 1967)
Weasel, *Mustela nivalis* (Uttendörfer 1930; Hens 1954; Schmidt & Szabo 1981)

Birds

Storm Petrel, *Hydrobates pelagicus* (Lockley 1938; Hayden 2004)
Manx Shearwater, *Puffinus puffinus* (Hayden 2004)
Sparrowhawk, *Accipiter nisus* (Schönn *et al.* 1991)
Kestrel, *Falco tinnunculus* (Schönn *et al.* 1991)
Grey Partridge, *Perdix perdix* (Schönn *et al.* 1991)
Red-legged Partridge, *Alectoris rufa* (Calderon 1977; Blache 2001)
Quail, *Coturnix coturnix* (Calderon 1977; Blache 2001)
Pheasant, *Phasianus colchicus* (Hibbert-Ware 1937/1938)
Moorhen, *Gallinula chloropus* (Barber 1925; Haverschmidt 1940)

Lapwing, *Vanellus vanellus* (Hibbert-Ware 1937/1938)
Jack Snipe, *Lymnocryptes minimus* (Schönn *et al.* 1991)
Snipe, *Gallinago gallinago* (Boyd 1940)
Black-tailed Godwit, *Limosa limosa* (Buker *et al.* 1984)
Common Sandpiper, *Actitis hypoleucos* (Sageder 1990)
Turnstone, *Arenaria interpres* (Freethy 1976)
Common Tern, *Sterna hirundo* (Buker *et al.* 1984; Pailley *et al.* 1991)
Little Tern, *Sterna albifrons* (Pailley *et al.* 1991)
Woodpigeon, *Columba palumbus* (Haverschmidt 1940; Glue 1972)
Turtle Dove, *Streptopelia turtur* (Manez 1983a)
Cuckoo, *Cuculus canorus* (Glue 1972; Stalling 1997)
Little Owl, *Athene noctua* (Schönn *et al.* 1991)
Hoopoe, *Upupa epops* (Heydt 1968)
Nightjar, *Caprimulgus europaeus* (Bisseling 1936)
Swift, *Apus apus* (Haensel & Walter 1966; Agelet 1979; Ciani 1988; Kuhn 1992; Blache 2001)
Wryneck, *Jynx torquilla* (Schönn *et al.* 1991; Génot & Bersuder 1995)
Grey-headed Woodpecker, *Picus canus* (Rockenbauch 1976)
Great Spotted Woodpecker, *Dendrocopos major*
Calandra Lark, *Melanocorypha calandra* (Manez 1983a)
Short-toed Lark, *Calandrella cinerea* (Manez 1983a)
Crested Lark, *Galerida cristata* (Simeonov 1968)
Wood Lark, *Lullula arborea* (Schönn *et al.* 1991)
Skylark, *Alauda arvensis*
Sand Martin, *Riparia riparia* (Hellebrekers 1928)
Swallow, *Hirundo rustica*
House Martin, *Delichon urbica* (Owen 1939; Sageder 1990)
Tree Pipit, *Anthus trivialis*
Meadow Pipit, *Anthus pratensis* (Hayden 2004)
Rock Pipit, *Anthus spinoletta* (Manez 1983a)
Pied Wagtail, *Motacilla alba* (Sageder 1990)
Yellow Wagtail, *Motacilla flava* (Schönn *et al.* 1991)
Wren, *Troglodytes troglodytes* (Manez 1983a)
Dunnock, *Prunella modularis* (Glue & Morgan 1977)
Robin, *Erithacus rubecula*
Nightingale, *Luscinia megarhynchos* (Schönn *et al.* 1991)
Black Redstart, *Phoenicurus ochruros*
Redstart, *Phoenicurus phoenicurus* (Schönn *et al.* 1991)
Whinchat, *Saxicola rubetra* (Génot & Bersuder 1995)
Wheatear, *Oenanthe oenanthe* (Madon 1933)
Blue Rock Thrush, *Monticola solitarius* (Manez 1983a)
Fieldfare, *Turdus pilaris* (Glue 1972)
Song Thrush, *Turdus philomelos*
Redwing, *Turdus iliacus*
Mistle Thrush, *Turdus viscivorus*
Reed Warbler, *Acrocephalus scirpaceus* (Haverschmidt 1940)
Lesser Whitethroat, *Sylvia curruca* (Schönn *et al.* 1991)
Whitethroat, *Sylvia communis* (Glue & Morgan 1977)

Garden warbler, *Sylvia borin* (Schönn *et al.* 1991)
Willow Tit, *Parus montanus*
Blue Tit, *Parus caeruleus*
Great Tit, *Parus major*
Nuthatch, *Sitta europea* (Schönn *et al.* 1991)
Golden Oriole, *Oriolus oriolus* (Schönn *et al.* 1991)
Red-backed Shrike, *Lanius collurio* (Simeonov 1968; Sageder 1990)
Great Grey Shrike, *Lanius excubitor* (Manez 1983a; Yosef 1993a)
Woodchat Shrike, *Lanius senator* (Schönn *et al.* 1991)
Jay, *Garrulus glandarius* (Hibbert-Ware 1937/1938)
Magpie, *Pica pica* (De la Hoz 1982; Manez 1983a)
Chough, *Pyrrhocorax pyrrhocorax* (Tella & Blanco 1993)
Jackdaw, *Corvus monedula* (Williams 1945; Kulczycki 1964)
Starling, *Sturnus vulgaris*
Spotless Starling, *Sturnus unicolor* (Manez 1983a)
House Sparrow, *Passer domesticus*
Spanish Sparrow, *Passer hispaniolensis* (Gusberti 1998)
Tree Sparrow, *Passer montanus*
Chaffinch, *Fringilla coelebs*
Serin, *Serinus serinus* (Manez 1983a)
Greenfinch, *Carduelis chloris*
Goldfinch, *Carduelis carduelis*
Linnet, *Carduelis cannabina*
Hawfinch, *Coccothraustes coccothraustes*
Yellowhammer, *Emberiza citrinella*
Corn Bunting, *Emberiza calandra*
Domestic birds as Pigeons, *Columba livia*, and Domestic Fowl, *Gallus gallus*, are also taken by the Little Owl (Grant-Ives 1936; Hibbert-Ware 1937/1938; Glue 1972; Gania & Zubkov 1975; Rule 1977; Schönn *et al.* 1991).

In New Zealand, the Little Owl can catch other owls such as the Morepork, *Ninox novaeseelandiae* (Rule 1977)

Reptiles

Moorish Gecko, *Tarentola mauritanica* (Moschetti & Manchini 1993)
Bedriaga's Skink, *Chalcides bedriagai* (Cramp 1985)
Three-toed Skink, *Chalcides chalcides* (Nappi & Mastrorilli 2003)
Spiny-footed Lizard, *Acanthodactylus erythrurus* (Manez 1983a)
Sand Lizard, *Lacerta agilis*
Green Lizard, *Lacerta viridis*
Erhard's Wall Lizard, *Podarcis erhardii* (Angelici *et al.* 1997)
Iberian Wall Lizard, *Podarcis hispanica* (Manez 1983a)
Common Wall Lizard, *Podarcis muralis*
Balkan Wall or Crimean Lizard, *Podarcis taurica* (Alivizatos *et al.* 2005)
Italian Wall Lizard, *Podarcis sicula* (Moschetti & Manchini 1993; Bux & Rizzi 2005)
Snake-eyed Lizard, *Ophisops elegans* (Simeonov 1968)
Large Psammodromus, *Psammodromus algirus* (Manez 1983a)

Slow Worm, *Anguis fragilis* (Madon 1933; Génot & Bersuder 1995)
Viperine Snake, *Natrix maura* (Bernis 1974; Manez 1983a)
Grass Snake, *Natrix natrix* (Buxton 1947; Nappi & Mastrorilli 2003)
Greek Tortoise, *Testudo graeca* (Mienis 1979)

Amphibians

Ribbet Newt, *Pleurodeles waltl* (Manez 1983a)
Spotted Salamander, *Salamandra salamandra* (Farsky 1928)
Bosca's Newt, *Triturus boscai* (Melendro & Gisbert 1978)
Smooth Newt, *Triturus vulgaris* (Trimnel 1945)
Painted Frog, *Discoglossus pictus* (Manez 1983a)
Western Spadefoot, *Pelobates cultripes* (Manez 1983a; Anthonioz *et al.* 2000)
Spadefoot, *Pelobates fuscus* (Marian & Schmidt 1967)
Common Toad, *Bufo bufo* (Raevel 1986)
Green Toad, *Bufo viridis* (Ille 1992)
Common Tree Frog, *Hyla arborea* (Thiollay 1966; Lovari 1974)
Stripeless Tree Frog, *Hyla meridionalis* (Barthelemy 2000)
Common Frog, *Rana temporaria*
Edible Frog, *Rana esculenta* (Farsky 1928; Cramp 1985)
Marsh Frog, *Rana ridibunda*

Fish

Carp, *Cyprinus carpio* (Manez 1983a; Cramp 1985)
Minnow, *Phoxinus phoxinus* (Cramp 1985)

A Goldfish was caught in a pond but fish are exceptional in the diet of the Little Owl.

Insects

Coleoptera

Carabidae: *Cicindella campestris, Calosoma maderae indigator, Calosoma sycophanta, Carabu auratus, Carabus auronitens, Carabus cancellatus, Carabus convexus, Carabus coriaceus, Carabu glabratus, Carabus granulatus, Carabus lineatus, Carabus lusitanicus, Carabus melancholicus Carabus nemoralis, Carabus problematicus, Carabus purpurascens, Carabus rossii, Carabu scabriusculus, Carabus scheidleri, Carabus ullrichi, Carabus violaceus, Cychrus rostratus, Leistu ferrugineus, Leistus spinibarbis, Nebria brevicollis, Nebria salina, Clivina fossor, Scarites laeviga tus, Scarites occidentalis, Scarites planus, Broscus cephalotes, Bembidion tetracolum, Bembidion andreae, Anisodactylus signatus, Anisodactylus binotatus, Diachromus germanus, Stenolophu teutonus, Ditomus clypeatus, Carterus tricuspidatus, Scybalicus oblongiusculus, Ophonus diffinis Ophonus griseus, Ophonus rufibarbis, Ophonus rufipes, Ophonus sabulicola, Harpalus affinis Harpalus attenuatus, Harpalus dimidiatus, Harpalus distinguendus, Harpalus punctatostriatus Harpalus rubripes, Harpalus sulphuripes, Poecilus crenatus, Poecilus cupreus, Poecilus nitidus Poecilus sericeus, Poecilus versicolor, Poecilus vicinus, Pterostichus aralensis, Pterostichus glo bosus ebenus, Pterostichus madidus, Pterostichus melanarius, Pterostichus melas italicus, Pteros tichus niger, Pterostichus nigrita, Pterostichus strenuus, Pterostichus vernalis, Pterostichus vul garis, Abax ater, Abax continuus, Percus bilineatus, Calathus ambiguus, Calathus fuscipes latus, Calathus melanocephalus, Pristonychus terricola, Agonum fuliginosum, Agonum mülleri, Agonum sordidum gridellii, Platynus dorsalis, Zabrus tenebrioides, Amara aenea, Amara apricaria, Amara eurynota, Amara fulva, Chlaenius circumscriptus, Chlaenius nitidulus, Chlaenius spoliatus, Oodes helopioides, Licinus cassideus, Licinus punctatulus, Dromius quadrimaculatus, Brachynus crepitans, Hydrophylus* sp.

Dytiscidae: *Dytiscus marginalis, Agabus bipustulatus, Agabus guttatus, Colymbetes fuscus, Cybister lateralimarginalis*

Hydrophilidae: *Helephorus aquaticus, Sphaeridium scarabaeoides, Hydrobius fuscipes, Hydrous piceus, Hydrous pistaceus*

Histeridae: *Dendrophilus punctatus, Gnathoncus buyssoni, Paralister bipustulatus, Paralister carbonarius, Hister unicolor*

Silphidae: *Necrophorus germanicus, Necrophorus humator, Necrophorus investigator, Necrophorus vespillo, Necrophorus vespilloides, Ablattaria laevigata, Blithophaga opaca, Phosphuga atrata, Silpha carinata, Silpha obscura, Silpha opaca, Silpha tristis, Thanatophilus rugosus, Thanatophilus sinuatus, Oeceoptoma thoracica*

Staphylinidae: *Oxytelus inustus, Paederus baudii, Philontus fuscipes, Philontus politus, Creophilus maxillosus, Ontholestes murinus, Ontholestes tessellatus, Trichoderma pubescens, Ocypus aeneocephalus, Ocypus ater, Ocypus fuscatus, Ocypus olens, Quedius fuliginosus*

Cleridae: *Thanasimus formicarius*

Elateridae: *Adelocera murina, Ampedus nigroflavus, Ampedus sanguineus, Agriotes lineatus, Agriotes obscurus, Agriotes sputator, Athous haemorrhoidalis, Athous niger, Athous vittatus, Melanotus rufipes, Ctenicera pectinicornis*

Buprestidae: *Capnodis* sp.

Byrrhidae: *Byrrhus pilula*

Dermestidae: *Dermestes* sp., *Attagenus pellio*

Coccinellidae: *Adalia bipunctata, Coccinella septempunctata*

Anobiidae: *Xestobium rufovillosum*

Ptinidae: *Ptinus* sp.

Anthicidae: *Anthicus antherinus*

Meloidae: *Meloe proscarabeus, Meloe rugosus, Meloe violaceus*

Melandryidae: *Melandrya caraboides*

Tenebrionidae: *Blaps halophila, Blaps mortisaga, Blaps mucronata, Tentyriab* sp., *Erodius* sp., *Asisda* sp., *Akis* sp., *Pimelia* sp., *Pedinus meridianus, Opatrum sabulosum, Alphitobius diaperinus, Tenebrio molitor, Stenomax aeneus, Nalassus laevioctostriatus*

Geotrupidae: *Geotrupes spiniger, Geotrupes stercorarius, Geotrupes stercorosus, Geotrupes vernalis, Typhaeus typhaeus*

Scarabaeidae: *Trox hispidus, Trox perlatus, Trox scaber, Bubas bison, Copris hispanus, Copris lunaris, Oniticellus fulvus, Onthophagus fracticornis, Onthophagus ovatus, Onthophagus taurus, Onthophagus vacca, Aphodius fimetarius, Aphodius fossor, Aphodius granarius, Aphodius paykulli, Aphodius prodromus, Aphodius rufipes, Aphodius rufus, Maladera holosericea, Amphimallon assimile, Amphimallon solstitiale, Rhizotrogus aequinoctialis, Rhizotrogus aestivus, Melolontha hippocastani, Melolontha melolontha, Melolontha papposa, Anomala vitis, Phylloperta horticola, Anisoplia austriaca, Hoplia caerulea, Hoplia philanthus, Pentodon idiota, Pentodon punctatus, Oryctes nasicornis, Phyllognatus excavatus, Cetonia aurata, Potosia cuprea, Potosia morio*

Lucanidae: *Lucanus cervus, Dorcus parallelopipedus, Sinodendron cylindrum*

Cerambycidae: *Prionus coriarius, Stenochorus meridianus, Leptura rubra, Rhagium bifasciatum, Rhamnusium bicolor, Cerambyx cerdo, Cerambyx scopoli, Clytus arietis, Dorcadion* sp., *Saperda carcharias, Saperda scalaris*

Chrysomelidae: *Lema lichenis, Oulema melanopus, Entomoscelis adonidis, Leptinotarsa decemlineata, Chrysolina haemoptera, Chrysolina marginata, Chrysolina oricalcia, Chrysolina staphylea, Chrysolina sturmi, Gastroidea polygoni, Phratora vittellinae, Timarcha coriaria, Timarcha nicaensis, Timarcha tenebricosa, Galerucella luteola, Agelastica alni, Cassida rubiginosa*

Scolytidae: *Hylesinus fraxini*

Curculionidae: *Otiorhynchus aurosparsus, Otiorhynchus picipes, Otiorhynchus porcatus, Otiorhynchus raucus, Otiorhynchus sulcatus, Otiorhynchus tenebricosus, Phyllobius pyri, Phyllobius viridicollis, Stophosoma faber, Stophosomas laterale, Barynotus obscurus, Liophloeus tessulatus, Barypeithes* sp., *Sitona humeralis, Tanymecus palliatus, Bothynoderes punctiventris, Cleonis piger, Larinus* sp., *Procas armillatus, Dorytomus longimanus, Rhytidoderes plicatus, Hylobius abietis, Hylobius transversovittatus, Minyops* sp., *Alophus triguttatus, Lepyrus* sp., *Liparus coronatus, Hypera zoilus, Tychius venustus*

Orthoptera

Gryllidae: *Gryllus campestris*

Tettigoniidae: *Tesellana tessellata, Tettigonia viridissima, Dectitus verrucivorus, Dectitus albifrons, Amphistris soetica, Ephippigerida* sp.

Gryllotalpidae: *Gryllotalpa vulgaris*

Acrididae: *Locusta* sp., *Calliptamis* sp., *Anacrydium aegyptium, Schistocerca gregaria*

Homoptera

Cicadidae: *Lyristes plebejus*

Dermaptera: *Forficula auricularia, Forficula lurida, Chelidurella acanthopygia, Apterygida media Labidura riparia*

Lepidoptera

Arctiidae: *Arctia caja*

Hepialidae: *Hepialus lupulinus*

Notodontidae: *Phalera bucephala*

Noctuidae: *Agrotis exclamationis, Agrotis segetum, Triphaena pronuba*

Lasiocampidae: *Lasiocampa quercus, Cosmotriche potatoria*

Sphingidae: *Smerinthus ocellata, Laothoe populi*

Tortricidae: *Tortrix viridana*

Hymenoptera

Ichneumonidae: *Ophion obscurus*

Vespidae: *Vespa crabro, Dolichovespula media, Polistes gallicus*

Formicidae: *Formica rufa, Messor barbara*

Dictyoptera: *Mantis religiosa*
Hemiptera: *Podops inuncta*
Chrysopa: undetermined

Diptera

Tipulidae: *Tipula paludosa, Tipula oleracea*
Syrphidae: *Cristalis tenax*
Calliphoridae: *Lucilla caesar*
Muscidae: *Cyclorrhapha* sp., *Nematocera* sp.
Odonata: *Lestes* sp., *Aeschna isosceles*
Millipede: *Julus* sp., *Polydesmus* sp., *Glomeris* sp.
Centipede: *Scolopendra cingulata, Geophilus* sp.
Woodlice: *Oniscus asellus, Porcellio scaber*
Scorpion: *Buthus occitanus, Mesobuthus gibbosus, Scorpio maurus, Scorpio fuscus*
Spider: *Nemesia* sp., *Pompilus* sp., *Calicurgus* sp., *Agenia* sp., *Pelopeus* sp., *Segestria perfida*

Other

Snail: *Helicella obvia, Helicella itala, Helix aspersa*
Mollusk: *Limax* sp., *Testacella haliotidea, Agriolimax agrestis*
Earthworm: *Lumbricus* sp.
Crab: *Carcinus maenas, Hemigrapsus penicillatus*

Information about insects and other invertebrates is from Rey (1910), Collinge (1922), Farsky (1928), Tinbergen (1932), Hibbert-Ware (1937/1938), Parigi-Bini (1948), Vachon (1954), Kumerloeve (1955), Lovari (1974), Ganya & Zubkov (1975), Libois (1977), Kerunian *et al.* (1982), Bameul (1982), De la Hoz (1982), Manez (1983a), Juillard (1984a), Popescu *et al.* (1986), Serrano *et al.* (1987), Bayle & Ziano (1989), Sageder (1990), Schönn *et al.* (1991), Ille (1992), Kuhn (1992), Moschetti & Mancini (1993), Moschetti & Mancini (1995), Guyot & Lecomte (1996), Angelici *et al.* (1997), Leveque (1997), Gusberti (1998), Bavoux *et al.* (2000), Fattorini *et al.* (2000), Vezinet (2003), Nappi (2005).

Vegetation includes mainly grass and sometimes leaves, but also small fruits, berries, and maize (*Zea*). In England, vegetable matter comprised 6.5% of the bulk of food from 212 stomachs collected throughout the year (Collinge 1922).

Eastern Palearctic

In the steppes and deserts of the Eastern Palearctic (Russia, Kazakhstan, Uzbekistan, Turkmenistan), the following specific prey composed the diet of the Little Owl (Sosnikhina 1950, Formozov & Osmolovskaya 1953, Osmolovskaya 1953, Gavrin 1962, Sukhinin *et al.* 1972, Kostin 1983, Sagitov 1990, Bel'skaya 1992, Mambetjumaev 1998, Tarkhidze & Loskutova 1999, Murzov & Berezovikov 2001).

Mammals

Mid-day Jird, *Meriones meridianus*
Great Gerbil, *Meriones opimus*
Red-tailed Gerbil, *Meriones erythrourus*
Libyan Jird, *Meriones libycus*
Persian Jird, *Meriones persicus*
Tamarisk Jird, *Meriones tamariscinus*
Small Five-toed Jerboa, *Allactaga elater*
Allactaga williamsi
Mongolian Five-toed Jerboa
Northern Three-toed Jerboa, *Dipus sagitta*
Lichtenstein's Jerboa, *Eremodipus lichtensteini*
Hairy-footed Jerboa
Thick-tailed Three-toed Jerboa
Little Earth Hare, *Pygeretmus pumilio*
Migratory Hamster, *Cricetulus migratorius*
Gray Dwarf Hamster
Thin-toed Souslik
Tolai Hare, *Lepus tolai*
Northern Mole Vole, *Ellobius talpinus*
Mole Vole, *Ellobius lutescens*
Afghan Vole, *Microtus afghanus*
Social Vole, *Microtus socialis*
Lesser White-toothed Shrew
Desert Eared Hedgehog, *Erinaceus auritus*
Bucharian Horseshoe Bat, *Rhinolophus bocharicus*
Hemprich's Long-eared Bat, *Otonycteris hemprichi*
Schreiber's Bat, *Miniopterus schreibersi*

Birds

Teal, *Anas crecca*
Rock Partridge, *Alectoris graeca*
Lesser Kestrel, *Falco naumanni*
Pallas's Sandgrouse, *Syrrhaptes paradoxus*
Caspian Plover, *Charadrius asiaticus*
Spotted Redshank, *Tringa erythropus*
Blue-cheeked Bee-eater, *Merops superciliosus*
Bee-eater, *Merops apiaster*
Roller, *Coracias garrulus*
Short-toed Lark, *Calandrella brachydactyla*
Bimaculated Lark, *Melanocorypha bimaculata*
Rufous Bush Robin, *Cercotrichas galactotes*
Isabelline Wheatear, *Oenanthe isabellina*
Pied Wheatear, *Oenanthe pleschanka*
Finsch's Wheatear, *Oenanthe finschii*
Black Wheatear, *Oenanthe leucora*

Black-throated Thrush, *Turdus ruficollis atrogularis*
Olivaceous Warbler, *Hippolais pallida*
Spotted Flycatcher, *Muscicapa striata*
Rock Nuthatch, *Sitta neumayer*
Rock Sparrow, *Petronia petronia*
Rose-colored Starling, *Sturnus roseus*
Trumpeter Finch, *Bucanetes githagineus*
Reed Bunting, *Emberiza schoeniclus*
Red-headed Bunting, *Emberiza bruniceps*

Reptiles

Stepperunner, *Eremias aruta*
Racerunner, *Eremias velox*
Scink Gecko, *Teratoscincus scincus*
Steppe Agama, *Agama sanguinolenta*
Toad-headed Agama, *Phrynocephalus mystaceus*
Dwarf Sand Boa, *Eryx miliaris*
Sand Boa, *Erix jaculus*

Fish

Varicorhinus capoeta sevangi

Insects

Coleoptera

Carabidae: *Scarites* sp., *Amara* sp., *Broscus cephalotes*, *Platysma* sp., *Taphoxenus* sp., *Cardioderuschloroticus*, *Dioctes* sp., *Chlaelnius spoliatus*, *Harpalus* sp., *Elytrodon bidentalus*

Tenebrionidae: *Ocnera* sp., *Trignoscelis* sp., *Cyphogenia* sp., *Tentrya* sp., *Pachyscelis musiva*, *Blaps* sp., *Pedinus* sp.

Scarabaeidae: *Scarabus sacer*, *Ceratophyrus polyceras*, *Aphodius* sp., *Oryctes* sp., *Chroneosoma* sp., *Rhyzotrogus* sp., *Polyphylla* sp., *Coptognathus* sp., *Trox* sp., *Epicometis* sp., *Oxythyria* sp.

Buprestidae: *Julodis variolaris*, *Dicera* sp., *Sphenoptera* sp.

Other countries

Jordan

In deserts and semi-desert habitats of Jordan, the diet of the Little Owl is composed of reptiles (Gekkonidae: *Stenodactylus grandiceps*, *Ptyodactylus puiseuxi*; Lacertidae: *Acanthodactylus* sp.; and Agamidae: *Pseudotrapelus sinaitus*, *Trapelus pallidus haasi*, *Uromastyx aegyptius microlepis*), mammals (Cricetidae: *Gerbillus dasyurus dasyurus*, *Meriones crassus crassus*; Muridae: *Acomys russatuslewisi*; and Dipodidae: *Jaculus jaculus vocator*), scorpions, beetles (Tenebrionidae and Buprestidae) and Orthoptera (Al-Mehim *et al.* 1997).

Egypt, Syria, and Iran

In the arid zones of Egypt, Syria, and Iran, the diet is composed of reptiles (Gekkonidae, Lacertidae, and Agamidae) birds (*Passer domesticus*, *Galerida cristata*, *Alauda arvensis*), mammals (*Gerbillus dasyurus*, *Gerbillus nanus*, *Gerbillus pyramidum*, *Gerbillus amoenus*, *Cricetulus migratorius*, *Meriones macedonicus*, *Rattus rattus*, *Mus musculus*, *Pipistrellus kuhlii*), poisonous sunspiders (Solifugae) and scorpions (Scorpionidae), earwigs (Dermaptera) and deserticolous tenebrionid beetles (*Pimelia* sp.), coprophagous and deserticolous species of Scarabeidae and Orthoptera (Obuch & Kristin 2004).

Israel

In two areas of Israel, the diet is composed as in Europe mainly by mammals in biomass (*Rattus rattus*, *Mus musculus*, *Pipistrellus kuhlii*, *Meriones tristrami*, *Microtus socialis*, *Suncus etruscus*, and *Crocidura suaveolens*) and less by birds (*Passer domesticus*) but also by insects in number (Dermaptera, Orthoptera, Coleoptera) and spiders. Latitudinal trends of birds and Orthoptera in the diet were found (Charter *et al.* 2006).

Portugal

In steppe areas in Portugal, the diet in spring is predominantly invertebrates (Dermaptera and Coleoptera): 98% by number and 71% by biomass (n = 1310 prey) (Tome unpublished).

Appendix B Little Owl bibliography

1. Abdusalyamov, I. A. 1971. Little Owl (*A. n. bactriana* Hutton). *Fauna Tadzhikskoi SSR* [*Fauna of Tajik SSR*]. Vol. XIX. Part 1. Ptitsy [Birds]. Dushanbe, Donish. [Little Owl data on pp. 332–336.]
2. Abreu, M. V. 1987. Décomptes de rapaces le long des routes au Portugal. Comparaison entre trois distritos. Rapaci Mediterranei III. Suppl. Ric. Biol. Selvaggina XII. [Little Owl data on pp. 295–300].
3. Acland, C. M. 1912. The food of the Little Owl. *British Birds* **6**: 66.
4. Adam, G. 1973. Steinkauz (*Athene noctua*) jagt zu Fuss nach Heuschrecken. *Ornithologische Mitteilungen* **25**: 249.
5. Aellen, E. 1939. Offene Bodenbrut des Steinkauzes. *Vögel der Heimat* **2**: 25–27.
6. Afanasova, L. V. & Khokhlov, A. N. 1995. Pitanie domovogo sycha v period razmnojeniya na zapade Stavropol'skogo kraya [The Little Owl feeding during breeding period in the West of Stavropol Territory]. Trudy Teberdinskogo gosudarstvennogo zapovednika [Proceedings of Teberda State Nature Reserve] **14**: 187–190.
7. AG zum Schutz bedrohter Eulen NRW, 1978. Steinkauz-Verbreitung. *Info* **7**: 5–6.
8. Agelet, A. 1979. Nota sobre la alimentacion del Mochuelo comun *Athene noctua* (*Aves, Strigidae*). *Miscellana Zoologica* **5**: 186–188.
9. Ainslie, D. 1907. Little Owl in Bedfordshire. *Zoologist* **11**: 353–354.
0. Alamany, O. & Tico, J. R. 1983. *In* Muntaner, J., Ferreri, X. & Martinez-Vilalta, A. (eds). *Atlas delo ocello nidificants de Catalunya i Andorra*. Ketres. Barcelona. [Little Owl data on pp. 135–136.]
1. Aleman, Y. & Dalmau, J. 1990. La Chouette chevêche (*Athene noctua*) dans les Pyrénées-Orientales. 1. Les sites de nidification. *Mélanocéphale* **7**: 16–27.
2. Alexander, C. J. 1911. Little Owl in Herefordshire and Worcestershire. *British Birds* **5**: 195.
3. Alexander, H. G. 1936. Sounds produced by Little Owl. *British Birds* **29**: 361.
4. Alexander, W. B. 1935. Note. *Bulletin British Ornithological Club* **55**: 60.
5. Alivizatos, H., Goutner, V. & Zogaris, S. 2005. Contribution to the study of the diet of four owl species (*Aves, Strigiformes*) from mainland and island areas of Greece. *Belgium Journal of Zoology* **135**: 109–118.
6. Allanazarova, N. A. 1988. Nekotorye dannye pitaniyu domovogo sycha v yugo-zapadnom Uzbekistane [Some data on the Little Owl feeding in Uzbekistan]. Ekologiya, okhrana i ratsionl'noe ispol'zovanie ptits Uzbekistana [Ecology, conservation and wise management of birds of Uzbekistan]. Materialy II Respublikanskoi ornitologicheskoi konferentsii [Materials of the 2nd Republican Orn. Conf.]. Tashkent, 23–25 November 1988. [Little Owl data on pp. 6–7.]
7. Alleinjn, F. 1966. De verspreiding van Velduil en Steenuil. *Vogeljaar* **14**: 180–181.
8. Allemand, G. 1995. La Chouette chevêche *Athene noctua* en Livradois-Forez. PNR Livradois-Forez. CORA Loire. CPIE Monts du Forez.
9. Al-Melhim, W. N., Amr, Z. S., Disi, A. M. & Katbeh-Bader, A. 1997. On the diet of the Little Owl, *Athene noctua*, in the Safawi area, eastern Jordan. *Zoology in the Middle East* **15**: 19–28.
0. Alvarez, C. J. & Jimenez, M. F. 1973. Estudio de Leucocytozoon en la sangre de mochuelo (*Athene noctua*). *Cuad. C. Biol.* **2**: 141–148.
1. Ammersbach, R. 1952. Zur "Kiewitt"-Frage (Steinkauz oder Waldkauz). *Ornithologische Mitteilungen* **4**: 184.
2. Ancelet, C. 1988. Ecologie et protection de la Chouette chevêche (*Athene noctua* Scop.). Contribution au programme de recherche inter-parcs naturels régionaux. BTSA PN.
3. Ancelet, C. 1994. Observations sur la répartition et l'habitat de la Chouette chevêche *Athene noctua* sur la bordure nord de la plaine de la Scarpe. Quel avenir face à l'évolution du milieu naturel? *Héron* **27**: 179–187.
4. Ancelet, C. 1994. Accouplement automnal chez la Chouette chevêche *Athene noctua*. *Alauda* **62**: 246.

25. Ancelet, C. 1995. La Chouette chevêche *Athene noctua* sur la bordure nord de la plaine de la Scarpe: état des lieux de l'habitat, quelles adaptations pour des mesures de "gestion". *Héron* **28**: 85–92.
26. Ancelet, C. 2001. Prises de bains par une jeune Chouette chevêche *Athene noctua*. *Héron* **34**: 105–107.
27. Ancelet, C. 2001. Observation d'une Chouette chevêche *Athene noctua* en forêt. *Héron* **34**: 23.
28. Ancelet, C. 2003. Exiguïté de la cavité de reproduction et mortalité juvénile chez la Chouette chevêche *Athene noctua*. *Héron* **36**: 160–165.
29. Ancelet, C. 2003. Reproduction au sol, dans un tas de bois, d'un couple de Chouette chevêche *Athene noctua* *Héron* **36**: 166.
30. Ancelet, C. 2003. Des Chouettes chevêches *Athene noctua* casanières, confiantes ou terrorisées? *Héron* **36** 167.
31. Ancelet, C. 2004. Parade et accouplement chez la Chevêche d'Athéna *Athene noctua*. *Alauda* **72**: 211–219.
32. Anderson, S. I. 1949. Little Owl raiding starlings' nest. *New Zealand Bird Notes* **3**: 110.
33. Andreotti, A. 1989. *In* Lega Italiana Protezione Uccelli. *Atlante degli uccelli nidificante in Liguria Regione Liguria*. Asserorato All'urbanistica Asserorato All'agricultura e Foreste. Catalogli dei beni naturali. Genova [Little Owl data on p. 71.]
34. Andris, K. 1965. Die Vogelwelt einer Kiefernaufforstungsfläche in der südbadischen Oberrheinebene. Mitt bad. Landesver. *Naturkunde und Naturschutz* **4**: 585.
35. Angelici, F. M., Latella, L., Luiselli, L. & Riga, F. 1997. The summer diet of the Little Owl (*Athene noctua* on the island of Astipalaia (Dodecanese, Greece). *Journal of Raptor Research* **31**: 280–282.
36. Angyal, L. & Konopka, H. P. 1975. Beobachtungen an einem zahmen Steinkauz. *Gefierdete Welt* **99**: 212–217
37. Angyal, L. & Konopka, H. P. 1977. Weitere Beobachtungen zum Verhalten des zahmen Steinkauz "Cheeta" *Gefierdete Welt* **101**: 66–68.
38. Anonymous, 1863. Mémoires de la Société d'agriculture du département de la Meuse. Laurent. Châlons-sur Marne. [Little Owl data on p. 134.]
39. Anonymous, 1918. Note. *Bulletin British Ornithological Club* **38**: 75.
40. Anonymous, 1922. The Little Owl. *Bird Notes and News* **10**: 21–23.
41. Anonymous, 1928. Notes. *Bird Notes and News* **13**: 34.
42. Anonymous, 1935. Note. *Bird Notes and News* **17**: 153–154.
43. Anonymous, 1936. The Little Owl. *Bird Notes and News* **17**: 16–17.
44. Anonymous, 1937. Little Owl taking spider. *British Birds* **31**: 126.
45. Anonymous, 1938. The Little Owl. *Bird Notes and News* **18**: 9.
46. Anonymous, 1974. Coléoptères trouvés dans les pelotes de Chevêches en Tunisie. *Alauda* **42**: 236.
47. Anonymous, 1974. Broedkasten voor steenuilen. *Vogeljaar* **22**: 803–805.
48. Anonymous, 1976. Nieuwe broedkast voor Steenuilen. *Vogeljaar* **24**: 323.
49. Anonymous, 1977. Schleiereulen und Steinkäuze sollen in Nordrhein-Westfalen nicht länger unter "Wohnungsnot" leiden. *Ornithologische Mitteilungen* **29**: 197.
50. Anonymous, 1985. Erfolgreiche Auswilderung von 7 Turmfalken, 7 Uhus und 9 Steinkäuzen. *Vivarium* **1** 14–15.
51. Anonymous, 1986. Atlas de la présence hivernale des oiseaux de Bretagne. *Ar Vran* **12**: 10–79.
52. Anonymous, 1987. Rapaces sur la plaine. *La vie secrète de la nature en France* **9**: 1951–1952.
53. Anonymous, 1987. Steinkauzbestand in der BRD. *D. B. V. AG zum Schutz bedrohter Eulen Informationsbla* **27**: 5.
54. Anonymous, 2004. Aktiver Steinkauzschutz in der Stadt Frankfurt. *Flieg und Flatter* **11**: 11.
55. Anthonioz, J., Bavoux, C. & Seguin, N. 2000. Prédation du Pélobate cultripède *Pelobates cultripes* pa la Chevêche d'Athéna *Athene noctua* dans l'île d'Oléron (Charente-Maritime). *Annales de la Société de Sciences Naturelles de la Charente-Maritime* **8**: 1097–1100.
56. Aplin, O. V. 1892. Blue-headed Wagtail and Little Owl in Oxfordshire. *Zoologist* **16**: 332.
57. Arenberg, Prince d' 1909. La Chouette chevêche (*Noctua minor*) est-elle nuisible? *Revue Français d'Ornithologie* **1**: 91.
58. Arias, J. M. 1994. Observation on food habits of the Little Owl (*Athene noctua* L., *Aves, Strigiformes*). Doñana *Acta Vertebrata* **21**: 183–185.
59. Arnhem, J. & Gregoire, A. 1962. Rapaces nocturnes de Belgique. *Naturalistes Belges* **43**: 401–402.
60. Arrigoni Degli Oddi, E. 1903. Letter on *Athene chiaradiae*. *Ibis* **3**: 140.
61. Arwentiev, W. 1938. Beitrag zur Nahrungsbiologie des Steinkauzes (*Athene noctua* (Scop.)) und Turmfalke (*Cerchneis tinnunculus*) in Rumänien. *Bulletin Section Science Académie de Roumanie* **21**: 81–84.
62. Ash, J. 1954. Comatose condition of Little Owl. *British Birds* **47**: 84.
63. Ash, J. 1955. Grass in stomachs of kestrel and Little Owl. *British Birds* **48**: 327.
64. Ashford, W. J. 1918. Little Owl breeding in Dorsetshire. *British Birds* **12**: 20.
65. Association Des Naturalistes Orleanais, 1995. *Découvrir les Oiseaux du Loiret*. [Little Owl data on p. 86].

6. Ataev, K. 1977. Novye dannye po ekologii domovogo sycha [New data on the Little Owl ecology]. *VII Vsesoyuznaya ornitologicheskaya konferentsiya [VIIth All-Union Orn. Conference]*. Kishinev, Naukova Dumka Press. [Little Owl data on pp. 196–197.]
7. Atchinson, G. T. 1912. The food of the Little Owl. *British Birds* **6**: 66.
8. Atchinson, G. T. 1917. Large clutch of eggs of Little Owl. *British Birds* **11**: 93.
9. Atelier d'Ecologie Rurale et Urbaine, 1988. Observatoire Doller. Suivi écologique année 1984. Etude SETRA. Cete de l'est. [Little Owl data on pp. 64–67.]
0. Attlee, H. G. 1917. Unusual birds in Oxfordshire in 1915. *British Birds* **10**: 270.
1. Audenaert, T. 2003. Voorkomen en verspreiding van de Steenuil *Athene noctua* te Sint-Pauwels tijdens het broedseizoen 2003. *Groene Waasland* **132**: 10–13.
2. Augst, U. & Manka, G. 1997. Vorkommen, Verbreitung und Bestandsentwicklung von Steinkauz (*Athene noctua*), Sperlingskauz (*Glaucidium passerinum*) und Rauhfusskauz (*Aegolius funereus*) in der Sächsischen Schweiz. Naturschutz report. Rauhfusskauz und Sperlingskauz in Deutschland. Thüringer Ministerium für Landwirtschaft, Naturschutz und Umwelt. [Little Owl data on pp. 122–131.]
3. Baboss, J. 1955. Ornithological observations in Eastern Hungary. *Aquila* **30**: 465.
4. Bacmeister, W. 1922. Die Vogelwelt Strassburgs und seiner Umgebung. *Mitteilungen über die Vogelwelt* **21**: 108.
5. Bacmeister, W. & Kleinschmidt, O. 1918. Zur Ornithologie von Nordost Frankreich. *Journal für Ornithologie* **3**: 266–267.
6. Baer, W. & Uttendörfer, O. 1897. Auf den Spuren gefierdeter Räuber Studien zweier Waldpolizisten. *Ornithologische Monatsschrift* **22**: 81.
7. Baer, W. & Uttendörfer, O. 1898. Auf den Spuren gefierdeter Räuber. *Ornithologische Monatsschrift* **23**: 250.
8. Bagnoli, P., Fontanesi, G., Casini, G. & Porciatti, V. 1990. Binocularity in the Little Owl, *Athene noctua* 1: anatomical investigation of the Thalamo Wulst Pathway. *Brain Behavior and Evolution* **35**: 31.
9. Bailleul, F. 1999. Etude d'une population de Chouettes chevêches en relation avec les modifications de l'habitat. Stage de maîtrise.
0. Bakaev, S. 1974. K biologii razmnozheniya domovogo sycha (*A. noctua bactriana* Hutton) v nizov'yakh reki Zerafshan [On breeding biology of the Little Owl (*A. noctua bactriana* Hutton) in Lower Zerafshan River]. *Trudy Samarkandskogo Universiteta [Proceedings of Samarkand University]* **211**: 35–39.
1. Baker, E. C. S. 1920. Description of a new subspecies of Little Owl (*Carine brema fryi*) from Madras, with remarks on the distribution of the species. *Bulletin British Ornithologist Club* **40**: 60–61.
2. Baldacchino, A. E. 1981. Le statut des rapaces à Malte. Rapaces méditerranéens. Parc naturel régional de Corse. CROP. [Little Owl data on pp. 17–18.]
3. Balducci, E. 1903. Osservazioni sullo sterno dell'*Athene chiaradiae. Archivio Zoologico Italiano* **1**: 375–380.
4. Balducci, E. 1905. Osservazioni e considerazioni sulla pigmentazione dell'iride dell'*Athene chiaradiae* Gigl. *Monitore Zoologico Italiano* **18**: 258–272.
5. Balten, B., De Bruin, B., Hulscher, J. B. & Koks, B. 1993. Avifauna Groningen 1968–1993. *De Grauwe Gors* **21**: 165.
6. Bameul, F. 1982. *Hydrophilus piceus* L. captured by an owl. *Balfour Browne Club Newsletter* **25**: 14.
7. Bang, J. & Rosendahl, S. 1972. Nogle oplysninger om kirkeuglens forekomst i Danmark 1960–1972 (*Athene noctua* (Scopoli.)). *Danske Fugle* **24**: 249–257.
8. Bangs, O. & Peters, J. L. 1928. *Athene noctua impasta. Bulletin of the Museum of Comparative Zoology (Harvard University)* **68**: 330–331.
9. Banks, R. C. 1918. Little Owl in Monmouthshire. *British Birds* **12**: 162.
0. Banks, R. C. 1919. Little Owl in Monmouthshire. *British Birds* **12**: 210.
1. Bannermann, D. A. 1955. *The Birds of the British Isles. 4.* Edinburgh. London: Oliver and Boyd. [Little Owl data on pp. 198–207.]
2. Banzhaf, W. 1938. Von unserem heimischen Eulen. *Natur und Volk* **68**: 92–95.
3. Barataud, M. 1977. Les rapaces nicheurs de Leycuras. *Bulletin de la Société pour l'Etude et la Protection des Oiseaux en Limousin* **7**: 45.
4. Barber, W. E. 1925. Increase of Little Owl in north Lancashire. *British Birds* **19**: 26.
5. Barbieri, F., Bogliani, G., Cesaris, C., Fasola, M. & Prigioni, C. 1978. Indicazioni sal censimento dell alloco, *Strix aluco*, e della civetta, *Athene noctua. Avocetta* **2**: 49–50.
6. Barbu, P. 1978. Contribution à l'écologie des petits mammifères du Sud de la R. S. Roumanie. *Analele Universitatii Bucuresti Biologie* **27**: 101–104.
7. Barbu, P. & Sorescu, C. 1970. Contribution concernant la nourriture de la Chouette (*Athene noctua noctua* Scop.). *Analele Universitatii Bucuresti Biologie* **19**: 67–72.
8. Barrow, W. H. 1919. Status of the Little Owl in Leicestershire. *British Birds* **13**: 30.

99. Barthelemy, E. 2000. L'avifaune du massif du Garlaban et de sa périphérie. Suivie de notes sur les mammifères, les reptiles et les amphibiens. *Faune de Provence. Bulletin du Conservatoire et Etudes des Ecosystèmes de Provence* **20**: 29–65.
100. Barthelemy, E. & Bertrand, P. 1997. Recensement de la Chevêche d'Athéna *Athene noctua* dans le massif du Garlaban (Bouches-du-Rhône). *Faune de Provence. Bulletin du Conservatoire et Etudes des Ecosystèmes de Provence* **18**: 61–66.
101. Basenina, N. V. 1968. Geographische Variation der Greifvögel und ihre Nahrung. *Ornitologija* **9**: 49–57.
102. Bashta, T. V. 1994. Do ekologii zhyvlennya khatnjogo sycha [On the Little Owl feeding ecology]. Materialy 1-j konferentsii molodykhoritologiv Ukrainy (Lutsk, 4–6 bereznya 1994). [Materials of the 1st Conference of Young Ornithologists of Ukraine, March 4 1994, Lutsk.] Chernivtsi. [Little Owl data on pp. 126–127.]
103. Battaglia, A., Ghidini, S., Campanini, G. & Spaggiari, R. 2004. Heavy metal contamination in Little Owl (*Athene noctua*) and common buzzard (*Buteo buteo*) from northern Italy. *Ecotoxicology and Environmental Safety* **60**: 61–66.
104. Batten, L. A. 1971. Bird population changes on farmland and in woodland for the years 1968–69. *Bird Study* **18**: 6.
105. Baudvin, H. 1974. Le surmulot (*Rattus norvegicus*) proie de la Chevêche (*Athene noctua*). *Jean-le-Blanc* **13**: 25.
106. Baudvin, H. 1983. Le régime alimentaire de la Chouette effraie (*Tyto alba*). *Jean-le-Blanc* **22**: 44–45.
107. Bauer, H.-G. & Berthold, P. 1996. Die Brutvögel Mitteleuropas. Bestand und Gefährdung. Aula-Verlag, Wiesbaden. [Little Owl data on pp. 254–256.]
108. Bauer, S. & Thielcke, G. 1982. Gefährdete Brutvogelarten in der Bundesrepublik Deutschland und im Land Berlin: Bestandsentwicklung, Gefährdungsursachen und Schutzmassnahmen. *Vogelwarte* **31**: 265–267.
109. Baum, H. G. & Grimm, H. 1993. Zur Situation des Steinkauzes (*Athene noctua*) in Thüringen. *Landschaftspflege und Naturschutz in Thüringen* **30**: 79–81.
110. Baumann, M. 1980. Chronique ornithologique d'Alsace du 1er Novembre 1978 au 31 Décembre 1979. *Ciconia* **4**: 188.
111. Baumgart, W. 1980. Wodurch ist der Steinkauz bedroht? *Falke* **27**: 228–229.
112. Baumgart, W. 1991. Gegenwärtiger Status und Gefährdungsgrad von Greifvögeln und Eulen in Syrien. *Birds of Prey Bulletin* **4**: 119–131.
113. Baumgart, W., Simeonov, S. D., Zimmermann, M. *et al.* 1973. An Horsten des Uhus (*Bubo bubo*) in Bulgarien.I. Der Uhu im Iskerdurchbruch (Westbalkans). *Zoologische Abhandlungen des Staatlichen Museum für Tierkunde Dresden* **32**: 203–247.
114. Baur, P. 1980. Steinkäuze in der Region Basel 1980. *Vögel der Heimat* **3**: 60–61.
115. Baur, P. 1981. Steinkäuze in der Region Basel 1981. *Vögel der Heimat* **5**: 106–108.
116. Baur, P. 1982. Steinkäuze in der Region Basel 1982. *Vögel der Heimat* **4**: 82–83.
117. Baur, P. 1982. Kritisches zur Steinkauzröhre. *Vögel der Heimat* **4**: 84–86.
118. Baur, P. 1983. Steinkäuze in der Region Basel 1983. *Vögel der Heimat* **3**: 54–55.
119. Baur, P. 1985. Steinkäuze in der Region Basel 1984. *Vögel der Heimat* **9**: 188–189.
120. Baur, P. 1986. Steinkäuze in der Region Basel 1985. *Vögel der Heimat* **56**: 62–63.
121. Baur, P. 1987. Steinkäuze in der Region Basel 1986. *Vögel der Heimat* **57**: 118–119.
122. Baur, P. 1988. Steinkäuze in der Region Basel 1987. *Vögel der Heimat* **58**: 162–163.
123. Baur, P. 1988. Steinkäuze in der Region Basel 1988. *Vögel der Heimat* **59**: 42–43.
124. Bauschmann, G. 1985. Der Steinkauz (*Athene noctua*) als Waldbewohner. *Vogel und Umwelt* **3**: 377.
125. Bavoux, C. & Burneleau, G. 1983. Statut des rapaces nocturnes à l'Ile d'Oléron. *Bulletin du Groupe Ornithologique Aunis-Saintonge* **13**: 64–68.
126. Bavoux, C., Burneleau, G. & Seguin, N. 2000. Consommation de crabes par la Chevêche d'Athéna *Athene noctua* dans l'île d'Oléron (Charente-Maritime). *Alauda* **68**: 329–330.
127. Bavoux, C., Faux, E., Mimaud, L. & Seguin, N. 2002. Capture d'écrevisses par la Chevêche d'Athéna *Athene noctua* dans le marais de Brouage (Charente-Maritime, France). *Alauda* **70**: 225–226.
128. Bayle, P. 1980. Rapport intermédiaire sur l'impact de la déviation routière de Sélestat sur les populations de rapaces diurnes et nocturnes. Service d'Etudes Techniques des Routes et Autoroutes. [Little Owl data on pp. 28–29.]
129. Bayle, P. 1981. Enquête sur la mortalité des oiseaux de proie en Alsace en 1980. *In* Le 10 ème colloque régional d'ornithologie et de mammalogie. Strasbourg 1980. *Ciconia* **5**: 61–62.
130. Bayle, P. & Ziano, M.-T. 1989. Les limaciens, proies potentielles des rapaces nocturnes en Provence et dans les Alpes du sud. *Faune de Provence. Bulletin du Conservatoire et Etudes des Ecosystèmes de Provence* **10**: 23–29.
131. Baziz, B., Sekour, M., Doumandji, S. *et al.* 2005. [Data on the diet of the Little Owl (*Athene noctua*) in Algeria.] *Aves* **42**: 149–155.

132. Bednarek, W., Hausdorf, W., Jörissen, U., Schulte, E. & Wegener, H. 1975. Über die Auswirkung der chemischen Umwelt belastung auf Greifvögel in zwei Probeflächen Westfalen. *Journal für Ornithologie* **116**: 184–185.
133. Beersma, P. & Beersma, W. 2000. Temperatuurmetingen in nestkasten voor Steenuilen *Athene noctua*. *Hel Vogeljaar* **48**: 269–271.
134. Beersma, P. & Beersma, W. 2001. Little Owls (*Athene noctua*) and biocides: reasons for concern? *In* Van Nieuwenhuyse, D., Leysen, M. & Leysen, K. (eds). *Oriolus* **67**: 94–99.
135. Beersma, P. & Beersma, W. 2003. Vliegen met elke kleine prooi? *Athene Nieuwsbrief STONE* **8**: 16.
136. Beersma, P. & Beersma, W. 2005. Hoe is het grondstadium van uitvliegende pullen te voorkomen? *Athene Nieuwsbrief STONE* **10**: 14–15.
137. Beersma, P. F. M., Beersma, W. E. & Van Den Burg, A. 2007. Steenuilen. Uitgeverij Roodbont, Zutphen, Netherlands.
138. Beesau, H., Besnault, J., Collette, J. *et al.* 1986. Oiseaux nicheurs du Parc Naturel Régional Normandie-Maine. PNR Normandie-Maine. [Little Owl data on pp. 92–93.]
139. Bekaert, M. 2002. Habitatvoorkeur en populatiedichtheden van de Steenuil *Athene noctua* in Vlaanderen. *Oriolus* **68** (4): 181–190.
140. Bekaert, M. 2006. Onderzoek naar de invloed van intrinsieke en extrinsieke limiterende factoren op het reproductief succes van de Steenuil (*Athene noctua*). Masters thesis, Universiteit Gent.
141. Beley, C. & Beley, J. J. 1974. Nos Oiseaux. *Bihoreau* **3**: 26.
142. Belik, V. P. 1997. Sovremennoe sostoyanie avifauny stepnogo Podon'ya [Modern status of avifauna of the steppe Cis-Don River area]. *Russkiy Ornitologicheskiy Zhurnal [Russian Journal of Ornithology] (Express issue)* **29**: 20–38.
143. Belko, N. G., Ivabchev, V. P., Priklonskiy, S. G. *et al.* 1998. Redkie, malochislennye i maloizuchennye vidy sokolooobraznykh i sov Yugo-Vostochnoi Meshchery [Rare, little-number and little-studied *Falconiformes* and *Strigiformes* species in South-Eastern Meshchera]. [Little Owl data on pp. 159–162.]
144. Bel'skaya, G. S. 1992. *Osobennosti biologii ptits v aridnykh usloviyakh [Peculiarities of Bird Biology in Arid Conditions]*. Asgabat: Ylym Press. [Little Owl data on pp. 222–248.]
145. Bel'skaya, G. S. & Zhernonov, I. V. 1962. Pitanie domovogo sycha v Zaunguzskikh i Tsentral'nykh Karakumakh [The Little Owl feeding in Zaunguz and Central Karakum]. Tez. 1 i Resp. konf. zoologov Turkmenistana [Abstracts of the 1st Rep. Conf. of Zoologists of Turkmenistan]. Ashkhabad. [Little Owl data on pp. 8–9.]
146. Benda, P. & Marek, J. 2001. Abundance of Little Owl (*Athene noctua*) in Decin, Czech Republic, in 2000. *Buteo* **12**: 135–138.
147. Benoist, O., Grandpierre, J.-L. & Pouneau, J. 1986. Avifaune de l'estuaire Seine, bilan des connaissances (1972–1986). *Cormoran* **30**: 462.
148. Bentham, H. 1923. Little Owls in north Devon. *British Birds* **16**: 309.
149. Benussi, E. 1997. Indagine su una popolazione di rapaci notturni (*Strigiformes*) dell'Italia nord-orientale. *Falco* **12**: 5–12.
150. Benussi, E. Galeotti, P. & Gariboldi, A. 1997. La comunità di Strigiformi della Val Rosandra nel Carso Triestino. *Annales, Koper, Series Historia Naturalis* **11**: 85–92.
151. Berezovikov, N. N., Gubin, B. M., Etrokhov, S. N., Karpov, F. F. & Kovalenko, A. V. 1999. Ptitsy pustyni Taukumy i ravniny Zhusandala (yuzhnoe Pribalkash'ye) [The birds of Taukumy desert and Zhusandala plain (Lake Balkhash region). Part 1]. *Russkiy Ornitologiches kiy Zhurnal [Russian Journal of Ornithology] (Express issue)* **73**: 3–22.
152. Berg, H.-M. 1992. Status und Verbreitung der Eulen (*Strigiformes*) in Österreich. *Egretta* **35**: 4–8.
153. Bergerhausen, W. 1992. Versuch einer bundesweiten Erhebung zur Verbreitung und Siedlungsdichte des Steinkauzes. *D. B. V. AG zum Schutz bedrohter. Eulen Informationsblatt* **38**: 2–6.
154. Bergerhausen, W. & Breuer, W. 1994. Quo vadis Steinkauz? *Eulen Rundblick* **40**: 39–43.
155. Berghmans, H. 2001. Little Owl *Athene noctua* preying on dead day-old chickens. *Oriolus* **68**: 191.
156. Berghmans, H. 2001. Adult Little Owl *Athene noctua* sharing nestbox with three Jackdaw *Corvus monedula* chicks. *Oriolus* **68**: 191.
157. Berghmans, H. 2001. Nestbox with two incubating female Little Owls *Athene noctua* and nine chicks: a case of bigamy? *Oriolus* **68**: 192–193.
158. Berghmans, H. 2001. Shell of snail *Cepaea nemoralis* in pellet of Little Owl *Athene noctua*. *Oriolus* **68**: 194.
159. Bergier, P. 1980. L'avifaune nicheuse des Alpilles. *Bulletin du Centre de Recherches Ornithologiques de Provence* **3**: 27.
160. Berndt, R. K., Koop, B. & Struwe-Juhl, B. 2002. *Vogelwelt Schleswig-Holstein.* Band 5: Brutvogelatlas. Ornithologische Arbeitsgemeinschaft für Schleswig-Holstein und Hamburg e.V. Wachholtz Verlag Neumünster. [Little Owl data on pp. 235–237.]

161. Bernis, F. 1974. Algunos datos de alimentacion y depredacion de falconiformes y estrigiformes Ibericas. *Ardeola* **19**: 245–246.
162. Bersuder, D. 1985. Contribution à l'étude du régime alimentaire de la Chouette chevêche en Alsace. *In* Colloque interrégional d'Ornithologie et de mammalogie. Saint-Dié 1984. *Ciconia* **9**: 50–51.
163. Bettmann, H. 1951. Waldkauz oder Steinkauz. *Ornithologische Mitteilungen* **3**: 132–133.
164. Beven, G. 1979. Little Owl's method of catching cockchafers. *British Birds* **72**: 594–595.
165. Bezzel, E. 1967. Besonders gefährdete Vogelarten in Bayern. *Landesverband für Vogelschutz in Bayern* **4**: 6.
166. Bezzel, E. 1982. *Vögel in der Kulturlandschaft*. Verlag Eugen Ulmer Stuttgart. [Little Owl data on pp. 50–55.]
167. Bezzel, E. 1985. *Kompendium der Vögel Mitteleuropas. Nonpasseriformes. Nichtsingvögel*. Aula-Verlag. Wiesbaden. [Little Owl data on pp. 648–651.]
168. Bienaime, E. & Chaib, F. 1980. La Chouette chevêche (*Athene noctua*) dans la région de Hazebrouck (Nord). *Héron* **4**: 75–79.
169. Birdlife International, 2004. *Birds in Europe: Population Estimates, Trends and Conservation Status*. Cambridge. UK. BirdLife Conservation Series No. 12. [Little Owl data on p. 163.]
170. Bisseling, G. A. L. 1933. Een partieel albinistische Steenuil, *Athene noctua vidalii* A. E. Brehm. *Orgaan Club Nederlandse Vogelkunde* **5**: 170.
171. Bisseling, G. A. L. 1936. Nachtzwaluwen, *Caprimulgus eu. europaeus* L., als prooi van Steenuil, *Athene noctua vidalii* A. E. Brehm. *Orgaan Club Nederlandse Vogelkunde* **9**: 139–141.
172. Blaauw, F. E. 1906. On an albino specimen of the Little Owl. *Bulletin British Ornithological Club* **16**: 41.
173. Blaauw, F. E. 1908. Note. *Bulletin British Ornithological Club* **16**: 41.
174. Blache, S. 2001. Etude du régime alimentaire de la Chevêche d'Athéna (*Athene noctua* Scop.) en période de reproduction en zone agricole intensive dans le sud-est de la France. *In* Génot, J.-C., Lapios, J.-M., Lecomte, P. & Leigh, R. S. (eds). *Ciconia* **25**: 77–94.
175. Blache, S. 2004. La Chevêche (*Athene noctua*) en zone d'agriculture intensive (plaine de Valence; Drôme): habitat, alimentation, reproduction. Ecole Pratique des Hautes Etudes. Sciences de la Vie et de la Terre. Laboratoire Biogéographie et Ecologie des Vertebrés. Montpellier.
176. Blache, S., Iborra, O. & Ulmer, A. 2003. Chevêche d'Athéna. *Atlas des Oiseaux Nicheurs de Rhône-Alpes*. [Little Owl data on p. 141.]
177. Blagosklonov, C. 1968. *Guide de la Protection des Oiseaux Utiles*. MIR Moscou. [Little Owl data on p. 246.]
178. Blanc, T. 1958. Au garde-manger hivernal de la Chevêche. *Nos Oiseaux* **24**: 321.
179. Blanke, E. 2003. Steenuilen in de gemeente Raalte in 2003. *Athene Nieuwsbrief STONE* **8**: 43–44.
180. Blanke, E. 2004. Steenuilen en steenmarters: een probleem? Oproep! *Athene Nieuwsbrief STONE* **9**: 13.
181. Blanke, E. 2005. Steenuilen en steenmarters: een probleem? *Athene Nieuwsbrief STONE* **10**: 7–13.
182. Blathwayt, F. I. 1902. Little Owl and Shore-Lark in Lincolnshire. *Zoologist* **6**: 112.
183. Blathwayt, F. L. 1904. Little Owl (*Athene noctua*) and Waxwing (*Ampelis garrulus*) in Lincolnshire. *Zoologis* **1904**: 74–75.
184. Bloem, H., Boer, K., Groen, N., van Harxen, R. & Stroeken, P. 2001. De Steenuil in Nederland. Handleiding voor onderzoek en bescherming.
185. Blondel, J. & Badan, O. 1976. La biologie du Hibou grand-duc en Provence. *Nos Oiseaux* **33**: 212–213.
186. Blum, V. 1991. *In* Kilzer, R. & Blum, V. (eds). *Atlas der Brutvögel Vorarlbergs*. Österreich Gesellschaft fü Vogelkunde, Landesstelle Vorarlberg. Bregenz. [Little Owl data on p. 130.]
187. Blyth, E. 1847. *In* Hutton T. Rough notes on the ornithology of Candahar and its neighbourhood. *Journal o the Asiatic Society of Bengal* **16**: 776–777.
188. Bobkov, Y. V., Zhukov, V. S., Kan, V. & Nikolaev, V. V. 1997. Materialy po nekotorym zimuiyushchin ptitsam Novosibirskoi oblasti [Materials on some wintering birds of Novosibirsk Region]. Materialy k rasprostraneniyu ptits na Urale, v Priural'ye I Zapadnoi Sibi ri [Materials on bird distribution in the Ural, i Cis-Uralia and West Siberia]. Ekaterinburg, Inst. of Ecology of Plants and Animals. [Little Owl data on pp 9–12.]
189. Bock, H. 1988. Projekt Steinkauz, erster Erfolg bei der Wiedereinbürgerung. Vogelschutz Report, Jahres bericht Landesbund für Vogelschutz-Bayreuth. pp. 30–35.
190. Boetticher, von H. 1927. Kurze Uebersicht über die Raubvögel und Eulen Bulgariens. *Verhandlunge Ornithologie Gesellschaft Bayern* **17**: 548.
191. Boie, H. 1822. *Athene*. Isis oder Encyclopaedische Zeitung, von Oken 1817–1848 Jena. [Little Owl data on pp. 10–11; col. 549.]
192. Boitier, E. 2003. Inventaire de la population de Chouette chevêche sur la commune de Bort-l'Etang en 2002 Report. Parc naturel régional Livradois-Forez.
193. Bolam, G. 1911. Little Owl and Wood-Sandpiper in Lincolnshire. *Zoologist* **15**: 432.
194. Bolam, G. 1920. A new bird for the local lists. *Vasculum Hexham* **6**: 53.
195. Boley, A. & Frey, H. 1934. Fernfunde beringter Steinkäuze (*Athene noctua*). *Vogelzug* **5**: 150.

96. Bon, M., Ratti, E. & Sartor, A. 2001. Variazione stagionale della dieta della Civetta *Athene noctua* (Scopoli, 1769) in una localita agricola della gronda lagunare Venezia. *Bollettino del Museo Civico di Storia Naturale di Venezia* **52**: 193–212.
97. Bonin, B. 1983. *In* De Beaufort, F. (ed). Livre rouge des espèces menacées en France. Tme 1: Vertébrés. Secrétariat de la Faune et de la Flore. Paris. [Little Owl data on pp. 145–147.]
98. Bonin, B. & Strenna, L. 1986. Sur la biologie du Faucon crécerelle, *Falco tinnunculus*, en Auxois. *Alauda* **54**: 251.
99. Bönsel, A. 2001. Erste Ergebnisse zum Wiederansiedlungsprojekt des Steinkauzes (*Athene noctua*) in Mecklenburg-Vorpommern. Landesverband Eulen-Schutz im Schleswig-Holstein. [Little Owl data on pp. 29–31.]
00. Bonvicini, P. & Maino, M. 1993. Predazione di barbagianni *Tyto alba* su Civetta *Athene noctua* in provincia di Arezzo. *Picus* **19**: 69.
01. Borgsteede, H. M., Okulewicz, A., Zoun, P. E. F. & Okulewicz, J. 2003. The helminth fauna of birds of prey (*Accipitriformes, Falconiformes* and *Strigiformes*) in the Netherlands. *Acta Parasitologica* **48**: 200–207.
02. Borodin, O. V. 1994. Konspekt fauny ptits Ul'yanovskoi oblasti [Conspectus of the ornithological fauna of Ul'yanovsk Region]. Ul'yanovsk, Branch of Moscow State University.
03. Bösiger, E. & Faucher, P. 1958. *Les Oiseaux de la Nuit.* Flammarion. [Little Owl data on pp. 48–53.]
04. Botta, E. & Ravasini, M. 1983. Andar per borre. *Uccelli* **18**: 11.
05. Böttcher, M. 1981. Endoscopy of birds of prey in chemical veterinary practice. Recent advances in the study of raptor diseases. Proceedings of the International Symposium on Diseases of Birds of Prey. 1st–3rd July 1980. London. [Little Owl data on pp. 101–104.]
06. Bouchy, P. 2004. Analyse démo-génétique de viabilité dans le cas de petites populations fragmentées: applications à la Chouette chevêche (*Athene noctua*). Ph.D. thesis, Université Paris 6 Pierre et Marie Curie.
07. Boudeau, D. 1998. Etude et protection de la Chouette chevêche dans le Gâtinais du Loiret. *Recherches Naturalistes en Région Centre* **2**: 31–37.
08. Bouisset, L. & Ruffie, J. 1956. Sur un exemple d'*Athene noctua* polyparasitée. *Bulletin de la Société Histoire Naturelle de Toulouse* **91**: 160–162.
09. Boutet, J.-Y. & Petit, P. 1987. *Atlas des Oiseaux Nicheurs d'Aquitaine 1974–1984.* Centre Régional Ornithologique Aquitaine-Pyrénées. [Little Owl data on p. 100.]
10. Boutinot, S. 1980. Etude écologique de l'avifaune du Vermandois. Structure, dynamique et évolution des populations depuis 1950. Masters thesis. Université de Reims. [Little Owl data on p. 178.]
11. Boutrouille, C. 2001. Un site de nidification insolite pour un couple de Chouette chevêche *Athene noctua. Héron* **34**: 24–25.
12. Boyd, A. W. 1940. Little Owl preying on birds. *British Birds* **34**: 18.
13. Boyd, A. W. & Leach, E. P. 1931. Ringed birds in pellets and nests of owls and hawks. Little Owl. *British Birds* **24**: 292–293.
14. Božič, L. & Vrezec, A. 2000. Sove Pohorja [Owls of the Pohorje Mountains]. *Acrocephalus* **21** (98/99): 47–53.
15. Bracker, M. 1950. Erlebnis mit Steinkäuzen. *Die Heimat Verein zur Pflege der Natur und Landeskunde in Schleswig-Holstein* **57**: 313–315.
16. Bradshaw, G. W. 1901. Little Owl at Henley. *Zoologist* **5**: 476.
17. Brandenburg, E. 2004. Friese Steenuil op leeftijd. *Athene Nieuwsbrief STONE* **9**: 19.
18. Brands, R. 1980. Beiträge zur Brutbiologie des Steinkauzes (*Athene noctua*). Masters thesis, Universität Köln.
19. Braun, M. & Simon, L. 1981. Rote Liste der bestandsgefährdeten Vogelarten in Rheinland-Pfalz. Stand 1.1.1981. *Naturschutz und Ornithologie in Rheinland-Pfalz* **2**: 61–70.
20. Braun, M. & Simon, L. 1983. Rote Liste der bestandsgefährdeten Vogelarten in Rheinland-Pfalz. Stand 31.8.1983. *Naturschutz und Ornithologie in Rheinland-Pfalz* **2**: 583–592.
21. Braun, M., Keil, W., Simon, L. & Viertel, K. 1985. Vögel in Rote Liste der bestandsgefährdeten Wirbeltiere in Rheinland-Pfalz. Ministerium für Soziales, Gesundheit und Umwelt Rheinland-Pfalz. [Little Owl data on p. 24.]
22. Brehm, C. L. 1823. Lehrbuch der Naturgeschichte aller europäischen Vögel. A. Schmidt. Jena. [Little Owl data on pp. 73–74.]
23. Brehm, C. L. 1855. Steinkauz, *Athene*, Boje. Vogelfang. [Little Owl data on pp. 36–38.]
24. Brehm, C. L. 1858. Die Steinkäuze. *Athene*, Boje, *Noctua*, Cuv. (*Strix psilodactyla*, L. *Strix passerina*, Gm., L. *Strix noctua*, Retz.). *Naumannia* **8**: 221–230.
25. Brentjes, B. 1967. Zur Rolle der Eulen im Alten Orient. *Beiträge zur Vogelkunde* **13**: 72–80.
26. Bretagnolles, V., Bavoux, C., Burneleau, G. & Van Nieuwenhuyse, D. 2001. Abondance et distribution des Chevêches d'Athéna: approche méthodologique pour des enquêtes à grande échelle en plaine céréalière. *In* Génot, J.-C., Lapios, J.-M., Lecomte, P. & Leigh, R. S. (eds). *Ciconia* **25**: 173–184.

227. Breuer, W. 1996. Rechtsschutz für Steinkäuze von der Veranwortung der Städte und Gemeinden für Natur und Landschaft in der Bauleitplanung. *Eulen Rundblick* **44**: 3–15.
228. Breuer, W. 1998. Kaüze, Klüngel, kommunale Nachsicht. Vom Wohl und Wehe der Steinkäuze in Flächennützungensplänen. *Eulen-Rundblick* **47**: 3–10.
229. Breuer, W. 1998. Berücksichtigung von Steinkauzehabitaten in der Flächennutzungsplanung am Beispiel von drei nordrhein-westfälischen Gemeinden. *Natur und Landschaft* **73**: 175–180.
230. Broekhuizen, S., Van' t Hoff, S., Jansen, L. A. & Niewold, F. J. J. 1980. Application of radio tracking in wildlife research in the Netherlands. *In* Amlaner, C. J. & Macdonald, D. W. (eds). *A Handbook on Biotelemetry and Radio Tracking*. Oxford: Pergamon Press. [Little Owl data on pp. 65–84.]
231. Broggi, M. F. & Willi, G. 1985. Rote Liste der gefährdeten und seltenen Vogelarten des Füstentums Liechtenstein. *Naturkundliche Forschung im Füstentum Liechtenstein* **5**: 13.
232. Brouwer, G. A. 1938. Over het bidden van *Athene noctua vidalii* A. E. Brehm. *Ardea* **27**: 260–261.
233. Brown, L. H. 1977. The status of, and threats to, diurnal and nocturnal birds of prey in east Africa and Ethiopia. International Council for Bird Preservation World Conference on Birds of Prey. Vienna 1976 p. 27.
234. Brücher, H. & Kostrzewa, A. 1984. Rasterkartierung der Greifvögel und Eulen im Naturpark Kottenforst-Ville und in den angrenzenden Gebieten. *Charadrius* **20**: 130–136.
235. Bruderer, B. & Thönen, W. 1977. Rote Liste der gefährdeten und seltenen Vogelarten der Schweiz. *Ornithologische Beobachter* **3**: 20.
236. Brugiere, D. 1988. Evolution de l'avifaune reproductrice des départements de l'Allier, du Puy-de-Dome de la Haute-Loire, du Cantal et de la Lozère au cours des quinze dernières années. Mise au point sur cette avifaune. *Grand-Duc* **33**: 47.
237. Brugiere, D. & Duval, J. 1983. Annales ornithologiques du Parc National des Cévennes 1970–1983. *Grand-Duc* **23**: 26.
238. Brüll, H. 1968. *Greifvögel und Eulen Mitteleuropas*. Minden.
239. Büchi, O. 1952. Forte mortalité de rapaces nocturnes dans le canton de Fribourg. *Nos Oiseaux* **21**: 224.
240. Bugnet, C. 2003. La Chouette Chevêche d'Athéna et les Vergers Traditionnels. Stage BTS. GPN.
241. Buker, J. B., De Wit, J. N. & Van Zuilen, W. A. 1984. Forse prooien van Steenuil *Athene noctua*. *Limosa* **57**: 118.
242. Bultot, J. 1990. Construction d'un nouveau type de nichoir pour chouette chevêche (*Athene noctua*) et premiers résultats. *Bulletin à l'Usage du Bagueur Ornithologique* **5**: 56–59.
243. Bultot, J. 1991. Nouveau type de nichoir pour Chouette Chevêche. *L'Homme et l'Oiseau* **29**: 258–262.
244. Bultot, J. 1996. Opération Chevêche. *L'Homme et l'Oiseau* **34**: 101–107.
245. Bultot, J. 1996. Gestion de l'habitat et protection de la Chouette chevêche. Les Aménagements écologique et les rapaces. Septembre 14–15, 1995. AFIE PNRVN. [Little Owl data on pp. 45–52.]
246. Bultot, J. 2000. *Noctua*. qu'est-ce que c'est que ça? *Noctuathene* **4**: 19.
247. Bultot, J. 2000. Le coin brico. Nichoirs pour chouette chevêche, type caisse à vin. *Noctuathene* **4**: 20–21.
248. Bultot, J., Marie, P. & Van Nieuwenhuyse, D. 2001. Population dynamics of Little Owl *Athene noctua* in Wallonia and its driving forces. Evidence for density-dependence. *In* Van Nieuwenhuyse, D., Leysen, M. & Leysen, K. (eds). *Oriolus* **67**: 110–125.
249. Bünger, 1893. Ein 18 Jahre in Gefangenschaft lebender Steinkauz. *Ornithologische Monatsschrift* **18**: 472.
250. Bunting, W. 1952. The Little Owl. *Bird Notes and News* **25**: 176–180.
251. Burbach, K. 1997. *In* Hessische Gesellschaft für Ornithologie und Naturschutz (ed.). Steinkauz – *Athene noctua*. *Avifauna von Hessen* **3**: 1–16.
252. Burget, J.-P. & Jenn, J.-P. 1984. La faune en danger. *Lien Ornithologique d'Alsace* **40**: 7.
253. Burton, J. A. 1970 Little Owl. *In* Gooders, J. (ed.). *Birds of the World*. London: [Little Owl data on pp. 1336–1339.]
254. Burton, J. A. 1985. *Eulen der Welt*. Melsungen.
255. Burton, P. 1993. Birds of prey on farmland. The 1993 breeding season. *The Raptor* **21**: 34–36.
256. Burton, P. J. K. 1983. Little Owl raiding Starling's nest. *British Birds* **76**: 314–315.
257. Busse, H. 1983. Der Steinkauz. *Falke* **30**: 34–35.
258. Bux, M. & Rizzi, V. 2005. *In* Mastrorilli, M., Nappi, A. & Barattieri, M. (eds). Dieta della Civetta, *Athene noctua*, in una salina dell'Italia Meridionale. Atti I Convegno italiano sulla Civetta. Osio Sotto (BG), 2 Marzo 2004. Gruppo Italiano Civette. pp. 50–52.
259. Buxton, E. J. M. 1947. Little Owl taking grass-snake. *British Birds* **40**: 55.
260. Buxton, P. A. 1907. Spread of the Little Owl in Herts. *Zoologist* **11**: 430.
261. Calderon, J. 1977. El papel de la Perdiz roja (*Alectoris rufa*) en la dieta de los predatores ibericos. Doñana *Acta Vertebrata* **4**: 77–78.
262. Calvet, A. 2002. La chouette chevêche (*Athene noctua* Scop.) dans la vallée du Thore: évaluation de la population. Propositios d'actions de conservation. Rapport. Parc naturel régional du Haut-Languedoc.

263. Calvi, G., De Carli, E., Buvoli, L. & Fornasari, L. 2005. La Civetta nel progetto MITO2000 dal 2000 al 2002. *In* Mastrorilli, M., Nappi, A. & Barattieri, M. (eds). Atti I Convegno italiano sulla Civetta. Osio Sotto (BG), 21 Marzo 2004. Gruppo Italiano Civette. pp. 13–17.
264. Camarda, A., De Paolis, S. & Di Modugno, D. 2003. Aspetti di patologie oculari in rapaci diurni e notturni. *Avocetta* **27**: 149.
265. Camberlein, G. & Petit, J. 1992. Statut de la Chouette chevêche dans les Cotes d'Armor. Premier bilan Déc. 1991. *Le Fou* **26**: 25–31.
266. Cameron-Brown, F. A. 1975. Little Owl behaviour. *Gloucestershire Naturalists' Society Journal* **26**: 83–84.
267. Camuzat, S. & Faucon, L. 1997. Recensement et cartographie des chouettes chevêches dans le Boulonnais. Maîtrise de Biologie des populations et des écosystèmes. Rapport. Parc naturel régional du Boulonnais. Université des Sciences et Techniques de Lille.
268. Canteneur, R. 1964. Les oiseaux sauvages victimes de la circulation routière dans l'Est de la France. *L'Oiseau et Revue Française d'Ornithologie* **34**: 254.
269. Capanna, E., Civitelli, M. V. & Martinico, E. 1987. I cromosomi degli uccelli. Citotassonomia ed evoluzione cariotipica. *Avocetta* **11**: 101–123.
270. Capizzi, D. & Luiselli, L. 1995. Comparison of the trophic niche of four sympatric owls (*Asio otus, Athene noctua, Strix aluco* and *Tyto alba*) in Mediterranean Central Italy. *Ecologia Mediterranea* **21**: 13–20.
271. Capizzi, D. & Luiselli, L. 1996. Feeding relationships and competitive interactions between phylogenetically unrelated predators (owl and snakes). *Acta Oecologica* **17**: 265–284.
272. Carabantes, F. 1970. The white Little Owls, *Athene noctua*, at Jerez Zoo. *International Zoo Yearbook* **10**: 33.
273. Carmona, P. 1981. Mettez une Chevêche dans votre nichoir. *Lien Ornithologique d'Alsace* **33**: 3–4.
274. Casalini, R. 2005. Attivita di mobbing nei confronti di *Athene noctua* (Scopoli, 1769) in un ambiente mediterraneo. Atti I Convegno italiano sulla Civetta. Osio Sotto (BG), 21 Marzo 2004. Gruppo Italiano Civette. 41–44.
275. Casini, G., Porciatti, V., Fontanesi, G. & Bagnoli, P. 1992. Wulst efferents in the Little Owl *Athene noctua* an investigation of projections to the optic tectum. *Brain Behavior and Evolution* **39**: 101–115.
276. Casini, L. 1987. *In* Foschi, U. F. & Gellini, S. *Atlante degli uccelli nidificanti in provincia di Forli (1982–1986)*. Maggioli Editori. Rimini. [Little Owl data on p. 68.]
277. Casini, L. 1993. *In* Meschini, E. & Frugis, S. *Atalante degli uccelli nidificante in Italia. Supplemento alle Ricerche di Biologia della Selvaggina* **20**: 148.
278. Castaldi, A. & Guerrieri, G. 2001. Rete viaria e mortalità di *Strigiformes* nella Riserva naturale Statale del Litorale Romano. *Uccelli d'Italia* **26**: 59–67.
279. Castellucci, M. & Zavalloni, D. 1989. Predazione di barbagianni, *Tyto alba*, su civetta, *Athene noctua*. *Rivista Italiana di Ornithologia* **59**: 282–283.
280. Caterini, F. 1938. Catture rare ed interessanti. *Rivista Italiana di Ornithologia* **8**: 87–94.
281. Caupenne, M. & Prevost, O. 1983. Synthèse des observations du 1.8.81 au 31.7.82. Outarde. *Bulletin du Groupe Ornithologique de la Vienne* **15**: 77.
282. Centili, D. 1995. Dati preliminari sulla Civetta *Athene noctua* in un'area dei Monti della Tolfa (Roma). *Avocetta* **19**: 113.
283. Centili, D. 1996. Censimento, distribuzione e habitat della Civetta *Athene noctua* in un'area dei monti della Tolfa. Masters thesis, Universita "La Sapienza". Roma.
284. Centili, D. 2001. A Little Owl population and its relationships with humans in central Italy. *In* Génot, J.-C., Lapios, J.-M., Lecomte, P. & Leigh, R. S. (eds). *Ciconia* **25**: 153–158.
285. Centili, D. 2001. Playback and Little Owls *Athene noctua*: preliminary results and considerations. *In* Van Nieuwenhuyse, D., Leysen, M. & Leysen, K. (eds). *Oriolus* **67**: 88–93.
286. Centre D'Etudes Ornithologiques D'Alsace, 1989. Livre Rouge des Oiseaux nicheurs d'Alsace. *Ciconia* (spécial) **13**: 99–102.
287. Centre Ornithologique Auvergne, 1983. *Atlas des Oiseaux Nicheurs du Département de l'Allier 1972–1982*. [Little Owl data on pp. 100–101.]
288. Centre Ornithologique Champagne-Ardenne, 1991. *Les Oiseaux de Champagne-Ardenne*. COCA. Bar-sur-Aube. [Little Owl data on pp. 172–173.]
289. Centre Ornithologique Rhone-Alpes, 1986. Synthèse des observations de l'année 1983. *Niverolle* **10**: 61.
290. Cesaris, C. 1988. Populazioni di Alloco *Strix aluco* e di Civetta *Athene noctua* in un'area del Parco Lombardo della Valle del Tiano. *Avocetta* **12**: 115–118.
291. Ceska, V. 1978. Wie lange können Eulen hungern? *Nationalpark Sonderheft* **5**: 17–19.
292. Ceska, V. 1980. Untersuchungen zu Nahrungsverbrauch, Nahrungsnutzung und Energiehaushalt bei Eulen. *Journal of Ornithology* **121**: 186–199.
293. Chaline, J., Baudvin, H., Jammot, D. & Saint Girons, M.-C. 1974. *Les Proies des Rapaces*. Paris: Doin. [Little Owl data on p. 17].

294. Chappell, B. M. A. 1950. Little Owl attacking Jackdaw. *British Birds* **43**: 123.
295. Chappuis, C. 1979. Emissions vocales nocturnes des oiseaux d'Europe. *Alauda* **47**: 285.
296. Charter, M., Leshem, Y., Izhaki, I., Guershon, M. & Kiat, Y. 2006. The diet of the Little Owl, *Athene noctua*, in Israel. *Zoology in the Middle East* **39**: 31–40.
297. Charteris, G. 1923. Ferret seizing a Little Owl. *British Birds* **16**: 309.
298. Chartier, A. 2001. La Chouette chevêche (*Athene noctua*) dans le Parc Naturel Régional des Marais du Cotentin et du Bessin (Manche-Calvados). Groupe Ornithologique Normand. PNR des Marais du Cotentin et du Bessin. DIREN Basse Normandie.
299. Chavigny, D. 1984. Avifaune en Sologne du Loiret. Région de Vannes-sur-Cosson. *Naturalistes Orléanais* **3**: 25.
300. Cheylan, G. 1974. Biogéographie d'une montagne méditerranéenne: la Sainte Victoire (Bouches-du-Rhône). *Alauda* **42**: 63.
301. Cheylan, G. 1986. Inventaire ornithologique préliminaire des îles de Marseille. *Faune de Provence. Bulletin du Conservatoire et Etudes des Ecosystèmes de Provence* **7**: 30–38.
302. Choussy, M. 1971. *Les rapaces*. Parc Naturel Régional des Volcans d'Auvergne. Coll. Découverte de la Nature. ARPEGE. Clermont-Ferrand. [Little Owl data on pp. 48–50.]
303. Christie, H. H. V. 1931. Buzzard killing Little Owl. *British Birds* **24**: 367.
304. Ciani, C. 1988. Predazione di Civetta, *Athene noctua*, su Rondone, *Apus apus*. *Rivista Italiana di Ornithologia* **58**: 87–88.
305. Clark, I. 1926. Little Owl. *Trans. Natural History Society Northumberland. Durham. Newcastle-upon-Tyne* **6**: 241–242.
306. Clark, R. J., Smith, D. G. & Kelso, L. H. 1978. *Working Bibliography of Owls of the World*. Washington DC: National Wildlife Federation.
307. Clark, S. V. 1912. Little Owl breeding in Middlesex and Sussex. *British Birds* **6**: 18–19.
308. Clech, D. 1993. La Chouette chevêche *Athene noctua* en Bretagne. *Ar Vran* **4**: 5–34.
309. Clech, D. 1994. La Chouette chevêche *Athene noctua* en Bretagne. Deuxième partie. *Ar Vran* **5**: 10–37.
310. Clech, D. 1994. Une régression inquiétante. La Chouette chevêche en Bretagne. *Revue du Fonds d'Intervention pour les Rapaces* **25**: 10–11.
311. Clech, D. 1996. La chouette chevêche: chronique d'une mort annoncée? *Penn ar Bed* **163**: 31–43.
312. Clech, D. 1997. *In* Groupe Ornithologique Breton, (eds). *Les Oiseaux Nicheurs de Bretagne 1980/1985* Groupe Ornithologique Breton. [Little Owl data on p. 149.]
313. Clech, D. 2001. Impact de la circulation routière sur la Chevêche d'Athéna, *Athene noctua*, par l'étude de la localisation de ses sites de reproduction. *Alauda* **69**: 255–260.
314. Clech, D. 2001. Etude d'une population de Chevêche d'Athéna dans le Haut-Léon (Bretagne–France). *In* Génot, J.-C., Lapios, J.-M., Lecomte, P. & Leigh, R. S. (eds). *Ciconia* **25**: 119–128.
315. Cleine, H. 1976. Onbekend gedrag van Steenuil. *Vogeljaar* **24**: 326.
316. Clever, K. H. 1981. Massnahmen zur Erhaltung des Steinkauzes (*Athene noctua*) in Hessen. *Vogel und Umwelt* **1**: 302–306.
317. Clitherow, C. E. S. 1911. Little Owls in Lincolnshire. *British Birds* **5**: 51.
318. Clotuche, E. 1981. Chronique ornithologique Décembre 1980, Janvier et Février 1981. *Aves* **18**: 165.
319. Clous, G., Clous, P. & Paineau, G. 1986. Compte rendu des observations du 1 Mars au 31 Août 1986. *Bulletin du Groupe Sarthois Ornithologique* **19**: 26.
320. Čmelík, T. 1995. Několik poznámek k hnízdění sýčka obecného (*Athene noctua*) na Kyjovsku. [On the Little Owl (*Athene noctua*) breeding in the Kyjov region.] *Crex* **6**: 33.
321. Coburn, F. 1909. Little Owl in Warwickshire and Worcestershire. *British Birds* **2**: 344.
322. Codd, R. B. 1950. Dusting of owls. *British Birds* **43**: 338.
323. Cohen, E. 1960. Little Owl taking Nuthatch. *British Birds* **53**: 574.
324. Collectif D'Associations "Cheveche-Ajoie" & Office Des Eaux Et De La Protection De La Nature, 2003. Plan d'action pour la Chevêche d'Athéna en Ajoie (JU). Report.
325. Collinge, W. E. 1919. The Little Owl. *Bird Notes and News* **8**: 51.
326. Collinge, W. E. 1920. On the proposed new subspecies of the Little Owl (*Carine noctua* Scopoli). *Scottish Naturalist* **1920**: 69–70.
327. Collinge, W. E. 1922. The food and feeding habits of the Little Owl. *Journal Min. Agric.* **28**: 1022–1031.
328. Collinge, W. E. 1922. The Little Owl. *Bird Notes and News* **10**: 21–23.
329. Collinge, W. E. 1923. A local investigation of the food of the Little Owl. *Journal Ministery of Agriculture* **29**: 750–752.
330. Commency, X. & Sueur, F. 1983. Avifaune de la Baie de Somme et de la plaine maritime Picarde. GEPOP Amiens. [Little Owl data on pp. 154–155.]
331. Congreve, W. M. 1923. Little Owl in Pembrokeshire. *British Birds* **17**: 22.

332. Connolly, C. 1964. Birds of prey in Ireland – their status and control. ICBP Working Conference on Birds of Prey and Owls. Caen. [Little Owl data on p. 119.]
333. Conrad, B. 1976. Die Belastung der freilebenden Vogelwelt der Bundesrepublik Deutschland mit Chlorierten Kohlenwasserstoffen und PCB und deren mögliche Auswirkungen. Ph.D. thesis.
334. Conrad, B. 1977. *Die Giftbelastung der Vogelwelt Deutschlands*. Kilda-Verlag Greven.
335. Conrad, B. 1981. Zur Situation der Pestizid belastung bei Greifvögeln und Eulen in der Bundesrepublik Deutschland. *Ökologie der Vögel* **3**: 161–167.
336. Contoli, L., Aloise, G. & Filipucci, M. G. 1988. Sulla diversificazione trofica di Barbagianni *Tyto alba* e Civetta *Athene noctua* in rapporto al livello diagnostico delle preda. *Avocetta* **12**: 21–30.
337. Conzemius, T. 1993. Wetteharte Steinkäuze im luxemburgischen Ösling. *Eulen Rundblick* **39**: 12.
338. Cooke, A. S., Bell, A. A. & Haas, M. B. 1982. Predatory birds, pesticides and pollution. *The Cambrian News*. Aberystwyth: Institute of Terrestrial Ecology. [Little Owl data on pp. 53–54.]
339. Cooper, A. D. 1898. Little Owl and Great Spotted Woodpecker. *Nat. Journal Guide* **7**: 181.
340. Coppée, J.-L., Bultot, J. & Hanus, B. 1995. Population et reproduction de la Chouette chevêche (*Athene noctua*) en Hainaut. Protection et restauration de ses habitats. *Aves* **32**: 73–99.
341. Cora Drome, 2003. *Oiseaux de la Drôme. Atlas des Oiseaux Nicheurs de la Drôme*. Cora Drome. [Little Owl data on p. 100.]
342. Cornulier, T. & Bretagnolle, V. 2006. Assessing the influence of environmental heterogeneity on bird spacing patterns: a case study with two raptors. *Ecography* **29**: 240–250.
343. Corradetti, A. & Neri, I. 1956. Plasmodium subpraecox Grassi e Feletti, 1892 ceppo di Plasmodium precox Grassi e Faletti, 1890 adattato a vivere nella civetta *Carine noctua*. *Rivista di Parasitologia* **17**: 165–169.
344. Corradetti, A., Cavallucci, S. & Crescenzi, R. 1941. Incidenza degli ematozoi delle civette nella campagna romana all'inizio dell'autunno. *Rivista di Parasitologia* **5**: 251–252.
345. Corti, U. A. 1962. *Juravögel. Die Brutvögel des schweizerischen Jura*. Verlag Bischofberger. Chur. [Little Owl data on pp. 97–98.]
346. Cosson, F. 1985. Some recent data of the raptors of Rhodes (Greece). *Bulletin of World Working Group on Birds of Prey* **2**: 59.
347. Costaglia, A., Lepore, G., De Falco, G. & Cecio, A. 1981. Neuroistologia comparata dell'esofago in uccelli granivori onnivori e carnivori. *Atti della Societa Italiana die Scienze Veterinarie* **35**: 323–324.
348. Courtois, J.-M. 1988. Deux lépidoptères commensaux de la Chouette chevêche (*Athene noctua*). *Ciconia* **12**: 177.
349. Coward, T. A. 1912. A note on the Little Owl: *Carine noctua* (Scop.) and its food. *Memoirs of the Literary and Philosophical Society of Manchester* **56**: 1–11.
350. Cox, A. H. M. 1912. Little Owl in south Devon. *British Birds* **5**: 333.
351. Craighed, J. C. & Craighed, F. C. 1969. *Hawks, Owls and Wildlife*. New York.
352. Cramp, S. 1963. Toxic chemicals and birds of prey. *British Birds* **56**: 129–130.
353. Cramp, S. 1964. Predators and toxic chemicals. ICBP Working Conference on Birds of Prey and Owls. Caen. [Little Owl data on pp. 53–58.]
354. Cramp, S. (ed.) 1985. *Handbook of the Birds of Europe, the Middle East and North Africa. The Birds of the Western Palearctic*. Vol. 4. New York: Oxford University Press. [Little Owl data on pp. 514–525.]
355. Crelier, D. 1999. Aperçu du projet régional de sauvegarde de la Chevêche d'Athéna *Athene noctua*. *Héron* **199**: 9–10.
356. Crespon, A. & Chretienne, M. 2001. La Chouette chevêche sur le territoire du Parc naturel régional du Perche. Bilan des prospections 1999–2000 et campagne 2001 de pose de nichoirs. Rapport. Parc naturel régional du Perche.
357. Cresson, J. 1840. *Ornithologie du Gard et des Pays Inconvoisins*. Nismes. [Little Owl data on pp. 53–54.]
358. Csermely, D., Casagrande, S. & Sponza, S. 2002. Adaptative details in the comparison of predatory behaviour of four owl species. *Italian Journal of Zoology* **69**: 239–243.
359. Cugnasse, J.-M. 1984. Contribution à l'étude des rapaces du département du Tarn. *Bulletin de l'Association Régionale Ornithologique du Midi et des Pyrénées* **8**: 41.
360. Cugnasse, J.-M. 1985. Prédation d'un poussin de Chouette effraye *Tyto alba* par une fouine *Martes foina*. *Guepier* **2**: 88–91.
361. Cummings, B. F. 1913. Little Owl in north Devon. *Zoologist* **17**: 37.
362. Cungs, J. & Waltener, M. 1987. Obstanbau gestern und heute. *Regulus* **2**: 36–40.
363. D'Agostino, M. 1995. Bilan de la population de la Chouette chevêche (*Athene noctua*) en 1995, au sein du Parc naturel régional de Lorraine. LPO. PNRL.
364. Dalbeck, L., Bergerhausen, W. & Hachtel, M. 1999. Habitatpräferenzen des Steinkauzes *Athene noctua* Scopoli, 1769 im ortsnahen Grünland. *Charadrius* **35**: 100–115.
365. Dalbeck, L., Bergerhausen, W. & Hachtel, M. 1999. Habitatpräferenzen des Steinkauzes (*Athene noctua*) im ortsnahen Grünland. *Eulen Rundblick* **48/49**: 3–15.

366. Danko, S. 1985. Report on activities of the group for research and protection of birds of prey and owls in Czechoslovakia in 1982. *Bulletin of World Working Group on Birds of Prey* **2**: 48.
367. Danko, S. 1985. Report on activities of the group for research and protection of birds of prey and owls in Czechoslovakia in 1983. *Bulletin of World Working Group on Birds of Prey* **2**: 50.
368. Danko, S. 1986. Report on activities of the group for research and protection of birds of prey and owls in Czechoslovakia in 1986. *Buteo* **1**: 22.
369. Danko, S. 1989. Report on activities of the group for research and protection of birds of prey and owls in Czechoslovakia in 1989. *Buteo* **4**: 1–28.
370. Danko, S., Chavko, J. & Karaska, D. 1995. Report on the activity of the Group on Protection of Birds of Prey and Owls in the Slovak Republic in 1993. *Buteo* **7**: 109–121.
371. Danko, S., Chavko, J. & Karaska, D. 1995. Report on the activity of the Group on Protection of Birds of Prey and Owls in the Slovak Republic in 1994. *Buteo* **7**: 132–148.
372. Danko, S., Darolova, A. & Kristin, A. 2002. *Bird Distribution in Slovakia*. Bratislava: Veda, vydavatelstv SAV.
373. Dathe, H. 1966. Steinkauz, *Athene noctua*, Brutvogel auf Kreta. *Beiträge zur Vogelkunde* **12**: 116.
374. Daufresne, T., Lecomte, P. & Lapios, J.-M. 1998. La Chevêche d'Athéna en France. Bilan, enquête, analyse et propositions pour sa conservation.
375. Davies, W. 1916. Little Owl in Staffordshire. *British Birds* **10**: 120.
376. Daws, W. 1915. Little Owl in north Nottinghamshire. *Zoologist* **19**: 395.
377. Dawson-Smith, F. 1913. British owls. *Bird Notes and News* **4**: 169.
378. De Jong, J. 2002. De Steenuil *Athene noctua* in Fryslân: plan van aanpak.
379. De Juana, E. 1980. *Atlas Ornitologico de la Rioja*. Servicio de Cultura de la Exema Diputacion Provincial. Logiono. [Little Owl data on pp. 266–268.]
380. De La Hoz, M. 1982. Algunos datos sobre la alimentacion del Mochuelo comun (*Athene noctua*) en Asturias *Boletín de Ciencias Naturales, Instituto de Estudios Asturianos* **29**: 113–120.
381. De Ruffray, P. 1981. Enquête sur la mortalité des oiseaux de proie en Alsace. *Lien Ornithologique d'Alsace* **33**: 32.
382. De Smet, A. 2002. Onderzoek naar het voorkomen van de Steenuil in West-Zeeuws-Vlaanderen 2000–2001 Natuurbes chermingsvereniging 't Duumpje.
383. De Tiersant, M.-P., Majorel, L., Loose, D. & Gagne, A. 2001. La situation de la Chevêche d'Athéna en Isère *Nouv'ailes Cora Isère* **134**: 7–12.
384. Deboulonne, A. 1985. Connaître et protéger la Chouette chevêche. *Héron* **1**: 122–124.
385. Deboulonne, A. 2000. Un exemple de nichoir pour la Chevêche d'Athéna *Athene noctua*. *Héron* **33**: 196–198
386. Debout, G. 1984. Chronique ornithologique. Septembre 1982–Février 1983. *Cormoran* **26**: 80.
387. Degland, C. D. & Gerbe, Z. 1867. *Ornithologie Européenne ou Catalogue Descriptif, Analytique et Raisonné des Oiseaux Observés en Europe*. Baillière. Paris. [Little Owl data on pp. 121–123.]
388. Delafosse, W. 1954/1955. Les Oiseaux du Pays Messin. *Mém. Acad. Nat. Metz* **3**: 129.
389. Delamain, J. 1938. *Portraits d'Oiseaux*. Paris: Stock. [Little Owl data on pp. 59–62.]
390. Delamain, J. 1980. *Pourquoi les Oiseaux Chantent*. Paris: Stock/Nature. [Little Owl data on pp. 10–40.]
391. Delaporte, P. 1987. Un nichoir pour Chevêche. *Oiseau Magazine* **6**: 28–29.
392. Delibes, M., Brunet-Lecomte, P. & Manez, M. 1983. Datos sobre la alimentacion de la Lechuza comun (*Tyto alba*), el Buho Chico (*Asio otus*) y el Mochuelo (*Athene noctua*) en una misma localidad de Castilla la Vieja *Ardeola* **30**: 57–63.
393. Della Pietà, C. 1997. L'astuta Civetta. *Airone* **194**: 153–154.
394. Delmee, E. 1988 *Athene noctua*. *In* Devillers, P., Roggeman, W., Tricot, J. *et al.* (eds). *Atlas des Oiseaux Nicheurs de Belgique*. Bruxelles: Institut Royal des Sciences Naturelles de Belgique. [Little Owl data on p. 172.]
395. Delogu, M., Delgado Montero, M. L. & De Marco, M. 2000. Epatosplenite infettiva degli Strigiformi (HSIS.) aspetti istopatologici ed ecografici a confronto. *Selezione Veterinaria* **8**: 857–860.
396. Dementiew, G. 1931. Sur quelques points de systématique et de nomenclature. *Alauda* **3**: 258–259.
397. Dementiew, G. & Gladkov, N. A. 1951. *Ptitsy Sovetskogo Soyusa*. [*Birds of the Soviet Union*.] 1. Moskva [Little Owl data on pp. 440–447.]
398. Demyanchik, V. T. & Ol'Gomets, A. I. 1992. Statsialnoe raspredelenie domovogo sycha na zapade Belarus [Microhabitat distribution of the Little Owl in the west of Belarus]. Nauka miru, mir nauke: Tez. i soobshch nauchn. prakt. konf. Brestskgo ped. in ta [Science for peace, peace for science: Abstracts and reports of scient. prakt. conf. of Brest Ped. Inst.] Brest. [Little Owl data on p. 17.]
399. Denac, D., Marčič, M., Radolič, P. & Tomažič, A. 2002. Sove v cerkvah, gradovih in drugih objektih na obmocju Vipavske doline [Owls in churches, castles and other buildings in the Vipava valley and the Kars (SW Slovenia)]. *Acrocephalus* **23**: 91–95.
400. Dent, G. 1940. Habits of the Little Owl in Essex. *Essex Naturalist* **26**: 247–248.

401. Deppe, H. J. 1975. Zur Ernährung durchziehender Greifvögel und Eulen im Naturschutzgebiet Geltinger-Birk (Landkreis Flensburg). *Ornithologische Mitteilungen* **27**: 202.
402. Deroussen, F. 1984. Une saison de nidification au bois de Vincennes. *Passer* **21**: 195.
403. Desmots, D. 1988. Prédation de Moineaux domestiques (*Passer domesticus*) par une Chouette chevêche (*Athene noctua*). *Biotopes* **54**(6): 44–45.
404. Destre, R. 2000. *Faune Sauvage de Lozère. Les Vertébrés*. Association lozérienne pour l'étude et la protection de l'environnement. [Little Owl data on p. 132.]
405. Deutsche Sektion Des Internationalen Rates Fur Vogelschutz, 1972. Die in der Bundesrepublik Deutschland gefährdeten Vogelarten (Rote Liste). Stand 31.12.1972. *Berichte der Deutschen Sektion des Internationalen Rates für Vogelschutz* **12**: 8–15.
406. Deutsche Sektion Des Internationalen Rates Fur Vogelschutz, 1974. Die in der Bundesrepublik Deutschland gefährdeten Vogelarten (Rote Liste). Stand 30.11.1974. *Berichte der Deutschen Sektion des Internationalen Rates für Vogelschutz* **14**: 7–19.
407. Deutsche Sektion Des Internationalen Rates Fur Vogelschutz, 1975. Die in der Bundesrepublik Deutschland gefährdeten Vogelarten (Rote Liste). *Vogelwelt* **96**: 193–198.
408. Deutsche Sektion Des Internationalen Rates Fur Vogelschutz, 1976. Rote Liste der in der Bundesrepublik Deutschland und in Westberlin gefährdeten Vogelarten (4. Fassung Stand 1.1.1977). *Berichte der Deutschen Sektion des Internationalen Rates für Vogelschutz* **16**: 12–19.
409. Deutsche Sektion Des Internationalen Rates Fur Vogelschutz, 1981. Rote Liste der in der Bundesrepublik Deutschland und in Berlin (West) gefährdeten Vogelarten Stand 1.1.1982. *Berichte der Deutschen Sektion des Internationalen Rates für Vogelschutz* **21**: 15–30.
410. Deutscher Bund Fur Vogelschutz, 1984. Steinkauzbestand in der BRD. *DBV AG zum Schutz bedrohter. Eulen Informationsblatt* **21**: 2.
411. Deutscher Bund Fur Vogelschutz, 1988. Ergebnisse der Bestandserhebung Steinkauz (*Athene noctua*). *DBV AG zum Schutz bedrohter. Eulen Informationsblatt* **29**: 3.
412. Dewitte, T. 1996. Enquête sur un hôte prestigieux de nos vieux vergers. La Chouette Chevêche *Athene noctua*. *Le Viroinvol* **3**: 29–36.
413. Dewitte, T. 2000. Un hôte prestigieux de nos vieux vergers. La chouette chevêche *Athene noctua*. *Noctuathene* **4**: 13–18.
414. Dewulf, H. 1998. Relations prédateur-proies-paysage chez la Chevêche d'Athéna (*Athene noctua* Scop.): Etude de faisabilité d'une évaluation de la richesse et de la quantité des proies principales de l'espèce dans différents types paysagers. Mémoire d'études. ENESAD.
415. Di Pietro, A., Mastrorilli, M., Pavesi, C. & Sangiovanni, M. 1998. Analisi e considerazioni sui rapaci recuperati dal WWF di Crema negli anni 1996 e 1997. *Pianura-Science e storia dell'ambiente padano* **10**: 19–26.
416. Dicerbo, E. 1962. Little Owl in Dumfriesshire. *Scottish Birds* **2**: 248.
417. Dicerbo, E. 1963. Little Owl in Dumfriesshire. *Scottish Birds* **3**: 314.
418. Dickinson, H. 1969. Little Owl in Lanarkshire. *Scottish Birds* **5**: 390.
419. Diehl, O. 1988. Lebensraum Obstwiese-Gefährdung und Massnahmen zur Erhaltung. *Vogelwelt* **109**: 141–144.
420. Dietrich, F. 1928. *Hamburgs Vogelwelt*. [Little Owl data on pp. 256–257.]
421. Dietzen, C., Folz, H.-G. & Henss, E. 2003. Ornithologische Sammelbericht 2003 für Rheinland-Pfalz. *Fauna Flora Rheinland-Pfalz* **32**: 137.
422. Dinetti, M. & Fraissinet, M. 2001. *Ornitologia Urbana*. [Little Owl data on pp. 158–159.]
423. Dini, G. 1937. Dalla invasione di Civette (*Carine noctua*) nidificanti, alla completa sparizione delle medesime. *Rivista Italiana di Ornithologia* **7**: 301.
424. Divilek, T. 1997. K problematice ochrany sýcka obecného (*Athene noctua*). [The Little Owl (*Athene noctua*) protection]. *Crex* **9**: 27–39.
425. Dobie, W. H. 1900. The Little Owl in north Wales. *Zoologist* **1900**: 556.
426. Dobie, W. H. 1917. Little Owl in Cheshire. *British Birds* **10**: 271–272.
427. Dobinson, H. M. & Richards, A. J. 1964. The effects of the severe winter of 1962/63 on birds in Britain. *British Birds* **57**: 391–392.
428. Doby, J. M. & Boisseau-Lebreuil, M. T. 1971. Rôle possible des animaux suavages carnassiers (Oiseaux et mammifères) dans la dissémination mécanique de l'adiaspiromycose par *Emmonsia crescens* Emmons et Jellison, 1960, dans la nature. *Compte Rendu de Séance de Biologie* **165**: 1119–1122.
429. Dombrowski, A., Fronczak, J., Kowalski, M. & Lippoman, T. 1991. Liczebnosc i preferencje srodowiskowe sow *Strigiformes* na terenach rolniczych Niziny Mazowieckiej [Population density and habitat preferences of owls *Strigiformes* on agricultural areas of Marowsz e Lowland (Central Poland)]. *Acta Ornitholecologica* **26**: 39–53.

430. Dooly, T. L. S. 1921. Little Owl in Lancashire. *British Birds* **15**: 45.
431. Dornbusch, M. 1987. Bestand und Schutz vom Aussterben bedrohter Tierarten in der DDR. *Arch. Nat. Schutz Landsch. Forsch.* **3**: 161–169.
432. Doublet, M. 1987. Synthèse des observations ornithologiques en Eure-et-Loir pour 1986. *Bulletin de la Société des Amis du Museum de Chartres Naturalistes d'Eure-et-Loir* **7**: 37.
433. Drost, R. & Schüz, E. 1940. Folgen des harten Winters 1939–1940. *Vogelzug* **11**: 178–179.
434. Drtilek, R. 1961. [An uncommon manner of breeding of Little Owl *Athene noctua*. Scop.] *Zprávy* **7**: 33.
435. Duarte, P. 1994. La Chouette chevêche (*Athene noctua*) en Livradois-Forez. CPIE Monts du Forez.
436. Dulphy, J.-P., Guelin, F. & Guelin, R. 1989. *Atlas des Oiseaux Nicheurs du Puy-de-Dôme 1980–1985.* Centre Ornithologique Auvergne. Clermond-Ferrand. [Little Owl data on p. 76.]
437. Dunaeva, T. N. & Kucheruk, V. V. 1938. Osobennosti pitaniya domovogo sycha v svyazi s geograficheskimi i statsionalnymi usloviyami i sezonami goda [Feeding peculiarities of the Little Owl in connection with geographical and microhabitat conditions and seasons of year]. *Zoologicheskij Zhurnal* [*Zoological Journal*] **17**: 1080–1090.
438. Duncan, A. B. 1951. Little Owl in Dumfriesshire. *Scottish Naturalist* **63**: 189.
439. Dunthorn, A. A. & Errington, F. P. 1964. Casualties among birds along a selected road in Wiltshire. *Bird Study* **11**: 175.
440. Dupond, C. 1943. *Les Oiseaux de la Belgique.* Patrimoine du Musée Royal d'Histoire naturelle de Belgique. Bruxelles. [Little Owl data on pp. 71–72.]
441. Durand, J. & Goedert, A. 1992. Impact des poteaux métalliques non obturés sur l'avifaune en France. France Telecom. LPO.
442. Dyczkowski, J. & Yalden, D. W. 1998. An estimate of the impact of predators on the British field vole *Microtus agrestis* population. *Mammal Review* **28**: 165–184.
443. Eakle, W. L. 1994. A raptor survey in western Turkey and eastern Greece. *Journal of Raptor Research* **28**: 186–191.
444. Eble Vicomte, 1946. La situation d'après-guerre dans un coin du Finistère. *Alauda* **14**: 150.
445. Eck, S. & Busse, H. 1973. *Eulen.* Die Neue Brehm-Bücherei. A. Ziemsen Verlag. Wittenberg Lutherstadt. [Little Owl data on pp. 148–151.]
446. Ehrle, W. & Mohr, H. 1989. Nistkasten-Einstreu beim Steinkauz. *Vögel der Heimat* **59**: 86.
447. Eick, M. J. 2003. Habitatnutzung und Dismigration des Steinkauzes *Athene noctua.* Masters thesis, Universität Hohenheim.
448. Eijkman, C. 1935. De nederlandsche uilen en roofvogels. *Levende Natuur* **38**: 314–316.
449. Elliott, J. S. 1907. Little Owl (*Athene noctua*) in Bedfordshire. *Zoologist* **11**: 384–385.
450. Elliott, J. S. 1940. Little Owl preying on birds. *British Birds* **33**: 312–313.
451. Elliott, W. J. 1914. Food of the Little Owl. *Zoologist* **18**: 274.
452. Ellis, J. C. S. 1946. Little Owl's method of taking pheasant chick by daylight. *British Birds* **39**: 217.
453. Ellis, J. C. S. 1949. Singular behaviour of Little Owl. *British Birds* **42**: 152.
454. Ellison, N. F. 1943. Little Owl in Wirrel. *North Western Nat.* **18**: 210.
455. Elton, H. B. 1947. Little Owl in Breconshire. *British Birds* **40**: 216.
456. Elvers, H. & Witt, K. 1973. Rote Liste der gefährdeten Brutbiotope und rote Liste der Brutvögel in Berlin (West). *Berichte der Deutschen Sektion des Internationalen Rates für Vogelschutz* **13**: 41–43.
457. Enehjelm af, C. 1957. The breeding of the Little Owl (*Athene noctua*). *The Avicultural Magazine* **63**: 44–45.
458. Enehjelm af C. 1969. Breeding owls. *The Avicultural Magazine* **72**: 56.
459. Engel, R. 1997. Steinkauz-Niströhre mit neuartigem Schutz gegen Marder. Zeitschrift für Vogelkunde und Naturschutz in Hessen. *Vogel und Umwelt* **9**: 163–167.
460. Engländer, H. & Weitz, H. 1984. Ornithologischer Sammelbericht für Westfalen 1.3.83–31.8.83. *Charadrius* **20**: 99.
461. Epple, W. & Hölzinger, J. 1987 *In* Hölzinger, J. (ed). Die Vögel Baden-Württembergs. Gefährdung und Schutz, Artenhilfsprogramme. *Avifauna Baden-Württemberg* **1**: 1085–1095.
462. Eppler, G. 1984. Der Steinkauz zwischen Gernsheim und Lampertheim. Bericht 1983. *DBV AG zum Schutz bedrohter. Eulen Informationsblatt* **20**:
463. Erard, C., Guillou, J.-J., Meininger, D. & Vielliard, J. 1968. Contribution à l'étude des oiseaux du Nord-Est de la France. *Alauda* **36**: 167.
464. Erkert, H. 1967. Beleuchtungsabhängige Aktivitätsoptima bei Eulen und circadiane Regel. *Naturwissenschaften* **54**: 231–232.
465. Erlanger, C. F. von, 1904. Beiträge zur Vogelfauna Nordostafrikas. *Journal für Ornithologie* **52**: 238–239.
466. Erlinger, G. 1968. Meine erfolgreiche Steinkauz-Zucht. *Gefierdete Welt* **92**: 232–234.
467. Erz, W. 1970. Eulen und ihr Schutz. *Charadrius* **6**: 144.

468. Erz, W. 1975. Die im Rheinland gefährdeten Vogelarten "Rote Liste" (Stand 1.1.1975). *Charadrius* **11**: 28–36.
469. Erz, W. 1979. Rote Liste der in Nordrhein-Westfalen gefährdeten Vogelarten. Schriftenreihe der Landesanstalt für Ökologie, *Landschaftsentwicklung und Forstplanung N. W.* **4**: 40.
470. Esnouf, S. 1998. Contribution à la sauvegarde de la population de Chouette chevêche du Causse Méjean. Université de Rennes I. Maîtrise des Sciences et Techniques "Aménagement et Mise en Valeur des Régions" Parc National des Cévennes.
471. Estoppey, F. 1992. Une densité élevée de Chouettes chevêches, *Athene noctua*, dans la plaine du Pô en Italie. *Nos Oiseaux* **41**: 315–319.
472. Etchecopar, R. D. & Hüe, F. 1964. *Les Oiseaux du Nord de l'Afrique, de la Mer Rouge aux Canaries*. Boubée. Paris. [Little Owl data on pp. 328–329.]
473. Etchecopar, R. D. & Hüe, F. 1978. *Les Oiseaux de Chine non passereaux*. Editions du Pacifique. Papeete. [Little Owl data on pp. 479–482.]
474. Etienne, P. 2003. La reproduction de la Chouette chevêche *Athene noctua*. Parades et occupations de l'espace. *Avocette* (numéro spécial): 113–116.
475. Etienne, P., Robert, J.-C. & Triplet, P. 1991. Avifaune nicheuse du Marquenterre (deuxième partie). *Picardie Écologie* **6**: 28.
476. Etoc, G. 1907. *Les Oiseaux de Loir-et-Cher*. Société d'Histoire Naturelle de Loir-et-Cher. Blois. [Little Owl data on pp. 6–7.]
477. Evans, H. W. 1920. Little Owl in Pembrokeshire and Flintshire. *British Birds* **13**: 297.
478. Everett, M. & Sharrock, J.T. R. 1980. The European atlas: owls. *British Birds* **73**: 246–248.
479. Exo, K.-M. 1979. Der Rückgang der Kopfbäume gefährdet den Steinkauz. *Kalender für das Klever Land* **1979**: 51–55.
480. Exo, K.-M. 1980. Habitatstruktur, Brutbiologie und Bestandsentwicklung einer niederrheinischen Steinkauzpopulation (*Athene noctua*). Masters thesis, Universität Köln.
481. Exo, K.-M. 1980. Wo und wieviele Steinkauz-Brutröhren. *Wir und die Vögel* **12**: 20.
482. Exo, K.-M. 1981. Zur Nistökologie des Steinkauzes (*Athene noctua*). *Vogelwelt* **102**: 161–180.
483. Exo, K.-M. 1982. Habitatstruktur, Brutbiologie und Bestandsentwicklung einer Steinkauzpopulation. *Journal of Ornithology* **123**: 346.
484. Exo, K.-M. 1983. Habitat, Siedlungsdichte und Brutbiologie einer niederrheinischen Steinkauzpopulation (*Athene noctua*). *Ökologie der Vögel* **5**: 1–40.
485. Exo, K.-M. 1984. Die akustische Unterscheidung von Steinkauz-männchen und-weibchen (*Athene noctua*). *Journal für Ornithologie* **125**: 94–97.
486. Exo, K.-M. 1986. [First results on territorial and nutritional searching methods of owlets]. *Journal für Ornithologie* **127**: 381–382.
487. Exo, K.-M. 1986. Jahres- und tageszeitliche Aktivitäts-muster beim Steinkauz. *Journal für Ornithologie* **127**: 387.
488. Exo, K.-M. 1987. Das Territorialverhalten des Steinkauzes (*Athene noctua*). Eine verhaltensökologie Studie mit Hilfe der Telemetrie. Ph.D. thesis, Universität Köln.
489. Exo, K.-M. 1988. Themenheft "Biologie und Schutz von Eulen". *Vogelwelt* **109**: 134.
490. Exo, K.-M. 1988. Radiotelemetrische Untersuchungen zum Territorialverhalten des Steinkauzes (*Athene noctua*). *Vogelwelt* **109**: 182.
491. Exo, K.-M. 1988. Jahreszeitliche ökologische Anpassungen des Steinkauzes (*Athene noctua*). *Journal für Ornithologie* **129**: 393–415.
492. Exo, K.-M. 1989. Wie markieren Steinkäuze ihre Reviere? *Journal of Ornithology* **130**: 132.
493. Exo, K.-M. 1989. Tagesperiodische Aktivitätmuster des Steinkauzes (*Athene noctua*). *Vogelwarte* **35**: 94–114.
494. Exo, K.-M. 1990. Geographische Variation des Reviergesangs beim Steinkauz (*Athene noctua*) – ein Vergleich des gesangs nordwestdeutscher und ostenglischer Vögel. *Vogelwarte* **35**: 279–286.
495. Exo, K.-M. 1991. Der Untere Niederrhein – ein Verbreitungsschwerpunkt des Steinkauzes (*Athene noctua*) in Mitteleuropa. *Natur und Landschaft* **66**: 156–159.
496. Exo, K.-M. 1992. Population ecology of Little Owl *Athene noctua* in Central Europe: a review. *In* Galbraith, C. A., Taylor, I. R. & Percival, S. (eds). *The Ecology and Conservation of European Owls*. UK Nature Conservation No. 5. Peterborough: Joint Nature Conservation Committee. pp. 64–75.
497. Exo, K.-M. 2001. The challenges of studying Little Owls at the edge of the 20th century. *In* Van Nieuwenhuyse, D., Leysen, M. & Leysen, K. (eds). The Little Owl in Flanders in its international context. Proceedings of the 2nd International Little Owl Symposium, 16–18 March 2001. *Oriolus* **67**(2–3): 5–7.
498. Exo, K.-M. & Hennes, R. 1977. Empfehlungen zur Methodik von Siedlungsdichte Untersuchungen vom Steinkauz (*Athene noctua*). *DBV AG zum Schutz bedrohter. Eulen Informationsblatt* **1**: 1–8.

499. Exo, K.-M. & Hennes, R. 1978. Empfehlungen zur Methodik von Siedlungsdichte Untersuchungen vom Steinkauz (*Athene noctua*). *Vogelwelt* **99**: 137–141.
500. Exo, K.-M. & Hennes, R. 1978. Ringfunde des Steinkauzes (*Athene noctua*). *Auspicium* **6**: 363–374.
501. Exo, K.-M. & Hennes, R. 1980. Beitrag zur Populationsökologie des Steinkauzes (*Athene noctua*)-eine Analyse deutscher und niederländischer Ringfunde. *Vogelwarte* **30**: 162–179.
502. Exo, K.-M. & Scherzinger, W. 1989. Stimme und Lautrepertoire des Steinkauzes (*Athene noctua*): Beschreibung, Kontext und Lebensraumanpassung. *Ökologie der Vögel* **11**: 149–187.
503. Faggio, G. 2005. La Civetta *Athene noctua* in Corsica. *In* Mastrorilli, M., Nappi, A. & Barattieri, M. (eds). Atti I Convegno italiano sulla Civetta. Osio Sotto (BG), 21 Marzo 2004. Gruppo Italiano Civette. pp. 35–36.
504. Faivre, B. 1995. Suivi de la reproduction de la Chouette chevêche *Athene noctua* en bordure des Vosges du Nord. BTS Protection de la Nature. Option Gestion des Espaces Naturels.
505. Fajardo, I., Pividal, V., Trigo, M. & Jimenez, M. 1998. Habitat selection, activity peaks and strategies to avoid road mortality by the Little Owl *Athene noctua*. A new methodology on owl research. *Alauda* **66**: 49–60.
506. Falcone, S. 1987. Rapaci abbattuti illegalmente in Sicilia. Rapaci Mediterranei III. Suppl. Ric. Biol. Selvaggina XII (Atti del IV Colloquio, Sant'Antioco, Cagliari, 11–13 Ottobre 1984. [Little Owl data on pp. 93–95.]
507. Farman, C. 1869. On some of the birds of prey of central Bulgaria. *Ibis* **11**: 204.
508. Farsky, O. 1928. De l'utilité de quelques oiseaux de proie et corvidés déterminée par l'examen de leurs aliments. Ph.D. thesis, Université de Nancy. [Little Owl data on pp. 99–103.]
509. Fattorini, S. 2001. Temporal and spatial variations in darkling beetle predation by kestrels and other raptors in a Mediterranean urban area. *Biologia* **56**: 165–170.
510. Fattorini, S., Manganaro, A. & Salvati, L. 1999. Variations in the winter Little Owl *Athene noctua* diet along an urbanization gradient: a preliminary study. *Avocetta* **23**: 189.
511. Fattorini, S., Manganaro, A., Piattella, E. & Salvati, L. 1999. Role of the beetles in raptor diets from a Mediterranean urban area (*Coleoptera*). *Fragmenta Entomologica* **31**: 57–69.
512. Fattorini, S., Manganaro, A., Ranazzi, L., Cento, M. & Salvati, L. 2000. Insect predation by Little Owl, *Athene noctua*, in different habitats of Central Italy. *Rivista Italiana di Ornithologia* **70**: 139–142.
513. Federation Rhone-Alpes De Protection De La Nature-Section Isere, 1983. Les poteaux PTT métalliques creux, pièges mortels pour la faune. *Courrier de la Nature* **88**: 32–33.
514. Fellenberg, W. 1981. Ornithologischer Sammelbericht für Westfalen 13.3.79–31.8.80. *Charadrius* **17**: 35.
515. Fellenberg, W. 1985. Ornithologischer Sammelbericht für Westfalen 1.9.83–29.2.84. *Charadrius* **21**: 43.
516. Fellenberg, W. 1987. Ornithologischer Sammelbericht für Westfalen (Zeitabschnitt 1.9.1986–29.2.1987) *Charadrius* **23**: 160–161.
517. Fennel, C. M. 1960. Recent records of the Little Owl and Water Rail in Korea. *Condor* **62**: 409.
518. Fequant, G. 1984. Nos dernières Chevêches. *Courrier de la Nature* **91**: 23–28.
519. Ferguson-Lees, I. J. 1964. Summary of the present status of birds of prey and owls in western Europe. ICBP Working Conference on Birds of Prey and Owls. Caen. [Little Owl data on pp. 132–140.]
520. Ferlier, M. & Zakeossian, D. 1998. Situation de la Chouette chevêche dans le Parc naturel régional de la Haute Vallée de Chevreuse. Stage de 2ème année INA-PG.
521. Ferrer, D., Molina, R., Castella, J. & Kinsella, J. M. 2004. Parasitic helminths in the digestive tract of six species of owls (*Strigiformes*) in Spain. *The Veterinary Journal* **167**: 181–185.
522. Ferrus, L. 1997. Influence de l'organisation des paysages sur la répartition de la Chouette chevêche (*Athene noctua* Scop.). ENESAD. INRA. PNRVN. Université de Bourgogne.
523. Ferrus, L., Génot, J.-C., Topin, F., Baudry, J. & Giraudoux, P. 2002. Répartition de la Chevêche d'Athéna (*Athene noctua* Scop.) et variation d 'échelle d'analyse des paysages. *Terre et Vie* **57**: 39–51.
524. Festetics, A. 1952. Observations on the Barn Owl and Little Owl's life habitats. *Aquila* **60**: 452–453.
525. Festetics, A. 1959. Gewölluntersuchungen an Steinkauzen der Camargue. *Terre et Vie* **106**: 121–127.
526. Finch, O.-D. 1995. Zur Verbreitung nachtaktiver Eulen im Landkreis Wesermarsch unter besonderer Berücksichtigung von Schleiereule *Tyto alba* und Steinkauz *Athene noctua*. *Jahresbericht Ornithologische Arbeitsgemeinschaft Oldenburg* **13**: 22–33.
527. Finck, P. 1989. Variabilität des Territorialverhaltens beim Steinkauz (*Athene noctua*). Ph.D. thesis, Universität Köln.
528. Finck, P. 1990. Seasonal variation of territory size with the Little Owl (*Athene noctua*). *Oecologia* **83**: 68–75.
529. Finck, P. 1993. Territoriengrösse beim Steinkauz (*Athene noctua*): Einfluss der Dauer der Territorienbesetzung. *Journal für Ornithologie* **134**: 35–42.
530. Fischer, L. 1897. *Katalog der Vögel Badens*. Karlsruhe. [Little Owl data on pp. 14–15.]
531. Flint, P. R. & Stewart, P. F. 1983. *The Birds of Cyprus*. London: British Ornithologists' Union. [Little Owl data on p. 86.]

532. Flint, V. E., Boehme, R. L., Kostin, Y. V. & Kuznetsov, A. A. 1984. *A Field Guide to Birds of the USSR*. Princeton: Princeton University Press. [Little Owl data on pp. 176–177.]

533. Flipo, S. 2003. Résultats du suivi pendant six années (1994 à 1999) d'une population de Chevêche d'Athéna *Athene noctua* dans un secteur bocager de la plaine maritime picarde. *Avocette (numéro spécial)*: 105–112.

534. Fonds D'Intervention Pour Les Rapaces D'Alsace, 1983. *Ces Oiseaux de Proie qu'on appelle Rapaces*. Strasbourg: Nouvel Alsacien. [Little Owl data on p. 31.]

535. Fontaneto, D. & Guidali, F. 2001. Resti vegetali nelle borre di civetta *Athene noctua* (Scopoli, 1769). *Avocetta* **25**: 321–323.

536. Ford, H. 1962. Current notes. *Scottish Birds* **2**: 260.

537. Formon, A. 1969. Contribution à l'étude d'une population de Faucons pèlerins *Falco peregrinus* dans l'est de la France. *Nos Oiseaux* **30**: 136.

538. Formozov, A. N. & Osmolovskaya, V. I. 1953. Chislennost nekotorykh landshaftnykh zhivotnykh tsentral'nogo Kazakhstana po dannym analiza dobychi, vylovlennoi khishchnymi ptitsami [Number of some landscape animals of the central Kazakhstan on the base of analysis of prey, caught by birds of prey]. Trudy Instituta geografii [Proceedings of Institute of Geography]. *AN SSSR [Academy of Science of the USSR]* **54**: 329–350.

539. Forrest, H. E. 1900. The Little Owl in Flintshire? *Zoologist* **4**: 482.

540. Forrest, H. E. 1912. Spread of the Little Owl in Yorkshire. *British Birds* **5**: 245.

541. Forrest, H. E. 1914. Little Owl in Shropshire. *British Birds* **8**: 18.

542. Forrest, H. E. 1916. Little Owl in Shropshire. *British Birds* **10**: 43.

543. Forrest, H. E. 1919. Little Owl in Montgomeryshire. *British Birds* **13**: 196.

544. Forrest, H. E. 1919. Little Owl breeding in Shropshire and Radnorshire. *British Birds* **13**: 30.

545. Fournel, D. H. L. 1836. *Faune de la Moselle ou Manuel de Zoologie Contenant la Description des Animaux Libres ou Domestiques Observés dans le Département de la Moselle*. Veronnais. Metz. [Little Owl data on pp. 99–100.]

546. Fournier, M. 1988. Répartition des rapaces pendant l'hiver 1986–87 dans un secteur cultivé de l'Avesnois. *Héron* **21**: 56.

547. Fowler, W. W. 1909. Little Owl in northwest Oxfordshire. *British Birds* **2**: 280.

548. Francois, R. 1996. La Chouette chevêche *Athene noctua* dans le bocage des franges normandes de l'Oise et de la Somme. *L'Avocette* **20**: 25–28.

549. Franke, C. 1968. Steinkauz als Mäusefänger im Zimmer. *Wild und Hund* **70**: 555.

550. Freethy, R. 1976. Little Owl flying at Dunlin. *British Birds* **69**: 272.

551. Fremaux, S. 2002. La Chevêche d'Athéna *Athene noctua* en Haute-Garonne et Tarn année 2000. *Le Pistrac* **18**: 11–25.

552. Friedrich, H. 1990. Die Entwicklung der Steinkauz-Population im Landkreis-Weilburg von 1978–1989. *Vogel und Umwelt* **6**: 59–61.

553. Frier, J. 1984. Chronique ornithologique 1.9.83–31.8.84. *Ardèche Nature* **12**: 13.

554. Frionnet, C. 1925. *Les Oiseaux de la Haute-Marne*. Société des Sciences Naturelles de Haute-Marne. [Little Owl data on pp. 317–319.]

555. Frolov, V. V. & Korkina, S. A. 1997. O statuse redkikh vidov ptits Penzenskoi oblasti na primere nevorob'inykh [On status of rare bird species of the Penza Region on the example of Non-Passerines]. Fauna, ekologiya i okhrana redkikh ptits Srednego Povolzh'ya [Fauna, ecology and conservation of rare birds of the Middle Volga River area]. Saransk. [Little Owl data on pp. 46–49.]

556. Frolov, V. V. & Rodionov, E. V. 1991. Sovy Penzenskoi oblasti [Owls of the Penza Region]. [Little Owl data on pp. 273–274.]

557. Frolov, V. V., Korkina, S. A., Frolov, A. V. *et al.* 2001. Analiz sostoyaniya fauny nevorob'inykh ptits yuga lesostepnoi zony pravoberezhnogo Povolzh'ya v XX veke. [Analysis of the state of fauna of non-passerine birds in south of the forest-steppe zone of the Right-bank Volga region in XX century]. *Berkut* **10**: 156–183.

558. Fronczak, J. & Dombrowski, A. 1991. Owls *Strigiformes* in the agricultural and forest landscapes of South Bialystok Lowlands. *Acta Ornitholecologica* **26**: 55–61.

559. Fuchs, E. 1976. Bedrohtes Vogelleben. Bericht 1976 der Schweizerische Vogelwarte Sempach. [Little Owl data on p. 7.]

560. Fuchs, E. & Schifferli, L. 1981. Sommerbestand von Waldkauz *Strix aluco* und Waldohreule *Asio otus* im aargauischen Reusstal. *Ornithologische Beobachter* **78**: 88.

561. Fuchs, P. 1982. Hoogstamboomgaarden en Steenuilen. *Vogeljaar* **30**: 241–250.

562. Fuchs, P. 1982. Analysis of factors which determine dispersion, population density and reproduction of the Little Owl. *Annual Report Research Institute for Nature Management* **1981**: 51–52.

563. Fuchs, P. 1986. Structure and functioning of a Little Owl *Athene noctua* population. *Annual Report Research Institute for Nature Management* **1985**: 113–126.

564. Fuchs, P. & Gussinklo, D. J. 1977. *The Status of Birds of Prey in the Netherlands*. [Little Owl data on pp. 139–143.]
565. Fuchs, P. & Thissen, J. B. M. 1981. Die Pestizid-und PCB-Belastung bei Greifvögeln und Eulen in den Niederlanden nach gesetzlich verordneten Einschränkungen im Gebrauch der Chlorierten Kohlenwasserstoffpestizide. *Ökologie der Vögel* **3**: 189–195.
566. Führer, von L. 1904. Über zwei neue palearktische Formen. *Ornithologisches Jahrbuch* **15**: 56.
567. Furrington, H. 1979. Eine Röhre schütz vor dem Marder. *Wir und die Vögel* **11**: 20–22.
568. Furrington, H. 1979. Nisthilfe für Steinkäuze. Eine Röhre schützt vor dem Marder. *Wir und die Vögel* **11**: 20–22.
569. Furrington, H. 1987. Steinkauz-Bestandsentwicklung auf einer ca. 750 km^2 grossen Fläche im Landkreis Heilbronn von 1973–1987. *DBV AG zum Schutz bedrohter. Eulen Informationsblatt* **27**: 2–4.
570. Furrington, H. 1994. Das Rauf und Runter als Zitterpartie einer kleinen Population des Steinkauzes (*Athene noctua*) im Landkreis Heilbronn. *Eulen Rundblick* **40**: 37–39.
571. Furrington, H. 1996. 25 Jahre Artenhilfsprogramm für den Steinkauz (*Athene noctua*) im Stadt-und Landkreis Heilbronn mit Randgebiet in Nordwürttemberg. *Jahresbericht zum Monitoring Greifvögel Eulen Europas* **8**: 97–100.
572. Furrington, H. 1998. 27 Jahre Steinkauz-Schutz im Stadt-und Landkreis Heilbronn mit Randgebieten auf einer Kontrollfläche von ca.750 km^2. *Ornithologische Schnellmitteilungen Baden-Württemberg* **59**: 14–16.
573. Furrington, H. 1998. 27 Jahre Steinkauz-Schutz im Raum Heilbronn/Nordwürttemberg. *Eulen Rundblick* **47**: 17.
574. Furrington, H. & Exo, K.-M. 1985. Schaffung und Erhaltung von Steinkauzbrutplätze. Deutsche Bund für Vogelschutz Jahresheft.
575. Gaedechens, E. 1938. *Die Vogelwelt im Westen Hamburgs*. Hamburg. [Little Owl data on p. 99.]
576. Gaffney, M. F. & Hodos, W. 2003. The visual activity and refractive state of the American kestrel (*Falco sparverius*). *Vision Research* **43**: 2053–2059.
577. Gager, L., Gelinaud, G., Henry, J. *et al.* 1985. Actualités ornithologiques du 16.7.84 au 15.11.84. *Ar Vran* **12**: 45.
578. Gailliez, D. 1993. *La Chouette Chevêche en Avesnois: Habitat et Population*. Espace Naturel Régional. Groupe des Naturalistes de l'Avesnois.
579. Gailliez, D. 1994. Un bastion d'importance nationale. Haut-Avesnois: recensements des chouettes chevêches. *Revue du Fonds d'Intervention pour les Rapaces* **25**: 11.
580. Galeotti, P. 1989. Tavola rotonda: metodi di censimento per gli strigiformi. *In* Fasola, M. (ed.). Atti II Seminario Italiano Censimenti Faunistici dei Vertebrati. Suppl. Ric. Biol. Selvaggina, XVI. [Little Owl data on pp. 437–445.]
581. Galeotti, P. & Morimando, F. 1991. Dati preliminari sul censimento della Civetta in ambiente urbano. *Suppl. Ric. Biol. Selvaggina* **16**: 349–351.
582. Galeotti, P. & Sacchi, R. 1996. Owl census project in the Lombardy region: preliminary data on the Tawny Owl (*Strix aluco*), the Little Owl (*Athene noctua*) and the Long-eared Owl (*Asio otus*), populations. Abstracts II Intern. Conference on Raptor. Raptor Research Foundation e Università di Urbino. pp. 79–80.
583. Gallego, A., Baron, M. & Gayoso, M. 1975. Organization of the outer plexiform layer of the diurnal and nocturnal bird retinae. *Vision Research* **15**: 1027–1028.
584. Gallego, A., Baron, M. & Gayoso, M. 1975. Horizontal cells of the avian retina. *Vision Research* **15**: 1029–1030.
585. Galuppo, C. & Borgo, E. 2001. Primi dati sull'alimentazione di Civetta *Athene noctua* e Allocco *Strix aluco* a Genova. *Avocetta* **25**: 210.
586. Ganya, I. M. & Zubkov, N. I. 1975. Pitanie domovogo sycha (*Athene noctua* Scop.) in srednei chasti Moldavii [The Little Owl (*Athene noctua* Scop.) feeding at the middle part of Moldavia]. Ekologiya ptits im mlekopitayushchikh Moldavii [*Ecology of Birds and Mammals of Moldavia*]. Kishinev: Shiintsa Press [Little Owl data on pp. 63–72.]
587. Garcia-Fernandez, A. J., Motas-Guzman, M., Navas, I. *et al.* 1997. Environmental exposure and distribution of lead in four species of raptors in Southeastern Spain. *Archives of Environmental Contamination and Toxicology* **33**: 76–82.
588. Garguil, P. 1982. *La Vie des Rapaces*. Ouest France. Rennes. [Little Owl data on pp. 63–65.]
589. Garnett, R. O. B. & Garnett, J. M. R. 1975. Little Owl behaviour. *Gloucestershire Naturalists' Society Journal* **26**: 84.
590. Garzon, H. J. 1968. Las rapaces y otras *Aves* de la Sierra de Gata. *Ardeola* **14**: 120.
591. Garzon, H. J. 1977. Birds of prey in Spain, the present situation. ICBP World Conference on Birds of Prey Vienna. [Little Owl data on pp. 162–163.]

592. Gassmann, H. 1987. Rettet die Obstwiesen! Ein Beitrag zum Schutz des Steinkauzes und seiner Lebensraüme. *Rheinische Heimatpflege* **24**: 14–19.
593. Gassmann, H. & Bäumer, B. 1993. Zur Populationsökologie des Steinkauzes (*Athene noctua*) in der westlichen Jülicher Börde. Erste Ergebnisse einer 15jährigen Studie. *Vogelwarte* **37**: 130–143.
594. Gassmann, H., Bäumer, B. & Glasner, W. 1994. Faktoren der Steuerung des Bruterfolges beim Steinkauz *Athene noctua*. *Vogelwelt* **115**: 5–13.
595. Gavrin, V. F. 1962. *Otryad Sovy Striges. Ptitsy Kazakhstana [Birds of Kazakhstan]*. Vol. 2. Alma-Ata. Publishing House of Academy of Sciences of the Kazakh SSR. [Little Owl data on pp. 708–779.]
596. Gelinaud, G. 1986. Actualités ornithologiques du 16 Mars 1985 au 15 Juillet 1985. *Ar Vran* **13**: 65–66.
597. Gengler, J. 1910. Ornithologische Beobachtungen in und um Metz. Natur und Offenbarung. [Little Owl data on p. 350.]
598. Génot, J.-C. 1986. Ecologie et protection de la Chouette chevêche (*Athene noctua*). Première phase: Répartition, densité, habitat. Parc Naturel Régional des Vosges du Nord.
599. Génot, J.-C. 1986. Répartition, densité et habitat de la Chouette chevêche en bordure des Vosges du Nord. In Colloque interrégional d'ornithologie et de mammalogie. Saint-Dié. 1986. *Ciconia* **10**: 164.
600. Génot, J.-C. 1988. Ecologie et protection de la Chouette chevêche (*Athene noctua* Scop.). Deuxième partie: Habitat, reproduction, régime alimentaire. Parc Naturel Régional des Vosges du Nord.
601. Génot, J.-C. 1989. Suivi des populations de Chouettes chevêches dans six Parcs Naturels Régionaux. Parc Naturel Régional des Vosges du Nord. Direction de la Nature et des Paysages. Ministère de l'Environnement.
602. Génot, J.-C. 1989. Bibliographie mondiale *Athene noctua*. Fondation Suisse pour les Rapaces. Miécourt.
603. Génot, J.-C. 1989. Répartition et habitat de la Chouette chevêche (*Athene noctua*) dans cinq Parcs Naturels Régionaux français. *Aves (numéro spécial)* **26**: 125–132.
604. Génot, J.-C. 1989. Ecology and Protection of the Little Owl *Athene noctua* in France. Raptors in the Modern World. Proceedings of the III World Conference on Birds of Prey and Owls. Eilat, Israel, 22–27 March 1987. pp. 433–434.
605. Génot, J.-C. 1990. La Chouette chevêche en France: constat d'une régression. *L'Oiseau Magazine* **18**: 22–27.
606. Génot, J.-C. 1990. Régression de la Chouette chevêche, *Athene noctua* SCOP., en bordure des Vosges du Nord. *Ciconia* **14**: 65–84.
607. Génot, J.-C. 1990. Habitat et sites de nidification de la Chouette chevêche, *Athene noctua* SCOP., en bordure des Vosges du Nord. *Ciconia* **14**: 85–116.
608. Génot, J.-C. 1991. La chouette chevêche *In* Baudvin, H., Génot, J.-C. & Muller, Y. (eds). *Les Rapaces Nocturnes*. Paris: Sang de la Terre. [Little Owl data on pp. 69–113.]
609. Génot, J.-C. 1991. Statut des repaces dans le Parc Natural Régional des Vosges du Nord. *Birds of Prey Bulletin* **4**: 149–154.
610. Génot, J.-C. 1991. Chouette chevêche, *In* Yeatman-Berthelot, D. (ed.). *Atlas des Oiseaux de France en Hiver.* Paris: Société Ornithologique de France. [Little Owl data on pp. 318–319.]
611. Génot, J-C. 1991. Mortalité de la Chouette chevêche, *Athene noctua*, en France. *In* Juillard, M., Bassin, P., Baudvin, H. *et al.* (eds). Rapaces Nocturnes Actes du 30e colloque interrégional d'ornithologie Porrentruy (Suisse): 2–4 Novembre 1990. *Nos Oiseaux*. pp. 139–148.
612. Génot, J.-C. 1992. Observatoire de la Chouette chevêche. Suivi inter-parcs 1991. Parc Naturel Régional des Vosges du Nord. Direction de la Nature et des Paysages. Ministère de l'Environnement.
613. Génot, J.-C. 1992. Contribution à l'écologie de la Chouette chevêche, *Athene noctua* (Scop.), en France. Ph.D. thesis, Université de Dijon.
614. Génot, J.-C. 1992. Biologie de reproduction de la Chouette chevêche *Athene noctua*, en France. *L'Oiseau et Revue Française d'Ornithologie* **62**: 309–319.
615. Génot, J.-C. 1992. Bilan de l'enquête Chouette chevêche du CEOA. *Cigogneau* **34**: 19–21.
616. Génot, J.-C. 1992. Biologie de reproduction de la Chouette chevêche *Athene noctua* SCOP., en bordure des Vosges du Nord. *Ciconia* **16**: 1–18.
617. Génot, J.-C. 1993. La Chouette chevêche: causes de régression et protection. *Lien Ornithologique d'Alsace* **58**: 7–9.
618. Génot, J.-C. 1994. *La Chouette Chevêche*. Eveil Editeur.
619. Génot, J.-C. 1994. Breeding biology of the Little Owl *Athene noctua* in France. *In* Meyburg, B.-U. & Chancellor, R. D. (eds). [Little Owl data on pp. 511–520.]
620. Génot, J.-C. 1994. *Chouette Chevêche ou Chevêche d'Athéna. In* Yeatman-Berthelot, D. & Jarry, G. (eds). [Little Owl data on pp. 398–401.]
621. Génot, J.-C. 1995. Données complémentaires sur la population de Chouettes chevêches, *Athene noctua*, en déclin en bordure des Vosges du Nord. *Ciconia* **19**: 145–157.
622. Génot, J.-C. 1995. Chouette la Chevêche. *Combat Nature* **111**: 5–6.

623. Génot, J.-C. 1996. Observatoire de la Chouette chevêche. Suivi inter-parcs 1996. Ministère de l'Environnement. Direction de la Nature et des Paysages.
624. Génot, J.-C. 1997. Monitoring studies of the Little Owl in France. *The Raptor* **1996–1997**: 24–28.
625. Génot, J.-C. 1999. Chevêche d'Athéna *In* Rocamora, G. & Yeatman-Berthelot, D. 1999. *Oiseaux Menacés et à Surveiller en France Listes Range et Recherche de Priorités. Populations Tendances. Menaces. Conservation.* Société d'Etudes Ornithologiques de France. Ligue pour le Protection des Oiseaux. [Little Owl data on pp. 302–303.]
626. Génot, J.-C. 2000. Observatoire de la Chouette chevêche. Suivi inter-parcs 2000. Parc Naturel Régional des Vosges du Nord. Direction de la Nature et des Paysages. Ministère de l'Aménagement du Territoire et de l'Environnement.
627. Génot, J.-C. 2001. Etat des connaissances sur la Chevêche d'Athéna, *Athene noctua*, en bordure des Vosges du Nord (Nord-Est de la France) de 1984 à 2000. *In* Génot, J.-C., Lapios, J.-M., Lecomte, P. & Leigh, R. S. (eds). *Ciconia* **25**: 109–118.
628. Génot, J.-C. 2001. Overview of Little Owl *Athene noctua* literature. *In* Van Nieuwenhuyse, D., Leysen, M. & Leysen, K. (eds). *Oriolus* **67**: 84–87.
629. Génot, J.-C. 2004. Observatoire de la Chevêche d'Athéna. Suivi inter-parcs 2004. Parc Naturel Régional des Vosges du Nord. Direction de la Nature et des Paysages. Ministère de l'Ecologie et du Développement Durable.
630. Génot, J.-C. 2006. La Chevêche d'Athéna, *Athene noctua*, dans la Réserve de la Biosphère des Vosges du Nord de 1984 à 2004. *Ciconia* **29**: 1–272.
631. Génot, J.-C. & Bersuder, D. 1995. Le régime alimentaire de la Chouette chevêche, *Athene noctua*, en Alsace-Lorraine. *Ciconia* **19**: 35–51.
632. Génot, J.-C. & Lecomte, P. 1998. Essai de synthèse sur la population de Chevêche d'Athéna *Athene noctua* en France. *Ornithos* **5**: 124–131.
633. Génot, J.-C. & Lecomte, P. 1998. Statut de la population française. *Bulletin du Fonds d'Intervention pour les Rapaces* **32**: 9.
634. Génot, J.-C. & Lecomte, P. 2002. *La Chevêche d'Athéna*. Delachaux et Niestlé.
635. Génot, J.-C. & Lecomte, P. 2004. *El Mochuelo*. Ediciones Omega.
636. Génot, J.-C. & Sturm, F. 1995. Chouette chevêche. Expérience de renforcement des populations. *Bulletin du Fonds d'Intervention pour les Rapaces* **27**: 6–7.
637. Génot, J.-C. & Sturm, F. 1996. Du renfort pour les chevêches. *L'Oiseau Magazine* **42**: 16–18.
638. Génot, J.-C. & Sturm, F. 2001. Biologie de reproduction de la Chevêche d'Athéna (*Athene noctua*) en captivité. *Ciconia* **25**: 219–230.
639. Génot, J.-C. & Sturm, F. 2003. Bilan de l'expérience de renforcement des populations de Chevêche d'Athéna *Athene noctua* dans le Parc naturel régional des Vosges du Nord. *Alauda* **71**: 175–178.
640. Génot, J.-C. & Van Nieuwenhuyse, D. 2002. *Athene noctua* Little Owl. *BWP Update* **4**: 35–63.
641. Génot, J.-C. & Wilhelm, J.-L. 1992. Modes d'occupation et d'utilisation du milieu développés par la Chouette chevêche dans le Parc Naturel Régional des Vosges du Nord. Contrat SRETIE/MERE No. 89300. Ministère de l'Environnement. Parc Naturel Régional des Vosges du Nord.
642. Génot, J.-C. & Wilhelm, J.-L. 1992. Domaine vital de la Chouette chevêche (*Athene noctua*) dans la Réserve de la Biosphère des Vosges du Nord. *Annales Scientifiques de la Réserve de Biosphère des Vosges du Nord* **2**: 33–52.
643. Génot, J.-C. & Wilhelm, J.-L. 1993. Occupation et utilisation de l'espace par la Chouette chevêche, *Athene noctua*, en bordure des Vosges du Nord. *Alauda* **61**: 181–194.
644. Génot, J.-C., Wilhelm, J.-L. & Loukianoff, S. 1992. Modes d'occupation et d'utilisation du milieu développés par la Chouette chevêche, *Athene noctua*, dans le Parc naturel régional des Vosges du Nord. XVe Colloque Francophone de Mammalogie: Les Carnivores. [Little Owl data on pp. 55–57.]
645. Génot, J.-C., Lecci, D., Bonnet, J., Keck, G. & Venant, A. 1995. Quelques données sur la contamination chimique de la Chouette chevêche, *Athene noctua*, et de ses oeufs en France. *Alauda* **63**: 105–110.
646. Génot, J.-C., Juillard, M. & Van Nieuwenhuyse, D. 1997. Little Owl. *In* Hagemeijer, W. J. M. & Blair M. J. (eds). *The EBCC Atlas of European Breeding Birds: Their Distribution and Abundance*. London: T. & A. D. Poyser. [Little Owl data on pp. 408–409.]
647. Génot, J.-C., Sturm, F., Pfitzinger, H. *et al.* 2000. Expérience de renforcement des populations de Chevêche d'Athéna (*Athene noctua* Scop.) dans la Réserve de Biosphère des Vosges du Nord. *Annales Scientifiques de la Réserve de Biosphère Transfrontalière Vosges du Nord-Pfälzerwald* **8**: 31–51.
648. Génot, J.-C., Lapios, J.-M., Lecomte, P. & Leigh, R. S. (eds). 2001. Chouette chevêche et territoires. Actes du Colloque International de Champ-sur-Marne, 25 et 26 Novembre 2000. ILOWG. *Ciconia* **25**: 61–204.
649. Gepp, J. 1983. Rote Listen gefährdeter Tiere Österreichs. Im Auftrag des Bundesministeriums für Gesundheit und Umweltschutz. Wien. [Little Owl data on pp. 39–40.]

50. Gerdol, R., Mantovani, E. & Perco, F. 1982. Indagine preliminare comparata sulle abitudini alimentari di tre strigiformi nel Carso triestino. *Rivista Italiana di Ornithologia* **52**: 55–60.

51. Geroudet, P. 1964. L'hiver rigoureux de 1962–1963 en Suisse romande. *Nos Oiseaux* **27**: 222.

52. Geroudet, P. 1965. *Les Rapaces Diurnes et Nocturnes d'Europe*. 3e édition entièrement nouvelle. Delachaux et Niestlé. Neuchâtel. [Little Owl data on pp. 369–377.]

53. Geroudet, P., Guex, C. & Maire, M. 1983. *Les Oiseaux Nicheurs du Canton de Genève*. Museum de Genève. [Little Owl data on p. 105.]

54. Gesellschaft Zur Erhaltung Der Eulen, 2003. Steinkauz in NRW immer seltener. *Naturschutz und Landschaftsplanung* **35**: 28.

55. Geslin, E. 1994. Etude de la Chouette chevêche, *Athene noctua*, dans un secteur favorable de la vallée de la Loire. Mémoire de maîtrise des Sciences de l'Environnement. Université d'Angers.

56. Geslin, E. 1994. Recensement de la Chouette chevêche *Athene noctua* dans un secteur favorable de la vallée de la Loire. *Bulletin du Groupe Angevin Etudes Ornithologiques* **22**: 61–68.

57. Geyr Von Schweppenburg, H. 1906. Untersuchungen über die Nahrung einiger Eulen. *Journal für Ornithologie* **54**: 540.

58. Giannotti, M., Balestrieri, R., Gori, F. *et al.* 2005. *In* Mastrorilli, M., Nappi, A. & Barattieri, M. (eds). Dati considerazioni sulla distribuzione della Civetta *Athene noctua* nei Campi Flegrei (Na). Atti I Convegno italiano sulla Civett a. Osio Sotto (BG), 21 Marzo 2004. Gruppo Italiano Civette. pp. 27–28.

59. Giglioli, E. H. 1900. Intorno ad una presunta nuova specie di *Athene* trovata in Italia. *Avicula* **4**: 57–60.

60. Giglioli, E. H. 1901. Notaintorno ad una presunta nuova specie di *Athene* trovata in Italia. *Ornis* **11**: 237–242.

61. Giglioli, E. H. 1902. L'*Athene chiaradiae*, sp. n. *Ornis* **11**: 237–242.

62. Giglioli, E. H. 1903. The strange case of *Athene chiaradiae*. *Ibis* **3**: 1–18.

63. Gil Lleget, A. 1944. Base para un método de estudio cientifico de alimentación en las avec y resultado del análisis de 400 estómagos. *Boletin de la Real Sociedad Espãnola de Historia Natural* **42**: 177–206.

64. Gilard, B. 1996. Chouette chevêche (*Athene noctua*), Rossignol philomèle (*Luscinia megarhynchos*) et Hypolais polyglotte (*Hippolais polyglotta*) présents à plus de 1050 m d'altitude sur la commune de Landos (Haute-Loire). *Autres données semblables en Auvergne. Grand-Duc* **48**: 17–23.

65. Ginter, F. 1971. [Contribution to the knowledge of the food of the Long eared Owl (*A. o.*) and Little Owl (*A. n.*) in the southwest Slovakia]. *Polovnícky Zborník* **1**: 55–68.

66. Giuseppe, A. 2005. La Civetta *Athene noctua* nella provincial di Cosenza (Calabria). *In* Mastrorilli, M., Nappi, A. & Barattieri, M. (eds). Atti I Convegno italiano sulla Civetta. Osio Sotto (BG), 21 Marzo 2004. Gruppo Italiano Civette. pp. 27–28.

67. Gladstone, H. S. 1922. Abundance of Little Owl in Norfolk. *British Birds* **16**: 190–191.

68. Glayre, D. & Magnenat, D. 1984. Oiseaux nicheurs de la Haute Vallée de l'Orbe. *Nos Oiseaux* No. 398 (fascicule spécial du vol.) **37**: 125.

69. Glue, D. 1969. Owl pellets. *In* Gooders, J. (ed.). *Birds of the World*. London. [Little Owl data on pp. 1368–1370.]

70. Glue, D. E. 1971. Ringing recovery circumstances of small birds of prey. *Bird Study* **18**: 137–146.

71. Glue, D. E. 1971. Avian predator pellet analysis and the mammalogist. *Mammal Review* **1**: 56–57.

72. Glue, D. E. 1972. Bird prey taken by British owls. *Bird Study* **19**: 91–95.

73. Glue, D. E. 1973. Seasonal mortality in four small birds of prey. *Ornis Scandinavica* **4**: 97–102.

74. Glue, D. E. 1986. *In* Lack, P. (ed.). *The Atlas of Wintering Birds in Britain and Ireland*. Calton: T. & A. D. Poyser. [Little Owl data on pp. 272–273.]

75. Glue, D. E. 2002. Little Owl *Athene noctua*. *In* Wernham, C. V., Toms, M. P., Clark, G. M. *et al.* (eds). *The Migration Atlas: Movements of the Birds of Britain and Ireland*. London: T. & A. D. Poyser. [Little Owl data on pp. 429–431.]

76. Glue, D. E. & Morgan, R. 1977. Recovery of bird rings in pellets and other prey traces of owls, hawks and falcons. *Bird Study* **24**: 111–113.

77. Glue, D. E. & Scott, D. 1980. Breeding biology of the Little Owl. *British Birds* **73**: 167–180.

78. Glutz Von Blotzheim, U. N. 1962. *Die Brutvögel der Schweiz*. Verlag Aargauer Tagblatt A. G. Aarau. [Little Owl data on pp. 319–321.]

79. Glutz Von Blotzheim, U. N. & Bauer, K. M. 1980. *Handbuch der Vögel Mitteleuropas*. Bd 9 Akad. Verlagsges. Wiesbaden. [Little Owl data on pp. 501–530.]

80. Godin, J. & Loison, M. 1975. Observations et baguage de rapaces nocturnes à Saint-Aybert (Nord-France)-Hensies (Hainaut-Belgique) de 1967 à 1970. *Aves* **12**: 57–71.

81. Goemare, B. 2005. Plus tard, je deviendrai un "rocker punk". Récit de Spartacus, une jeune Chevêche d'Athéna. *L'Homme et l'Oiseau* **43**: 40–47.

82. Golubeva, T. B., Chernyl, A. G. & Il'Chev, V. D. 1970. Summary responses of the auditory nerve in relation to acoustic signal parameters in the owls *Asio otus* and *Athene noctua*. *Journal of Evolutionary Biochemistry and Physiology* **6**: 169–175.

683. Gonzales, B. & Junco, O del, 1968. Notas sobre *Aves* de la provincia de Cadiz. *Ardeola* **12**: 214–217.
684. Gooch, G. B. 1940. The bill-snapping of a Little Owl. *British Birds* **33**: 316.
685. Goodman, S. M. 1988. The food habits of the Little Owl inhabiting Wadi el Natrun (Egypt). *Sandgrouse* **10**: 100–104.
686. Gorman, G. 1995. The status of owls (*Strigiformes*) in Hungary. *Buteo* **7**: 95–108.
687. Görner, M. 1982. Zur Ökologie unserer heimischen Eulen und Massnahmen zu ihrem Schutz. *Landschaftspflege und Naturschutz in Thüringen* **19**: 1–17.
688. Goszcynski, J. 1981. Comparative analysis of food of owls in agrocenoses. *Ekologia Polska* [*Polish Journal of Ecology*] **29**: 431–439.
689. Gotta, A. & Pigozzi, G. 1997. Trophic niche of the barn owl and Little Owl in a rice field habitat in northern Italy. *Italian Journal of Zoology* **64**: 55–59.
690. Goulliart, A. 1988. La Chouette chevêche *Athene noctua*, grande consommatrice de géotrupes. *Bulletin de la Société Entomologique du Nord de la France* **246**: 10.
691. Goutner, V. & Alivizatos, H. 2003. Diet of the Barn Owl (*Tyto alba*) and Little Owl (*Athene noctua*) in wetlands of northeastern Greece. *Belgian Journal of Zoology* **133**: 15–22.
692. Grachev, V. A. 2000. Ptitsy okrestnostei Aral'ska i ozera Kamyshlybash (po nablyudeniyam 1951–1954) Nevorob'inye. [Birds of Aral'sk town vicinities and Kamyshlybash Lake (By observations 1951–1954) Non-Passerines]. *Selevinia* **1–4**: 95–104.
693. Grant, C. H. B. & Hackworth-Praed, C. W. 1937. Note. *Bulletin British Ornithological Club* **57**: 158.
694. Grant-Ives, D. M. 1936. Observation on the Little Owl. *The Avicultural Magazine* **7**: 198.
695. Granval, P. & Aliaga, R. 1988. Analyse critique des connaissances sur les prédateurs de lombriciens. *Gibier Faune Sauvage* **5**: 71–94.
696. Graubitz, G. 1983. Domovyi sych [Little Owl]. Ptitsy Latvii. Territorialnoe razmeshchenie i chislennos [Birds of Latvia. Territorial distribution and number]. Riga, "Zinatne" Publ. House. [Little Owl data on pp 122–123.]
697. Greenwood, J. J. D, Crick, H. Q. P. & Bainbridge, I. P. 2003. Numbers and international importance of raptors and owls in Britain and Ireland. *In* Thompson, D. B. A., Redpath, S. M., Fielding, A. H., Marquiss, M. & Galbraith, C. A. (eds). *Birds of Prey in a Changing Environment.* Edinburgh: TSO Scotland. [Little Owl data on pp. 25–49.]
698. Gregory, T. C. 1944. Little Owl hovering. *British Birds* **38**: 55–56.
699. Greschik, E. 1911. Magen-und Gewölluntersuchungen unserer einheimischen Raubvögel. *Aquila* **28**: 169-173.
700. Gretz, M. & Bayle, P. 1984. Contamination par le plomb et le cadmium des oiseaux de proie dans l'est de la France. *Bulletin de l'Association Philomatique Alsace-Lorraine* **20**: 135–161.
701. Grimm, H. 1985. Zum Vorkommen und Schutz des Steinkauzes (*Athene noctua*) in Thüringen. *Veröffentlichungen Museum Gera Naturwissenschaftliche* **11**: 83–89.
702. Grimm, H. 1985. Über Nistplatz-und Raumkonkurrenten des Steinkauzes (*Athene noctua*) im Thüringer Becken. *Acta Ornitholecologica* **1**: 95–96.
703. Grimm, H. 1986. Zur Strukturierung zweier Graslandhabitate und deren potentielles Nahrungsangebot für den Steinkauz (*Athene noctua*) in Thüringer Becken. *Landschaftspflege und Naturschutz in Thüringen* **22**: 94–104.
704. Grimm, H. 1988. Wiesenpflege als Voraussetzung zur Erhaltung des Lebensraumes des Steinkauzes (*Athene noctua*). *Veröffentlichungen Museum Gera Naturwissenschaftliche* **15**: 74–76.
705. Grimm, H. 1989. Die Erhaltung und Pflege von Streuobstwiesen unter dem Aspekt des Steinkauzschutzes (*Athene noctua*). *Abhandlungen und Berichte des Museums der Natur Gotha* **15**: 103–107.
706. Grimm, H. 1991. Zur Ernährung Thüringischer Steinkäuze (*Athene noctua*). *Acta Ornitholecologica* **2**: 277–284.
707. Grimm, P. 1991. Rauschschwalbe (*Hirundo rustica*) imitierte Rufe des Steinkauzes (*Athene noctua*). *Beiträge zur Vogelkunde* **37**: 191.
708. Grishanov, G. V. 1994. Gnezdyashchiesya ptitsy Kaliningradskoi oblasti: territorialnoe razmeshchenie dinamikachislennosti v XIX–XX vv. I. Non-Passeriformes [Breeding birds of Kaliningrad Region: territorial distribution and dynamics of number in XIX–XXth century. I. Non-Passeriformes]. *Russkiy Ornitologicheskiy Zhurnal* [*Russian Journal of Ornithology*] **3**: 83–116.
709. Groen, N., Boudewijn, T. & De Jonge, J. 2000. De effecten van overstroming van de uiterwaarden op de Steenuil. *De Levende Natuur* **101**: 143–148.
710. Groen, N., Van Harxen, R. & Stroeken, P. 2001. Effects of pollution on birds in river ecosystems: a case study on the Little Owl (*Athene noctua*) in the Netherlands. *In* Génot, J.-C., Lapios, J.-M., Lecomte, P. & Leigh, R. S. (eds). *Ciconia* **25**: 141–146.
711. Grosch, K., Natterer, S. & Schepperle, K. 1995. Zur Nahrung eines Luwigsburger Steinkauzpaares. *Ornithologische Schnellmitteilungen Baden-Württemberg* **47**: 41–42.

12. Groupe Ornithologique De L'Yonne, 1994. *Atlas des Oiseaux Nicheurs de l'Yonne 1979–1992*. [Little Owl data on pp. 99–100.]
13. Groupe Ornithologique De La Vienne, 1991. *Les Oiseaux Nicheurs de la Vienne*. [Little Owl data on p. 38.]
14. Groupe Ornithologique Des Avaloirs & Groupe Ornithologique Normand, 1990. Protection de la Chouette chevêche sur le territoire du Parc. Information-Mise en place de nichoirs. Rapport Parc Naturel Régional de Normandie-Maine.
15. Groupe Ornithologique Normand & Groupe Ornithologique Des Avaloirs, 1997. Suivi des populations de chouettes chevêches. Dénombrement des mâles chanteurs au printemps 1997. Rapport Parc Naturel Régional de Normandie-Maine.
16. Groupe Sarthois Ornithologique, 1991. *Les Oiseaux Nicheurs de la Sarthe*. Le Mans. [Little Owl data on p. 72.]
17. Grzywaczewski, G. 1998. Rozmieszczenie sow (*Strigiformes*) w wojewodztwie Chelmskim [Distribution of *Strigiformes* in province of Chelm]. *In* Sobisz, Z. & Wolk, E. (eds). Materialy II Przegladu Dzialalnosci Kol Naukowych Przyrodnik WSP, Slupsk: 123–127.
18. Grzywaczewski, G. 2000. Wystepowanie pojdzki *Athene noctua* (Scop. 1769) w krajobrazie rolniczym okolic Chelma [Occurence of Little Owl *Athene noctua* (Scop. 1769) in agricultural landscape near Chelm city (Eastern Poland). The problems of conservation and using of the rural areas of great natural values.] *In* Radwan, S. & Lorkiewicz, Z. (eds). Lublin: Marie Curie-Sklodowska University Press. pp. 371–376.
19. Grzywaczewski, G. & Kitowski, I. 2000. Owls (*Strigiformes*) in cities of the Lublin region (SE Poland). Conference Birds of Prey and Owls, Czech Society for Ornithology. Mikulov, 24–26 November 2000. [Little Owl data on pp. 22–23.]
20. Grzywaczewski, G., Kitowski, I. & Scibior, R. 2006. Diet of Little Owl *Athene noctua* during breeding in the central part of Lublin region (SE Poland). *Acta Zoologica Sinica* **52**: 1155–1161.
21. Gubin, B. M. 1998. Gnezdyashchiesya ptitsy vostochnoi kromki peskov Kyzylkum [Breeding birds of eastern edge of Kyzylkum sands]. *Russkiy Ornitologicheskiy Zhurnal* [*Russian Journal of Ornithology*] (express issue) **50**: 3–23.
22. Gubin, B. M. 1999. Ptitsy vostochnogo Priaralya [Birds of Eastern Cis-Aral Sea area]. *Russkiy Ornitologicheskiy Zhurnal* [*Russian Journal of Ornithology*] (express issue) **80**: 3–16.
23. Guenaux, G. 1920. *Oiseaux Utiles et Nuisibles à l'Agriculture*. Paris: Baillières. [Little Owl data on pp. 78.]
24. Guermeur, Y. & Monnat, J. Y. 1980. *Histoire et Géographie des Oiseaux Nicheurs de Bretagne*. SEPNB Centrale Ornithologique Ar Vran. Ministère de l'Environnement et du cadre de Vie. Direction de la Protection de la Nature. ARPEGE Clermont-Ferrand. [Little Owl data on pp. 103–104.]
25. Gunston, D. 1948. Little Owl as prey of Tawny Owl. *British Birds* **41**: 388.
26. Günther, R. 1982. Zur Bestandsituation des Steinkauzes, *Athene noctua* (Scopoli), im Bezirk Gera. *Thüringer Ornithologische Mitteilungen* **28**: 39–42.
27. Gurney, C. J. 1913. Little Owl breeding in Essex. *British Birds* **7**: 85.
28. Gurney, J. H. 1919. Ornithological notes from Norfolk for 1919. *British Birds* **13**: 259.
29. Gurston, D. 1948. Little Owl as prey of tawny owl. *British Birds* **41**: 388.
30. Gurtner, W. 1984. Steinkauz im Gürbetal. *Vögel der Heimat* **54**: 231.
31. Gusberti, V. 1998. Evaluation des facteurs limitants et des exigences minimales pour la chevêche d'Athéna (*Athene noctua*, Scop. 1769) au Tessin: le cas de la population du Mendrisiotto. Université de Neuchâtel. Institut de Zoologie.
32. Gusev, V. M. 1952. Oznachenii glubiny snezhnogo pokrova dlya ptits, pitayushchikhsya myshevidnymi gryzunami [On the significance of the snow cover depth for the birds, feeding by mouse-looking rodents]. *Zoologicheskij Zhurnal* [*Zoological Journal*] **31**: 471–473.
33. Gusev, V. M. 1956. [On winter food of *Athene noctua* in Apsheron peninsula]. *Zoologicheskij Zhurnal* [*Zoological Journal*] **35**: 300–303.
34. Guyot, H. & Lecomte, P. 1996. La Chouette chevêche et les insectes. *Insectes* **102**: 22–24.
35. Guyot, H. & Lecomte, P. 1996. La Chouette chevêche et les insectes. *Insectes* **101**: 8–10.
36. Gyllin, R. 1968. Ryttling som jaktmethod hos minervauggla (*Athene noctua*) [Hovering as hunting method of Little Owl *Athene noctua*]. *Var Fagelvarld* **27**: 172–173.
37. Haarmann, K. 1976. Die im Hamburger Raum gefährdeten Vogelarten "Rote Liste" (Stand 1.1.1976). *Hamburger Avifaunistische Beiträge* **14**: 1–16.
38. Haase, P. 1993. Zur Situation und Brutbiologie des Steinkauzes *Athene n. noctua* SCOP., 1769 im Westhaveland. *Naturschutz und Landschatspflege in Brandenburg* **2**: 29–37.
39. Haase, P. 2001. Steinkauz – *Athene noctua*. *In* Arbeitsgemeinschft Berlin-Brandenburgischer Ornithologen. Die Vogelwelt von Brandenburg und Berlin Natur & Text. Rangsdorf. [Little Owl data on pp. 382–385.]
40. Haensel, J. & Walter, H. J. 1966. Beitrag zur Ernährung der Eulen im Nordharz-Vorland unter besonderer Berücksichtigung der Insektennahrung. *Beiträge zur Vogelkunde* **11**: 345–358.
41. Hague, J. B. & Hague, A. E. 1969. Some observations on copulation and associated behaviour and food searching of the Little Owl. *Naturalist* **911**: 115–116.

742. Hahn, E. 1984. Welche der mitteleuropäischen Eulenarten eignet sich als Biomonitor? Diplomarbeit, Universität des Saarlandes. Saarbrücken.
743. Hainard, R. 1955. Accouplement "inutile" et attitude de chant chez la Chouette chevêche. *Nos Oiseaux* **23**: 48–49.
744. Haller, W. 1951. *Unsere Vögel*. Verlag der AZ Aarau. [Little Owl data on p. 71.]
745. Hamonville, D'L, 1895. Les oiseaux de la Lorraine. *Mémoires de la Société Zoologique de France* **8**: 257.
746. Hand, R. & Heine, K. H. 1984. Vogelfauna des Regierungsbezirkes Trier. Faunistische und ökologische Grundlagenstudien sowie Empfehlungen für Schutzmassnahmen. *Pollichia* **6**: 127–128.
747. Hansen, W., Synnatzschke, J. & Oelke, H. 1988. Bestimmungsbuch für Rupfungen und Mauserfedern. *Beiträge zur Naturkunde Niedersachsens* **41**: 371–373.
748. Hanson, D. E. 1973. X-ray photographs of Little Owl pellets. *British Birds* **66**: 33.
749. Harbodt, A. & Pauritsch, G. 1987. Lebensraum "Streuobstwiese" Programme und gesetzliche Schutzmöglichkeiten. *In* Keil, W. (ed.). Staatliche Vogelschutzwarte für Hessen, Rheinland-Pfalz und Saarland, 1937–1987. Festschrift. Institut für Angewandte Vogelkunde, 81–91. [Little Owl data on pp. 81–91.]
750. Hardouin, L. 2002. Communication acoustique et reconnaissance individuelle chez la Chevêche d'Athéna CEPE. CNRS. DEA. Strasbourg: Université Louis Pasteur.
751. Hardouin, L. 2006. Communication acoustique et territorialité chez les rapaces nocturnes. Ph.D. thesis, Université Louis Pasteur. Strasbourg.
752. Hardouin, L. A., Tabel, P. & Bretagnolles, V. 2006. Neighbour–stranger discrimination in the Little Owl, *Athene noctua*. *Animal Behaviour* **72**: 105–112.
753. Hardy, M. & Lemaitre, T. 2002. La Chouette chevêche (*Athene noctua* Scopoli, 1769) dans la région de Bréhal (50) au printemps 2002.
754. Harisson, T. H. 1936. Little Owl. *Listener* **766**: 768.
755. Harrison, D. P. 1916. Little Owl breeding in Wiltshire. *British Birds* **10**: 120.
756. Harrison, J. M. 1957. Exhibition of a new race of the Little Owl from the Iberian peninsula. *Bulletin British Ornithological Club* **77**: 2–3.
757. Harrison, J. M. & Hovel, H. 1964. On the taxonomy of *Athene noctua* in Israel. *Bulletin British Ornithological Club* **84**: 91–94.
758. Hartert, E. 1913. Die Vögel der paläarktischen Fauna. Systematisch Übersicht der in Europa, Nord Asien under der Mittelmeerregion vorkommende Vögel. 2. Friedländer und Sohn. Berlin. [Little Owl data on pp. 999–1006.]
759. Hartert, E. 1924. Ornithological results of Captain Buchanan's second Sahara expedition. *Novitates Zoologicae* **31**: 18.
760. Harthan, A. J. 1948. Fledging-period of Little Owl. *British Birds* **41**: 22.
761. Hartmann-Müller, B. 1973. Beitrag zur Mauser des Steinkauzes, *Athene noctua* (Scop.). I. Der Handflügel Philippia. *Abh. Ber. Naturkde. Mus. Kassel* **1**: 286–295.
762. Hartmann-Müller, B. 1974. Beitrag zur Mauser des Steinkauzes, *Athene noctua* (Scop.). II. Armflügel und Schwanzfedern. Philippia. *Abh. Ber. Naturkde. Mus. Kassel* **2**: 182–184.
763. Harvey, G. H. 1923. Little Owl in Cornwall. *British Birds* **17**: 90.
764. Haverschmidt, F. 1939. Beobachtungen über die Begattung beim Steinkauz (*Athene noctua*). *Beiträge zur Fortpflanzungsbiologie der Vögel* **15**: 236–239.
765. Haverschmidt, F. 1940. Prooiresten bij Steenuilenbroedsels. *Ardea* **29**: 56–57.
766. Haverschmidt, F. 1946. Observations on the breeding habits of the Little Owl. *Ardea* **34**: 214–246.
767. Hayden, J. 2004. The diet of the Little Owl on Skomer Island NNR 1998–2003. CCW Contract Science Report 673.
768. Heatwole, H. & Muir, R. 1982. Population densities, biomass and trophic relations of birds in the pre-Saharan steppe of Tunisia. *Journal of Arid Environments* **5**: 145–167.
769. Heckenroth, H. 1985. *Atlas der Brutvögel Niedersachsens 1980 und des Landes Bremen mit Ergänzungen aus den Jahren 1976–1979*. Naturschutz und Landschaftspflege in Niedersachsen. Hannover. [Little Owl data on p. 187.]
770. Heckenroth, H. & Laske, V. 1997. *Atlas der Brutvögel Niedersachsens 1981–1995 und des Landes Bremen*. Naturschutz und Landschaftspflege in Nierdersachsens. [Little Owl data on p. 173.]
771. Hegger, H. L. 1977. Steinkauz, Waldkauz und Waldohreule als Brutvögel im Kempener Land. *Heimatbuch des Kreises Viersen* **1977**: 58–63.
772. Heilig, D. & Stahlheber, K.-H. 2004. Die Entwicklung des Steinkauzbestandes (*Athene noctua*) 1987–2004 einer Untersuchungsfläche in der Südpfalz (Rheinland-Pfalz). *Faune Flora Rheinland-Pfalz* **10**: 425–437.
773. Heim De Balzac, H. & Mayaud, N. 1962. *Les Oiseaux du Nord-ouest de l'Afrique*. Paris: Lechevalier. [Little Owl data on pp. 184–185.]
774. Heinroth, O. & Heinroth, M. 1926. *Die Vögel Mitteleuropas*, Vol. 2. Berlin: Hugo Bermühler Verlag. [Little Owl data on pp. 15–17.]

775. Heitmann, U. 1984. Gute Erfahrungen mit Nisthilfe für Schleiereule (*Tyto alba*) und Steinkauz (*Athene noctua*) im Raum Dinklage. *Jahresbericht Ornithologische Arbeitsgemeinschaft Oldenburg* **8**: 68–69.
776. Helbig, A. 1981. Auswirkungen des strengen Winters 1978–79 auf die Vogelwelt in Westfalen. *Charadrius* **17**: 82–103.
777. Hell, P. 1964. Prispevok k poznaniu potravy niektorych dracov a sov mimoriadne ukej zime 1962–1963. *Zoologicke Listy* **13**: 207–220.
778. Hellebrekers, W. P. J. 1928. Aantal eieren bij *Asio otus* (L.), *Athene noctua* A. E. Brehm en *Riparia riparia* (L.). *Ardea* **17**: 165–166.
779. Hellwege, B. 1968. Diskussionsbeitrag zu Siedlungsdichte-Untersuchungen an Eulen. *Ornithologische Mitteilungen* **20**: 147–148.
780. Hendy, E. W. 1921. Little Owl breeding in Cheshire. *British Birds* **15**: 141.
781. Henrioux, P. 1980. Nidification simultanée de l'Etourneau et de la Chouette chevêche dans le même nichoir. *Nos Oiseaux* **35**: 343.
782. Henry, J.-M. 1992. Suivi de la population de Chouette chevêche sur un secteur témoin situé dans le Parc Naturel Régional de Brotonne. Rapport Groupe Ornithologique Normand. Parc Naturel Régional de Brotonne.
783. Hens, P. A. 1940. Driest optreden van Steenuilen, *Athene noctua vidalii* A. E. Brehm, tijdens de strenge koude. *Limosa* **13**: 103.
784. Hens, P. A. 1954. Een Wezel (*Mustela nivalis*) als prooi van de Steenuil (*Athene noctua*). [Weasel *Mustela nivalis* as a prey of Little Owl *Athene noctua*]. *Limosa* **27**: 63.
785. Hernandez, A., Sanchez, A. J. & Alegre, J. 1987. Datos sobre el regimen alimenticio del mochuelo (*Athene noctua*) y la lechuza comun (*Tyto alba*) en habitats esteparios en la cuenca del duero (Leon y Zamora, Espana). Actas I Congress Internacional de *Aves* Este pasias Leon. pp. 183–193.
786. Hernandez, A., Alegre, J., Salgado, J. M. & Gutierrez, A. 1991. The role of coleopterans in the diet of some vertebrates in northwestern Spain. *Elytron Supplement* **5**: 231–237.
787. Hernandez, M. 1988. La mortalidad del Mochuelo en carreteras. *Panda* **24**: 4–7.
788. Hernandez, M. 1988. Road mortality of the Little Owl – *Athene noctua* – in Spain. *Journal Raptor Research* **22**: 81–84.
789. Herremans, M. & Van Nieuwenhuyse, D. 2004. De Steenuil als indicator voor landschappelijke kwaliteit. Natuurpunt.
790. Herren, H. 1977. The situation regarding birds of prey in Switzerland in 1975. ICBP World Conference on Birds of Prey. Vienna. [Little Owl data on p. 181.]
791. Herrera, C. M. 1973. La captura de carnivores por las *Strigiformes*. *Ardeola* **19**: 441.
792. Herrera, C. M. & Hiraldo, F. 1976. Food-niche and trophic relationships among European owls. *Ornis Scandinavica* **7**: 29–41.
793. Herroelen, P. 1992. Chouettes chevêches menacées? *L'Homme et l'Oiseau* **30**: 38–42.
794. Herweijer, P. 1942. Enquête betreffende de Steenuil (*Athene noctua* Scop.). *Ardea* **31**: 303.
795. Hessel, K. 2004. Onderzoek naar de ruimtelijke verspreiding van vogels, in het bijzonder de Steenuil *Athene noctua*, aan de hand van een aantal statistische en cartografische methodes. Batchelors thesis, Erasmus Hogeschool, Brussels.
796. Heuer, J. 1980. Zum Vorkommen des Steinkauzes (*Athene noctua*) im Braunschweigerhügelland. *Vogelkunde Berichte Niedersachsen*: pp. 3–5.
797. Heuglin, T. von, 1863. Beiträge zur Ornithologie Nord-Ost Afrika's. *Journal für Ornithologie* **11**: 14.
798. Heussler, W. & Heussler, T. 1896. Die Vögel der Rheinpfalz und der unmittelbar angrenzenden Gebiete. *Ornis* **8**: 488.
799. Hewson, R. 1972. Changes in the number of stoats, rats and Little Owl in Yorkshire as shown by tunnel trapping. *Journal of Zoology* **168**: 427–429.
800. Hey, E.-O. 1980. Der Steinkauz (*Athene noctua*)-Brutvogel in der Wesermarsch zwischen Intschede und Bollen. Mitteilungen 1979 DBV Kreisgruppe Verden e.V. pp. 24–25.
801. Heyberger, M. 1979. *Villages et Cultures*. Editions Mars et Mercure. Wettolsheim. [Little Owl data on pp. 84–85.]
802. Heyberger, M. 1985. La Chouette chevêche, une espèce qui était commune. *Saisons d'Alsace* **41**: 13–15.
803. Heyder, R. 1952. *Die Vögel des Landes Sachsen*. Akademische Verlagsgesellschaft. Leipzig. [Little Owl data on pp. 259–260.]
804. Heyder, R. 1962. Nachträge zur sächsischen Vogelfauna. *Beiträge zur Vogelkunde* **8**: 60.
805. Heyder, R. 1967. Der Vogelname "Wichtel". *Beiträge zur Vogelkunde* **13**: 198–204.
806. Heydt, J. G. 1968. Las rapaces y ostras *Aves* de la Sierra de Gata. *Ardeola* **14**: 120.
807. Heyne, K. H. 1978. Beitrag zur Bedeutung der Streuobstwiesen, insbesondere für gefährdete Vogelarten. *Dendrocopos* **5**: 10–15.
808. Hibbert-Ware, A. 1936. Report of an investigation of the food of captive Little Owls. *British Birds* **29**: 302–305.

809. Hibbert-Ware, A. 1936. Sounds produced by Little Owl. *British Birds* **29**: 332.
810. Hibbert-Ware, A. 1937/1938. Report of the Little Owl Food Inquiry. 1936–37. *British Birds* **31**: 162–187.
811. Hibbert-Ware, A. 1937/1938. Report of the Little Owl Food Inquiry. 1936–37. *British Birds* **31**: 205–229.
812. Hibbert-Ware, A. 1937/1938. Report of the Little Owl Food Inquiry. 1936–37. *British Birds* **31**: 249–264.
813. Hibbert-Ware, A. 1938. The Little Owl. Is it injurious? *School Nature Study*: 1–5.
814. Hibbert-Ware, A. 1938. The food habits of the Little Owl (*Carine noctua vidalii*). *Annals of Applied Biology* **25**: 218–220.
815. Hoblingre, F. 2001. Réseau Chevêche-Bilan 1996. *Milvus* **30**: 49–52.
816. Hodson, N. L. 1962. Some notes on the causes of bird road casualties. *Bird Study* **9**: 169.
817. Holandre, J. J. J. 1836. *Faune du département de la Moselle, Animaux vertébrés, Mammifères, Oiseaux, Reptiles et Poissons*. Editions Thiel. Metz. [Little Owl data on p. 52.]
818. Holder, F. W. 1921. Little Owl in Lancashire. *British Birds* **15**: 63.
819. Holupirek, H. 1970. Die Vögel des hohen Mittelgebirges. *Beiträge zur Vogelkunde* **15**: 136.
820. Hölzinger, J. 2001. *Athene noctua* (Scopoli, 1769) Steinkauz. *In* Hölzinger, J. & Mahler, U. (eds). *Die Vögel Baden-Württembergs Band* 2.3: Nicht-Singvögel 3. Verlag Eugen Ulmer. [Little Owl data on pp. 195–210.]
821. Hony, G. B. 1916. Status of Little Owl in Wiltshire. *British Birds* **9**: 210.
822. Hope, L. E. 1924. Little Owl in Cumberland. *British Birds* **18**: 23.
823. Hosking, E. J. & Newberry, C. W. 1945. *Birds of the Night*. London: Collins. [Little Owl data on pp. 42–52.]
824. Hothum, G. 1990. Eulenvorkommen im Odenwaldkreis. *Vogel und Umwelt* **6**: 83–85.
825. Hounsome, T., O'Mahony, D. & Delahay, R. 2004. The diet of Little Owls *Athene noctua* in Gloucestershire, England. *Bird Study* **51**: 282–284.
826. Hovorka, W. & Ille, R. 1999. Das niederösterreichische Artensicherungsprogramm für den Steinkauz 1996–1998. *Egretta* **42**: 156–163.
827. Hoz, de la M. 1982. Algunos datos sobre la alimentacion del Mochuelo comun (*Athene noctua*) en Asturias. *Boletín de Ciencias Naturales, Instituto de Estudios Asturianos* **29**: 113–120.
828. Hubl, H. 1952. Beiträge zur Kenntnis der Vorhaltensweisen junger Eulenvögel in Gefangenschaft: (Schleiereule, *Tyto alba*; Steinkauz, *Athene noctua* und Waldkauz, *Strix aluco*). *Zeitschrift für Tierpsychologie* **9**: 102–119.
829. Hüe, F. & Etchecopar, R. D. 1970. *Les Oiseaux du Proche et du Moyen Orient*. Boubée et Cie. Paris. [Little Owl data on pp. 411–412.]
830. Hugues, A. 1924. La diminution de la Chevêche. *Revue Française d'Ornithologie* **8**: 320–321.
831. Hugues, S. W. M. 1977. Little Owl dead with head stuck in nest-hole of Great Tit. *British Birds* **70**: 501–502.
832. Hulten, M. 1955. Gewöllenuntersuchungen. *Regulus* **35**: 45.
833. Hulten, M. & Wassenich, V. 1961. Die Vogelfauna Luxemburgs.2. Teil. *Bulletin de la Société des Naturalistes Luxembourgeois* **66**: 446–447.
834. Hurstel, A. 1986. Les proies de la Chouette chevêche. *Lien Ornithologique d'Alsace* **43**: 10–11.
835. Hurstel, A. 1987. Les proies de la Chouette chevêche dans le Haut-Rhin, en 1986. *Lien Ornithologique d'Alsace* **45**: 22–23.
836. Hurstel, A. 1991. La Chouette chevêche, *Athene noctua* Scop. dans le Haut-Rhin. *Ciconia* **15**: 99–110.
837. Igalffy, K. 1949. Jedan par cukova obienih (*Athene noctua*, Scop.) istrijebio lastavice. *Larus* **3**: 370–371.
838. Ille, R. 1983. Ontogenese des Beutefangverhaltens beim Steinkauz (*Athene noctua*). *Journal für Ornithologie* **124**: 133–146.
839. Ille, R. 1992. Zur Biologie und Ökologie des Steinkauzes (*Athene noctua*) im Marchfeld: Aktuelle Situation und mögliche Schutzmassnahmen. *Egretta* **35**: 49–57.
840. Ille, R. 1995. Ergebnisse einer Bestandserhebung beim Steinkauz (*Athene noctua*) im Nordburgenland zwischen 1992–1994. *Biologisches Forschungsinstituts Burgenland-Berichte* **83**: 23–29.
841. Ille, R. 1996. Zur Biologie und Ökologie zweier Steinkauzpopulationen in Österreich. *Abhandlungen Der Zoologisch-Botanischen Gesellschaft in Österreich* **129**: 17–31.
842. Ille, R. & Grinschgl, F. 2001. Little Owl (*Athene noctua*) in Austria. Habitat characteristics and population density. *In* Genot, J.-C., Lapios, J.-M., Lecomte, P. & Leigh, R. S. (eds). *Ciconia* **25**: 129–140.
843. Illner, H. 1979. Eulenbestandsaufnahmen auf den M. T. B. Werl von 1974–1978. *DBV. AG zum Schutz bedrohter Eulen Informationsblatt* **9**: 5–6.
844. Illner, H. 1979. Erfahrungsbericht über Steinkauzbruten in Niströhren. *DBV. AG zum Schutz bedrohter Eulen Informationsblatt* **9**: 6.
845. Illner, H. 1981. Populationsentwicklung der Eulen (*Strigiformes*) auf einer Probefläche Mittelwestfalen 1974–1979 und bestandsbeeinflussende Faktoren, insbesondere anthropogener Art. *Ökologie der Vögel* **3**: 301–310.
846. Illner, H. 1988. Langfristiger Rückgang von Schleiereule *Tyto alba*, Waldohreule *Asio otus*, Steinkauz *Athene noctua* und Waldkauz *Strix aluco* in der Agrarlandschaft Mittelwestfalens 1974–1986. *Vogelwelt* **109**: 145–151.

847. Illner, H. 1990. Sind durch Nistkasten-Untersuchungen verlässliche Populationstrends zu ermitteln? Eine Fallstudie am Steinkauz (*Athene noctua*). *Vogel und Umwelt* **6**: 47–57.
848. Illner, H. 1991. Influence d'un apport de nourriture supplémentaire sur la biologie de reproduction de la Chouette chevêche, *In* Juillard, M., Bassin, P., Baudvin, H. *et al.* (eds). *Athene noctua*. 153–157.
849. Illner, H. 1992. Road deaths of Westphalian owls: methodological problems, influence of road type and possible effects on population levels. *In* Galbraith, C. A., Taylor, I. R. & Percival, S. (eds). *Nature Conservation* **5**: 94–100.
850. Illner, H. 1995. Strassentod westfälicher Eulen (*Strigiformes*) und Vorschläge zur Vermeidung. *Eulen Rundblick* **42**: 18–20.
851. Illner, H., Lederer, W. & Loske, K.-H. 1989. *Atlas der Brutvögel des Kreises Soest/Mittelwestfalen 1981–1986*. Arbeitsgemeinschaft Biologischer Umweltschutz im Kreis Soest Lohne. [Little Owl data on pp. 132–133.]
852. Il'Yukh, M. P. 2002. Gnezdovaya biologiya domogovo sycha Predkavkazie [Breeding biology of the Little Owl in Front-Caucasian area]. Ptitsy Yuzhnoi Rossii [Birds of south Russia]. [Materials of Intern. Conf. "Results and prospects of ornithology development on the Northern Caucasus in the 21st century" devoted to twenty-year anniversary of activity of the North Caucasian Ornithological Group. October 24–27 2002]. Rostov-on-Don. *Proc.* **31**: 113–118.
853. Il'Yukh, M. P. & Khokhlov, A. N. 1999. Kladki i razmery yaits ptits Tsentral'nogo Predkavkaz'ya [Clutches and sizes of bird eggs of the Central Front-Caucasian area]. Stavropol.
854. Ingendahl, D. & Tersteegen, H.-J. 1992. Bestand von Steinkauz und Schleiereule im Sudkreis Kleve (Unterer Mittelrhein). Ein Bericht der Eulenschutzgruppe Issum. *DBV. AG zum Schutz bedrohter Eulen Informationsblatt* **38**: 11–13.
855. Isenmann, P. & Moali, A. 2000. *Oiseaux d'Algérie*. SEOF. [Little Owl data on pp. 190–191.]
856. Ishunin, G. I. 1965. Pitanie domovogo sycha v zapovednike Aral-Paigambar (Uzbekistan) [The Little Owl feeding in Aral-Paigambar Nature Reserve (Uzbekistan)]. *Ornitologiya* **7**: 471–472.
857. Ishunin, G. I. & Pavlenko, T. A. 1966. Materialy po ekologii zhivotnykh pastbishch Kyzylkuma [Materials on animal ecology of the pastures of Kyzylkum]. Pozvonochnye zhivotnye Azii [Vertebrates of the Middle Asia]. Tashkent. [Little Owl data on pp. 28–66.]
858. Ivanov, A. I. 1976. *Katalog Ptits SSSR*. [*Catalogue of the Birds of the USSR*]. Leningrad: Nauka Press.
859. Izmailov, I. V. & Borovitskaya, G. K. 1973. Ptitsy *Yugo-Zapadnogo Zabaikal'ya* [*Birds of the South-Western Trans-Baikalia*]. Vladimir, Vladimir State Pedagogical Institute Press. [Little Owl data on pp. 296–303.]
860. Jablonski, B. 1976. Estimation of birds abundance in large areas. *Acta Ornithologica* **16**: 23–62.
861. Jacob, J. & Hoerschelmann, H. 1984. Chemotaxonomische Untersuchungen an Eulen (*Strigiformes*). *Funktionelle Biologie & Medizin* **3**: 56–61.
862. Jacob, J. P. 1979. Chronique ornithologique Mars–Avril–Mai 1979. *Aves* **16**: 149.
863. Jacob, J. P. 1982. Chronique ornithologique Juin-Juillet-Août 1981. *Aves* **19**: 156.
864. Jacobs, F. 2003. Steenuilinventarisarie midden-Betuwe 2002. *Athene Nieuwsbrief STONE* **7**: 11–16.
865. Jacobs, F. 2003. Succes van het verplaatsen van eieren of jongen van Steenuilen. *Athene Nieuwsbrief STONE* **8**: 20–23.
866. Jacobsen, L. B. 2006. Ynglebestanden af Kirkeugle *Athene noctua* i Vendsyssel og Himmerland 1981–2000. [The Little Owl in northern Jutland, Denmark]. *Dansk Ornitologisk Forenings Tidsskrift* **100**: 35–43.
867. Jacobsen, L. B. & Dabelsteen, T. 2006. Individuel variation i Kirkeuglens *Athene noctua* territoriekald. [Individual variation in the territorial call of the Little Owl *Athene noctua*.] *Dansk Ornitologisk Forenings Tidsskrift* **100**: 51–56.
868. Jacoby, H., Knötzsch, G. & Schuster, S. 1970. Die Vögel des Bodenseegebietes. *Ornithologische Beobachter* **67**: 177.
869. Jacquat, B. 1978. Schweizerische Ringfundmeldung für 1975 und 1976 (67 Ringfundbericht). *Ornithologische Beobachter* **75**: 151.
870. Jagoš, B. 1995. K rozšíření sýcka obecného (*Athene noctua*) ve východní cásti okresu Hodonín. [The abundance of Little Owl (*Athene noctua*) in east part of the Hodonin district]. *Crex* **6**: 33–34.
871. Jaksic, F. M. & Marti, C. D. 1981. Trophic ecology of *Athene* owls in mediterranean-type ecosystems: a comparative analysis. *Canadian Journal of Zoology* **59**: 2331–2340.
872. Janes, H. R. 1935. The Little Owl. *Bird Notes and News* **16**: 177.
873. Janossy, D. & Haraszthy, L. 1985. The status of birds of prey in Hungary 1982. *Bulletin of World Working Group on Birds of Prey* **2**: 45.
874. Jarry, G. 1983. Fluctuations des aires de reproduction et des densités de peuplement chez les oiseaux d'Europe Occidentale sous l'influence directe ou indirecte de l'homme. *Compte Rendu de la Société de Biogéographie* **59**: 95.
875. Jaschke, M. 1978. Herbstbalz beim Steinkauz (*Athene noctua*). *Charadrius* **14**: 23–24.
876. Jaschke, M. 1989. Hilfe für einen Steinkauz. *Falke* **36**: 238.

877. Jaspers, V., Covaci, A., Maervoet, J. *et al.* 2005. Brominated flame retardants and organochlorine pollutants in eggs of little owls (*Athene noctua*) from Belgium. *Environmental Pollution* **136**: 81–88.
878. Jay, M. 1993. *Athene noctua In* Centre Ornithologique Du Gard. *Oiseaux Nicheurs du Gard. Atlas Biogéographique 1985–1993.* [Little Owl data on p. 145.]
879. Jegen, H. 1997. Ergebnisse einer Steinkauzkartierung in der Verbandsgemeinde Kyllburg 1996 zusammengestellt. *Dendrocopos* **24**: 32–37.
880. Jennings, M. C. 1981. *Birds of the Arabian Gulf.* London: George Allen & Unwin Ltd. [Little Owl data on p. 73.]
881. Jermaczek, A., Czwalga, T., Jermaczek, D. *et al.* 1995. *Birds of the Gorzów Wielkopolski Region-faunistic monograph.* Publishers of Naturalists of the Gorzów Wielkopolski Region, Swiebodzin.
882. Jespersen, P. 1937. De forskellige uglers udbredelse og forekomst i Danmark. *Dansk Ornitologisk Forenings Tidsskrift* **31**: 115–120.
883. Jöbges, M. 2004. Der Steinkauz (*Athene noctua*). *In* Gedeon, K., Mitschke, A. & Sudfeldt, C. (eds). *Brutvögel in Deutschland.* Hohenstein-Ernsthal. pp. 22–23.
884. Johansen, H. 1956. Die Vogelfauna Westsiberiens. *Journal für Ornithologie* **97**: 213–214.
885. Joiris, C. & Delbecke, K. 1981. Rückstände chlororganischer Pestizide und PCBs in belgischen Greifvögeln. *Ökologie der Vögel* **3**: 173–180.
886. Joiris, C. & Martens, P. 1971. Teneur en pesticides d'oeufs de rapaces récoltés en Belgique en 1969. *Aves* **8**: 8.
887. Joiris, C., Lauwereys, M. & Vercruysse, A. 1977. PCB and organochlorine pesticides residues in eggs of birds of prey collected in Belgium in 1972, 1973, 1974. *Gerfaut* **67**: 447–458.
888. Joiris, C., Dejaegher, J. & Delbecke, K. 1979. Changes of eggshell thickness in Belgian birds of prey. *Gerfaut* **69**: 165–210.
889. Joiris, C., Delbecke, K., Martens, E., Lauwereys, M. & Vercruysse, A. 1979. PCB and organochlorine pesticides residues in eggs of birds of prey found dead in Belgium from 1973 to 1977. *Gerfaut* **69**: 319–337.
890. Jones, C. G. 1981. Abnormal and maladaptive behaviour in captive raptors. Recent advances in the study of raptor diseases. Proceedings of the International Symposium on Diseases of Birds of Prey. 1st–3rd 1980 London. [Little Owl data on p. 57.]
891. Jones, R. W. 1920. Little Owl in Caernarvonshire. *British Birds* **14**: 135.
892. Jones, W. M. & Hodgson, T. V. 1920. Little Owl in Cardiganshire, Cornwall and Montgomeryshire. *British Birds* **13**: 274–275.
893. Jong, de J. 1976. De Steenuil. *Onze Vogels* **37**: 468.
894. Jong, de J. 1996. Weinig broedgewallen van de Steenuil. *Vanellus* **5**: 153.
895. Jongbloed, R. H., Traas, T. P. & Luttik, R. 1996. A probabilistic model for deriving soil quality criteria based on secondary poisoning of top predators. II. Calculations for dichlorodiphenyltrichlorethane (DDT) and cadmium. *Ecotoxicity and Environmental Safety* **34**: 279–306.
896. Jonkers, D. A. & De Vries, G. W. 1977. Verkeersslachtoffers onder de fauna. [Little Owl data on p. 73.]
897. Jordans, A. von & Steinbacher, J. 1941. Beiträge zur Avifauna der Iberischen Halbinsel. *Annalen Naturhistorischen Museum Wien* **52**: 200–244.
898. Jorek, N. 1975. Die im Westfalen gefährdeten Vogelarten "Rote Liste Stand 1.1.1975". *Alcedo* **2**: 82–87.
899. Jouanin, C. 1963. La destruction des prétendus nuisibles. *Nos Oiseaux* **27**: 132–136.
900. Jouard, H. 1937. Sur un cas d'accouplement "pour le plaisir" chez des Mésanges. *Alauda* **9**: 230–231.
901. Joubert, B. 1992. *Oiseaux du Massif Central.* Une avifaune de Haute-Loire. C. P. I. E. du Velay. [Little Owl data on pp. 177–178.]
902. Jourdain, F. C. R. 1909. Little Owls in Anglesey. *British Birds* **3**: 126–127.
903. Jourdain, F. C. R. 1915. Status of the Little Owl in Staffordshire. *British Birds* **9**: 250–251.
904. Jourdain, F. C. R. 1917. Spread of the Little Owl in south Oxfordshire and north Berkshire. *British Birds* **10**: 271.
905. Joveniaux, A. 1988. Atlas des oiseaux nicheurs du département du Jura. *Nos Oiseaux* **39**: 289–293.
906. Joveniaux, A. 1991. La Chouette chevêche. Biologie, comportement, habitat. *Jura Nature* **47**: 10–12.
907. Juillard, M. 1974. La Chouette chevêche *Athene noctua* (Scop.) en Ajoie. *Bulletin Association pour la Défense des Intérêts du Jura (ADI Journal)* **45**: 245–253.
908. Juillard, M. 1979. La croissance des jeunes Chouettes chevêches, *Athene noctua*, pendant leur séjour au nid. *Nos Oiseaux* **35**: 113–124.
909. Juillard, M. 1979. Quelques remarques sur l'écologie du verger. *Bulletin Association pour la Défense des Intérêts du Jura (ADI Journal)* **50**: 283–287.
910. Juillard, M. 1980. Répartition, biotopes et sites de nidification de la Chouette chevêche, *Athene noctua*, en Suisse. *Nos Oiseaux* **35**: 309–337.
911. Juillard, M. 1980. Chouette chevêche *In* Schifferli, A., Geroudet, R., Winkler, R. *et al.* (eds). *Atlas des Oiseaux Nicheurs de Suisse.* Station ornithologique de Suisse. Sempach. [Little Owl data on pp. 182–183.]
912. Juillard, M. 1981. Trois malformations anatomiques apparentes chez la Chouette chevêche, *Athene noctua*. *Nos Oiseaux* **36**: 121–125.

13. Juillard, M. 1981. Zur Kontamination von Greifvögeln aus der Französischen Schweiz mit chlororganischen Verbindungen mit PCBs und einigen Schwermetallen. *Ökologie der Vögel* **3**: 169–170.
14. Juillard, M. 1981. Nichoir pour Chouette chevêche (*Athene noctua*). *Nos Oiseaux* **36**: 40–42.
15. Juillard, M. 1983. La photographie sur pellicule infrarouge: une méthode pour l'étude du régime alimentaire des oiseaux cavicoles. *Terre et Vie* **37**: 267–285.
16. Juillard, M. 1983. Der Schutz der jurassischen Obstgärten. *Vögel der Heimat* **2**: 35–36.
17. Juillard, M. 1984. *La Chouette Chevêche.* Nos Oiseaux. Prangins.
18. Juillard, M. 1984. Contribution à la connaissance éco-éthologique de la Chouette chevêche, *Athene noctua* (Scop.), en Suisse. Ph.D. Thesis, Université de Neuchâtel.
19. Juillard, M. 1985. A propos des habitats de la Chouette chevêche, *Athene noctua*, dans les régions méditerranéennes. *Nos Oiseaux* **38**: 121–132.
20. Juillard, M. 1987. Chevêche la chouette. *Quatre Saisons du Jardinage* **42**: 26–32.
21. Juillard, M. 1989. The decline of the Little Owl *Athene noctua* in Switzerland. *In* Meyburg, B.-U. & Chancellor, R. D. (eds). *Raptors in the Modern World.* Bevlin: W. W. G. B. P. pp. 435–439.
22. Juillard, M. 1990. Le déclin de la Chouette chevêche (*Athene noctua*), en Suisse *In* Colloque interrégional d'ornithologie et de mammalogie, Strasbourg 1989. *Ciconia* **14**: 55–57.
23. Juillard, M. 1997. Les vergers de la Chouette. *Pro Natura Magazine* **5**: 6–8.
24. Juillard, M. 2000. Appel d'aide pour le verger jurassien d'arbres à hautes tiges et pour la Chevêche, suite au passage de l'ouragan "Lothar". *Nos Oiseaux* **47**: 86.
25. Juillard, M. & Jacquat, B. 1982. A propos du verger jurassien. Bulletin de l'Association pour la Sauvegarde du Patrimoine Rural Jurassien (ASPRU Journal) *L'Hôta* **5**: 4–11.
26. Juillard, M., Praz, J.-C., Etournaud, A. & Beaud, P. 1978. Données sur la contamination des rapaces de Suisse romande et de leurs oeufs par les biocides organochlorés, les PCB et les métaux lourds. *Nos Oiseaux* **34**: 189–206.
27. Juillard, M., Baudvin, H., Bonnet, J., Génot, J.-C. & Teyssier, G. 1990. Sur la nidification en altitude de la Chouette chevêche, *Athene noctua*. Observations dans le Massif central (France). *Nos Oiseaux* **40**: 267.
28. Juillard, M., Baudvin, H., Bonnet, J. & Génot, J.-C. 1992. Habitat et sites de nidification de la Chouette chevêche (*Athene noctua*) sur le Causse Méjean (Lozère). *Nos Oiseaux* **41**: 415–440.
29. Kaeser, G. 1979. Tröstliches von beinahe Kandidaten, bzw. Kandidaten der roten Liste. *Vögel der Heimat* **49**: 247.
30. Kahmann, H. 1953. Das Ergebnis der Zergliederung von Eulengewöllen und seine wissenschaftliche Verwertung. *Ornithologische Mitteilungen* **5**: 201–206.
31. Kämpfer, A. & Lederer, W. 1988. Dismigration des Steinkauzes *Athene noctua* in Mittelwestfalen. *Vogelwelt* **109**: 155–164.
32. Kämpfer-Lauenstein, A. & Lederer, W. 1991. Zur Dismigration und Populationsdynamik des Steinkauzes (*Athene noctua*) in Mittelwestfalen. *In* Stubbe, M. (ed.). Populationsökologie von Greifvogel und Eulenarte 2, Wissenschaftliche Beitrage Universität Halle 1991: 479–491.
33. Kämpfer-Lauenstein, A. & Lederer, W. 1995. Bestandsentwicklung einer Steinkauzpopulation in Mittelwestfalen (1974–1994). *Charadrius* **4**: 211–216.
34. Karyakin, I. V. 1998. Konspekt fauny ptits Respubliki Bashkortostan [Conspectus of bird fauna of Republic of Bashkortostan]. Perm: Centre of Field Investigations of Uralian Animal Protection Union.
35. Karyakin, I. V. 1998. Pernatye khishchniki Ural'skogo regiona. [Birds of prey and owls of the Ural region]. Perm: Centre of Field Researches of Ural Animal Conservation Union.
36. Karyakin, I. V. & Kozlov, A. A. 1999. Predvaritelínyikasdastr ptits Chelyabinskoi oblasti [Preliminary cadastre of birds of Chelyabinsk Region (Oblast)]. Novosibirsk: Manuscript Press. [Little Owl data on pp. 221–223.]
37. Kaus, D. 1978. Obstbaumprogram-ökologische Forderung zur Erhaltung des Streuobstbaus. *Vogelschutz* **4**: 13–16.
38. Kaus, D. 1981. Steinkauz in Franken. Vortrag auf der Tagung der AG zum Schutz bedrohter Eulen, Giessen: 6–8.
39. Kayser, Y. 1995. Regime alimentaire inhabituel de la Chouette chevêche (*Athene noctua*) dans les salins de Thyna, Tunisie. *Alauda* **63**: 152–153.
40. Kehrer, S. 1972. *Der Steinkauz* (*Athene noctua*). Stuttgart: DBV Verlag.
41. Keil, D. 1984. Die Vögel des Kreises Hettstedt. *Apus* **5**: 173.
42. Kemp, A. C. 1989. Estimation of biological indices for little-known African owls. *In* Meyburg, B.-U. & Chancellor, R. D. (eds). *Raptors in the Modern World.* Berlin: W. W. G. B. P. pp. 441–449.
43. Kempeneers, Y. 1996. Recensement de la Chouette chevêche sur les communes de Siberet et Hompé. *Aves Feuille de Contact* **6**: 266.
44. Kempf, C. 1973. Les rapaces nocturnes d'Alsace. *Alauda* **41**: 415–416.
45. Kempf, C. 1976. *Oiseaux d'Alsace.* Strasbourg: Istra. [Little Owl data on pp. 147–148.]

946. Keraultret, L. 1981. Liste rouge des espèces d'oiseaux nicheurs menacées et rares dans le Nord de la France (Départements du Nord et du Pas-de-Calais). *Héron* **4**: 6.
947. Kesteloot, E. 1977. Present situation of birds of prey in Belgium. ICBP World Conference on Birds of Prey. Vienna. 1975. [Little Owl data on p. 86.]
948. Keve, A. & Kohl, S. 1961. A new race of the Little Owl from Transylvania. *Bulletin British Ornithological Club* **81**: 51–52.
949. Keve, A., Kohl, I., Matousek, F., Mosansky, A. & Rucner-Kroneisl, R. 1962. Über die taxonomische Stellung der Südesteuropäischen Steinkäuze, *Athene noctua* (Scop.). *Larus* **14**: 26–74.
950. Key, A. S. & Gribble, F. C. 1951. Observations on a pair of Little Owls. *Bedfordshire Naturalist* **5**: 29–32.
951. Khokhlov, A. N. 1992. Osobennosti ekologii sov vantropogennykh landshaftakh Tsentral'nogo Predkavkz'ya [Ecological characters of Owls in the anthropogenic landscapes of the central Pre-Caucasus]. Sovremennatya ornitologiya 1991 [Modern ornithology 1991]. Nauka. [Little Owl data on pp. 85–95.]
952. Khokhlov, A. N. 1995. Ornitologicheskie nablyudeniya v Zapadnoi Turkmenii [Ornithological observations in Western Turkmenia]. Stavropol, Stavropol State Pedagogical University.
953. Khokhlov, A. N. & Kulikov, V. T. 1991. Letnyaya ornitofauna Severnogo Stavropol'ya [Summer ornithofauna of the Northern part of Stavropol Territory]. Fauna, naselenie i ekologiya ptits Severnogo Kavkaza [Fauna, population and ecology of birds of North Caucasia]. Stavropol. [Little Owl data on pp. 107–112.]
954. Khokhlov, A. N., Tel'Pov, V. A. & Bitarov, V. N. 1991. Zimnyaya avifauna g.Kislovodska i ego okrestnostei (Stavropolskiy krai) [Winter avifauna of Kislovodsk town and its vicinities (Stavropol Territory)]. [Little Owl data on pp. 123–135.]
955. Khokhlov, A. N., Il'Yukh, M. P., Emel'Yanov, S. A. *et al.* 1998. K letnei ornitofaune nizoviy reki Kumy i prilezhashchikh territoriy [On the summer ornithofauna of lower parts of Kuma River and adjoining territories]. [Little Owl data on pp. 135–143.]
956. Kimmel, O. 1983. Erfahrungen mit der mardersicheren Steinkauzröhre an 27 Standorten im Raum Ibbenbüren. *AG zum Schutze bedrohter Eulen. Info* **16**: 2–3.
957. Kimmel, O. 1999. Zur Brutplatzwahl des Steinkauzes. *Eulen Rundblick* **48/49**: 23.
958. Kimmel, O. & Radler, K. 1999. Niströhrenstandort und Ansiedlungserfolg beim Steinkauz (*Athene noctua*). *Eulen Rundblick* **48/49**: 21–23.
959. Kinsky, F. C. 1973. The subspecific status of the New Zealand population of the Little Owl, *Athene noctua* (Scopoli, 1769). *Notornis* **20**: 9–13.
960. Kirchberger, K. 1988. Artenschutzmöglichkeiten beim Steinkauz und Schwarzmilan. Vogelschutz in Österreich. *Mitteilungen der Österreichischen Gesellschaft für Vogelkunde* **2**: 52–55.
961. Kirmann, F. B. & Jourdain, F. C. R. 1966. *British Birds*. London and Edinburgh: Nelson. [Little Owl data on p. 85.]
962. Kiselev, F. A. & Ovchinnikova, E. N. 1953. Domovoi sych kakistrebitel myshevidnykh gryzunov v stepnom Krymu [Little Owl as small rodent predator in steppe zone of Crimea]. *Proceedings of the Crimean Branch of USSR Academy of Science* **3**: 51–52.
963. Kislenko, G. S. & Erokhin, V. B. 1998. Novye svedeniya o rasprostranenii i ekologii redkikh vidov ptit Moskovskoi oblasti [New data on distribution and ecology of rare bird species of Moscow Region]. Redkie vidy ptits Nechernozmengo tsentra Rossi [Rare bird species of non-black-earth centre of Russia]. Moscow. [Little Owl data on pp. 74–79.]
964. Kitowski, I. 2002. Coexistence of owl species in the farmland of southeastern Poland. *Acta Ornithologica* **37**: 121–124.
965. Kitowski, I. 2003. Distribution of Little Owl *Athene noctua* and Barn Owl *Tyto alba* in the Zamosc Region (SE Poland) in the light of atlas studies. *Ornis Hungarica* **12–13**: 271–274.
966. Kitowski, I. & Grzywaczewski, G. 2001. Monitoring Little Owl *Athene noctua* (Scop. 1769) in the town of Chelm (SE Poland) from 1998–2000. Bird numbers. 15th International Conference of the EBCC 26th–31th March, Nyiregyhaza, Hungary.
967. Kitowski, I. & Grzywaczewski, G. 2003. The monitoring of Little Owl *Athene noctua* in Chelm (SE Poland) in 1998–2000. *Ornis Hungarica* **12–13**: 279–282.
968. Kitowski, I. & Wojtak, E. 2001. Distribution of Little Owl *Athene noctua* (Scop. 1769) and Barn Owl *Tyto alba* (Scop. 1769) in the Zamosc region (SE Poland) in the light of atlas studies. Bird numbers. 15th International Conference of the EBCC 26th–31th March, Nyiregyhaza, Hungary.
969. Klaas, C. 1963. Vom Steinkauz und seinem Beutetieren. *Natur und Museum* **93**: 79–84.
970. Klaehn, D. 1983. Steinkauz *In* Grosskopf, G. & Klaehn, D. (eds). *Die Vogelwelt des Landkreises Stade*. Verlag Friedrich Schwaumburg. Stade. [Little Owl data on pp. 285–286.]
971. Klein, H. P. 1984. Bestandserfassung ausgewählter gefährdeter Brutvogelarten im Kreis Viersen. *Charadrius* **20**: 12–16.
972. Kleinschmidt, O. 1906. *Strix Athene*. Steinkauz. Berajah.

973. Kleinschmidt, O. 1907. Zum geographischen Variieren von *Strix Athene*. *Falco* **3**: 63–67.
974. Kleinschmidt, O. 1909. Beschreibung neuer Formen (*Strix Athene saharae* nov. spp.). *Falco* **5**: 19–20.
975. Kleinschmidt, O. 1934. *Die Raubvögel der Heimat*. Leipzig: Verlag Quelle & Meyer. [Little Owl data on p. 15.]
976. Kleinschmidt, O. 1958. *Raubvögel und Eulen der Heimat*. Wittenberg-Lutherstadt.
977. Kneule, W. 1996. Wiederfund eines gepflegten Steinkauzes (*Athene noctua*). *Ornithologische Schnellmitteilungen Baden-Württemberg* **50**: 70.
978. Kneule, W. & Michels, H. 1994. Populationsentwicklung des Steinkauzes im Raum Nürtingen und Filderstadt, Landkreis Esslingen. *Ornithologische Schnellmitteilungen Baden-Württemberg* **42**: 39–41.
979. Knopfli, W. 1971. Die Vogelwelt der Limmathal und Zürichseeregion. *Ornithologische Beobachter* **68**: 157–159.
980. Knötzsch, G. 1978. Ansiedlungsversuche und Notizen zur Biologie des Steinkauzes (*Athene noctua*). *Vogelwelt* **99**: 41–54.
981. Knötzsch, G. 1985. Zwölfjährige Populationsuntersuchungen an einem isolierten Vorkommen des Steinkauzes im Bodenseegebiet. Bundesweiter Tagung der AG zum Schutz bedrohter Eulen am 2/3 März 1985 in Grävenwiesbach. DBV Zusammenfassung der Vorträge. [Little Owl data on pp. 3–4.]
982. Knötzsch, G. 1988. Bestandsentwicklung einer Nistkasten-Population des Steinkauzes *Athene noctua* am Bodensee. *Vogelwelt* **109**: 164–171.
983. Koblik, E. A. 2001. Raznoobrazie ptits (po materialam ekspozitsii Zoologicheskogo Muzeya MGU) [Bird diversity (on materials of exposition of the Zoological Museum of Moscow State University)]. Moscow, Moscow Univ. Press. **3**: 360.
984. Kochan, W. 1979. Materialy do skladu pokarmu ptakow drapieznych i sow. *Acta Zoologica Cracoviensia* **23**: 244–246.
985. Koebel, E. 1923. Ueber die Vögel der Stadt Stuttgart und ihrer Umgebung. *Mitteilungen über die Vogelwelt* **2**: 163.
986. Komarov, Y. E. 1998. Ptitsy seliskikh naselennykh punktov respubliki Severnaya Ossetia-Alania [Birds of countryside settlement of Republic of Northern Ossetia-Alania]. *Kavkazskiy Ornithologicheskiy Vestnik* [*Caucasian Ornithological Bulletin*] **10**: 65–74.
987. König, C. 1968. Siedlungsdichte-Untersuchungen an Eulen. *Ornithologische Mitteilungen* **20**: 145–147.
988. König, C. 1969. Extrem kurze Brutdauer beim Steinkauz (*Athene noctua*). *Vogelwelt* **90**: 66–67.
989. König, C., Weick, F. & Becking, J.-H. 1999. *Owls: A Guide to the Owls of the World*. New Haven, Connecticut: Yale University Press.
990. Koop, M.-H. 1996. Freilandökologischer Vergleich der Kleinsäuger-, Laufkäfer-und Regenwurm-Abundanz mit deren Nutzung als Beute durch den Steinkauz *Athene noctua* (Scopoli, 1769) in drei Steinkauzrevieren der Mechernicher Voreifel. Masters thesis.
991. Kopystynska, K. 1962. Investigations on the vision of infra-red in animals. Part III. Preliminary experiments on the Little Owl *Athene noctua* (Scop.). *Prace Zoologiczne* **7**: 95–107.
992. Korovin, V. A. 1997. Ptitsy yuzhnoi okonechnosti Chelyabinskoi oblasti [Birds of the southern edge of Chelyabinsk Region]. [Little Owl data on pp. 74–97.]
993. Kostin, A. B. & Rozovskaya, T. A. 1998. Redkie vidy ptits yugo-vostoka Moskovskoi oblasti [Rare bird species of the Moscow Region]. [Little Owl data on pp. 91–93.]
994. Kostin, Y. V. 1983. *Ptitsy Kryma [Birds of the Crimea]*. Moscow: Nauka. [Little Owl data on pp. 144–145.]
995. Kostrzewa, A., Ferrer-Lerin, F. & Kostrzewa, R. 1986. Abundance, status and vulnerability of raptors and owls in parts of the Spanish Pyrennees. *Bulletin of World Working Group on Birds of Prey* **3**: 186–187.
996. Kotkova, L. I. & Smogorzhevsky, L. O. 1971. [Diet of Little Owl (*Athene noctua* indigena) in Chernomorsky Nature Reserve and neighbouring territories]. Visnik Kiivskogo Universitetu. *Seriya Biologichna* [*Arch. Kiev Univ. Ser. Biol.*] **13**: 117–120.
997. Kouzmanov, G., Todorov, R. & Stoyanov, G. 1995. Informations sur la répartition et la situation des rapaces nocturnes en Bulgarie. *Circulaire du GTMR* **21**: 14–17.
998. Köves, E. 1965. Steinkauz als Haustier. *Das Tier Berlin Monatschrift* **5**: 24–25.
999. Kovshar, A. F. & Levin, A. S. 1993. Ptitsy pustyni Betpak-Dala (Letniy aspect) [The birds of Betpak-Dala desert (summer aspect)]. Fauna i biologiya ptits Kazakhstana [Fauna and biology of birds of Kazakhstan]. Almaty, Inst. of Zool. [Little Owl data on p. 104.]
1000. Kowalski, M., Lippoman, T. & Oglecki, P. 1991. Liczebnosc sow *Strigiformes* we wschodniej czesci Puszczy Kampinoskiej [Census of owls *Strigiformes* in the Eastern part of Kampinos National Park (Central Poland)]. *Acta Ornitholecologica* **26**: 23–29.
1001. Krahe, R. 1970. Über das Beuteschlagen und den Sinn des Abdeckens von Beutetieren bei Eulen. *Gefierdete Welt* **94**: 107–109.

1002. Kramer, H. & Wolters, H. E. 1964. Zur Taxonomie europäischer, in Neuseeland eingebürgerter Vögel. *Journal of Ornithology* **112**: 202–226.
1003. Kraus, E. 1988. Steinkauz *In* Spitzenberger, F. (ed.). Artenschutz in Österreich. Besonders gefährdete Säugetiere und Vögel Österreichs und ihre Lebensräume. Bundesministerium für Umwelt, Jugend und Familie. Wien. [Little Owl data on pp. 275–276.]
1004. Krause, F. 1999. The conservation of the Barn (*Tyto alba*) and Little (*Athene noctua*) Owls in Germany. *Crex* **14**: 101–103.
1005. Krause, F. 1999. The conservation of the Barn (*Tyto alba*) and Little (*Athene noctua*) Owls in the Netherlands. *Crex* **14**: 99–100.
1006. Krause, F. 2000. Owls in the Breclav region in 1999. *Crex* **15**: 45–47.
1007. Krause, F. 2001. Owls in the Breclav region in 2000. *Crex* **16**: 63–64.
1008. Krause, F. 2002. Raptors and owls in the agricultural landscape in the Breclav region in 2001. *Crex* **18**: 108–109.
1009. Krischer, O. 1990. Rückgang der Steinkauzpopulation in der Region Basel. *Mitglieder-Informationen AG zum Schutz bedrohter Eulen* **35**: 3–4.
1010. Kroener, C. A. 1865. *Aperçu des Oiseaux de l'Alsace et des Vosges*. Derivaux. Strasbourg. [Little Owl data on p. 5.]
1011. Krohn, H. 1915. Vogelgewichte. *Ornithologische Monatsber* **23**: 148.
1012. Krohn, H. 1924. *Die Vogelwelt Schleswig-Holsteins*. Im Sonneschein-Verlag. Hamburg. [Little Owl data on pp. 259–260.]
1013. Krzanowski, A. 1971. Numerical comparison of Vespertilionidae and Rhinolophidae (Chiroptera: Mammalia) in the owl pellets. *Acta Zoologica Cracoviensia* **18**: 133–140.
1014. Kübel, M. & Ullrich, B. 2004. Mäusebusard (*Buteo buteo*) erbeutet adulten Steinkauz (*Athene noctua*). *Ornithologische Jahreshefte für Baden-Württemberg* **20**: 201.
1015. Kuhn, M. & Dewitz, von W. 1979. Steinkauz (*Athene noctua*) brütet in Industriegerät. *Charadrius* **15**: 90–91.
1016. Kuhn, W. 1992. Analyse von Steinkauzgewöllen mit besonderer Berücksichtigung der Koleopteren. Master thesis, Ludwig-Maximilian Universität.
1017. Kuhn, W. 1995. Struktur und jahreszeitliche Verteilung von Käfern in Steinkauzgewöllen. *Eulen Rundblick* **42**: 12–15.
1018. Kulaeva, T. M. 1977. *Sovoobraznye [Strigiformes]. Ptitsy Volzhsko-Kamskogo Kraya [Birds of the Volga Kama Territory]*. Moscow: Nauka Press. [Little Owl data on pp. 239–257.]
1019. Kulczycki, A. 1964. Badania na skladem pokarmu sow z Beskidu Niskiego. *Acta Zoologica Cracoviensia* **9**: 535–558.
1020. Kumerloeve, H. 1955. Spalax und Skorpione als Steinkauz-Nahrung. *Vogelwelt* **76**: 110.
1021. Kutzer, E. H., Frey, H. & Nöbauer, N. 1982. Zur Parasitenfauna österrechischer Eulenvögel (*Strigiformes*). *Angewandte Parasitologie* **23**: 190–197.
1022. Kwast, F. 1978. La Chouette chevêche. *Lien Ornithologique d'Alsace* **28**: 12.
1023. L'Hardy, J.-P. 1985. Bilan de l'enquête sur la mise à jour de l'atlas des oiseaux nicheurs. *Bulletin du Groupe Sarthois Ornithologique* **14**: 5–15.
1024. L'Hermitte, J. 1921. Aberration chez la Chevêche. *Revue Française d'Ornithologie* **7**: 12–13.
1025. Labes, R. & Patzer, J. 1987. Steinkauz – *Athene noctua*. *In* Klafs, G. & Stübs, J. (eds). *Die Vogelwelt Mecklenburgs*, 3rd edition. Jena. pp. 247–248.
1026. Labitte, A. 1951. Notes biologiques sur la Chouette chevêche, *Carine noctua vidalii* A. E. Brehm, 1857. *L'Oiseau et Revue Française d'Ornithologie* **21**: 120–126.
1027. Lack, D. 1946. Competition for food by birds of prey. *Journal of Animal Ecology* **15**: 128.
1028. Lacordaire, L. 1877. *Catalogue des Oiseaux Observés de 1845 à 1874, dans les Départements du Doubs et de la Haute-Saône*. Besançon: Dodivers. p. 25.
1029. Laiu, L. & Murariu, D. 1997. Nourriture de la Chouette Chevêche (*Athene noctua* Scop., 1769) (*Aves Strigiformes*) pendant l'été, dans une dépression sous-carpatique de Moldavie-Roumanie. *Travaux* **37**: 319–326.
1030. Lakmann, G. 1988. Zur Avifauna des Naturschutzgebietes "Rabbruch" (Salzkotten, Kr. Paderborn). *Bericht des Naturwissenschaftlichen Vereins für Bielefeld* **29**: 121–175.
1031. Lallemant, J.-J. 1996. Essai d'estimation de la population de Chouette chevêche (*Athene noctua*) dans le département du Puy-de-Dôme. *Grand-Duc* **49**: 12–13.
1032. Lambert, G. C. 1918. Exhibition of a collection of small beetle's wings from the nest of the little owl. *Bulletin British Ornithological Club* **38**: 75.
1033. Lambrecht, K. 1917. Az europai madarvilag kialakulasa. *Aquila* **24**: 203.
1034. Lampe, A. 2005. Caractérisation de l'habitat de la Chevêche d'Athéna (*Athene noctua*) dans un secteur du Livradois-Forez (Puy-de-Dôme). Propositions de mesures de gestion. Masters thesis, Université de Perpignan.

1035. Lancum, F. H. 1925. Further observations on the food of the Little Owl. *Journal Min. Agric.* **32**: 170–173.

1036. Lancum, F. H. 1936. British owls. *Journal Min. Agric.* **32.**

1037. Lange, H. 1942. Kirkeuglen og dens fode. *Naturens Verden* **26**: 108–120.

1038. Lange, W. 1992. Meine Steinkäuze. *Gefierdete Welt* **116**: 278–279.

1039. Lathbury, G. 1970. A review of the birds of Gibraltar and its surrounding waters. *Ibis* **112**: 35.

1040. Laursen, J. T. 1981. Kirkeuglens, *Athene noctua*, fodevalg i Ostjylland. *Dansk Ornitologisk Forenings Tidsskrift* **75**: 105–110.

1041. Lazar, P. 1983. Data to the small mammal fauna from the environs of Surneg obtained from owl pellet investigations. *Folia Musei Historico-Naturalis Bakonyiensis* **2**: 217–228.

1042. Lebreton, P. 1977. Atlas ornithologique Rhône-Alpes. *Les Oiseaux Nicheurs Rhônalpins.* CORA. Lyon. [Little Owl data on pp. 154–155.]

1043. Lebreton, P. 1980. Atlas ornithologique Rhône-Alpes-compléments 1976–1979. *Bièvre* **2**: 29.

1044. Lecomte, P. 1994. Ile-de-France: sur la piste de la chouette aux yeux d'or. *Courrier de la Nature* **147**: 23–27.

1045. Lecomte, P. 1995. La Chouette chevêche autour de Paris. *Circulaire du GTMR* **21**: 12–13.

1046. Lecomte, P. 1995. Le statut de la Chouette chevêche *Athene noctua* en Ile-de-France. Evolution et perspectives. *Alauda* **63**: 43–50.

1047. Lecomte, P. 2000. Balade en Aubrac. Sur la piste de la chouette aux yeux d'Or. *Courrier de la Nature* **186**: 33–37.

1048. Lecomte, P., Lapios, J.-M. & Génot, J.-C. 2001. Plan de restauration des populations de Chevêches d'Athéna en France. *In* Génot, J.-C., Lapios, J.-M., Lecomte, P. & Leigh, R. S. (eds). *Ciconia* **25**: 159–171.

1049. Lecorre, M. 1987. La Chouette chevêche (*Athene noctua*) dans le bocage herbignaçais. Premiers résultats. 1986. *Bulletin du Groupe Ornithologique Loire Atlantique* **7**: 58–65.

1050. Ledant, J.-P., Jacob, J.-P. & Devillers, P. 1982. Enquête sur les espèces de vertébrés menacées de disparition en Wallonie. Les oiseaux menacés de disparition en Wallonie. Tome 2. Ministère de la Région Wallonne pour l'Eau, l'Environnement et la Vie Rurale. [Little Owl data on pp. 465–470.]

1051. Lederer, W. & Kämpfer-Lauenstein, A. 1996. Einfluss der Witterung auf die Brutbiologie einer Steinkauzpopulation (*Athene noctua*) in Mittelwestfalen. *Populationsökologie Greifvögel und Eulenarten* **3**: 353–360.

1052. Lees, A. E. 1923. Early nesting of Little Owl in Huntingdonshire. *British Birds* **16**: 329.

1053. Lefevre, T. 1992. Etude et protection de la Chouette chevêche dans le Domfrontais. Inventaire des populations de mâles chanteurs et couples nicheurs au cours de l'année 1992. Groupe Ornithologique Normand. Parc Naturel Régional Normandie-Maine.

1054. Lefranc, N. 1979. *Les Oiseaux des Vosges.* Kruch. Raon l'Etape. [Little Owl data on p. 130.]

1055. Legeleux, C. 1999. A l'aide des Chouettes chevêches (*Athene noctua*). *Cormoran* **11**: 46–48.

1056. Lei, F.-M. 1995. A study on diet of the Little Owl (*Athene noctua plumipes*) in Qishan, Shaanxii Province, China. *Wuyi Science Journal* **12**: 136–142.

1057. Lei, F. M., Cheng, T. H. & Yin, Z. H. 1997. On distribution, habitat and the clinal variations of the subspecies of the Little Owl *Athene noctua* in China (*Strigiformes*: *Strigidae*). *Acta Zootaxonomical Sinica* **22**: 327–334.

1058. Leibak, E., Lilleleht, V. & Veromann, H. 1994. *Birds of Estonia. Status, Distribution and Numbers.* Tallinn: Estonian Academy Publishers.

1059. Leicht, U. 1992. Erfahrungen mit der Steinkauzzucht und der Auswildeung. *Naturschutzzentrum Wasserschloss Midwitz* **2**: 35.

1060. Leigh, A. G. 1908. Little Owl in Warwickshire. *British Birds* **2**: 240.

1061. Leigh, A. G. 1909. Little Owl in Anglesey and Warwickshire. *British Birds* **3**: 127.

1062. Leigh, A. G. 1910. Little Owl in Staffordshire. *British Birds* **3**: 307.

1063. Leigh, A. G. 1911. Little Owl in Warwickshire. *British Birds* **4**: 287.

1064. Leigh, R. S. 1993. Little Owls on Cheshire farmland. *The Raptor* **4**: 83–85.

1065. Leigh, R. S. 1995. Operation Little Owl – an ecological project. *The Raptor* **6**: 1–4.

1066. Leigh, R. S. 2001. The breeding dynamics of Little Owls (*Athene noctua*) in North West England. *In* Génot, J.-C., Lapios, J.-M., Lecomte, P. & Leigh, R. S. (eds). *Ciconia* **25**: 67–76.

1067. Leigh, R. S. 2001. *In* Van Nieuwenhuyse, D., Leysen, M. & Leysen, K. (eds). Working structure of the European conservation plan for the Little Owl *Athene noctua*: the way forward. *Oriolus* **67**: 142–146.

1068. Lembach, J. 1993. Obstwiesen als Lebensräume für Höhlenbrüter. *Falke* **7**: 234–240.

1069. Lemoine, O. 1986. Ecologie d'une population de Chouettes chevêches dans la région du Parc Naturel Régional de Brotonne. Approche biogéographique: densités, répartition, étude de l'habitat. DEA d'écologie. Université Paris XI.

1070. Lemoine, O. 1988. La Chouette chevêche, écologie d'une population de Chouettes chevêches dans la région du Parc Naturel de Brotonne. SRETIE. Rapport.

1071. Lemoine, O. 1989. Un oiseau des bocages du Parc: la Chouette chevêche. Dénombrements. Analyse des biotopes. Mesures de protection. Parc Naturel Régional Normandie-Maine. Rapport.

1072. Lengagne, T. 2000. La Chouette chevêche dans les Ardennes. *Bulletin Annuel du Regroupement des Naturalistes Ardennais*: **64**.

1073. Letty, J., Génot, J.-C. & Sarrazin, F. 2001. Viabilité de la population de Chevêche d'Athéna *Athene noctua* dans le Parc naturel régional des Vosges du Nord. *Alauda* **69**: 359–372.

1074. Letty, J., Génot, J.-C. & Sarrazin, F. 2001. Analyse de viabilité de la population de Chevêche d'Athéna (*Athene noctua*) dans le Parc naturel régional des Vosges du Nord. *In* Génot, J.-C., Lapios, J.-M., Lecomte, P. & Leigh, R. S. (eds). *Ciconia* **25**: 147–152.

1075. Leveque, G. 1997. Reproduction et régime alimentaire de la population de Chouette chevêche (*Athene noctua*) du Causse Méjean. Rapport Stage.

1076. Lewis, S. 1914. The Little Owl breeding in Somerset. *Zoologist* **18**: 112–113.

1077. Leysen, M., Van Nieuwenhuyse, D. & Steenhoudt, K. 2001. The Flemish Little Owl project: data collection and processing methodology. *In* Van Nieuwenhuyse, D., Leysen, M. & Leysen, K. (eds). *Oriolus* **67**: 22–31

1078. Lheritier, J.-N. 1974. L'avifaune nicheuse de la Limagne Brivadoise (Haute-Loire). *Alauda* **42**: 96–99.

1079. Lheritier, J.-N. 1977. Présentation de l'atlas des oiseaux nicheurs du Massif Central. *Grand-Duc* **10**: 7–85

1080. Lheritier, J.-N., Debussche, M. & Lepart, J. 1979. L'avifaune nicheuse des reboisements de Pin noir du causse Méjean. *L'Oiseau et Revue Française d'Ornithologie* **49**: 185–211.

1081. Libois, R. 1977. Contribution à l'étude du régime alimentaire de la Chouette chevêche (*Athene noctua*) en Belgique. *Aves* **14**: 165–177.

1082. Liebe, K. T. 1893. Sand- und Staubbäder der Raubvögel und Eulen. *Ornithologische Monatsschrift* **18** 6–9.

1083. Lipej, L. & Gjerkes, M. 1992. Bats in the diet of owls in NW Istra. *Myotis* **30**: 133–138.

1084. Lippens, L. & Wille, H. 1972. *Atlas des Oiseaux de Belgique et d'Europe Occidentale*. Lannoo Tielt. [Little Owl data on pp. 500–501.]

1085. Lipsberg, Y. K. 1985. Domovyi sych [Little Owl]. *Sarkana gramata Latvijas PSR* [*Red Data Book of the Latvian SSR*]. Riga: "Zinatne" Publ. House. [Little Owl data on pp. 452–453.]

1086. Lo Verde, G. & Massa, B. 1988. Abitudini alimentari della Civetta (*Athene noctua*) in Sicilia. *Naturalista Siciliano* **12**: 145–149.

1087. Lobachev, V. S. & Schenbrol, G. I. 1974. Pitanie domovogo syca v Severnom Priarale. [Food of the Little Owl in north Priarale]. *Ornitologiya* **11**: 382–390.

1088. Lockley, R. M. 1928. Status of the Little Owl in Pembrokeshire. *British Birds* **21**: 200.

1089. Lockley, R. M. 1938. The Little Owl inquiry and the Skokholm-Petrels. *British Birds* **31**: 278–279.

1090. Lodge, R. B. 1912. Little owl breeding in Middlesex. *British Birds* **6**: 18–19.

1091. Loewis, von O. 1883. Livlands Eulen, wildlebende Hühnerarten und Watvögel. *Zoologischer Beobachter* **24**: 115–116.

1092. Loose, D. 1989. Aider la Chouette chevêche en Isère. *Courrier du Hérisson* **88**: 23–25.

1093. Loose, D. 1989. L'étude et la protection de la Chouette chevêche. *Niverolle*, No. spécial: 89–91.

1094. Loose, D. 1990. La protection et l'étude de la Chouette chevêche (*Athene noctua*) dans le département de l'Isère. Premier bilan. *Niverolle* **12**: 47–54.

1095. Lord, J. & Ainsworth, G. H. 1945. Fight between Little Owls and Barn Owls. *British Birds* **38**: 275.

1096. Lorge, P. 2000. Der Steinkauz, eine bedrohte Vogelart unserer Dörfer. *Regulus* **4**: 12.

1097. Lorge, P. 2006. Gehört der Steinkauz *Athene noctua* in Luxemburg bald zum alten Eisen? *Regulus Wissenschaftliche Berichte* **21**: 54–58.

1098. Loske, K.-H. 1977. Der hohle Baum mit Innenleben. *Wir und die Vögel* **6**: 18–21.

1099. Loske, K.-H. 1978. Hilfe für den Steinkauz. *Ornithologische Mitteilungen* **30**: 19–21.

1100. Loske, K.-H. 1978. Gezielte Massnahmen zur Bestandserhaltung bzw Vermehrung des Steinkauzes (*Athene noctua*) in Mittelwestfalen. *Vogelwelt* **99**: 226–229.

1101. Loske, K.-H. 1978. Pflege, Erhaltung und Neuanlage von Kopfbäumen. *Natur und Landschaft* **53**: 279–281.

1102. Loske, K.-H. 1984. Zerstörung von Steinkauzrevieren. *DBV AG zum Schutz bedrohter Eulen Informationsblatt* **20**: 4–5.

1103. Loske, K.-H. 1985. Erhaltung, Pflege und Neuanlage von Kopfbäumen. *DBV AG zum Schutz bedrohter Eulen Informationsblatt* **2**.

1104. Loske, K.-H. 1986. Zum Habitat des Steinkauzes (*Athene noctua*) in der Bundesrepublik Deutschland *Vogelwelt* **107**: 81–101.

1105. Loske, K.-H. & Loske, R. 1981. Quantitative Erfassung von Biotopverlusten dargestellt am Beispiel des Langeneicker Bruches (Kreis Soast). *Natur und Landschaft Westfalen* **17**: 79–82.

1106. Loudon, H. 1906. Über des Vorkommen des Steinkauzes, *Carine noctua* Retz, in den Ostseeprovinzen *Ornithologische Monatsber* **14**: 190–191.

1107. Lovari, S. 1973. Shooting and preservation of wildlife in Italy. *Biological Conservation* **5**: 235–236.

1108. Lovari, S. 1974. The feeding habits of four raptors in central Italy. *Raptor Research* **8**: 45–57.

109. Lovaty, F. 1990. Distribution et densités des oiseaux reproducteurs sur les pelouses des causses de la région de Mende (Lozère). *L'Oiseau et Revue Française d'Ornithologie* **60**: 10–15.
110. Loyd, L. R. W. 1912. The food of the Little Owl. *British Birds* **6**: 65–66.
111. Loyd, L. R. W. 1919. Little Owl in south Devon. *British Birds* **13**: 164.
112. Lübcke, W. 1979. Fang von Steinkäuzen durch Rufkontakt mit Volierenvögeln. *Vogelkundliche Hefte* **5**: 85–86.
113. Lucan, V., Nitsche, L. & Schumann, G. 1974. *Vogelwelt des Land-und Stadtkreises Kassel*. Kassel. [Little Owl data on pp. 147–148.]
114. Lucas, P. 1996. Comparaison entre deux populations de Chouette chevêche (*Athene noctua*) en province de Liège. *Aves* **33**: 85–92.
115. Luder, R. & Stange, C. 2001. Entwicklung einer Population des Steinkauzes *Athene noctua* bei Basel 1978–1993. *Ornithologische Beobachter* **98**: 237–248.
116. Luginbuhl, J. 1908. Etwas vom Steinkauz. *Ornithologische Beobachter* **6**: 166–167.
117. Lundberg, A. 1986. Adaptive advantages of reversed sexual size dimorphism in European owls. *Ornis Scandinavica* **17**: 133–140.
118. Lüps, P., Hauri, R., Herren, H., Märki, H. & Ryser, R. 1978. Die Vogelwelt des Kantons Bern. *Ornithologische Beobachter* **75**: 136–137.
119. Lutzau, von R. & Metze, H. 1937. Neuer Fernfund eines beringten Steinkauz (*Athene noctua*). *Vogelzug* **8**: 29.
120. Lynden, A. J. H. 1981. Vogels van Zeeland, vroeger en nu. *Zeeuws Nieuws* **6**: 75.
121. Lynes, H. 1910. Nesting of the Little Owl in Hampshire. *British Birds* **3**: 336–338.
122. Lysaght, W. R. 1919. Little Owl breeding in Monmouthshire. *British Birds* **12**: 237–238.
123. Mädlow, W. & Mayr, C. 1996. Die Bestandsentwicklung ausgewählter gefährdeter Vogelarten in Deutschland 1990–1994. *Vogelwelt* **117**: 249–260.
124. Madon, P. 1933. Les rapaces d'Europe, leur régime, leurs relations avec l'agriculture et la chasse. Toulon. [Little Owl data on pp. 97–113.]
125. Makatsch, W. 1950. *Die Vogelwelt Macedoniens*. Leipzig: Geest & Portig. [Little Owl data on pp. 243–244.]
126. Makatsch, W. 1976. *Die Eier der Vögel Europas*. Band 2. Verlag J. Neumann. Neudamn. Melsungen. Berlin. Basel. Wien. [Little Owl data on pp. 43–45.]
127. Makatsch, W. 1981. *Verzeichnis der Vögel der Deutschen Demokratisches Republik*. Neumann Verlag. Leipzig. Rodebeul. [Little Owl data on pp. 90–91.]
128. Malmstigen, J. E. 1970. Ryttling hos minervauggla, *Athene noctua*. [Hovering in Little Owl *Athene noctua*]. *Var Fagelvärld* **29**: 231.
129. Mambetjumaev, A. M. 1998. Pitanie khishchnykh ptits i sov v nijnem i srednem techenii Amudar'i na Ustyurte i v Kyzylkumakh [Diet of birds of prey and owls in Lower and Middle Amu Darya, Ustyurt and Kyzyl Kum]. *Russkiy Ornitologicheskiy Zhurnal* [*Russian Journal of Ornithology*] **47**: 6–16.
130. Mammen, U. 1997. Eulen-Brutsaison 1995 und 1996. *Eulen Rundblick* **46**: 24–28.
131. Mammen, U. 1998. Eulen-Brutsaison. *Eulen Rundblick* **47**: 22–25.
132. Mammen, U. & Stubbe, M. 1996. Monitoring of raptors and owls in Europe: annual report 1995. *Jahresbericht zum Monitoring Greifvögel und Eulen Europas* **8**: 1–92.
133. Mammen, U. & Stubbe, M. 1999. Monitoring of raptors and owls in Europe; annual report 1998. *Jahresbericht zum Monitoring Greifvögel und Eulen Europas* **11**: 1–108.
134. Mammen, U. & Stubbe, M. 2005. Zur Lage der Greifvögel und Eulen in Deutschland 1999–2002. *Vogelwelt* **126**: 53–65.
135. Mandrillon, L. & Renaudier, A. 1989. Chronique ornithologique départementale. Effraie. *CORA Section Rhône* **7**: 12–53.
136. Manez, M. 1982. Geographical and seasonal variation in the diet of the Little Owl in Spain. Abstr. XVIII Congr. Int. Orn. Moscow 1982. p. 274.
137. Manez, M. 1983. Espectro alimentacio del Mochuelo comun (*Athene noctua*) en Espana. *Alytes* **1**: 275–290.
138. Manez, M. 1983. Variaciones geograficas y estacionales en la dieta del Mochuelo comun (*Athene noctua*) en Espana. *XV Congr. Int. Fauna Cinegetica y Silvestre*. Trujillo, 1981. pp. 617–634.
139. Manez, M. 1994. Little Owl *Athene noctua In* Tucker, G. M. & Heath, M. F. (eds). *Birds in Europe. Their Conservation Status*. BirdLife International. [Little Owl data on pp. 330–331.]
140. Manganaro, A., Ranazzi, L., Ranazzi, R. & Sorace, A. 1990. La dieta dell' alloco, *Strix aluco*, nel parco du villa Doria Pamphili (Roma). *Rivista Italiana di Ornithologia* **60**: 37–52.
141. Manganaro, A., Natalini, R., Demartini, L., Salvati, L. & Ranazzi, L. 1997. Il sistema trofico Barbagianni-Civetta/Vertebrati nella tenuta di Castelporziano (Roma). *Avocetta* **21**: 95.
142. Manganaro, A., Ranazzi, L. & Salvati, L. 2001. Diet overlap of Barn Owl (*Tyto alba*) and Little Owl (*Athene noctua*) in Mediterranean urban areas. *Buteo* **12**: 67–70.
143. Manson-Bahr, P. 1950. Dusting of owls. *British Birds* **43**: 19.
144. Maples, S. 1907. Little Owl (*Athene noctua*) in Hertfordshire. *Zoologist* **11**: 353.

1145. Marchant, J. 1980. La Chouette chevêche. *Journal des Oiseaux* **137**: 2.
1146. Marchant, J. H., Hudson, R., Carter, S. P. & Whittington, P. 1990. *Population Trends in British Breeding birds*. Nature Conservancy Council, BTO. [Little Owl data on pp. 114–116.]
1147. Marian, M. & Schmidt, E. 1967. Adatok a kurvik (*Athene noctua* (Scop.)) gerinces taplalenak ismeretehez Magyarorszagon. *Mora Ferenc Muzeum Evkonyve*: 271–275.
1148. Marie, P. & Leysen, M. 2001. Contribution to the design of an anti-marten *Martes foina* system to limit predation in Little Owl *Athene noctua* nest boxes. *In* Van Nieuwenhuyse, D., Leysen, M. & Leysen, K. (eds). *Oriolus* **67**: 126–131.
1149. Marples, B. J. 1942. A study of the Little Owl (*Athene noctua*) in New Zealand. *Transactions of the Royal Society of New Zealand* **72**: 237–252.
1150. Marriage, A. W. 1909. Little Owl in Hampshire. *British Birds* **2**: 310.
1151. Martin, R. & Rollinat, R. 1914. *Description et Moeurs des Mammifères, Oiseaux, Reptiles, Batraciens et Poissons de la France Centrale*. Paris: Paul Lechevalier. [Little Owl data on pp. 97–98.]
1152. Martin, R. & Rollinat, R. 1982. *Les oiseaux du département de l'Indre à la fin du 19ème siècle*. Texte intégral du chapitre ornithologique de l'ouvrage "Les vertébrés sauvages du département de l'Indre". Groupe d'Etude de l'Avifaune de l'Indre. [Little Owl data on pp. 17–18.]
1153. Martinez, J., Tomas, G., Merino, S., Arriero, E. & Moreno, J. 2003. Detection of serum immunoglobulins in wild birds by direct ELISA: a methodological study to validate the technique in different species using antichicken antibodies. *Functional Ecology* **17**: 700–706.
1154. Martínez, J. A. & Zuberogoitia, I. 2003. La urbanización litoral de Alicante reduce la población de mochuelos. *Quercus* **203**: 52.
1155. Martínez, J. A. & Zuberogoitia, I. 2004. Habitat preferences for Long-eared Owls *Asio otus* and Little Owl *Athene noctua* in semi-arid environments at three spatial scales. *Bird Study* **51**: 163–169.
1156. Martínez, J. A. & Zuberogoitia, I. 2004. Effects of habitat loss on perceived and actual abundance of the Little Owl *Athene noctua* in eastern Spain. *Ardeola* **51**: 215–219.
1157. Martínez, J. A., Sanchez, M. A., Carmona, D. *et al.* 1992. The Ecology and Conservation of the Eagle Owl *In* Galbraith, C. A., Taylor, I. R. & Percival, S. (eds). *Bubo bubo* in Murcia, south-east Spain. pp. 84–88.
1158. Martínez, J. A., Zuberogoitia, I. & Alonso, R. 2002. *Rapaces Nocturnas*: *Guia para la Determinacion de la Edad y el sexo de las Estrigiformes Ibericas*. Madrid: Monticola Ed. [Little Owl data on pp. 86–95.]
1159. Martínez, J. A., Zuberogoitia, I. & Zabala, J. 2007. Las rapaces nocturnas y los mosaicos agroforestales *Quercus* **251**: 18–23.
1160. Martiško, J. 1994. Návrh koncepce ochrany sovy pálené a sýčka obecného v české republice. [Plan of Barn Owl and Little Owl conservation in the Czech Republic]. Czech Union for the Nature Protection – branch Brno.
1161. Martiško, J. 1994. Monitoring of the abundance of Barn Owl (*Tyto alba*) and Little Owl (*Athene noctua*) in 1993. *Crex* **2**: 21.
1162. Martiško, J. 1995. K rozšíření sýčka obecného (*Athene noctua*) v okolí Brna. [The abundance of the Little Owl (*Athene noctua*) in Brno region]. *Crex* **6**: 34.
1163. Martiško, J. 1995. 2. setkání ochránců sovy pálené a sýčka obecného. (The second meeting of Barn Owl (*Tyto alba*) and Little Owl (*Athene noctua*) conservationists.). *Crex* **6**: 38–39.
1164. Martiško, J. 1995. Několik poznámek k praktické ochraně sýčka obecného (*Athene noctua*). [Comment to the practical conservation activities of Little Owl (*Athene noctua*).]. *Crex* **6**: 39–40.
1165. Martiško, J. 1995. K záchranným chovům sovy pálené (*Tyto alba*) a sýčka obecného (*Athene noctua*) u nás. [Comments to the rearing in captivity of Barn Owl (*Tyto alba*) and Little Owl (*Athene noctua*)]. *Crex* **6**: 40–42.
1166. Martiško, J. *et al.* 1995. *Ochrana ptáků I – sova pálená a sýček obecný*. [*Bird Protection I – Barn Owl and Little Owl*]. EkoCentrum Brno.
1167. Martorelli, G. 1902. Nota ornithologica ulteriori osservazioni sull *Athene chiaradiae* Giglioli. *Atti della Societa Italiana die Scienze Naturale* **40**: 325–338.
1168. März, R. 1949. Der Raubvogel-und Eulenbestand einer Kontrollfläche des Elbsandsteingebirges in den Jahre 1932–1940. *Beiträge zur Vogelkunde* **1**: 138.
1169. März, R. 1955. Eifrige Helfer gegen die Mäuseplatz. *Falke* **2**: 151–153.
1170. März, R. 1958. Eulen als Fledermausfänger. *Beiträge zur Vogelkunde* **6**: 87–96.
1171. März, R. & Weglau, I. 1957. Rupfungs-und Gewöllaufsammlung bei Darfel/Westfalen. *Vogelwelt* **78**: 110–112.
1172. Masefield, J. R. B. 1916. Little Owl in Staffordshire. *British Birds* **9**: 250.
1173. Masefield, J. R. B. 1927. Little owls nesting in railway point box. *British Birds* **16**: 95–96.
1174. Massa, B. 1981. Le régime alimentaire de quatorze espèces de rapaces en Sicile. Rapaces méditerranéens Parc Naturel Régional de Corse. Centre de Recherches Ornithologiques de Provence. [Little Owl data on p. 120.]

175. Mastrorilli, M. 1997. Popolazioni di Civetta (*Athene noctua*) e selezioni dell'habitat in un'area di pianura della provincia di Bergamo. *Rivista Museo civico Scienze Naturali "E.Caffi"* **19**: 15–19.
176. Mastrorilli, M. 1998. Sistema trofico Civetta (*Athene noctua*) – Vertebrati in un'area di pianura della provincia di Bergamo. *Bubo* **2**: 19–23.
177. Mastrorilli, M. 1999. Caratteristiche dei ricoveri e dei siti riproduttivi di Civetta *Athene noctua* nelle province di Bergamo e Cremona. *Avocetta* **23**: 163.
178. Mastrorilli, M. 1999. Rapaci pervenuti alla sede LIPU di Bergamo, dal 1990 al 1996, con particolare riguardo al periodo ed alle cause di ritrovamento. *Picus* **25**: 35–39.
179. Mastrorilli, M. 1999. Ripetuta ed inusuale osservazione di Civetta *Athene noctua* in un dormitorio invernale di Gufo comune *Asio otus*. *Picus* **25**: 47.
180. Mastrorilli, M. 2000. Seconda nidificazione di Civetta, *Athene noctua*, in Corsica. *Rivista Italiana di Ornithologia* **70**: 89–91.
181. Mastrorilli, M. 2001. Presence and breeding of Little Owl *Athene noctua* in Orobie Alps (North of Italy, Lombardia, Bergamo district). *In* Génot, J.-C., Lapios, J.-M., Lecomte, P. & Leigh, R. S. (eds). *Ciconia* **25**: 199–204.
182. Mastrorilli, M. 2001. Little Owl *Athene noctua* status and habitat selection in the town of Bergamo (Lombardy, Northern Italy). *In* Van Nieuwenhuyse, D., Leysen, M. & Leysen, K. (eds). *Oriolus* **67**: 136–141.
183. Mastrorilli, M. 2005. *La Civetta in Italia*. Brescia: Araspix Editrice.
184. Mastrorilli, M. & Nappi, A. 2003. Predazioni inusuali da parte degli Strigiformi in Italia: Uccelli e Mammiferi. *Avocetta* **27**: 91.
185. Mastrorilli, M., Barbagallo, A. & Bassi, E. 1999. Dati sulla nicchia trofica invernale del Gufo comune *Asio otus* in provincia di Bergamo. *Avocetta* **23**: 54.
186. Mastrorilli, M., Sacchi, R. & Gentili, A. 2001. Importanza dell'erpetofauna nella dieta degli Strigiformi italiani. *Pianura-Scienze e storia dell'ambiente padano* **13**: 339–342.
187. Mastrorilli, M., Nappi, A. & Barattieri, M. 2005. Atti I Convegno italiano sulla Civetta. Osio Sotto (BG), 21 Marzo 2004. Gruppo Italiano Civette.
188. Mastrorilli, M., Barattieri, M., Calvi, G. & Fasano, D. 2005. Le popiolazioni di Civetta *Athene noctua* nei comprensori urbani d'Italia: proposta di standardizzazione del metodo di raccolta. *In* Mastrorilli, M., Nappi, A. & Barattieri, M. (eds). Atti I Convegno italian o sulla Civetta. Osio Sotto (BG), 21 Marzo 2004. Gruppo Italiano Civette. pp. 20–23.
189. Mathes, P. 1992. Modes d'occupation et d'utilisation du milieu par la Chouette chevêche (*Athene noctua*) dans le Parc Naturel Régional des Vosges du Nord. Stage BTS Protection de la Nature. Neuvic.
190. Maurer, A. 1986. Premières données sur la Chouette chevêche (*Athene noctua*). *Lien Ornithologique d'Alsace* **43**: 6–9.
191. Maurer, A. 1987. Campagne de protection "Chouette chevêche". Bilan 1986. *Lien Ornithologique d'Alsace* **45**: 19–21.
192. Maurer, A. 1988. Campagne de protection "Chouette chevêche". Bilan 1987. *Lien Ornithologique d'Alsace* **49**: 3–5.
193. Maurer, A. & Hurstel, A. 1989. La Chouette chevêche (*Athene noctua*). Notes sur son régime alimentaire en Alsace. *Lien Ornithologique d'Alsace* **50**: 10–13.
194. Maury-Peratony, J. L. 2002/2003. Densités et habitats de la Chevêche d'Athéna (*Athene noctua*) dans trois secteurs représentatifs de la Plaine du Roussillon (Pyrénées Orientales). Groupe Ornithologique du Roussillon. Institut Universitaire Professionnalisé: Génie des Territoires et de l'Environnement.
195. Mauvieux, S. 1992. Prospection de la Chouette chevêche sur deux communes des Cotes d'Armor (printemps 1991). *Le Far* **26**: 32–35.
196. Mayenne Nature Environnement, 1991. *Atlas des Oiseaux Nicheurs de la Mayenne 1984–1988. Les Oiseaux de la Mayenne*. Editions Rives reines. Laval. [Little Owl data on p. 75.]
197. Meade-Waldo, E. G. B. 1900. The Little Owl (*Carine noctua*). *Zoologist* **4**: 556.
198. Meade-Waldo, E. G. B. 1912. The food of the Little Owl. *British Birds* **6**: 64–65.
199. Meares, D. H. 1917. Little Owl breeding in Essex. *British Birds* **10**: 271.
200. Mebs, T. 1966. *Eulen und Käuze*. Stuttgart: Kosmos. [Little Owl data on pp. 73–78.]
201. Mebs, T. 1968. Zum Kennenlernen der Eulen und zur Linderung ihrer Wohnungsnot. *Ornithologische Mitteilungen* **20**: 217–218.
202. Mebs, T. 1987. *Eulen un Käuze*. Stuttgart: Granchk'sche Verlagshandlung, W. Keller & Co. [Little Owl data on pp. 52–59.]
203. Mebs, T. 1989. *Guide des Rapaces Nocturnes. Chouettes et Hiboux*. Neuchâtel et Paris: Delachaux & Niestlé. [Little Owl data on pp. 52–59.]
204. Mebs, T. 1992. Ergebnisse einer Umfrage zu Methoden der Bestandserfassung beim Steinkauz (*Athene noctua*). *DBV AG zum Schutz bedrohter Eulen Informationsblatt* **38**: 7–10.

1205. Mebs, T. 2002. *In* Nordrhein-Westfäliche Ornithologen-Gesellschaft. *Die Vögel Westfalens. Ein Atlas der Brutvögel von 1989 bis 1994. beiträge zur Avifauna Nordrhein-Westfalens.* Bd 37. Bonn: Nordrhein-Westfäliche Ornithologengesellschaft. [Little Owl data on pp. 138–139.]
1206. Meinertzhagen, R. 1951. On the genera *Athene* Boie 1822 (genotype *Athene noctua* (Scopoli)) and *Speotyto* Gloger 1842 (genotype *Strix cunicularia* Molina). *Bulletin British Ornithological Club* **70**: 8–9.
1207. Meininger, P. L. 1984. Inventarisatie van een aantal soorten broedvogels op Walcheren in 1981 en 1982. *Vogeljaar* **2**: 62–71.
1208. Meisser, C. 1995. Programme de protection et d'étude de la Chouette chevêche (*Athene noctua*) dans le canton de Genève. *Nos Oiseaux* **43**: 193–201.
1209. Meisser, C. 1998. Suivi et protection de la Chouette chevêche (*Athene noctua*) dans le canton de Genève, Suisse. Aperçu de la période d'étude 1984–1997. *Nos Oiseaux-Genève*: 1–4.
1210. Meisser, C. 2003. Plan d'action pour la Chevêche d'Athéna en Ajoie. Collectif d'associations "Chevêche – Ajoie" (ASPO/BirdLife Suisse, Pro Natura Jura, Nos Oiseaux, SSNPP, ASB), Canton et République du Jura (OEPN).
1211. Meisser, C. 2005. Plan d'action pour la Chevêche d'Athéna en Ajoie. Bilan 2003–2004. Collectif d'associations "Chevêche – Ajoie" (ASPO/BirdLife Suisse, Pro Natura Jura, Nos Oiseaux, SSNPP, ASB) Canton et République du Jura (OEPN).
1212. Meisser, C. & Albrecht, P. 2001. Suivi et protection de la Chevêche d'Athéna (*Athene noctua*) dans le canton de Genève, Suisse (période 1984–2000). *In* Génot, J.-C., Lapios, J.-M., Lecomte, P. & Leigh, R. S (eds). *Ciconia* **25**: 191–197.
1213. Meklenburtsev, R. N. 1993. Ob izmeneniyakh prirodnoi obstanovki i sostava naseleniya ptits v okrestnoste Tashkenta [The changes in environment and avifauna of surroundings of Tashkent]. *Russkiy Ornitologicheskiy Zhurnal* [*Russian Journal of Ornithology*] **2**: 29–36.
1214. Melchior, E., Mentgen, E., Peltzer, R., Schmitt, R. & Weiss, J. 1987. *Atlas des Oiseaux Nicheurs du Grand-Duché de Luxembourg*. Lëtzelbuerger Natur-a Vulleschutzliga. [Little Owl data on pp. 122–123.]
1215. Melendro, J. & Gisbert, J. 1978. Notas ornithologicas breves. *Ardeola* **24**: 261.
1216. Melliger, I. 1987. Steinkauzfamilie in Naturhöhle. *Vögel der Heimat* **57**: 116–117.
1217. Mendelssohn, H. 1972. The impact of pesticides on bird life in Israel. *Bulletin of the International Council for Bird Preservation* **11**: 97.
1218. Mendelssohn, H. & Marder, U. 1970. Problems of reproduction in birds of prey in captivity. *International Zoo Yearbook* **10**: 6–11.
1219. Meng, J.-P. 1995. La Chouette chevêche (*Athene noctua*) en Livradois-Forez. Parc Naturel Régional du Livradois-Forez. CPIE des Monts du Forez. CORA Loire. Rapport.
1220. Menzel, F. 1909. Die Vogelwelt von Helmstedt und Umgebung. *Ornithologisches Jahrbuch* **20**: 100.
1221. Mercier, A. 1921. Valeur économique de quelques rapaces. *Gerfaut* **11**: 65.
1222. Meyer de Schauensee, R. 1984. *The Birds of China*. Oxford: Oxford University Press. [Little Owl data on p. 270.]
1223. Meyknecht, J. 1941. Farbensehen und Helligkeitsunterscheidung beim Steinkauz (*Athene noctua vidalii* A. E. Brehm). *Ardea* **30**: 129–170.
1224. Michelucci, S. & Lupetti, M. 1969. Contributo allo studio del corpo ultimobranchiale con ricerche morfologiche esperimentoli in *Athene noctua*. *Archivio Italiano di Anatomia e di Embriologia* **73**: 263–284.
1225. Mienis, H. K. 1971. *Thebe pisana* in pellets of an Israelian owl. *Basteria* **35**: 73–75.
1226. Mienis, K. H. 1979. Predation on *Testudo graeca* by the Little Owl in Israel. *Salamandra* **15**: 107–108.
1227. Mikiara, Š. 1990. Skoré zahniezdenie kuvika obycajného (*Athene noctua*). [Early breeding of Little Owl (*Athene noctua*).]. *Buteo* **5**: 91–92.
1228. Mikkola, H. 1976. Owls killing and killed by other owls and raptors in Europe. *British Birds* **69**: 144–154.
1229. Mikkola, H. 1983. *Owls of Europe*. Calton: Poyser. [Little Owl data on pp. 126–135.]
1230. Mildenberger, H. 1967/68. Untersuchungen über die Bestandsdichte der Eulen im Amtsbezirk Schermbeck, Niederrhein. *Jahrbuch* **10**: 85–86.
1231. Mildenberger, H. 1984. *Die Vögel des Rheinlandes*. 2. Kilda. Verlag Greven. Bonn. [Little Owl data on pp. 66–71.]
1232. Mills, D. G. H. 1981. Cannibalism among Little Owls. *British Birds* **74**: 354.
1233. Minchin, E. A. & Woodcock, H. M. 1911. Observations on the trypanosome of the Little Owl (*Athene noctua*) with remarks on other protozoan parasites occuring in this bird. *Quarterly Journal of Microscopical Science* **57**: 141–185.
1234. Ministere De L'Environnement, 1996. La diversité biologique en France. Programme d'action pour la faune et la flore sauvages. [Little Owl data on p. 225.]
1235. Minoranskiy, V. A., Dobrinov, A. V., Markitan, L. V. & Podgornaya, Y. Y. 1998. Materialy po chiclennost ptits v del'te Dona [Materials on bird numbers in Don River delta]. [Little Owl data on pp. 86–96.]
1236. Misonne, X. 1948. Le régime de la Chouette chevêche *Athene noctua vidalii* A. E. Brehm. *Gerfaut* **38**: 89

237. Misra, M. & Srivastava, M. D. 1974. The W-chromosome in two species of *Strigiformes*. *Chromosome Inf. Serv.* **1974**: 28–29.
238. Mitschke, A. & Baumung, S. 2001. *Brutvogel-Atlas Hamburg*. Hamburger avifaunistische Beiträge Arbeitskreis an der Staalichen Vogelschutzwarte Hamburg. [Little Owl data on pp. 146–147.]
239. Mlíkovský, J. 1996. Zimní potrava sýčka obecného (*Athene noctua*) v Jablonné. [Winter food of the Little Owl at Jablonná, Central Bohemia.]. *Buteo* **8**: 113–114.
240. Mlíkovský, J. & Piechocki, R. 1983. Biometrische Untersuchungen, zum Geschlechtsdimorphismus einiger mitteleuropäische Eulen. *Beiträge zur Vogelkunde* **29**: 1–11.
241. Mnatsekanov, R. A., Tilba, P. A. & Solovyev, S. A. 1990. [On the breeding of Kestrel and Little Owl in eastern Azov Sea Region]. Tezisy Dokladov Konferentsii Aktualnuye Voprosy Ekologii I Okhrany Prirody Azovskogo Morya i Vostochnogo Priazovya. Krasnodar: Kub ansky Gosudarstvenny Universitet [Abst. Conf. Actual Probl. of Ecol. & Nature Conserv. of Azov Sea and East. Azov. See Reg.]. [Little Owl data on pp. 180–183.]
242. Mohr, H. 1989. Steinkäuze brüten wieder in Schwaben. *Gefierdete Welt* **113**: 308–309.
243. Mohr, H. 1990. Der Steinkauz. eine der ammeisten "vom Aussterben bedrohten Vogelart". Arbeitsgemeinschaft Naturschutz Lkr. Biberach.
244. Mohr, H. 1991. Mit dem Steinkauz geht es weiter bergab! *Gefierdete Welt* **115**: 40.
245. Mohr, R. 1983. Bemerkenswerter Fund eines beringten Steinkauzes (*Athene noctua*). *Vogel und Umwelt* **2**: 359.
246. Mohr, R. 1990. Bemerkenswertes aus dem Leben hessischer Steinkäuze (*Athene noctua*). *Vogel und Umwelt* **6**: 63–69.
247. Möller, B. 1993. Erste Ergebnisse zur Wiedereinbürgerung des Steinkauzes (*Athene noctua*) in den Landkreisen Hildesheim und Peine. *Beiträge zur Naturkunde Niedersachsens* **46**: 72–81.
248. Moltoni, E. 1937. Osservazioni bromatologische sugli uccelli rapaci italiani. *Rivista Italiana di Ornithologia* **7**: 29.
249. Moltoni, E. 1948. Ulteriori osservazioni bromatologiche sugli uccelli rapaci italiani. *Rivista Italiana di Ornithologia* **18**: 101–125.
250. Moltoni, E. 1949. Alcuni dati sul peso e sulla longevita degli uccelli rapaci italiani. *Rivista Italiana di Ornithologia* **19**: 120–121.
251. Monsara, P. 1999. Etude du régime alimentaire de la chouette chevêche en période de reproduction. CORA Drôme. Université de droit d'économie et des sciences d'Aix-Marseille III. Maîtrise de biologie des populations et des écosystèmes.
252. Montel, F. 1981. La vallée de la Bouvaque. *Avocette* **5**: 7.
253. Moore, N. W. 1965. Environmental contamination by pesticides. Ecology and the Industrial Society. Symposium of the British Ecological Society. Swansea. 1964. Goodmann *et al.* Oxford. [Little Owl data on p. 228.]
254. Moore, N. W. & Walker, C. H. 1964. Organic chlorine insecticide residues in wild birds. *Nature* **201**: 1072–1073.
255. Morbach, J. 1939. Zum Nahrungsregim unserer Eulenvögel. *Bulletin de la Ligue Luxembourgeoise pour l'Etude et la Protection des Oiseaux* **10**: 17–18.
256. Morbach, J. 1939. Was lehren uns die Gewölleuntersuchungen? *Bulletin de la Ligue Luxembourgeoise pour l'Etude et la Protection des Oiseaux* **9**: 4–9.
257. Moreau, R. E. 1923. Little Owl in Cornwall. *British Birds* **17**: 42.
258. Moreau, W. M. 1947. Kestrel robbing Little Owl. *British Birds* **40**: 216–217.
259. Morillo, C. 1976. *Guida de las Rapaces Ibericas*. Instituto Nacional para la conservacion de la naturaleza. Madrid. [Little Owl data on pp. 154–157.]
260. Moronvalle, J. & Moronvalle, P. 1992. Recensement de la Chouette chevêche *Athene noctua* dans le nord amiénois. *Avocette* **16**: 23–32.
261. Morris, R. 1916. The Little Owl in Sussex. *Zoologist* **20**: 112.
262. Mörzer Bruijns, M. F. 1962. De massasterfte van vogels in Nederland door vergiftiging met bestrijdingsmiddelen in het voorjaar van 1960. *Landbouwkundig Tijdschrift* **74**: 583.
263. Moschetti, G. & Mancini, D. 1993. Dieta della Civetta *Athene noctua* (Scopoli) e sue variazioni stagionali in un parco urbano in ambiente mediterraneo. *Gli Uccelli d'Italia* **28**: 3–12.
264. Moschetti, G. & Mancini, D. 1995. Variazioni stagionali nella dieta della Civetta *Athene noctua* (Scopoli) in un parco urbano in ambiente mediterraneo. *Supplemento alle Ricerche di Biologia della Selvaggina* **22**: 261–263.
265. Mostini, L. & Piccolino, D. 2002. Indagine sull'alimentazione della Civetta *Athene noctua* nella pianura novarese. *Rivista Piemontese di Storia Naturale* **23**: 227–232.
266. Mourgues, J. C. 1985. De l'utilité des centres de sauvetage de la faune sauvage. Union Nationale des Centres de Soins. Ministère de l'Environnement.

1267. Mouton, J. 1982. L'avifaune des prés Duhem-Armentières (Nord). *Héron* **4**: 79.
1268. Moyne, G. 1991. La Chouette chevêche. Causes de régression et mesures de protection. *Jura Nature* **47**: 13–15.
1269. Müller, B. 1993. Erste Ergebnisse zur Wiedereinbürgerung des Steinkauzes (*Athene noctua*) in den Landkreisen Hildesheim und Peine. *Beiträge zur Naturkunde Niedersachsens* **46**: 72–81.
1270. Müller, H. H. 1968. Neunachweise von Beutetieren des Steinkauzes (*Athene noctua* (Scop. 1769)). Dortmunder Beiträge zur Landeskunde.
1271. Müller, W. 1976. Zum Vorkommen des Steinkauzes (*Athene noctua*) am unteren Niederrhein. Eine Bestandsaufnahme im südlichen Kreis Wesel. *Rheinische Heimatpflege* **13**: 288–289.
1272. Müller, W. 1983. Die Vogelwelt der Obstgärten. *Vögel der Heimat* **2**: 26–30.
1273. Müller, W. 1999. Untersuchungen zum Auftreten alternativer Fortpflanzungsstrategien und zum Geschlechterverhältnis beim Steinkauz (*Athene noctua*). Masters thesis, Universität Bonn.
1274. Müller, W., Epplen, J. T. & Lubjuhn, T. 2001. Genetic paternity analyses in Little Owls (*Athene noctua*): does the high rate of paternal care select against extra-pair young? *Journal für Ornithologie* **142**: 195–203
1275. Mullie, W. C. & Meininger, P. L. 1985. The decline of bird of prey populations in Egypt. *Conservation Studies on Raptors ICBP* **5**: 61–82.
1276. Murray, W. 1958. Little Owl nesting in Scotland. *Scottish Birds* **1**: 37–38.
1277. Murzov, V. N. & Berezovikov, N. N. 2001. Pitanie domovogo sycha *Athene noctua* v pustynnykh landshaftakh Yuzhnogo Pribalkashya (Yugo-Vostochnyi Kazakhstan) [The food of the Little Owl *Athene noctua* in the deserts of southern part of Balkash region (South-Eastern Kazakhstan)]. *Russkiy Ornitologicheskiy Zhurnal* [*Russian Journal of Ornithology*] **134**: 176–178.
1278. Muselet, D. 1984. L'avifaune de la vallée de la Corrie (Eure et Loir). *Passer* **21**: 156.
1279. Nankinov, D. N. 2002. Sovremennoe sostoyanie populyatsiy sov Bolgarii [Present situation of population of owls in Bulgaria]. *Berkut* **11**: 48–60.
1280. Nappi, A. & Mastrorilli, M. 2003. The Little Owl *Athene noctua* diet in Italy. III International Symposium on Wild Fauna. Ischia 24–28 May 2003. Luigi Esposito & Bianca Gasparrini. [Little Owl data on pp 403–406.]
1281. Nappi, A. & Mastrorilli, M. 2003. Predazioni inusuali da parte degli Strigiformi in Italia: Invertebrati, Pesci Anfibi, Rettili. *Avocetta* **27**: 90.
1282. Nappi, A. & Mastrorilli, M. 2003. Predazioni inusuali da parte degli Strigiformi in Italia: Invertebrati, Pesci Anfibi, Rettili. *Avocetta* **27**: 90.
1283. Nash, E. G. 1925. Little Owl attacking cat. *British Birds* **19**: 25–26.
1284. Navarro, J., Minguez, E., Garcia, D. *et al.* 2005. Differential effectiveness of playbacks for Little Owl (*Athene noctua*) surveys before and after sunset. *Journal of Raptor Research* **39**: 454–457.
1285. Navarro, J., Sanchez-Zapata, J. A., Carrette, M. *et al.* 2003. Diet of three sympatric owls in steppe habitat of Eastern Kazakhstan. *Journal Raptor Research* **37**: 256–258.
1286. Neeracher, H. 1975. Erfahrung mit Steinkauzbrutröhren. *Vögel der Heimat* **46**: 63.
1287. Negro, J.-J., De La Riva, M. & Hiraldo, F. 1990. Daytime activity of Little Owls (*Athene noctua*) in southwestern Spain. *Journal Raptor Research* **24**: 72–74.
1288. Neri, F. 1990. Bilan du premier printemps de prospection sur la population de Chouette chevêche du Parc Naturel Régional du Haut-Languedoc 1990. Parc Naturel Régional du Haut-Languedoc.
1289. Newstead, R. 1951. Notes on the menu of Little Owl. *North Western Nat.* **23**: 156.
1290. Newton, I. 2002. Population limitation in Holarctic owls. *In* Newton, I., Kavanagh, R., Olsen, J. & Taylor, I. (eds). *Ecology and Conservation of Owls*. Australia: CSIRO Publishing. pp. 3–29.
1291. Nichols, W. B. & Glegg, W. E. 1915. Little Owls in Essex. *British Birds* **8**: 197.
1292. Nicolai, B. 1994. Steinkauz Information. Artenhilfsprogramm des Landes Sachsen-Anhalt. Ministerium für Umwelt und Naturschutz des Landes Sachsen-Anhalt.
1293. Nicolai, B. 2000. Bestandsentwicklung und Situation des Steinkauzes (*Athene noctua*) in Sachsen-Anhalt. *Apus* **10**: 55–64.
1294. Nie, G. J. van, 1976. Steenuil voert regenwormen aan jongen. *Vogeljaar* **24**: 326.
1295. Niehuis, M. 1982. Anderungen in der Vogelfauna von Rheinland-Pfalz. *Pfälzer Heimat* **33**: 105.
1296. Niethammer, G. 1943. Beiträge zur Kenntnis der Brutvogel des Peloponnes. *Journal für Ornithologie* **91**: 224–225.
1297. Niethammer, G. 1971. Zur Taxonomie europäischer in Neuseeland eingebürgeter Vögel. *Journal für Ornithologie* **112**: 221–223.
1298. Nikiforov, M. E., Kozulin, A. V., Grichik, V. V. & Tishechkin, A. K. 1997. *Ptitsy Belarusi na rubezhe XXI veka* [*Birds of Belarus at the edge of XXI century*]. Minsk: N. A. Korolev Producer.
1299. Nikiforov, M. E., Yaminskiy, B. V. & Shklyarov, L. P. 1989. *Belorussii* [*Birds of Belarus*]. Spravochnik opredelitel gnedz i yaits [Manual-guide of nests and eggs]. Minsk: Vysheishaya Shkola Press.

1300. Nitsche, G. & Plachter, H. 1987. *Atlas der Brutvögel Bayerns 1979–1983*. München: Bayerisches Landesamt für Umweltschutz. [Little Owl data on p. 126.]

1301. Nöbauer, H. 1982. Beitrag zur Parasitenfauna einheimischer Eulenvögel. Masters thesis, Universität Wien.

1302. Noble, H. 1912. Spread of the Little Owl in Berkshire, Norfolk and Lincolnshire. *British Birds* **5**: 245.

1303. Nore, T. 1977. L'autour et l'épervier en Limousin. *Bulletin de la Société pour l'Etude et la Protection des Oiseaux en Limousin* **7**: 40.

1304. Nore, T. 1979. Rapaces diurnes communs en Limousin pendant la période de nidification (II: autour, épervier et faucon crecerelle). *Alauda* **47**: 265.

1305. Normand, N. & Lesaffre, G. 1976. *Les Oiseaux de la Région Parisienne et de Paris*. Paris: L'Association Parisienne Ornithologique. [Little Owl data on p. 93.]

1306. Noval, A. 1975. *Aves de Presa*. Naranco-Bilbao. [Little Owl data on pp. 323–328.]

1307. Nowicki, F. 1996. Contribution à la connaissance des habitats de la Chouette chevêche: étude des couples recensés dans le secteur de Vigy. Stage première année Maîtrise Science et Techniques. UFR Metz. LPO Lorraine.

1308. O'Brien, S. & Julian, S. 1993. The diet of Little Owls. *The Raptor* **21**: 80–82.

1309. Obuch, J. & Kristin, A. 2004. Prey composition of the Little Owl *Athene noctua* in an arid zone (Egypt, Syria, Iran). *Folia Zoologica* **53**: 65–79.

1310. Obuch, J. & Kürthy, A. 1995. The diet of three owls species commonly roosting in buildings (*Tyto alba, Athene noctua, Strix aluco*). *Buteo* **7**: 27–36.

1311. Oehme, H. 1961. Vergleichend-histologische Untersuchungen an der Retina von Eulen. *Zoologische Jahrbücher Anatomie* **79**: 439–478.

1312. Oeser, R. 1978. Zum Vorkommen des Steinkauzes, *Athene noctua*, im Erzgebirge. *Beiträge zur Vogelkunde* **24**: 103–104.

1313. Olea, P. P. 1997. Mochuelo comun. *Athene noctua*. *In* Purroy, J. (ed.). *Atlas de las Aves de Espana (1975–1995)*. [Little Owl data on pp. 260–261.]

1314. Oles, T. 1961. Observations on the food habits in Little Owl. *Przeglad Zoologiczny* **4**: 377–378.

1315. Olioso, G. 1996. *Oiseaux de Vaucluse et de la Drôme Provençale*. CROP & CEEP et SEOF [Little Owl data on pp. 86.]

1316. Olioso, G., Volot, R. & Gallardo, M. 1982. Contributions à l'étude des vertébrés du Sud Vaucluse. *Bulletin du Centre de Recherches Ornithologiques de Provence* **4**: 13.

1317. Opluštil, L. 2004. Zkušenosti z praktické ochrany sýčků obecných (*Athene noctua*) na jižní Moravě. [Practical experience in Little Owl (*Athene noctua*) conservation in South Moravia.]. *Crex* **22**: 40–49.

1318. Opluštil, L. & Krause, F. 2005. Výskyt a hnízdění sýčků obecných (*Athene noctua*) na jižní a východní Moravě v roce 2005. [The records and breeding of the Little Owl (*Athene noctua*) in South and East Moravia in 2005.]. *Crex* **25**: 93–98.

1319. Ordelman, G. J. 1982. Jonge Steenuilen en Zwarte Kraaien. *Vogeljaar* **30**: 101.

1320. Ordelman, G. J. 1985. Belevenissen rond uilebroedsels. *Vogeljaar* **33**: 155–157.

1321. Orf, M. 2001. Habitatnutzung und Aktionsraumgrösse des Steinkäuze *Athene noctua* im Main-Taunus-Kreis. Masters thesis, Johann Wolgang Goethe-Universität, Frankfurt am Main.

1322. Orlova, E. A. & Ilyashenko, V. Y. 1978. Materialy po pitaniyu nekotorykh dnevnykh khishchnykh ptits I sov yugo-vostochnogo Ataya [Notes on the feeding of some birds of prey and owls of the South-eastern Altai]. Sistematika i biologiya redkikh i maloizuchenn ykh ptits. *Trudy Zool. Instituta* (*Sys. Trudy Zoologicheskogo Instituta*) [*Proceedings of Zoological Institute*] **76**: 94–100.

1323. Orsini, P. 1985. Le régime alimentaire du Hibou grand-duc *Bubo bubo* en Provence. *Alauda* **53**: 11–28.

1324. Osmolovskaya, V. I. 1953. Geograficheskoe raspredelenie khishchnykh ptits ravninnogo Kazakhstana i ikh znachenie v istreblenii vreditelei [Geographical distribution of the birds of prey of the plain of Kazakhstan and their importance in extermination of pests]. Trudy Instituta geografii [Proceedings of Institute of Geography]. *AN SSSR* [*Academy of Science of the USSR*] **54**: 219–307.

1325. Otto, A. & Ullrich, T. 2000. Schutz des Steinkauzes (*Athene noctua*) in der südlichen Ortenau und in angrenzenden Gebieten. *Naturschutz am südlichen Oberrhein* **3**: 49–54.

1326. Owen, D. F. 1951. Nightjars mobbing owls. *British Birds* **44**: 324.

1327. Owen, J. H. 1912. Little Owls breeding in Essex. *British Birds* **6**: 63–64.

1328. Owen, J. H. 1919. On the procuring of food by the male for the female among birds of prey. *British Birds* **13**: 29.

1329. Owen, J. H. 1939. Little Owl taking House-Martin. *British Birds* **33**: 111.

1330. Paevskiy, V. A. 1993. Vstrecha domovogo sycha *Athene noctua* v Luzhskom raione Leningradskoi oblasti [The record of the Little Owl *Athene noctua* in the Luga District, Leningrad Region]. *Russkiy Ornitologicheskiy Zhurnal* [*Russian Journal of Ornithology*] **2**: 5.

1331. Pailley, M. & Pailley, P. 1994. Evolution des effectifs de la Chouette chevêche *Athene noctua* dans une commune de Maine-et-Loire entre 1975 et 1989. *Bulletin du Groupe Angevin Etudes Ornithologiques* **22**: 69.
1332. Pailley, M., Pailley, P. & Preau, L.-M. 1991. Prédation de la Chouette chevêche (*Athene noctua*) sur des populations nicheuses de Sternes pierregarins (*Sterna hirundo*) et de Sternes naines (*S. albifrons*) en Maine-et-Loire. *L'Oiseau et Revue Française d'Ornit Hologie* **61**: 337–338.
1333. Palmer, P. 1989. Little Owl and grey squirrel in same nest-box. *British Birds* **82**: 221.
1334. Panzke C. 1986. Steinkauz-*Athene noctua*. *In* Zang, H. & Heckenroth, H. (eds). *Die Vögel Niedersachsens*. Naturschutz und Landschaftspflege in Niedersachsen.
1335. Pardieu, Marquis de, 1927. Nidification précoce d'une Chevêche. *Revue Française d'Ornithologie* **11**: 118.
1336. Parigi-Bini P. 1948. Strano reperto nel continuto gastrico di una Civetta. *Rivista Italiana di Ornithologia* **18**: 146–147.
1337. Paris, P. 1909. Anomalies observées chez quelques oiseaux de la Côte d'Or. *Revue Française d'Ornithologie* **1**: 104.
1338. Paris, P. 1909. Notes pour servir l'ornithologie de l'Aube. *Revue Française d'Ornithologie* **1**: 242.
1339. Parker, A. 1970. Some observations on the Little Owl (*Athene noctua*) and Barn Owl (*Tyto alba*) at their nests, near Romford, Essex, 1968. *London Bird Report* **33**: 81–87.
1340. Parker, T. H. 1945. Some owl notes. *Birds Notes and News* **6**: 87–88.
1341. Parslow, J. L. F. 1967. Changes in status among breeding birds in Britain and Ireland. *British Birds* **60**: 190–200.
1342. Patten, C. J. 1930. The Little Owl and little rabbits. *Irish Naturalists' Journal* **3**: 92–94.
1343. Patzer, J. 1979. Steinkauz *In* Klafs, G. & Stübs, J. (eds). *Die Vogelwelt Mecklenburgs*. VEB Gustav Fischer Verlag. Jena. [Little Owl data on pp. 206–207.]
1344. Paulus, M. 1943. Notes sur des contenus stomacaux d'oiseaux. *Bulletin du Musee d'Histoire Naturelle de Marseille* **3**: 44.
1345. Pavlov, A. I. 1965. Pitanie domovogo sycha na Mangyshlake [The Little Owl role in Mangyshlak]. Materialy 4-i nauch. Konf. po prirodnoi ochagovosti i profilaktike chumy [Materials of 4th scient. conf. on natural nidusness and prophylaxis of plague]. Alma-Ata. [Little Owl data on pp. 184–185.]
1346. Pavlov A. N. 1962. O khozyaistvennom znachenii domovogo sycha v polupustyne Severo Zapadnogo Ptikaspiya [On practical importance of the Little Owl in semidesert of the north western Cis Caspian Sea area]. *Zoologicheskij Zhurnal* [*Zoological Journal*] **41**: 1898–1901.
1347. Paz, U. 1987. *The Birds of Israel*. London: Christopher Helin. [Little Owl data on pp. 148–149.]
1348. Peachell, F. H. 1897. A Little Owl. *Nat. Journal Guide* **6**: 38.
1349. Pearson, C. E. 1913. Little Owl breeding in Nottinghamshire. *British Birds* **7**: 55.
1350. Pecherand, R. 1961. *Chouettes et Hiboux*. Paris: Crépin-Leblond et Cie. [Little Owl data on pp. 41–44.]
1351. Pedrini, P. 1984. Osservazioni sugli Strigiformi del Trentino. *Natura Alpina* **35**: 1–10.
1352. Pedrocchi-Renault, C. 1987. Fauna Ornitica del Alto Aragon Occidental. Consejo Superior de Investigaciones Cientificas. Monograficas del Instituto Pirenaico de ecologico. JACA. [Little Owl data on pp. 61–62.]
1353. Peirce, M. A. 1981. Current knowledge of the haematozoa of raptors. Recent advances in the study of raptor diseases. Proceedings of the International Symposium on Diseases of Birds of Prey 1st–3rd July 1980.
1354. Peitzmeier, J. 1952. Zur Ernährung der Brut des Steinkauzes. *Vogelwelt* **73**: 135.
1355. Peitzmeier, J. 1952. Langsamer Ausgleich der Winterluste beim Steinkauz. *Vogelwelt* **73**: 136.
1356. Peklo, A. M. 1982. [On the diet of Little Owl in Chernomorsky Nature Reserve]. Tezisy Dokladov Pribaltiyskoy Konferentsii Molodykh Ornitologov Posvyashchennoy 100-letiyu so Dnya Rozhdeniya Professora T. Ivanauskasa. Kaunas: Gosudarstvenny Komitet Lit.SSR p o Okhrane Prirody. [*Abst. E Baltic Conf Young Ornithologists Dedicated*] **100**: 65–67.
1357. Penpeny, M. 2005. La Chevêche d'Athena (*Athene noctua*) dans le Vexin français: problématique et enjeux. *Courrier Scientifique du Parc Naturel Régional du Vexin Français* **1**: 34–38.
1358. Pepin, D. 1987. Quelques notes sur la répartition de la Chouette chevêche en France-Comté. *Falco* **22**: 22–25.
1359. Pepin, D. 1989. Bilan quantitatif d'une enquête pas comme les autres: l'enquête "Chouette chevêche" Novembre 1988. *Falco* **23**: 244–246.
1360. Pepin, D. 1989. Dans l'intimité des petites chouettes aux yeux d'or. *L'Oiseau Magazine* **16**: 9–11.
1361. Percy Harrisson, D. 1916. Little Owl breeding in Wiltshire. *British Birds* **10**: 120.
1362. Perthuis, A. 1974. Inventaire ornithologique de la région Centre. *Naturalistes Orléanais* **11**: 19.
1363. Perthuis, A. 1983. *Les Oiseaux Nicheurs du Perche en Loir-et-Cher*. PUF Vendôme. [Little Owl data on pp. 41.]
1364. Peter, W. 1990. "Steinkauz-Fernwanderer" aus dem Main-Kinzig-Kreis. *Vogel und Umwelt* **6**: 133.
1365. Peter, W. 1999. Schutzmassnahmen für den Steinkauz (*Athene noctua*) und dessen Bestandsentwicklung im Main-Kinzig-Kreis. *Eulen Rundblick* **48/49**: 24–25.

366. Petzold, H. & Raus, T. 1973. Steinkauz (*Athene noctua*) – Bestandsaufnahmen in Mittelwestfalen. *Anthus* **10**: 25–38.
367. Pfister, O. 2001. Owls in Ladakh. *Oriental Bird Club Bulletin* **29**: 22–29.
368. Pharisat, A. 1978. Observations sur le régime alimentaire de la Chouette chevêche à Pierrefontaine-les-Varans (Doubs). *Bulletin de la Société Histoire Naturelle du Pays de Montbéliard.* [Little Owl data on pp. 38–43.]
369. Pharisat, A. 1995. Nouvelles observations sur le contenu des pelotes de régurgitation de la Chouette chevêche (*Athene noctua*) à Pierrefontaine-les-Varans (Doubs). *Bulletin de la Société Histoire Naturelle du Pays de Montbéliard.* pp. 191–192.
370. Piciocchi, S. 1989. *Athene noctua In* Fraissinet, M. & Kalby, M. (eds). Atlante degli uccelli nidificanti in Campania (1983–1987). Assoziane Studi Ornitologici Italia Meridionale. Salerno. [Little Owl data on pp. 69–70.]
371. Picciocchi, S. & Mastronardi, D. 2003. Atlante degli uccelli rapaci diurni e notturni nidificanti in Campania: risultati dei primi due anni di studio. *Avocetta* **27**: 114.
372. Piechocki, R. 1960. über die Winterverluste der Schleiereule (*Tyto alba*). *Vogelwarte* **29**: 277–280.
373. Piechocki, R. 1968. Die Grossfieder-Mauser des Steinkauzes (*Athene noctua*). *Journal für Ornithologie* **109**: 30–36.
374. Piotte P. 1984. *Atlas des Oiseaux Nicheurs de Franche-Comté.* Groupe Naturaliste de France-Comté. [Little Owl data on p. 84.]
375. Pirogov, N. G. 1991. Sovy Chernomorskogo zapovednika [Owls of the Black Sea Nature Reserve]. Materialy 10-i Vsesoyuznoi ornitologicheskoi konferentsii [Materials of the 10th All-Union Ornith. Conference]. Minsk, Navuka i tekhnika press. Book 2. Part 2: 148–149.
376. Pirotte, S. 2005. Etat des vieux vergers sur la commune de Theux et étude de leur intérêt ornithologique. Rapport de graduation en agronomie. Haute Ecole de la Province de Liège.
377. Pirovano, A. & Galeotti, P. 1999. Territorialismo intra- e interspecifica delle Civette *Athene noctua* in provincia di Pavia. *Avocetta* **23**: 139.
378. Pitt, F. 1919. Little Owl breeding in Shropshire. *British Birds* **13**: 163.
379. Pitzer, W. 1995. Schleiereulen-und Steinkauzvorkommen am südlichen Rand des Ruhrgebiets. *Falke* **4**: 121–124.
380. Plantinga, J-E. 1999. Plan van aanpak Steenuil. Actierapport Vogelbescherming Nederland 14, Zeist.
381. Pompidor, J.-P. 1989. Rapport sur la destruction des rapaces dans les Pyrénées orientales. *Bulletin du Fonds d'Intervention pour les Rapaces* **16**: 8–9.
382. Popescu, A. & Blidarescu, S. 1983. Date referitoare la hrana devara a cucuvelei (*Athene noctua noctua* Scop.). *Analele Universitatii Bucuresti Biologie* **32**: 77–82.
383. Popescu, A. & Negrea, I. 1987. Date privind hrana unor pasari rapitoare de noapte (*Athene noctua noctua* Scop. *Strigiformes*) din judetul timis. *Analele Universitatii Bucuresti Biologie* **36**: 67–70.
384. Popescu, A. & Savu, V. 1981. Date privind hrana unor pasarilor rapitoare de noapte (*Athene noctua noctua* Scop.si *Strix aluco aluco* L.). *Analele Universitatii Bucuresti Biologie* **30**: 63–68.
385. Popescu, A., Nitu, E. & Negrea, I. 1986. Insects, an important component part of the food of the Little Owl (*Athene noctua noctua* Scop.) in summer time. *Analele Universitatii Bucuresti Biologie* **35**: 17–20.
386. Popham, H. L. 1914. The Little Owl breeding in Somerset. *Zoologist* **18**: 150.
387. Popov, V. V. & Verzhutsky, D. B. 1990. Wintering of raptors and Owls in south-western Tuva. *Bull. Moscow. Soc. Naturalists. Sera Biol.* **95**: 66–69.
388. Porciatti, V., Fontanesi, G. & Bagnoli, P. 1989. The electroretinogram of the Little Owl (*Athene noctua*). *Vision Research* **29**: 1693–1698.
389. Porciatti, V., Fontanesi, G., Raffaelli, A. & Bagnoli, P. 1990. Binocularity in the Little Owl, *Athene noctua* II. Properties of visually evoked potentials from the wulst in response to monocular and binocular stimulation with sine wave gratings. *Brain Behavior Evolution* **35**: 40–48.
390. Portal, M. 1928. Little Owl in Northumberland. *British Birds* **22**: 170.
391. Potters, H. 2002. [Possible case of bigamy by Little Owl *Athene noctua.*] *Limosa* **75**: 33.
392. Poulsen, C. M. 1940. Kirkeuglene og Vinberen. 1940. *Flora og Fauna* **46**: 123–126.
393. Poulsen C. M. 1957. Massedodsfald blandt kirkeugler (*Athene noctua* (Scop.)). *Dansk Ornitologisk Forenings Tidsskrift* **51**: 40–41.
394. Poulter, D. 1953. Grass found on stomach of Little Owl. *British Birds* **46**: 414.
395. Pozio, E., Goffredo, M., Fico, R. & La Rosa, G. 1999. *Trichinella pseudospiralis* in sedentary night birds of prey from central Italy. *Journal of Parasitology* **85**: 759–761.
396. Prestt, I. 1977. A review of the status of birds of prey in Great Britain. ICBP World Conference on Birds of Prey. Vienna 1975. [Little Owl data on p. 117.]
397. Prestt, I. & Ratcliffe, D. A. 1972. Effects of organochlorine insecticides on European birdlife. [Little Owl data on pp. 486–513.]

1398. Prestt, I., Jefferies, D. J. & Moore, N. W. 1970. Polychlorinated biphenyls in wild birds in Britain and their avian toxicity. *Environmental Pollution* **1**: 7.
1399. Prestwich, A. A. 1950. *Records of Birds of Prey in Captivity*. London: Prestwich. [Little Owl data on p. 22.]
1400. Profira, B. & Sorescu, C. 1970. Contributions concernant la nourriture de la Chouette (*Athene noctua noctua* Scop.). *Analele Universitatii Bucuresti: Biologie Animalia* **19**: 67–72.
1401. Przygodda, W. 1964. Rodenticides and birds of prey. ICBP World Conference on Birds of Prey and Owls. Caen 1964. [Little Owl data on p. 59.]
1402. Pukinskiy, Y. B. 1977. *Zhizn' sov* [*Life of Owls*]. Leningrad: Leningrad Univ. Press. [Little Owl data on pp. 175–184.]
1403. Pulsford, A. H. 1989. Redstarts attack Little Owl. *Cheshire and Wirral Bird Report* [*Cheshire and Wirral Ornithological Society*] **1988**: 88.
1404. Pykal, J. Krafka, Z., Klimes, Z. *et al.* 1994. Populacni hustota sycka obecneho (*Athene noctua*) ve vybranych oblastech jiznich Cech. [Population density of Little Owl (*Athene noctua*) in selected regions of Southern Bohemia (Czech Republic)]. *Sylvia* **30**: 59–63.
1405. Qu, Y.-H., Lei, F.-M. & Yin, Z.-H. 2002. A study on genetic polymorphisms within *Athene noctua*. *In* Lian, Z. M., Xi, G. S., Huang, Y. *et al.* (ed). *Animal Science*. Shaanxi. Normal University Press. Xian. pp. 189–195.
1406. Quadrelli, G. 1985. Pesenza di fibre vegetali nelle borre di civetta *Athene noctua* (Scopoli). *Picus* **11**: 69–71.
1407. Quepat, N. 1899. *Ornithologie du Val de Metz*. Vanière. Metz. [Little Owl data on pp. 11–12.]
1408. Rabacchi, R. 1983. Offresi casa per rapaci notturni. *Uccelli* **18**: 10.
1409. Rabacchi, R. 1985. Osservazioni comportamentali insolite riscontrate in civeti *Athene noctua* e merlo *Turdus merula*. *Picus* **11**: 109–111.
1410. Rabosee, D. 1995. *Atlas des Oiseaux Nicheurs de Bruxelles 1989–1991*. *Aves* [Little Owl data on pp. 80–81.]
1411. Radu, D. 1990. Cucuveaua (*Athene noctua*) reprodusa in captivitate pentu prima data in Romania. *Ocrot. Nat. Med. Inconj.* [*Environmental and Nature Protection Journal Romania*] **34**: 29–38.
1412. Raevel, P. 1986. Concentration ponctuelle de rapaces nocturnes sur un lieu de passage de crapauds communs (*Bufo bufo*). *Héron* **4**: 52.
1413. Raevel, P. 1987. L'avifaune nicheuse d'un secteur bocager de Flandre intérieure. *Héron* **20**: 17–26.
1414. Raffaele, G. 1931. Il Plasmodium della civetta (*Athene noctua*). *Rivista di Malariologia* **10**: 684–688.
1415. Raffaele, G. 1934. Un ceppo italiano di *Plasmodium elongatum*. *Rivista di Malariologia* **13**: 332–337.
1416. Rakhimov, I. I. 1995. Sych domovyi [Little Owl]. *Krasnaya kniga Respubliki Tatarstan* [*Red Data Book of Republic of Tatarstan*]. Kazan, Publishing House "Priroda /Nature", Publisher TOO "Star" [Little Owl data on pp. 78–79].
1417. Rakhimov, I. I. & Pavlov, Y. I. 1999. Sych domovyi [Little Owl]. *Khishchnye ptitsy i sovy Tatarstana* [*Birds of Prey and Owls of Tatarstan*]. Kazan', "Tatpoligraf" Press. [Little Owl data on pp. 76–78.]
1418. Randi, E., Lorenzini, R. & Fusco, G. 1991. Biochemical variability in four species of *Strigiformes*. *Biochemical Systematics and Ecology* **19**: 13–16.
1419. Randla, T. 1976. *Eesti röövlinnud Kullilised ja kakulised*. Talinn: Valgus. [Little Owl data on pp. 158–159.]
1420. Rappe, A. 1979. Pesticides et oiseaux, quelques données récentes. *Aves* **16**: 130.
1421. Redon, M. 2002. La Chouette chevêche sur le Causse Méjean. Bilan des suivis 2001. Etude de l'habitat et propositions pour le maintien de l'espèce. BTS GPN Option Gestion des Espaces Naturels. Parc National des Cévennes. Rapport.
1422. Reichenow, A. 1904. *Die Vögel Afrikas*. 3. Neudamm. Verlag von J. Neumann. [Little Owl data on p. 822.]
1423. Reinard, M. 1977. Steinkauz zieht seine Jungen in einer Erdhöhle gross. *Regulus* **12**: 234–236.
1424. Reisinger, L. 1926. Hypnose des Steinkauzes. *Biologisches Zentralblatt* (*Leipzig*) **46**: 630–631.
1425. Reisinger, L. 1926. Ein interessanter Fehler am Auge eines Steinkauzes. *Biologisches Zentralblatt* (*Leipzig*) **46**: 632.
1426. Remmert, H. 1953. Zur Ernährung des Steinkauzes. *Vogelwelt* **74**: 217.
1427. Renner, M. 1986. Chouette chevêche *Athene noctua*. Synthèse 1986–1987. *Milvus* **21**: 64–66.
1428. Renner, M. 1987. La Chouette chevêche en Lorraine. *Milvus* **22**: 59–60.
1429. Renner, M. 1992. Chouette chevêche (*Athene noctua*). Suivi des populations dans le Parc Naturel Régional de Lorraine en 1991. Parc Naturel Régional de Lorraine. LPO. Rapport.
1430. Renson, D. 1998. Enquête Chouette chevêche. EPOPS Liaison. *La revue des naturalistes du Limousin* 1: 24–33.
1431. Renzoni, A. & Vegni Talluri, M. 1966. The karyogram of some *Falconiformes* and *Strigiformes*. *Chromosoma* **20**: 133–150.
1432. Reske, E. 1969. Die Aachener Vogelwelt. *Charadrius* **5**: 90.
1433. Rey, E. 1908. Mageinhalt einiger Vögel. *Ornithologische Monatsschrift* **33**: 193.
1434. Rey, E. 1910. Mageinhalt einiger Vögel. *Ornithologische Monatsschrift* **35**: 227.

435. Rheinwald, G. 1975. *Atlas der Brutverbreitung Westdeutscher Vogelarten.* Dachverbandes Deutscher Avifaunisten.
436. Rheinwald, G. 1980. Verbreitung der Eulen in Europa. *Vogelwelt* **101**: 114–118.
437. Richetti, F., Vittoria, A., Costagliola, A., De Falco, G. & Lepore, G. 1980. Prime osservazioni istomorfologiche sull'apparato digerente di alcuni uccelli rapaci diurni (*Buteo buteo*) e noturni (*Athene noctua*). *Atti della Societa Italiana die Scienze Veterinarie* **34**: 125.
438. Richter, B. 1973. Der Steinkauz (*Athene noctua*) im Hamburger Raum. *Hamburger Avifaunistische Beiträge* **11**: 1–16.
439. Richter, T. & Gerlach, H. 1981. The bacterial flora of the nasal mucosa of birds of prey. [Little Owl data on pp. 11–14.]
440. Riedel, W. 1996. Gewölleuntersuchungen von Steinkäuzen (*Athene noctua*) aus dem mittleren Neckarraum. *Ornithologische Jahreshefte für Baden-Württemberg* **12**: 313–318.
441. Ringleben, H. 1958. Kopfüber hängender Steinkauz (*Athene noctua*). *Vogelring* **27**: 22.
442. Rischer, F. 1978. Fluchthemmung beim Steinkauz (*Athene noctua*)? *Ornithologische Mitteilungen* **30**: 128–129.
443. Risdon, D. H. S. 1949. Little Owls, *Athene noctua. The Avicultural Magazine* **55**: 90–92.
444. Risenthal, von D. 1879. Verkannte und Missachtete. Das Steinkäuzchen. *Ornithologische Monatsschrift* **4**: 119–124.
445. Ristow, D. 1968. Beobachtungen über die Brutvögel der Siegniederung. *Charadrius* **4**: 90.
446. Ritter, F. 1974. Steinkauz – *Athene noctua* (Scopoli). *Berichte zur Avifauna des Bezirkes Gera Loseblattsammlung*: 3–4.
447. Ritzel, L. 1978. Fluchthemmung beim Steinkauz (*Athene noctua*)? *Ornithologische Mitteilungen* **30**: 128–129.
448. Ritzel, L. & Wulf, J. 1978. Elstern (*Pica pica*) attackiert Steinkauz (*Athene noctua*). *Ornithologische Mitteilungen* **30**: 255.
449. Robert, J.-C. 1979. Compte rendu ornithologique de la Baie de Somme. Automne–Hiver 1974–1975. *Documents Zoologiques. Université de Picardie.* **2**: 23.
450. Robert, J.-C. 1979. Oiseaux du Parc Naturel Régional Normandie-Maine. *Cormoran* **26**: 87–90.
451. Robiller, F. 1987. Über den Bodenbelag in Bruthöhlen des Steinkauzes (*Athene noctua*). *Acta Ornithol-ecologica* **1**: 299–301.
452. Robiller, F. 1993. Zur lokalen Bestandsentwicklung des Steinkauzes, *Athene noctua*, am Stadtrand von Weimar. *Beiträge zur Vogelkunde* **39**: 1–3.
453. Robiller, F. & Günzler, W. 1991. Erfolgreiche Bruten eines ausgewilderten Steinkauzes (*Athene noctua*). *Artenschutzreport* **1**: 51–52.
454. Robiller, F. & Robiller, F. C. 1992. Gelungene Ansiedlung des Steinkauzes (*Athene noctua*) am Stadtrand von Weimar. *Beiträge zur Vogelkunde* **38**: 1–5.
455. Robiller, F. & Robiller, M. 1986. Ein Beitrag zur Brutbiologie des Steinkauzes (*Athene noctua*). *Beiträge zur Vogelkunde* **32**: 161–174.
456. Robinson, H. W. 1924. Little Owl in Lancashire. *British Birds* **18**: 22–23.
457. Robinson, H. W. 1925. More Little Owls in north Lancashire. *British Birds* **18**: 267.
458. Robinson, J. 1967. Residues of organochlorine insecticides in dead birds in the United Kingdom. *Chemistry and Industry* **86**: 1978–1981.
459. Rochon Duvigneaud, A. 1934. Notes d'ophtalmologie comparée. Les yeux de la Chouette chevêche (*Athene noctua*). *Bulletin de la Société Zoologique de France* **59**: 224–226.
460. Rockenbauch, D. 1976. Ergänzungen zur Nahrungsbiologie einiger Eulenarten. *Anzeiger der Ornithologischer Gesellschaft in Bayern* **15**: 78–84.
461. Rode, P. 1930. Observations sur le phénomène d'immobilisation réflexe chez la Chouette chevêche (*Athene noctua* Scop.). *Bulletin de la Société Zoologique de France* **55**: 451–454.
462. Rodts, J. 1994. La faune et le trafic routier. *L'Homme et l'Oiseau* **2**: 109–118.
463. Rogacheva, E. V. 1988. *Ptitsy Srednei* Sibiri [*Birds of the Middle Siberia*]. Moscow: Nauka.
464. Roger, M. 2001. Etude de l'effectif d'une espèce en régression: la Chouette chevêche. Plan de restauration de la qualité de l'habitat de la campagne littorale de Guidel-Ploemeur (56). 1. UFR Sciences de la Vie et de l'Environnement. Masters thesis, Université de Rennes.
465. Rollinat, R. 1909. Sur la Chouette chevêche et la Chouette hulotte. *Revue Française d'Ornithologie* **1**: 167.
466. Romanowski, J. 1988. Trophic ecology of *Asio otus* (L) and *Athene noctua* (Scop.) in the suburbs of Warsaw. *Polish Ecological Studies* **14**: 223–234.
467. Rooth, J. & Bruijns, M. F. M. 1964. Birds of prey and owls in the Netherlands. Report on ICBP Working Conference on Birds of Prey and Owls. [Little Owl data on pp. 107–108.]

1468. Rörig, G. 1900. Mageuntersuchungen land-und forstwirtschaftlich wichtiger Vögel. *Arb. Biol. Abt. Land u. Forstwirtschaft Kaiserl. Gesundheitsamte*. Berlin **1**: 15–17.
1469. Rosendahl, S. 1973. *Ugler I Danmark*. DOCS Forlag. Skjern. [Little Owl data on pp. 24–29.]
1470. Rosendahl, S. & Skovgaard, P. 1972. Nogle meldinger af kirkeugle (*Athene noctua* (Scopoli)). *Danske Fugle* **24**: 248.
1471. Roth, H. 1956. Knochenverdauung bei einem jungen Steinkauz (*Athene noctua* Scop.). *Journal für Ornithologie* **97**: 90–91.
1472. Roth, N. & Barth, R. 1984. Zusammenstellung der ornith. Einzelbeobachtungen des Jahres 1983 für das Saarland. *Lanius* **23**: 164.
1473. Roth, N. & Hahn, E. 1983. Arbeitsanleitung zur saarländischen Steinkauz-und Schleiereulenkartierung 1983/84 der Eulen-Arbeitsgemeinschaft Saar.
1474. Roth, N., Nicklaus, G. & Weyers, H. 1990. *Die Vogel des Saarlandes. Eine Übersicht*. Ornithologische Beobachtung Saar. Homburg. [Little Owl data on p. 136.]
1475. Royer, P. 1983. Les rapaces dans le département de la Somme. Réflexions sur les causes de leur raréfaction. Master thesis, Université d'Amiens. [Little Owl data on pp. 72–74.]
1476. Rubinic, B. 2000. Mortality rate of owls *Strigiformes* on motorways between Bologna and Monfalcone (Italy) in the winter of 1998–99. *Acrocephalus* **21**: 67–70.
1477. Rufino, R. 1989. *Atlas das Aves que Nidifican em Portugal Continental*. Centro de Estudos de Migroçoes e Protecçao de *Aves*. Servicio Nacional de Parques Reservas e Conservaçao da Natureza. Lisboa. [Little Owl data on p. 97.]
1478. Rufino, R., Araujo, A. & Abreu, M. V. 1985. Breeding raptors in Portugal: distribution and population estimates. *Conservation Studies on Raptors ICBP* **5**: 15–28.
1479. Ruge, K. 1977. Wie wir Eulen helfen können. *Wir und die Vögel* **1**: 5.
1480. Rüger, A. 1976. In Schleswig-Holstein gefährdete sowie seltene Vogelarten und deren Lebensräume "Rote Liste". *Corax* **5**: 156.
1481. Rule, M. 1977. Diet of nesting Little Owl. *Notornis* **24**: 40.
1482. Runte, P. 1951. Zur "kuwitt Frage" (Steinkauz oder Waldkauz?). *Ornithologische Mitteilungen* **3**: 133–136
1483. Rusch, W. 1988. Zwei Brutnachweise des Steinkauzes (*Athene noctua*) in der Nähe eines Waldkauz-Brutplatzes (*Strix aluco*). *Kiebitz* **8**: 145.
1484. Rusch, W. 1989. Schleiereulen-und Steinkauzbestandserhebung im Kreis Coesfeld 1988. *Kiebitz* **9** 34–36.
1485. Rusch, W. 1991. Zum Bestand von Steinkauz und Schleiereule im Kreis Coesfeld. *Mitglieder-Informationen AG zum Schutz bedrohter Eulen* **36**: 9–11.
1486. Ryabitsev, V. K. 2001. *Ptitsy Urala, Priural'ya i Zapadnoi Sibiri* [*Birds of the Ural, Ural Area and West Siberia*]. Ekaterinburg: Uralian University Press.
1487. Rydzewski, W. 1978. The longevity of ringed birds. *Ring* **97**: 239.
1488. Ryslavy, T. 2001. Zur Bestandssituation ausgewählter Vogelarten in Brandenburg Jahresbericht 1999 *Naturschutz und Landschaftspflege in Brandenburg* **10**: 4–16.
1489. Ryslavy, T. 2002. Zur Bestandssituation ausgewählter Vogelarten in Brandenburg Jahresbericht 2000 *Naturschutz und Landschaftspflege in Brandenburg* **11**: 183–197.
1490. Ryslavy, T. 2004. Zur Bestandssituation ausgewählter Vogelarten in Brandenburg Jahresbericht 2002 *Naturschutz und Landschaftspflege in Brandenburg* **13**: 147–155.
1491. Sageder, G. 1990. Nahrungsspektrum und Mechanismen der Entstehung der Beutewahl beim Steinkauz eine vergleichende Freiland-und Laboruntersuchung. Ph.D. thesis, Universität Wien.
1492. Sagitov, A. K. 1990. *Otryad Sovy Striges. Ptitsy Uzbekistan*. [*Birds of Uzbekistan*]. Vol. 2. Tashkent. [Little Owl data on pp. 225–244.]
1493. Sagot, C. 1999. *Athene noctua In* Couloumy, C. (ed.) *Faune Sauvage des Alpes du Haut-Dauphiné. Atlas des Vertébrés. Tome 2. Les Oiseaux*. Parc National des Ecrins. Centre de Recherches Alpin sur les Vertébrés [Little Owl data on p. 104.]
1494. Salasse, J.-P. 1972. Inventaire et distribution des rapaces dans le Cantal. *Grand-Duc* **4**: 33–34.
1495. Salasse, J.-P. 1974. Régime alimentaire des rapaces dans le Cantal. *Grand-Duc* **6**: 21–23.
1496. Salek, M. 2004. Vyvoj populace sykha obecneho *Athene noctua* na Ceskobudejovecku a Pisecku. *Sluka Holysov* **1**: 87–88.
1497. Salek, M. & Berec, M. 2001. Distribution and biotope preferences of the Little Owl (*Athene noctua*) in selected areas of the Southern Bohemia (Czech Republic). *Buteo* **12**: 127–134.
1498. Salikhbaev, Kh. S. & Ostapenko, M. M. 1964. *Ptitsy* [*Birds*]. *Ekologiya i khozyaistvennoe znacheni pozvonochnykh zhivotnykh yuga Uzbekistana (bassein Surkhandar'i)* [*Ecology and Practical Importance of Vertebrates of the South of Uzbekistan (SurkhanDarya Basin)*]. Tashkent. [Little Owl data on pp. 98–99.]
1499. Salvan, J. 1983. *L'Avifaune du Gard et du Vaucluse*. Nîmes: Société d'Etude des Sciences Naturelles de Nîmes et du Gard. [Little Owl data on p. 108.]

1500. Salvati, L. Manganaro, A. & Ranazzi, L. 2002. Little Owl *Athene noctua* density and habitat preferences in urban Rome, Italy. *Vogelwelt* **123**: 155–160.

1501. Samigullin, G. M. 1989. Sovinye Orenburgskoi oblasti [*Strigiformes* of the Orenburg Region]. *Rasprostranenie i fauna ptits Urala* [*Distribution and Fauna of birds of Ural*]. Sverdlovsk. [Little Owl data on pp. 92–93.]

1502. Samoilov, B. & Morozova, G. 1998. *Redkie ptitsy Tsentral'noi Rossii na territorii Moskvy* [*Rare Birds of Central Russia in the Territory of Moscow*]. [Little Owl data on pp. 125–132.]

1503. Sancho, J. S. D. & Ruiz, O. C. 1984. Algunos datos sobre status, distribucion y alimentacion del Buho real (*Bubo bubo*) en Navarra. [Little Owl data on p. 249.]

1504. Sane, R. 1996. La Chouette chevêche dans le Haut-Rhin. Résultats de deux saisons de prospection (1994 et 1995). *Cigogneau* **54**: 17–26.

1505. Sane, R., Hurstel, A., Sane, F. & Jaegly, E. 1996. La Chouette chevêche, *Athene noctua* SCOP., dans le Haut-Rhin en 1994 et 1995. *Ciconia* **20**: 81–92.

1506. Sanmartin, M. L., Alvarez, F., Barreiro, G. & Leiro, J. 2004. Helminth fauna of Falconiform and Strigiform birds of prey in Galicia, Northwest Spain. *Parasitol. Res.* **92**: 255–263.

1507. Sara, M. & Zanca, L. 1989. Considerazioni sul censimento degli Strigiformi. *Rivista Italiana di Ornithologia* **59**: 3–16.

1508. Sara, M. & Zanca, L. 1990. Aspetti della nicchia ecologica degli Strigiformi in Sicilia. *Naturalista Siciliano* **14**: 109–122.

1509. Sardenne, J.-P. 1991. *Les Oiseaux de Charente*. Angoulême: Charente Nature. [Little Owl data on p. 94.]

1510. Sauer, T. 1988. Hat der Steinkauz noch eine Zukunft? *Falke* **35**: 234–235.

1511. Savasta, F. 1992. La chouette chevêche. *Courrier des Epines Dromoises* **47**: 13–16.

1512. Savean, G. 1984. Premières données ornithologiques sur La Puisaye du Loiret en période de nidification. *Naturalistes Orléanais* **3**: 60–62.

1513. Scaravelli, D., Delogu, M. & De Marco, A. 2003. Note di morfometria, patologia comparata e costruzione del nido in *Athene noctua* nord italiane. *Avocetta* **27**: 104.

1514. Schaaf, R. 2005. Der Steinkauz von Crespina. *Kauzbrief* **17**: 14–24.

1515. Schäfer, E. 1938. Ornithologische Ergebnisse zweier Forschungsreisen nach Tibet. *Journal für Ornithologie* **86**: 170–172.

1516. Schaub, M., Ullrich, B., Knötzsch, G., Albrecht, P. & Meisser, C. 2006. Local population dynamics and the impact of scale and isolation: a study on different little owl populations. *Oikos* **115**: 389–400.

1517. Scheifler, R., Deforet, T., Coeurdassier, M. *et al.* 1999. Prospection de la Chevêche d'Athéna *Athene noctua* dans le nord de la Haute-Saône: méthodologie et résultats des enquêtes 1994 et 1995. *Falco* **32**: 3–10.

1518. Schepers, F. 1976. Steenuil op insektenjacht. *Vogeljaar* **24**: 329.

1519. Scherzinger, W. 1971. Zum Feindverhalten einiger Eulen (*Strigidae*). *Zeitschrift für Tierpsychologie* **29**: 165–174.

1520. Scherzinger, W. 1971. Beobachtungen zur Jugendentwicklung einiger Eulen (*Strigidae*). *Zeitschrift für Tierpsychologie* **28**: 494–504.

1521. Scherzinger, W. 1975. Breeding european owls. *The Avicultural Magazine* **81**: 7–71.

1522. Scherzinger, W. 1980. Verhalten und Stimme des Steinkauzes. *In* Glutz Von Blotzheim, V. N. & Bauer, K. M. (eds). *Handbuch der Vögel Mitteleuropas*. Vol. 9. Wiesbaden.

1523. Scherzinger, W. 1981. Vorkommen und Gefährdung der vier kleinen Eulenarten in Mitteleuropa. *Ökologie der Vögel* **3**: 283–292.

1524. Scherzinger, W. 1986. Kontrastzeichnungen im Kopfgefieder der Eulen (*Strigidae*)-als visuelle Kommunikationsmittel. *Annalen Naturhistorischen Museum Wien* **89**: 37–56.

1525. Scherzinger, W. 1988. Vergleichende Betrachtung der Lautrepertoires innerhalb der Gattung *Athene* (*Strigiformes*). Proc. Int. 100. DO-G Meeting, Current Topics Avian Biol. Bonn. [Little Owl data on pp. 89–96.]

1526. Schmid, P. 2003. Gewöllenanalyse bei einer Population des Steinkauzes *Athene noctua* im Grossen Moos, einer intensive genutzen Agrarlandschaft des schweizerischen Mittellandes. *Ornithologische Beobachter* **100**: 117–126.

1527. Schmidt, E. 1976. Kleinsäugerfaunistische Daten aus Eulengewöllen in Ungarn. *Aquila* **82**: 119–143.

1528. Schmidt, E. & Szabo, L. V. 1981. Data on the small mammal fauna of the Hortobagy based on owl pellet examinations. *Natural Hist. Ntn. Pks. Hung.* **1**: 409–411.

1529. Schmidt, P. 2003. Gewöllanalyse bei einer Population des Steinkauzes *Athene noctua* im Grossen Moos, einer intensiv genutzen Agrarlandschaft des schweizerischen Mittellandes. *Ornithologische Beobachter* **100**: 117–126.

1530. Schmidt-Bey, W. 1925. Die Vögel der Rheinebene zwischen Karlsruhe und Basel. *Ornithologische Monatsschrift* **50**: 136.

1531. Schmitt, B. 1966. Les oiseaux de Strasbourg. *Lien Ornithologique d'Alsace* **4**: 6.

1532. Schmitz-Scherzer, J. & Dorn, C. 1984. Untersuchungen an Gewöllen von Schleiereulen (*Tyto alba*) und Steinkäuzen (*Athene noctua*). *Ornithologische Mitteilungen* **36**: 17.
1533. Schnabel, E. 1926. Erster Bericht über die Tätigkeit der Beringungstelle Unterfranken. *Verhandlungen Ornithologie Gesellschaft Bayern* **17**: 82.
1534. Schnabel, E. 1934. Wiederfunde beringter fränkischer Vögel. *Verhandlungen Ornithologie Gesellschaft Bayern* **20**: 443–453.
1535. Schnurre, O. 1921. *Die Vögel der Deutschen Kulturlandschaft.* Marburg.
1536. Schnurre, O. 1935. Rüttelnde Eulen. *Beiträge zur Fortpflanzungsbiologie der Vögel* **11**: 143.
1537. Schnurre, O. 1940. Drei Jahre aus dem Leben eines Berliner Waldkauzes. *Beiträge zur Fortpflanzungsbiologie der Vögel* **16**: 84.
1538. Schnurre, O. 1956. Ernährungsbiologische Studien an Raubvögeln und Eulen der Darsshablinsel (Mecklenburg). *Beiträge zur Vogelkunde* **4**: 211–245.
1539. Schönn, S. 1980. Käuze als Feinde anderer Kauzarten und Nisthilfen für höhlenbrütende Eulen. *Falke* **27**: 294–299.
1540. Schönn, S. 1983. Ökologie und Schutz des Steinkauzes, *Athene noctua*, in der DDR. *Bulletin of World Working Group on Birds of Prey* **1**: 210–211.
1541. Schönn, S. 1986. Zu Status, Biologie, ökologie und Schutz des Steinkauzes (*Athene noctua*) in der DDR. *Acta Ornitholecologica* **1**: 103–133.
1542. Schönn, S. Scherzinger, W., Exo, K.-M. & Ille, R. 1991. *Der Steinkauz.* Die Neue Brehm-Bücherei. Wittenberg Lutherstadt: A. Ziemsen Verlag.
1543. Schoop, G., Siegert, R., Galassi, D. & Klöppel, G. 1955. Newcastle-Infektionen beim Steinkauz (*Athene noctua*), Hornraben (*Bucorvus* sp.), Seeadler (*Haliaetus albicilla*) und Rieseneisvogel (*Dacelo gigas*). *Mh Praktischer Tierheilkunde* **7**: 223–235.
1544. Schot, K. 1980. Little Owl. *Blijdorp Geluiden* **28**: 6.
1545. Schröpfer, L. 1995. Sledování početnosti a rozšíření sýčka obecného (*Athene noctua*) v Ceské republice [Monitoring of the abundance and distribution of the Little Owl (*Athene noctua*) in the Czech Republic. *Buteo* **7**: 195–196.
1546. Schröpfer, L. 1996. Sýček obecný (*Athene noctua*) v České republice-početnost a rozšíření v letech 1993-1995. [The Little Owl (*Athene noctua*) in the Czech Republic-abundance and distribution in the years 1993–1995.] *Buteo* **8**: 23–38.
1547. Schröpfer, L. 1996. Ein Beitrag zur Verbreitung des Steinkauzes *Athene noctua* (Scopoli) in Westböhmen. *Populationsökologie Greifvögel und Eulenarten* **3**: 361–364.
1548. Schröpfer, L. 1997. Abundance and distribution of the Little Owl (*Athene noctua*) in the Czech Republic – to be continued in 1998–1999. *Buteo* **9**: 123–124.
1549. Schröpfer, L. 2000. Sledování pocetnosti a rozšíření sýčka obecného (*Athene noctua*) v České republice - pokračování v letech 1998 a 1999. [The Little Owl (*Athene noctua*) in the Czech Republic – abundance and distribution in the years 1998–1999.] *Buteo* **11**: 161–174.
1550. Schröpfer, L. 2000. Ergebnisse des Programms "Häufigkeit und Verbreitung des Steinkauzes (*Athene noctua*) in der Tscheschichen Republik 1998–99" – vorläufiger Stand 1998. *Populationsökologie Greifvögel und Eulenarten* **4**: 517–522.
1551. Schuster, S., Blum, V., Jacoby, H. *et al.* 1983. *Die Vögel des Bodenseegebietes.* Ornithologische Arbeitsgemeinschaft Bodensee. Konstanz. [Little Owl data on p. 221.]
1552. Schwab, E. 1972. Massnahmen zur Erhaltung des Steinkauzes – *Athene noctua* – im Beobachtungsgebiet Rodgau und Dreieich. *Luscinia* **41**: 272–276.
1553. Schwarz, M. 1971. L'avifaune de la petite Camargue alsacienne (40 années d'observations). *Bulletin de la Société Industrielle de Mulhouse* **4**: 72.
1554. Schwarzenberg, L. 1970. Hilfe unserem Steinkauz. *Deutsche Bund für Vogelschutz Jahresheft*: **20–23**.
1555. Schwarzenberg, L. 1980. Steinkauzpflege: Vögel ziehen um. *Wir und die Vögel* **12**: 18–19.
1556. Schwarzenberg, L. 1981. Nisthilfen für Steinkauz, Schleiereule und Turmfalken. *Ökologie der Vögel* **3**: 349–353.
1557. Schwarzenberg, L. 1984. Kritisches zur Steinkauzröhre: Modell 1983-ein Ausweg! Eulen Arbeitsgemeinschaft Saar.
1558. Schwarzenberg, L. 1985. Die mardersichere Steinkauzröhre durch chemischen Abwehrstoff. *DBV AG zur Schutz bedrohter Eulen Informationsblatt* **23**: 1–4.
1559. Schwarzenberg, L. 1985. Kritisches zur Steinkauzröhre: Modell 1983 – ein Ausweg! *Thüringer Ornithologische Mitteilungen* **33**: 19–28.
1560. Schwarzenberg, L. 1985. Kritisches zur Steinkauzröhre: Modell 1983-ein Ausweg! Bundesweite Tagung der AG zum Schutz bedrohter Eulen am 2–3 März 85 in Grävenwiesbach DBV Zusammenfassungen der Vorträge. Thüringer Ornithologische Mitteilungen. pp. 6–7.

561. Schwarzenberg, L. 1986. Die mardersichere Steinkauzröhre durch chemischen Abwehrstoff. Eulen Arbeitsgemeinschaft Saar.
562. Schwarzenberg, L. 1986. Schaukelröhre. Nisthilfe für den Steinkauz. Eulen Arbeitsgemeinschaft Saar.
563. Schwarzenberg, L. 1990. Bauanleitung zum Steinkauz-Innenkasten. Eulen Arbeitsgemeinschaft Saar.
564. Scopoli, G. A. 1769. *Athene noctua*. Annus I-(V) Historico-Naturalis (Annus I Hist.-Nat.). [Little Owl data on p. 22.]
565. Scott, D. 1980. Owls allopreening in the wild. *British Birds* **73**: 436–439.
566. Selmi, E., Ronchetti, G., Zoboli, A. & Conventi, L. 2005. Posatoi di Civetta *Athene noctua* in un'area rurale della Provincia di Modena. *In* Mastrorilli, M. Nappi, A. & Barattieri, M. 2005. Atti I Convegno italiano sulla Civetta. Osio Sotto (BG), 21 Marzo 2004. Gruppo Italiano Civette. pp. 33–36.
567. SEPOL, 1993. *Atlas Des Oiseaux Nicheurs en Limousin*. SEPOL. Editions Lucien Souny. [Little Owl data on p. 79.]
568. Sermet, E. & Ravussin, P.-A. 1996. *Les Oiseaux du Canton de Vaud*. Nos Oiseaux. [Little Owl data on p. 234].
569. Serrano, J., De Los Santos, A. & Manez, M. 1987. Los caraboidea de Donana y zonas adyacentes (*Coleoptera*). *Graellsia* **43**: 39–48.
570. Sgorlon, G. 2003. Densita e spaziatura dei siti di nidificazione di Civetta *Athene noctua* in un ambiente urbano del Veneto Orientale. *Avocetta* **27**: 88.
571. Sgorlon, G. 2005. Exitus di Civetta *Athene noctua* dopo interazione con il Barbagianni *Tyto alba*. *In* Mastrorilli, M., Nappi, A. & Barattieri, M. (eds). Atti I Convegno italiano sulla Civetta. Osio Sotto (BG), 21 Marzo 2004. Gruppo Italiano Civette. p. 49.
572. Sharrock, J. T. R. 1976. *The Atlas of Breeding Birds in Britain and Ireland*. Berkhamsted: Poyser. [Little Owl data on pp. 254–255.]
573. Shaw, A. C. 1919. Little Owl in the Isle of Wight. *British Birds* **12**: 186–187.
574. Shnitnikov, V. N. 1949. *Ptitsy Semirech'ya* [*Birds of Semirechie*]. M.; L., Publ. House of Ac. of Sc. of the USSR. [Little Owl data on p. 280.]
575. Shtegman, B. K. 1960. K biologii domovoga sycha v yuzhnom Pribalkhashie [On the Little Owl biology in southern Pribalkhash'ye (Cis-Balkhash Lake Area)]. *Ornitologiya* **3**: 315–319.
576. Siblet, J.-P. 1988. *Les Oiseaux du Massif de Fontainebleau et des Environs*. Lechevallier et R. Chabaud. [Little Owl data on pp. 145.]
577. Sick, M. 1987. Auch Käuzen kann geholfen werden. Hans Mohr startet Hilfsprogramm für den Steinkauz. *Gefierdete Welt* **5**: 130–132.
578. Siegmund, B. 1914. Der Steinkauz in der Stadt. *Ornithologische Beobachter* **11**: 175–178.
579. Siegmund, B. 1915. Der Steinkauz in der Stadt. *Ornithologische Beobachter* **12**: 104–105.
580. Signoret, F. 2002. La Chouette chevêche *Athene noctua* en marais Breton Vendéen: inventaire 1999–2000. *Gorgebleue* **17**: 25–30.
581. Sill, K. & Ullrich, B. 2005. Reproduktive Leistung eines über zwölf Jahre brütend kontrollierten Steinkauzweibchens *Athene noctua*. *Vogelwarte* **43**: 43–45.
582. Silloway, P. M. 1901. A handsome little owl. *Zoologist* **18**: 85–87.
583. Sim, G. 1902. The Little Owl in Kincardineshire. *Annals of Scottish Natural History*. p. 119.
584. Simakin, L. V. 2000. Zimnyaya fauna i naselenie ptits Badkhyza (Yugo-vostochnaya Turkmenia) [Winter fauna and bird population of Badkhyz]. *Ornitologiya* **29**: 87–92.
585. Simeonov, S. D. 1968. Materialen über die Nahrung des Steinkauzes (*Athene noctua* Scopoli) in Bulgarien. *Fragmenta Balcanica. Skopje, Musei Macedonici Scientiarum Naturalium* **6**: 157–165.
586. Simeonov, S. D. 1983. New data on the diet of the Little Owl (*Athene noctua* (Scop.)) in Bulgaria. *Ekologiya Sofiya* **11**: 53–60.
587. Simeonov, S. D. 1988. A study of the nutritive spectrum of the Eagle Owl (*Bubo bubo* L.) in Bulgaria. *Ekologiya Sofiya* **21**: 47–56.
588. Simon, L. 1979. Bestandserfassung des Steinkauzes – *Athene noctua* – in einem Gebiet der Südpfalz. *Naturschutz und Ornithologie in Rheinland-Pfalz* **1**: 187–202.
589. Simon, L. 1979. Arbeitsbericht der AG-Eulen der Gesellschaft für Naturschutz und Ornithologie Rheinland-Pfalz e.v. (GNOR). *Naturschutz und Ornithologie in Rheinland-Pfalz* **1**: 343–349.
590. Smeets, W. 2000. De "Thoonen-steenuilenkast". *Vogeljaar* **48**: 55–58.
591. Smets, M. 1981. Belevenissen bij het Steenuiltje (*Athene noctua*). *Wielewaal* **47**: 326–328.
592. Smets, P. 2004. Onderzoek naar het broedsucces van de Steenuil *Athene noctua* in het Hageland (Vlaanderen). *Athene Nieuwsbrief STONE* **9**: 11–12.
593. Snouckaert Van Schauburg, R. 1914. De steenuil van Palestina. *Jaarbericht Club Nederlandse Vogelkunde* **4**: 62–63.
594. Snow, D. W. & Perrins, C. M., 1998. *The Birds of the Western Palearctic. Concise Edition*. Volume 1. *Non-Passerines*. Oxford: Oxford University Press. [Little Owl data on pp. 903–906.]

1595. Soares, A. A. 1971. Rapinaceos de Portugal. II. *Strigiformes. Arquivos do Museu Bocage* **3**: 109–114.
1596. Soares, A. A. 1972. Sur la non validité de *Athene noctua grüni* Jordans & Steinbacher et de *Athene noctua cantabriensis* Harrisson (*Aves-Strigiformes*). *Arquivos do Museu Bocage* **3**: 355–365.
1597. Soba Nature Nievre & Camosine, 1994. *Atlas des Oiseaux Nicheurs de la Nièvre.* Soba Nature Nièvre. Station ornithologique du Bec d'Allier. Camosine. Nevers. [Little Owl data on p. 84.]
1598. Soler, M. & Soler, J. J. 1992. Latitudinal trends in clutch size in single brooded hole nesting bird species: a new hypothesis. *Ardea* **80**: 293–300.
1599. Solymosy, L. 1951. A kuvik teli költese. [Little Owl breeding in winter.] *Aquila* **56**: 242.
1600. Sopyiev, O. 1982. On some adaptations of birds due to the anthropogenic effects on the Kara-Kum desert. Abstracts XVIII Cong. Int. Orn. Moscow. [Little Owl data on pp. 1060–1061.]
1601. Sosnikhina, T. M. 1950. Khozyaistvennoe znachenie domovogo sycha v usloviyakh polupustyni yuga Armyanskoi SSR. [Practical importance of the Little Owl in conditions of semidesert of the south of Armenian SSR.] *Proceedings of the Academy of Science of Armenian SSR* **3**: 95–100.
1602. Souvairan, P. 1963. Quelques observations concernant la densité et le comportement ornithologique d'une commune de Saône et Loire. *Jean-le-Blanc* **2**: 25.
1603. Speicher, K. 1970. Greifvogelschutz im Saarland. *Faunistisch-floristische Notizen aus dem Saarland* **3**: 18
1604. Spencer, K. G. 1965. Avian casualties on railways. *Bird Study* **12**: 257.
1605. Spenlehauer, T. 2004. La Chevêche d'Athéna. *Ligue de Protection des Oiseaux, Alsace* **32**: 22–23.
1606. Sponza, S. & Csermely, D. 1995. Analisi del comportamento predatorio in due specie di Strigiformi Barbagianni *Tyto alba* e Civetta *Athene noctua. Avocetta* **19**: 108.
1607. Staatliche Vogelschutz Fur Hessen, Rheinland-Pfalz Und Saarland & Hessische Gesellschaft Fur Ornithologie Und Naturschutz EV. 1987. Rote Liste des bestandsgefährdeten Vogel arten in Hessen-7. Fassung, Stand 1. Januar 1988. Zeitschrift für Vogelkunde und Naturschutz in Hessen. *Vogel und Umwelt* **4**: 335–344.
1608. Staats von Wacquant Geozelles, 1890. Zur Lebensweise des Steinkauzes (*Athene noctua* Retz). *Ornithol ogische Monatsschrift* **15**: 194–202.
1609. Staats von Wacquant Geozelles, 1893. Auch der Steinkauz besitzt Nachahmungstalent. *Ornithologische Monatsschrift* **18**: 471–472.
1610. Stadler, H. 1932. La voix des chouettes d'Europe moyenne. *Alauda* **4**: 271–278.
1611. Stadler, H. 1945. Die Stimmen der mitteleuropäischen Eulen. *Vögel Heimat* **14–16**: 17–23.
1612. Stahl, D. 1982. Zucht und Auswilderung des Steinkauzes (*Athene noctua*). *Voliere* **5**: 178–180.
1613. Stalla, F. & Stoltz, M. 2004. Die Vogelwelt des Naturparks Pfälzerwald. Pollichia. [Little Owl data on pp. 211–213.]
1614. Stalling, T. 1997. Zur Nahrung des Steinkauzes (*Athene noctua*). *Ornithologische Schnellmitteilungen Baden-Württemberg* **56**: 30.
1615. Stam, F. 2003. De overlevingskans en levensverwachting van steenuilen in het werkgebied van de vogelw erkgroep Stad en Ambt Doesborgh. *Athene Nieuwsbrief STONE* **8**: 9–11.
1616. Stam, F. M. 1988. Lichtsluis in steenuilkast succesvol. *Vogels* **48**: 250.
1617. Stange, C. 1999. Die Steinkäuze kommen. *Ornis* **4**: 29–31.
1618. Stange, C. 1999. Steinkauz – Artenschutztagung in Niedersachsen vom 16–18 April 1999. *Eulen Rundblick* **48/49**: 45–49.
1619. Stankovics, B. 1997. Kukumavka (*Athene noctua*) u urbanoj sredini Jagodine [Little Owls (*Athene noctua*) living in an urban area of Jagodina town.] *Ciconia* **6**: 91–92.
1620. Starikov, S. V. 2002. Materialy k ornitofaune severo-vostochnoi chasti Alakolskoi kotloviny (Vostochny Kazakhstan) [On avifauna of north-eastern part of Alakol' basin (Eastern Kazakhstan)]. *Russkiy Ornitologicheskiy Zhurnal [Russian Journal of Ornithology*] (*express issue*) **178**: 187–213.
1621. Starley, J. R. 1913. Little Owl breeding in Warwickshire. *British Birds* **7**: 118.
1622. Stastny, K., Bejcek, V. & Hudec, K. 1996. *Atlas hnizdniho rozireni ptaku v Ceske republice 1985–1989* Jinocany. H & H. [Little Owl data on pp. 188–189.]
1623. Staton, J. 1947. Little Owl as prey of Tawny Owl. *British Birds* **40**: 279.
1624. Steffen, J. 1958. Chevêches et Friquets. *Nos Oiseaux* **24**: 321.
1625. Stegman, B. K. 1960. K biologii domovogo syca v Juznom Pribalchase [On the biology of the Little Owl south of the Balchasch Lake.] *Ornithologija* **3**: 315–318.
1626. Steiner, J. 1988. Artenschutzmassnahmen für Schleiereule und Steinkauz im Seewinkel. Vogelschutz in österreich. *Mitteilungen der österreichischen Gesellschaft für Vogelkunde* **2**: 56.
1627. Stelzl, W. 1998. Zum Vorkommen und Bruterfolg einiger heimischen Eulenarten im Jahr 1997. *OBS-Info* **14**: 5–6.
1628. Stepanyan, L. S. 1990. *Konspekt ornitologicheskoi fauny SSSR* [*Conspectus of the Ornithological Fauna of the USSR.*]. Moscow: Nauka.
1629. Stevenson, H. 1862. Note on the Shore Lark (*Alauda alpestris*) and Little Owl (*Strix passerina*) in Norfolk *Zoologist* **20**: 7931.

1630. Stickroth, H., Schlumprecht, H. & Achtziger, R. 2004. Zielwerte für den "Nachhaltigkeitsindikator für die Artenvielfalt" – Messlatte für eine nachhaltige Entwicklung in Deutschland aus Sicht des Natur- und Vogelschutzes. *Berichte zum Vogelschutz* **41**: 78–98.
1631. Stocker, L. 1983. The comical Little Owl. *Countryside* **1028**: 247–248.
1632. Stracey Clitherow, C. E. 1911. Little Owl in Lincolnshire. *British Birds* **5**: 51.
1633. Stresemann, E. & Stresemann, V. 1966. Die Mauser der Vögel. *Journal für Ornithologie* **107**: 357–362.
1634. Stroeken, P. & Van Harxen, R. 2003. Steenuil bereikt leeftijd van 15 kalenderjaren. *Athene Nieuwsbrief STONE* **8**: 12–16.
1635. Stroeken, P. & Van Harxen, R. 2003. Verslag van het broedbiologisch onderzoek in de Zuidoost-Achterhoek in 2003. *Athene Nieuwsbrief STONE* **8**: 29–31.
1636. Stroeken, P. & Van Harxen, R. 2003. Resten van gewervelde prooidieren bij steenuilnesten. *Athene Nieuwsbrief STONE* **8**: 39–42.
1637. Stroeken, P. & Van Harxen, R. 2005. Overschatting broedsucces Steenuil – het effect van controles na het ringbezoek op de berekening van het broedsucces. *Athene Nieuwsbrief STONE* **10**: 38–43.
1638. Stroeken, P., Bloem, H., Van Harxen, R., Groen, N. & Boer, K. 2001. The Little Owl in the Netherlands: distribution, breeding densities, threats and protection measurements. *In* Génot, J.-C., Lapios, J.-M., Lecomte, P. & Leigh, R. S. (eds). *Ciconia* **25**: 185–190.
1639. Strucker, R. & Verkerk, J. 2000. Sterke achteruitgang van de Steenuil *Athene noctua* in het Noordelijk Deltagebied. *Vogeljaar* **48**: 49–54.
1640. Stubbe, H. 1973. *Buch der Hege*. Band 2. Federwild. Berlin. [Little Owl data on pp. 230–235.]
1641. Sturm, F. & Génot, J.-C. 1994. Renforcement des populations de Chouettes chevêches dans le Parc Naturel Régional des Vosges du Nord. Bilan 1993–1994. Ministère de l'Environnement. DNP. Parc Naturel Régional des Vosges du Nord. Rapport.
1642. Sturm, F., Faivre, B. & Génot, J.-C. 1996. Renforcement des populations de Chouettes chevêches dans le Parc Naturel Régional des Vosges du Nord. Bilan 1995. Ministère de l'Environnement. DNP. Parc Naturel Régional des Vosges du Nord. Rapport.
1643. Sturm, F., Duchamp, L. & Génot, J.-C. 1997. Renforcement des populations de Chouettes chevêches dans le Parc Naturel Régional des Vosges du Nord. Bilan 1996. Ministère de l'Environnement. DNP. Parc Naturel Régional des Vosges du Nord. Rapport.
1644. Sturm, F., Duchamp, L., Brignon, E., Pfitzinger, H. & Génot, J.-C. 1997. Renforcement des populations de Chouettes chevêches dans le Parc Naturel Régional des Vosges du Nord. Bilan 1997. Ministère de l'Environnement. DNP. Parc Naturel Régional des Vosges du Nord. Rapport.
1645. Sukhinin, A. N., Bel'Skaya, G. S. & Zhernov, I. V. 1972. Pitanie domovogo sycha v Turkmenii [The Little Owl feeding in Turkmenia]. *Ornitologiya* **10**: 216–227.
1646. Sutcliffe, S. J. 1989. The diet of Little Owls on Skomer. Pembs. *Bird Rep.* **1989**: 24–26.
1647. Swinhoe, R. 1870. *Athene noctua plumipes*. Proceedings of the Zoological Society of London.
1648. Tahon, F. 1986. Recensement de rapaces nocturnes dans la région de Gembloux. Résumé de communication. Journée d'Etude d'Aves du 23 Novembre 1986.
1649. Tarres, M. A. G., Baron, M. & Gallego, A. 1986. The horizontal cells in the retina of the owl, *Tyto alba*, and owlet, *Carine noctua*. *Experimental Eye Research* **42**: 315–321.
1650. Tattorini, S., Manganaro, A. & Salvati, L. 2001. Insect identification in pellet analysis: implications for the foraging behaviour of Raptors. *Buteo* **12**: 61–66.
1651. Tavistock, Lord 1915. My Little Owls. *Bird Notes and News* **6**: 154–159.
1652. Tavistock, Lord 1915. My Little Owls. *Bird Notes and News* **6**: 183–186.
1653. Tavistock, Lord 1915. My Little Owls. *Bird Notes and News* **6**: 224–227.
1654. Tavistock, Lord 1915. My Little Owls. *Bird Notes and News* **6**: 243–249.
1655. Tavistock, Lord 1917. Moult of owl's beaks. *Ibis* **5**: 639.
1656. Tayler, A. G. 1944. Little Owl hovering. *British Birds* **37**: 178.
1657. Teixera, R. M. 1979. *Atlas van de Nederlandse Broedvogels*. De Lange van Leer. Deventer. [Little Owl data on pp. 196–197.].
1658. Tella, J.-L. & Blanco, G. 1993. Possible predation by Little Owl *Athene noctua* on nestling Red-billed Choughs *Pyrrhocorax pyrrhocorax*. *Butlletí del Grup Catalá D'anellament* **10**: 55–57.
1659. Tennent, J. R. M. 1949. Little Owl attacking Moorhen. *British Birds* **42**: 329.
1660. Terrasse, J.-F. 1977. The situation of birds of prey in France in 1975. ICBP World Conference on Birds of Prey. Vienna. 1975. [Little Owl data on p. 106].
1661. The Cyprus Ornithological Society, 1984. Tenth bird report 1979. Nicosia Cyprus. [Little Owl data on p. 20.]
1662. The Cyprus Ornithological Society, 1985. Eleventh bird report 1980. Nicosia Cyprus. [Little Owl data on p. 26.]

1663. The Cyprus Ornithological Society, 1986. Twelfth bird report 1981. Nicosia Cyprus. [Little Owl data on p. 18.]
1664. The Cyprus Ornithological Society, 1986. Thirteenth bird report 1981. Nicosia Cyprus. [Little Owl data on p. 21.]
1665. The Ornithological Society of Turkey, 1969. Bird report 1966–1967. Nicosia Cyprus. [Little Owl data on p. 88.]
1666. The Ornithological Society of Turkey, 1978. Bird report 1974–1975. Nicosia Cyprus. [Little Owl data on p. 113.]
1667. Thebault, F. 1988. Les parasites des oiseaux de proies. Etude necropsique des helminthes rencontrés chez les rapaces de la région toulousaine. Masters Thesis, Université de Toulouse.
1668. Thévenot M., Bergier, P. & Beaubrun, P. 1983. Répartition actuelle et statut des rapaces nocturnes au Maroc. *Bièvre* **5**: 34–35.
1669. Thibault, J.-C. 1983. *Les Oiseaux de la Corse*. Histoire et répartition aux XIXe et XXe siècles. Parc Naturel Régional de Corse. De Gerfau. Paris. [Little Owl data on pp. 148–149.].
1670. Thibault, J.-C. & Bonaccorsi, G. 1999. The Birds of Corsica. An annotated checklist. British Ornithologist Union. Checklist No. **17**: 64.
1671. Thibault, J.-C., Delaugere, M. & Noblet, J.-F. 1984. *Livre Rouge des Vertébrés Menacés de la Corse*. Parc Naturel Régional de Corse. [Little Owl data on p. 43.]
1672. Thiede, W. 1986. Bemerkenswerte faunistische Feststellungen 1982/83 in Europa. *Vogelwelt* **107**: 191–198.
1673. Thienemann, J. 1918. XVII Jahresbericht der Vogelwarte Rossitten. *Journal für Ornithologie* **66**: 371–372.
1674. Thienemann, J. 1924. XXI Jahresbericht der Vogelwarte Rossitten. *Journal für Ornithologie* **72**: 216.
1675. Thienemann, J. 1926. XXIII und XXIV Jahresbericht der Vogelwarte Rossitten. *Journal für Ornithologie* **74**: 77.
1676. Thiollay J.-M. 1963. Les pelotes de quelques rapaces. *Nos Oiseaux* **27**: 130–131.
1677. Thiollay, J.-M. 1966. Note sur le régime de *Tyto alba* et *Athene noctua* en Corse. *L'Oiseau et Revue Française d'Ornithologie* **36**: 282–283.
1678. Thiollay. J.-M. 1968. Le régime alimentaire de nos rapaces: quelques analyses françaises. *Nos Oiseaux* **29**: 252–253.
1679. Thiollay, J.-M. 1969. Essai sur les rapaces du midi de la France: Distribution. Ecologie. *Alauda* **37**: 20.
1680. Thobias, G. 1948. A kuvik baromfipusztitasa [Little Owl destroying poultry]. *Aquila* **55**: 242.
1681. Thomas, J. F. 1939. The food of the Little Owl. Derived from pellets, February 1936 to January 1937 *Discovery New Series* **2**: 94–99.
1682. Thompson, R. 1947. Little Owl breeding in Anglesey. *British Birds* **40**: 216.
1683. Thorpe, W. H. & Griffin, D. R. 1962. The lack of ultrasonic components in the flight noise of owls compared with birds. *Ibis* **104**: 256–257.
1684. Ticehurst, N. F. 1912. A further extension of the breeding range of the Little Owl. *British Birds* **6**: 188–189
1685. Tinbergen, L. & Broekhuysen, G. J. 1934. Ein Beitrag zur Kenntnis der Ernährung der Steinkauzjungen (*Athene noctua vidalii* A. E. Brehm). *Beiträge zur Fortpflanzungsbiologie der Vögel* **10**: 17–20.
1686. Tinbergen, L. & Tinbergen, N. 1932. über die Ernährung einer Steinkauzbrut (*Athene noctua vidalii* A. E Brehm). *Beiträge zur Fortpflanzungsbiologie der Vögel* **8**: 11–14.
1687. Tinbergen, N. 1932. Eine Beobachtung über die Ernährung des Steinkauzes (*Athene noctua vidalii* A. E Brehm). *Ardea* **21**: 74–75.
1688. Tischler, F. 1941. *Die Vögel Ostpreussens und seiner Nachgebiete* 1. Ost Europa Verlag. Königsberg und Berlin. [Little Owl data on pp. 553–555.]
1689. Toffoli, R. & Beraudo, P. 2005. Considerazioni sulla densita della Civetta *Athene noctua* in Provincia di Cuneo. *In* Mastrorilli, M., Nappi, A. & Barattieri, M. 2005. Atti I Convegno italiano sulla Civetta. Osio Sotto (BG), 21 Marzo 2004. Gruppo Italiano Civette. pp. 24–26.
1690. Tomassi, R., Piattella, E., Manganaro, A. *et al.* 1999. Primi dati su dieta e densità della Civetta *Athene noctua* nella Tenuta Presidenziale di Castelporziano (Roma). *Avocetta* **23**: 159.
1691. Tombal, J.-C. 1979. Synthèse des observations de l'hiver 1978/79. *Héron* **3**: 71.
1692. Tombal, J.-C. 1984. Synthèse des observations de l'hiver 1983–1984. Héron **3**: 46–47.
1693. Tombal, J.-C., 1996. Oiseaux nicheurs dans l'arrondissement de Kortrijk (West-Vlaanderen, Belgique) et dans la région lilloise et ses abords (Nord, France): comparaison de deux atlas récents. *Héron* **29**: 338–360
1694. Tombal, J.-C. 1996. Les Oiseaux de la Région Nord-Pas-de-Calais. Effectifs et distribution des espèce nicheuses: période 1985–1995. *Héron* **29**: 90–91.
1695. Tombal, J.-C. & Boutrouille, C. 1985. Synthèse des observations du printemps 1984. *Héron* **1**: 48.
1696. Tome, D. 1996. Owls in Slovenia – present status and distribution. *In* Stubbe, M. & Stubbe, A. (eds) *Populationsökologie Greifvogel-und Eulenarten*, 3. Wittenberg. [Little Owl data on pp. 343–351.]
1697. Tome, D. 1996. Višinska razširjenost sov v Sloveniji. [Vertical distribution of owls in Slovenia.] *Acrocephalus* **17**: 2–3.

1698. Tome, R., Bloise, C. & Korpimäki, E. 2004. Nest-site selection and nesting success of Little Owls (*Athene noctua*) in Mediterranean woodlands and open habitats. *Journal Raptor Research* **38**: 35–46.
1699. Tome, R., Santos, N., Cardia, P., Ferrand, N. & Korpimäki, E. 2005. Factors affecting the prevalence of blood parasites of Little Owls *Athene noctua* in southern Portugal. *Ornis Fennica* **82**: 63–72.
1700. Tomialojc, L. 1976. *Birds of Poland: Ptaki Polski*. Warsaw. [Little Owl data on p. 123.]
1701. Toms, M. P., Crick, H. Q. P. & Shawyer, C. R. 2000. *Project Barn Owl Final Report*. British Trust for Ornithology, Research Report 197. Thetford, Norfolk: British Trust for Ornithology.
1702. Topin, F. 1996. Effet de la structure du paysage sur la Chouette chevêche, sa distribution et son abondance. Stage DESS Espace rural et environnement. Université de Bourgogne. Parc Naturel Régional des Vosges du Nord.
1703. Törmälä T. 1978. Minervanpöllö, *Athene noctua*, tavattu Suomena. *Ornis Fennica* **55**: 43.
1704. Torregiani, F. 1981. Osservazione dell'accopiamento della civetta, *Athene noctua*. *Gli Uccelli d'Italia* **6**: 178–179.
1705. Toschki, A. 1999. Bestand und Verbreitung des Steinkauzes (*Athene noctua*) in Aachen. *Eulen Rundblick* **48/49**: 16–20.
1706. Townsend, G. 1924. Little Owl in Lancashire. *British Birds* **18**: 80.
1707. Transehe, von N. 1965. *Die Vogelwelt Lettlands*. Hannover: Verlag Hans von Hirschheydt. [Little Owl data on p. 73.]
1708. Transon, P. 1992. Enquête Chouette chevêche sur le secteur des Alpes Mancelles et de la région de Carrouges effectuée par le groupe ornithologique dse Avaloirs. Parc Naturel Régional Normandie-Maine. Rapport.
1709. Trap-Lind, I. 1963. Kirkeuglen. *Vor Viden* **7**: 210–219.
1710. Tricot, J. 1968. A propos de la capture des lombrics par la Chouette chevêche (*Athene noctua*). *Aves* **5**: 11.
1711. Trimnell, H. C. 1945. Little Owl feeding young on newts. *British Birds* **38**: 174.
1712. Tristan De Marquis, 1931. *La Faune Ornithologique de la Région Orléanaise et en Particulier de la Sologne*. Mémoires de la Société d'Agriculture, Sciences, Belles-lettres et Arts d'Orléans. Orléans. [Little Owl data on p. 13.]
1713. Trouche, L. 1957. Contribution à l'étude du bocage normand et du Département de la Manche. *Alauda* **25**: 60.
1714. Tubbs, C. R. 1953. Little Owl "smoke-bathing". *British Birds* **46**: 377.
1715. Tully, H. 1936. The Little Owl in Northumberland and Durham. *Vasculum* **22**: 41–45.
1716. Turchin, V. G. 1995. Stanovlenie ornitokompleksa khishchnykh ptits i sov Kamennoi Stepi [Formation of ornithocomplex of birds of prey and owls in Kamennaya Steppe]. *Voprosy estestvoznaniya* [*Problems of Natural History Sciences*]. **2**: 73–74.
1717. Turchin, V. G. 1999. Fauna khishchnykh ptits i sov Kamennoi stepi i eyo istoricheskie izmeneniya [Fauna of birds of prey and owls of the Kamennaya [Stone] Steppe and its historical changes]. *Berkut* **8**: 141–146.
1718. Turchin, V. G. 2000. Annotirovannyi spisok vodov vesenne-letnei ornithofauny Kamennoi stepi [Review of bird species of the Stone Steppe]. *Berkut* **9**: 1–8.
1719. Tutis, V., Bartovsky, V., Radovic, D. & Susic, G. 1991. Les rapaces nicheurs en Yougoslavie. *In* Juillard, M., Bassin, P., Baudvin, H. *et al.* (eds). Rapaces Nocturnes Actes du 30e colloque inter-régional d'orn thologie Porrenbury (Suisse): 2–4 Novembre 1990. *Nos Oiseaux*: 283–286.
1720. Tyurckhodzaev, Z. M. & Zaleskij, A. N. 1983. [The influence of the Little Owl on rodent populations on crops of Kzyl-Ordinsk region]. *In* Nurmuratov, T. N. (ed). Biologicheskie Metody Zashchity Selskokhozyaystvennykh Kultur v Kazakhstane. Alma-Ata: Vostochn oe Otdelenie VASHNIL. [Biol. Methods of Crop Protection in Kazakhstan. AA: East. Div. of All-Union Agric. Acad. Sci.]. [Little Owl data on pp. 105–110.]
1721. Ullrich, B. 1970. Ersatzbrut und Mauserbeginn beim Steinkauz (*Athene noctua*). *Vogelwelt* **91**: 28–29.
1722. Ullrich, B. 1973. Beobachtungen zur Biologie des Steinkauzes (*Athene noctua*). *Anzeiger der Ornithologischer Gesellschaft in Bayern* **12**: 163–175.
1723. Ullrich, B. 1975. Zu Legeabstand, Brutbeginn, Schlupffolge und Brutdauer beim Steinkauz (*Athene noctua*). *Journal für Ornithologie* **116**: 324–325.
1724. Ullrich, B. 1975. Bestandsgefährdung von Vogelarten im ökosystem "Streuobstwiese" unter besonderer Berücksichtigung von Steinkauz und den einheimischen Würgerarten der Gattung Lanius. *Beih. Veröff. Naturschutz u. Landschaftspflege Baden-Württemberg* **7**: 90–110.
1725. Ullrich, B. 1980. Zur Populationsdynamik des Steinkauzes (*Athene noctua*). *Vogelwarte* **30**: 179–198.
1726. Ullrich, B. 1985. Gibt es Steinkauz einen Geschlechtsdimorphismus in der Körpergrösse (Gewicht, Klügellänge, Schnabellänge)? Bundesweite Tagung der AG zum Schutz bedrohter Eulen am 2–3 März 85 in Grävenwiesbach. DBV Zusammenfassungen der Vorträge. [Little Owl data on pp. 9–11.]
1727. Uspensky, G. A. 1977. O gnezdovanii pustel'gi obyknovennoi, kobchika i domovogo sycha v gnezdakh vranovykh ptits [On the breeding of Kestrel, Red-footed Falcon and Little Owl in corvids' nests]. [Abstr. VII All. Union Orn. Conf. Part 1]. Kiev: Naukova Dumka. pp. 330–331.

1728. Uttendörfer, O. 1930. Studien zur Ernährung unserer Tagraubvögel und Eulen. *Abhandlungen der Naturforschenden Gesellschaft zu Görlitz* **31**: 136–141.
1729. Uttendörfer, O. 1931. Beobachtungen über die Ernährung unserer Tagraubvögel und Eulen im Jahr 1930. *Journal für Ornithologie* **79**: 305.
1730. Uttendörfer, O. 1939. *Die Ernährung der deutschen Raubvögel und Eulen und ihre Bedeutung in der heimischen Natur*. Verlag J. Neumann. Neudamm. [Little Owl data on pp. 257–266.].
1731. Uttendörfer, O. 1952. *Neue Ergebnisse über die Ernährung der Greifvögel und Eulen*. Eugen. Ulmer. Stuttgart. [Little Owl data on pp. 123–128.]
1732. Vachon, M. 1954. Remarques sur les ennemis des scorpions. A propos de la présence de restes de scorpions dans l'estomac de la Chouette *Athene noctua*. *L'Oiseau et Revue Française d'Ornithologie* **24**: 171–174.
1733. Vagliano, C. 1977. The status of birds of prey in Greece. ICBP World Conference on Birds of Prey. Vienna. 1975. [Little Owl data on p. 121.]
1734. Vagliano, C. 1981. Contribution au statut des rapaces diurnes et nocturnes en Crète. Rapaces méditerranéens. Parc Naturel Régional de Corse. Centre de Recherches Ornithologique de Provence. [Little Owl data on pp. 14–16.]
1735. Vagliano, C. 1985. The Little Owl (*Athene noctua*) in Crete (Greece). *Bulletin of World Working Group on Birds of Prey* **2**: 107–108.
1736. Vallon, G. 1901. Nota intorno alla nuova specie di Civetta scoperta nella provincia del Friuli. *Atti dell'Accademia di scienze, lettere e arti di Udine* **8**: 101–117.
1737. Vallon, G. 1901. Über *Athene chiaradiae* Giglioli in Friaul. *Ornithologisches Jahrbuch* **12**: 217–219.
1738. Vallon, G. 1902. Note ornitologiche per la provincia del Friuli durante l'anno 1902 (dal 1 gennaio al 1 agosto) (continuazione e fine). *Avicula* **6**: 108–117.
1739. Vallon, G. 1902. Note ornitologiche per la provincia del Friuli durante l'anno 1902 (dal 1 gennaio al 1 agosto) (continuazione e fine). *Avicula* **6**: 126–130.
1740. Vallon, G. 1907. Sulla nuova opera ornitologica "Berajah", Zoografia infinita di O. Kleinschmidt. *Bollettino della Societa Zoologica Italiana* **16**: 259–264.
1741. Van Ballegoie, A. 2003. Mandarijneend als prooidier van Steenuil. *Athene Nieuwsbrief STONE* **8**: 8.
1742. Van De Reep, M. 2004. Zijn de steenuilen in Zuid-Holland intelligenter? *Athene Nieuwsbrief STONE* **9**: 18.
1743. Van De Velde, E. & Mannaert, P. 1980. Steenuil (*Athene noctua*) eet wegslachtoffers. *Veldornithologisch Tijdschrift* **3**: 137–139.
1744. Van Den Brink, N. W., Groen, N. M., De Jonge, J. & Bosveld, A. T. C. 2003. Ecotoxicological suitability of floodplain habitats in The Netherlands for the Little Owl (*Athene noctua vidalii*). *Environmental Pollution* **122**: 127–134.
1745. Van Den Burg, A., Beersma, P. & Beersma, W. 2003. De temperatuur van nest-en roestplaatsen van de Steenuil *Athene noctua*. *Vogeljaar* **51**: 147–152.
1746. Van Den Burg, A. B., Beersma, P. F. & Beersma-Both, W. E. 2003. De temperatuur van nest-en roestplaatsen van de Steenuil. *Athene Nieuwsbrief STONE* **8**: 24–28.
1747. Van Dijk, A. J. 2003. Vliegende start monitoring Steenuil in 2003. *Athene Nieuwsbrief STONE* **8**: 4–7.
1748. Van Dijk, A. J. & Ottens, H.-J. 2001. Actuele verspreiding van de Steenuil en van Steenuilonderzoekers in Nederland. SOVON-informatierapport 2001/15 uitgevoerd in opdracht van Vogelbescherming Nederland.
1749. Van Dijk, A. J. & Van Turnhout, C. 2003. Monitoring van de Steenuil. *Athene Nieuwsbrief STONE* **7**: 3–8.
1750. Van Dijk, T. 1973. A comparative study of hearing in owls of the family *Strigidae*. *Netherlands Journal of Zoology* **23**: 131–167.
1751. Van Harxen, R. 2004. Broedseizoen 2004: een eerste indruk. *Athene Nieuwsbrief STONE* **9**: 14–17.
1752. Van Harxen, R. & Stroeken, P. 2003. Prooiaanvoer bij een steenuilenbroedpaar. *Vanellus* **56**: 144–154.
1753. Van Harxen, R. & Stroeken, P. 2003. Prooiaanvoer bij een steenuilenbroedpaar. *Athene Nieuwsbrief STONE* **7**: 17–28.
1754. Van Harxen, R. & Stroeken, P. 2003. Bijzondere schuilplaats van uitgevlogen Steenuiljongen. *Athene Nieuwsbrief STONE* **8**: 45–46.
1755. Van Leeuwen, M. 2003. Steenuilen binnen het Utrechtse project Uilen & Zwaluwen. *Athene Nieuwsbrief STONE* **8**: 17–19.
1756. Van Nieuwenhuyse, D. 2004. Steenuil. *In* Vermeersch, G., Anselin, A., Devos, K. *et al. Atlas van de Vlaamse Broedvogels 2000–2002*. Mededelingen van het Instituut voor Natuurbehoud 23, Brussel. [Little Owl data on pp. 256–257.]
1757. Van Nieuwenhuyse, D. & Bekaert, M. 2001. Modèle de régression logistique de prédiction de la qualité de l'habitat de la Chevêche d'Athéna (*Athene noctua*) à Herzele, Flandre Orientale (Nord-Ouest de la Belgique). *In* Génot, J.-C., Lapios, J.-M., Lecomte, P. & Leigh, R. S. (eds). *Ciconia* **25**: 95–102.

1758. Van Nieuwenhuyse, D. & Bekaert, M. 2001. Study of Little Owl *Athene noctua* habitat preference in Herzele (East-Flanders, Northern Belgium) using the median test. *In* Van Nieuwenhuyse, D., Leysen, M. & Leysen, K. (eds). *Oriolus* **67**: 62–71.

1759. Van Nieuwenhuyse, D. & Bekaert, M. 2002. An (auto)logistic regression model for prediction of Little Owl (*Athene noctua*) suitability of landscapes in East Flanders. Evidence for socially induced distribution patterns of Little Owl. Raptors in the new millenium. *In* Yosef, R. (ed.). Proceedings of the World Conference on Birds of Prey and Owls. Eilat, Israel 2–8 April 2000. International Birding and Research Center. Eilat. pp. 80–90.

1760. Van Nieuwenhuyse, D. & Leysen, M. 2001. Habitat typologies of Little Owl *Athene noctua* territories in Flanders (Northern Belgium). *In* Van Nieuwenhuyse, D., Leysen, M. & Leysen, K. (eds). *Oriolus* **67**: 72–83.

1761. Van Nieuwenhuyse, D. & Leysen, M. 2004. Distribution of Little Owl (*Athene noctua*) in Flanders (northern Belgium), in relation to environment: spatial modeling through GIS data and logistic regression. *In* Rodríguez, R. E. & Bojórquez Tapia, L., *Spatial Analysis in Raptor Ecology and Conservation*. Centro de Investigaciones Biologicas del Noroeste/Comision Nacional para el Conocimiento y Uso de la Biodiversidad. pp. 75–109.

1762. Van Nieuwenhuyse, D. & Nollet, F. 1990. Een onderzoek naar het verspreidingspatroon van de Steenuil *Athene noctua* in relatie met enkele landschappelijke kenmerken binnen de gemeente Meulebeke (West-Vlaanderen). *Oriolus* **56**: 50–55.

1763. Van Nieuwenhuyse, D. & Nollet, F. 1991. Biotoopstudie van de Steenuil *Athene noctua* met behulp van het clusteringsprogramma TWINSPAN. *Oriolus* **57**: 57–61.

1764. Van Nieuwenhuyse, D., Bekaert, M., Steenhoudt, K. & Nollet, F. 2001. Longitudinal analysis of habitat selection and distribution patterns in Little Owls *Athene noctua* in Meulebeke (West-Flanders), Northern Belgium. *In* Van Nieuwenhuyse, D., Leysen, M. & Leysen, K. (eds). *Oriolus* **67**: 52–61.

1765. Van Nieuwenhuyse, D., Leysen, M. & Steenhoudt, K. 2001. Analysis and spatial prediction of Little Owl *Athene noctua* distribution in relation to its environment in Flanders (Northern Belgium). *In* Van Nieuwenhuyse, D., Leysen, M. & Leysen, K. (eds). *Oriolus* **67**: 32–51.

1766. Van Nieuwenhuyse, D., Leysen, M., De Leenheer, I. & Bracquene, J. 2001. Towards a Conservation Strategy for Little Owl *Athene noctua* in Flanders. *In* Van Nieuwenhuyse, D., Leysen, M. & Leysen, K. (eds). *Oriolus* **67**: 12–21.

1767. Van Nieuwenhuyse, D., Belis, W. & Bodson, S. 2002. Une méthode standardisée pour l'inventaire des Chevêches d'Athéna (*Athene noctua*). *Aves* **39**: 179–190.

1768. Van Orden, C. & Paklina, N. V. 2003. An association between Tibetan Owlets *Athene noctua ludlowi* and Himalayan Marmots *Marmota himalayana* in West-Tibet and Southeast Ladakh. *De Takkeling* **11**: 80–85.

1769. Van Rijen, T., Schriks, H. & Van Leest, M. 2004. Steenuileninventarisatieproject West-Brabant afgerond. *Athene Nieuwsbrief STONE* **9**: 7–10.

1770. Van't Hoff, J. 1999. Nu het nog kan . . . Een soortenbeschermingsplan voor de Steenuil in Groningen. Steenuilwerkgroep Groningen.

1771. Van't Hoff, J. 2001. Balancing on the edge. The critical situation of the Little Owl *Athene noctua* in an intensive agricultural landscape. *In* Van Nieuwenhuyse, D., Leysen, M. & Leysen, K. (eds). *Oriolus* **67**: 100–109.

1772. Van't Hoff, J., 2003. Die Steinkauz-Arbeitsgemeinschaft Groningen. *Kauzbrief* **15**: 21–27.

1773. Van't Hoff, J. 2003. Uilenboekje over Nederlandse uilen. Steenuilenwerkgroep Groningen.

1774. Van't Hoff, J. 2004. Een sterk verhaal? *Athene Nieuwsbrief STONE* **9**: 21.

1775. Vanden Wyngaert, L. 2005. Steenuilproject Dijleland, Verslag 2004.

1776. Vanderlee, B. 1984. Poging tot vergelijkend voedselonderzoek aan de hand van braakballen van enkele roofvogels. *Wielewaal* **50**: 232–240.

1777. Vaurie, C. 1960. Systematic notes on palearctic birds. No. 42 *Strigidae*: The genus *Athene*. *American Museum Novitates* **2015**: 1–21.

1778. Vaurie, C. 1965. *The Birds of the Palearctic Fauna. A Systematic Reference. Non Passeriformes*. London: Witherby Limited. [Little Owl data on pp. 607–613.]

1779. Veit, W. 1988. Die Bestandsentwicklung zweier Steinkauzpopulationen in den Kreis Limburg-Weilburg, Lahn-Dill und Giessen von 1978–1987. *Vogel und Umwelt* **5**: 87–91.

1780. Veit, W. 1990. Brutzeitbeobachtungen von Eulen in den Kreisen Lahn-Dill, Giessen und Limburg-Weilburg 1988 und 1989. *Vogel und Umwelt* **6**: 133–134.

1781. Veit, W. 2001. Ornithologische Jahresbericht für Hessen 2 (2000). Zeitschrift für Vogelkunde und Naturschutz in Hessen. *In* Korn, M., Kreuziger, J., Norgall, A., Roland, H.-J. & Stübing, S. (eds). *Vogel und Umwelt* **12**: 101–213.

1782. Venant, A., Richou-Bac, L., Gleizes, E. *et al.* 1984. Contamination des oeufs de Rapaces par les Hydrocarbures Organochlorés entre 1974 et 1980. *Environmental Pollution* **7**: 179–191.

1783. Vendramin, E. & Marchesi, L. 2003. Densità e dispersione territoriale della Civetta *Athene noctua* in lessinia (VR). *Avocetta* **27**: 185.
1784. Vercauteren, P. 1986. Reactie op het artikel: "Balts en copulatie bij de Steenuil in November". *Ornis Flandriae* **5**: 107–108.
1785. Vercauteren, P. 1989. Steenuil *Athene noctua*. Vogels in Vlaanderen, Voorkomen en verspreiding. IMP. [Little Owl data on pp. 248–249.]
1786. Verhaeghe, K., Van Nieuwenhuyse, & Nollet, F. 1996. Studie van Steenuilen *Athene noctua* in Meulebeke (West-Vlaanderen). Inventarisatie, nestkastenonderzoek en analyse van de in-en uitgaande bewegingen in een nestkast. *Oriolus* **62**: 1–14.
1787. Verheyden, C. 1991. Une nouvelle méthode pour évaluer les densités de rapaces nocturnes et leur utilisation de l'habitat. *L'Oiseau et Revue Française d'Ornithologie* **61**: 17–26.
1788. Verwaerde, J. 1998. Wat wil de Steenuil? *Wielewaal* **64**: 2–3.
1789. Verwaerde, J., Van Nieuwenhuyse, D., Nollet, F. & Bracquene, J. 1999. Standaardprotocol voor het inventariseren van Steenuilen *Athene noctua* Scop. (*Strigidae*) in de West-Europese Laagvlatke. *Oriolus* **65**: 109–114.
1790. Vezinet, P. 2003. A propos du comportement alimentaire de la Chevêche d'Athéna dans notre région. La chouette d'Eoures. *Bulletin Association La Chevêche* **45**: 4.
1791. Vickers, H. S. 1934. Little Owl's nest under railway. *British Birds* **28**: 84–85.
1792. Vidal, P. 1986. Avifaune des îles d'Hyères (Var). Faune de Provence. *Bulletin du Conservatoire et Etudes des Ecosystèmes de Provence* **7**: 55.
1793. Vieron, J.-P. & Savasta, F. 1992. Survivantes. *Courrier des Epines Dromoises* **50**: 38.
1794. Vignes, J.-C. 1984. Les oiseaux victimes de la circulation routière au Pays basque français. *L'Oiseau et Revue Française d'Ornithologie* **54**: 139.
1795. Vilkov, E. V. 2003. Status, dinamika chislennosti i osobennosti ekologii sov Dagestana [Status, number dynamics and peculiarities of ecology of owls of Dagestan].
1796. Visser, D. 1977. De Steenuil-in het Rijk van Nijmegen. *De Mourik* **3**: 13–27.
1797. Vivier, C. & Telle, A.-S. 1997. Inventaire des populations et des sites de reproduction de la Chouette chevêche dans l'Audomarois. Maîtrse de Biologie des Populations et des Ecosystèmes. Université des Sciences et Technologies de Lille. Parc naturel régional Audomarois. Rapport.
1798. Vogrin, M. 1997. Little Owl (*Athene noctua*): a highly endangered species in NE Slovenia. *Buteo* **9**: 99–102.
1799. Vogrin, M. 2000. Owls of the Lower Savinja valley. *Acrocephalus* **21**: 43–45.
1800. Vogrin, M. 2001. Little Owl *Athene noctua* in Slovenia: an overview. *In* Van Nieuwenhuyse, D., Leysen, M. & Leysen, K. (eds). *Oriolus* **67**: 132–135.
1801. Volkemer, A. 1969. Die Brutvögel der Westeifel in den Kreisen Daun und Prüm. *Charadrius* **5**: 8.
1802. Vondracek, J. 1981. [On the presence of the Little Owl *Athene noctua* in northern Bohemia]. *Fauna Bohemiae Septentrionalis* 5–6.
1803. Voous, K. H. 1962. *Die Vogelwelt Europas und ihre Verbreitung*. Hamburg und Berlin: Verlag P. Parey. [Little Owl data on p. 160.]
1804. Voous, K. H. 1986. *Roofvogels en Uilen van Europa*. E. J. Brill. Dr. W. Backhuys. Leiden. [Little Owl data on p. 177.]
1805. Vorisek, P. 1995. Report of the activity of the Working Group on Protection and Research of Birds of Prey and Owls in the Czech Republic in 1993. *Buteo* **7**: 122–131.
1806. Vorisek, P. 1995. Report of the activity of the Working Group on Protection and Research of Birds of Prey and Owls in the Czech Republic in 1994. *Buteo* **7**: 149–158.
1807. Voronetskiy, V. I. 1998. Sipukha (*Tyto alba guttata* Brehm) i zapadnyi domovyi sych (*Athene noctua noctua* Scop.) – kandidaty na vklyuchenie v Krasnuyu knigu Rossii. [The Barn Owl (*Tyto alba guttata* Brehm) and the western Little Owl (*Athene noctua noctua* Scop.) as a candidate to the Red Data Book of Russia] *Ornitologiya* [*Ornithology*]. Moscow: Moscow Univ. Press. **28**: 136–139.
1808. Wagner, P. 1981. Der Steinkauz. *Regulus* **6**: 100–101.
1809. Wahlsted, J. 1971. Jaktmethod hos minervauggla *Athene noctua* [Hunting method of Little Owl *Athene noctua*]. *Var Fagelvarld* **30**: 46–47.
1810. Walker, C. H., Hamilton, G. A. & Harrisson, R. B. 1967. Organochlorine insecticide residues in wild birds. *Journal of the Science of Food and Agriculture* **18**: 123–129.
1811. Walter, G. & Hudde, H. 1987. Die Gefiederfliege *Carnus hemapterus* (*Milichiidae, Diptera*), ein Ektoparasit der Nestlinge. *Journal für Ornithologie* **128**: 251–255.
1812. Ward, H. L. 1964. Acanthocephala from the Little Owl, *Athene noctua*, in Egypt. *Journal of the Tennessee Academy of Science* **39**: 83–85.
1813. Wardhaugh, A. A. 1984. Wintering strategies of British owls. *Bird Study* **31**: 76–77.
1814. Ware, A. H. 1935. The food of the Little Owl. *Cambridge Bird Club Report* **1934**: 23–25.
1815. Warham, J. 1949. Photographing the Little Owl. *Country Life* **1949**: 1358–1359.

1816. Warham, J. 1955. Experiments with owls. *Country Life*: 1550–1551.
1817. Wassenich, V. 1956. Der Steinkauz in der Ofenröhre. *Regulus* **36**: 37.
1818. Wassmer, B. & Wilhelm, J.-L. 1983. Chouette. *Description des Différentes Espèces* **3**: 1710–1712.
1819. Waterson, G. 1951. Little Owl in east Lothian. *Scottish Naturalist* **63**: 189.
1820. Waterson, G. 1961. Little owl in Midlothian. *Scottish Birds* **1**: 453.
1821. Watson, J. B. 1922. Little Owl breeding in Merioneth. *British Birds* **16**: 84.
1822. Wayre, P. 1969. Breeding the Little Owl (*Athene noctua vidalii*). *The Avicultural Magazine* **75**: 96–97.
1823. Wayre, P. 1970. Breeding birds of prey and owls in the Norfolk wildlife park. *International Zoo Yearbook* **10**: 5–6.
1824. Weber, B. 1973. Beitrag zur Kenntnis der Ernährung der Eulen in der Magdeburger Börde, im Gebiet zwischen Ohre und oberer Aller in der Altmark. *Beiträge zur Vogelkunde* **19**: 363–375.
1825. Weesie, P. D. M. 1982. A Pleistocene endemic island form within the genus *Athene*: *Athene cretensis* nov. spp. (*Aves Strigiformes*) from Crete. *Proceedings of the Koninklijke Nederlandse Akademie van Wetenschappen* B 85: 323–336.
1826. Weigold, H. 1924. VIII. Bericht der Vogelwarte der Staatl. Biologischen Anstalt auf Helgoland. *Journal für Ornithologie* **72**: 54.
1827. Weiss, J. 1988. Rote Liste der Brutvögel Luxemburgs. *Regulus* **1**: 7–9.
1828. Weitnauer, E. & Bruderer, B. 1987. Veränderungen der Brutvogel Fauna der Gemeinde Oltingen in den Jahren 1935–1985. *Ornithologische Beobachter* **84**: 1–9.
1829. Weller, L. G. 1947. Little Owl nesting in cornstacks. *British Birds* **40**: 181.
1830. Wells, C. H. 1909. Little Owl breeding in Derbyshire. *British Birds* **3**: 84.
1831. Wemer, P. 1910. Etwas vom Steinkauz (*Athene noctua*). *Zoologischer Beobachter* **51**: 137–141.
1832. Wendland, V. 1967. Zischen und Fauchen bei Vögeln, insbesondere bei Eulen. *Beiträge zur Vogelkunde* **12**: 406.
1833. Wendt, E. 1978. Brutröhre für Steinkäuze. *Wir und die Vögel* **10**: 24.
1834. Wendt, E. 2004. Igel als Beute beim Steinkauz? *Ornithologische Schnellmitteilungen Baden-Württemberg* **74/75**: 49.
1835. Wenner, M. V. 1927. Little Owl in Derbyshire. *British Birds* **20**: 275.
1836. Werner, L. G. 1932. L'utilité et la protection des oiseaux. *Bulletin de la Société Industrielle de Mulhouse* **98**: 40.
1837. Werner, P. 1990. "Steinkauz-Fernwanderer" aus dem Main-Kinzig-Kreis. *Vogel und Umwelt* **6**: 133.
1838. Wernet, A. 1998. Une Chevêche d'Athéna (*Athene noctua*) attrape une proie au vol. *Ciconia* **22**: 40.
1839. Westerhof, G. 1981. Vogelasiel "De Hamrik" wil met Steenuil fokken. *Vogeljaar* **29**: 328.
1840. Weyer, B. V. 1911. Probable breeding of the Little Owl in Berkshire. *British Birds* **5**: 138.
1841. Weyers, H. 1982. Rote Liste der im Sarland bestandsgefährdeten Vogelarten. 4. Fassung. DBV Saarland. [Little Owl data on pp. 63–67.]
1842. Wheatley, I. S. 1969. Little Owl nesting in Midlothian. *Scottish Birds* **5**: 287.
1843. Whitaker, F. 1897. Little Owl near Newark on Trent. *Zoologist* **1**: 329.
1844. Wigger, B. 1910. Zur Naturgeschichte des Käuzchens *Athene noctua* (Retz). *Münster Jahresberichte Prov. Ver. Wiss.* **38**: 56–58.
1845. Wiglesworth, J. 1917. The Little Owl (*Athene noctua*) in Somerset. *Proceedings of the Somerset Archaeological and Natural History Society* **63**: 152–161.
1846. Wigman, A. B. 1937. De Steenuil, *Athene noctua vidalii* A. E. Brehm, als bodembroeder. *Limosa* **10**: 126–127.
1847. Wilhelm, J.-L., Loukianoff, S. & Génot, J.-C. 1991. Modes d'occupation et d'utilisation du milieu développés par la Chouette chevêche dans le Parc Naturel Régional des Vosges du Nord. SRETIE Ministère de l'Environnement. Parc Naturel Régional des Vosg es du Nord. Rapport.
1848. Willems, F. & Majoor, F. 2005. Veranderingen in het broedsucces van Steenuilen. *SOVON-Nieuws* **18**: 10–11.
1849. Willems, F., Van Harxen, R., Stroeken, P. & Majoor, F. 2004. Reproductie van de Steenuil in Nederland in de periode 1977–2003. *Athene Nieuwsbrief STONE* **9**: 32–70.
1850. Williams, P. D. 1920. Little Owl in Cornwall. *British Birds* **13**: 219.
1851. Williams, R. 1945. Little Owl attacking and carrying off Jackdaw. *British Birds* **38**: 194–195.
1852. Winck, M. 1987. *Die Vögel des Rheinlandes. 3.* Rheinischer Landwirschafts Verlag. Bonn. [Little Owl data on pp. 182–183.]
1853. Winde, H. 1970. Osteologische Untersuchungen an einigen deutschen Eulenarten (*Aves, Strigidae*). *Zoologische Abhandlungen des Staatlichen Museums für Tierkunde Dresden* **30**: 149–157.
1854. Wink, M., Sauer-Gürth, H. & Fuchs, M. 2004. Phylogenetic relationships in owls based on nucleotide sequences of mitochondrial and nuclear marker genes. *In* Chancellor, R. D. & Meyburg, B.-U. (eds). Raptors Worldwide. Proceedings of the VI World Conference on Birds of Prey and Owls. Budapest,

Hungary May 18–23, 2003. World Working Group on Birds of Prey and Owls (WWGBP). [Little Owl data on pp. 517–526.]
1855. Witherby, H. F. 1920. Note. *Bulletin British Ornithological Club* **40**: 118.
1856. Witherby, H. F. 1920. The Dutch and British Little Owls. *British Birds* **13**: 283.
1857. Witherby, H. F. & Ticehurst, N. F. 1908. The spread of the Little Owl from the chief centres of its introduction. *British Birds* **1**: 335–342.
1858. Witherby, H. F., Jourdain, F. C. R., Ticehurst, N. F. & Tucker, B. W. 1938. *The Handbook of British Birds*. 2. London: Witherby Ltd. [Little Owl data on pp. 322–327.]
1859. Witt, K. 1983. Bestandserfassung einiger ausgewählter gefährdeter Vogelarten (1977–1980). *Vogelwelt* **104**: 237–239.
1860. Witt, K. 1986. Bestandserfassung einiger ausgewählter gefährdeter Vogelarten (1982–1984) in der Bundesrepublik Deutschland. *Vogelwelt* **107**: 234–237.
1861. Wolf, H. 1981. Die Senne als Lebensraum für Greifvögel (*Falconiformes*) und Eulen (*Strigiformes*). *Berichte des Naturwissenschaftlichen Vereins für Bielefeld* **3**: 207–208.
1862. Woltered, K. L. 1895. Der Steinkauz (*Carine noctua* Retz) in der Gefangenschaft. *Ornithologische Monatsschrift* **20**: 252–255.
1863. Wuczynski, A. 1994. Population numbers and distribution of owls *Strigiformes* in the agricultural landscape of the Dzierzonów Dale. *Birds of the Silesian Region* **10**: 118–121.
1864. Wüst, W. 1970. *Die Brutvögel Mitteleuropas*. Bayerischer Schulbuch Verlag. [Little Owl data on pp. 237–239.]
1865. Wüst, W. 1986. *Avifauna Bavariae. Band 2*. Ornitholgische Gesellschaft Bayern. München. [Little Owl data on pp. 797–804.]
1866. Yeatman, L. 1970. *Histoire des Oiseaux d'Europe*. Bordas. [Little Owl data on pp. 232.]
1867. Yeatman, L. 1976. *Atlas des Oiseaux Nicheurs de France*. Paris: Soc. Orn. de France. [Little Owl data on pp. 128–129.]
1868. Yew, D. T. 1980. Neuronal cells and types of contacts in the owl's retina. *Anatomischer Anzeiger* **147**: 255–259.
1869. Yin, X. C. 1964. Daytime activity of the Ural Wood Owl *Strix uralensis* and Little Owl *Athene noctua*. *Chinese Journal of Zoology* **6**: 172.
1870. Yosef, R. 1993. Effects of Little Owl predation on Northern Shrike postfledging success. *Auk* **110**: 396–398.
1871. Zabala, J., Zuberogoita, I., Martinez-Climent, J. A. *et al.* 2006. Occupancy and abundance of Little Owl *Athene noctua* in an intensively managed forest area in Biscay. *Ornis Fennica* **83**: 97–107.
1872. Zaccaroni, A., Amorena, M., Naso, B. *et al.* 2003. Cadmium, chromium and lead contamination of *Athene noctua*, the little owl, of Bologna and Parma, Italy. *Chemosphere* **52**: 1251–1258.
1873. Zarkhidze, V. A. & Loskutova, E. A. 1999. Dinamika troficheskikh svyazei khishchnykh ptits i melkikh mlekopitayushchikh v severozapadnoi Turkmenii [Relationships between the birds of prey and small mammals in north-western Turkmenistan and their dynamics]. *Russkiy Ornitologicheskiy Zhurnal* [*Russian Journal of Ornithology*] **82**: 3–17.
1874. Zav'Yalov, E. V., Tabachishin, V. G., Shlyakhtin, G. V., Yakushev, N. N., Kochetova, I. B. 2000. Sovy Saratovskoi oblasti [Owls of the Saratov Region]. *Berkut* **9**: 73–74.
1875. Zav'Yalov, E. V., Tabachishin, V. G., Shlyakhtin, G. V., Yakushev, N. N. & Kochetova, I. B. 2000. Sovy Saratovski oblasti [Owls of the Saratov region]. *Berkut* **9**: 74–81.
1876. Zens, K.-W. 1992. Ökologische Studien an einer Population des Steinkauzes (*Athene noctua* SCOP. 1769) in der Mechernicher Voreifel unter Einbeziehung der radiotelemetrischen Methode. Masters thesis, Rheinischen Friedrich-Wilhelms-Universität.
1877. Zens, K.-W. 2005. Langstudie (1987–1997) zur Biologie, ökologie und Dynamik einer Steinkauzpopulation (*Athene noctua* SCOP. 1769) im Lebensraum der Mechenicher Voreifel. Ph.D. thesis, Rheinischen Friedrich-Wilhelm-Universität Bonn.
1878. Zerunian, S., Franzini, G. & Sciscione, L. 1982. Little Owls and their prey in a Mediterranean habitat. *Bollettino di Zoologia* **49**: 195–206.
1879. Zhalakevichius, M., Paltanavichius, S., Shvazhas, S., & Stanevichius, V. 1995. Lietuvos Paukshchiai [Birds of Lithuania]. *Acta Ornithologica Lituanica* (special issue) **11**: 59.
1880. Zhitkov, B. M. & Buturlin, S. A. 1906. Materialy dlya ornitofauny Simbirskoi gubernii [Materials for ornithofauna of the Simbirsk Province]. S.-Petersburg. *Zapiski Imperatorskogo Russkogo geograficheskogo obshchestva po obshchei geografii* [*Proceedings of the Imperial Russian Geographical Society on General Geography*]. **49**: 270.
1881. Ziesemer, F. 1978. Die Eulen in Schleswig-Holstein-ein Beitrag zur Verbreitung und Siedlungsdichte (*Athene noctua*) in Schleswig-Holstein. *Zoologischer Anzeiger* **207**: 323–334.
1882. Ziesemer, F. 1981. Zur Situation der Eulen (*Strigiformes*) in Schleswig-Holstein. *Ökologie der Vögel* **3**: 311–316.

1883. Ziesemer, F. 1981. Zur Verbreitung und Siedlungsdichte des Steinkauzes (*Athene noctua*) in Schleswig-Holstein. *Zoologischer Anzeiger Jena* **207**: 323–334.

1884. Zimmerli, E. 1975. Dreizehenspecht und Steinkauz. *Vögel der Heimat* **45**: 230–238.

1885. Zimmerli, E. 1982. Steinkauz und Obstgärten. *Vögel der Heimat* **53**: 81.

1886. Zinner, H. 1978. Observations on Little Owls feeding on snails in the Negev Argamon, Israel. *Israel Journal of Malacology* **6**: 57–60.

1887. Zmihorski, M., Altenburg-Bacia, D., Romanowski, J., Kowalski, M. & Osojca, G. 2006. Long-term decline of the Little Owl (*Athene noctua* Scop. 1769) in Central Poland. *Polish Journal of Ecology* **54**: 321–324.

1888. Zoest, van J. G. A. & Fuchs, P. 1988. Jaaggedrag en prooiaanvoer van een Steenuil *Athene noctua* broedpaar. *Limosa* **61**: 105–112.

1889. Zuberogoitia, I. 2002. Ecoetologia de las rapaces nocturnas de Bizkaia. Ph.D. thesis, Universidad del Pais Vasco.

1890. Zuberogoitia, I. & Campos, L. F. 1997. Intensive census of nocturnal raptors in Biscay. *Munibe* **49**: 117–127.

1891. Zuberogoitia, I. & Campos, L. F. 1998. Censuring owls in large areas: a comparison between methods. *Ardeola* **45**: 47–53.

1892. Zuberogoitia, I. & Martinez-Climent, J. A. 2001. The Little Owl in the "Proyecto *Noctua*". *In* Génot, J.-C., Lapios, J.-M., Lecomte, P. & Leigh, R. S. (eds). *Ciconia* **25**: 103–108.

1893. Zuberogoitia, I. & Torres, J. J. 1997. *Aves Rapaces de Bizkaia*. Temas Vizcainos. BBK. Bilbao.

1894. Zuberogoitia, I., Campos, L. F., Crespo, T. & Ocio, G. 1994. Situacción y datos sobre la reproducción de las rapaces nocturnas en Bizkaia. Actas de las XII Jornadas Ornitológicas Españolas. Almerimar (El Ejido-Almería), 15 a 19 de Septiembre de 1994: 297–305.

1895. Zuberogoitia, I., Martínez, J. A., Zbala, J. & Martinez, J. E. 2005. Interspecific aggression and nest-site competition in a European owl community. *Journal of Raptor Research* **39**: 156–159.

1896. Zubkov, N. I. 1978. Temperature conditions of incubation by Little Owl in Moldavia. *Proceedings of the Moldavian SSR Academy of Science* **4**: 88–90.

1897. Zubkov, N. I. & Muntyanu, A. I. 1981. Znachenie v pitanii nekotorykh vidov sov Moldavii myshevidnykh gryzunov [Significance of the mouse-looking rodents in some owl species feeding in Moldavia]. Ecology and conservation of birds. Kishinev: Shtiintsa Press.

1898. Zvářal, K. 2002. Can "architectural traps" be the cause of the critical decrease of Little Owl (*Athene noctua*)? *Crex* **18**: 94–99.

1899. Zverzhanovskii, M. 1967. Pitanie domovogo sycha v Stavropol'ye [Food of the little owl (*Athene noctua*) in Stavropol]. *Prirodo Sev. Kavkaza i oxrana* **1967**: 133–136.

1900. Zwiesele, D. 1906. Die Eulen Württembergs. *Ornithologische Beobachter* **5**: 6.

Appendix C Keyword index for Little Owl bibliography

Census

2, 79, 95, 104, 146, 168, 169, 192, 197, 207, 219, 220, 221, 226, 231, 234, 235, 240, 251, 262, 267, 283, 285, 286, 290, 298, 356, 374, 407, 414, 431, 468, 469, 470, 471, 498, 505, 533, 548, 551, 578, 580, 581, 606, 608, 615, 623, 624, 626, 627, 629, 648, 649, 655, 715, 731, 753, 770, 782, 794, 815, 836, 847, 848, 860, 864, 879, 898, 943, 946, 958, 965, 966, 968, 987, 1016, 1053, 1071, 1077, 1114, 1115, 1130, 1131, 1156, 1181, 1212, 1260, 1288, 1293, 1359, 1366, 1377, 1413, 1421, 1429, 1430, 1464, 1473, 1480, 1496, 1504, 1505, 1517, 1548, 1550, 1580, 1588, 1639, 1648, 1671, 1703, 1705, 1708, 1749, 1757, 1758, 1760, 1762, 1764, 1765, 1767, 1771, 1788, 1789, 1798, 1827, 1841, 1890, 1891.

Competition

200, 672, 1184, 1185, 1258, 1319, 1448, 1621, 1851, 189.

Conservation

23, 25, 108, 117, 134, 154, 184, 227, 229, 243, 244, 245, 246, 247, 262, 311, 316, 340, 355, 374, 378, 383, 385, 391, 413, 419, 462, 467, 470, 533, 552, 567, 592, 593, 605, 625, 634, 635, 648, 714, 739, 742, 746, 749, 772, 826, 839, 876, 914, 916, 920, 923, 924, 937, 1003, 1048, 1050, 1071, 1092, 1093, 1094, 1096, 1098, 1099, 1100, 1104, 1115, 1139, 1197, 1201, 1208, 1212, 1231, 1234, 1242, 1243, 1244, 1268, 1380, 1408, 1464, 1479, 1510, 1539, 1552, 1557, 1558, 1561, 1562, 1563, 1577, 1612, 1617, 1638, 1748, 1766, 1770, 1772, 1793, 1808, 1833, 1839.

Contamination

03, 134, 334, 335, 338, 352, 353, 565, 609, 645, 648, 709, 710, 742, 885, 886, 887, 888, 889, 895, 913, 917, 918, 926, 1050, 1253, 1254, 1262, 1341, 1397, 1401, 1420, 1458, 475, 1744, 1872, 1882.

Density

, 10, 18, 66, 119, 139, 146, 166, 168, 179, 192, 226, 244, 245, 248, 262, 267, 282, 283, 286, 290, 298, 309, 311, 314, 318, 340, 356, 370, 404, 412, 413, 422, 435, 470, 471, 484, 95, 496, 499, 505, 546, 548, 551, 561, 581, 598, 599, 601, 603, 604, 606, 612, 613, 618, 620, 623, 624, 626, 627, 629, 646, 648, 652, 653, 655, 674, 678, 679, 715, 719, 724, 771, 82, 801, 807, 820, 826, 842, 843, 845, 846, 851, 860, 864, 901, 911, 943, 944, 945, 966, 68, 986, 987, 1031, 1044, 1053, 1069, 1070, 1099, 1114, 1118, 1182, 1200, 1203, 1212, 229, 1231, 1260, 1271, 1288, 1343, 1365, 1366, 1377, 1413, 1427, 1464, 1492, 1494, 496, 1497, 1517, 1550, 1552, 1570, 1572, 1580, 1588, 1602, 1622, 1638, 1657, 1668, 693, 1708, 1757, 1762, 1796, 1798, 1818, 1860, 1867, 1876, 1890.

Distribution

, 9, 10, 12, 17, 18, 23, 33, 56, 59, 64, 65, 70, 71, 82, 87, 89, 90, 94, 98, 108, 109, 110, 12, 138, 141, 147, 148, 152, 160, 167, 179, 186, 193, 194, 209, 214, 216, 226, 236, 237, 51, 260, 265, 276, 277, 286, 287, 288, 289, 299, 300, 305, 307, 308, 309, 311, 312, 321, 30, 331, 332, 337, 341, 345, 350, 354, 355, 356, 357, 361, 363, 370, 371, 372, 375, 376, 78, 379, 382, 383, 386, 388, 393, 394, 397, 398, 401, 402, 404, 410, 411, 420, 421, 422,

Food

General

Genetics

Habitat

1, 11, 16, 23, 25, 27, 28, 29, 34, 79, 82, 87, 139, 146, 166, 168, 192, 209, 210, 211, 226, 227, 229, 240, 244, 245, 262, 276, 282, 283, 284, 297, 300, 302, 309, 311, 314, 337, 340, 345, 354, 362, 364, 365, 378, 386, 388, 393, 394, 397, 399, 402, 412, 414, 419, 422, 463, 470, 473, 476, 482, 484, 507, 509, 518, 522, 531, 532, 533, 534, 551, 559, 561, 575, 577, 578, 581, 588, 590, 593, 596, 597, 599, 600, 603, 604, 607, 613, 618, 619, 620, 627, 628, 643, 646, 648, 652, 653, 664, 677, 678, 679, 703, 718, 730, 731, 742, 746, 749, 753, 771, 773, 798, 807, 820, 826, 832, 836, 842, 855, 868, 880, 891, 895, 901, 910, 911, 916, 917, 918, 919, 923, 927, 928, 937, 941, 958, 967, 970, 980, 994, 1003, 1015, 1028, 1042, 1045, 1055, 1069, 1070, 1071, 1077, 1078, 1088, 1089, 1098, 1099, 1100, 1102, 1104, 1115, 1118, 1121, 1125, 1151, 1152, 1155, 1156, 1181, 1200, 1203, 1209, 1215, 1216, 1219, 1222, 1231, 1260, 1271, 1292, 1306, 1307, 1321, 1343, 1350, 1354, 1366, 1370, 1374, 1377, 1386, 1407, 1421, 1423, 1445, 1450, 1464, 1469, 1473, 1481, 1484, 1494, 1497, 1499, 1551, 1567, 1573, 1575, 1578, 1579, 1588, 1589, 1594, 1600, 1617, 1632, 1638, 1639, 1657, 1661, 1662, 1663, 1664, 1666, 1668, 1669, 1671, 1692, 1696, 1702, 1724, 1733, 1734, 1757, 1758, 1759, 1760, 1762, 1763, 1764, 1768, 1771, 1791, 1796, 1801, 1817, 1828, 1829, 1840, 1846, 1852, 1858, 1865, 1876, 1882, 1895.

Methodology

56, 98, 139, 174, 184, 223, 226, 361, 447, 477, 490, 499, 501, 506, 528, 541, 542, 549, 550, 576, 613, 641, 643, 644, 731, 813, 822, 860, 915, 931, 987, 991, 1060, 1112, 1148, 1153, 1189, 1251, 1321, 1366, 1430, 1473, 1588, 1606, 1629, 1659, 1752, 1753, 1767, 1789, 1847, 1850, 1876, 1888, 1891.

Monitoring

518, 1130, 1131, 1234, 1749, 1872.

Morphology

29, 78, 81, 170, 205, 354, 464, 505, 638, 652, 679, 806, 908, 912, 942, 991, 1012, 1125, 1126, 1158, 1222, 1223, 1229, 1389, 1513, 1598, 1655, 1726.

Mortality

29, 64, 67, 69, 89, 90, 119, 122, 128, 129, 161, 167, 239, 248, 289, 305, 313, 314, 321, 350, 376, 415, 426, 433, 439, 441, 467, 484, 496, 501, 505, 506, 513, 540, 543, 594, 600, 609, 614, 618, 651, 667, 670, 677, 679, 709, 772, 776, 787, 788, 790, 794, 816, 818, 842, 844, 849, 850, 862, 892, 896, 899, 903, 917, 918, 941, 944, 945, 953, 979, 981, 982, 1050, 1051, 1060, 1061, 1062, 1063, 1066, 1076, 1118, 1144, 1150, 1172, 1178, 1185, 1229, 1231, 1266, 1292, 1327, 1341, 1355, 1378, 1390, 1423, 1427, 1428, 1449, 1456, 1457, 1462, 1513, 1573, 1604, 1621, 1664, 1688, 1700, 1725, 1779, 1801, 1818, 1843.

Nestboxes

114, 115, 116, 117, 133, 184, 199, 207, 243, 245, 246, 248, 316, 340, 363, 364, 365, 378, 383, 385, 391, 446, 460, 462, 482, 504, 514, 515, 552, 559, 567, 569, 592, 714, 738, 742, 781, 844, 847, 910, 914, 940, 957, 958, 980, 982, 1055, 1093, 1096, 1098, 1100, 1115,

Nesting sites

Parasitology

Population dynamics

Predation

Prey

Reinforcement

Statistics

Taxonomy

Voice

Weather

Index